Wilson and Walker's
Principles and Techniques of

BIOCHEMISTRY AND MOLECULAR BIOLOGY

Bringing this best-selling textbook right up-to-date, the new edition uniquely integrates the theories and methods that drive the fields of biology, biotechnology and medicine, comprehensively covering both the techniques students will encounter in lab classes and those that underpin current key advances and discoveries. The contents have been updated to include both traditional and cutting-edge techniques most commonly used in current life science research. Emphasis is placed on understanding the theory behind the techniques, as well as analysis of the resulting data. New chapters cover proteomics, genomics, metabolomics and bioinformatics, as well as data analysis and visualisation. Using accessible language to describe concepts and methods, and with a wealth of new in-text worked examples to challenge students' understanding, this textbook provides an essential guide to the key techniques used in current bioscience research.

ASSOCIATE PROFESSOR ANDREAS HOFMANN is the Structural Chemistry Program Leader at Griffith University's Eskitis Institute and an Honorary Senior Research Fellow at the Faculty of Veterinary and Agricultural Sciences at The University of Melbourne. He is also Fellow of the Higher Education Academy (UK). Professor Hofmann's research lab focusses on the structure and function of proteins in infectious and neurodegenerative diseases, and also develops computational tools. He is the author of *Methods of Molecular Analysis in the Life Sciences* (2014) and *Essential Physical Chemistry* (2018).

DR SAMUEL CLOKIE is the Principal Clinical Bioinformatician at the Birmingham Women's and Children's Hospital, UK, where he leads the implementation of bioinformatic techniques to decipher next-generation sequencing data in clinical diagnostics. He is also Honorary Senior Research Fellow at the Institute of Cancer and Genomic Sciences at The University of Birmingham and before that was a research fellow at the National Institutes of Health in Bethesda, USA.

Wilson and Walker's
Principles and Techniques of
Biochemistry and Molecular Biology

Edited by

ANDREAS HOFMANN
Griffith University, Queensland

SAMUEL CLOKIE
Birmingham Women's and Children's Hospital

CAMBRIDGE
UNIVERSITY PRESS

Shaftesbury Road, Cambridge CB2 8EA, United Kingdom

One Liberty Plaza, 20th Floor, New York, NY 10006, USA

477 Williamstown Road, Port Melbourne, VIC 3207, Australia

314–321, 3rd Floor, Plot 3, Splendor Forum, Jasola District Centre, New Delhi – 110025, India

103 Penang Road, #05–06/07, Visioncrest Commercial, Singapore 238467

Cambridge University Press is part of Cambridge University Press & Assessment,
a department of the University of Cambridge.

We share the University's mission to contribute to society through the pursuit of
education, learning and research at the highest international levels of excellence.

www.cambridge.org
Information on this title: www.cambridge.org/9781107162273
DOI: 10.1017/9781316677056

First published 2018 (version 4, February 2023)

Printed in the United Kingdom by Bell & Bain Ltd, Glasgow, February 2023

A catalogue record for this publication is available from the British Library
Names: Wilson, Keith, 1936– editor. | Walker, John M., 1948– editor. | Hofmann, Andreas, editor.
| Clokie, Samuel, editor.
Title: Wilson and Walker's principles and techniques of biochemistry and molecular biology /
edited by Andreas Hofmann, Griffith University, Queensland, Samuel Clokie, Birmingham
Women's and Children's Hospital.
Other titles: Principles and techniques of biochemistry and molecular biology.
Description: [2018 edition]. | Cambridge : Cambridge University Press, 2018. | Includes
bibliographical
references and index.
Identifiers: LCCN 2017033452 | ISBN 9781107162273 (alk. paper)
Subjects: LCSH: Biochemistry – Textbooks. | Molecular biology – Textbooks.
Classification: LCC QP519.7 .P75 2018 | DDC 572.8–dc23
LC record available at https://lccn.loc.gov/2017033452

ISBN 978-1-107-16227-3 Hardback
ISBN 978-1-316-61476-1 Paperback

Additional resources for this publication at www.cambridge.org/hofmann

CONTENTS

FOREWORD TO THE EIGHTH EDITION

When the first edition of our book was published in 1975, biochemistry was emerging as a new discipline that had the potential to unify the previously divergent -ologies that constituted the biological sciences. It placed emphasis on the understanding of the structure, function and expression of individual proteins and their related genes, and relied on the application of analytical techniques such as electrophoresis, chromatography and various forms of spectrometry. In the ensuing 40 years the completion of the Human Genome Project has confirmed the central role of DNA in all of the activities of individual cells and the emergence of molecular biology as the means of understanding complex biological processes. This in turn has led to the new disciplines of bioinformatics, chemoinformatics, proteomics and metabolomics. The succeeding six editions of our book have attempted to reflect this evolution of biochemistry as a unifying discipline. All editions have placed emphasis on the experimental techniques that undergraduates can expect to encounter during the course of their university studies and to this end we have been grateful for the excellent feedback we have received from the users of the book.

The point has now been reached where we believed that it was appropriate for us to hand over the direction and academic balance of future editions to a new editorial team and to this end we are delighted that Andreas Hofmann and Samuel Clokie have agreed to take on the role. We wish them well and look forward to the continuing success of the book.

KEITH WILSON AND JOHN WALKER

PREFACE

It has been a tremendous honour being asked by Keith Wilson, John Walker and Cambridge University Press to take on the role of editorship for this eighth edition of *Principles and Techniques*. In designing the content, we extend the long and successful tradition of this text to introduce relevant methodologies in the biochemical sciences by uniquely integrating the theories and practices that drive the fields of biology, biotechnology and medicine. Methodologies have improved tremendously over the past ten years and new strategies and protocols that once were just applied by a few pioneering laboratories are now applied routinely, accompanied by ever more fluent boundaries between core disciplines – leading to what is now called the life sciences.

In this eighth edition, all core methodologies covered in the previous edition have been kept and appropriately updated and consolidated. New chapters have been added to address the requirements of today's students and scientists, who operate in a much broader area than did the typical biochemists one or two decades ago. The contents of this text are thus structured into six areas, all of which play pivotal roles in current research: basic principles, biochemistry and molecular biology, biophysical methods, information technology, -omics methods and chemical biology. Of course, due to space restraints that help to keep this text accessible, the addition of new material required us to consolidate and carefully select content to be kept from the previous edition. These decisions have been guided by the over-arching theme of this text, namely to present principles and techniques. Importantly, our experience as teachers has always been that many undergraduates are challenged by quantitative calculations based on these principles, and hence we have continued the tradition of this text by including relevant mathematical and numerical tools as well as examples.

Indeed, new chapters on data processing, visualisation and Python have example applications that can be accessed from the CUP website to aid understanding.

Sadly, two authors of previous editions have passed away: John Fyffe died shortly after the seventh edition was launched and Alastair Aitken passed away in 2014. Also, Keith Wilson, John Walker and Robert Burns decided to retire, and we thus invited several new authors to contribute to this eighth edition.

We are grateful to the authors and publishers who have granted us permission to reproduce and adapt their figures. We further wish to express our gratitude to Sonja Biberacher, Madeleine Dallaston, Emma Klepzig, Hannah Leeson and Megan Cross who have helped tremendously with preparing figures and photographs. The manuscript and figures for this book have been compiled entirely with open-source and academic software under Linux, and we would like to acknowledge the efforts of software developers and programmers who make their products freely available.

Finally, our sincere thanks also extend to Katrina Halliday and her team at Cambridge University Press whose invaluable support has made it possible to produce this new edition.

We welcome constructive comments from all students who use this book within their studies, and from teachers and academics who adopt the book to complement their courses.

ANDREAS HOFMANN AND SAMUEL CLOKIE

CONTRIBUTORS

PARISA AMANI
Structural Chemistry Program, Eskitis Institute for Drug Discovery,
Griffith University,
Nathan, Queensland, Australia

ANWAR R. BAYDOUN
School of Life Sciences, University of Hertfordshire,
Hatfield, Hertfordshire, UK

DAVID CAMP
School of Environment, Griffith University,
Nathan, Queensland, Australia

CINZIA CANTACESSI
Cambridge Veterinary School,
University of Cambridge,
Cambridge, UK

JEAN-BAPTISTE CAZIER
Centre for Computational Biology,
University of Birmingham,
Edgbaston, Birmingham, UK

SAMUEL CLOKIE
Bioinformatics Department,
West Midlands Regional Genetics Laboratory
Birmingham Women's and Children's NHS Foundation Trust,
Birmingham, UK

MEGAN CROSS
Structural Chemistry Program, Eskitis Institute for Drug Discovery,
Griffith University,
Nathan, Queensland, Australia

KATJA FISCHER
Scabies Laboratory, QIMR Berghofer,
Locked Bag 2000, Royal Brisbane Hospital,
Herston, Queensland, Australia

ROBIN B. GASSER
Faculty of Veterinary and Agricultural Sciences,
The University of Melbourne,
Parkville, Victoria, Australia

JOHN GRAINGER
Faculty of Biology, Medicine and Health,
University of Manchester,
Manchester, UK

STEPHEN E. HARDING
School of Biosciences,
University of Nottingham,
Sutton Bonington, UK

SONJA HESS
Proteome Exploration Laboratory, Beckman Institute,
California Institute of Technology,
Pasadena, USA

ANDREAS HOFMANN
Structural Chemistry Program, Eskitis Institute for Drug Discovery,
Griffith University,
Nathan, Queensland, Australia
and
Faculty of Veterinary and Agricultural Sciences,
The University of Melbourne,
Parkville, Victoria, Australia

JOANNE KONKEL
Faculty of Biology, Medicine and Health,
University of Manchester,
Manchester, UK

PASI K. KORHONEN
Faculty of Veterinary and Agricultural Sciences,
The University of Melbourne,
Parkville, Victoria, Australia

JOANNE MACDONALD
School of Science and Engineering,
University of the Sunshine Coast,
Maroochydore DC, Queensland, Australia
and

Department of Medicine,
Columbia University,
New York, NY, USA

JAMES I. MACRAE
Metabolomics Unit,
The Francis Crick Institute,
London, UK

KAY OHLENDIECK
Department of Biology,
National University of Ireland, Maynooth,
Co. Kildare, Ireland

STEPHEN W. PADDOCK
Howard Hughes Medical Institute,
Department of Molecular Biology,
University of Wisconsin,
Madison, Wisconsin, USA

ANNA V. PROTASIO
Wellcome Trust Sanger Institute,
University of Cambridge,
Hinxton, Cambridge, UK

RALPH RAPLEY
Department of Biosciences,
University of Hertfordshire,
Hatfield, Hertfordshire, UK

GILL RUMSBY
Clinical Biochemistry,
HSL Analytics LLP,
London, UK

ANNE SIMON
Université Claude Bernard Lyon 1,
Bâtiment Curien, Villeurbanne,
France
and
Laboratoire Chimie et Biologie des Membranes et des Nanoobjets,
Université de Bordeaux, Pessac,
France

ROBERT J. SLATER
Royal Society of Biology,
Charles Darwin House,
London, UK

TIM J. STEVENS
MRC Laboratory of Molecular Biology,
University of Cambridge,
Cambridge, UK

PAUL TAYLOR
School of Biological Sciences,
University of Edinburgh,
Edinburgh, Scotland, UK

MICHAEL WEISS
Phase I Pilot Consortium,
Children's Oncology Group,
Monrovia, CA, USA

0 Tables and Resources

Table 0.1 **Abbreviations**

ADP	adenosine 5'-diphosphate
AMP	adenosine 5'-monophosphate
ATP	adenosine 5'-triphosphate
bp	base pairs
cAMP	cyclic AMP
CAPS	N-cyclohexyl-3-aminopropanesulfonic acid
CHAPS	3-[(3-cholamidopropyl)dimethylamino]-1-propanesulfonic acid
cpm	counts per minute
CTP	cytidine triphosphate
DDT	2,2-bis-(p-chlorophenyl)-1,1,1-trichloroethane
DMSO	dimethylsulfoxide
DNA	deoxyribonucleic acid
e^-	electron
EDTA	ethylenediaminetetra-acetate
ELISA	enzyme-linked immunosorbent assay
FACS	fluorescence-activated cell sorting
FAD	flavin adenine dinucleotide (oxidised)
$FADH_2$	flavin adenine dinucleotide (reduced)
FMN	flavin mononucleotide (oxidised)
$FMNH_2$	flavin mononucleotide (reduced)
GC	gas chromatography
GTP	guanosine triphosphate
HAT	hypoxanthine, aminopterin, thymidine medium
HEPES	4(2-hydroxyethyl)-1-piperazine-ethanesulfonic acid
HPLC	high-performance liquid chromatography
IMS	industrial methylated spirit
kb	kilobase pairs

Table 0.1 (cont.)

M_r	relative molecular mass
MES	4-morpholine-ethanesulfonic acid
min	minute
MOPS	4-morpholine-propanesulfonic acid
NAD^+	nicotinamide adenine dinucleotide (oxidised)
NADH	nicotinamide adenine dinucleotide (reduced)
$NADP^+$	nicotinamide adenine dinucleotide phosphate (oxidised)
NADPH	nicotinamide adenine dinucleotide phosphate (reduced)
PIPES	1,4-piperazinebis(ethanesulfonic acid)
ppb	parts per billion
ppm	parts per million
RNA	ribonucleic acid
rpm	revolutions per minute
SDS	sodium dodecyl sulfate
TAPS	N-[tris-(hydroxymethyl)]-3-aminopropanesulfonic acid
TRIS	2-amino-2-hydroxymethylpropane-1,3-diol
v/v	volume per volume
w/v	weight per volume

Table 0.2 Biochemical constants

Unit	Symbol	SI equivalent
Atomic mass unit	u	1.661×10^{-27} kg
Avogadro constant	L or N_A	6.022×10^{23} mol^{-1}
Faraday constant	F	9.648×10^4 C mol^{-1}
Planck constant	h	6.626×10^{-34} J s
Universal or molar gas constant	R	8.314 J K^{-1} mol^{-1}
Molar volume of an ideal gas at standard conditions[a]	V_m	22.41 dm^3 mol^{-1}
Velocity of light in a vacuum	c	2.997×10^8 m s^{-1}

Note: [a]Standard conditions: $p^0 = 1$ bar, $\theta_{normal} = 25\ ^\circ C$.

Table 0.3 SI units: basic and derived units

Quantity	SI unit	Symbol (basic SI units)	Definition of SI unit	Equivalent in SI units
Basic units				
Length	metre	m		
Mass	kilogram	kg		
Time	second	s		
Electric current	ampere	A		
Temperature	kelvin	K		
Luminous intensity	candela	cd		
Amount of substance	mole	mol		
Derived units				
Area	square metre	m^2		
Concentration	mole per cubic metre	$mol\ m^{-3}$		
Density	kilogram per cubic metre	$kg\ m^{-3}$		
Electric charge	coulomb	C	$1\ A\ s$	$1\ J\ V^{-1}$
Electric potential difference	volt	V	$1\ kg\ m^2\ s^{-3} A^{-1}$	$1\ J\ C^{-1}$
Electric resistance	ohm	Ω	$1\ kg\ m^2\ s^{-3} A^{-2}$	$1\ V\ A^{-1}$
Energy, work, heat	joule	J	$1\ kg\ m^2\ s^{-2}$	$1\ N\ m$
Enzymatic activity	katal	katal	$1\ mol\ s^{-1}$	$60\ mol\ min^{-1}$
Force	newton	N	$1\ kg\ m\ s^{-2}$	$1\ J\ m^{-1}$
Frequency	hertz	Hz	$1\ s^{-1}$	
Magnetic flux density	tesla	T	$1\ kg\ s^{-2}\ A^{-1}$	$1\ V\ s\ m^{-2}$
Molecular mass	atomic mass units (dalton)	u	$1\ u = 1\ Da$	
Molar mass	gram per mole	$g\ mol^{-1}$	$1\ g\ mol^{-1}$	
Power, radiant flux	watt	W	$1\ kg\ m^2\ s^{-3}$	$1\ J\ s^{-1}$
Pressure	pascal	Pa	$1\ kg\ m^1\ s^{-2}$	$1\ N\ m^{-2}$
Volume	cubic metre	m^3		

Table 0.4 Conversion factors for non-SI units

Unit	Symbol	SI equivalent
Energy		
calorie	cal	4.184 J
erg	erg	10^{-7} J
electron volt	eV	1.602×10^{-19} J
Pressure		
atmosphere	atm	1.013×10^5 Pa
bar	bar	10^5 Pa
millimetres of mercury	mm Hg (Torr)	133.322 Pa
pounds per square inch	psi	6.895×10^4 Pa
Temperature		
degree Celsius	°C	$\left[\dfrac{\theta}{1\,^{\circ}\text{C}} + 273.15 \right] \text{K}$
degree Fahrenheit	°F	$\left[\left(\dfrac{t}{1\,^{\circ}\text{F}} - 32 \right) \times \dfrac{5}{9} + 273.15 \right] \text{K}$
Length		
ångstrøm	Å	10^{-10} m
inch	in	0.0254 m
Mass		
pound	lb	0.4536 kg

Table 0.5 Common unit prefixes associated with quantitative terms

Multiple	Prefix	Symbol	Multiple	Prefix	Symbol
10^{24}	yotta	Y	10^{-1}	deci	d
10^{21}	zetta	Z	10^{-2}	centi	c
10^{18}	exa	E	10^{-3}	milli	m
10^{15}	peta	P	10^{-6}	micro	µ
10^{12}	tera	T	10^{-9}	nano	n
10^{9}	giga	G	10^{-12}	pico	p
10^{6}	mega	M	10^{-15}	femto	f
10^{3}	kilo	k	10^{-18}	atto	a
10^{2}	hecto	h	10^{-21}	zepto	z
10^{1}	deca	da	10^{-24}	yocto	y

Table 0.6 Interconversion of non-SI and SI units of volume

Non-SI unit		Non-SI subunit		SI subunit		SI unit
1 litre (l)	=	10^3 ml	=	1 dm^3	=	10^{-3} m^3
1 millilitre (ml)	=	1 ml	=	1 cm^3	=	10^{-6} m^3
1 microlitre (μl)	=	10^{-3} ml	=	1 mm^3	=	10^{-9} m^3
1 nanolitre (nl)	=	10^{-6} ml	=	1 nm^3	=	10^{-12} m^3

Table 0.7 Interconversion of mol, mmol and μmol in different volumes to give different concentrations

Molar (M)		Millimolar (mM)		Micromolar (μM)
1 mol dm^{-3}	=	10^3 mmol dm^{-3}	=	10^6 μmol dm^{-3}
=		=		=
1 mmol cm^{-3}	=	10^3 μmol cm^{-3}	=	10^6 nmol cm^{-3}
=		=		=
1 μmol mm^{-3}	=	10^3 nmol mm^{-3}	=	10^6 pmol mm^{-3}

Table 0.8 pK_a values of some acids and bases that are commonly used as buffer solutions

Acid or base	pK_a
Acetic acid	4.8
Barbituric acid	4.0
Carbonic acid	6.1, 10.2
CAPS[a]	10.4
Citric acid	3.1, 4.8, 6.4
Glycylglycine	3.1, 8.1
HEPES[a]	7.5
MES[a]	6.1
MOPS[a]	7.1
Phosphoric acid	2.0, 7.1, 12.3
Phthalic acid	2.8, 5.5
PIPES[a]	6.8
Succinic acid	4.2, 5.6
TAPS[a]	8.4
Tartaric acid	3.0, 4.2
TRIS[a]	8.1

Note: [a]See list of abbreviations (Table 0.1).

Table 0.9 **Abbreviations for amino acids and average molecular masses, free and within peptides (residue mass)**

Amino acid	Three-letter code	One-letter code	Molecular mass M Nominal mass (Da)	Residue mass $M\text{-}M(H_2O)$ Average mass (Da)	Monoisotopic mass (Da)	Side Chain
Alanine	Ala	A	89	71.08	71.037114	
Arginine	Arg	R	174	156.19	156.10111	
Asparagine	Asn	N	132	114.10	114.04293	
Aspartic acid	Asp	D	133	115.09	115.02694	
Asparagine or aspartic acid	Asx	B				
Cysteine	Cys	C	121	103.15	103.00919	
Glutamine	Gln	Q	146	128.13	128.05858	
Glutamic acid	Glu	E	147	129.12	129.04259	
Glutamine or glutamic acid	Glx	Z				
Glycine	Gly	G	75	57.05	57.021464	
Histidine	His	H	155	137.14	137.05891	
Isoleucine	Ile	I	131	113.16	113.08406	
Leucine	Leu	L	131	113.16	113.08406	

Table 0.9 (cont.)

Amino acid	Three-letter code	One-letter code	Molecular mass M Nominal mass (Da)	Residue mass M-$M(H_2O)$ Average mass (Da)	Monoisotopic mass (Da)	Side Chain
Lysine	Lys	K	146	128.17	128.09496	
Methionine	Met	M	149	131.20	131.04048	
Met-sulfoxide[a]	MetSO	MSO	165	147.20	147.03540	
Phenylalanine	Phe	F	165	147.18	147.06841	
Proline	Pro	P	115	97.12	97.052764	
Serine	Ser	S	105	87.08	87.032029	
Threonine	Thr	T	119	101.11	101.04768	
Tryptophan	Trp	W	204	186.21	186.07931	
Tyrosine	Tyr	Y	181	163.18	163.06333	
Valine	Val	V	117	99.13	99.068414	

Notes: Although some amino acids are similar in mass, they can be distinguished by modern high-resolution mass spectrometry. [a]This is a frequently found modification in mass spectrometric investigation of proteins and peptides.

Table 0.10 Ionisable groups found in amino acids

Amino acid	pK$_{a1}$ α-carboxyl group	pK$_{a2}$ α-ammonium ion	pK$_{a3}$ side-chain group	pI
Alanine	2.3	9.7	–	6.0
Arginine	2.2	9.0	12.5	10.8
Asparagine	2.0	8.8	–	5.41
Aspartic acid	1.9	9.6	3.7	2.8
Cysteine	2.0	8.2	8.4	5.1
Glutamine	2.2	9.1	–	5.7
Glutamic acid	2.2	9.7	4.3	3.2

Amino acid					
Glycine	2.3	9.6		–	6.0
Histidine	1.8	9.2	(equilibrium structure, imidazolium ⇌ imidazole + H⁺)	6.0	7.6
Isoleucine	2.4	9.6		–	6.0
Leucine	2.4	9.6		–	6.0
Lysine	2.2	9.0	(equilibrium structure, –NH₃⁺ ⇌ –NH₂ + H⁺)	10.5	9.7
Methionine	2.3	9.2		–	5.7
Phenylalanine	1.8	9.1		–	5.5
Proline	2.0	10.6	(equilibrium structure, –NH₂⁺ ⇌ –NH + H⁺)	–	6.3
Serine	2.2	9.2		–	5.7
Threonine	2.1	9.1		–	5.6
Tryptophan	2.8	9.4		–	5.9
Tyrosine	2.2	9.1	(equilibrium structure, phenol ⇌ phenolate + H⁺)	10.1	5.7
Valine	2.3	9.6		–	6.0

Table 0.11 **The genetic code: triplet codons and their corresponding amino acids**

	U		C		A		G		
U	UUU	Phe	UCU	Ser	UAU	Tyr	UGU	Cys	U
	UUC	Phe	UCC	Ser	UAC	Tyr	UGC	Cys	C
	UUA	Leu	UCA	Ser	UAA	Stop	UGA	Stop	A
	UUG	Leu	UCG	Ser	UAG	Stop	UGG	Trp	G
C	CUU	Leu	CCU	Pro	CAU	His	CGU	Arg	U
	CUC	Leu	CCC	Pro	CAC	His	CGC	Arg	C
	CUA	Leu	CCA	Pro	CAA	Gln	CGA	Arg	A
	CUG	Leu	CCG	Pro	CAG	Gln	CGG	Arg	G
A	AUU	Ile	ACU	Thr	AAU	Asn	AGU	Ser	U
	AUC	Ile	ACC	Thr	AAC	Asn	AGC	Ser	C
	AUA	Ile	ACA	Thr	AAA	Lys	AGA	Arg	A
	AUG	Met	ACG	Thr	AAG	Lys	AGG	Arg	G
G	GUU	Val	GCU	Ala	GAU	Asp	GGU	Gly	U
	GUC	Val	GCC	Ala	GAC	Asp	GGC	Gly	C
	GUA	Val	GCA	Ala	GAA	Glu	GGA	Gly	A
	GUG	Val	GCG	Ala	GAG	Glu	GGG	Gly	G

1 Biochemical and Molecular Biological Methods in Life Sciences Studies

SAMUEL CLOKIE AND ANDREAS HOFMANN

1.1 FROM BIOCHEMISTRY AND MOLECULAR BIOLOGY TO THE LIFE SCIENCES

Biochemistry is a discipline in the natural sciences that is chiefly concerned with the chemical processes that take place in living organisms. Starting in the 1950s, a new stream evolved from traditional biochemistry, which, until then, mainly investigated bulk behaviour and macroscopic phenomena. This new stream focussed on the molecular basis of biological processes and since it put the biologically important molecules into the spotlight, the term molecular biology was coined. Importantly, molecular biology goes beyond the mere characterisation of molecules. It includes the study of interactions between biologically relevant molecules with the clear goal to reveal insights into functions and processes, such as replication, transcription and translation of genetic material.

Major technological and methodological advances made during the 1980s enabled the development and establishment of several specialised areas in the life sciences. These include structural biology (in particular the determination of three-dimensional structures), genetics (for example DNA sequencing) and proteomics. The refinement and improvement of methodologies, as well as the development of more efficient software (in line with more powerful computing resources), contributed substantially to specialist techniques becoming more accessible to researchers in neighbouring disciplines. What once had been the task of a highly specialised scientist who had been extensively trained in that particular area has consistently been transformed into a routinely applied methodology. These tendencies have pushed the feasibility of cross-disciplinary studies into an entirely new realm, and in many contemporary laboratories and research groups, methods originally at home in different basic disciplines are frequently used next to each other.

It is thus not surprising, that in the past 15–20 years, the term 'life sciences' has been used to describe the general nature of studies and research areas of a scientist working on studies related to living organisms.

1.2 THE EDUCATION OF LIFE SCIENTISTS

The life sciences embrace different fields of the natural and health sciences, all of which involve the study of living organisms, from microorganisms, plants and animals to human beings. Importantly, even satellite areas that are methodologically rooted outside natural or health sciences, for example bioethics, have become a part of the life sciences. And with neuroscience and artificial intelligence being two major current areas of interest, one might expect further subjects to be included under the life sciences umbrella.

The growing complexity of scientific studies requires ever-increasing **cross-disciplinarity** when it comes to particular methodologies utilised in the quest to reach the goals of these studies. This poses entirely new problems when it comes to the education of future scientists and researchers. At the core, this requires an appreciation or even acquisition of mindsets from other disciplines, as opposed to a mere concatenation of methods specific to one discipline. Therefore, the achievement of true cross-disciplinarity poses particular challenges since, according to Simon Penny, it requires deep professional humility, intellectual rigour and courage. At the same time, due to time and practical constraints, educational programmes that teach life science studies often focus on a select group of individual disciplines, rather than attempting to include every subject that could be included under the global term 'life sciences'.

In order to successfully embark on a contemporary life science study or contribute to large cross-disciplinary teams, scientists also need to possess knowledge and skills in a diverse range of areas. For example, in order to screen a small-molecule compound library against a target protein, the chemist needs to understand the nature and behaviour of proteins. Vice versa, if a cell biologist wants to screen a small-molecule library to identify novel effectors for a pathway of interest, they need to deal with the logistics and characteristics of small molecules.

One relatively new aspect that the life scientist is faced with is the large amount of data generated by improved instrumentation and methodologies. The term 'bioinformatics' has been coined that loosely describes the processing of biological data. It is now considered a stand-alone discipline that includes methods on processing '**big data**' that are commonplace in the field of genomics. Such data-processing techniques are ubiquitous in the life sciences, and bioinformatics serves as an example that spans almost all the life science disciplines.

Life scientists work in many diverse areas, including hospitals, academic teaching and research, drug discovery and development, agriculture, food institutes, general education, cosmetics and forensics. Aside from their specialist knowledge of the interfacing of core disciplines, they also require a solid basis of transferable skills, such as analytical and problem-solving capabilities, and written and verbal communication, as well as planning, research, observation and numerical skills.

1.3 AIMS OF LIFE SCIENCE STUDIES

Studies in the life sciences ultimately aspire to an advanced understanding of the nature of life in molecular and mechanistic terms. Biochemistry still constitutes the core discipline of the experimental life sciences, and involves the study of the chemical processes that occur in living organisms. Such studies rely on the application of appropriate techniques to advance our understanding of the nature, and relationships between, biological molecules, especially proteins and nucleic acids in the context of cellular function.

The huge advances made in the past 10–15 years, with the **Human Genome Project** being a particular milestone, have stimulated major developments in our understanding of many diseases and led to identification of strategies that might be used to combat these diseases. Such progress was accompanied – and enabled – by substantial developments in technologies, data acquisition and data mining. For example, the genome of any living organism includes coding regions that are transcribed into messenger RNAs (mRNAs), which are subsequently translated into proteins. In vitro, mRNAs are reverse-transcribed (Section 4.10.4), resulting in stable complementary DNA, which is traditionally sequenced using a DNA polymerase, an oligonucleotide primer and four deoxyribonucleotide triphosphates (dNTPs) to synthesise the complementary strand to the template sequence (see Section 20.2.1). The development of high-throughput sequencing ('next-generation sequencing') technologies that can produce millions or billions of sequences concurrently, has made the sequencing of entire genomes orders of magnitude faster and, at the same time, less expensive. This particular development has made it possible to sequence many thousands of human genomes, making it possible to truly understand **population genetics**. Such information can be used to aid and improve genetic diagnosis of common and rare human diseases.

The combination of molecular biology and genomics applied to the benefit of humankind can be best illustrated by the invention and recent improvement of genome editing techniques, such as the CRISPR/Cas method (see Section 4.17.6). The potential impact on **health economics** is substantial; individually unpleasant and costly conditions, genetically or environmentally acquired, can be addressed and potentially be eradicated.

Similar developments have occurred in many other disciplines. In structural biology, robotics, especially at synchrotron facilities, have drastically reduced the time required for handling individual samples. Plate readers that perform particular spectroscopic applications (see Chapter 13) in a multi-well format are now ubiquitously present in laboratories and enable medium and **high throughput** for many standard assays (see also Section 24.5.3).

All these developments are accompanied by a massive increase in (digital) data generated, which opens an avenue for entirely new types of studies that are more or less exclusively concerned with data mining and analysis.

This text aims to cover the principles and methodologies underpinning life science studies and thus address the requirements of today's students and scientists who operate in a much broader area than the typical biochemists one or two decades ago.

The contents of this text are therefore structured around six different disciplines or methodologies, all of which play pivotal roles in current research:

- Basic Principles (Chapter 2)
- Biochemistry and Molecular Biology (Chapters 3–10)
- Biophysics (Chapters 11–15)
- Information Technology (Chapters 16–19)
- 'omics Methods (Chapters 20–22)
- Chemical Biology (Chapters 23–24)

The Basic Principles chapter introduces some important general concepts surrounding biologically relevant molecules, as well as their handling in aqueous solution. It further highlights fundamental considerations when designing and conducting experimental research.

Information technology has become an integral part of scientific research. The ability to handle, analyse and visualise data has always been a core skill of science and gained even more importance with the advent of 'big data' on the one hand, and the fact that data acquisition, processing and communication is done entirely in digital format on the other. Furthermore, in the areas of bio- and chemoinformatics, standardisation of data formats has resulted in the availability of unprecedented volumes of information in databases that require an appropriate understanding in order to fully utilise the data resource.

Among the core methodologies of biochemistry and molecular biology are techniques to culture living cells and microorganisms, and the preparation and handling of DNA, as well as the production and purification of proteins. Such experimental work is frequently accompanied by analytical or preparative gel electrophoresis, the use of antibodies (immunochemistry) or radio isotopes. In the medical sciences, a diagnostic test requires the application of biochemical or molecular techniques in a regulated environment, comprising the area of clinical biochemistry (discussed in Chapter 10).

The characterisation of cells and molecules, as well as their interactions and processes, involves an array of biophysical techniques. Such techniques apply gravitational (centrifugation) or electrical forces (mass spectrometry), as well as interactions with light over a broad range of energy.

The neologism 'omics is frequently being used when referring to methodologies that characterise large pools of biologically relevant molecules, such as DNA (genome), mRNA (transcriptome), proteins (proteome) or metabolic products (metabolome). The first application of these methodologies were mainly concerned with the acquisition and collection of large datasets specific to one experiment or biological question. However, several areas of the life sciences now leave behind the phase of pure observation and are increasingly applied to study the dynamics of an organism, a so-called systems biology approach.

A hallmark of chemical biology is the use of small molecules, either purified or synthetically derived natural products or purpose-designed chemicals, to study the modulation of biological systems. The use of small-molecule compounds can be either exploratory in nature (probes) or geared towards therapeutic use where the desired activities of the compounds are to either activate or inhibit a target protein, which is typically, but not necessarily, an enzyme. Due to the specific role of a target protein in a given cellular pathway, the molecular interaction will ideally lead to modulation of processes involved in pathogenic situations. Even if a small molecule has a great number of side effects, it can be used as a probe and be useful to delineate molecular pathways in vitro.

Given the breadth of topics, methodologies and applications, selections as to the individual contents presented in this text had to be made. The topics selected for this text have been carefully chosen to provide undergraduate students and non-specialist researchers with a solid overview of what we feel are the most relevant and fundamental techniques.

Methods and techniques form the tool set of an experimental scientist. They are applied in the context of studies which, in the life sciences, address questions of the following nature:

- the structure and function of the total protein component of the cell (proteomics) and of all the small molecules in the cell (metabolomics)
- the mechanisms involved in the control of gene such expression
- the identification of genes associated with a wide range of diseases and the development of gene therapy strategies for the treatment of diseases
- the characterisation of the large number of 'orphan' receptors, whose physiological role and natural agonist are currently unknown, present in the host and pathogen genomes and their exploitation for the development of new therapeutic agents
- the identification of novel disease-specific markers for the improvement of clinical diagnosis
- the engineering of cells, especially stem cells, to treat human diseases
- the understanding of the functioning of the immune system in order to develop strategies for protection against invading pathogens
- the development of our knowledge of the molecular biology of plants in order to engineer crop improvements, pathogen resistance and stress tolerance
- the discovery of novel therapeutics (drugs and vaccines) to the nature and treatment of bacterial, fungal and viral diseases.

1.4 PERSONAL QUALITIES AND SCIENTIFIC CONDUCT

The type of tasks in scientific research and the often long-term goals pursued by science require, very much like any other profession, particular attributes of people working in this area:

- Quite obviously, a substantial level of intelligence is required to grasp the scientific concepts in the area of study. The ability to think with clarity and logically, and to transform particular observations (low level of abstraction) into concepts (high level of abstraction) is also required, as is a solid knowledge of the basic mathematical syllabus for any area of the life sciences.

- For pursuing longer-term goals it is also necessary to possess stamina and persistence. Hurdles and problems need to be overcome, and failures need to be coped with. Often, experimental series can become repetitive and it is important to not fall into the trap of boredom (and then become negligent).
- A frequently underestimated quality is attention to detail. Science and research is about getting things right. At the time when findings from research projects are written up for publication, or knowledge is summarised for text books such as this one, the readership and the public expect that all details are correct. Of course, mistakes can and do happen, but they should not happen commonly, and practices need to be in place that prevent mistakes being carried over to the next step. Critical self-appraisal and constant attention to detail is probably the most important element in this process.
- Communication skills and the ability to describe and visualise fairly specialist concepts to non-experts is of great importance as well. The best set of data is not put to good use if it is not presented to the right forum in the right fashion. Likewise, any set of knowledge acquired cannot be taught effectively to others without such skills.
- Lastly, curiosity and the willingness to explore are a requirement if an independent career in the life sciences is being sought. Just having excellent marks in science subjects does not automatically make a scientist if one needs to be told every single next step through a research project.

Since science does not happen in an isolated situation but a community (the scientific community on the one hand, but also society as a whole on the other), norms need to be put in place that define and guide what is acceptable and unacceptable behaviour. Beset by the occasional fraudulent study and scientific misconduct case, more attention has been paid to ethical conduct in science in the recent decade (see also Section 2.8.2).

Despite ethical frameworks and policies put in place to varying degrees, ultimately, the responsibility of ethical conduct rests with the individual. And while many of the ethical norms are geared towards the interactions of scientists when it comes to specific scientific tasks or procedures, there are certainly elements that apply to any (scientific or non-scientific) situation:

- Critically assess potential conflicts of interest. In a surprisingly large number of situations, any individual might find themselves playing multiple roles, and the objectives of each of the roles may be in conflict with each other. Where conflicts of interest arise, the individual should withdraw from the decision-making process.
- Respect confidentiality and privacy. This not only applies to ongoing research work, results, etc., but also to conversations and advice. When approached for advice or with a personal conversation, most parties expect this to be treated in confidence and not made available to the public.
- Follow informed consent rules and discuss intellectual property frankly. In order to avoid disagreements about who should get credit and for which aspects, talk about these issues at the beginning of a working relationship.
- Engage in a sharing culture. Resources, methodologies, knowledge and skills that have been established should be shared where reasonably possible. This fosters positive

interactions, contributes to transparency and frequently leads to new collaborations, all of which advance science.

- Lead by example. Not just in a teaching situation, but in most other settings, too, general principles and cultural norms are only effectively imprinted into the environment if the rules and guidelines are lived. In many instances, rules and protocols are 'implemented' by institutions and exist mainly as a ticking-the-box exercise. Such policies do not address the real point as they should and are largely ineffective.

1.5 SUGGESTIONS FOR FURTHER READING

1.5.1 Experimental Protocols

Holmes D., Moody P. and Dine D. (2010) *Research Methods for the Biosciences, 2nd Edn.*, Oxford University Press, Oxford, UK.

1.5.2 General Texts

Smith D. (2003) Five principles for research ethics. *Monitor on Psychology* 34, 56.

1.5.3 Review Articles

Duke C.S. and Porter J.H. (2013) The ethics of data sharing and reuse in biology. *BioScience* 63, 483–489.

Puniewska M. (2014) Scientists have a sharing problem. *The Atlantic*, 15 Dec 2014, www.theatlantic.com/health/archive/2014/12/scientists-have-a-sharing-problem/383061/ (accessed April 2017).

Taylor P.L. (2007) Research sharing, ethics and public benefit. *Nature Biotechnology* 25, 398–401.

1.5.4 Websites

Mike Brotherton's Blog: Five qualities required to be a scientist www.mikebrotherton.com/2007/11/05/five-qualities-required-to-be-a-scientist/ (accessed April 2017)

Simon Penny: Rigorous Interdisciplinary Pedagogy simonpenny.net/texts/rip.html (accessed April 2017)

Andy Polaine's Blog: Interdisciplinarity vs Cross-Disciplinarity www.polaine.com/2010/06/interdisciplinarity-vs-cross-disciplinarity/ (accessed April 2017)

2 | Basic Principles

PARISA AMANI AND ANDREAS HOFMANN

2.1 BIOLOGICALLY IMPORTANT MOLECULES

Molecules of biological interest can be classified into ions, small molecules and macromolecules. Typical organic small molecules include the ligands of enzymes, substrates such as adenosine triphosphate (ATP) and effector molecules (inhibitors, drugs). Ions such as Ca^{2+} play a key role in signalling events. Biological macromolecules are polymers which, by definition, consist of covalently linked monomers, the building blocks. The four types of biologically relevant polymers are summarised in Table 2.1.

2.1.1 Proteins

Proteins are formed by a condensation reaction of the α-amino group of one amino acid (or the imino group of proline) with the α-carboxyl group of another. Concomitantly, a water molecule is lost and a **peptide bond** is formed. The peptide bond possesses partial double-bond character and thus restricts rotation around the C–N bond. The progressive condensation of many amino acids gives rise to an unbranched polypeptide chain. Since biosynthesis of proteins proceeds from the N- to the C-terminal amino acid, the N-terminal amino acid is taken as the beginning of the chain and the C-terminal amino acid as the end. Generally, chains of amino acids containing fewer than 50 residues are referred to as peptides, and those with more than 50 are referred to as proteins. Most proteins contain many hundreds of amino acids; ribonuclease, for example, is considered an extremely small protein with only 103 amino-acid residues. Many biologically active peptides contain 20 or fewer amino

Table 2.1 **The four types of biologically relevant polymers**

Polymer	Monomers	Monomer details
Ribonucleic acid (RNA)	4 Bases	Adenine (A), uracil (U), cytosine (C), guanine (G)
Deoxyribonucleic acid (DNA)	4 Bases	Adenine (A), thymine (T), cytosine (C), guanine (G)
Protein	20 Amino acids	Ala, Cys, Asp, Glu, Phe, Gly, His, Ile, Lys, Leu, Met, Asn, Pro, Gln, Arg, Ser, Thr, Val, Trp, Tyr
Polysaccharide	Monosaccharides	Trioses, tetroses, pentoses, hexoses, heptoses

acids, such as the mammalian hormone oxytocin (nine amino-acid residues) which is clinically used to induce labour since it causes contraction of the uterus, and the neurotoxin apamin (18 amino-acid residues) found in bee venom.

2.2 THE IMPORTANCE OF STRUCTURE

Three main factors determine the three-dimensional structure of a macromolecule:

- allowable backbone angles
- interactions between the monomeric building blocks
- interactions between solvent and macromolecule.

The solvent interactions can be categorised into two types: binding of solvent molecules (solvation) and hydrophobic interactions. The latter arise from the inability or reluctance of parts of the macromolecule to interact with solvent molecules (hydrophobic effect), which, as a consequence, leads to exclusive solvent–solvent interactions. Phenomenologically, a collection of molecules that cannot be solvated will stick close to one another and minimise solvent contact.

The interactions between the building blocks of the macromolecule comprise negative interactions (by avoiding atomic clashes) and positive interactions, which may be provided by hydrogen bonds, electrostatic interactions and van der Waals interactions (see also Section 17.4). Hydrogen bonds are the main constitutive force of backbone interactions in proteins, but can also be observed between residue side chains. Electrostatic interactions in proteins occur between residue side chains that possess opposite charges (arginine, lysine and aspartate, glutamate). The van der Waals attraction is a weak short-range force that occurs between all molecules. It becomes particularly important if two molecules possess highly complementary shapes. Thus, van der Waals interactions are responsible for producing complementary surfaces in appropriate regions of macromolecules.

The allowable backbone angles provide a framework of geometric constraints and balance attractive interactions and geometrical/steric tension within the macromolecule.

2.2.1 Conformation

The structural arrangement of groups of atoms is called conformation (see also Section 17.1.1) and the **conformational isomerism** of molecules describes isomers that can be inter converted exclusively by rotations about formally single bonds. The rotation about a single bond is restricted by a rotational energy barrier that must be overcome; the individual isomers are called **rotamers**. Conformational isomerism arises when the energy barrier is small enough for the interconversion to occur. The angle describing the rotation around a bond between two atoms is called the **dihedral** (or torsion) angle.

The protein **backbone** (Figure 2.1) is geometrically defined by three dihedral angles, namely Φ (N-Cα), Ψ (Cα-C) and ω (C-N); the last angle can take only two values, $0°$ and $180°$, due to the partial double-bond character of the peptide bond. The conformation of the protein backbone is therefore determined by the Φ and Ψ torsion angles. The values these angles assume determine which type of secondary structure (see below) a certain consecutive region in the protein will adopt.

Many organic small molecules possess cyclic aromatic structures and are thus planar. However, there are also many non-aromatic cyclic structures, in particular **carbohydrates (sugars)**, which are of great importance in biochemical processes. Notably, many sugars exist in aqueous solution as both open-chain and cyclic forms. Figure 2.2 illustrates this using the example of *D*-glucose. It is obvious from the open-chain form, that there are a number of stereogenic centres in *D*-glucose. If the positions of the hydroxyl groups are changed to the opposite enantiomer for each stereogenic carbon, the resulting molecule is *L*-glucose. Upon ring closure of the **open-chain** form, atoms or groups bonded to tetrahedral ring carbons are either pointing up or down, as indicated by the use of dashed or solid wedges when drawing the two-dimensional structures. If two neighbouring hetero-atom substituents (e.g. hydroxyl groups) on the ring are both pointing in the same direction, this conformation is called *cis*; if they point in opposite directions, they are said to be in the *trans* conformation. The open-chain form is characterised by an aldehyde function which, upon ring closure, is converted to a **hemi-acetal** (comprising $R_1C(OR_2)(OH)H$; see the carbon

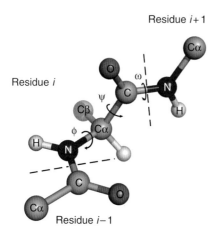

Residue *i*+1

Residue *i*

Residue *i*−1

Figure 2.1 Polypeptide chain comprising three amino acids (numbered *i*−1, *i*, *i*+1). The limits of a single residue (*i*) are indicated by the dashed lines. The torsion angles Φ, Ψ and ω describe the bond rotations around the N-Cα, Cα-C and C-N bonds, respectively.

surrounded by the blue and red highlighted oxygen atoms in Figure 2.2). Formation of the hemi-acetal can result in either α-D-glucose or β-D-glucose, depending on the position of the hydroxyl group in the anomeric position (highlighted in blue).

In five- and six-membered non-aromatic ring structures, the bond angles are close to tetrahedral (109.5°) giving rise to a ring shape that is not flat. The lowest energy (and thus most stable) ring conformation is the so-called chair conformation (Figure 2.3). An alternate chair conformation is obtained through a process called ring inversion, if one of the 'up' carbon atoms moves downwards and one of the 'down' carbon atoms upwards. Another ring conformation of six-membered rings is the so-called boat conformation, where the substituents at the 'bow' and the 'stern' are close enough to each other to cause van der Waals repulsion. Therefore, the boat conformation possesses higher energy and is hence less favoured than the chair conformation.

2.2.2 Folding and Structural Hierarchy

Macromolecular molecules, due to their physical extents, have the ability to bend and therefore bring various regions of their atomic arrangements into contact with each other. This process and the resulting three-dimensional structure will be driven

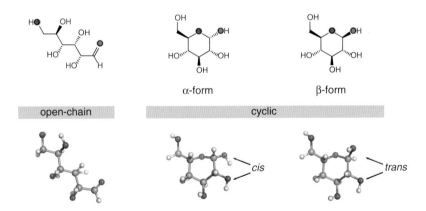

Figure 2.2 The open-chain and the cyclic forms of D-glucose shown as two-dimensional drawings as well as with their three-dimensional atomic structures. Ring closure of the open-chain form can result in either α-D-glucose or β-D-glucose, which differ in the position of the hydroxyl group in the anomeric position (highlighted in blue).

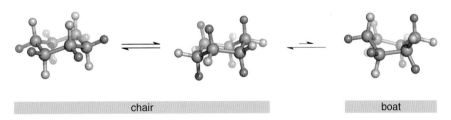

Figure 2.3 Different ring conformations observed with six-membered non-aromatic rings. In the left panel, the yellow substituents are in axial position and the violet substituents are in equatorial position. Upon ring inversion (centre panel), these positions change such that the yellow substituents become equatorial and the violet substituents axial. In the boat conformation (right panel), the axial/equatorial distinction is obsolete.

by possible attractive interactions as well as repulsions between different parts of the large molecule.

In order to distinguish the local three-dimensional structure of atoms bonded directly to each other from structures appearing at a more macroscopic scale, a structural hierarchy is required to distinguish the different types of structures when describing the shape and fabric of molecules (Figure 2.4). We will limit our discussion of structural hierarchy to proteins, where the concept of folding is a dominant aspect of characterisation; however, the concept can equally be applied to any type of macromolecule comprising a set of building blocks, such as carbohydrates as well as nucleic acids (Section 4.2) or synthetic polymers.

The primary structure of a protein is defined by the sequence of the amino acid residues, and is thus naturally dictated by the base sequence of the corresponding gene. From the primary structure, information about the amino acid composition (which of the possible 20 amino acids are actually present) as well as the content (the relative proportions of the amino acids present) is immediately available.

Secondary structure defines the localised folding of a polypeptide chain due to hydrogen bonding. It includes structures such as the α-helix and β-strand, known as regular (ordered) secondary structure elements, as opposed to the conformation of a random coil. Some proteins have up to 70% ordered secondary structure, but others have none. The super-secondary structure refers to a specific combination of particular secondary structure elements, such as β-α-β units or a helix-turn-helix motif. Such assemblies are often referred to as structural motifs.

The tertiary structure defines the overall folding of a polypeptide chain. It is stabilised by electrostatic attractions between oppositely charged ionic groups ($-NH_3^+$, $-COO^-$), by weak van der Waals forces, by hydrogen bonding, by hydrophobic interactions and, in some proteins, by disulfide ($-S-S-$) bridges formed by the oxidation of spatially adjacent sulfydryl groups ($-SH$) of cysteine residues. The three-dimensional folding of soluble proteins (i.e. those that are typically surrounded by an aqueous environment) is such that the interior comprises predominantly non-polar, hydrophobic amino-acid residues such as valine, leucine and phenylalanine. The clustering of non-polar amino acids, which may also include stacking of aromatic side chain rings) in the interior is a result of minimisation of contact of these residues with polar solvent – a phenomenon we introduced earlier as the hydrophobic effect. However, electrostatic interactions between adjacent side chain residues (colloquially termed salt bridges) are also regularly found in the interior of soluble proteins. The surface of soluble proteins is typically decorated with polar, ionised, hydrophilic residues, since they are compatible with the aqueous environment. But again, some proteins also have hydrophobic residues located on their outer surface that may give rise to hydrophobic clefts or patches.

In enzymes, the specific three-dimensional folding of the polypeptide chain(s) results in the juxtaposition of certain amino acid residues that constitute the active site or catalytic site. At the level of tertiary structure, the active site is often located in a cleft that is lined with hydrophobic amino-acid residues, but that contains some polar residues. The binding of the substrate at the catalytic site and the subsequent conversion of substrate to product involves different amino-acid residues (Section 23.4.2).

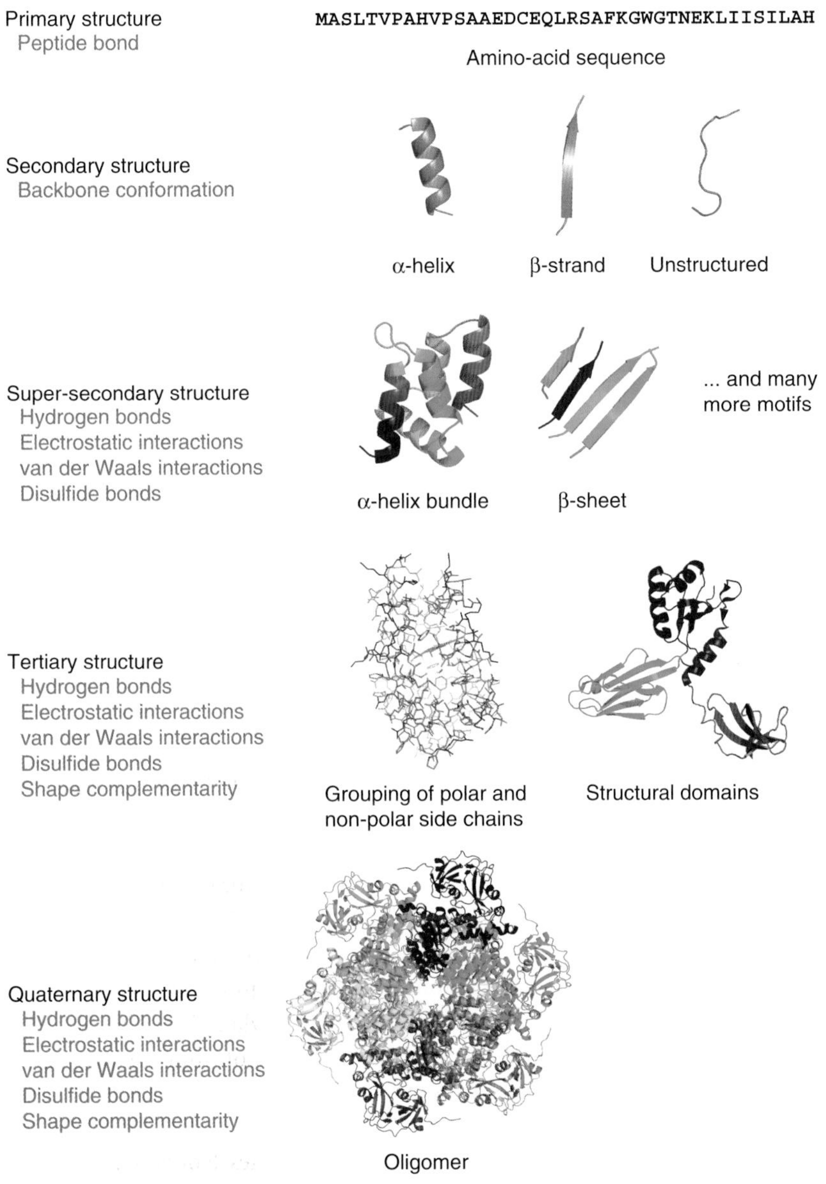

Primary structure
Peptide bond

MASLTVPAHVPSAAEDCEQLRSAFKGWGTNEKLIISILAH

Amino-acid sequence

Secondary structure
Backbone conformation

α-helix   β-strand   Unstructured

Super-secondary structure
Hydrogen bonds
Electrostatic interactions
van der Waals interactions
Disulfide bonds

... and many
more motifs

α-helix bundle   β-sheet

Tertiary structure
Hydrogen bonds
Electrostatic interactions
van der Waals interactions
Disulfide bonds
Shape complementarity

Grouping of polar and
non-polar side chains

Structural domains

Quaternary structure
Hydrogen bonds
Electrostatic interactions
van der Waals interactions
Disulfide bonds
Shape complementarity

Oligomer

Figure 2.4 The fold of a protein molecule can be described at different levels of hierarchy and is illustrated in this tier system. The primary structure arises from the amino-acid sequence of the protein and comprises the sequential arrangement of individual amino acids covalently linked through peptide bonds. Secondary structure refers, in its most basic form, to three different backbone conformations: α-helix, β-strand and unstructured coil. The super-secondary structure arises from the assembly of these three secondary-structure elements into particular motifs. The figure above illustrates this for a helix-bundle and a β-sheet, but there are many more motifs found in proteins, such as β-α-β (e.g. Rossmann fold), helix-loop-helix (e.g. EF-hand) and others. The tertiary structure comprises the folding of a protein chain into a more or less compact shape driven by the grouping of non-polar and polar side chains. In a soluble protein, the non-polar side chains (shown in brown) are located predominantly in the interior, whereas polar (orange) and charged (red, blue) side chains are located predominantly on the surface. This grouping may also lead to the formation of domains. At the level of quaternary structure, one distinguishes between monomeric and oligomeric (dimer, trimer, etc.; a hexamer is shown here) proteins. The formation of oligomers is driven by shape complementarity between the individual monomers and the interaction interfaces are most commonly stabilised by non-covalent interactions, but inter-molecular disulfide bonds are also possible.

Particular consecutive regions of a protein, often with an assembly of super-secondary structures, can form a compact three-dimensional entity that is stable on its own, i.e. can fold independently. Such independent folding units are called **domains** and can evolve function and exist independently of a particular protein. Domains may therefore be found as conserved parts in different proteins that have evolved by molecular evolution. Typically, individual domains comprise less than 200 amino-acid residues, but extreme examples ranging from 36 residues (E-selectin) to 692 residues (lipoxygenase-1) are also known. When comparing two domains that possess the same three-dimensional folds (structural homology), it is most commonly observed that they possess similar (not necessarily identical) amino acid sequences, since protein folds have been conserved throughout evolution, despite changes of the primary structure. This type of structural homology is based on an ancestral protein where subsequent evolution introduced individual differences. However, there are also cases where structural homology is shared between two domains that possess very different amino-acid sequences. In such cases of low sequence homology, the structural homology arose out of convergent evolution, where nature has deployed a particular structural fold on more than one occasion.

At the level of **quaternary** structure, the association of two or more macromolecules is described; with polypeptides, this term is therefore exclusively used for oligomeric proteins. Importantly, the interactions between the individual monomeric molecules is based only on electrostatic attractions, hydrogen bonding, van der Waals forces (all non-covalent) and occasionally disulfide bridges (covalent). An individual polypeptide chain in an oligomeric protein is referred to as a **subunit**. The subunits in a protein may be identical or different: for example, haemoglobin consists of two α- and two β-chains, and lactate dehydrogenase of four (virtually) identical chains.

A special mechanism for achieving formation of quaternary structure arises from the phenomenon of **domain swapping**. In the overall arrangement of the individual domains of an individual protein, one of the domains can be provided by a second monomer. The very domain of the first monomer, in turn, takes the appropriate place in the second monomer. Notably, this mechanism is not confined to entire domains, but may also be established by exchange of one or a few secondary-structure elements. This interchange of structural elements then leads to formation of a protein dimer.

Domain swapping is an important evolutionary mechanism for functional adaptation by oligomerisation, e.g. oligomeric enzymes that have their active site at subunit interfaces.

2.2.3 Classification of Protein Structure

With experimental determination of ever more three-dimensional structures of proteins, collectively assembled in the **Protein Data Bank** (PDB), it is possible to classify the various observed folds, compare the folds of two different proteins, and develop methods to predict the likely fold of a protein whose three-dimensional structure has not yet been determined.

Historically, two types of classification systems have been established, the SCOP and the CATH systems. Both systems use hierarchical descriptors ('domain', 'family', 'superfamily', 'fold', 'class', 'topology', 'architecture') for the various levels of structural features, but apply different principles, and hence result in different classifications (Table 2.2). Whereas the CATH Protein Structure Classification is a semi-automatic, hierarchical classification of protein domains, SCOP relies substantially on human expertise to decide whether certain proteins are evolutionarily related and therefore should be assigned to the same 'superfamily', or their similarity is a result of structural constraints and therefore should belong to the same 'fold'.

With availability of the first algorithm (DALI) that allowed automated comparison of protein structures and thus an analysis of their similarity, a database (FSSP: Families of Structurally Similar Proteins) was established that stored the results of structural similarity assessment of all structures available in the PDB at the time. FSSP was purely automatically generated (including regular automatic updates), but offered no classification, thus leaving it to the user to draw their own conclusion as to the significance of structural relationships based on the pairwise comparisons of individual protein structures.

Table 2.2 **Comparison of the SCOP and CATH protein structure classification systems**

SCOP			CATH		
Descriptor	**Explanation**	**Level of Hierarchy**	**Level of Hierarchy**	**Explanation**	**Descriptor**
Domain	Compact structural subunit	1	1	Compact structural subunit	Domain
Family	Sets of domains, grouped into families of homologues (sequences imply common evolutionary origin)	2			
Superfamily	Families that share common structure and function	3	2	Indicative of a demonstrable evolutionary relationship	Homologous Superfamily
Fold	Superfamilies that share a common folding motif	4	3	High structural similarity, but no evidence of homology	Topology
			4	Large-scale grouping of topologies that share particular structural features	Architecture
Seven classes	All α, all β, α/β, α+β, multi-domain, membrane proteins, small proteins	5	5	Mainly α, mainly β, α/β, few secondary structure elements	Four classes

More recent developments have focussed on novel algorithms to automatically assess structural similarities of proteins and gave rise to the popular SSM (Secondary Structure Match) algorithm, which has found its way into various crystallographic and modelling software, and is also available in protein structure search engines such as PDBeFold.

The accumulated knowledge of protein sequence (primary structure) data and experimental three-dimensional structures after decades of research now allows, in many cases, inference of the family membership of a protein just based on its primary structure. A popular database in this context is the Pfam protein families database where each protein family is represented by multiple sequence alignments and hidden Markov models (Section 16.4.4). In the absence of experimental three-dimensional structures, models can in many cases be generated by using methods of molecular modelling (Section 17.4).

2.2.4 Macroscopic Parameters Describing Macromolecules

The size of a macromolecule may be described by its molecular mass M, but also in terms of its spatial expansion. Ideally, in a sample of a purified biological macro-molecule, all molecules have the same value of M; such samples are called mono-disperse. If species with different molecular masses are present, the sample is called polydisperse, and the molecular mass obtained by experimental techniques is neces-sarily an average value.

The spatial expansion of a macromolecule is also important and typically described by the end-to-end distance and the radius of gyration. The end–to–end distance (L) is a useful parameter for molecules with regular, mostly linear shape (e.g. rods), such as DNA molecules. It is the average separation between the two ends of the molecule and depends on the molecular mass as well as the degree of flexibility. Entirely flexible molecules are called random coils and in such cases the end-to-end distance (now called h) can be calculated by random walk statistics:

$$h = \sqrt{L \times l} = \sqrt{N} \times l \qquad \text{(Eq 2.1)}$$

where l is the effective segment length of the macromolecule and L is the contour length (total length). So if the macromolecule consists of N monomers, the contour length is $L = N \times l$. The effective length l can be compared to the linear dimensions of the monomeric building block. If the values differ substantially, the macromolecule is rigid (very long and thin); however, if they are similar, the macromolecule is flex-ible. The end-to-end distance h can be determined from sedimentation experiments (Section 12.5) and l be calculated from Equation 2.1. For flexible macromolecules such as polyethylene, $l \approx 3$ Å (which agrees with the length of the methylene group). In contrast, for DNA, values of about 1000 Å are observed, which is much larger than the length of the phospho-sugar unit, thus indicating a high rigidity.

The radius of gyration (R_g) is a quantity that can be used to estimate the physical extent of a macromolecule. It is defined as the root mean square average of the dis-tances r_i of all mass elements m_i from the centre of mass of the entire macromolecule and calculated as

$$R_g^2 = \frac{\sum m_i \times r_i^2}{\sum m_i}$$

(Eq 2.2)

where i is the running index over all individual mass elements in the macromolecule. Whereas the direct physical significance of R_g is not obvious, its relevance stems from the fact that it can be determined experimentally by scattering techniques (static light scattering, small-angle scattering; Section 14.3.5). One may visualise R_g for a given macromolecule by a hypothetical hard sphere centred at the centre of gravity of the molecule (Figure 2.5). For select geometric shapes, the radius of gyration can be calculated (Table 2.3). In contrast, a similar quantity, the hydrodynamic radius R_{hydro}, can be interpreted more readily. R_{hydro} can be measured with dynamic light scattering (Section 14.3.5) and is based on the diffusional properties of the molecule in solution. Therefore, a hard sphere with radius R_{hydro} is indicative of the apparent size of the dynamic solvated particle. In contrast to the geometric radius of the macromolecule, which arises from rotation of the protein around its centre of gravity as a rigid body, the radius of gyration reflects the variable (not necessarily spherical) shape of the molecule. The hydrodynamic radius takes into account the shape, but also the solvation shell in solution.

Table 2.3 **Calculation of the radius of gyration for select geometrical shapes**

Shape	R_g
Sphere	$\sqrt{\dfrac{3}{5}} \times r$
Rod	$\sqrt{\dfrac{r^2}{2} + \dfrac{L^2}{12}}$
Random coil	$\sqrt{\dfrac{N}{6}} \times l$

Notes: r: radius; L: length of rod; N: number of units of length l.

Radius of gyration   Hydrodynamic radius   Geometric radius

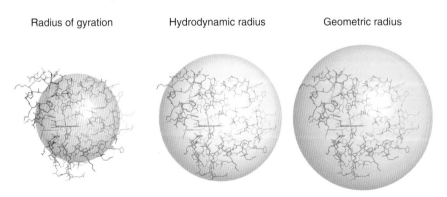

Figure 2.5 Comparison of the radius of gyration, the hydrodynamic radius and the geometric radius for the protein myoglobin. The protein is depicted as an all atom stick model and spheres with the different radii are superimposed; the centre of each sphere is at the centre of gravity of the protein. For explanation of the radius of gyration and hydrodynamic radius see main text. The geometric radius arises from rotation of the protein around its centre of gravity as a rigid body.

Example 2.1   **RADIUS OF GYRATION**

The experimentally determined radius of gyration can be used to assess the shape of macromolecules, especially when comparing the measured value with that expected for a sphere. Some examples are listed in Table A.

Table A

Molecule	M (g mol^{-1})	$R_{g,obs}$ (Å)	$R_{g,sphere}$ (Å)	$R_{g,obs}$ / $R_{g,sphere}$
Lysozyme	13 930	14.3	12.2	1.17
Myoglobin	16 890	16.0	13.2	1.21
tRNA	26 600	21.7	13.8	1.57
Myosin	493 000	468	45.2	10.4
DNA	4×10^6	1170	74	15.8

These data show that lysozyme and myoglobin may be approximated by roughly spherical shapes, since the deviation of their radii of gyration from that of a spherical molecule of the same mass is relatively small. Myosin and DNA, in contrast, show a large deviation from the radius of gyration expected for a sphere, suggesting that they adopt rod-like shapes.

2.3 PARAMETERS OF BIOLOGICAL SAMPLES

Water is the essential ingredient of life; therefore, biological samples are typically handled in aqueous environments. This is especially true for samples containing partially or fully purified molecules of biological interest.

When characterising and describing macromolecules, in the majority of cases it is implied that the macromolecule exists in its **native structure**. This may refer to the macromolecule:

● existing in its natural environment
● existing as an isolated molecule that retains its enzymatic activity
● adopting the same secondary structure as in its natural environment.

Importantly, for isolated macromolecules, the question of whether they indeed adopt their native fold and conformations cannot be experimentally answered with absolute certainty. If the protein in question is an enzyme, the native fold is typically assumed if the expected enzymatic activity can be observed.

The term **denatured structure** is defined as the opposite of native structure and typically means a form of the macromolecule that possesses less secondary structure than the native form. For proteins, this is typically a (near) random coil conformation, for DNA, whose native state is the double-stranded form, the denatured state denotes the single-stranded form.

2.3.1 Quantification

In order to quantify the amount of a particular component in a given sample, its concentration needs to be specified. Table 2.4 summarises the most commonly used measures of concentration for solution samples.

Historically, a few different preferences have developed, and sometimes different terms are used to describe a particular quantity (e.g. 'molarity' for 'molar concentration'). We therefore highlight a few general concepts:

- The molecular mass is the sum of the atomic masses of the constituent atoms.
- The term 'molecular mass' is preferred to the older term 'molecular weight'.
- When describing molar concentrations, some textbooks and journals tend to use the volume unit of litre and its subunits (Table 0.6) rather than cubic decimetres (Table 0.7).
- Atomic and molecular masses are both expressed in daltons (Da) or kilodaltons (kDa), where one dalton is an atomic mass unit equal to one-twelfth of the mass of one atom of the ^{12}C isotope. However, many biochemists prefer to use the term relative molecular mass (M_r). This is defined as the molecular mass of a substance relative to one-twelfth of the atomic mass of the ^{12}C isotope. M_r therefore has no units. Thus the relative molecular mass of sodium chloride is 23 (Na) plus 35.5 (Cl), which yields 58.5; one mole of NaCl thus has a mass of 58.5 grams. If 1 mol of NaCl (i.e. 58.5 g) was dissolved in 1 dm^3 (= 1 l) of water, the concentration of this solution would be 1 molar (1 M).

Table 2.4 **Different types of concentration measures for solution samples.**

Concentration measure	Definition	Units
Molar concentration	$c = \dfrac{n(\text{solute})}{V(\text{solution})}$, note: $\quad [\text{solute}] = \dfrac{c(\text{solute})}{1\text{ M}}$	$[c] = 1\dfrac{\text{mol}}{\text{dm}^3} = 1\text{ M}$
Mass concentration	$\rho^* = \dfrac{m(\text{solute})}{V(\text{solution})}$	$[\rho^*] = 1\dfrac{\text{mg}}{\text{cm}^3}$
Mass ratio	$w = \dfrac{m(\text{solute})}{m(\text{solution})}$	$[w] = 1 = 100\%$
	$w = \dfrac{m(\text{solute})}{V(\text{solution}) \times \rho(\text{solvent})}$	$[w] = 1 = 100\%$
Molality	$b = \dfrac{n(\text{solute})}{m(\text{solvent})}$	$[b] = 1\dfrac{\text{mol}}{\text{kg}}$
Mole fraction	$x = \dfrac{n(\text{component})}{n(\text{total})}$	$[x] = 1$

2.3.2 Properties of Aqueous Solutions

Electrolytes

Many molecules of biochemical importance are weak electrolytes in that they are acids or bases that are only partially ionised in aqueous solution. Examples include amino acids, peptides, proteins, nucleosides, nucleotides and nucleic acids. Of particular importance are the reagents used in the preparation of buffers such as acetic and phosphoric acids. The biochemical function of many of these molecules is dependent upon their state of ionisation at the prevailing environmental pH. The catalytic sites of enzymes, for example, contain functional carboxyl and amino groups, from the side chains of constituent amino acids in the protein chain (Section 23.4.2), which need to be in a specific ionised state to enable the catalytic function of the enzyme to be realised. Before the ionisation of these compounds is discussed in detail, it is necessary to appreciate the importance of the ionisation of water.

Ionisation of weak acids and bases

One of the most important weak electrolytes is water, since it ionises to a small extent to give hydrogen ions and hydroxyl ions. In fact there is no such species as a free hydrogen ion in aqueous solution as it reacts with water to give a hydronium ion (H_3O^+):

$$H_2O \rightleftharpoons H^+ + OH^-$$

$$H_2O + H^+ \rightleftharpoons H_3O^+ \tag{Eq 2.3}$$

However, for the calculation of the equilibrium constant (K_{eq}) for the ionisation of water only the first reaction in Equation 2.3 is considered. K_{eq} has a value of 1.8×10^{-16} at 24 °C:

$$K_{eq} = \frac{[H^+] \times [OH^-]}{[H_2O]} = 1.8 \times 10^{-16} \tag{Eq 2.4}$$

The molar concentration of pure water is $[H_2O] = (55.6\ M)/(1\ M) = 55.6$. This can be incorporated into a new constant, K_w:

$$K_w = K_{eq} \times [H_2O] = 1.8 \times 10^{-16} \times \frac{55.6\ M}{1\ M} = 1.0 \times 10^{-14} \tag{Eq 2.5}$$

K_w is known as the autoprotolysis constant of water. Its numerical value of exactly 10^{-14} relates specifically to 24 °C. At 0 °C, K_w has a value of 1.14×10^{-15} and at 100 °C a value of 5.45×10^{-13}. The stoichiometry in Equation 2.5 shows that hydrogen ions and hydroxyl ions are produced in a 1 : 1 ratio; hence both of them must be present at a concentration of 1.0×10^{-7}. According to the Sörensen definition, the pH is equal to the negative logarithm of the hydrogen ion concentration; it follows that the pH of pure water is 7.0. This is the definition of pH neutrality.

Amino acids, the building blocks of peptides and proteins, possess carboxyl (-COOH) as well as amine (-NH$_2$) groups, which can donate or accept hydrogen ions, respectively. These groups confer the features of weak acids and bases and therefore amino acids share the principles of ionisation with those of pH buffers, which also constitute either weak acids or bases.

The tendency of a weak acid, generically represented as HA, to ionise is expressed by the equilibrium reaction:

HA $\rightleftharpoons$ $H^+ + A^-$ (Eq 2.6)

weak acid   conjugate base

The ionisation of a weak acid results in the release of a hydrogen ion and the conjugate base of the acid, both of which are ionic in nature. This reversible reaction can be represented by an equilibrium constant, K_a, known as the acid dissociation constant (Equation 2.7). Numerically, it is very small.

$$K_a = \frac{[H^+] \times [A^-]}{[HA]} = \frac{[H^+] \times [\text{conjugate base}]}{[\text{weak acid}]}$$ (Eq 2.7)

Similarly, amines as weak bases can exist in ionised and unionised forms and the concomitant ionisation process is represented by an equilibrium constant, K_b (Equation 2.9). In this case, the non-ionised form of the base abstracts a hydrogen ion from water to produce the conjugate acid that is ionised:

$RNH_2 + H_2O$ $\rightleftharpoons$ $OH^- + RNH_3^+$ (Eq 2.8)

weak base   conjugate acid

$$K_b = \frac{[OH^-] \times [RNH_3^+]}{[RNH_2]} = \frac{[OH^-] \times [\text{conjugate acid}]}{[\text{weak base}]}$$ (Eq 2.9)

A specific and simple example of the ionisation of a weak acid is that of acetic acid, $H_3C\text{-}COOH$:

$H_3C\text{-}COOH$ $\rightleftharpoons$ $H^+ + H_3C\text{-}COO^-$ (Eq 2.10)

acetic acid   acetate anion

Acetic acid and its conjugate base, the acetate anion, are known as a conjugate acid–base pair. The acid dissociation constant can be written in the following way:

$$K_a = \frac{[H^+] \times [H_3C\text{-}COO^-]}{[H_3C\text{-}COOH]} = \frac{[H^+] \times [\text{conjugate base}]}{[\text{weak acid}]}$$ (Eq 2.11)

For acetic acid, K_a has a value of 1.75×10^{-5}. In practice, however, it is far more common to express the K_a value in terms of its negative logarithm (i.e. $-\lg K_a$) referred to as pK_a. Thus, in this case, pK_a is equal to 4.75. It can be seen from Equation 2.7 that pK_a is numerically equal to the pH at which 50% of the acid is protonated (unionised) and 50% is deprotonated (ionised).

It is possible to write an expression for K_b of the acetate anion as a conjugate base:

$H_3C\text{-}COO^- + H_2O$ $\rightleftharpoons$ $OH^- + H_3C\text{-}COOH$ (Eq 2.12)

acetate anion   acetic acid

$$K_b = \frac{[OH^-] \times [H_3C\text{-}COOH]}{[H_3C\text{-}COO^-]} = \frac{[OH^-] \times [\text{weak acid}]}{[\text{conjugate base}]} \tag{Eq 2.13}$$

For acetic acid, K_b has a value of 1.77×10^{-10}, hence its pK_b (i.e. $-\lg K_b$) is 9.25. Importantly, multiplying K_a and K_b yields the autoprotolysis constant of water, as introduced in Equation 2.5:

$$K_a \times K_b = [H^+] \times [OH^-] = K_w = 1.0 \times 10^{-14} \text{ at } 24\ °C$$

and hence

$$pK_a + pK_b = pK_w = 14 \tag{Eq 2.14}$$

This relationship holds for all acid–base pairs and enables one pK_a value to be calculated from knowledge of the other. Biologically important examples of conjugate acid–base pairs are lactic acid/lactate, pyruvic acid/pyruvate, carbonic acid/bicarbonate and ammonium/ammonia.

In the case of the ionisation of weak bases, the most common convention is to quote the K_a or the pK_a of the conjugate acid rather than the K_b or pK_b of the weak base itself. Examples of the pK_a values of some weak acids and bases are given in Table 0.8. Remember that the smaller the numerical value of pK_a, the stronger the acid (more ionised) and the weaker its conjugate base. Weak acids will be predominantly unionised at low pH values and ionised at high values. In contrast, weak bases will be predominantly ionised at low pH values and unionised at high values. This sensitivity to pH of the state of ionisation of weak electrolytes is important both physiologically and in in vitro biochemical studies where analytical techniques such as electrophoresis and ion-exchange chromatography are employed.

Ionisation of Polyprotic Weak Acids and Bases

Polyprotic weak acids and bases are capable of donating or accepting more than one hydrogen ion. Each ionisation stage can be represented by a K_a value using the convention that K_{a1} refers to the acid with the most ionisable hydrogen atoms and K_{an} the acid with the least number of ionisable hydrogen atoms. One of the most important biochemical examples is phosphoric acid, H_3PO_4, as it is widely used as the basis of a buffer in the pH region of 6.7:

$$H_3PO_4 \rightleftharpoons H^+ + H_2PO_4^- \qquad pK_{a1} = 2.0$$
$$H_2PO_4^- \rightleftharpoons H^+ + HPO_4^{2-} \qquad pK_{a2} = 6.7 \tag{Eq 2.15}$$
$$HPO_4^{2-} \rightleftharpoons H^+ + PO_4^{3-} \qquad pK_{a3} = 12.3$$

A special form of polyprotic acid/base behaviour arises if acid and base groups are present simultaneously in one molecule, as is the case in amino acids (see Section 2.6 and Figure 2.7).

Example 2.2 **CALCULATION OF THE EXTENT OF IONISATION OF A WEAK ELECTROLYTE AND PH OF THE SOLUTION**

Question Calculate the pH of a 0.01 M solution of acetic acid and its fractional ionisation, given that its K_a is 1.75×10^{-5}.

Answer To calculate the pH we remember the definition of the acid dissociation constant K_a

$$K_a = \frac{[\text{acetate}] \times [\text{H}^+]}{[\text{acetic acid}]} = 1.75 \times 10^{-5}$$

and attempt to determine the numerical value of the equilibrium concentration of hydrogen ions, which will allow us to compute the pH.

Acetate and hydrogen ions are produced in equal quantities. If we set $x = [\text{H}^+]$, then we know that $[\text{acetate}] = x$. The equilibrium concentration of non-dissociated acetic acid is then $0.01 - x$. The above equation therefore becomes:

$$\frac{x \times x}{0.01 - x} = 1.75 \times 10^{-5}$$

$$x^2 = 1.75 \times 10^{-7} - 1.75 \times 10^{-5} \times x$$

This equation can be solved either by use of the quadratic formula or, more easily, by neglecting the term $x \times 1.75 \times 10^{-5}$, since it is so small. Adopting the latter alternative gives:

$$x^2 = 1.75 \times 10^{-7}, \text{ and thus}$$

$$x = 4.18 \times 10^{-4}, \text{hence}$$

$$[\text{H}^+] = 4.18 \times 10^{-4} \text{ and pH} = 3.4$$

The fractional ionisation (α) of acetic acid is defined as the fraction of the acetic acid that is in the form of acetate and is therefore given by the equation:

$$\alpha = \frac{[\text{acetate}]}{[\text{acetate}] + [\text{acetic acid}]}$$

$$\alpha = \frac{4.18 \times 10^{-4}}{4.18 \times 10^{-4} + (0.01 - 4.18 \times 10^{-4})}$$

$$\alpha = \frac{4.18 \times 10^{-4}}{0.01}$$

$$\alpha = 4.18 \times 10^{-2} = 4.2\%$$

Thus, the majority of the acetic acid is present in the non-ionised form. If the pH is increased above 3.4, the proportion of acetate present will increase in accordance with the Henderson–Hasselbalch equation.

Ionic Strength

Solutions of biochemical samples are not only affected by the pH, but also by the ionic strength. The ionic strength describes the concentration of all ions in solution and is defined as

$$I = \frac{1}{2}\sum_{i=1}^{N} c_i \times z_i^2 \qquad \text{(Eq 2.16)}$$

where c_i is the molar concentration of ion species i and z_i is the charge number of ion species i. The sum is over all N ion species in solution. The ionic strength takes into account that salts (electrolytes) dissociate into individual ions when dissolved and thus increase the effective number of particles in the solution. For a 1:1 electrolyte (e.g. NaCl), the ionic strength is equal to the molar concentration, but for salts with multivalent ions (e.g. $MgSO_4$) the ionic strength is four times higher.

Since proteins possess charged residue side chains, the presence of positively and negatively charged counterions are important to shield those charged side chains from each other. The charged side chains therefore can carry a so-called counterion atmosphere, the extent of which depends on the amount of available ions in solution, and hence on the ionic strength. Through these effects, the salt concentration in a buffer can affect the solubility of proteins (and other charged biological molecules). When increasing the salt concentration of a solution, two effects are possible:

- salting in: interactions between charged residue side chains are shielded and aggregation of protein molecules is prevented, so the solubility is increased;
- salting out: the large number of ions require large numbers of water molecules to satisfy their hydration spheres and water molecules that would otherwise hydrate the protein are removed, leading to precipitation of the protein.

2.4 MEASUREMENT OF THE pH: THE pH ELECTRODE

The pH electrode is an example of an ion-selective electrode that responds to one specific ion in solution, in this case the hydrogen ion. The electrode consists of a thin glass porous membrane, sealed at the end of a hard glass tube containing 0.1 M hydrochloric acid (see Figure 2.6). A silver wire coated with silver chloride is immersed into the HCl solution and acts as an internal reference that generates a constant potential. The porous membrane is typically 0.1 mm thick, the outer and inner 10 nm consisting of a hydrated gel layer containing exchange binding sites for hydrogen or sodium ions. On the inside of the membrane, the exchange sites are predominantly occupied by hydrogen ions from the hydrochloric acid, whilst on the outside, the exchange sites are occupied by sodium and hydrogen ions. The bulk of the membrane is a dry silicate layer in which all exchange sites are occupied by sodium ions. Most of the coordinated ions in both hydrated layers are free to diffuse into the surrounding solution. Hydrogen ions in the test solution can diffuse in the opposite direction and replace bound sodium ions in a process called ion-exchange equilibrium. Any other types of cation present in the test solution are unable to bind to the exchange sites,

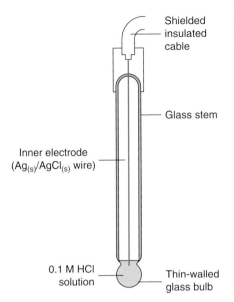

Shielded
insulated
cable

Glass stem

Inner electrode
$(Ag_{(s)}/AgCl_{(s)}$ wire$)$

0.1 M HCl
solution

Thin-walled
glass bulb

Figure 2.6 The most common form of the pH electrode
is the glass electrode.

which ensures the high specificity of the electrode. Importantly, hydrogen ions can not diffuse across the dry glass layer, but sodium ions can. Effectively, this establishes a membrane system consisting of two hydrated layers separated by a sodium ion transport system; the two hydrated layers possess different hydrogen ion activities (note: activity describes an effective concentration, based on the chemical potential of a species).

The principle of operation of the pH electrode rests on the fact that a gradient of hydrogen ion activity across the membrane will generate an electric potential E, the size of which is determined by the hydrogen ion gradient across the membrane. Moreover, since the hydrogen ion concentration (and thus activity) on the inside is constant, due to the use of 0.1 M hydrochloric acid, the observed potential is directly dependent upon the hydrogen ion concentration of the test solution. In practice, however, a small junction or asymmetry potential (E^*) is also created as a result of linking the glass electrode to a reference electrode. The observed potential across the membrane is therefore given by the equation:

$$E = E^* + (0.0592 \text{ V}) \times \text{pH} \qquad \text{(Eq 2.17)}$$

For each tenfold change in the hydrogen ion concentration across the membrane (equivalent to a pH change of 1 in the test solution) there will be a potential difference change of 59.2 mV across the membrane. The sensitivity of pH measurements is influenced by the prevailing absolute temperature. Since the precise composition of the porous membrane varies with time, so too does the asymmetry potential E^*. This contributes to the need for frequent recalibration of the electrode, which is commonly achieved by using two standard buffers of known pH.

2.5 BUFFERS

2.5.1 pH Buffers

A buffer solution is one that resists a change in pH on the addition of either acid or base. They are of enormous importance in practical biochemical work, as many biochemical molecules are weak electrolytes whose ionic status varies with pH. In order to provide constant environmental conditions, there is a need to stabilise the ionic status during the course of a practical experiment. In practice, a buffer solution consists of an aqueous mixture of a weak acid and its conjugate base. The conjugate base component will neutralise any hydrogen ions generated during an experiment whilst the unionised acid will neutralise any base generated (see Figure 2.7). The Henderson–Hasselbalch equation is of central importance in the preparation of buffer solutions, since it allows estimation of the pH of a buffer solution, based on the acid dissociation constant K_a, and the molar concentrations of the components of the buffer pair:

$$pH = pK_a + lg \frac{[conjugate\ base]}{[weak\ acid]} \qquad\qquad (Eq\ 2.18)$$

For a buffer based on the conjugate acid of a weak base:

$$pH = pK_a + lg \frac{[weak\ base]}{[conjugate\ acid]} \qquad\qquad (Eq\ 2.19)$$

Table 0.8 lists some weak acids and bases commonly used in the preparation of buffer solutions. Phosphate, HEPES and PIPES are commonly used because their optimum pH is close to 7.4. The buffer action and pH of blood is illustrated in Example 2.3 and the preparation of a phosphate buffer is given in Example 2.4.

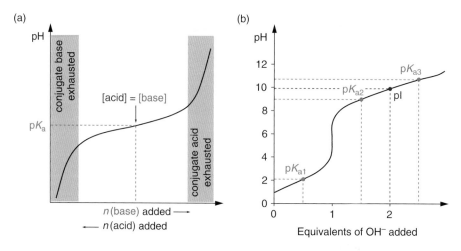

Figure 2.7 (a) Buffer titration. The buffer solution has the strongest ability to protect against pH changes when pH = pK_a, because the conjugate acid and conjugate base molecules are available in equal amounts and can thus neutralise added base or acid. If the pH of the solution is too far from the buffer pK_a, there are too few remaining conjugate acid/base molecules to neutralise added base/acid. (b) Titration curve for the amino acid lysine as an example of a polyprotic acid/base. See also Example 2.5.

| Example 2.3 | **BUFFER ACTION AND THE pH OF BLOOD** |

The normal pH of blood is 7.4 and is maintained at this value by buffers, in particular by the buffering action of HCO_3^- resulting from gaseous CO_2 dissolved in blood and the resulting ionisation of carbonic acid:

$$CO_2 + H_2O \rightleftharpoons H_2CO_3$$

$$H_2CO_3 \rightleftharpoons H^+ + HCO_3^-$$

It is possible to calculate an overall equilibrium constant (K_{eq}) for these two consecutive reactions and to incorporate the concentration of water (55.6 M) into the value:

$$K_{eq} = \frac{[H^+] \times [HCO_3^-]}{[CO_2]} = 7.95 \times 10^{-7}, \text{ and therefore } pK_{eq} = 6.1$$

Rearranging this equation yields:

$$[H^+] = K_{eq} \times \frac{[CO_2]}{[HCO_3^-]} \text{ and thus}$$

$$-\lg[H^+] = -\lg K_{eq} - \lg \frac{[CO_2]}{[HCO_3^-]}, \text{ which resolves to}$$

$$pH = pK_{eq} + \lg \frac{[HCO_3^-]}{[CO_2]}$$

We recognise this as the Henderson–Hasselbalch equation for a buffer based on a weak acid (carbonic acid, i.e. carbon dioxide) and its conjugate base (bicarbonate).

When the pH of blood falls due to the metabolic production of H^+, these equilibria shift in favour of increased production of H_2CO_3, which in turn results in higher levels of CO_2 that is then expired. In contrast, when the pH of blood rises, more HCO_3^- is produced, and breathing is adjusted to retain more CO_2 in the blood, thus maintaining blood pH.

Some disease states may change this pH causing either **acidosis** (e.g. obstructive lung disease) or **alkalosis** (e.g. hyperventilation), which may cause serious problems and, in extreme cases, death. Clinical biochemists routinely monitor a patient's acid–base balance in the blood, in particular the ratio of HCO_3^- and CO_2. Reference ranges for these at pH 7.4 are $c(HCO_3^-)$ = 22.0–26.0 mM and pCO_2 = 4.6–6.9, which corresponds to $c(CO_2)$ of the order of 1.20 mM.

Question   A patient suffering from acidosis had a blood pH of 7.15 and a CO_2 concentration of 1.15 mM. What was the patient's bicarbonate (HCO_3^-) concentration and what are the implications of this value to the buffer capacity of the blood?

Answer   Applying the above equation we get:

$$pH = pK_{eq} + \lg \frac{[HCO_3^-]}{[CO_2]}$$

$$7.15 = 6.1 + \lg \frac{[HCO_3^-]}{1.15 \times 10^{-3}}$$

$$10^{1.05} = \frac{[HCO_3^-]}{1.15 \times 10^{-3}}$$

and therefore $[HCO_3^-] = 12.9 \times 10^{-3}$, i.e. the concentration of bicarbonate is 12.9 mM.

This indicates that the bicarbonate concentration in the patient's blood had decreased by 11.1 mM, i.e. 46%, thereby severely reducing the buffer capacity of the blood so that any further significant production of acid would have serious implications for the patient.

2.5.2 pH Buffer Capacity

It can be seen from the Henderson–Hasselbalch equations that when the concentrations (or more strictly the activities) of the weak acid and base are equal, their ratio is 1 and their logarithm 0, so that $pH = pK_a$. The ability of a buffer solution to resist a change in pH on the addition of strong acid or alkali is expressed by its buffer capacity (β). This is defined as the molar amount of acid or base required to change the pH by one unit, i.e.:

$$\beta = \frac{dn(\text{base})}{d(\text{pH})} = -\frac{dn(\text{acid})}{d(\text{pH})} \qquad \text{(Eq 2.20)}$$

where dn(base) and dn(acid) are the amounts of base and acid, respectively, and d(pH) is the resulting change in pH. In practice, β is largest within the pH range of $pH = pK_a \pm 1$.

2.5.3 Selection of a Biochemical Buffer

When selecting a buffer for a particular experimental study, several factors should be taken into account:

- Select the one with a pK_a as near as possible to the required experimental pH and within the range $pK_a \pm 1$.

Example 2.4 **PREPARATION OF A PHOSPHATE BUFFER**

Question How would you prepare 1 dm³ of 0.1 M phosphate buffer, pH 7.1, given that pK_{a2} for phosphoric acid is 6.7 and that the atomic masses for Na, P and O are 23, 31 and 16 Da, respectively?

Answer The buffering action will be based on the ionisation:

$$H_2PO_4^- \rightleftharpoons H^+ + HPO_4^{2-} \qquad pK_{a2} = 6.7$$

and will therefore involve the use of solid sodium dihydrogen phosphate (NaH_2PO_4) and disodium hydrogen phosphate (Na_2HPO_4).

Applying the appropriate Henderson–Hasselbalch equation (Equation 2.18) gives:

$$7.1 = 6.7 + \lg\frac{\left[HPO_4^{2-}\right]}{\left[H_2PO_4^-\right]}$$

$$0.4 = \lg\frac{\left[HPO_4^{2-}\right]}{\left[H_2PO_4^-\right]}$$

$$2.51 = \frac{\left[HPO_4^{2-}\right]}{\left[H_2PO_4^-\right]}$$

Since the total concentration of the two species needs to be 0.1 M, it follows that the concentration of HPO_4^{2-} must be 72 mM and the concentration of $H_2PO_4^-$ 28 mM. Their molecular masses are 142 and 120 Da, respectively, hence the mass of each required is 0.072 mol x 142 g mol⁻¹ = 10.2 g (Na_2HPO_4) and 0.028 mol x 120 g mol⁻¹ = 3.36 g (NaH_2PO_4). These masses would be dissolved in approximately 800 cm³ pure water, the pH measured and adjusted as necessary, and the volume finally made up to 1 dm³.

- Ensure that the selected buffer does not form insoluble complexes with any anions or cations essential to the reaction being studied (phosphate buffers tend to precipitate polyvalent cations, for example, and may be a metabolite or inhibitor of the reaction).
- If required, ensure that the proposed buffer has other desirable properties such as being non-toxic, able to penetrate membranes and not absorbing in the visible or ultraviolet region.
- Identify the buffering species. For example, in the case of an acetate buffer, the buffering species are $H_3C\text{-}COOH$ (acid) and $H_3C\text{-}COONa$ (base). In a phosphate buffer at pK_{a2}, the buffer species are NaH_2PO_4 (acid) and Na_2HPO_4 (base).

- Choose the correct acid/base for adjusting the pH. For example, an acetate buffer must not be made from H_3C-COONa titrated with HCl, as this would result in formation of the unwanted salt NaCl. Instead, the buffer should be made from either H_3C-COONa titrated with acetic acid, or alternatively, from acetic acid titrated with NaOH, so that no non-buffer ions are introduced into the system.
- Select an appropriate concentration of buffer to have adequate buffer capacity for the particular experiment. Buffers are most commonly used in the range 0.02–0.1 M.
- It is possible, but not required, to make up separate solutions of the two buffer constituents. Typically, buffers are made up by weighing out one component, dissolving in a volume short of the final volume, titrating to the right pH, and making up to volume.

2.6 IONISATION PROPERTIES OF AMINO ACIDS

We have previously mentioned the behaviour of polyprotic acids and bases. Amino acids are molecules where an acid and a base function is present at the same time (so-called amphoteric substances). Their ionisation behaviour therefore includes cationic species at low pH and anionic species at high pH. Inevitably, in the course of the pH change, a particular intermediate species exists where the molecule carries no net charge, but is still ionised; this is called a zwitterion. It has been shown that, in the crystalline state and in solution in water, amino acids exist predominantly as this zwitterionic form. This confers upon them physical properties characteristic of ionic compounds, i.e. high melting and boiling points, water solubility and low solubility in organic solvents such as ether and chloroform. The pH at which the zwitterion predominates in aqueous solution is referred to as the isoionic point, because it is the pH at which the number of negative charges on the molecule produced by ionisation of the carboxyl group is equal to the number of positive charges acquired by proton acceptance by the amino group (see Figure 2.7). In the case of amino acids this is equal to the isoelectric point (pI), since the molecule carries no net charge and is therefore electrophoretically immobile. The numerical value of the pI for a given amino acid is related to its acid strength (pK_a values) by the equation:

$$pI = \frac{pK_{a1} + pK_{a2}}{2} \tag{Eq 2.21}$$

where pK_{a1} and pK_{a2} are equal to the negative logarithm of the acid dissociation constants, K_{a1} and K_{a2} (Section 2.3.2).

In the case of glycine, pK_{a1} and pK_{a2} are 2.3 and 9.6, respectively, so that the isoelectric point is 6.0. At pH values below this, the cation and zwitterion will coexist in equilibrium in a ratio determined by the Henderson–Hasselbalch equation (Section 2.5.1), whereas at higher pH values the zwitterion and anion will coexist in equilibrium. For acidic amino acids such as aspartic acid, the ionisation pattern is different owing to the presence of a second carboxyl group (Example 2.5). Basic amino acids, such as lysine, possess yet another ionisation pattern (Example 2.5).

Example 2.5 **THE IONISATION OF ASPARTIC ACID AND LYSINE**

Question What are the isoelectric (isoionic) points of the amino acids aspartic acid and lysine?

Answer

| Cation | Zwitterion pH 3.0 (isoionic point) | Anion | Anion |

For aspartic acid, the zwitterion will predominate in aqueous solution at a pH determined by pK_{a1} and pK_{a2}. According to Equation 2.21, the isoelectric point is the mean of pK_{a1} and pK_{a2} and thus pI = 3.0.

| Cation | Cation | Zwitterion pH 9.8 (isoionic point) | Anion |

In the case of lysine, the ionisation pattern is different, and the isoelectric point is the mean of pK_{a2} and pK_{a3}, therefore pI = 9.8.

Beside amino or carboxyl groups, an amino acid side chain (R) may contain a different chemical group that is also capable of ionising at a characteristic pH. Such groups include a phenolic group (tyrosine), guanidino group (arginine), imidazolyl group (histidine) and sulfydryl group (cysteine) (Table 0.10). It is clear that the state of ionisation of these groups of amino acids (acidic, basic, neutral) will be grossly different at a particular pH. Moreover, even within a given group there will be minor differences due to the precise nature and environment of the R group. These differences are exploited in the electrophoretic and ion-exchange chromatographic separation of mixtures of amino acids such as those present in a protein hydrolysate (Section 8.4.2).

Notably, since proteins are formed by the condensation of the α-amino group of one amino acid with the α-carboxyl of the adjacent amino acid (Section 2.1.1), with the exception of the two terminal amino acids, all amino and carboxyl groups are involved in peptide bonds and are no longer ionisable in the protein. Therefore, only appropriate functional groups on amino acid side chains groups are free to ionise and, as introduced above, there will be many of these. The number of surface-accessible positively and negatively charged groups in a protein molecule influence aspects of its physical behaviour, such as solubility and electrophoretic mobility (Section 6.5).

The isoionic point of a protein and its isoelectric point, unlike that of an amino acid, are generally not identical. This is because, by definition, the isoionic point is the pH at which the protein molecule possesses an equal number of positive and negative groups.

In contrast, the isoelectric point is the pH at which the protein is electrophoretically immobile. In order to determine electrophoretic mobility experimentally, the protein must be dissolved in a buffered medium containing anions and cations, of low relative molecular mass, that are capable of binding to the multi-ionised protein. Hence the observed balance of charges at the isoelectric point could be due in part to there being more bound mobile anions (or cations) than bound cations (anions) at this pH. This could mask an imbalance of charges on the actual protein.

In practice, protein molecules are always studied in buffered solutions, so it is the isoelectric point that is important. It is the pH at which, for example, the protein has minimum solubility, since it is the point at which there is the greatest opportunity for attraction between oppositely charged groups of neighbouring molecules and con-sequent aggregation and easy precipitation. The isoelectric point of a protein with known primary structure (amino-acid sequence) can be estimated by using the known tabulated pK_a values of the individual amino acids (Table 0.10). Different software in the form of stand-alone programs as well as web services (e.g. EMBOSS) have implemented this feature, although it is important to remember that the calculated pI of a peptide sequence remains an estimate, since the true pI depends on a number of factors discussed above, including the protein fold and the buffer environment.

2.7 QUANTITATIVE BIOCHEMICAL MEASUREMENTS

All analytical methods can be characterised by a number of performance indicators that define how the selected method performs under specified conditions. Knowledge of these performance indicators allows the analyst to decide whether or not the method is acceptable for a particular application. Performance indicators are estab-lished by the use of well-characterised test and reference analyte samples. The order in which they are evaluated will depend on the immediate analytical priorities, but initially the three most important are specificity, detection limit and analytical range. Once a method is in routine use, the question of assuring the quality of analytical data by the implementation of quality assessment procedures comes into play. The major performance indicators are:

- Precision (also called imprecision, variation or variability). This is a measure of the reproducibility of a particular set of analytical measurements on the same sample of test analyte. If the replicated values agree closely with each other, the measurements are said to be of high precision. Clearly, the aim is to develop or devise methods that have as high a precision as possible, but it is common to observe variation over the analytical range (see below) and over periods of time (especially between batches). A result that is of high precision may nevertheless be a poor estimate of the 'true' value (i.e. of low accuracy or high bias), because of the presence of unidentified errors. Methods for assessment of the precision of a dataset are discussed in Section 19.3.3.

- Accuracy (also called trueness, bias or inaccuracy). This describes the difference between the mean of a set of analytical measurements on the same sample of test analyte and the 'true' value for the test sample. As pointed out above, the 'true' value is normally unknown, except in the case of standard measurements. In other cases, accuracy has to be assessed indirectly by use of an internationally agreed reference method and/or by the use of external quality assessment schemes and/or by the use of population statistics (see Section 19.3.4).
- Detection limit (also called sensitivity). This is the smallest concentration of the test analyte that can be distinguished from zero with a defined degree of confidence. Concentrations below this limit should be reported as 'less than the detection limit'. All methods have their individual detection limits for a given analyte and this may be one of the factors that influence the choice of a specific analytical method for a given study. In clinical biochemical measurements, sensitivity is often defined as the ability of the method to detect the analyte without giving false negatives (see Section 10.1.2).
- Analytical range. This is the range of concentrations of the test analyte that can be measured reproducibly. The lower end of this range is the detection limit. In most cases the analytical range is defined by an appropriate calibration curve. As mentioned above, the precision of the method may vary across the analytical range.
- Analytical specificity (also called selectivity). This is a measure of the extent to which other substances that may be present in the sample may interfere with the analysis and therefore lead to a falsely high or low value. A simple example is the ability of a method to measure glucose in the presence of other hexoses such as mannose and galactose. In clinical biochemical measurements, selectivity is an index of the ability of the method to give a consistent negative result for known negatives (see Section 10.1.2).
- Analytical sensitivity. This is a measure of the change in response of the method to a defined change in the quantity of analyte present. In many cases analytical sensitivity is expressed as the slope of a linear calibration curve.
- Robustness. This is a measure of the ability of the method to give a consistent result, in spite of small changes in experimental parameters such as pH, temperature and amount of reagents added. For routine analysis, the robustness of a method is an important practical consideration.

2.8 EXPERIMENT DESIGN AND RESEARCH CONDUCT

Since experiments are the key element of scientific research, their careful design, execution and data analysis are of utmost importance in addressing specific questions or hypotheses. Well-informed experimental design involves a discrete number of compulsory stages listed in Table 2.5.

The results of well-designed and analysed studies are finally published in the scientific literature after being subjected to independent peer review. One of the major

Table 2.5 Sequence of tasks for proper experiment design

Subject	Identify the subject for experimental investigation.
Background	Critically appraise the current state of knowledge (the 'literature') of the chosen subject. Note the strengths and weaknesses of the methodologies previously applied and the new hypotheses that emerged from the studies.
Hypothesis	Formulate the question or hypothesis to be addressed by the planned experiment.
System	Carefully select the biological system (species, in vivo or in vitro) to be used for the study.
Variables	Identify the variable that is to be studied; other variables will need to be controlled so that the selected variable is the only factor that will determine the experimental outcome.
Protocol	Design the experimental protocol, including the statistical analysis of the results, careful evaluation of the materials and apparatus to be used and consider the safety aspects of the study.
Execution	Execute the experiment, including appropriate calibrations and controls, with a carefully written record of the outcomes.
Replication	Repeat the experiment as necessary for unambiguous analysis of the outcomes.
Evaluation	Assess the outcomes and apply the appropriate statistical tests to quantitative data.
Conclusion	Formulate the main conclusions that can be drawn from the results.
Extrapolation	Formulate new hypotheses and future experiments that emerge from the study.

challenges faced by life scientists is keeping abreast of current advances in the literature. On the one hand, the advent of the web has made access to the literature easier than it was decades ago; on the other, the sheer number of published studies and the fact that contemporary life science research operates in an inter-disciplinary area into which several different traditional disciplines feed can still pose a significant challenge.

2.8.1 Health and Safety

General
Safe working practices in the laboratory are of utmost importance for the safety of oneself, colleagues and the environment. Unfortunately, in many instances, the effective care applied can fall short of what is required, and health and safety matters have become a heavily bureaucracy-driven exercise. It is thus important to remember that any particular task in the laboratory should be critically assessed as to what the potential consequences of that task are and whether one possesses the required training and confidence to execute the task.

Before commencing experimental work, the following aspects should always be assessed:

- What potential risks does the task involve, have those risks been outlined (risk assessment) and what are the procedures to minimise those risks?

- What type of personal protective equipment is required?
- Is the equipment safe to use, in terms of physical state as well as training?
- What are the correct waste disposal procedures?
- Are there any adverse medical conditions of staff to be considered?
- What are the emergency and evacuation procedures?
- What are the standard operating procedures, rules in the safety manuals, etc?

General good housekeeping, clean and tidy work benches, as well as labelling are a mandatory requirement of good laboratory practice.

Working with Microbiological Organisms

Genetic material and, typically, microbiological organisms are frequently handled in biochemical and molecular biology studies. Inevitably, much of this work will give rise to genetically modified organisms (GMOs). A GMO is a viable biological entity capable of reproduction or transferring genetic material that has had its genetic material modified by any technique, unless it arose from sexual reproduction, homologous recombination or somatic cell nuclear transfer.

Permission to work with GMOs is typically given by an institutional biosafety committee (IBC) to which the proposed type of work needs to be submitted in advance. IBCs are primarily responsible for ensuring compliance with government legislation that regulates work with potentially hazardous biological agents.

Work with GMOs is generally classified into different biosafety levels which describe the containment precautions required to isolate biological agents in an enclosed laboratory facility. Table 2.6 summarises the different levels of containment (level 1: lowest–level 4: highest) and their implementation in a few select countries. At the lowest level of containment (level 1), the containment zone may just be a chemical fume hood. The highest level of containment (level 4) involves isolation of an organism by means of building systems, sealed rooms and containers, positive pressure personnel suits ('space suits'), elaborate procedures for entering the room and decontamination procedures for leaving the room.

Laminar flow clean benches protect the work from contamination by the environment; they do not protect the worker, as the air passes unfiltered onto the worker and into the laboratory. In contrast, biological safety cabinets are used when infectious aerosols need to be handled. In such cabinets, the contaminated air is passed through a high-efficiency particulate air (HEPA) filter. There are three types of biosafety cabinets (see also Section 3.2):

- Class I: inward flow of air away from the operator. The air is passed through a HEPA filter before being discharged from the cabinet.
- Class II: an air barrier protects the operator and a flow of filtered air is passed over the work to prevent it from becoming contaminated. The air is passed through a HEPA filter before being discharged from the cabinet.
- Class III: completely enclosed unit with built-in air locks for introducing and removing materials. Both incoming and outgoing air passes through HEPA filters. Class III cabinets are intended for use with highly hazardous microorganisms.

Table 2.6 Overview of biosafety levels and administration for handling GMOs in research laboratories in select countries

Country	United States	UK	Germany	Australia[a]
Controlling department	Centers for Disease Control and Prevention (CDC)	Health and Safety Executive	State Agencies for Consumer Safety	Office of the Gene Technology Regulator
Main applicable law or regulation	Coordinated Framework for Regulation of Biotechnology, 1986	Control of Substances Hazardous to Health Regulations 2002 (COSHH)	Genetic Engineering Act 1990, Amendment 1994	Gene Technology Act 2000, Gene Technology Amendment Act 2007, Biosecurity Act 2015
Well-characterised agents not known to consistently cause disease in healthy adult humans (e.g. *E. coli*)	BSL–1	Class 1	S1	PC1
Agents of moderate potential hazard to personnel and the environment (e.g. hepatitis A)	BSL–2	Class 2	S2	PC2
Infectious microorganisms that pose a high individual risk and a limited to moderate community risk (e.g. *M. tuberculosis*)	BSL–3	Class 3	S3	PC3
Dangerous and exotic agents that pose a high individual risk (e.g. Ebola virus)	BSL–4	Class 4	S4	PC4

Note: [a] In addition to GMOs, Australia also regulates goods subject to biosecurity (previously called quarantine) control for which a Department of Agriculture and Water Resources import permit is required prior to importation. The biosecurity containment requirements closely relate to the OGTR physical containment requirements and conditions, albeit with some differences.

Ethics Approvals

The use of animals in research is subject to scrutiny with respect to ethical conduct and thus requires ethics approval. For animal work this is to ensure that all facets of animal care and use meets the requirements of the applicable laws and regulations. Similar to the procedures for work with GMOs, ethics approvals need to be applied for and are considered by an institutional ethics committee. If research on animals involves surveys to collect data from farmers, pet owners or other members of the public, a human ethics approval may be required to collect and use such data.

Generally, all research involving human participants requires approval from the appropriate responsible ethics review body. Such research is defined as being conducted with or about people, their biological materials or data about them, and thus involves a broad variety of activities, including:

- being observed by researchers
- taking part in surveys, interviews or focus groups
- access to information as part of an existing database
- access to personal documents or other materials
- undergoing psychological, physiological or medical testing or treatment
- the collection and use of biological materials (body organs, tissues, fluids or exhaled breath).

2.8.2 Ethical Conduct in Research

Ethical conduct typically defines norms that distinguish between acceptable and unacceptable ways of behaviour. Since many ethical norms frequently apply throughout all types of interactions in society, they are often regarded as common sense. Within professional and academic disciplines, ethical norms are implemented either as methods, procedures, or perspectives in analysis and decision-making processes of problems and issues. Examples of some major guiding principles include honesty, objectivity, integrity, carefulness and openness, to name just a few.

In order to support the aims of research (e.g. knowledge, truth), the accountability of researchers to the public and the essential values for collaborative work, ethical norms also apply to people who conduct scientific research and other scholarly or creative activities. These norms are typically summarised as research ethics. Given the rooting of ethical conduct in the very fabric of society, it is obvious that any research activity has to adhere to research ethics. Promoting this cause, many institutions, learned societies and professional organisations have defined codes and policies for research ethics. In many cases, authors of scientific articles also need to confirm that their scientific conduct within the submitted study complies with particular rules of research ethics.

Frequently, the rules defined in a code of conduct can be applied straightforwardly to make a particular decision, but there are also activities that most researchers in the scientific community would regard as misconduct, despite there being either no written code or a written policy would not classify this activity as unethical. Some frequent examples of this include:

- discussing confidential data from a paper that is under peer review with a journal;
- submitting the same paper to different journals without telling the editors;

- overworking, neglecting or exploiting graduate or post-doctoral students;
- failing to keep acceptable research records;
- making derogatory comments about and personal attacks on reviewers;
- rejecting a manuscript for publication without even reading it.

Like every regulatory policy, the codes and policies developed in research ethics do not cover each and every situation. There may even be conflicts, and, in many cases the rules require substantial interpretation. It is therefore an important aspect of any research training to raise awareness of the applicable research ethics, and to learn how difficult situations are assessed and decisions made to resolve potential ethical issues and avoid scientific misconduct.

2.9 SUGGESTIONS FOR FURTHER READING

2.9.1 Experimental Protocols

Beran J.A. (2013) *Laboratory Manual for Principles of General Chemistry*, 10th Edn., Wiley, Chichester, UK.

Hofmann A. (2016) *Methods in Structural Chemistry: A Lab Manual*, 3rd Edn., Structural Chemistry Program, Griffith University, Brisbane, Australia.

Krissinel E. and Henrick K. (2004) Secondary-structure matching (SSM), a new tool for fast protein structure alignment in three dimensions. *Acta Crystallographica D* 60, 2256–2268.

Westermeier R., Naven T. and Höpker H.-R. (2008) *Proteomics in Practice: A Guide to Successful Experimental Design*, 2nd Edn., Wiley-Blackwell, New York, USA.

2.9.2 General Texts

Beynon R. and Easterby J. S. (2003) *Buffer Solutions (The Basics)*, Taylor & Francis, New York, USA.

Boyer R.F. (2011) *Biochemistry Laboratory: Modern Theory and Techniques*, 2nd Edn., Prentice Hall, Upper Saddle River, NJ, USA.

Creighton T.E. (2010) *The Biophysical Chemistry of Nucleic Acids and Proteins*, Helvetian Press, Eastbourne, UK.

Lesk A.M. (2010) *Introduction to Protein Science: Architecture, Function, and Genomics*, 2nd Edn., Oxford University Press, Oxford, UK.

Whitford D. (2005) *Proteins: Structure and Function*, 1st Edn., Wiley, Chichester, UK.

2.9.3 Review Articles

Umscheid C.A., Margolis D.J. and Grossman C.E. (2011) Key concepts of clinical trials: a narrative review. *Postgraduate Medicine* 123, 194–204.

2.9.4 Websites

CATH: Classification of Protein Structures
www.cathdb.info/ (accessed April 2017)

DALI and FSSP: Families of Structurally Similar Proteins
ekhidna.biocenter.helsinki.fi/dali (accessed April 2017)

EMBOSS tools for peptide sequence analysis
www.ebi.ac.uk/Tools/emboss/ (accessed April 2017)

PDBeFold: Structure Similarity
www.ebi.ac.uk/msd-srv/ssm/ (accessed April 2017)

Pfam: protein families represented by multiple sequence alignments and hidden
Markov models
pfam.xfam.org/ (accessed April 2017)

Protein Data Bank
www.rcsb.org/pdb (accessed April 2017)

SCOP: Structural Classification of Proteins
scop.mrc-lmb.cam.ac.uk/scop/ (accessed April 2017)

Structure Function Linkage Database
sfld.rbvi.ucsf.edu/django/ (accessed April 2017)

What is Ethics in Research & Why is it Important?
www.niehs.nih.gov/research/resources/bioethics/whatis/ (accessed April 2017)

3 Cell Culture Techniques

ANWAR R. BAYDOUN

3.1 INTRODUCTION

Cell culture is a technique that involves the isolation and in vitro maintenance of cells isolated from tissues or whole organs derived from animals, microbes or plants. In general, animal cells have more complex nutritional requirements and usually need more stringent conditions for growth and maintenance than microbial cells. By comparison, microbes and plants require less rigorous conditions and grow effectively with the minimum of needs. Regardless of the source of material used, practical cell culture is governed by the same general principles, requiring a sterile pure culture of cells, the need to adopt appropriate aseptic techniques and the utilisation of suitable conditions for optimal viable growth of cells.

Once established, cells in culture can be exploited in many different ways. For instance, they are ideal for studying intracellular processes including protein synthesis, signal transduction mechanisms and drug metabolism. They have also been widely used to understand the mechanisms of drug actions, cell-to-cell interactions and genetics. Additionally, cell culture technology has been adopted in medicine, where genetic abnormalities can be determined by chromosomal analysis of cells derived, for example, from expectant mothers. Similarly, viral infections can be assayed both qualitatively and quantitatively on isolated cells in culture. In industry as well as academic drug discovery, cultured cells are used routinely to test both the pharmacological and toxicological effects of pharmaceutical compounds. This technology thus provides a valuable tool to scientists, offering a user-friendly system that is relatively cheap to run and the exploitation of which avoids the legal, moral and ethical questions generally associated with animal experimentation. More importantly, cell culture also presents a tremendous potential for future exploitation in disease treatment, where, for instance, defective or malfunctioning genes could be corrected in the host's own cells and transplanted back into the host to treat a disease. Furthermore, successful development of culture techniques for stem cells provides a much needed cell-based strategy for treating diseases where organ transplant is currently the only available option.

In this chapter, fundamental information required for standard cell culture, together with a series of principles and protocols used routinely in growing animal and bacterial cells are discussed. Additionally, a section has been dedicated to human embryonic stem cell culture, outlining techniques that are now becoming routine for stem cell culture. This should provide the basic knowledge for those new to cell culture and act as a revision aid for those with some experience in the field. Throughout the chapter, particular attention is paid to the importance of the work environment, outlining safety considerations, together with adequate description and hints on the essential techniques required for cell culture work.

3.2 THE CELL CULTURE LABORATORY AND EQUIPMENT

3.2.1 The Cell Culture Laboratory

The design and maintenance of the cell culture laboratory is perhaps the most important aspect of cell culture, since a sterile surrounding is critical for handling of cells and culture media, which should be free from contaminating microorganisms. Such organisms, if left unchecked, would outgrow the cells being cultured, eventually killing them due to the release of toxins and/or depletion of nutrients from the culture medium. Even low-level contamination must be avoided, as it can potentially negatively influence the experiment, for example by causing cellular stress.

Where possible, a cell culture laboratory should be designed in such a way that it facilitates preparation of media and allows for the isolation, examination, evaluation and maintenance of cultures under controlled sterile conditions. In an ideal situation, there should be a room dedicated exclusively to cell culture. However, many cell culture facilities, especially in academia, form part of an open-plan laboratory. This is not a serious problem, as long as good aseptic techniques (discussed below) and a few basic guidelines are adopted. All surfaces within the culture area should be non-porous to prevent adsorption of media and other materials that may provide a good breeding ground for microorganisms, leading to infection of the cultures. Surfaces should also be easy to clean and all waste generated should be disposed of immediately. The disposal procedure may require prior autoclaving of the waste, which can be carried out using pressurised steam at 121 °C under 105 kPa for a defined period of time. These conditions are required to destroy microorganisms. For smooth running of the facilities, daily checks should be made of the temperature in incubators, and of the gas supply to the incubators by checking the CO_2 cylinder pressure. Water baths should be kept clean at all times and areas under the work surfaces of the flow cabinets cleaned of any spills.

Several pieces of equipment are essential. These include a tissue culture hood, incubator(s), autoclave and microscope. A brief description will be given of these and other essential equipment items.

3.2.2 Cell Culture Hoods

The cell culture hood is the central piece of equipment where all the cell handling is carried out and is designed not only to protect the cultures from the operator, but in some cases to protect the operator from the cultures. These hoods are generally referred to as laminar flow hoods as they generate a smooth uninterrupted streamlined flow (laminar flow) of sterile air that has been filtered through a high-efficiency particulate air (HEPA) filter. There are two types of laminar flow hood classified as either vertical or horizontal. The horizontal hoods allow air to flow directly at the operator and as a result are generally used for media preparation or when working with non-infectious materials, including those derived from plants. The vertical hoods (also known as biosafety cabinets) are best for working with hazardous organisms, since air within the hood is filtered before it passes into the surrounding environment.

Currently, there are at least three different classes of hood used, which all offer various levels of protection to the cultures, the operator or both, and these are described below.

Class I Hoods

These hoods, as with the class II type, have a screen at the front that provides a barrier between the operator and the cells, but allows access into the hood through an opening at the bottom of the screen (Figure 3.1). This barrier prevents too much turbulence

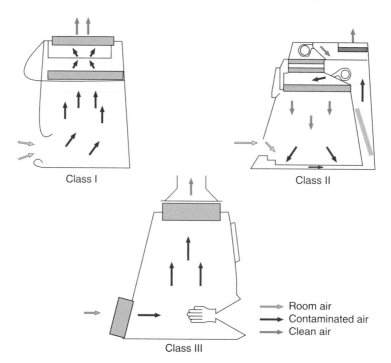

Figure 3.1 Schematic representation of tissue culture cabinets.

to air flow from the outside and, more importantly, provides good protection for the operator. Cultures are also protected, but to a lesser extent when compared to the class II hoods as the air drawn in from the outside is sucked through the inner cabinet to the top of the hood. These hoods are suitable for use with low-risk organisms and when operator protection only is required.

Class II Hoods

Class II hoods are the most common units found in tissue culture laboratories. These hoods offer good protection to both the operator and the cell culture. Unlike class I hoods, air drawn from the outside is passed through the grill in the front of the work area and filtered through the HEPA filter at the top of the hood before streaming down over the tissue culture (Figure 3.1). This mechanism protects the operator and ensures that the air over the cultures is largely sterile. These hoods are adequate for animal cell culture, which involves low to moderate toxic or infectious agents, but are not suitable for use with high-risk pathogens, which may require a higher level of containment.

Class III Hoods

Class III safety cabinets are required when the highest levels of operator and product protection are required. These hoods are completely sealed, providing two glove pockets through which the operator can work with material inside the cabinet (Figure 3.1). Thus the operator is completely shielded, making class III hoods suitable for work with highly pathogenic organisms, including tissue samples carrying known human pathogens.

Practical Hints and Safety Aspects of Using Cell Culture Hoods

All hoods must be maintained in a clutter-free and clean state at all times as too much clutter may affect air flow and contamination will introduce infections. Thus, as a rule of thumb, put only items that are required inside the cabinet and clean all work surfaces before and after use with industrial methylated spirit (IMS). The latter is used at an effective concentration of 70% (prepared by mixing IMS and ultrapure water in v/v ratio of 70:30), which acts against bacteria and fungal spores by dehydrating and fixing cells, thus preventing contamination of cultures.

Some cabinets may be equipped with a short-wave ultraviolet light that can be used to irradiate the interior of the hood to kill microorganisms. When present, switch on the ultraviolet light for at least 15 min to sterilise the inside of the cabinet, including the work area. Note, however, that ultraviolet radiation can cause damage to the skin and eyes. Precautions should be taken at all times to ensure that the operator is not directly exposed to the light when using this option to sterilise the hood. Once finished, ensure that the front panel door (class I and II hoods) is replaced securely after use. In addition, always turn the hood on for at least 10 min before starting work to allow the flow of air to stabilise. During this period, monitor the air flow and check all dials in the control panel at the front of the hood to ensure that they are within the safe margin.

3.2.3 CO$_2$ Incubators

Water-jacketed incubators are required to facilitate optimal cell growth under strictly maintained and regulated conditions, normally requiring a constant temperature of 37 °C and an atmosphere of 5–10% CO$_2$ plus air. The purpose of the CO$_2$ is to ensure that the culture medium is maintained at the required physiological pH (usually pH 7.2–7.4). This is achieved by the supply of CO$_2$ from a gas cylinder into the incubator through a valve that is triggered to draw in CO$_2$ whenever the level falls below the set value of 5% or 10%. The CO$_2$ that enters the inner chamber of the incubator dissolves into the culture medium, which contains bicarbonate. The latter reacts with H$^+$ (generated from cellular metabolism), forming carbonic acid, which is in equilibrium with water and CO$_2$, thereby maintaining the pH in the medium at approximately pH 7.2.

$$HCO_{3(aq)}^- + H_{(aq)}^+ \rightleftharpoons H_2CO_{3(aq)} \rightleftharpoons CO_{2(g)} + H_2O_{(l)} \qquad \text{(Eq 3.1)}$$

These incubators are generally humidified by the inclusion of a tray of sterile water on the bottom deck. This prevents evaporation of medium from the cultures by creating a highly humidified atmosphere in the incubator.

An alternative to humidified incubators is a dry, non-gassed unit that relies on the use of alternative buffering systems such as 4-(2-hydroxyethyl)-1-piperazine-ethanesulfonic acid (HEPES) or morpholinopropane sulfonic acid (MOPS) for maintaining a balanced pH within the culture medium. The advantage of this system is that it eliminates the risk of infection that can be posed by the tray of water in the humidified unit. The disadvantage, however, is that the culture medium will evaporate rapidly, thereby stressing the cells. One way round this problem is to place the cell cultures in a sandwich box containing little pots of sterile water. With the sandwich box lid partially closed, evaporation of water from the pots creates a humidified atmosphere within the box, thus reducing the risk of evaporation of medium from the cultures.

Practical Hints and Safety Aspects of Using Cell Culture Incubators

The incubator should be maintained at 37 °C and supplied with 5% CO$_2$ at all times. CO$_2$ levels inside the unit can be monitored and adjusted using a gas analyser such as the Fyrite Reader. Regular checks should also be made on the levels of CO$_2$ in the gas cylinders that supply CO$_2$ to the incubators and these should be replaced when levels are low. Most incubators are designed with an inbuilt alarm that sounds when the CO$_2$ level inside the chamber drops. At this point the gas cylinder must be replaced immediately to avoid stressing or killing the cultures. Typically, two gas cylinders are connected to a cylinder changeover unit that switches automatically to the second source of gas supply when the first is empty.

When using a humidified incubator, it is essential that the water tray is maintained and kept free from microorganisms. This can be achieved by adding various agents to the water, such as the antimicrobial agent Roccal-D® at a concentration of 1% (w/v). Other products such as Thiomersal®, SigmaClean® or just copper sulfate can also be used. Proper care and maintenance of the incubator should, however, include regular cleaning of the interior of the unit using any of the above reagents then swabbing with 70% IMS. Alternatively, copper-coated incubators may be used which, due to the antimicrobial properties of copper, are reported to reduce microbial contamination.

3.2.4 Microscopes

Inverted phase contrast microscopes (see Section 11.2.5) are routinely used for visualising cells in culture. These are expensive, but easy to operate, with a light source located above and the objective lenses below the stage on which the cells are placed. Visualisation of cells by microscopy can provide useful information about the morphology and state of the cells. Early signs of cell stress may be easily identified and appropriate action taken to prevent loss of cultures.

3.2.5 Other General Equipment

Several other pieces of equipment are required in cell culture. These include a centrifuge to separate cells from culture medium, a water bath for thawing frozen samples of cells and warming media to 37 °C before use, and a fridge and freezer for storage of media and other materials required for cell culture. Some cells need to attach to a surface in order to grow and are therefore referred to as adherent. These cells are cultured in non-toxic polystyrene plastics that contain a biologically inert surface on which the cells attach and grow. Various types of plastics are available for this purpose and include Petri dishes, multi-well plates (with either 96, 24, 12 or 6 wells per plate) and screwcap flasks classified according to their growth surface areas: T-25, T-75, T-225 (cm^2 of surface area). A selection of these plastics is shown in Figure 3.2.

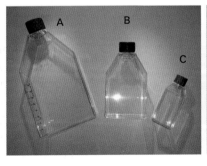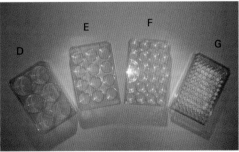

Figure 3.2 Tissue culture plastics used generally for cell culture. (A–C) T-flasks; (D–G) representative multi-well plates. (A) T–225 (225 cm^2), (B) T-75 (75 cm^2), (C) T-25 (growth area: 25 cm^2), (D) 6-well plate, (E) 12-well plate, (F) 24-well plate and (G) 96-well plate.

3.3 SAFETY CONSIDERATIONS IN CELL CULTURE

Safety in the cell culture laboratory must be of major concern to the operator. This is particularly the case when working with pathogenic microbes, fresh primate or human tissues, or with cells that may contain agents that use humans as hosts. One very good example of this would be working with fresh human lymphocytes, which may contain infectious agents such as the human immunodeficiency virus (HIV) and/or hepatitis B virus. Thus, when one is working with fresh human tissue, it is essential that the infection status of the donor is determined in advance of use and all necessary precautions taken to eliminate or limit the risks to which the operator is exposed. A recirculation class II cabinet is a minimum requirement for this type of cell culture work and the operator should be provided with protective clothing, including latex gloves and a face mask if required. In the UK, such work should also be carried out under the guidelines laid down by the UK Advisory Committee on Dangerous Pathogens (ACDP).

Apart from the risks posed by the biological material being used, the operator should also be aware of his or her work environment and be fairly conversant with the equipment being used, as these may also pose a serious hazard. The culture cabinet should be serviced routinely and checked (approximately every 6 months) to ensure its safety to the operator. Additionally, the operator must ensure his or her own safety by adopting some common precautionary measures, such as refraining from eating or drinking whilst working in the cabinet and using a pipette aid as opposed to mouth pipetting to prevent ingestion of unwanted substances. Gloves and adequate protective clothing, such as a clean laboratory coat, should be worn at all times and gloves must be discarded after handling of non-sterile or contaminated material.

3.4 ASEPTIC TECHNIQUES AND GOOD CELL CULTURE PRACTICE

3.4.1 Good Practice

In order to maintain a clean and safe culture environment, adequate aseptic or sterile technique should be adopted at all times. This simply involves working under conditions that prevent contaminating microorganisms from the environment from entering the cultures. Part of the precautions taken involves washing hands with antiseptic soap and ensuring that all work surfaces are kept clean and sterile by swabbing with 70% IMS before starting work. Moreover, all procedures, including media preparation and cell handling, should be carried out in a cell culture cabinet that is maintained in a clean and sterile condition.

Other essential precautions should include avoiding talking, sneezing or coughing into the cabinet or over the cultures. A clean pipette should be used for each different procedure and under no circumstance should the same pipette be used between different bottles of media, as this will significantly increase the risk of cross-contamination. All spillages must be cleaned quickly to avoid contamination from microorganisms that may be present in the air. Failing to do so may result in infection of the cultures, which may be reduced by using antibiotics. However, this is not always guaranteed and good aseptic techniques should eliminate the need for antibiotics. In the event of cultures becoming contaminated, they should be removed immediately from the laboratory, disinfected and autoclaved to prevent the contamination spreading. Under no circumstance can an infected culture be opened inside the cell culture cabinet or incubator. All waste generated must be disposed of in accordance with the national legislative requirements, which state that cell culture waste including media be inactivated using a disinfectant before disposal and that all contaminated materials and waste be autoclaved before being discarded or incinerated.

The risk from infections is the most common cause for concern in cell culture. Various factors can contribute to this, including poor work environment, poor aseptic techniques and indeed poor hygiene of the operator. The last is important, since most of the common sources of infections such as bacteria, yeast and fungus originate from the worker. Maintaining a clean environment, and adopting good laboratory practice and aseptic techniques, should, therefore, help to reduce the risk of infection. However, should infections occur, it is advisable to address this immediately and eradicate the problem. To do this, it helps to know the types of infection and what to look for.

In animal cell cultures, bacterial and fungal infections are relatively easy to identify and isolate. The other most common contamination originates from mycoplasma. These are the smallest (approximately 0.3 mm in diameter) self-replicating prokaryotes in existence. They lack a rigid cell wall and generally infect the cytoplasm of mammalian cells. There are at least five species known to contaminate cells in culture: *Mycoplasma hyorhinis*, *Mycoplasma arginini*, *Mycoplasma orale*, *Mycoplasma fermentans* and *Acholeplasma laidlawii*. Infections caused by these organisms are more problematic and not easily identified or eliminated. Moreover, if left unchecked, mycoplasma contamination will cause subtle but adverse effects on cultures, including changes in metabolism, DNA, RNA and protein synthesis, morphology and growth. This can lead to non-reproducible, unreliable experimental results and unsafe biological products.

3.4.2 Identification and Eradication of Bacterial and Fungal Infections

Both bacterial and fungal contaminations are easily identified as the infective agents are readily visible to the naked eye, even in the early stages. This is usually made noticeable by the increase in turbidity and the change in colour of the culture medium owing to the change in pH caused by the infection. In addition, bacteria can be easily identified under microscopic examination as motile round bodies. Fungi, on the other hand, are distinctive by their long hyphal growth and by the fuzzy colonies they form in the medium. In most cases, the simplest solution to these infections is to remove and dispose of the contaminated cultures. In the early stages of an infection, attempts can be made to eliminate the infecting microorganism using repeated washes and incubations with antibiotics or antifungal agents. This is, however, not advisable as handling infected cultures in the sterile work environment increases the chance of the infection spreading.

As part of good laboratory practice, sterile testing of cultures should be carried out regularly to ensure that they are free from microbial organisms. This is particularly important when preparing cell culture products or generating cells for storage. Generally, the presence of these organisms can be detected much earlier and necessary precautions taken to avoid a full-blown contamination crisis in the laboratory. The testing procedure usually involves culturing a suspension of cells or products in an appropriate medium such as tryptone soya broth (TSB) for bacterial or thioglycollate medium (TGM) for fungal detection. The mixture is incubated for up to 14 days, but examined daily for turbidity, which is used as an indication of microbial growth. It is essential that both positive and negative controls are set up in parallel with the sample to be tested. For this purpose a suspension of bacteria such as *Bacillus subtilis* or fungus such as *Clostridium sporogenes* is used instead of the cells or product to be tested. Flasks containing only the growth medium are used as negative controls. Any contamination in the cell cultures will result in the broth appearing turbid, as would the positive controls. The negative controls should remain clear. Infected cultures should be discarded, whilst clear cultures would be safe to use or keep.

3.4.3 Identification of Mycoplasma Infections

Mycoplasma contaminations are more prevalent in cell culture than many workers realise. The reason for this is that mycoplasma contaminations are not evident under light microscopy, nor do they result in turbid growth in culture. Instead, the changes induced are more subtle and manifest themselves mainly as a slowdown in growth and in changes in cellular metabolism and functions. However, cells generally return to their native morphology and normal proliferation rates relatively rapidly after eradication of mycoplasma.

The presence of mycoplasma contamination in cultures has in the past been difficult to determine and samples had to be analysed by specialist laboratories. This is no longer the case and techniques such as enzyme-linked immunosorbent assay (ELISA), polymerase chain reaction (PCR) and DNA staining using the fluorescent dye Hoechst 33258 are now available for detection of mycoplasma in cell culture laboratories. The gold standard, however, is the microbiological culture technique in which cells in suspension are inoculated into liquid broth and then incubated under aerobic conditions at 37 °C for 14 days. A non-inoculated flask of broth is used as a negative control. Aliquots of broth are taken every 3 days and inoculated onto an agar plate, which is incubated anaerobically as above. All plates are then examined under an inverted microscope at a magnification of 300× after 14 days of incubation. Positive cultures will show typical mycoplasma colony formation, which has an opaque granular central zone surrounded by a translucent border, giving a 'fried egg' appearance (Figure 3.3). It may be necessary to set up positive controls in parallel, in which case plates and broth

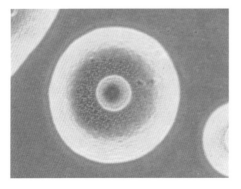

Figure 3.3 Photograph of mycoplasma, showing the characteristic opaque granular central zone surrounded by a translucent border, giving a 'fried egg' appearance.

should be inoculated with a known strain of mycoplasma such as *Mycoplasma orale* or *Mycoplasma pneumoniae*.

The DNA binding method offers a rapid alternative for detecting mycoplasma and works on the principle that Hoechst 33258 fluoresces under ultraviolet light once bound to DNA. Thus, in contaminated cells, the fluorescence will be fairly dispersed in the cytoplasm of the cells owing to the presence of mycoplasma. In contrast, uncontaminated cells will show localised fluorescence in their nucleus only. The Hoechst 33258 assay, although rapid, is of relatively low sensitivity when compared with the culture technique described above. For this assay, an aliquot of the culture to be tested is placed on a sterile coverslip in a 35-mm culture dish and incubated at 37 °C in a cell culture incubator to allow cells to adhere. The coverslip is then fixed by adding a fixative consisting of one part glacial acetic acid and three parts methanol, prepared fresh on the day. A freshly prepared solution of Hoechst 33258 stain is added to the fixed coverslip, incubated in the dark at room temperature to allow the dye to bind to the DNA and then viewed under ultraviolet fluorescence at 1000×. All positive cultures will show fluorescence of mycoplasma DNA, which will appear as small cocci or filaments in the cytoplasm of the contaminated cells (Figure 3.4b). Negative cultures will show only fluorescing nuclei of uncontaminated cells against a dark cytoplasmic background (Figure 3.4a). However, this technique is prone to errors, including false-negative results. To avoid the latter, cells should be cultured in antibiotic-free medium for two to three passages before being used. A positive control using a strain of mycoplasma seeded onto a coverslip is essential. Such controls should be handled away from the cell culture laboratory to avoid contaminating clean cultures of cells. It is also important to ensure that the fluorescence detected is not due to the presence of bacterial contamination or debris embedded into the plastics during manufacture. The former normally appear larger than the

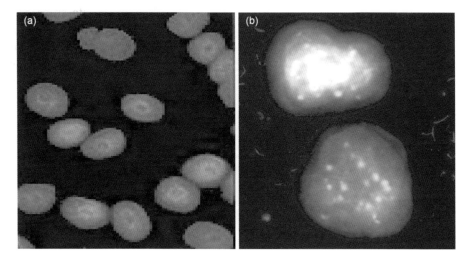

Figure 3.4 Hoechst 33258 staining of mycoplasma in cells. (a) A Hoechst-negative stain, with the dye staining cellular DNA in the nucleus and thus showing nuclear fluorescence. (b) A Hoechst-positive stain, showing staining of mycoplasma DNA in the cytoplasm of the cells.

fluorescing cocci or filaments of mycoplasma. Debris, on the other hand, would show a non-uniform fluorescence owing to the variation in size of the particles usually found in plastics.

ELISA detection (see Section 7.3.1) of mycoplasma is also commonly used and can be carried out using specifically designed kits following the manufacturer's protocol and reagents supplied. In this assay, 96-well plates are coated with the antibodies against different mycoplasma species. Each plate is then incubated at 37 °C for 2 h with the required antibody or antibodies before blocking with the appropriate blocking solution and incubating with the test sample(s). A negative control, which is simply media with sample buffer, and a positive control normally provided with the kits, should also be included in each assay. A detection antibody is subsequently added to the samples, incubated for a further 2 h at 37 °C before washing and incubating with a streptavidin solution for 1 h at 37 °C. Each plate is then tested for mycoplasma by adding the substrate solution and reading on a plate reader at 405 nm after a further 30 min incubation at room temperature. This method is apparently suitable for detecting high levels of mycoplasma and could also be used to identify several species in one assay.

As with the ELISAs, commercial PCR kits are also available that contain the required primers, internal control template, positive control template and all the relevant buffers. Samples are generated and set up in a reaction mix as instructed in the manufacturer's protocol. The PCR is performed according to manufacturer's protocols and the products generated analysed by electrophoresis on a high-grade 2% agarose gel. Typically, cell lines are checked every 1–3 months by ELISA and PCR.

3.4.4 Eradication of Mycoplasma

Until recently, the most common approach for eradicating mycoplasma has been the use of antibiotics such as gentamycin. This approach is, however, not always effective, as not all strains of mycoplasma are susceptible to this antibiotic. Moreover, anti-biotic therapy does not always result in long-lasting successful elimination and many antibiotics can be cytotoxic to the cell culture. A new generation of bactericidal antibiotic preparation referred to as Plasmocin™ is effective against mycoplasma even at relatively low, non-cytotoxic concentrations. The antibiotics contained in this product are actively transported into cells, thus facilitating killing of intracellular mycoplasma without any adverse effects on actual cellular metabolism.

Apart from antibiotics inhibiting growth, other products have been introduced to eradicate mycoplasma efficiently and quickly without causing any adverse effects to eukaryotic cells or viruses. One such product is Mynox®, a biological agent that integrates into the membrane of mycoplasma, compromising its integrity and eventually initiating its disintegration, thereby killing mycoplasma. This process apparently occurs within an hour of application and may have the added advantage that it will not lead to the development of resistant strains. It is safe to cultures and eliminated once the medium has been replaced. Moreover, this reagent is highly sensitive, detecting as little as 1–5 fg of mycoplasma DNA, which corresponds to two to five mycoplasma cells per sample, and is effective against many of the common mycoplasma contaminations encountered in cell culture.

3.5 TYPES OF ANIMAL CELLS, CHARACTERISTICS AND MAINTENANCE IN CULTURE

The cell types used in cell culture fall into two categories, generally referred to as either a primary culture or a cell line.

3.5.1 Primary Cell Cultures

Primary cultures are cells derived directly from tissues following enzymatic dissociation or from tissue fragments referred to as explants. These are usually the cells of preference, since it is argued that primary cultures retain their characteristics and reflect the true activity of the cell type in vivo. The disadvantage of using primary cultures, however, is that their isolation can be labour-intensive and may produce a heterogeneous population of cells. Moreover, primary cultures have a relatively limited lifespan and can be used over only a limited period of time in culture.

Primary cultures can be obtained from many different tissues and the source of tissue used generally defines the cell type isolated. For instance, cells isolated from the endothelium of blood vessels are referred to as endothelial cells, whilst those isolated from the medial layer of the blood vessels and other similar tissues are smooth muscle cells. Although both can be obtained from the same vessels, endothelial cells are different in morphology and function, generally growing as a single monolayer characterised by a cobble-stoned morphology. Smooth muscle cells, in contrast, are elongated, with spindle-like projections at either end, and grow in layers, even when maintained in culture. In addition to these cell types there are several other widely used primary cultures derived from a diverse range of tissues, including fibroblasts from connective tissue, lymphocytes from blood, neurons from nervous tissues and hepatocytes from liver tissue.

3.5.2 Continuous Cell Lines

Cell lines consist of a single cell type that has gained the ability for infinite growth. This usually occurs after transformation of cells by one of several means that include treatment with carcinogens or exposure to viruses such as the monkey simian virus 40 (SV40), Epstein–Barr virus (EBV) or Abelson murine leukaemia virus (A-MuLV), amongst others. These treatments cause the cells to lose their growth regulation ability. As a result, transformed cells grow continuously and, unlike primary culture, have an infinite lifespan (become 'immortalised'). The drawback to this is that transformed cells generally lose some of their original in vivo characteristics. For instance, certain established cell lines do not express particular tissue-specific genes. One good example of this is the inability of liver cell lines to produce clotting factors. Continuous cell lines, however, have several advantages over primary cultures, not least because they are available without requiring labour-intensive isolation. In addition, they require less serum for growth, have a shorter doubling time and can grow without necessarily needing to attach or adhere to the surface of the flask.

Many different cell lines are currently available from various cell banks, which makes it fairly easy to obtain these cells without having to generate them. One of the largest organisations that supplies cell lines is the **European Collection of Animal Cell Cultures** (ECACC) based in Salisbury, UK. A selection of the different cell lines supplied by this organisation is listed in Table 3.1.

Table 3.1 **Examples of cell lines available from commercial sources**

Cell line	Morphology	Species	Tissue origin
BAE-1	Endothelial	Bovine	Aorta
BHK-21	Fibroblast	Syrian hamster	Kidney
CHO	Fibroblast	Chinese hamster	Ovary
COS-1/7	Fibroblast	African green monkey	Kidney
HeLa	Epithelial	Human	Cervix
HEK-293	Epithelial	Human	Kidney
HT-29	Epithelial	Human	Colon
MRC-5	Fibroblast	Human	Lung
NCI-H660	Epithelial	Human	Lung
NIH/3T3	Fibroblast	Mouse	Embryo
THP-1	Monocytic	Human	Blood
V-79	Fibroblast	Chinese hamster	Lung
HEP1	Hepatocytes	Human	Liver

3.5.3 Cell Culture Media and Growth Requirements for Animal Cells

The cell culture medium used for animal cell growth is a complex mixture of nutrients (amino acids, a carbohydrate such as glucose, and vitamins), inorganic salts (e.g. containing magnesium, sodium, potassium, calcium, phosphate, chloride, sulfate and bicarbonate ions) and broad-spectrum antibiotics. In certain situations, it may be essential to include a fungicide such as amphotericin B, although this may not always be necessary. For convenience and ease of monitoring, the status of the medium, the pH indicator phenol red may also be included. This will change from red at pH 7.2–7.4 to yellow or fuchsia as the pH becomes either acidic or alkaline, respectively.

The other key basic ingredient in the cell culture medium is serum, usually bovine or fetal calf. This is used to provide a buffer for the culture medium, but, more importantly, enhances cell attachment and provides additional nutrients and hormone-like growth factors that promote the healthy growth of cells. An attempt to culture cells in the absence of serum does not usually result in successful or healthy cultures, even though cells can produce growth factors of their own. However, despite these benefits, the use of serum is increasingly being questioned, not least because of many of the other unknowns that can be introduced, including infectious agents such as viruses and mycoplasma. 'Mad cow disease' (bovine spongiform encephalitis) has introduced an additional drawback, posing a particular risk for the cell culturist, and has increased the need for alternative products. In this regard, several cell culture reagent manufacturers have now developed serum-free media supplemented with various components including albumin, transferrin, insulin, growth factors and other essential elements required for optimal cell growth. These is proving very useful, particularly for the pharmaceutical and biotechnology companies involved in the manufacture of drugs or biological products for human and animal consumption.

3.5.4 Preparation of Animal Cell Culture Medium

Preparation of the culture medium is perhaps taken for granted as a simple straight-forward procedure that is often not given due care and attention. As a result, most infections in cell culture laboratories originate from infected media. Following the simple yet effective procedures outlined in Section 3.4.1 should prevent or minimise the risk of infecting the medium during the preparation process.

Preparation of the medium itself should also be carried out inside the culture cabinet and usually involves adding a required amount of serum together with antibiotics to a fixed volume of medium. The amount of serum used will depend on the cell type, but usually varies between 10% and 20%. The most common antibiotics used are penicillin and streptomycin, which inhibit a wide spectrum of Gram-positive and Gram-negative bacteria. Penicillin acts by inhibiting the last step in bacterial cell wall synthesis, whilst streptomycin blocks protein synthesis.

Once prepared, the mixture, which is referred to as complete growth medium, should be kept at 4 °C until used. To minimise wastage and risk of contamination, it is advisable to make just the required volume of medium and use this within a short period of time. As an added precaution, it is also advisable to check the clarity of the medium before use. Any infected medium, which will appear cloudy or turbid, should be discarded immediately. In addition to checking the clarity, a close eye should also be kept on the colour of the medium, which should be red at physiological pH, owing to the presence of phenol red. Media where the pH indicator suggests acidic (yellow) or alkaline (fuchsia) conditions should be discarded, as these extremes will affect the viability and thus growth of the cells.

3.5.5 Subculture of Cells

Subculturing is the process by which cells are harvested, diluted in fresh growth medium and replaced in a new culture flask to promote further growth. This process, also known as passaging, is essential if the cells are to be maintained in a healthy and viable state, otherwise they may die after a certain period in continuous culture. The reason for this is that adherent cells grow in a continuous layer that eventually occupies the whole surface of the culture dish and at this point they are said to be confluent. Once confluent, the cells stop dividing and go into a resting state where they stop growing (senesce) and eventually die. Thus, cells must be subcultured before they reach full contact inhibition and ideally just before they reach a confluent state.

Cells can be harvested and subcultured using one of several techniques. The precise method used is dependent to a large extent on whether the cells are adherent or in suspension.

Subculture of Adherent Cells

Adherent cells can be harvested either mechanically, using a rubber spatula (also referred to as a 'rubber policeman') or enzymatically using proteolytic enzymes. Cells in suspension are simply diluted in fresh medium by taking a given volume of cell suspension and adding an equal volume of medium.

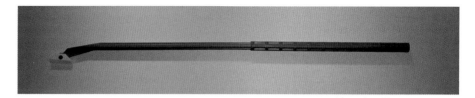

Figure 3.5 Cell scraper.

Harvesting of Cells Mechanically

This method is simple and easy. It involves gently scraping cells from the growth surface into the culture medium using a rubber spatula that has a rigid polystyrene handle with a soft polyethylene scraping blade (Figure 3.5). This method is not suitable for all cell types, as the scraping may result in membrane damage and significant cell death. Before adopting this approach, it is important to carry out some test runs, where cell viability and growth are monitored in a small sample of cells following harvesting.

Harvesting of Cells Using Proteolytic Enzymes

Several different proteolytic enzymes can be exploited, including trypsin, a proteolytic enzyme that destroys proteinaceous connections between cells, and between cells and the surface of the flask in which they grow. As a result, harvesting of cells using this enzyme results in the release of single cells, which is ideal for subculturing, as each cell will then divide and grow, thus enhancing the propagation of the cultures.

Trypsin is commonly used in combination with EDTA, which enhances the action of the enzyme by sequestering divalent cations that inhibit trypsin activity. EDTA alone can also be effective in detaching adherent cells as it chelates the Ca^{2+} required by some adhesion molecules (integrins) that facilitate cell–cell or cell–matrix interactions. Although EDTA alone is much gentler on the cells than trypsin, some cell types may adhere strongly to the plastic, requiring trypsin to detach.

The standard procedure for detaching adherent cells using trypsin and EDTA involves making a working solution of 0.1% trypsin plus 0.02% EDTA in Ca^{2+}/Mg^{2+}-free phosphate-buffered saline (PBS). The growth medium is aspirated from confluent cultures and washed at least twice with PBS to remove traces of serum that may inactivate the trypsin. The trypsin-EDTA solution (approximately 1 cm^3 per 25 cm^2 of surface area) is then added to the cell monolayer and swirled around for a few seconds. Excess trypsin-EDTA is aspirated, leaving just enough to form a thin film over the monolayer. The flask is then incubated at 37 °C in a cell culture incubator for 2–5 min, but monitored under an inverted light microscope at intervals to detect when the cells are beginning to round up and detach. This is to ensure that the cells are not overexposed to trypsin, as this may result in extensive damage to the cell surface, eventually resulting in cell death. It is therefore important that the proteolysis reaction is quickly terminated by the addition of complete medium containing serum to inactivate the trypsin. The suspension of cells is collected into a sterile centrifuge tube and spun at 1000 rpm for 10 min to pellet the cells, which are subsequently resuspended in a known volume of fresh complete culture medium to give a required density of cells per cubic centimetre volume.

As with all tissue culture procedures, aseptic techniques should be adopted at all times. This means that all the above procedures should be carried out in a tissue culture cabinet under sterile conditions. Other precautions worth noting include the handling of the trypsin stock. This should be stored frozen at −20 °C and, when needed, placed in a water bath just to the point where it thaws. Any additional time in the 37 °C water bath will inactivate the enzymatic activity of the trypsin. The working solution should be kept at 4 °C once made and can be stored for up to 3 months.

Subculture of Cells in Suspension

For cells in suspension, it is important initially to examine an aliquot of cells under a microscope to establish whether cultures are growing as single cells or clumps. If cultures are growing as single cells, an aliquot is counted as described in Section 3.5.6 and then reseeded at the desired seeding density in a new flask by simply diluting the cell suspension with fresh medium, provided the original medium in which the cells were growing is not spent. However, if the medium is spent and appears acidic, then the cells must be collected by centrifugation at 1000 rpm for 10 min, resuspended in fresh medium and transferred into a new flask. Similarly, cells that grow in clumps are collected by centrifugation and resuspended in fresh medium as single cells using a glass Pasteur or fine-bore pipette.

3.5.6 Cell Quantification

It is essential that when cells are subcultured they are seeded at the appropriate seeding density that will facilitate optimum growth. On the one hand, if cells are seeded at a lower seeding density they may take longer to reach confluency and some may expire before getting to this point. On the other hand, if seeded at a high density, cells will reach confluency too quickly, resulting in irreproducible experimental results. This is because trypsin can digest surface proteins, including receptors for drugs, and these will need time (sometimes several days) to renew. Failure to allow these proteins to be regenerated on the cell surface may result in variable responses to drugs specific for such receptors.

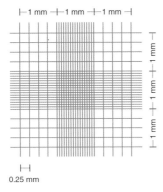

Figure 3.6 Haemocytometer.

Several techniques are now available for quantification of cells and of these, the most common method involves the use of a **haemocytometer** which is fairly simple and inexpensive to use. The haemocytometer itself is a thickened glass slide that has a small chamber of grids cut into the glass. The chamber has a fixed volume and is etched into nine large squares, of which the large corner squares contain 16 small squares each. Each large square measures 1 mm × 1 mm and is 0.1 mm deep (see Figure 3.6).

Thus, with a coverslip in place, each square represents a volume of 0.1 mm^3 (1.0 mm^2 area × 0.1 mm depth) or 10^{-4} cm^3. Knowing this, the cell concentration (and the total number of cells) can therefore be determined and expressed per cubic centimetre. The general procedure involves loading approximately 10 µl of a cell suspension into a clean haemocytometer chamber and counting the cells within the four corner squares with the aid of a microscope set at 20× magnification. The count is mathematically converted to the number of cells per cm^3 of suspension.

To ensure accuracy, the coverslip must be firmly in place. This can be achieved by moistening a coverslip with exhaled breath and gently sliding it over the haemocytometer chamber, pressing firmly until Newton's refraction rings (usually rainbow-like) appear under the coverslip. The total number of cells in each of the four 1 mm^3 corner squares should be counted, with the proviso that only cells touching the top and left borders, but not those touching the bottom and right borders, are counted. Moreover, cells outside the large squares, even if they are within the field of view, should not be counted. When present, clumps should be counted as one cell. Ideally ~100 cells should be counted to ensure a high degree of accuracy in counting. If the total cell count is less than 100 or if more than 10% of the cells counted appear to be clustered, then the original cell suspension should be thoroughly mixed and the counting procedure repeated. Similarly, if the total cell count is greater than 400, the suspension should be diluted further to get counts of between 100 and 400 cells.

Since some cells may not survive the trypsinisation procedure, it is usually advisable to add an equal volume of the dye Trypan Blue to a small aliquot of the cell suspension before counting. This dye is excluded by viable cells but taken up by dead cells. Thus, when viewed under the microscope, viable cells will appear as bright translucent structures, while dead cells will stain blue (see Section 3.5.12). The number of dead cells can therefore be excluded from the total cell count, ensuring that the seeding density accurately reflects viable cells.

Calculating Cell Number

Cell number is usually expressed per cm^3 ('cell density') and is determined by multiplying the average of the number of cells counted by a conversion factor of 10^4 to get the cell density.

Thus:

$$\text{Cell density} = \frac{\text{Number of cells}}{\text{cm}^3}$$

$$= \frac{\text{Number of cells counted}}{\text{Number of squares counted}} \times \text{Conversion factor} \times \text{cm}^{-3} \qquad \text{(Eq 3.2)}$$

If the cells were diluted before counting, then the dilution factor should also be taken into account; therefore:

$$\text{Cell density} = \frac{\text{Number of cells}}{\text{cm}^3}$$

$$= \frac{\text{Number of cells counted}}{\text{Number of squares counted}} \times \text{Conversion factor} \times \text{Dilution factor} \times \text{cm}^{-3} \qquad \text{(Eq 3.3)}$$

To get the total number of cells harvested, the cell density needs to be multiplied by the original volume from which the cell sample was removed, i.e.:

$$\text{Total number of cells} = \text{Cell density} \times \text{Total volume of cell suspension} \qquad \text{(Eq 3.4)}$$

Example 3.1 **CALCULATION OF CELL NUMBER**

Question 200 cells were counted in four squares of a haemocytometer. Calculate the total number of cells suspended in a final volume of 5 ml, taking into account that the cells were diluted 1:1 before counting.

Answer Using Equation 3.3, the cell density can be calculated as:

$$\text{Cell density} = \frac{200}{4} \times 10^4 \times 2 \times \text{cm}^{-3} = 1\ 000\ 000\ \text{cm}^{-3}$$

Thus, using Equation 3.4, in a final volume of 5 ml (= 5 cm³) the total number of cells present is:

Total number of cells = 10^6 cm^{-3} × 5 cm³ = 5 000 000.

Alternative Methods for Determining Cell Number

Several other methods are available for quantifying cells in culture, including direct measurement using an electronic Coulter counter. This is an automated method of counting and measuring the size of microscopic particles. The instrument itself consists of a glass probe with an electrode that is connected to an oscilloscope (Figure 3.7). The probe has a small aperture of fixed diameter near its bottom end. When immersed in a solution of cell suspension, cells are flushed through the aperture, causing a brief increase in resistance owing to a partial interruption of current flow. This will result in spikes being recorded on the oscilloscope and each spike is counted as a cell. One disadvantage of this method, however, is that it does not distinguish between viable and dead cells.

Indirectly, cells can be counted by determining total cell protein and using a standard curve relating protein content to cell number. However, protein content per cell can vary during culture and may not give a true reflection of cell number. Alternatively, the DNA content of cells may be used as an indicator of cell number, since the DNA content of diploid cells is usually constant. However, the DNA content of cells may change during the cell cycle and therefore not give an accurate estimate of cell number.

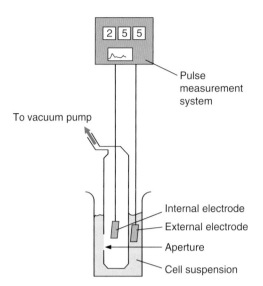

Figure 3.7 Coulter counter. Cells entering the aperture create a pulse of resistance between the internal and external electrodes that is recorded on the oscilloscope.

3.5.7 Seeding Cells onto Culture Plates

Once counted, cells should be seeded at a density that promotes optimal cell growth. It is therefore essential that when cells are subcultured, they are seeded at the appropriate seeding density (see Section 3.5.6). The seeding density will vary, depending on the cell type and on the surface area of the culture flask into which the cells will be placed. These factors should therefore be taken into account when deciding on the seeding density of any given cell type and the purpose of the experiments carried out.

3.5.8 Maintenance of Cells in Culture

It is important that after seeding, flasks are clearly labelled with the date, cell type and the number of times the cells have been subcultured or passaged. Moreover, a strict regime of feeding and subculturing should be established that permits cells to be fed at regular intervals without allowing the medium to be depleted of nutrients or the cells to overgrow or become super confluent. This can be achieved by following a routine procedure for maintaining cells in a viable state under optimum growth conditions. In addition, cultures should be examined daily under an inverted microscope, looking particularly for changes in morphology and cell density. Cell shape can be an important guide when determining the status of growing cultures. Round or floating cells in subconfluent cultures are not usually a good sign and may indicate distressed or dying cells. The presence of abnormally large cells can also be useful in determining the well-being of the cells, since the number of such cells increases as a culture ages or becomes less viable. Extremes in pH should be avoided by regularly replacing spent medium with fresh medium. This may be carried out on alternate days until the cultures are approximately 90% confluent, at which point the cells are either used for experimentation or trypsinised and subcultured following the procedures outlined in Section 3.5.5.

The volume of medium added to the cultures will depend on the confluency of the cells and the surface area of the flasks in which the cells are grown. As a guide, cells which are under 25% confluency may be cultured in approximately 1 cm³ of medium per 5 cm² and those between 25% and 40–45% confluency should be supplemented with 1.5 cm³ and 2 cm³ culture medium per 5 cm², respectively. When changing the medium, it is advisable to pipette the latter onto the sides or the opposite surface of the flask to where the cells are attached. This is to avoid making direct contact with the monolayers, as this will damage or dislodge the cells.

3.5.9 Growth Kinetics of Animal Cells in Culture

When maintained under optimum culture conditions, cells follow a characteristic growth pattern (Figure 3.8), exhibiting an initial lag phase in which there is enhanced cellular activity, but no apparent increase in cell growth. The duration of this phase is dependent on several factors, including the viability of the cells, the density at which the cells are plated and the media component.

The lag phase is followed by a log phase, in which there is an exponential increase in cell number with high metabolic activity. These cells eventually reach a stationary phase where there is no further increase in growth due to depletion of nutrients in the medium, accumulation of toxic metabolic waste or a limitation in available growth space. If left unattended, cells in the stationary phase will eventually begin to die, resulting in the decline phase on the growth curve.

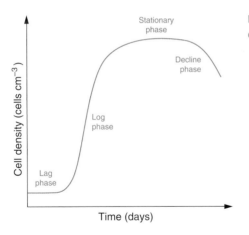

Figure 3.8 Growth curve showing the phases of cell growth in culture.

3.5.10 Cryopreservation of Cells

Cells can be preserved for later use by freezing stocks in liquid nitrogen. This process is referred to as cryopreservation and is an efficient way of sustaining stocks. Indeed, it is advisable that, when good cultures are available, aliquots of cells should be stored in the frozen state. This provides a renewable source of cells that could be used in the future without necessarily having to culture new batches from tissues. Freezing can, however, result in several lethal changes within the cells, including formation of ice crystals and changes in the concentration of electrolytes and in pH. To minimise these risks, a cryoprotective agent such as DMSO is usually added to the cells prior to freezing in order to lower the freezing point and prevent ice crystals from forming inside the cells. The actual cryogenic procedure is itself relatively straightforward. It involves harvesting cells as described in Section 3.5.5 and resuspending them in 1 cm^3 of freezing medium, which is basically culture medium containing 40% serum. The cell suspension is counted and appropriately diluted to give a final cell count of between 10^6 and 10^7 cells cm^{-3}. A 0.9 cm^3 aliquot is transferred into a cryogenic vial labelled with the cell type, passage number and date harvested. This is then made up to 1 cm^3 by adding 0.1 cm^3 of DMSO to give a final concentration of 10%. The freezing process is carried out in stages, allowing the cells initially to cool down slowly from room temperature to −80 °C at a rate of 1–3 K min^{-1}. This initial stage can be carried out using a freezing chamber or alternatively a cryo freezing container filled with isopropanol, which provides the critical repeatable −1 K min^{-1} cooling rate required for successful cell cryopreservation. When this process is complete, the cryogenic vials, which are polypropylene tubes that can withstand temperatures as low as −190 °C, are removed and immediately placed in a liquid nitrogen storage tank, where they can remain for an indefinite period or until required.

All procedures should be carried out under sterile conditions to avoid contaminating cultures, as this will appear once the frozen stocks are recultured. As an added precaution, it is advisable to replace the growth medium in the 24 h period prior to harvesting cells for freezing. Moreover, cells used for freezing should be in the log phase of growth and not too confluent in case they are already in growth arrest.

3.5.11 Resuscitation of Frozen Cells

When required, frozen stocks of cells may be revived by removing the cryogenic vial from storage in liquid nitrogen and placing in a water bath at 37 °C for 1–2 min or until the ice crystals melt. It is important that the vials are not allowed to warm up to 37 °C as this may cause the cells to rapidly die. The thawed cell suspension may then be transferred into a centrifuge tube, to which fresh medium is added and centrifuged at 1000 rpm for 10 min. The supernatant should be discarded to remove the DMSO used in the freezing process and the cell pellet resuspended in 1 cm^3 of fresh medium, ensuring that clumps are dispersed into single cells or much smaller clusters using a glass Pasteur pipette. The required amount of fresh pre-warmed growth medium is placed in a culture flask and the cells pipetted into the flask, which is then placed in a cell culture incubator to allow the cells to adhere and grow.

Practical Hints and Tips for Resuscitation of Frozen Cells
It is important to handle resuscitated cells delicately after thawing as these may be fairly fragile and could degenerate quite readily if not treated correctly. In addition, it is important to dilute the freezing medium immediately after thawing to reduce the concentration of DMSO or freezing agent to which the cells are exposed.

3.5.12 Determination of Cell Viability

Determination of cell viability is extremely important, since the survival and growth of the cells may depend on the density at which they are seeded. The degree of viability is most commonly determined by differentiating living from dead cells using the dye exclusion method. Basically, living cells exclude certain dyes that are readily taken up by dead cells. As a result, dead cells stain in the colour of the dye used whilst living cells remain unstained owing to the inability of the dye to penetrate into the cytoplasm. One of the most commonly used dyes in such assays is Trypan Blue. This is incubated at a concentration of 0.4% with cells in suspension and counted using a haemocytometer as described in Section 3.5.6, keeping separate counts for viable and non-viable cells.

The total number of cells is calculated using equation 3.3, and the percentage of viable cells determined using the following formula:

$$\text{Viability} = \frac{\text{Number of unstained cells counted}}{\text{Total number of cells counted}} \times 100\% \qquad \text{(Eq 3.5)}$$

To avoid underestimating cell viability, it is important that the cells are not exposed to the dye for more than 5 min before counting. This is because uptake of Trypan Blue is time sensitive and the dye may be taken up by viable cells during prolonged incubation periods. Additionally, Trypan Blue has a high affinity for serum proteins and as such may produce high background staining. The cells should therefore be free from serum, which can be achieved by washing the cells with PBS before counting.

3.6 STEM CELL CULTURE

Pluripotent stem cells are unspecialised cells that have the ability to undergo self-renewal, replicating many times over prolonged periods, thereby generating new unspecialised cells. More importantly, pluripotent stem cells have the potential to give rise to specialised cells with specific functions by the process of differentiation. Because of this property, these cells are now being developed and exploited for cell-based therapies in various disease states. It has therefore become essential to be able to isolate, maintain and grow these cells in culture. This section of the chapter will focus on techniques that are now becoming routine for stem cell culture, focussing essentially on human embryonic stem cells (hESCs). The latter are cells derived from the inner cell mass of the blastocyst, which is a hollow microscopic ball made up of an outer layer of cells (the trophoblast), a fluid-filled cavity (the blastocoel) and a cluster of inner cell mass.

Culturing of hESCs can be carried out in a standard cell culture laboratory using equipment already described earlier in the chapter. As with normal cell culture, the important criteria are that good aseptic techniques are adopted, together with good laboratory practice. Unlike normal specialised cells, however, culture of hESCs requires certain conditions specifically aimed at maintaining these cells in a viable undifferentiated state. Historically, hESCs, and indeed other stem cells, have been cultured on feeders, which act to sustain growth and maintain cells in the undifferentiated state without allowing them to lose their pluripotency (i.e. ability to differentiate, when needed, into specialised cell types of the three germ layers). The most common feeder cells used are fibroblasts derived from mouse embryos. The methodology for this, together with other techniques for successful maintenance and propagation of hESCs are described below. Other protocols such as freezing and resuscitation of frozen cells are similar to those already described and the reader is therefore referred to the relevant sections above.

3.6.1 Preparation of Mouse Embryonic Fibroblasts

Typically, fibroblasts are isolated under sterile conditions in a tissue culture cabinet from embryos obtained from mice at 13.5 days of gestation. Each embryo is minced into very fine pieces using sterile scissors and incubated in a cell culture incubator at 37 °C with trypsin/EDTA (0.25% (w/v)/5 mM) for 20 minutes. The mixture is then pipetted vigorously using a fine-bore pipette until it develops a sludgy consistency. This process is repeated, returning the digest into the incubator if necessary, until the embryos have been virtually digested. The trypsin is subsequently neutralised with culture medium containing 10% serum, ensuring that the volume of medium is at least twice that of the trypsin used. The minced tissue is plated onto a tissue culture flask and incubated overnight at 37 °C in a tissue culture incubator. The medium is subsequently removed after 24 h and the cell monolayer washed to remove any tissue debris and non-adherent cells. Adherent cells are cultured to 80–90% confluency for the first phase of propagation to give passage zero (P0), but subsequent cultures should only be allowed to reach 70–80% confluency prior to passaging, as described in Section 3.5.5. If needed, the trypsinised cells can be propagated, otherwise they should be frozen, especially at P0, as described in Section 3.5.10 and used as stock.

Practical Hints and Tips for Using Fibroblast Feeders

Embryos should be pooled for more consistent batches of feeders and to avoid variations between embryos. Additionally, each batch of feeder should be tested for its ability to support cells in an undifferentiated state. Feeders should also be used between passages three and five to ensure support for the growth of undifferentiated cells. After passage five, feeder cells may begin to senesce and may fail to maintain stem cells in the undifferentiated state.

3.6.2 Inactivation of Fibroblast Cells for Use as Feeders

Isolated fibroblasts should be inactivated before they can be used as feeders in order to prevent their proliferation and expansion during culture. This can be achieved using one of two protocols, which include either irradiation or treatment with mitomycin C, an antibiotic DNA cross-linker. With the former, cells in suspension are exposed to 30 Gy of irradiation using a caesium-source γ-irradiator (see Chapter 9). This is the dose of irradiation normally used for mouse fibroblasts; however, the radiation dose and exposure time may vary between species and batches of fibroblasts. As a result, a dose curve should be performed to determine the effective irradiation that is sufficient to stop cell division without cellular toxicity. Once irradiated, cells are subjected to centrifugation at 1000 rpm before resuspending the pellet using the appropriate medium and at the appropriate density for freezing or plating on gelatin-coated plates.

With the mitomycin procedure, cells are normally incubated with the compound at a concentration of 10 µg cm^{-3} for 2–3 h at 37 °C in a cell culture incubator. The mitomycin solution is aspirated and the cells washed several times with PBS or serum-free culture medium to ensure that there are no trace amounts of mitomycin that could affect the stem cells. The cells are then trypsinised, neutralised with serum-containing medium, centrifuged and re-plated onto gelatin-coated dishes at the appropriate cell density.

Practical Hints and Tips for Feeders

Of the two methods, exposure of cells to γ-irradiation is the much preferred methodology because this gives a more consistent and reliable inactivation of cells. More importantly, mitomycin can be harmful and toxic, with embryonic cells showing particular sensitivity to this compound. Use of mitomycin-inactivated fibroblasts should therefore generally be avoided if irradiated feeders can be obtained. If frozen stocks of inactivated feeders are required, these can be prepared as described in Section 3.5.10. It is, however, important to ensure that stocks are not kept frozen for periods exceeding 4 months at −80 °C, to avoid degeneration of cells. In addition, once plated, feeders should be used for stem cell culture within 24 h or no longer than 4 days after plating. It is important to always check the feeder cell density and morphology before use.

3.6.3 Plating of Feeder Cells

As with standard cell culture, fibroblast feeders are plated on tissue-culture-grade plastics, but usually in the presence of a substrate such as gelatin, to provide the extra-cellular matrix component needed for cell attachment of the inactivated fibroblasts. In brief, the plates or flasks are incubated for 1 h at room temperature or overnight at 4 °C with the appropriate volume of 0.1% sterile gelatin. Excess gelatin is subsequently removed and the feeder cells plated at the appropriate density for each cell line, e.g. 3.5 × 10^5 cells per 25-cm^2 flask. Feeders should be ready for use after 5–6 h, but are best left to establish overnight for better results.

Practical Hints and Tips for Plating Feeders

It is important to ensure that the seeding density is optimal for each cell line, otherwise feeders may fail to maintain the hESCs in the undifferentiated state. If frozen stocks of feeders are used for plating, these should be resuscitated, re-suspended in fresh growth medium and plated on gelatin-coated plates, as described in Section 3.5.11. Again the density of post-thaw feeders required to support the cells in an undifferentiated state should be established for each   batch of frozen feeders, since there is cell loss during the freeze–thaw process.

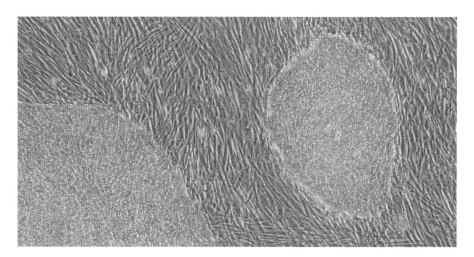

Figure 3.9 Undifferentiated hESCs on mouse feeder cells.

3.6.4 Culture of Human Embryonic Stem Cells

Once the feeders are ready, hESCs can be plated directly onto the feeder layer and the dishes placed in a cell culture incubator. Any non-adherent cells are removed during the first culture medium change, once the cells have attached and established. The cells are monitored and fed on a daily basis until the colonies are ready to be passaged. This can take up to 2 weeks, depending on the conditions of growth.

Practical Hints and Tips for hESC Culture

It is important to ensure that the colonies do not grow too large or to the point where adjacent colonies touch each other as this will initiate their differentiation. Similarly, the seeding density should be high enough to sustain growth, otherwise sparsely plated colonies will grow very slowly and may never establish fully.

Colonies should be plated on healthy feeders that are not more than 4 days old. More importantly, only tightly packed colonies containing cells with the typical hESC morphology should be passaged (see Figure 3.9). Any colony that has a less-defined border (see Figure 3.10) at the periphery, with loose cells spreading out or cells with atypical morphology, should not be passaged because these characteristics are evidence of cell differentiation. Should cells differentiate, these should be excised or aspirated before passaging the undifferentiated cells. Alternatively, if the majority of the colonies appear differentiated and no colonies display the characteristic morphology of undifferentiated cells, then it is advisable to discard the cultures and start with a new batch of undifferentiated hESCs.

3.6.5 Enzymatic Subculture of hESCs

One of the most commonly used enzymes for subculturing hESCs is collagenase. When employed, hESC colonies are washed with PBS and then incubated for 8–10 min with collagenase IV made up in serum-free medium at a concentration of 1 mg cm^{-3}. When ready to harvest, colonies begin to curl at the edges and curled-up colonies can

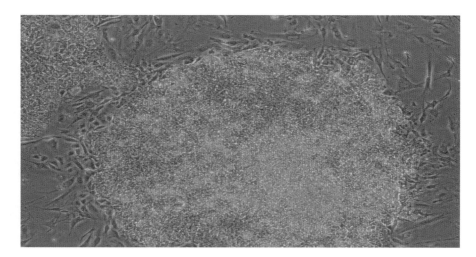

Figure 3.10 Partially differentiated hESCs on mouse feeder cells.

then be dislodged with gentle pipetting using a 5 ml pipette to break large clumps. Alternatively, colonies can be fragmented using glass beads. These are then washed with culture medium to remove the enzyme, which may otherwise impair the attachment and growth of the cells, thus reducing the plating efficiency. hESCs can be washed by allowing the colonies to sediment slowly over 5–10 mins, leaving any residual feeder cells in the supernatant, which is removed by aspiration. The colonies are subsequently resuspended in growth medium and are usually plated at a ratio of between 1:3 and 1:6. Alternatively, fragmented colonies could be frozen as described in Section 3.5.10 and stored for later use.

3.6.6 Mechanical Subculture of hESCs

An alternative to the enzymatic method of subculturing hESCs is to manually cut colonies into appropriate-size fragments using a fine-bore needle or a specially designed cutter such as the STEMPRO® EZPassage™ disposable stem cell passaging tool from ThermoFisher Scientific. To do this, the dish of hESCs is placed under a dissecting microscope in a tissue culture hood. Undifferentiated colonies are identified by their morphology and then cut into grids (see Figure 3.11) by scoring across and perpendicular to the first cut. Using a 1 ml pipette or Pasteur pipette, the cut segments are transferred to dishes containing fresh feeders and culture medium. The colony fragments are placed evenly across the feeders (see Figure 3.12) to avoid the colonies clumping together and attaching to the dish as one mass of cells. The dishes are then carefully transferred to a tissue culture incubator and left undisturbed until attached, before replacing the spent medium with fresh. Established colonies are then fed every day until subcultured.

3.6.7 Feeder-Free Culture of hESCs

Although culture of hESCs on feeders has been extensively used, there have been concerns over this procedure when stem cells are being considered for clinical use in

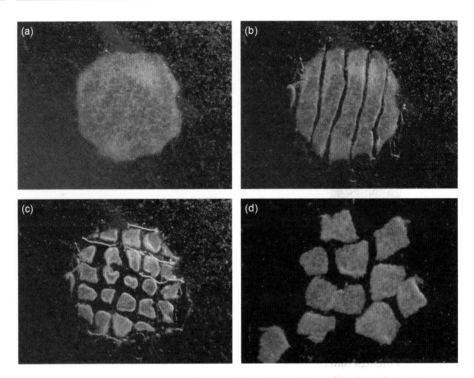

Figure 3.11 Mechanically harvested hESCs: (a) stem cell monolayer; (b) cut stem cell monolayer; (c) segments cut from stem cell monolayer; (d) isolated segments of cut stem cell monolayer.

humans. One of the main drawbacks of using feeders is the concern over potential transmission of animal pathogens to humans and the possibility of expression of immunogenic antigens. Feeders are also inconvenient, expensive and time-consuming to generate and inactivate. As a result of these limitations, there has been a drive towards developing a feeder-free culture system using feeder-conditioned media or media supplemented with different growth factors and other signalling molecules essential for sustaining growth. The conditioned medium can be generated by incubating normal growth medium with feeder cells for 24 h before use.

Feeder-free culture of hESCs is often carried out on tissue culture plastics coated with Matrigel™, a substrate derived from mouse tumour and rich in extra-cellular matrix proteins such as laminin, collagen and heparan sulfate proteoglycan. It is also rich in growth factors such as basic fibroblast growth factor (bFGF), which can help to sustain and promote stem cell growth, whilst maintaining them in an undifferentiated state.

Practically, dishes are coated with 1:30 dilution of Matrigel™ made up in culture medium. Just prior to use, the Matrigel™ is removed and replaced with culture medium before plating cells. The hESCs, subcultured from feeders or obtained from frozen stocks, are resuspended in conditioned medium often supplemented with bFGF at a concentration of 4 ng ml^{-1} before seeding. An alternative is to use mTeSR™1, a highly specialised serum-free medium that is reported to support the growth of homogeneous undifferentiated cultures.

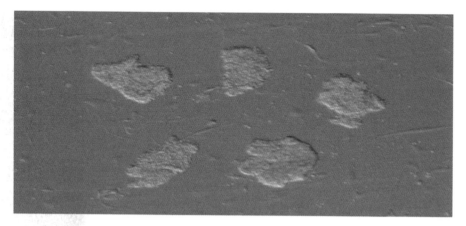

Figure 3.12 Plating of hESCs onto feeder layer.

Once established, hESCs are fed every day with fresh growth medium. Colonies on Matrigel™ tend to show a different morphology to those on feeders, tending to be larger and less packed initially.

Practical Hints and Tips for Using Matrigel™

All work with Matrigel™, other than plating of the hESCs, should be carried out at 4 °C. Thus, when coating tissue culture plastics with Matrigel™, all the plates and pipette tips should be kept on ice and used cold to prevent the Matrigel™ solidifying. Stock Matrigel™ is usually in the solid form and should be placed on ice or in the fridge at 4 °C overnight until it liquefies. Once liquefied, the Matrigel™ should be diluted in ice-cold culture medium at a final concentration of 1:30 or as recommended. Each plate should have a smooth even layer of Matrigel™ and if this is not the case, the plates should be incubated at 4 °C until the Matrigel™ liquefies and settles as a uniform layer. Once coated, Matrigel™ plates should be used within 7 days of preparation.

3.7 BACTERIAL CELL CULTURE

As with animal cells, pure bacterial cultures (cultures that contain only one species of organism) are cultivated routinely and maintained indefinitely using standard sterile techniques that are now well defined. However, since bacterial cells exhibit a much wider degree of diversity in terms of both their nutritional and environmental requirements, conditions for their cultivation are diverse and the precise requirements highly dependent on the species being cultivated. Outlined below are general procedures and precautions adopted in bacterial cell culture.

3.7.1 Safety Considerations for Bacterial Cell Culture

Culture of microbial cells, like that involving cells of animal origin, requires care and sterile techniques, not least to prevent accidental contamination of pure cultures with other organisms. More importantly, the utmost care should be given towards protecting the operator, especially from potentially harmful organisms. Aseptic techniques and safety conditions described for animal cell culture should be adopted at all times. Additionally, instruments used during the culturing procedures should be sterilised before and after use by heating in a Bunsen burner flame. Moreover, to avoid spread of bacteria, areas of work must be decontaminated after use using germicidal sprays and/or ultraviolet radiation. This is to prevent airborne bacteria from spreading rapidly. In line with these precautions, all materials used in microbial cell culture work must be disposed of appropriately; for instance, autoclaving of all plastics and tissue culture waste before disposal is usually essential.

3.7.2 Nutritional Requirements of Bacteria

The growth of bacteria requires much simpler conditions than those described for animal cells. However, due to their diversity, the composition of the medium used may be variable and largely determined by the nutritional classification of the organisms to be cultured. These generally fall into two main categories, classified as either autotrophs (self-feeding organisms that synthesise their own food in the form of sugars) or heterotrophs (non-self-feeding organisms that derive chemical energy by breaking down organic molecules consumed). These in turn are subgrouped into chemo- or photoautotrophs or -heterotrophs. Both chemo- and photoautotrophs rely on carbon dioxide as a source of carbon, but derive energy from completely different sources, with the chemoautotrophs utilising inorganic substances, whilst the photoautotrophs use light. Chemoheterotrophs and photoheterotrophs both use organic compounds as the main source of carbon with the photoheterotrophs using light for energy and the chemo subgroup getting their energy from the metabolism of organic substances.

3.7.3 Culture Media for Bacterial Cell Culture

Several different types of medium are used to culture bacteria, and these can be categorised as either complex or defined. The former usually consists of natural substances, including meat and yeast extract, and as a result is less well defined since the precise composition is largely unknown. Such media are, however, rich in nutrients and therefore generally suitable for culturing fastidious organisms that require a mixture of nutrients for growth. Defined media, by contrast, are relatively simple. These are usually designed to the specific needs of the bacterial species to be cultivated and as a result are made up of known components put together in the required amounts. This flexibility is usually exploited to select or eliminate certain species by taking advantage of their distinguishing nutritional requirements. For instance, bile salts may be included in media when selective cultivation of enteric bacteria (rod-shaped Gram-negative bacteria such as *Salmonella* or *Shigella*) is required, since growth of most other Gram-positive and Gram-negative bacteria will be inhibited.

3.7.4 Culture Procedures for Bacterial Cells

Bacteria can be cultured in the laboratory using either liquid or solid media. Liquid media are normally dispensed into flasks and inoculated with an aliquot of the organism to be grown. This is then agitated continuously on a shaker that rotates in an orbital manner, mixing and ensuring that cultures are kept in suspension. For such cultures, sufficient space should be allowed above the medium to facilitate adequate diffusion of oxygen into the solution. Thus, as a rule of thumb, the volume of medium added to the flasks should not exceed more than 20% of the total volume of the flask. This is particularly important for aerobic bacteria and less so for anaerobic microorganisms. At a medium scale, bacterial cultures are grown in culture flasks mounted on orbital shakers. Best results are obtained when the culture volume is approximately 10% to 25% of the flask volume; this is because a smaller volume provides a comparatively larger surface area exposed to air, thus enabling optimum aeration (i.e. dissolved oxygen). In large-scale culture, fermenters or bioreactors equipped with stirring devices for improved mixing and gas exchange may be used. The device (Figure 3.13) is usually fitted with probes that monitor changes in pH, oxygen concentration and temperature. In addition, most systems are surrounded by a water jacket with fast-flowing cold water to reduce the heat generated during fermentation. Outlets are also included to release CO_2 and other gases produced by cell metabolism.

When fermenters are used, precautions should be taken to reduce potential contamination with airborne microorganisms when air is bubbled through the cultures. Sterilisation of the air may therefore be necessary and can be achieved by introducing a filter (pore size of approximately 0.2 µm) at the point of entry of the air flow into the chamber.

Solid medium is usually prepared by solidifying the selected medium with 1–2% of the seaweed extract agar, which, although organic, is not degraded by most microbes, thereby providing an inert gelling medium on which bacteria can grow. Solid agar media are widely used to separate mixed cultures and form the basis for isolation of pure cultures of bacteria. This is achieved by streaking diluted cultures of bacteria onto the surface of an agar plate by using a sterile inoculating loop. Cells streaked across the plate will eventually grow into a colony, each colony being the product of a single cell and thus of a single species.

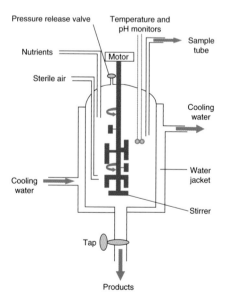

Figure 3.13 Schematic representation of a fermenter.

Once isolated, cells can be cultivated either in batch or continuous cultures. Of these, batch cultures are the most commonly used for routine liquid growth and entail inoculating an aliquot of cells into a sterile flask containing a finite amount of medium. Such systems are referred to as closed, since nutrient supply is limited to that provided at the start of culture. Under these conditions, growth will continue until the medium is depleted of nutrients or there is an excessive build-up of toxic waste products generated by the microbes. Thus, in this system, the cellular composition and physiological status of the cells will vary throughout the growth cycle.

In continuous cultures (also referred to as open systems), the medium is refreshed regularly to replace that spent by the cells. The objective of this system is to maintain the cells in the exponential growth phase by enabling nutrients, biomass and waste products to be controlled through varying the dilution rate of the cultures. Continuous cultures, although more complex to set up, offer certain advantages over batch cultures in that they facilitate growth under steady-state conditions in which there is tight coupling between cell division and biosynthesis. As a result, the physiological status of the cultures is more clearly defined, with very little variation in the cellular composition of the cells during the growth cycle. The main concern with the open system is the high risk of contamination associated with the dilution of the cultures. However, applying strict aseptic techniques during feeding or harvesting cells may help to reduce the risk of such contamination. In addition, the whole system can be automated by connecting the culture vessels to their reservoirs through solenoid valves that can be triggered to open when required. This minimises direct contact with the operator or outside environment and thus reduces the risk of contamination.

3.7.5 Determination of Growth of Bacterial Cultures

Several methods are available for determining the growth of bacterial cells in culture, including directly counting cells using a haemocytometer as described (Section 3.5.6). This is, however, suitable only for cells in suspension. When cells are grown on solid agar plates, colony counting can be used instead to estimate growth. This method assumes that each colony is derived from a single cell, which may not always be the case, since errors in dilution and/or streaking may result in clumps rather than single cells producing colonies. In addition, suboptimal culture conditions may cause poor growth, thus leading to an underestimation of the true cell count. When cells are grown in suspension, changes in the turbidity (see Section 13.2.2) of the growth medium could be determined using a spectrophotometer and the apparent absorbance value converted to cell number using a standard curve of optical density (see Section 13.2.2) versus cell number. For example, the optical density of E. coli at 600 nm (OD_{600}) can be measured in a plastic cuvette; an OD_{600} value of 1.0 is equal to 8×10^8 cells per ml. Such calibration curves should be constructed for each cell type by taking the readings of a series of known numbers of cells in suspension.

3.8 POTENTIAL USE OF CELL CULTURES

Cell cultures of various sorts from animal and microbes are increasingly exploited, not only by scientists for studying the physiology and biochemistry of cells in isolation, but also by various biotechnology and pharmaceutical companies for drug screening and for the production of valuable biological products, including viral vaccines (e.g. polio vaccine), antibodies (e.g. OKT3 used in suppressing immunological organ rejection in transplant surgery) and various recombinant proteins. The application of recombinant DNA techniques has led to an ever-expanding list of improved products, from both mammalian and bacterial cells, for therapeutic use in humans. These products include the commercial production of factor VIII for haemophilia, insulin for diabetes, interferon α and β for anticancer chemotherapy and erythropoietin for anaemia. Bacterial cultures have also been widely used for other industrial purposes, including the large-scale production of cell proteins, growth regulators, organic acids, alcohols, solvents, sterols, surfactants, vitamins, amino acids and many more products. In addition, degradation of waste products, particularly those from the agricultural and food industries is another important industrial application of microbial cells. They are also exploited in the bioconversion of waste to useful end products, and in toxicological studies where some of these organisms are rapidly replacing animals in preliminary toxicological testing of xenobiotics. The advent of stem cell culture now provides the possibility of treating diseases using cell-based therapy. This would be particularly important in regenerating diseased or damaged tissues by transplanting stem cells programmed to differentiate into a specific cell type specialised in carrying out a specific function.

3.9 ACKNOWLEDGEMENTS

Images courtesy of Lesley Young and Paula M Timmons, UK Stem Cell Bank, NIBSC, United Kingdom. Thanks also to Lyn Healy, UK Stem Cell Bank, NIBSC, United Kingdom for valuable comments and advice on stem cell culture.

3.10 SUGGESTIONS FOR FURTHER READING

3.10.1 Experimental Protocols

Freshney R.I. (2010) *Culture of Animal Cells: A Manual of Basic Technique and Specialized Applications*, 6th Edn., John Wiley & Sons, Inc., New York, USA.

HSE Advisory Committee on Dangerous Pathogens (2001) *The Management Design and Operation of Microbiological Containment Laboratories*, HSE books, Sudbury, UK.

Parekh S. R. and Vinci V.A. (2003) *Handbook of Industrial Cell Culture: Mammalian, Microbial, and Plant Cells*, Humana Press, Totowa, NJ, USA.

Peterson S. and Loring J. (2012) *Human Stem Cell Manual: A Laboratory Guide*, 2nd Edn., Academic Press, Cambridge, MA, USA.

3.10.2 General Texts

Ball A.S. (1997) *Bacterial Cell Culture: Essential Data*, John Wiley & Sons, New York, USA.

Davis J.M. (2011) *Animal Cell Culture: Essential Methods*, Wiley-Blackwell, Chichester, UK.

Furr A.K. (ed.) (2001) *CRC Handbook of Laboratory Safety*, 5th Edn., CRC Press, Boca Raton, FL, USA.

3.10.3 Review Articles

Geraghty R., Capes-Davis A., Davis J.M., *et al.* (2014) Guidelines for the use of cell lines in biomedical research. *British Journal of Cancer* 111, 1021–1046.

3.10.4 Websites

ATCC® Primary Cell Culture Guide
www.atcc.org/~/media/PDFs/Culture%20Guides/Primary_Cell_Culture_Guide.ashx (accessed April 2017)

Introduction to Cell Culture (ThermoFisher Scientific)
www.thermofisher.com/au/en/home/references/gibco-cell-culture-basics/introduction-to-cell-culture.html (accessed April 2017)

4 Recombinant DNA Techniques and Molecular Cloning

RALPH RAPLEY

4.1 INTRODUCTION

The human genome contains the blueprint for human development and maintenance, and may ultimately provide the means to understand human cellular and molecular processes in both health and disease. The genome is the full complement of DNA from an organism and carries all the information needed to specify the structure of every protein the cell can produce. The realisation that DNA lies behind all of the cell's activities led to the development of what is termed molecular biology. Rather than a discrete area of the biosciences, molecular biology is now accepted as a very important means of understanding and describing complex biological processes. The development of methods and techniques for studying processes at the molecular level has led to new and powerful ways of isolating, analysing, manipulating and exploiting nucleic acids. Moreover, to keep pace with the explosion in biological information, the discipline termed bioinformatics has evolved and provides a vital role in current biosciences. The completion of the human genome project and numerous other genome projects has allowed the continued development of new, exciting areas of biological sciences such as biotechnology, genome mapping, molecular medicine and gene therapy.

In considering the potential utility of molecular biology techniques it is important to understand the basic structure of nucleic acids and gain an appreciation of how this dictates the function in vivo and in vitro. Indeed, many techniques used in molecular biology mimic in some way the natural functions of nucleic acids such as replication and transcription. This chapter is therefore intended to provide an overview of the general features of nucleic acid structure and function, and describe some of the basic methods of recombinant DNA technology, together with an overview of molecular cloning and associated methods.

4.2 STRUCTURE OF NUCLEIC ACIDS

4.2.1 Primary Structure of Nucleic Acids

Deoxyribonucleic acid (DNA) and ribonucleic acid (RNA) are macromolecular structures composed of regular repeating polymers formed from nucleotides. These are the basic building blocks of nucleic acids and are derived from nucleosides, which are composed of two elements: a five-membered pentose carbon sugar (2-deoxyribose in DNA and ribose in RNA) and a nitrogenous base. The carbon atoms of the sugar are designated 'prime' (1', 2', 3', etc.) to distinguish them from the carbons of the

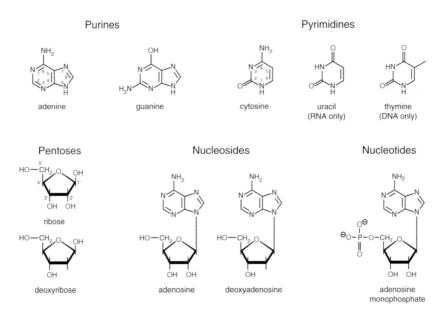

Figure 4.1 Structure of the building blocks of nucleic acids.

nitrogenous bases of which there are two types, purine and pyrimidine. A nucleotide, or nucleoside phosphate, is formed by the attachment of a phosphate to the 5′ position of a nucleoside by an ester linkage (Figure 4.1). Such nucleotides can be joined together by the formation of a second ester bond by reaction between the phosphate of one nucleotide and the 3′-hydroxyl of another, thus generating a 5′ to 3′ phosphodiester bond between adjacent sugars; this process can be repeated indefinitely to give long polynucleotide molecules (Figure 4.2). DNA has two such polynucleotide strands; however, since each strand has both a free 5′-hydroxyl group at one end and a free 3′-hydroxyl at the other, each strand has a polarity or directionality. The polarities of the two strands of the molecule are in opposite directions, and thus DNA is described as an **anti-parallel** structure (Figure 4.3).

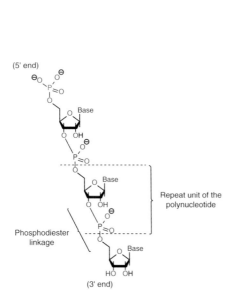

Figure 4.2 Polynucleotide structure.

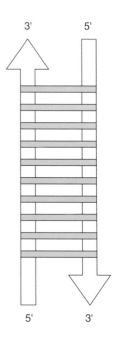

Figure 4.3 The anti-parallel nature of DNA. Two complementary strands of DNA in a double helix run in the opposite directions. The two strands are held together by hydrogen bonds between the bases.

The **purine bases** (composed of fused five- and six-membered rings), adenine (A) and guanine (G), are found in both RNA and DNA, as is the pyrimidine base (a single six-membered ring) cytosine (C). The other **pyrimidine bases** are each restricted to one type of nucleic acid: uracil (U) occurs exclusively in RNA, whilst thymine (T) is limited to DNA. Thus it is possible to distinguish between RNA and DNA on the basis of the presence of ribose and uracil in RNA, and deoxyribose and thymine in DNA. However, it is the sequence of bases along a molecule that distinguishes one DNA (or RNA) from another. It is conventional to write a nucleic acid sequence starting at the 5′ end of the molecule, using single capital letters to represent each of the bases, e.g. CGGATCT. Note that there is usually no point in indicating the sugar or phosphate groups, since these are identical throughout the length of the molecule. Terminal phosphate groups can, when necessary, be indicated by use of a 'p'; thus 5′-pCGGATCT-3′ indicates the presence of a phosphate on the 5′ end of the molecule.

4.2.2 Secondary Structure of Nucleic Acids

The two polynucleotide chains in DNA are usually found in the shape of a right-handed double helix, in which the bases of the two strands lie in the centre of the molecule, with the sugar–phosphate backbones on the outside. A crucial feature of this double-stranded structure is that it depends on the sequence of bases in one strand being complementary to that in the other. A purine base attached to a sugar residue on one strand is always linked by hydrogen bonds to a pyrimidine base attached to a sugar residue on the other strand. Moreover, adenine (A) always pairs with thymine (T) or uracil (U) in RNA, via two hydrogen bonds, and guanine (G) always pairs with cytosine (C) by three hydrogen bonds (Figure 4.4). When these conditions are met, a stable double helical structure results, in which the backbones of the two strands are, on average, a constant distance apart. Thus, if the sequence of one strand is known, that of the other strand can be deduced. The strands are designated as plus (+) and minus (−), and an RNA molecule complementary to the minus (−) strand is synthesised during transcription. The base sequence may cause significant local variations in the shape of the DNA molecule and these variations are vital for specific interactions between the DNA and its binding proteins to take

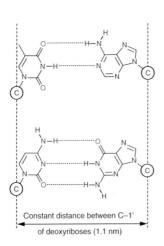

Figure 4.4 Base-pairing in DNA. 'C' in a circle represents carbon at the 1′ position of deoxyribose.

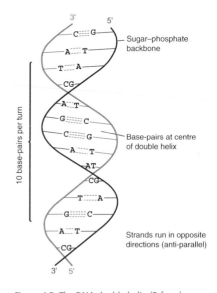

Figure 4.5 The DNA double helix (B form).

Table 4.1 **The various forms of DNA**

DNA form	% Humidity	Helix direction	No of bases per helix turn	Helix diameter (Å)
B	92%	RH	10	19
A	75%	RH	11	23
C	66%	RH	9.3	19
Z	$(Pu-Py)_n$	LH	12	18

Notes: RH, right-handed helix; LH, left-handed helix; Pu, Purine; Py, Pyrimidine. Different forms of DNA may be obtained by subjecting DNA fibres to different relative humidities. The B form is the most common form of DNA whilst the A and C forms have been derived under laboratory conditions. The Z form may be produced with a DNA sequence made up from alternating purine and pyrimidine nucleotides.

place. Although the three-dimensional structure of DNA may vary, it generally adopts a double helical structure termed the B form or B-DNA in vivo. There are also other forms of right-handed DNA such as A, C and Z, which are formed when DNA fibres are experimentally subjected to different relative humidities (Table 4.1). Experimental evidence suggests these alternative forms occur in both prokaryotic and eukaryotic cells.

The major distinguishing feature of B-DNA is that it has approximately 10 bases for one turn of the double helix (Figure 4.5); furthermore a distinctive major and minor groove may be identified. In the B form of DNA, the major groove is wider than the minor groove and therefore DNA-binding proteins such as transcription factors bind to the DNA molecule at the wider major groove. The minor groove is a binding site for the dye Hoechst 33258. In certain circumstances. In certain circumstances where repeated DNA sequences or motifs are found, the DNA may adopt a left-handed helical structure termed Z-DNA. This form of DNA was first synthesised in the laboratory and is thought not to exist in vivo. The various forms of DNA serve to show that it is not a static molecule, but dynamic and constantly in flux, and may be coiled, bent or distorted at certain times. Although RNA almost always exists as a single strand, it often contains sequences within the same strand that are self-complementary, and which can therefore base-pair if brought together by suitable folding of the molecule. A notable example is transfer RNA (tRNA) which folds up to give a clover-leaf secondary structure (Figure 4.6).

4.2.3 Separation of Double-Stranded DNA

The two anti-parallel strands of DNA are held together partly by the weak forces of hydrogen bonding between complementary bases and partly by hydrophobic interactions between adjacent, stacked base pairs, termed base-stacking. Little energy is needed to separate a few base pairs, and so, at any instant, a few short stretches of

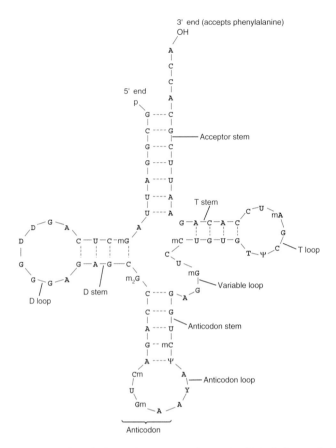

Figure 4.6 Clover-leaf secondary structure of yeast tRNA^{Phe}. A single strand of 76 ribonucleotides forms four double-stranded 'stem' regions by base-pairing between complementary sequences. The anticodon will base-pair with UUU or UUC (both are codons for phenylalanine); phenylalanine is attached to the 3′ end by a specific aminoacyl tRNA synthetase. Several 'unusual' bases are present: D, dihydrouridine; T, ribothymidine; ψ, pseudouridine; Y, very highly modified, unlike any 'normal' base. mX indicates methylation of base X (m$_2$X shows dimethylation); Xm indicates methylation of ribose on the 2′ position; p, phosphate.

DNA will be opened up to the single-stranded conformation. However, such stretches immediately pair up again at room temperature, so the molecule as a whole remains predominantly double-stranded.

If, however, a DNA solution is heated to approximately 90 °C or above, there will be enough kinetic energy to denature the DNA completely, causing it to separate into single strands. This is termed **denaturation** and can be followed spectrophotometrically by monitoring the absorbance of light at 260 nm (see Section 13.2.2). The stacked bases of double-stranded DNA are less able to absorb light than the less constrained bases of single-stranded molecules, and so the absorbance of DNA at 260 nm increases as the DNA becomes denatured, a phenomenon known as the **hyperchromic effect**.

The absorbance at 260 nm may be plotted against the temperature of a DNA solution, which will indicate that little denaturation occurs below approximately 70 °C,

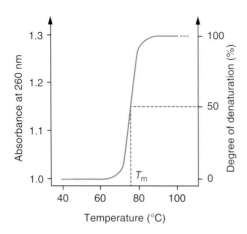

Figure 4.7 Melting curve of DNA.

but further increases in temperature result in a marked increase in the extent of denaturation. Eventually a temperature is reached at which the sample is totally denatured, or melted. The temperature at which 50% of the DNA is melted is termed the **melting temperature** (T_m), and this depends on the nature of the DNA (Figure 4.7). If several different samples of DNA are melted, it is found that the T_m is highest for those DNA molecules that contain the highest proportion of cytosine and guanine, and T_m can actually be used to estimate the percentage (C + G) in a DNA sample. This relationship between T_m and (C + G) content arises because cytosine and guanine form three hydrogen bonds when base-paired, whereas thymine and adenine form only two. Because of the differential numbers of hydrogen bonds between A–T and C–G pairs those sequences with a predominance of C–G pairs will require greater energy to separate or denature them. The conditions required to separate a particular nucleotide sequence are also dependent on environmental conditions such as salt concentration.

If melted DNA is cooled, it is possible for the separated strands to reassociate, a process known as **renaturation**. However, a stable double-stranded molecule will only be formed if the complementary strands collide in such a way that their bases are paired precisely, and this is an unlikely event if the DNA is very long and complex (i.e. if it contains a large number of different genes). Measurements of the rate of renaturation can give information about the complexity of a DNA preparation.

Strands of RNA and DNA will associate with each other, if their sequences are complementary, to give double-stranded, hybrid molecules. Similarly, strands of radioactively labelled RNA or DNA, when added to a denatured DNA preparation, will act as probes for DNA molecules to which they are complementary. This hybridisation of complementary strands of nucleic acids is very useful for isolating a specific fragment of DNA from a complex mixture. It is also possible for small single-stranded fragments of DNA (up to 40 bases in length), termed **oligonucleotides**, to hybridise to a denatured sample of DNA. This type of hybridisation is termed **annealing** and again is dependent on the base sequence of the oligonucleotide and the salt concentration of the sample.

4.3 GENES AND GENOME COMPLEXITY

4.3.1 Gene Complexity

Each region of DNA that codes for a single RNA or protein is called a gene, and the entire set of genes in a cell, organelle or virus forms its genome. Cells and organelles may contain more than one copy of their genome. Genomic DNA from nearly all prokaryotic and eukaryotic organisms is also complexed with protein and termed chromosomal DNA. Each gene is located at a particular position along the chromosome, termed the locus, whilst the particular form of the gene is termed the allele. In mammalian DNA, each gene is present in two allelic forms that may be identical (homozygous) or that may vary (heterozygous). There are approximately 27 000 genes present in the human genome, giving rise to a conservatively estimated ~60 000 known transcripts. Not all will be expressed in a given cell at the same time and it is more common for a particular cell type to express only a limited repertoire of genes. The occurrence of different alleles at the same site in the genome is termed polymorphism. In general, the more complex an organism the larger its genome, although this is not always the case, since many higher organisms have non-coding sequences, some of which are repeated numerous times and termed repetitive DNA. In mammalian DNA, repetitive sequences may be divided into low-copy-number and high-copy-number DNA. The latter is composed of repeat sequences that are dispersed throughout the genome and those that are clustered together. The repeat cluster DNA may be defined into so-called classical satellite DNA, minisatellite and microsatellite DNA (Table 4.2), the last being mainly composed of dinucleotide repeats. These sequences are termed polymorphic, collectively termed polymorphisms, and vary between individuals; they also form the basis of genetic fingerprinting.

Table 4.2 **Repetitive satellite sequences found in DNA and their characteristics**

Types of repetitive DNA	Repeat unit size (bp)	Characteristics/motifs
Satellite DNA	5–200	Large repeat unit range (Mb) usually found at centromeres
Minisatellite DNA		
Telomere sequence	6	Found at the ends of chromosomes. Repeat unit may span up to 20 kb G-rich sequence
Hypervariable sequence	10–60	Repeat unit may span up to 20 kb
Microsatellite DNA	1–4	Mononucleotide repeat of adenine dinucleotide repeats common (CA). Usually known as VNTR (variable number tandem repeat)

Notes: bp, base-pairs; kb, kilobase-pairs.

4.3.2 Single Nucleotide Polymorphisms (SNPs)

A further important source of polymorphic diversity known to be present in genomes is termed single nucleotide polymorphisms or SNPs (pronounced snips). SNPs are substitutions of one base at a precise location within the genome. Those that occur in coding regions are termed cSNPs. Estimates indicate that a SNP occurs every once in every 300 bases, which based on the size of the human genome (3×10^9 bases) would give 10 million SNPs. However, results from the 1000 Genomes Project suggest an average of 3.6 million SNPs per individual. Interest in SNPs lies in the fact that these polymorphisms may account for the differences in disease susceptibility, drug metabolism and response to environmental factors between individuals. Indeed, there are now a number of initiatives to identify SNPs and produce genomic SNP maps. One initiative is the international HapMap project, which is the production of a haplotype map of common sources of variations from groups of associated SNPs. This will potentially allow a set of so-called tag SNPs to be identified and potentially provide an association between the haplotype and a disease.

4.3.3 Chromosomes and Karyotypes

Higher organisms may be identified by using the size and shape of their genetic material at a particular point in the cell division cycle, termed the metaphase. At this point, DNA condenses to form a number of very distinct chromosome structures. Various morphological characteristics of chromosomes may be identified at this stage, including the centromere and the telomere. The array of chromosomes from a given organism may also be stained with dyes such as Giemsa stain and subsequently analysed by light microscopy. The complete array of chromosomes in an organism is termed the karyotype. In certain genetic disorders aberrations in the size, shape and number of chromosomes may occur and thus the karyotype may be used as an indicator of the disorder. Perhaps the most well-known example of such a correlation is the Down syndrome where three copies of chromosome 21 (trisomy 21) exist rather than two as in the normal state.

4.3.4 Renaturation Kinetics and Genome Complexity

When preparations of double-stranded DNA are denatured and allowed to renature, measurement of the rate of renaturation can give valuable information about the complexity of the DNA, i.e. how much information it contains (measured in base-pairs). The complexity of a molecule may be much less than its total length if some sequences are repetitive, but complexity will equal total length if all sequences are unique, appearing only once in the genome. In practice, the DNA is first cut randomly into fragments about 1 kb in length, and is then completely denatured by heating above its T_m. Renaturation at a temperature about 10 °C below the T_m is monitored either by decrease in absorbance at 260 nm (the hypochromic effect), or by passing samples at intervals through a column of hydroxylapatite, which will adsorb only double-stranded DNA, and measuring how much of the sample is bound. The degree of renaturation after a given time will depend on the concentration of double-stranded

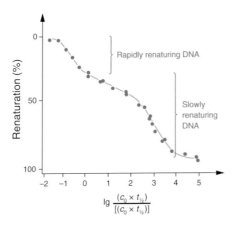

Figure 4.8 Cot curve of human DNA. DNA was allowed to renature at 60 °C after being completely dissociated by heat. Samples were taken at intervals and passed through a hydroxylapatite column to determine the percentage of double-stranded DNA present. This percentage was plotted against $\lg \{(c_0 \times t_{1/2})/[(c_0 \times t_{1/2})]\}$. (Because the argument of a logarithm cannot include units, $(c_0 \times t_{1/2})$ needs to be divided by its units, denoted as $[(c_0 \times t_{1/2})]$.)

DNA prior to denaturation (c_0; in nucleotides per unit volume), and t, the duration of the renaturation (in seconds).

For a given c_0, it should be evident that a preparation of bacteriophage λ DNA (genome size 49 kb) will contain many more copies of the same sequence per unit volume than a preparation of human DNA (haploid genome size 3×10^6 kb), and will therefore renature far more rapidly, since there will be more molecules complementary to each other per unit volume in the case of λ DNA, and therefore more chance of two complementary strands colliding with each other. In order to compare the rates of renaturation of different DNA samples, one usually measures c_0 and the time taken for renaturation to proceed halfway to completion, $t_{1/2}$, and then multiplies these values together to give a $c_0 \times t_{1/2}$ value. The larger this value, the greater the complexity of the DNA; hence λ DNA has a far lower $c_0 \times t_{1/2}$ than does human DNA.

In fact, the human genome does not renature in a uniform fashion. If the extent of renaturation is plotted against $\lg \{(c_0 \times t_{1/2})/[(c_0 \times t_{1/2})]\}$ – this is known as a cot curve – it is seen that part of the DNA renatures quite rapidly, whilst the remainder is very slow to renature (Figure 4.8). This indicates that some sequences have a higher concentration than others; in other words, part of the genome consists of repetitive sequences. These repetitive sequences can be separated from the single-copy DNA by passing the renaturing sample through a hydroxylapatite column early in the renaturation process, at a time which gives a low value of $c_0 \times t_{1/2}$. At this stage, only the rapidly renaturing sequences will be double-stranded, and they will therefore be the only ones able to bind to the column.

4.3.5 The Nature of the Genetic Code

DNA encodes the primary sequence of a protein by utilising sets of three nucleotides, termed a codon or triplet, to encode a particular amino acid. The four bases (A, C, G and T) present in DNA allow a possible 64 triplet combinations; however, since there are only 20 naturally occurring amino acids, more than one codon may encode an amino acid. This phenomenon is termed the degeneracy of the genetic code. With the exception of a limited number of differences found in mitochondrial DNA and one or two other species, the genetic code appears to be universal. In addition to coding for amino acids, particular triplet sequences also indicate the beginning (Start) and the end (Stop) of a particular gene. Only one start codon exists (ATG), which also codes for the amino acid methionine, whereas three dedicated stop codons are available (TAT, TAG and TGA) (Table 0.11). A sequence flanked by a start and a stop codon containing a number of codons that may be read in-frame to represent a continuous protein sequence is termed an **open reading frame** (ORF).

4.3.6 Epigenetics and the Nature of DNA

It has been well recognised for some time that DNA as a structure may be chemically modified without the underlying DNA sequence being altered. The most important modification is the addition of a methyl (CH_3) group to the 5′-carbon on cytosine to give methylcytosine (5mC) and results in what some describe as the fifth base in DNA. This is termed DNA methylation and catalysed by DNA methyltransferases. The DNA methyltransferases invariantly modify the cytosine in the pattern CpG (C-phosphate-G) and approximately 75% of CpGs in the human genome are methyl-ated, which equates to around 1.5% of human DNA (termed the **epigenome**).

The next most common DNA modifications are generated by the Ten-eleven trans-location (Tet) enzymes that catalyse the oxidation of 5mC to 5hmC (5-hydroxymethyl-cytosine). Further modification of 5hmC is achieved in a stepwise fashion to generate 5fC (5-formylcytosine) and 5caC (5-carboxycytosine), reflecting an active process of demethylation. However, rather than merely adducts produced in the pathway to demethylation, evidence is growing that these additional DNA modifications can also perform specific regulatory functions and global loss of 5hmC is a common hallmark of cancer cells. CpG sites are not evenly distributed in the genome; a background CpG level of 1/100 can be compared to 1/10 nucleotides in regions known as **CpG islands**. These high CpG density regions occur at 40% of gene promoter regions and methyla-tion of these sites inhibits gene expression. This feature of gene expression control is termed epigenetics and is a complex process which, in addition to DNA methylation, includes the modification of histone proteins and some small RNA molecules involved in gene expression control. Importantly, epigenetics appears to play a role in differ-entiation, disease states such as cancers (hyper- and hypomethylation) and certain neurological diseases which may lead to a new means of future treatment.

4.4 LOCATION AND PACKAGING OF NUCLEIC ACIDS

4.4.1 Cellular Compartments

The genetic information of cells and most viruses is stored in the form of DNA. This information is used to direct the synthesis of RNA molecules. In general, DNA in eukaryotic cells is confined to the nucleus and organelles such as mitochondria or chloroplasts, which contain their own genome. The predominant RNA species are, however, normally located within the cytoplasm and fall into three classes. Figure 4.9 indicates the locations of nucleic acids in eukaryotic cells.

- **Messenger RNA** (mRNA) contains sequences of ribonucleotides which code for the amino acid sequences of proteins. A single mRNA codes for a single polypeptide chain in eukaryotes, but may code for several polypeptides in prokaryotes.

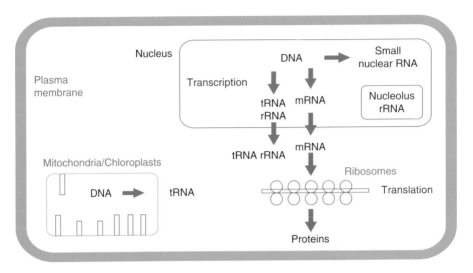

Figure 4.9 Location of DNA and RNA molecules in eukaryotic cells and the flow of genetic information.

- Ribosomal RNA (rRNA) forms part of the structure of ribosomes, which are the sites of protein synthesis. Each ribosome contains only three or four different rRNA molecules, complexed with a total of between 55 and 75 proteins.
- Transfer RNA (tRNA) molecules carry amino acids to the ribosomes, and interact with the mRNA in such a way that their amino acids are joined together in the order specified by the mRNA. There is at least one type of tRNA for each amino acid.

In eukaryotic cells alone a further group of RNA molecules termed small nuclear RNA (snRNA) is present, which functions within the nucleus and promotes the maturation of mRNA molecules. All RNA molecules are associated with their respective binding proteins and are essential for their cellular functions. In addition other species of RNA including micro-RNA and small interfering RNA are present and known to contribute to modulation of gene expression. Nucleic acids from prokaryotic cells are less well compartmentalised, although they serve similar functions.

4.4.2 The Packaging of DNA

The DNA in prokaryotic cells resides in the cytoplasm, although it is associated with nucleoid proteins, where it is tightly coiled and supercoiled by topoisomerase enzymes to enable it to physically fit into the cell. By contrast, eukaryotic cells have many levels of packaging of the DNA within the nucleus, involving a variety of DNA binding proteins.

First-order packaging involves the winding of the DNA around a core complex of four small proteins repeated twice, termed histones (H2A, H2B, H3 and H4). These are rich in the basic amino acids lysine and arginine, and form a barrel-shaped core octamer structure. Approximately 147 bp of DNA is wound on average 1.7 times around the structure, which together is termed a nucleosome. A further histone protein, H1, is found to associate with the outer surface of the nucleosome. The compacting effect of the nucleosome reduces the length of the DNA by a factor of six.

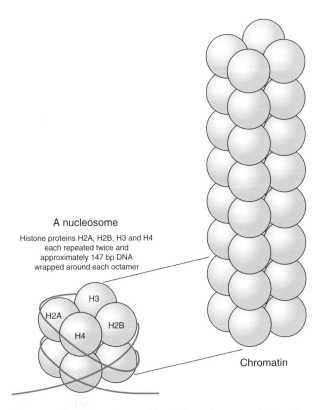

A nucleosome

Histone proteins H2A, H2B, H3 and H4
each repeated twice and
approximately 147 bp DNA
wrapped around each octamer

H3

H2A

H2B

H4

Chromatin

Figure 4.10 Structure and composition of the nucleosome and chromatin.

Nucleosomes also associate to form a second order of packaging termed the 30 nm chromatin fibre, thus further reducing the length of the DNA by a factor of seven (Figure 4.10). These structures may be further folded and looped through interaction with other non-histone proteins and ultimately form chromosome structures.

DNA is found closely associated with the nuclear lamina matrix, which forms a protein scaffold within the nucleus. The DNA is attached at certain positions within the scaffold, usually coinciding with origins of replication. Many other DNA binding proteins are also present, such as high mobility group (HMG) proteins, which assist in promoting certain DNA conformations during processes such as replication or active gene expression.

4.5 FUNCTIONS OF NUCLEIC ACIDS

4.5.1 DNA Replication

The double-stranded nature of DNA provides a means of replication during cell division since the separation of two DNA strands allows complementary strands to be synthesised upon them. Many enzymes and accessory proteins are required for in vivo replication, which in prokaryotes begins at a region of the DNA termed the **origin of replication.**

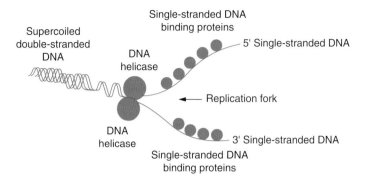

Figure 4.11 Initial events at the replication fork involving DNA unwinding.

DNA has to be unwound before any of the proteins and enzymes needed for replication can act, and this involves separating the double-helical DNA into single strands. This process is carried out by the enzyme DNA helicase. Furthermore, in order to prevent the single strands from reannealing, small proteins termed single-stranded DNA binding proteins (SSBs) attach to the single DNA strands (Figure 4.11).

On each exposed single strand, a short, complementary RNA chain termed a primer is first produced, using the DNA as a template. The primer is synthesised by an RNA polymerase enzyme known as a primase, which uses ribonucleoside triphosphates and itself requires no primer to function. Then DNA polymerase III (DNA pol III) also employs the original DNA as a template for synthesis of a DNA strand, using the RNA primer as a starting point. The primer is vital since it leaves an exposed 3′-hydroxyl group. This is necessary since DNA pol III can only add new nucleotides to the 3′ end and not the 5′ end of a nucleic acid. Synthesis of the DNA strand therefore occurs only in a 5′ to 3′ direction from the RNA primer. This DNA strand is usually termed the leading strand and provides the means for continuous DNA synthesis.

Since the two strands of double-helical DNA are anti-parallel, only one can be synthesised in a continuous fashion. Synthesis of the other strand must take place in a more complex way. The precise mechanism was worked out by Reiji Okazaki in the 1960s. Here the strand, usually termed the lagging strand, is produced in relatively short stretches of 1–2 kb, termed Okazaki fragments. This is still in a 5′ to 3′ direction, using many RNA primers for each individual stretch. Thus, discontinuous synthesis of DNA takes place and allows DNA pol III to work in the 5′ to 3′ direction. The RNA primers are then removed by DNA pol I, which possesses 5′ to 3′ exonuclease activity, and the gaps are filled by the same enzyme acting as a polymerase. The separate fragments are joined together by DNA ligase to yield a newly formed strand of DNA on the lagging strand (Figure 4.12).

The replication of eukaryotic DNA is less well characterised, involves multiple origins of replication and is certainly more complex than that of prokaryotes; however, in both cases the process involves 5′ to 3′ synthesis of new DNA strands. The net result of the replication is that the original DNA is replaced by two molecules, each containing one 'old' and one 'new' strand; the process is therefore known as semi-conservative replication. The ideas behind DNA synthesis, replication and the

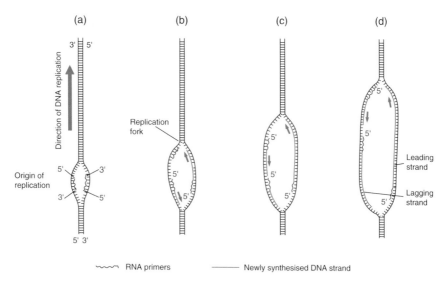

Figure 4.12 DNA replication. (a) Double-stranded DNA separates at the origin of replication. RNA polymerase synthesises short DNA primer strands complementary to both DNA strands. (b) DNA polymerase III synthesises new DNA strands in a 5' to 3' direction, complementary to the exposed, old DNA strands, and continuing from the 3'end of each RNA primer. Consequently, DNA synthesis is in the same direction as DNA replication for one strand (the leading strand) and in the opposite direction for the other (the lagging strand). RNA primer synthesis occurs repeatedly to allow the synthesis of fragments of the lagging strand. (c) As the replication fork moves away from the origin of replication, DNA polymerase III continues the synthesis of the leading strand, and synthesises DNA between RNA primers of the lagging strand. (d) DNA polymerase I removes RNA primers from the lagging strand and fills the resulting gaps with DNA. DNA ligase then joins the resulting fragments, producing a continuous DNA strand.

enzymes involved in them have been adopted in many molecular biology techniques and form the basis of many manipulations in genetic engineering.

4.5.2 DNA Protection and Repair Systems

Cellular growth and division require the correct and coordinated replication of DNA. Mechanisms that proofread replicated DNA sequences and maintain the integrity of those sequences are, however, complex and are only beginning to be elucidated for prokaryotic systems. Bacterial protection is afforded by the use of a restriction modification system based on differential methylation of host DNA, so as to distinguish it from foreign DNA, such as viruses. The most common is type II and consists of a host DNA methylase and **restriction endonuclease** that recognises short (4–6 bp) palindromic sequences and cleaves foreign unmethylated DNA at a particular target sequence. The enzymes involved in this process have been of enormous benefit for the manipulation and analysis of DNA.

Repair systems allow the recognition of altered, mispaired or missing bases in double-stranded DNA and invoke an excision repair process. The systems characterised for bacterial systems are based on the length of repairable DNA during either replication (**dam system**) or in general repair (**urr system**). In some cases, damage to DNA

activates a protein termed RecA to produce an SOS response that includes the activation of many enzymes and proteins; however, this has yet to be fully characterised.
The recombination–repair systems in eukaryotic cells share some common features
with prokaryotes, although the mechanisms in eukaryotic cells are more complex.
Defects in DNA repair may result in the stable incorporation of errors into genomic
sequences, which may underscore several genetic-based diseases.

4.5.3 Transcription of DNA

Expression of genes is carried out initially by the process of transcription, whereby
a complementary RNA strand is synthesised by an enzyme termed RNA polymerase
from a DNA template encoding the gene. Most prokaryotic genes are made up of three
regions. At the centre is the sequence that will be copied in the form of RNA, called
the structural gene. To the 5′ side (upstream) of the strand that will be copied (the
plus (+) strand) lies a region called the promoter, and downstream of the transcription unit is the terminator region. Transcription begins when DNA-dependent RNA
polymerase binds to the promoter region and moves along the DNA to the transcription unit. At the start of the transcription unit the polymerase begins to synthesise
an RNA molecule complementary to the minus (−) strand of the DNA, moving along
this strand in a 3′ to 5′ direction, and synthesising RNA in a 5′ to 3′ direction, using
ribonucleoside triphosphates. The RNA will therefore have the same sequence as the
+ strand of DNA, apart from the substitution of uracil for thymine. On reaching the
stop site in the terminator region, transcription is stopped, and the RNA molecule is
released. The numbering of bases in genes is a useful way of identifying key elements.
Point or base +1 is the residue located at the transcription start site; positive numbers
denote 3′ regions, whilst negative numbers denote 5′ regions (Figure 4.13).

   In eukaryotes, three different RNA polymerases exist, designated I, II and III.
Messenger RNA is synthesised by RNA polymerase II, while RNA polymerases I and III
catalyse the synthesis of rRNA (I), tRNA and snRNA (III).

4.5.4 Promoter and Terminator Sequences in DNA

Promoters are usually to the 5′ end or upstream of the structural gene and have been
best characterised in prokaryotes such as *Escherichia coli*. They comprise two highly
conserved sequence elements: the TATA box (consensus sequence 'TATAAT') which

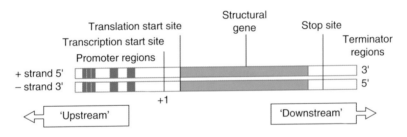

Figure 4.13 Structure and nomenclature of a typical gene.

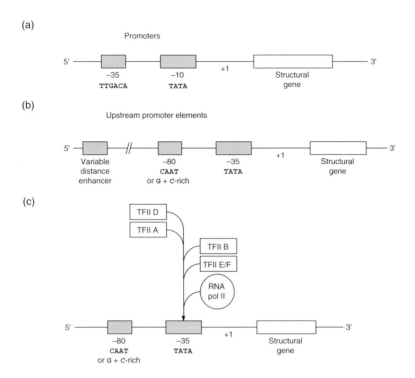

Figure 4.14 (a) Typical promoter elements found in a prokaryotic cell (e.g. *E. coli*). (b) Typical promoter elements found in eukaryotic cells. (c) Generalised scheme of binding of transcription factors to the promoter regions of eukaryotic cells. Following the binding of the transcription factors IID, IIA, IIB, IIE and IIF, a pre-initiation complex is formed. RNA polymerase II then binds to this complex and begins transcription from the start point +1.

is centred approximately 10 bp upstream from the transcription initiation site (−10 in the gene numbering system), and a 'GC-rich' sequence which is centred about −25 bp upstream from the TATA box. The GC element is thought to be important in the initial recognition and binding of RNA polymerase to the DNA, while the −10 sequence is involved in the formation of a transcription initiation complex (Figure 4.14a).

The promoter elements serve as recognition sites for DNA binding proteins that control gene expression, and these proteins are termed **transcription factors** or **trans-acting factors**. These proteins have a DNA binding domain for interaction with promoters and an activation domain to allow interaction with other transcription factors. A well-studied example of a transcription factor is TFIID which binds to the −35 promoter sequence in eukaryotic cells. Gene regulation occurs in most cases at the level of transcription, and primarily by the rate of transcription initiation, although control may also be by modulation of mRNA stability, or at other levels such as translation. Terminator sequences are less well characterised, but are thought to involve nucleotide sequences near the end of mRNA with the capacity to form a hairpin loop, followed by a run of U residues, which may constitute a termination signal for RNA polymerase.

In the case of eukaryotic genes, numerous short sequences spanning several hundred bases may be important for transcription, compared to normally less than 100 bp for prokaryotic promoters. Particularly critical is the TATA box sequence, located approximately −35 bp upstream of the transcription initiation point in the majority of genes (Figure 4.14b). This is analogous to the −10 sequence in prokaryotes. A number of other transcription factors also bind sequentially to form an initiation complex that includes RNA polymerase, subsequent to which transcription is initiated. In addition to the TATA box, a CAT box (consensus GGCCAATCT) is often located at about −80 bp, which is an important determinant of promoter efficiency. Many **upstream promoter elements** (UPEs) have been described that are either general in their action or tissue (or gene) specific. GC elements that contain the sequence GGGCG may be present at multiple sites, in either orientation, and are often associated with **housekeeping genes**, such as those encoding enzymes involved in general metabolism. Some promoter sequence elements, such as the TATA box, are common to most genes, while others may be specific to particular genes or classes of genes.

Of particular interest is a class of promoter first investigated in the virus SV40 and termed an **enhancer**. These sequences are distinguished from other promoter sequences by their unique ability to function over several kilobases either upstream or downstream of a particular gene in an orientation-independent manner. Even at such great distances from the transcription start point they may increase transcription by several hundred-fold. The precise interactions between transcription factors, RNA polymerase or other DNA binding proteins and the DNA sequences they bind to may be identified and characterised by a number of techniques, such as chromatin immunoprecipitation sequencing (ChIP-Seq; see Section 4.17.3) and electrophoretic mobility shift assay (EMSA; see also Section 4.17.3). For transcription in eukaryotic cells to proceed, a number of transcription factors need to interact with the promoters and with each other. This cascade mechanism is indicated in Figure 4.14c and is termed a **pre-initiation complex**. Once this has been formed around the −35 TATA sequence, RNA polymerase II is able to transcribe the structural gene and form a complementary RNA copy.

4.5.5 Transcription in Prokaryotes

Prokaryotic gene organisation differs from that found in eukaryotes in a number of ways. Prokaryotic genes are generally found as continuous coding sequences that are not interrupted. Moreover, they are frequently found clustered into **operons** that contain genes that relate to a particular function, such as the metabolism of a substrate or synthesis of a product. This is particularly evident in the best-known operon identified in *E. coli* termed the **lactose operon** where three genes *lacZ, lacY* and *lacA* share the same promoter and are therefore switched on and off at the same time. In this model the absence of lactose results in a repressor protein binding to an operator region upstream of the *Z, Y* and *A* genes and prevents RNA polymerase from transcribing the genes (Figure 4.15a). However, the presence of lactose requires the genes to be transcribed to allow its metabolism. Lactose binds to the repressor protein and causes a conformational change in its structure. This prevents it binding to the

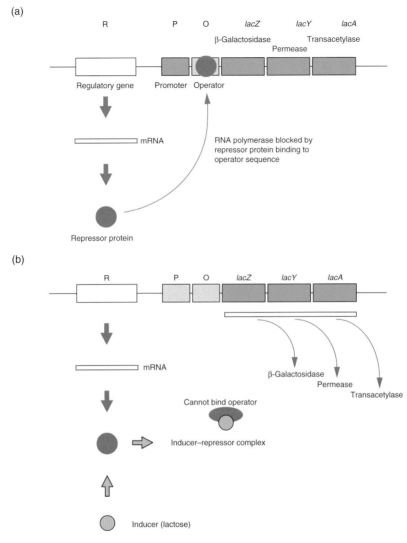

Figure 4.15 Lactose operon (a) in a state of repression (no lactose present) and (b) following induction by lactose.

operator and allows RNA polymerase to bind and transcribe the three genes (Figure 4.15b). Transcription and translation in prokaryotes is also closely linked or coupled, whereas in eukaryotic cells the two processes are distinct and take place in different cell compartments.

4.5.6 Post-Transcriptional Processing

Post-transcriptional gene regulation and its control is complex. One important control mechanism is through short ~22 nucleotide non-coding micro RNAs (miRNA). These are similar to siRNAs found in the RNA interference (RNAi; see Section 4.5.8)

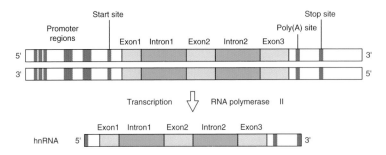

Figure 4.16 Transcription of a typical eukaryotic gene to form heterogeneous nuclear RNA (hnRNA).

pathways, but are genome encoded and fine tune expression by translational repression through binding to specific 3' untranslated regions (3' UTR) in the mRNA. In addition, mRNA may be targeted for destruction by deadenylation or decapping. There are estimated to be over 2500 miRNAs potentially controlling approximately two-thirds of human genes; importantly some have been proposed to be biomarkers for diseases such as cancer and heart disease.

Transcription of a eukaryotic gene results in the production of a heterogeneous nuclear RNA transcript (hnRNA), which faithfully represents the entire structural gene (Figure 4.16). Three processing events then take place. The first processing step involves the addition of a methylated guanosine residue (m7Gppp) termed a cap to the 5' end of the hnRNA. This may be a signalling structure or aid in the stability of the molecule (Figure 4.17). In addition, 150 to 300 adenosine residues, termed a poly(A) tail, are attached at the 3' end of the hnRNA by the enzyme poly(A) polymerase. The poly(A) tail allows the specific isolation of eukaryotic mRNA from total RNA by affinity chromatography; its presence is thought to confer stability on the transcript.

Unlike prokaryotic transcripts, those from eukaryotes have their coding sequence (expressed regions or exons) interrupted by non-coding sequences (intervening

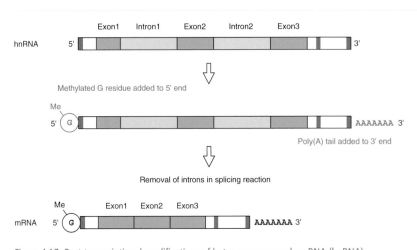

Figure 4.17 Post-transcriptional modifications of heterogeneous nuclear RNA (hnRNA).

regions or introns). Intron–exon boundaries are generally determined by the sequence GU–AG and need to be removed or spliced before the mature mRNA is formed (Figure 4.17). The process of intron splicing is mediated by small nuclear RNAs (snRNAs) that exist in the nucleus as ribonuclear protein particles. These are often found in a large nuclear structure complex termed the spliceosome where splicing takes place. Introns are usually removed in a sequential manner from the 5′ to the 3′ end and their number varies between different genes. Some eukaryotic genes such as histone genes contain no introns, whereas the gene for dystrophin, which when mutated is responsible for muscular dystrophy, contains over 250 introns. In some cases, however, the same hnRNA transcript may be processed in different ways to produce different mRNAs coding for different proteins in a process known as alternative splicing. Thus, a sequence that constitutes an exon for one RNA species may be part of an excised intron in another. The particular type or amount of mRNA synthesised from a cell or cell type may be analysed by a variety of molecular biology techniques.

4.5.7 Translation of mRNA

Messenger RNA molecules are read and translated into proteins by complex RNA–protein particles termed ribosomes. The ribosomes are termed 70S or 80S, depending on their sedimentation coefficient (see Section 12.5.2). Prokaryotic cells have 70S ribosomes, whilst those of the eukaryotic cytoplasm are 80S. Ribosomes are composed of two subunits that are held apart by ribosomal binding proteins until translation proceeds. There are sites on the ribosome for the binding of one mRNA and two tRNA molecules and the translation process is in three stages:

- Initiation: involving the assembly of the ribosome subunits and the binding of the mRNA.
- Elongation: where specific amino acids are used to form polypeptides, this being directed by the codon sequence in the mRNA.
- Termination: which involves the disassembly of the components of translation following the production of a polypeptide.

Transfer RNA (tRNA) molecules are also essential for translation. Each of these are covalently linked to a specific amino acid, forming an aminoacyl tRNA, and each has a triplet of bases exposed that is complementary to the codon for that amino acid. This exposed triplet is known as the anticodon, and allows the tRNA to act as an 'adapter' molecule, bringing together a codon and its corresponding amino acid. The process of linking an amino acid to its specific tRNA is termed charging and is carried out by the enzyme aminoacyl tRNA synthetase.

   In prokaryotic cells the ribosome binds to the 5′ end of the mRNA at a sequence known as a ribosome binding site or sometimes termed the Shine-Dalgarno sequence after its discoverers. In eukaryotes, the situation is similar, but involves a Kozak sequence (named after Marylin Kozak) located around the initiation codon. Following translation initiation, the ribosome moves towards the 3′ end of the mRNA, allowing an aminoacyl tRNA molecule to engage in base-pairing with each successive codon, thereby carrying in amino acids in the correct order for protein synthesis. There are two sites for tRNA molecules in the ribosome, the A site and the P site, and when these sites are occupied, directed by the sequence of codons in the mRNA, the ribosome allows the formation of a peptide bond between the amino acids. The process is also under the control of an enzyme, peptidyl transferase. When the ribosome encounters a termination codon (UAA, UGA or UAG), a release factor binds to the complex and translation stops, the polypeptide and its corresponding mRNA are released and the ribosome divides into its two subunits (Figure 4.18). A myriad of accessory initiation and elongation protein factors are involved in this process. In eukaryotic cells, the polypeptide may then be subjected to post-translational modifications such as glycosylation and by virtue of specific amino-acid signal sequences may be directed to specific cellular compartments or exported from the cell.

Since the mRNA base sequence is read in triplets, an error of one or two nucleotides in positioning of the ribosome will result in the synthesis of an incorrect polypeptide. Thus, it is essential for the correct reading frame to be used during translation. This is ensured in prokaryotes by base-pairing between the Shine–Dalgarno sequence (Kozak sequence in eukaryotes) and a complementary sequence of one of the ribosome's rRNAs, thus establishing the correct starting point for movement of the ribosome along the mRNA. However, if a mutation such as a deletion/insertion takes place within the coding sequence it will also cause a shift of the reading frame and result in an aberrant polypeptide.

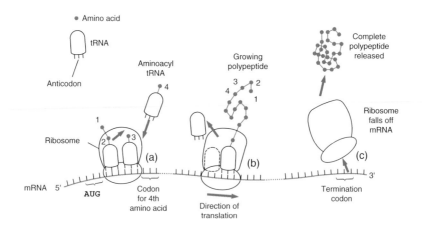

Figure 4.18 Translation. Ribosome (a) has moved only a short way from the 5′ end of the mRNA, and has built up a dipeptide (on one tRNA) that is about to be transferred onto the third amino acid (still attached to tRNA). Ribosome (b) has moved much further along the mRNA and has built up an oligopeptide that has just been transferred onto the most recent aminoacyl tRNA. The resulting free tRNA leaves the ribosome and will receive another amino acid. The ribosome moves towards the 3′ end of the mRNA by a distance of three nucleotides, so that the next codon can be aligned with its corresponding aminoacyl tRNA on the ribosome. Ribosome (c) has reached a termination codon, has released the completed polypeptide, and has fallen off the mRNA.

4.5.8 Control of Protein Production: RNA Interference

There are a number of mechanisms by which protein production is controlled; however, the control may be either at the gene level or at the protein level. Typically, this could include controlling levels of expression of mRNA, an increase or decrease in mRNA turnover, or controlling mRNA availability for translation. One important control mechanism that has also been adapted as a molecular biology technique to aid in the modulation of mRNA is termed RNA interference (RNAi). This involves the production of short 21–23 nucleotide-long double-stranded RNA molecules termed siRNA (small interfering RNA) from longer double-strand RNA (dsRNA) by an enzyme termed dicer, an endoribonuclease (Figure 4.19). Each siRNA is unwound into two single-stranded RNAs, the complementary (passenger) strand is degraded and the anti-sense (guide) strand is incorporated into an RNA-induced silencing complex (RISC). The guide strand pairs with a complementary sequence in mRNA and cleavage is induced through one of the components of the RISC. The RNAi process has been adapted into a vital research tool (see Section 4.17.5) and has a potential application in treating medical conditions where, for example, increased levels of specific mRNA molecules in certain cancers and viral infections may be reduced.

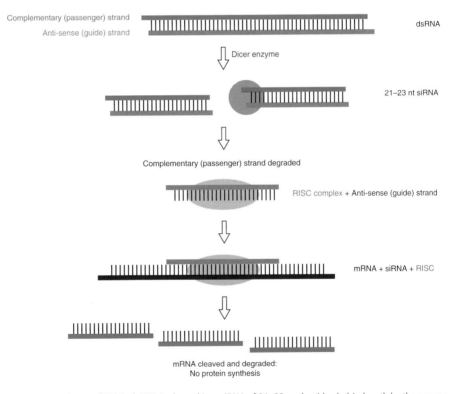

Figure 4.19 Scheme of RNAi: dsRNA is cleaved into siRNA of 21–23 nucleotides (nt) in length by the enzyme dicer. The RISC complexes with the guide strand and by complementarity binds with mRNA, which is degraded.

4.6 THE MANIPULATION OF NUCLEIC ACIDS: BASIC TOOLS AND TECHNIQUES

The discovery and characterisation of a number of key enzymes has enabled the development of various techniques for the analysis and manipulation of DNA. In particular, the enzymes termed type II restriction endonucleases have come to play a key role in all aspects of molecular biology. These enzymes recognise certain DNA sequences, usually 4–6 bp in length, and cleave them in a defined manner. The sequences recognised are palindromic or of an inverted repeat nature, that is they read the same in both directions on each strand. When cleaved they leave a flush-ended or staggered (also termed a cohesive-ended) fragment, depending on the particular enzyme used (Figure 4.20). An important property of staggered ends is that those produced from different molecules by the same enzyme are complementary (or 'sticky') and so will anneal to each other. The annealed strands are held together only by hydrogen bonding between complementary bases on opposite strands. Covalent joining of ends on each of the two strands may be brought about by the enzyme DNA ligase. This is widely exploited in molecular biology to enable the construction of recombinant DNA, i.e. the joining of DNA fragments from different sources. Approximately 500 restriction enzymes have been characterised that recognise over 100 different target sequences. A number of these, termed isoschizomers, recognise different target sequences, but produce the same staggered ends or overhangs. A number of other enzymes have proved to be of value in the manipulation of DNA, as summarised in Table 4.3, and are indicated at appropriate points within the text.

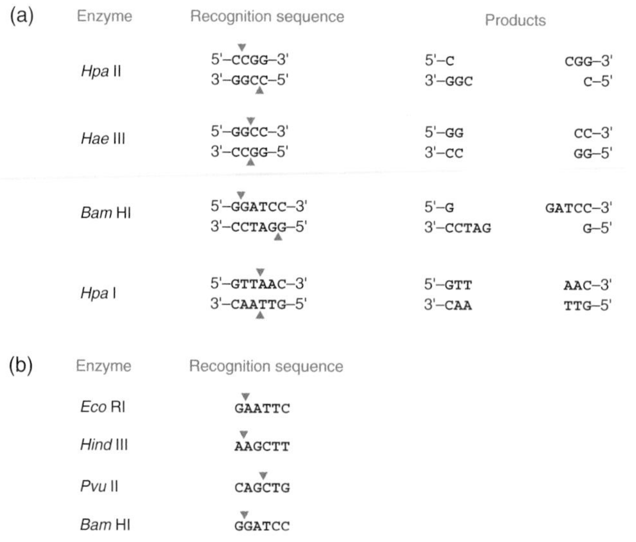

(a)

Enzyme	Recognition sequence	Products	
Hpa II	5'–CCGG–3' 3'–GGCC–5'	5'–C 3'–GGC	CGG–3' C–5'
Hae III	5'–GGCC–3' 3'–CCGG–5'	5'–GG 3'–CC	CC–3' GG–5'
Bam HI	5'–GGATCC–3' 3'–CCTAGG–5'	5'–G 3'–CCTAG	GATCC–3' G–5'
Hpa I	5'–GTTAAC–3' 3'–CAATTG–5'	5'–GTT 3'–CAA	AAC–3' TTG–5'

(b)

Enzyme	Recognition sequence
Eco RI	GAATTC
Hind III	AAGCTT
Pvu II	CAGCTG
Bam HI	GGATCC

Figure 4.20 Recognition sequences of some restriction enzymes showing (a) full descriptions and (b) conventional representations. Arrows indicate positions of cleavage. Note that all the information in (a) can be derived from knowledge of a single strand of the DNA. In (b) only one strand is shown, drawn 5' to 3'; this is the conventional way of representing restriction sites.

Table 4.3 **Types and examples of typical enzymes used in the manipulation of nucleic acids**

Enzyme	Specific example	Use in nucleic acid manipulation
DNA polymerases	DNA pol I	DNA-dependent DNA polymerase 5′→3′→5′ exonuclease activity
	Klenow	DNA pol I lacks 5′→3′ exonuclease activity
	T4 DNA pol	Lacks 5′→3′ exonuclease activity
	Taq DNA pol	Thermostable DNA polymerase used in PCR
	Tth DNA pol	Thermostable DNA polymerase with RT activity
	T7 DNA pol	Used in DNA sequencing
RNA polymerases	T7 RNA pol	DNA-dependent RNA polymerase
	T3 RNA pol	DNA-dependent RNA polymerase
	Qβ replicase	RNA-dependent RNA polymerase, used in RNA amplification
Nucleases	DNase I	Non-specific endonuclease that cleaves DNA
	Exonuclease III	DNA-dependent 3′→5′ stepwise removal of nucleotides
	RNase A	RNases used in mapping studies
	RNase H	Used in second strand cDNA synthesis
	S1 nuclease	Single-strand-specific nuclease
Reverse transcriptase	AMV-RT	RNA-dependent DNA polymerase, used in cDNA synthesis
Transferases	Terminal transferase (TdT)	Adds homopolymer tails to the 3′ end of DNA
Ligases	T4 DNA ligase	Links 5′-phosphate and 3′-hydroxyl ends via phosphodiester bond
Kinases	T4 polynucleotide kinase (PNK)	Transfers terminal phosphate groups from ATP to 5′-OH groups
Phosphatases	Alkaline phosphatase	Removes 5′-phosphates from DNA and RNA
Methylases	*Eco* RI methylase	Methylates specific residues and protects from cleavage by restriction enzymes

Notes: PCR, polymerase chain reaction; RT, reverse transcriptase; cDNA, complementary DNA; AMV, avian myeloblastosis virus.

4.7 ISOLATION AND SEPARATION OF NUCLEIC ACIDS

4.7.1 Isolation of DNA

The use of DNA for analysis or manipulation usually requires that it is isolated and purified to a certain extent. DNA is recovered from cells by the gentlest possible method of cell rupture to prevent the DNA from fragmenting by mechanical shearing. Therefore, the extraction process is usually done in the presence of EDTA, which chelates the Mg^{2+} ions needed for enzymes that degrade DNA (termed DNases). Ideally, cell walls, if present, should be digested enzymatically (e.g. lysozyme treatment of bacteria), and the cell membrane should be solubilised using detergent. If physical disruption is necessary, it should be kept to a minimum, and should involve cutting or squashing of cells, rather than the use of shear forces. Cell disruption (and most subsequent steps) should be performed at 4 °C, using glassware and solutions that have been autoclaved to destroy DNase activity.

After release of nucleic acids from the cells, RNA can be removed by treatment with ribonuclease (RNase) that has been heat-treated to inactivate any DNase contaminants; RNase is relatively stable to heat as a result of its disulfide bonds, which ensure rapid renaturation of the molecule upon cooling. The other major contaminant, protein, is removed by shaking the solution gently with water-saturated phenol, or with a phenol/chloroform mixture, either of which will denature proteins, but not nucleic acids. Centrifugation of the emulsion formed by this mixing produces a lower, organic phase, separated from the upper, aqueous phase by an interface of denatured protein. The aqueous solution is recovered and its protein content repeatedly depleted, until no more material is seen at the interface. Finally, the DNA preparation is mixed with two volumes of absolute ethanol, and the DNA allowed to precipitate out of solution in a freezer. After centrifugation, the DNA pellet is redissolved in a buffer containing EDTA to inactivate any DNases present. This solution can be stored at 4 °C for at least a month. DNA solutions can be stored frozen, although repeated freezing and thawing tends to damage long DNA molecules by shearing. The procedure described above is suitable for total cellular DNA. If the DNA from a specific organelle or viral particle is needed, it is best to isolate the organelle or virus before extracting its DNA, since the recovery of a particular type of DNA from a mixture is usually rather difficult. Where a high degree of purity is required, DNA may be subjected to density gradient ultracentrifugation through caesium chloride (see Section 12.4.2), which is particularly useful for the preparation of plasmid DNA. A flow chart of DNA extraction is shown in Figure 4.21.

It is possible to check the integrity of the DNA by **agarose gel electrophoresis** for qualitative and semi-quantitative analysis. The concentration of DNA in a solution can be determined by measuring the absorbance at 260 nm (A_{260}) in a spectrophotometer. Considering that one absorbance unit equates to 50 µg cm^{-3} of DNA, one obtains:

$$\rho^{\cdot}(DNA) = 50 \times A_{260} \frac{\mu g}{cm^3} \qquad\qquad (Eq\ 4.1)$$

Contaminants may also be identified by scanning **UV spectroscopy** from 200 nm to 300 nm (see Section 13.2.1). The ratio of absorbances at 260 nm and 280 nm of approximately 1.8 indicates that the sample is free of protein contamination (as proteins absorb strongly at 280 nm).

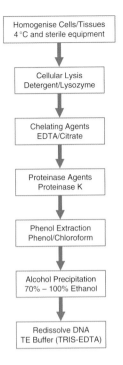

Figure 4.21 General steps involved in extracting DNA from cells or tissues.

4.7.2 Isolation of RNA

The methods used for RNA isolation are very similar to those described above for DNA; however, RNA molecules are relatively short, and therefore less easily damaged by shearing, so cell disruption can be rather more vigorous. RNA is, however, very vulnerable to digestion by RNases, which are present endogenously in various concentrations in certain cell types and exogenously on fingers. Gloves should therefore be worn, and a strong detergent should be included in the isolation medium to immediately denature any RNases. Subsequent removal of proteins should be particularly rigorous, since RNA is often tightly associated with proteins. DNase treatment can be used to remove DNA, and RNA can be precipitated by ethanol. One reagent which is commonly used in RNA extraction is guanidinium thiocyanate, which is both a strong inhibitor of RNase and a protein denaturant. A flow chart of RNA extraction is indicated in Figure 4.22. It is possible to check the integrity of an RNA extract by analysing it by agarose gel electrophoresis. The most abundant RNA species, the rRNA molecules 23S and 16S for prokaryotes, and 18S and 28S for eukaryotes, appear as discrete bands on the agarose gel and thus indicate that the other RNA components are likely to be intact. This is usually carried out under denaturing conditions to prevent secondary structure formation in the RNA. The concentration of the RNA may be estimated by using UV spectrophotometry. At 260 nm, one absorbance unit equates to 40 µg cm^{-3} of RNA and therefore:

$$\rho^{*}(\mathrm{RNA}) = 40 \times A_{260} \, \frac{\mu g}{cm^{3}} \hspace{3cm} \text{(Eq 4.2)}$$

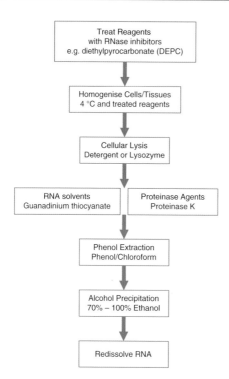

Figure 4.22 General steps involved in extracting RNA from cells or tissues.

Contaminants may also be identified in the same way as for DNA samples by scanning UV spectrometry; however, in the case of RNA, an absorbance ratio A(260 nm)/A(280 nm) ratio of approximately 2 would be expected for a sample containing no protein.

In many cases, it is desirable to isolate eukaryotic mRNA, which constitutes only 2–5% of cellular RNA, from a mixture of total RNA molecules that is almost entirely composed of ribosomal RNA. This may be carried out by affinity chromatography on oligo(dT)-cellulose columns. At high salt concentrations, the mRNA containing poly(A) tails binds to the complementary oligo(dT) molecules of the affinity column, and so mRNA will be retained; all other RNA molecules can be washed through the column by further high salt solution. Finally, the bound mRNA can be eluted using a low concentration of salt (Figure 4.23). Nucleic acid species may also be subfractionated by physical techniques such as electrophoretic or chromatographic separations based on differences in nucleic acid fragment sizes or physico-chemical characteristics. Nanodrop™ spectrometer systems have also aided the analysis of nucleic acids in recent years in allowing the full spectrum of information, whilst requiring only a very small (microlitre) sample volume.

4.7.3 Automated and Kit-Based Extraction of Nucleic Acids

Most of the current reagents used in molecular biology and the most common techniques can be found in kit form or can be automated, and the extraction of nucleic acids by these means is no exception. The advantage of the use of prefabricated kits or

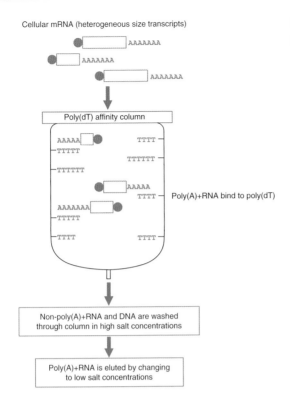

Cellular mRNA (heterogeneous size transcripts)

Figure 4.23 Affinity chromatography of poly(A)+RNA.

automated systems lies in the fact that the reagents are standardised and subjected to quality control, providing a high degree of reliability. For example, glass bead preparations for DNA purification, as well as small compact column-type preparations such as Qiagen columns, are used extensively in research and in routine DNA analysis. Essentially the same reagents for nucleic acid extraction may be used in a format that allows reliable and automated extraction. This is of particular use where a large number of DNA extractions are required. There are also many kit-based extraction methods for RNA; these in particular have overcome some of the problems of RNA extraction, such as RNase contamination. A number of fully automated nucleic acid extraction machines, such as the QIAsymphony system, are now employed in areas where high throughput is required, e.g. clinical diagnostic laboratories (see also Chapter 10). Here, the raw samples, such as blood specimens are placed in 96- or 384-well microtitre plates and subjected to a set computer-controlled processing pattern carried out by a robot. The samples are rapidly manipulated and extracted in approximately 45 min without any manual operations being undertaken.

4.7.4 Electrophoresis of Nucleic Acids

Electrophoresis in agarose or polyacrylamide gels is the usual way to separate DNA molecules according to size (see also Section 6.4). The technique can be used analytically or preparatively, and can be qualitative or quantitative. Large fragments of DNA,

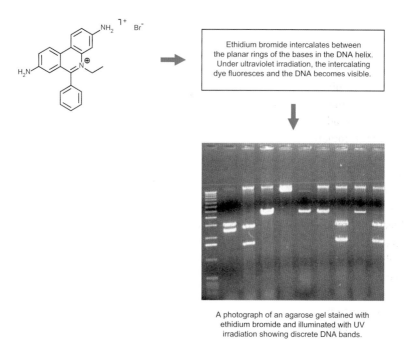

Ethidium bromide intercalates between the planar rings of the bases in the DNA helix. Under ultraviolet irradiation, the intercalating dye fluoresces and the DNA becomes visible.

A photograph of an agarose gel stained with ethidium bromide and illuminated with UV irradiation showing discrete DNA bands.

Figure 4.24 The use of ethidium bromide to detect DNA.

such as chromosomes, may also be separated by a modification of electrophoresis termed **pulsed–field gel electrophoresis** (PFGE; see Section 6.4.3). The easiest and most widely applicable method is electrophoresis in horizontal agarose gels, followed by staining with an intercalation dye (e.g. SYBR® Green or ethidium bromide). Such dyes bind to DNA by insertion between stacked base pairs (**intercalation**), and exhibit a strong fluorescence when illuminated with ultraviolet light (Figure 4.24). Very often, electrophoresis is used to check the purity and intactness of a DNA preparation or to assess the extent of an enzymatic reaction during, for example, the steps involved in the cloning of DNA. For such checks, 'minigels' are particularly convenient, since they need little preparation, use small sample volumes and give results quickly. Agarose gels can be used to separate molecules larger than about 100 bp. For higher resolution or for the effective separation of shorter DNA molecules, polyacrylamide gels are the preferred method.

When electrophoresis is used preparatively, the piece of gel containing the desired DNA fragment is physically removed with a scalpel. The DNA may be recovered from the gel fragment in various ways. This may include crushing with a glass rod in a small volume of buffer, using agarase to digest the agarose, leaving the DNA, or by the process of **electroelution**. In this last method, the piece of gel is sealed in a length of dialysis tubing containing buffer and is then placed between two electrodes in a tank containing more buffer. Passage of an electrical current between the electrodes causes DNA to migrate out of the gel piece, but it remains trapped within the dialysis tubing, and can therefore be recovered easily.

4.8 AUTOMATED ANALYSIS OF NUCLEIC ACID FRAGMENTS

Gel electrophoresis remains the established method for the separation and analysis of nucleic acids. However, a number of automated systems using pre-cast gels and standardised reagents are available that are now very popular. This is especially useful in situations where a large number of samples or high-throughput analysis is required. In addition, technologies such as the Agilent Bioanalyzer have been developed that obviate the need to prepare electrophoresis gels. These instruments employ microfluidic circuits constructed on small cassette units that contain interconnected microreservoirs. The sample is applied in one area and driven through microchannels under computer-controlled electrophoresis. The channels lead to reservoirs allowing, for example, incubation with other reagents such as dyes for a specified time. Electrophoretic separation is thus carried out in a microscale format. The small sample size minimises sample and reagent consumption and the units, being computer controlled, allow data to be captured within a very short time period. In addition, dedicated spectrophotometers, such as the Nanodrop, can provide nucleic acid concentrations quickly with limited sample volume. Alternative methods of analysis, including high-performance liquid chromatography (HPLC)-based approaches have gained in popularity, especially for DNA mutation analysis. Mass spectrometry methods traditionally used in protein analysis, such as MALDI-TOF are also used for nucleic acid analysis due to their rapidity and increasing reliability.

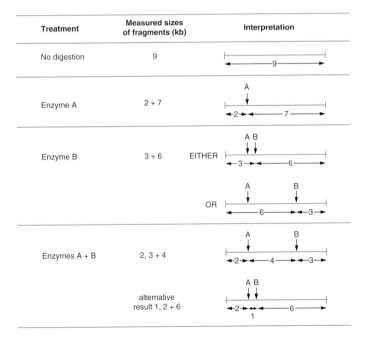

Figure 4.25 Restriction mapping of DNA. Note that each experimental result and its interpretation should be considered in sequence, thus building up an increasingly less ambiguous map.

4.9 MOLECULAR ANALYSIS OF NUCLEIC ACID SEQUENCES

4.9.1 Restriction Mapping of DNA Fragments

Restriction mapping involves the size analysis of restriction fragments produced by several restriction enzymes individually and in combination. The principle of this mapping is illustrated in Figure 4.25, in which the restriction sites of two enzymes, A and B, are being mapped. Cleavage with A gives fragments of 2 and 7 kb from a 9 kb molecule, hence we can position the single A site 2 kb from one end. Similarly, B gives fragments of 3 and 6 kb, so it has a single site 3 kb from one end; but it is not possible at this stage to say if it is near to A's site, or at the opposite end of the DNA. This can be resolved by a double digestion. If the resultant fragments are 2, 3 and 4 kb, then A and B cut at opposite ends of the molecule; if they are 1, 2 and 6 kb, the sites are near each other. Not surprisingly, the mapping of real molecules (which possess many different restriction sites) is rarely as simple as this, and bioinformatic analysis of the restriction fragment lengths is usually needed to construct a map.

4.9.2 Nucleic-Acid Blotting Methods

Electrophoresis of DNA restriction fragments allows separation based on size to be carried out; however, it provides no indication as to the presence of a specific, desired fragment among the complex sample. This can be achieved by transferring the DNA from the intact gel onto a piece of nitrocellulose or nylon membrane placed in contact with it. This provides a more permanent record of the sample since DNA begins to diffuse out of a gel that is left for a few hours. First, the gel is soaked in alkali to render the DNA single stranded. It is then transferred to the membrane so that the DNA becomes bound to it in exactly the same pattern as that originally on the gel. This transfer, named a **Southern blot** after its inventor Ed Southern, can be performed electrophoretically or by drawing large volumes of buffer through both gel and membrane, thus transferring DNA from one to the other by capillary action (Figure 4.26). The point of this operation is that the membrane can now be treated with a labelled DNA molecule, for example a **gene probe**. This single-stranded DNA probe will hybridise under the right conditions to complementary fragments immobilised onto the membrane. The conditions of hybridisation, including the temperature and salt concentration, are critical for this process to take place effectively. This is usually referred

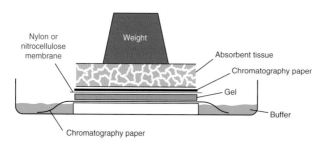

Figure 4.26 Southern blot apparatus.

to as the stringency of the hybridisation and it is particular for each individual gene probe and for each sample of DNA. A series of washing steps with buffer is then carried out to remove any unbound probe and the membrane is developed, after which the precise location of the probe and its target may be visualised. It is also possible to analyse DNA from different species or organisms by blotting the DNA and then using a gene probe representing a protein or enzyme from one of the organisms. In this way it is possible to search for related genes in different species. This technique is generally termed zoo blotting.

The same basic process of nucleic acid blotting can be used to transfer RNA from gels onto similar membranes. This allows the identification of specific mRNA sequences of a defined length by hybridisation to a labelled gene probe and is known as Northern blotting (a homage to Southern blotting). It is possible with this technique to not only detect specific mRNA molecules, but it may also be used to quantify the relative amounts of the specific mRNA. It is usual to separate the mRNA transcripts by gel electrophoresis under denaturing conditions since this improves resolution and allows a more accurate estimation of the sizes of the transcripts. The format of the blotting may be altered from transfer from a gel to direct application to slots on a specific blotting apparatus containing the nylon membrane. This is termed slot or dot blotting and provides a convenient means of measuring the abundance of specific mRNA transcripts without the need for gel electrophoresis; it does not, however, provide information regarding the size of the fragments.

4.9.3 Design and Production of Gene Probes

The availability of a gene probe is essential in many molecular biology techniques. A gene probe is defined as a single-stranded piece of DNA complementary to a desired target DNA sequence that is labelled with, for instance, a fluorescent dye or radioactive label.

The information needed to produce a gene probe may come from many sources. In some cases it is possible to use related genes, that is from the same gene family, to gain information on the most useful DNA sequence to use as a probe. In addition, protein or DNA sequences from different species may also provide a starting point with which to produce a so-called heterologous gene probe. However, the increasing number of DNA sequences in databases such as Genbank and the availability of bioinformatics resources has ensured that this is the usual starting point for gene probe design. Indeed, the method of choice of DNA production for probes is now almost exclusively by artificial chemical synthesis.

In this process, the chemical production of single-stranded DNA is undertaken by computer-controlled gene synthesiser machines. These link the monomers of DNA together based on the desired input sequence to form an oligonucleotide. However, the monomers used in chemical synthesis are so-called phosphoramidites, which are modified dNTPs (deoxyribonucleoside triphosphates) that are joined via the 5′ carbon; this is in the reverse direction to natural DNA synthesis (carried out by DNA polymerase), where monomers are added to the 3′ end. Oligonucleotide primers used in the PCR are all produced by this method. This process allows alternative nucleotides to be

Polypeptide		Phe	Met	Pro	Trp	His	
Corresponding nucleotide sequences	5'	T TTC	ATC	T CCC A G	TGG	T CAC	3'

Figure 4.27 Oligonucleotide probes. Note that only methionine and tryptophan have unique codons. It is impossible to predict which of the indicated codons for phenylalanine, proline and histidine will be present in the gene to be probed, so all possible combinations must be synthesised (16 in the example shown).

used in the synthesis, which may be useful for optimising protein production and in altering protein properties by protein engineering.

Advances in the technology of chemically synthesising DNA has also led to the production of larger custom-made whole-gene sequences. Small genomes and chromosomes such as chromosome 3 from the yeast *Saccharomyces cerevisiae* have all been produced and this process has given rise to the new field of synthetic biology, which is having a wide-ranging impact in biosciences and biotechnology.

Where little DNA information is available to prepare a gene probe, it is possible in some cases to use the knowledge gained from analysis of the corresponding protein. Thus it is possible to isolate and purify proteins and sequence part of the N-terminal end or an internal region of the protein. From our knowledge of the genetic code, it is possible to predict the various DNA sequences that could code for the protein, and then synthesise appropriate oligonucleotide sequences chemically. Due to the degeneracy of the genetic code, most amino acids are coded for by more than one codon, therefore there will be more than one possible nucleotide sequence that could code for a given polypeptide (Figure 4.27). The longer the polypeptide, the greater the number of possible oligonucleotides that must be synthesised. Fortunately, there is no need to synthesise a sequence longer than about 20 bases, since this should hybridise efficiently with any complementary sequences, and should be specific for one gene. Ideally, a section of the protein should be chosen that contains as many tryptophan and methionine residues as possible, since these have unique codons, and there will therefore be fewer possible base sequences that could code for that part of the protein. The synthetic oligonucleotides can then be used as probes in a number of molecular biology methods.

4.9.4 Labelling DNA Gene Probe Molecules

An essential feature of a gene probe is that it can be labelled and subsequently visualised by some means. This allows flagging or identification of any complementary sequence recognised by the probe.

The most common radioactive label is phosphorus-32 (^{32}P), although for certain techniques sulfur-35 (^{35}S) and tritium (^{3}H) are used (see also Chapter 9). These may be detected by the process of autoradiography (see Section 9.3.3) where the labelled probe molecule, bound to sample DNA, located for example on a nylon membrane, is placed in contact with a film sensitive to X-rays. Following exposure, the film is developed and fixed, just as a black-and-white negative. The exposed film reveals the precise location of the labelled probe and therefore the DNA to which it has hybridised.

Non-radioactive labels are increasingly being used to label DNA gene probes. Until recently, radioactive labels were more sensitive than their non-radioactive counterparts. However, recent developments have led to similar sensitivities that, when combined with their improved safety, have led to greater acceptance.

The labelling systems are termed either direct or indirect. Direct labelling allows an enzyme reporter such as alkaline phosphatase to be coupled directly to the DNA. Although this may alter the characteristics of the DNA gene probe, it offers the advantage of rapid analysis, since no intermediate steps are needed. However indirect labelling is at present more popular. This relies on the incorporation of a nucleotide that has a label attached. At present, three commonly used labels are biotin, fluorescein and digoxygenin. These molecules are covalently linked to nucleotides using a carbon spacer arm of 7, 14 or 21 atoms. Specific binding proteins may then be used as a bridge between the nucleotide and a reporter protein such as an enzyme. For example, biotin incorporated into a DNA fragment is recognised with a very high affinity by the protein streptavidin. This may either be coupled or conjugated to a reporter enzyme molecule such as alkaline phosphatase, which converts the colourless substrate p-nitrophenol phosphate (PNPP) into a yellow-coloured compound p-nitrophenol (PNP) and also offers a means of signal amplification. Alternatively, labels such as digoxygenin incorporated into DNA sequences may be detected by monoclonal antibodies (see Section 7.1.2), again conjugated to reporter molecules such as alkaline phosphatase. Thus, the detection of non-radioactive labels occurs by either chemiluminescence (see Section 13.6) or coupled reactions whose products allow for spectrophotometric detection. This has important practical implications since autoradiography may take 1–3 days, whereas colour and chemiluminescent reactions take minutes.

4.9.5 End-Labelling of DNA Molecules

The simplest form of labelling DNA is by 5′ or 3′ end-labelling. 5′ end-labelling involves a phosphate transfer or exchange reaction where the 5′-phosphate of the DNA to be used as the probe is removed and in its place a labelled phosphate, usually ^{32}P, is added. This is frequently carried out by using two enzymes. The first, alkaline phosphatase, is used to remove the existing phosphate group from the DNA. Subsequently, a second enzyme, polynucleotide kinase, is added that catalyses the transfer of a phosphate group (^{32}P-labelled) to the 5′ end of the DNA. The newly labelled probe is then purified by size-exclusion chromatography and may be used directly (Figure 4.28).

Using the other end of the DNA molecule, the 3′ end, is slightly less complex. Here, a new dNTP that is labelled (e.g. ^{32}P-αdATP or biotin-labelled dNTP) is added to the 3′ end of the DNA by the enzyme terminal transferase (Figure 4.29). Although this is a simpler reaction, a potential problem exists because a new nucleotide is added to the existing sequence and so the complete sequence of the DNA is altered, which may affect its hybridisation to its target sequence. A limitation of end-labelling methods arises from the fact that only one label is added to the DNA, so probes labelled in such a fashion are of a lower specific activity in comparison to probes that incorporate labels along the length of the DNA.

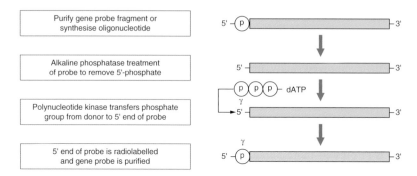

Figure 4.28 End-labelling of a gene probe at the 5′ end with alkaline phosphatase and polynucleotide kinase.

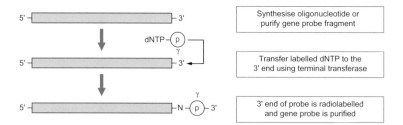

Figure 4.29 End-labelling of a gene probe at the 3′ end using terminal transferase. Note that the addition of a labelled dNTP at the 3′ end alters the sequence of the gene probe.

4.9.6 Random Primer Labelling of DNA

The DNA to be labelled is first denatured and then placed under renaturing conditions in the presence of a mixture of many different random sequences of hexamers or hexanucleotides. These hexamers will, by chance, bind to the DNA sample wherever they encounter a complementary sequence, and so the DNA will rapidly acquire an approximately random sprinkling of hexanucleotides annealed to it. Each of the hexamers can act as a primer for the synthesis of a fresh strand of DNA catalysed by DNA polymerase since it has an exposed 3′-hydroxyl group. The Klenow fragment of DNA polymerase is used for random primer labelling because it lacks a 5′ to 3′ exonuclease activity. The Klenow fragment is prepared by cleavage of DNA polymerase with subtilisin, giving a large enzyme fragment that has no 5′ to 3′ exonuclease activity, but which still acts as a 5′ to 3′ polymerase. Thus, when the Klenow fragment is mixed with the annealed DNA sample in the presence of dNTPs, including at least one that is labelled, many short stretches of labelled DNA will be generated (Figure 4.30). In a similar way to random primer labelling, the PCR may also be used to incorporate radioactive or non-radioactive labels.

A further traditional method of labelling DNA is by the process of nick translation. Low concentrations of DNase I are used to make occasional single-strand nicks in the double-stranded DNA that is intended as the gene probe. DNA polymerase then fills in the nicks, using an appropriate dNTP, at the same time making a new nick to the 3′ side of the previous one (Figure 4.31). In this way the nick is translated along the

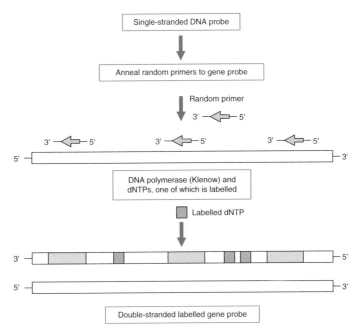

Figure 4.30 Random primer gene probe labelling. Random primers are incorporated and used as a start point for Klenow DNA polymerase to synthesise a complementary strand of DNA, whilst incorporating a labelled dNTP at complementary sites.

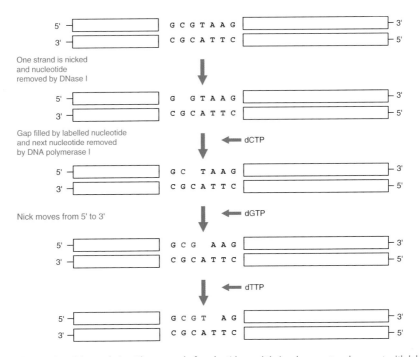

Figure 4.31 Nick translation. The removal of nucleotides and their subsequent replacement with labelled nucleotides by DNA polymerase I increase the label in the gene probe as nick translation proceeds.

DNA. If labelled dNTPs are added to the reaction mixture, they will be used to fill in the nicks, and so the DNA can be labelled to a very high specific activity.

4.9.7 Fluorescence-Based Probes

Molecular beacon probes (see Section 14.4.2) contain a fluorophore at one end of the probe and a quencher molecule at the other. The oligonucleotide has a stem–loop structure where the stems place the fluorophore and quencher in close proximity. The loop structure is designed to be complementary to the target sequence. When the stem–loop structure is formed, the fluorophore is quenched by fluorescence resonance energy transfer (FRET; see Section 13.5.3), i.e. the energy is transferred from the fluorophore to the quencher and ultimately released in a non-radiative fashion. The elegance of these types of probe lies in the fact that upon hybridisation to a target sequence, the stem and loop move apart, the quenching is then lost and emission of light occurs from the fluorophore upon excitation. These types of probe are used to detect DNA amplification accomplished by the polymerase chain reaction (PCR; see next section) and have the advantage that unhybridised probes need not be removed, therefore allowing uninterrupted monitoring of the reaction.

4.10 THE POLYMERASE CHAIN REACTION (PCR)

4.10.1 Basic concept of the PCR

The polymerase chain reaction or PCR is one of the mainstays of molecular biology. One of the reasons for the wide adoption of the PCR is the elegant simplicity of the reaction and relative ease of the practical manipulation steps. Indeed, combined with the relevant bioinformatics resources for its design and for determination of the required experimental conditions, it provides a rapid means for DNA identification and analysis. It has opened up the investigation of cellular and molecular processes to those outside the field of molecular biology.

The PCR is used to amplify a precise fragment of DNA from a complex mixture of starting material usually termed the template DNA and in many cases requires little DNA purification. It does require the knowledge of some DNA sequence information flanking the fragment of DNA to be amplified (target DNA). Using this information, two oligonucleotide primers may be chemically synthesised, each complementary to a stretch of DNA to the 3′ side of the target DNA, one oligonucleotide for each of the two DNA strands (Figure 4.32). It may be thought of as a technique analogous to the DNA replication process that takes place in cells since the outcome is the same: the generation of new complementary DNA stretches based upon the existing ones. It is also a technique that has replaced, in many cases, the traditional DNA cloning methods, since it fulfils the same function: the production of large amounts of DNA from limited starting material. However, this is achieved in a fraction of the time needed to clone a DNA fragment. Although not without its drawbacks, the PCR is a remarkable development that has been changing the approach of many scientists to the analysis of nucleic acids and continues to have a profound impact on core biosciences and biotechnology.

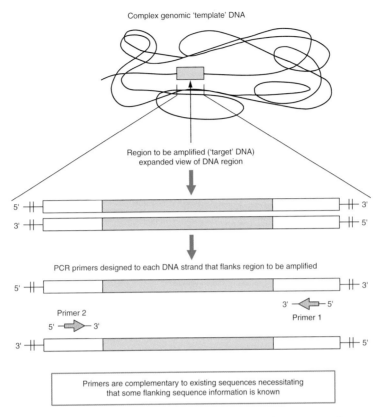

Figure 4.32 The location of PCR primers. PCR primers designed for sequences adjacent to the region to be amplified allow a region of DNA (e.g. a gene) to be amplified from a complex starting material of genomic template DNA.

4.10.2 Stages in the PCR

The PCR consists of three defined sets of times and temperatures, termed steps: (i) denaturation, (ii) annealing and (iii) extension. Each of these steps is repeated 30–40 times, each repetition being termed a cycle (Figure 4.33). In the first cycle, the double-stranded template DNA is denatured by heating the reaction mixture to above 90 °C. Upon denaturation, the region to be specifically amplified (target) is made accessible within the complex DNA. The reaction mixture is then cooled to a temperature between 40 and 60 °C. The precise temperature is critical and each PCR system has to be defined and optimised. One useful technique for optimisation is touchdown PCR, where a programmable thermal cycler is used to incrementally decrease the annealing temperature until the optimum is derived. Reactions that are not optimised may give rise to other DNA products in addition to the specific target or may not produce any amplified products at all. The annealing step allows the hybridisation of the two oligonucleotide primers, which are present in excess, to bind to their complementary sites that flank the target DNA. The annealed oligonucleotides act as primers for DNA synthesis, since they provide a free 3'-hydroxyl group for DNA polymerase. The DNA synthesis step is termed extension and is carried out by a thermostable DNA

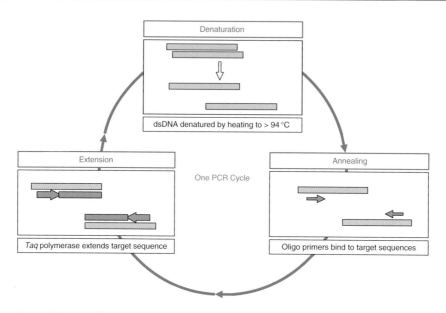

Figure 4.33 A simplified scheme of one PCR cycle that involves denaturation, annealing and extension. ds, double-stranded.

polymerase, most commonly *Taq* DNA polymerase. A range of commercially available DNA polymerases are available that improve on *Taq* polymerase; benefits include very high fidelity with less than 1 error in 200 000, owing to 3′→5′ exonuclease activity (also called proofreading activity) and an ability to tolerate GC-rich DNA sequences.

DNA synthesis proceeds from both of the primers until the new strands have been extended along and beyond the target DNA to be amplified. It is important to note that, since the new strands extend beyond the target DNA, they will contain a region near their 3′ ends that is complementary to the other primer. Thus, if another round of DNA synthesis is allowed to take place, not only will the original strands be used as templates, but also the new strands. Most interestingly, the products obtained from the new strands will have a precise length, delimited exactly by the two regions complementary to the primers. As the system is taken through successive cycles of denaturation, annealing and extension, all the new strands will act as templates and so there will be an exponential increase in the amount of DNA produced. The net effect is to selectively amplify the target DNA and the primer regions flanking it (Figure 4.34). Note that the original target DNA is also copied in a linear fashion, although this is outpaced by the exponential amplification from the new target strands.

One problem with early PCR reactions was that the temperature needed to denature the DNA also denatured the DNA polymerase. However, the availability of a thermo-stable DNA polymerase enzyme isolated from the thermophilic bacterium *Thermus aquaticus* found in hot springs provided the means to automate the reaction. *Taq* DNA polymerase has a temperature optimum of 72 °C and survives prolonged exposure to temperatures as high as 96 °C; it is therefore still active after each of the denaturation steps. The widespread utility of the technique is not least due to the ability to automate the reaction, and thermal cyclers have therefore been produced in which it is possible to program in the temperatures and times for a particular PCR reaction.

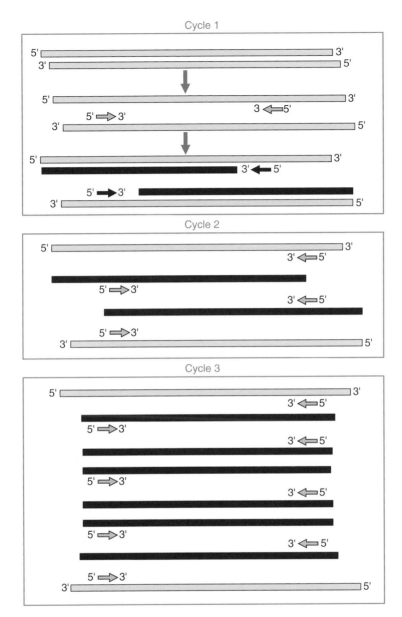

Figure 4.34 Three cycles in the PCR. As the number of cycles increases, the DNA strands that are synthesised and become available as templates are delimited by the ends of the primers. Thus, specific amplification of the desired target sequence flanked by the primers is achieved. Primers are denoted as 5′→3′.

4.10.3 PCR Primer Design and Bioinformatics

The specificity of the PCR lies in the design of the two oligonucleotide primers. These have to not only be complementary to sequences flanking the target DNA, but also must not be self-complementary or bind to each other to form dimers since either scenario prevents DNA amplification. They must also be matched in their GC content and have similar annealing temperatures in order to allow for a an optimum annealing temperature. The increasing use of bioinformatics resources (e.g. Primer3 or NCBI

Primer-BLAST) in the design of primers makes the design and selection of reaction conditions usually very straightforward. These resources allow input of the sequences to be amplified, primer length, product size, GC content, etc. and, following analysis, provide a choice of matched primer sequences. Indeed the initial selection and design of primers without the aid of bioinformatics would now be unnecessarily time-consuming.

It is also possible to design primers with additional sequences at their 5′ end, such as restriction endonuclease target sites or promoter sequences. However, modifications such as these require that the annealing conditions be altered to compensate for the areas of non-homology in the primers. A number of PCR methods have been developed where either one or both of the primers are random. This gives rise to arbitrary priming in genomic templates, but interestingly may give rise to discrete banding patterns when analysed by gel electrophoresis. In many cases, this technique may be used reproducibly to identify a particular organism or species. This is sometimes referred to as random amplified polymorphic DNA (RAPD) and has been used successfully in the detection and differentiation of a number of pathogenic strains of bacteria. In addition, primers can now be synthesised with a variety of labels, such as fluorophores, allowing easier detection and quantification using techniques such as quantitative PCR (qPCR).

4.10.4 PCR Amplification Templates

DNA from a variety of sources may be used as the initial source of amplification templates. It is also a highly sensitive technique and requires only one or two molecules for successful amplification. Unlike many manipulation methods used in conventional molecular biology, the PCR technique is sensitive enough to require very little template preparation. The extraction from many prokaryotic and eukaryotic cells may involve a simple boiling step. Indeed, the components of many chemical extraction techniques, such as SDS and proteinase K, may adversely affect the PCR. The PCR may also be used to amplify RNA, a process termed RT–PCR (reverse transcriptase-PCR). Initially, a reverse transcription reaction which converts the RNA to cDNA is carried out. This reaction normally involves the use of the enzyme reverse transcriptase, although some thermostable DNA polymerases used in the PCR (e.g. *Tth* DNA polymerase) have a reverse transcriptase activity under certain buffer conditions. This allows mRNA transcription products to be effectively analysed. It may also be used to differentiate latent viruses (detected by standard PCR) or active viruses that replicate and thus produce transcription products and are therefore detectable by RT-PCR (Figure 4.35). Additionally, the PCR may be extended to determine relative amounts of a transcription product. To produce long PCR products, so-called **long–range PCR** has been developed. Here, a cocktail of different DNA polymerases is used in the reaction and allows the production of PCR products of 20–30kb.

4.10.5 Sensitivity of the PCR

The exquisite sensitivity of the PCR system is also one of its main drawbacks, since the very large degree of amplification makes the system vulnerable to contamination. Even a trace of foreign DNA, such as that contained in dust particles, may be amplified to significant levels and may give misleading results. Hence cleanliness is paramount

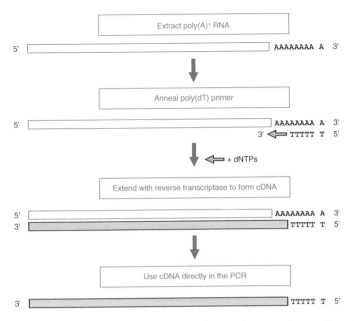

Figure 4.35 Reverse transcriptase-PCR (RT-PCR): mRNA is converted to complementary DNA (cDNA) using the enzyme reverse transcriptase. The cDNA is then used directly in the PCR.

when carrying out PCR, and dedicated equipment and in some cases dedicated laboratories are used. It is possible that amplified products may also contaminate the PCR, although this may be overcome by UV irradiation to damage already amplified products so that they cannot be used as templates. A further interesting solution is to incorporate uracil into the PCR and then treat the products with the enzyme **uracil** *N*-**glycosylase** (UNG), which degrades any PCR amplicons with incorporated uracil, rendering them useless as templates. Furthermore, most PCRs are now undertaken using a so-called **hotstart**. Here, the reaction mixture is physically separated from the template or the enzyme to avoid any mispriming; only when the reaction begins does mixing occur. The separation is achieved by means of a heat-labile chemical moiety or antibody binding to the DNA polymerase.

4.10.6 Alternative Amplification Methods

Many traditional methods in molecular biology have now been superseded by the PCR and the applications for the technique appear to be unlimited. Some of the key areas to which the PCR has been put to use are summarised in Table 4.4. The success of the PCR process has given impetus to the development of other amplification techniques that are based on either thermal cycling or non-thermal cycling (isothermal) methods. Indeed, the development of isothermal systems such as the **LAMP DNA amplification system** (loop-mediated isothermal amplification) do away with the need for a thermal cycler and have the advantage of being able to be used outside the laboratory. Two broad methodologies exist that either amplify the target molecules such as DNA and RNA, or detect the target and amplify a signal molecule bound to it (see Table 4.5).

Table 4.4 **Selected applications of the PCR. A number of the techniques are described in this chapter**

Field or area of study	Application	Specific examples or uses
General molecular biology	DNA amplification	Screening gene libraries
Gene probe production	Production/labelling	Use with blots/hybridisations
RNA analysis	RT–PCR	Active latent viral infections
Forensic science	DNA profiling	Analysis of DNA from blood
Infection/disease monitoring	Microbial detection	Strain typing/analysis RAPDs
DNA sequence analysis	NGS library preparation	Clinical mutation detection
Genome mapping studies	Referencing points in genome	Sequence-tagged sites (STS)
Gene discovery	mRNA analysis	Expressed sequence tags (EST)
Genetic mutation analysis	Detection of known mutations	Screening for cystic fibrosis
Quantification analysis	Quantitative PCR	5′ nuclease (TaqMan assay)
Genetic mutation analysis	Detection of unknown mutations	Gel-based PCR methods (DGGE)
Protein engineering	Production of novel proteins	PCR mutagenesis
Molecular archaeology	Retrospective studies	Dinosaur DNA analysis
Single-cell analysis	Sexing or cell mutation sites	Sex determination of unborn
In situ analysis	Studies on frozen sections	Localisation of DNA/RNA
Gene editing	Sequence modification	Novel proteins

Notes: RT, reverse transcriptase; RAPDs, rapid amplification polymorphic DNA; DGGE, denaturing gradient gel electrophoresis; NGS, next-generation sequencing.

Table 4.5 **Selected alternative amplification techniques to the PCR**

Technique	Type of assay	Specific examples or uses
Target amplification methods		
Ligase chain reaction (LCR)	Non-isothermal, employs thermostable DNA ligase	Mutation detection
Nucleic acid sequence based amplification (NASBA)	Isothermal, involving use of RNA, RNase H/ reverse transcriptase, and T7 DNA polymerase	Viral detection, e.g. HIV
Loop-mediated isothermal amplification (LAMP)	Isothermal, 4–6 primers, strand displacing, DNA polymerase	Viral, bacterial, parasitic disease
Signal amplification methods		
Branched DNA amplification (b-DNA)	Isothermal microwell format using hybridisation or target/capture probe and signal amplification	Mutation detection

Note: HIV, human immunodeficiency virus.

4.10.7 Quantitative PCR (qPCR)

One of the most useful PCR applications is quantitative PCR or qPCR. This allows the PCR to be used as a means of identifying the initial concentrations of DNA or cDNA template used. Early qPCR methods involved the comparison of a standard or control DNA template amplified with separate primers at the same time as the specific target DNA. However, these types of quantification rely on the fact that all the reactions are identical and so any factors affecting this may also affect the result. The introduction of thermal cyclers that incorporate the ability to detect the accumulation of DNA through fluorescent dyes binding to the DNA has rapidly transformed this area.

In its simplest form, a DNA-binding cyanine dye, such as SYBR® Green, is included in the PCR reaction. This dye binds to the minor groove of double-stranded DNA, but not single-stranded DNA; therefore, as amplicons accumulate during the PCR process, SYBR® Green binds the double-stranded DNA proportionally and fluorescence emission of the dye can be detected following excitation. Thus the accumulation of DNA amplicons can be followed in real time during the reaction run. In order to quantify unknown DNA templates, a standard dilution is prepared using DNA of known concentration. As the DNA accumulates during the early exponential phase of the reaction, an arbitrary point is taken where each of the diluted DNA samples cross. This is termed the crossing threshold (Ct) value. From the various Ct values, a log graph is prepared from which an unknown concentration can be deduced. Since SYBR® Green and similar DNA-binding dyes are non-specific, most qPCR cyclers have a built-in melting curve function in order to determine if a correctly sized PCR product is present. The cycler gradually increases the temperature of each tube until the double-stranded PCR product denatures or melts and allows a precise, although not definitive, determination of the product. Accurate confirmation of the product can be achieved by gel electrophoresis and DNA sequencing.

4.10.8 The TaqMan System

In order to make qPCR specific, a number of strategies may be employed that rely on specific hybridisation probes. One ingenious method is called the TaqMan assay or 5′ nuclease assay. Here, the probe consists of an oligonucleotide labelled with a fluorescent reporter at one end of the molecule and a quencher at the other.

During the PCR the oligonucleotide probe binds to the target sequence in the annealing step. As the *Taq* polymerase extends from the primer, its 5′ exonuclease activity degrades the hybridisation probe and releases the reporter from the quencher. Upon excitation of the reporter, a signal is thus generated that increases in direct proportion to the number of starting molecules, and fluorescence can be detected in real time as the PCR proceeds (Figure 4.36). Although relatively expensive in comparison to other methods for determining expression levels, the TaqMan approach is simple, rapid and reliable, and is now in use in many research and clinical areas. Further developments in probe-based PCR systems have also been used and include scorpion probe systems, amplifluor and real-time LUX probes. Quantification is generally undertaken against reference samples, usually housekeeping genes such as GAPDH, rRNA or actin. A further method of quantifying PCR products is digital PCR (dPCR).

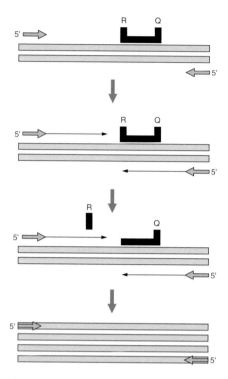

Figure 4.36 5′ nuclease assay (TaqMan assay). PCR is undertaken with a fluorescence reporter/quencher pair (called RQ probe). As the reporter and quencher are in close proximity, fluorescence is quenched. During extension by *Taq* polymerase, the probe is cleaved owing to the nuclease activity of the polymerase. Due to release of the reporter, there is a detectable increase in fluorescence, which is monitored in real-time PCR experiments.

Here, absolute quantification is achieved without the need for reference samples or standards. The initial DNA or cDNA is separated into many compartments or cells, so only a few molecules are present in each cell. These are then amplified by, for example, the TaqMan process. Compartments are then assessed for fluorescence and are either positive or negative, after which statistics are applied to generate the amount of DNA. This process is also useful for detection of rare alleles.

4.11 CONSTRUCTING GENE LIBRARIES

4.11.1 Digesting Genomic DNA Molecules

Following the isolation and purification of genomic DNA it is possible to specifically fragment it with enzymes termed restriction endonucleases. These enzymes are the key to molecular cloning because of the specificity they have for particular DNA sequences. It is important to note that every copy of a given DNA molecule from a specific organism will yield the same set of fragments when digested with a particular enzyme. DNA from different organisms will, in general, give different sets of fragments when treated with the same enzyme. By digesting complex genomic DNA from an organism, it is possible to reproducibly divide its genome into a large number of small fragments, each approximately the size of a single gene. Some enzymes cut straight across the DNA to give flush or blunt ends. Other restriction enzymes make staggered single-strand cuts, producing short single-stranded projections at each end of the digested DNA. These ends are not only identical, but complementary, and will base-pair with each other; they are therefore known as cohesive or sticky ends. In addition, the 5′ end projection of the DNA always retains the phosphate groups.

Over 600 enzymes, recognising more than 200 different restriction sites, have been characterised. The choice of which enzyme to use depends on a number of factors. For example, a recognition sequence of 6 bp will occur, on average, every 4096 (= 4^6) bases assuming a random sequence of each of the four bases. This means that digesting genomic DNA with *Eco* RI which recognises the sequence 5′-GAATTC-3′, will produce fragments, each of which is on average just over 4 kb. Enzymes with 8 bp recognition sequences produce much longer fragments. Therefore, very large genomes, such as human DNA, are usually digested with enzymes that produce long DNA fragments. This makes subsequent steps more manageable, since a smaller number of those fragments need to be cloned and subsequently analysed.

4.11.2 Ligating DNA Molecules

The DNA products resulting from restriction digestion forming sticky ends may be joined to any other DNA fragments treated with the same restriction enzyme. Thus, when the two sets of fragments are mixed, base-pairing between sticky ends will result in the annealing together of fragments that were derived from different starting DNA. There will, of course, also be pairing of fragments derived from the same starting DNA molecules, termed **reannealing**. All these pairings are transient, owing to the weakness of hydrogen bonding between the few bases in the sticky ends, but they can be stabilised by use of an enzyme called **DNA ligase** in a process termed **ligation**. This enzyme, usually isolated from bacteriophage T4 and termed T4 DNA ligase, forms a covalent bond between the 5′-phosphate at the end of one strand and the 3′-hydroxyl of the adjacent strand (Figure 4.37). The reaction, which is ATP dependent, is often carried out at 10 °C to lower the kinetic energy of the molecules, and so reduce the chances of base-paired sticky ends parting before they have been stabilised by ligation. However, long reaction times are needed to compensate for the low activity

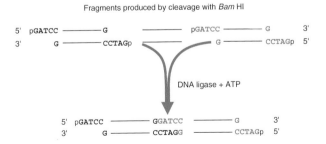

Figure 4.37 Ligation of molecules with cohesive ends. Complementary cohesive ends base-pair, forming a temporary link between two DNA fragments. This association of fragments is stabilised by the formation of 3′→5′ phosphodiester linkages between cohesive ends, a reaction catalysed by DNA ligase.

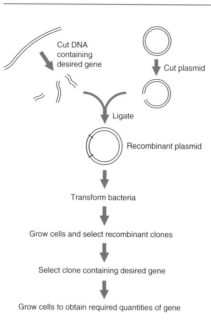

Figure 4.38 Outline of gene cloning.

of DNA ligase in the cold. It is also possible to join blunt ends of DNA molecules, although the efficiency of this reaction is much lower than sticky-ended ligations.

Since ligation reconstructs the site of cleavage, recombinant molecules produced by ligation of sticky ends can be cleaved again at the 'joins', using the same restriction enzyme that was used to generate the fragments initially. In order to propagate digested DNA from an organism it is necessary to join or ligate that DNA with a specialised DNA carrier molecule termed a vector. DNA fragments are thus inserted by ligation into the vector DNA molecule (Figure 4.38), which allows the whole recombined DNA to then be replicated indefinitely within microbial cells. In this way, a DNA fragment can be cloned to provide sufficient material for further detailed analysis, or for further manipulation. This method can be used to generate a collection of clones (gene library) if the DNA extracted from an organism is digested with a restriction enzyme and all fragments ligated into vector DNA.

4.11.3 Aspects of Gene Libraries

There are two general types of gene library: a genomic library that consists of the total chromosomal DNA of an organism and a cDNA library that represents only the mRNA from a particular cell or tissue at a specific point in time (Figure 4.39). The choice of the particular type of gene library depends on a number of factors, the most important being the final application of any DNA fragment derived from the library. If the ultimate aim is understanding the control of protein production for a particular gene or the analysis of its architecture, then genomic libraries must be used. However, if the goal is the production of new or modified proteins, or the determination of the tissue-specific expression and timing patterns, cDNA libraries are more appropriate. The main consideration in the construction of genomic or cDNA libraries is therefore

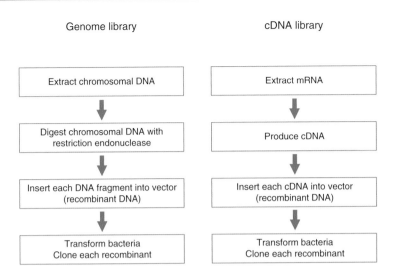

Figure 4.39 Comparison of the general steps involved in the construction of genomic and complementary DNA (cDNA) libraries.

the nucleic acid starting material. Since the genome of an organism is fixed, chromosomal DNA may be isolated from almost any cell type in order to prepare genomic libraries. In contrast, cDNA libraries only represent the mRNA being produced from a specific cell type at a particular time. Thus, it is important to consider carefully the cell or tissue type from which the mRNA is to be derived when constructing a cDNA library. There are a variety of cloning vectors available, many based on naturally occurring molecules, such as bacterial plasmids or bacteria-infecting viruses. The choice of vector depends on whether a genomic library or cDNA library is constructed.

4.11.4 Genomic DNA Libraries

Genomic libraries are constructed by isolating the complete chromosomal DNA from a cell, then digesting it into fragments of the desired average length with restriction endonucleases. This can be achieved by partial restriction digestion using an enzyme that recognises tetranucleotide sequences. Complete digestion with such an enzyme would produce a large number of very short fragments, but if the enzyme is allowed to cleave only a few of its potential restriction sites before the reaction is stopped, each DNA molecule will be cut into relatively large fragments (Table 4.6). The average fragment size will depend on the relative concentrations of DNA and restriction enzyme, and in particular, on the conditions and duration of incubation (Figure 4.40). It is also possible to produce fragments of DNA by physical shearing, although the ends of the fragments may need to be repaired to make them flush-ended. This can be achieved by using the Klenow fragment (see Section 4.9.6) which does not possess $5' \rightarrow 3'$ exonuclease activity and will fill in any recessed $3'$ ends on the sheared DNA using the appropriate dNTPs.

Table 4.6 **Numbers of clones required for representation of DNA in a genome library**

Species	Genome size (kb)	No. of clones required	
		17 kb fragments	35 kb fragments
Bacteria (*E. coli*)	4 000	700	340
Yeast	20 000	3 500	1 700
Fruit fly	165 000	29 000	14 500
Man	3 000 000	535 000	258 250
Maize	15 000 000	2 700 000	1 350 000

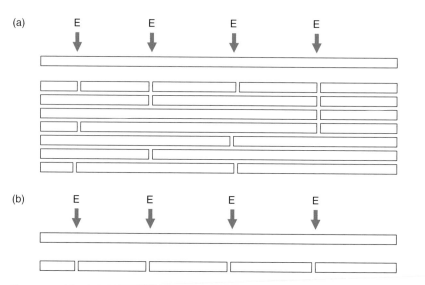

Figure 4.40 Comparison of (a) partial and (b) complete digestion of DNA molecules at restriction enzyme sites (E).

The mixture of DNA fragments is then ligated with a vector, and subsequently cloned. If enough clones are produced, there will be a very high chance that any particular DNA fragment, such as a gene, will be present in at least one of the clones. To keep the number of clones to a manageable size, fragments about 10 kb in length are needed for prokaryotic libraries, but the length must be increased to about 40 kb for mammalian libraries. It is possible to calculate the number of clones that must be present in a gene library to give a probability of obtaining a particular DNA sequence. This formula is:

$$N = \frac{\ln(1 - P)}{\ln(1 - f)} \qquad \text{(Eq 4.3)}$$

where N is the number of recombinants, P is the probability and f is the fraction of the genome in one insert. Thus for the *E. coli* DNA chromosome of 5×10^6 bp and an insert size of 20 kb ($f = 0.004$), the number of clones needed (N) would be 1×10^3, with a probability of $P = 0.99$.

4.11.5 cDNA Libraries

There may be several thousand different proteins being produced in a cell at any one time, all of which have associated mRNA molecules. To identify any one of those mRNA molecules, the clones of each individual mRNA have to be synthesised. Libraries that represent the mRNA in a particular cell or tissue are termed cDNA libraries. mRNA cannot be used directly in cloning since it is too unstable. However, it is possible to synthesise complementary DNA molecules (cDNAs) to all the mRNAs from the selected tissue. The cDNA may be inserted into vectors and then cloned. The production of cDNA (complementary DNA) is carried out using an enzyme termed reverse transcriptase, which is isolated from RNA-containing retroviruses.

Reverse transcriptase is an RNA-dependent DNA polymerase, and will synthesise a first-strand DNA complementary to an mRNA template, using a mixture of the four dNTPs. There is also a requirement (as with all polymerase enzymes) for a short oligonucleotide primer to be present (Figure 4.41). With eukaryotic mRNA bearing a poly(A) tail, a complementary oligo(dT) primer may be used. Alternatively, random hexamers may be used that randomly anneal to the mRNAs in the complex. Such primers provide a free 3'-hydroxyl group that is used as the starting point for the reverse transcriptase. Regardless of the method used to prepare the first-strand cDNA, one absolute requirement is high-quality non-degraded mRNA; the integrity of the RNA should always be checked by gel electrophoresis. In yet another approach, a

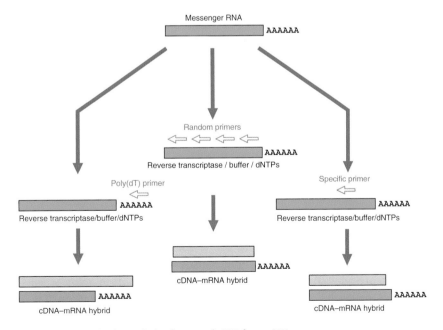

Figure 4.41 Strategies for producing first-strand cDNA from mRNA.

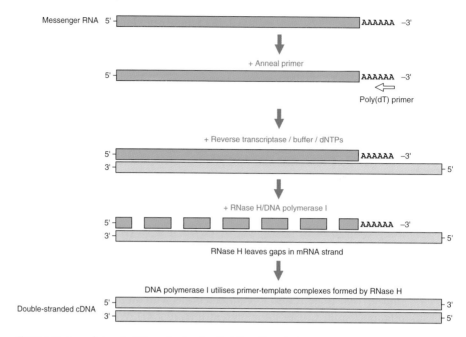

Figure 4.42 Second-strand cDNA synthesis using the RNase H method.

fraction of the extract may be used in a cell-free translation system, which, if intact mRNA is present, will direct the synthesis of proteins represented by the mRNA molecules in the sample.

Following the synthesis of the first DNA strand, a poly(dC) tail is added to its 3′ end, using terminal transferase and dCTP. This will also, incidentally, put a poly(dC) tail on the poly(A) of mRNA. Hydrolysis by alkali is then used to remove the RNA strand, leaving single-stranded DNA that can be used, like the mRNA, to direct the synthesis of a complementary DNA strand. The second-strand synthesis requires an oligo(dG) primer, base-paired with the poly(dC) tail, which is catalysed by the Klenow fragment of DNA pol I. The final product is double-stranded DNA, one of the strands being complementary to the mRNA. One further method of cDNA synthesis involves the use of RNase H. Here, the first-strand cDNA is produced as above with reverse transcriptase, but the resulting mRNA–cDNA hybrid is retained. RNase H is then used at low concentrations to nick the RNA strand. The resulting nicks expose 3′-hydroxyl groups that are used by DNA polymerase as a primer to replace the RNA with a second strand of cDNA (Figure 4.42).

4.11.6 Treatment of Blunt cDNA Ends

Ligation of blunt-ended DNA fragments is not as efficient as ligation of sticky ends, therefore additional procedures are undertaken before ligation when ligating with cDNA molecules with cloning vectors. One approach is to add small double-stranded molecules with one internal site for a restriction endonuclease, termed nucleic-acid linkers, to the cDNA. Numerous linkers are commercially available with internal

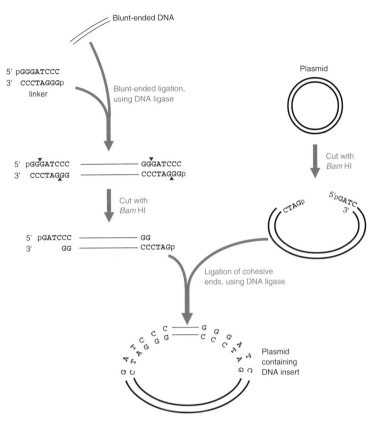

Figure 4.43 Use of linkers. In this example, blunt-ended DNA is inserted into a specific restriction site on a plasmid, after ligation to a linker containing the same restriction site.

restriction sites for many of the most commonly used restriction enzymes. Linkers are blunt-end ligated to the cDNA, but since they are added much in excess of the cDNA the ligation process is reasonably successful. Subsequently, the linkers are digested with the appropriate restriction enzyme, which provides the sticky ends for efficient ligation to a vector digested with the same enzyme. This process may be made easier by the addition of adaptors rather than linkers; adapters possess preformed sticky ends and so there is no need for restriction digestion following ligation (Figure 4.43).

4.11.7 Enrichment Methods for RNA

Frequently, an attempt is made to isolate the mRNA transcribed from a desired gene within a particular cell or tissue that produces the protein in high amounts. Thus, if the cell or tissue produces such a protein, a large fraction of the total mRNA will code for the protein. An example of this are the B cells of the pancreas, which contain high levels of pro-insulin mRNA. In such cases, it is possible to precipitate polysomes that are actively translating the mRNA, by using antibodies to the ribosomal proteins; the mRNA can then be dissociated from the precipitated ribosomes (ribosome profiling). More often, though, the mRNA required is only a minor component of the total

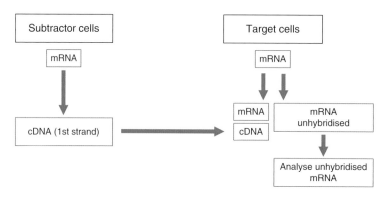

Figure 4.44 Scheme of analysing specific mRNA molecules by subtractive hybridisation.

cellular mRNA. In such cases, total mRNA may be fractionated by size using sucrose density gradient centrifugation (see Section 12.4.2). Then, each fraction is used to direct the synthesis of proteins using an in vitro translation system. An extension to this technique is to digest the ribosomal fractions with RNase and sequence the remaining short (~35 bases) pieces of protected RNA using next-generation sequencing (see Section 20.2.2). This allows for genome-wide assessment of actively translated mRNA.

4.11.8 Subtractive Hybridisation

It is often the case that genes are transcribed in a specific cell type or differentially activated during a particular stage of cellular growth, frequently at very low levels. It is possible to isolate those mRNA transcripts by subtractive hybridisation. The mRNA species common to the different cell types are removed, leaving the cell type or tissue-specific mRNAs for analysis (Figure 4.44). This may be undertaken by isolating the mRNA from the so-called subtractor cells and producing a first-strand cDNA. The original mRNA from the subtractor cells is then degraded and the mRNA from the target cells isolated and mixed with the cDNA. All the complementary mRNA–cDNA molecules common to both cell types will hybridise, thus rendering the unbound mRNA, which may be isolated and further analysed. A more rapid approach of analysing the differential expression of genes has been developed using PCR and is termed differential display.

4.11.9 Cloning PCR Products

Whereas PCR has to some extent replaced cloning as a method for the generation of large quantities of a desired DNA fragment, there is, in certain circumstances, still a requirement for the cloning of PCR-amplified DNA. For example, certain techniques, such as in vitro protein synthesis, are best achieved with the DNA fragment inserted into an appropriate plasmid or phage cloning vector. Cloning methods for PCR follow closely the cloning of DNA fragments derived from the conventional manipulation of DNA. The technique with which this may be achieved is through one of two ways,

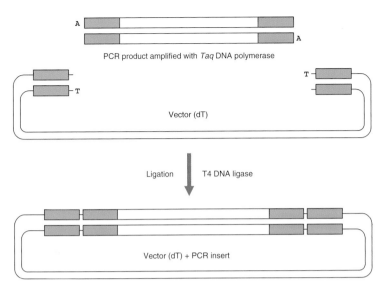

Figure 4.45 Cloning of PCR products using dA:dT cloning.

blunt-ended or cohesive-ended cloning. Certain thermostable DNA polymerases such as *Taq* DNA polymerase and *Tth* DNA polymerase give rise to PCR products having a 3′ overhanging A residue. It is possible to clone the PCR product into dT vectors termed dA:dT cloning. This makes use of the fact that the terminal additions of A residues may be successfully ligated to vectors prepared with T residue overhangs to allow ligation of the PCR product (Figure 4.45). The reaction is catalysed by DNA ligase as in conventional ligation reactions. The related method of TA-TOPO cloning is very similar and employs a precut vector with 5′-(C/T)CCTT-3′ at the linear ends, but with a linked topoisomerase enzyme that carries out the ligation of the PCR products to the vector.

It is also possible to carry out cohesive end cloning with PCR products. In this case, oligonucleotide primers are designed with a restriction endonuclease site incorporated into them. Since the complementarity of the primers needs to be absolute at the 3′ end, the 5′ end of the primer is usually the region for the location of the restriction site. This needs to be designed with care, since the efficiency of digestion with certain restriction endonucleases decreases if extra nucleotides, not involved in recognition, are absent at the 5′ end. In this case, the digestion and ligation reactions are the same as those undertaken for conventional reactions. An adaptation of this principle is used in the In-Fusion® cloning system, where homologous sequences in an amplified PCR product are joined with complementary sequences in a cloning vector.

4.11.10 Isothermal Assembly: Gibson Assembly Cloning System

One key method of efficiently cloning DNA that obviates the need for restriction enzymes is termed the Gibson assembly. This is a single-step technique based on three enzymes that allows assembly of multiple DNA fragments in an isothermal single-tube reaction (Figure 4.46). First, an exonuclease is used to create single-stranded

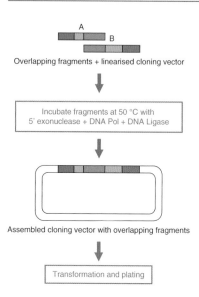

Figure 4.46 Gibson isothermal assembly system.

3' overhangs that allow annealing of the DNA fragments at the overlap region. A DNA polymerase is then used to fill in the gaps within each annealed fragment. Finally, a DNA ligase seals the nicks in the assembled DNA. A key feature of the method is the design of the primers requiring overlapping sequences with complementarity of 20–40 bases for the assembly of the construct. Up to six fragments of up to 12 kb can be assembled. Compared to conventional cloning, which involves restriction enzymes and a time-consuming protocol of many manipulation steps, the Gibson assembly is very efficient and provides DNA joining which is seamless. A number of bioinformatics resources are also available to assist in planning the sequences required for the process.

A number of alternative cloning methods are also available. For example, in GeneArt® Seamless Cloning the initial ligation step is omitted, yet it is possible to assemble up to four inserts and a linearised vector of up to 40 kb in a 30-minute room temperature reaction, with the ligation being carried out in E. coli. Other examples of non-traditional cloning systems include those based on Type IIS restriction enzymes (Golden Gate cloning) and ligation-independent cloning (LIC). Furthermore, it is also possible to source error-free synthesised gene sequences up to 2 Mbp in size from biotechnology companies.

4.12 CLONING VECTORS

For the cloning of any molecule of DNA it is necessary for that DNA to be incorporated into a cloning vector. These are DNA elements that may be stably maintained and propagated in a host organism for which the vector has replication functions. A typical host organism is a bacterium such as E. coli that grows and divides rapidly. Thus, any vector with a replication origin in E. coli will replicate (together with any incorporated DNA) efficiently. Concomitantly, any DNA cloned into a vector will enable the amplification of the inserted foreign DNA fragment and also allow any subsequent analysis to be undertaken. In this way, the cloning process resembles the

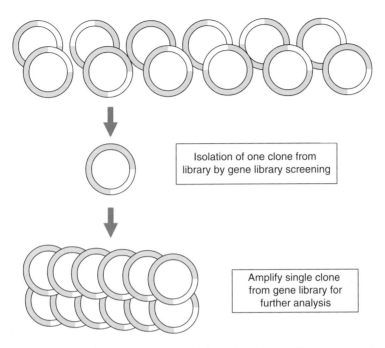

Stable gene bank (gene library),
each vector containing
a different foreign DNA fragment

Isolation of one clone from
library by gene library screening

Amplify single clone
from gene library for
further analysis

Figure 4.47 Production of multiple copies of a single clone from a stable gene bank or library.

PCR, although there are some major differences between the two techniques. By cloning, it is possible to not only store a copy of any particular fragment of DNA, but also produce virtually unlimited amounts of it (Figure 4.47).

The vectors used for cloning vary in their complexity, their ease of manipulation, their selection and the amount of DNA sequence they can accommodate (the insert capacity). Vectors have, in general, been developed from naturally occurring molecules such as bacterial plasmids, bacteriophages or combinations of the elements that make them up, such as cosmids (see Section 4.12.4). For gene library constructions there is a choice and trade-off between various vector types, usually related to the ease of the manipulations needed to construct the library and the maximum size of foreign DNA insert of the vector (Table 4.7). Thus, vectors with the advantage of large insert capacities are usually more difficult to manipulate, although there are many more factors to be considered, which are indicated in the following treatment of vector systems.

4.12.1 Plasmids

Many bacteria contain an extrachromosomal element of DNA, termed a plasmid, which is a relatively small, covalently closed circular molecule, carrying genes for antibiotic resistance, conjugation or the metabolism of 'unusual' substrates. Some

Table 4.7 **Comparison of vectors generally available for cloning DNA fragments**

Vector	Host cell
M13	*E. coli*
Plasmid	*E. coli*
Phage λ	*E. coli*
Cosmids	*E. coli*
BACs	*E. coli*
YACs	*S. cerevisiae*

Notes: BAC, bacterial artificial chromosome; YAC, yeast artificial chromosome.

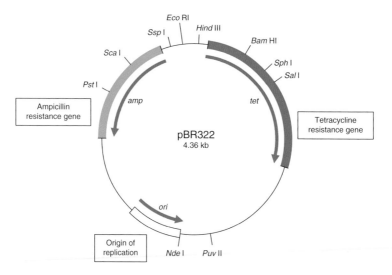

Figure 4.48 Map and important features of pBR322.

plasmids are replicated at a high rate by bacteria such as *E. coli* and so are excellent potential vectors. In the early 1970s, a number of natural plasmids were artificially modified and constructed as cloning vectors, by a complex series of digestion and ligation reactions. One of the most notable plasmids, termed pBR322 after its developers Bolivar and Rodriguez (pBR), was widely adopted and illustrates the desirable features of a cloning vector, as indicated below (Figure 4.48):

- The plasmid is much smaller than a natural plasmid, which makes it more resistant to damage by shearing, and increases the efficiency of uptake by bacteria, a process termed **transformation**.
- A bacterial **origin of DNA replication** (ori) ensures that the plasmid will be replicated by the host cell. Some replication origins display stringent regulation of replication, in which rounds of replication are initiated at the same frequency as cell division. Most

plasmids, including pBR322, have a relaxed origin of replication, whose activity is not tightly linked to cell division, and so plasmid replication will be initiated far more frequently than chromosomal replication. Hence a large number of plasmid molecules will be produced per cell.

- Two genes coding for resistance to antibiotics have been introduced. One of these allows the selection of cells that contain plasmid: if cells are plated on medium containing an appropriate antibiotic, only those that contain plasmid will grow to form colonies. The other resistance gene can be used, as described below, for detection of those plasmids that contain inserted DNA.
- There are single recognition sites for a number of restriction enzymes at various points around the plasmid, which can be used to open or linearise the circular plasmid. Linearising a plasmid allows a fragment of DNA to be inserted and the circle closed. The variety of sites not only makes it easier to find a restriction enzyme that is suitable for both the vector and the foreign DNA to be inserted, but, since some of the sites are placed within an antibiotic resistance gene, the presence of an insert can be detected by loss of resistance to that antibiotic. This is termed insertional inactivation.

Insertional inactivation is a useful selection method for identifying recombinant vectors with inserts. For example, a fragment of chromosomal DNA digested with *Bam* HI is isolated and purified. The plasmid pBR322 is also digested at a single site using *Bam* HI, and both samples are then subjected to conditions that denature all proteins (e.g. thermal denaturation), thus inactivating the restriction enzyme. *Bam* HI cleaves to give sticky ends, and so it is possible to obtain ligation between the plasmid and digested DNA fragments in the presence of T4 DNA ligase. The products of this ligation will include plasmids containing a single fragment of the DNA as an insert, but there will also be unwanted products, such as plasmid that has recircularised without an insert, dimers of plasmid, fragments joined to each other, and plasmid with an insert composed of more than one fragment. Most of these unwanted molecules can be eliminated during subsequent steps. The products of such reactions are usually identified by agarose gel electrophoresis.

The ligated DNA must now be used to transform *E. coli*. Bacteria do not normally take up DNA from their surroundings, but can be induced to do so by prior treatment with Ca^{2+} at 4 °C; they are then termed competent, since DNA added to the suspension of competent cells will be taken up during a brief increase in temperature termed heat shock. Small, circular molecules are taken up most efficiently, whereas long, linear molecules will not enter the bacteria.

After a brief incubation to allow expression of the antibiotic resistance genes, the cells are plated onto medium containing the antibiotic, e.g. ampicillin. Colonies that grow on these plates must be derived from cells that contain plasmid, since this carries the gene for resistance to ampicillin. It is not, at this stage, possible to distinguish between those colonies containing plasmids with inserts and those that simply contain recircularised plasmids. To do this, the colonies are replica plated, using a sterile velvet pad, onto plates containing tetracycline in their medium. Since the *Bam* HI site lies within the tetracycline resistance gene, this gene will be inactivated by the

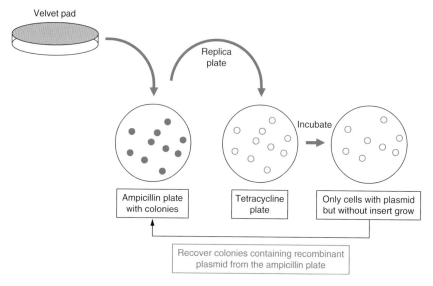

Figure 4.49 Replica plating to detect recombinant plasmids. A sterile velvet pad is pressed onto the surface of an agar plate, picking up some cells from each colony growing on that plate. The pad is then pressed on to a fresh agar plate, thus inoculating it with cells in a pattern identical to that of the original colonies. Clones of cells that fail to grow on the second plate (e.g. owing to the loss of antibiotic resistance) can be recovered from their corresponding colonies on the first plate.

presence of an insert, but will be intact in those plasmids that have merely recircularised (Figure 4.49). Thus colonies that grow on ampicillin but not on tetracycline must contain plasmids with inserts. Since replica plating gives an identical pattern of colonies on both sets of plates, it is straightforward to recognise the colonies with inserts, and to recover them from the ampicillin plate for further growth. This illustrates the importance of a second gene for antibiotic resistance in a vector.

Although recircularised plasmid can be selected against, its presence decreases the yield of recombinant plasmids containing inserts. If the digested plasmids are treated with the enzyme alkaline phosphatase prior to ligation, recircularisation will be prevented, since this enzyme removes the 5′-phosphate groups that are essential for ligation. Links can still be made between the 5′-phosphate of insert and the 3′-hydroxyl of the plasmid, so only recombinant plasmids and chains of linked DNA fragments will be formed. It does not matter that only one strand of the recombinant DNA is ligated, since the nick will be repaired by bacteria transformed with these molecules.

The valuable features of pBR322 have been enhanced by the construction of a series of plasmids termed pUC (produced at the University of California) (Figure 4.50). There is an antibiotic resistance gene for tetracycline and origin of replication for *E. coli*. In addition, the most popular restriction sites are concentrated into a region termed the **multiple cloning site** (MCS). In addition, the MCS is part of a gene in its own right and codes for a portion of a polypeptide called β-galactosidase. When the pUC plasmid has been used to transform the host cell *E. coli*, the gene may be switched on by adding the inducer IPTG (isopropyl-β-*D*-thiogalactopyranoside). Its presence causes the enzyme β-galactosidase to be produced. The functional enzyme is able

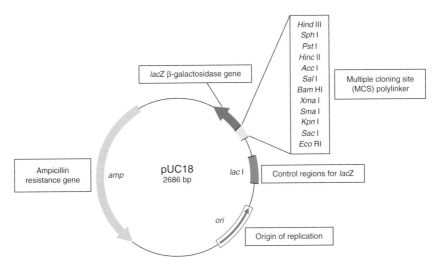

Figure 4.50 Map and important features of pUC18.

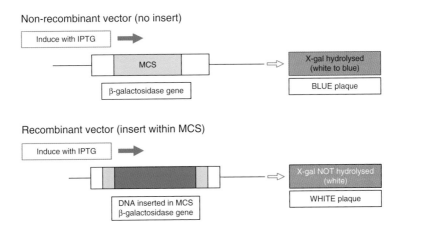

Figure 4.51 Principle of blue/white selection for the detection of recombinant vectors.

to hydrolyse a colourless substance called X-gal (5-bromo-4-chloro-3-indolyl-β-ga-lactopyranoside) into a blue insoluble material (5,5′-dibromo-4,4′-dichloro indigo) (Figure 4.51). However, if the gene is disrupted by the insertion of a foreign fragment of DNA, a non-functional enzyme results, which is unable to carry out hydrolysis of X-gal. Thus, a recombinant pUC plasmid can be easily detected since it is white or colourless in the presence of X-gal, whereas an intact non-recombinant pUC plasmid will be blue, since its gene is fully functional and not disrupted. This elegant system, termed **blue/white selection**, allows the initial identification of recombinants to be undertaken very quickly and has been included in a number of subsequent vector systems. This selection method and insertional inactivation of antibiotic resistance genes do not, however, provide any information on the character of the DNA insert, just the status of the vector.

4.12.2 Virus-Based Vectors

A useful feature of any cloning vector is the amount of DNA it may accept or have inserted before it becomes non-viable. Inserts greater than 5 kb increase plasmid size to the point at which efficient transformation of bacterial cells decreases markedly, and so bacteriophages (bacterial viruses) have been adapted as vectors in order to propagate larger fragments of DNA in bacterial cells. Cloning vectors derived from λ bacteriophage are commonly used, since they offer an approximately 16-fold advantage in cloning efficiency in comparison with the most efficient plasmid cloning vectors.

Phage λ is a linear double-stranded phage approximately 49 kb in length (Figure 4.52). It infects *E. coli* with great efficiency by injecting its DNA through the

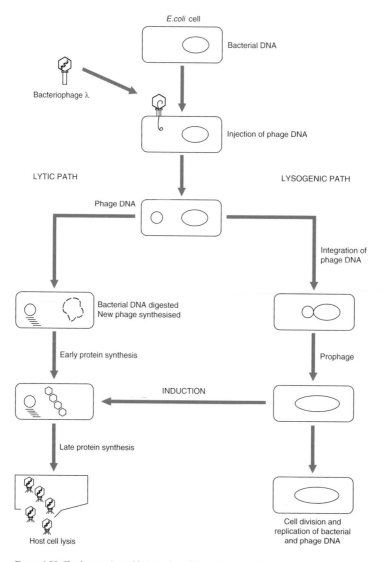

Figure 4.52 The lysogenic and lytic cycles of bacteriophage λ.

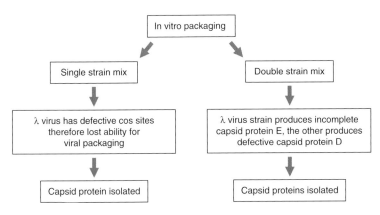

Figure 4.53 Two strategies for producing in vitro packaging extracts for bacteriophage λ.

cell membrane. In the wild-type phage λ, the DNA follows one of two possible modes of replication. Firstly, the DNA may either become stably integrated into the *E. coli* chromosome where it lies dormant until a signal triggers its excision; this is termed the lysogenic life cycle. Alternatively, it may follow a lytic life cycle where the DNA is replicated upon entry to the cell, phage head and tail proteins are synthesised rapidly and new functional phage assembled. The phage are subsequently released from the cell by lysing the cell membrane to infect further *E. coli* cells nearby. At the extreme ends of phage λ are 12 bp sequences termed *cos* (cohesive) sites. Although they are asymmetric, they are similar to restriction sites and allow the phage DNA to be circularised. Phage may be replicated very efficiently in this way, the result of which are concatemers of many phage genomes, which are cleaved at the *cos* sites and inserted into newly formed phage protein heads.

Much use of phage λ has been made in the production of gene libraries, mainly because of its efficient entry into the *E. coli* cell and the fact that larger fragments of DNA may be stably integrated. For the cloning of long DNA fragments, up to approximately 25 kb, much of the non-essential λ DNA that codes for the lysogenic life cycle is removed and replaced by the foreign DNA insert. The recombinant phage is then assembled into pre-formed viral protein particles, a process termed in vitro packaging. These newly formed phage are used to infect bacterial cells that have been plated out on agar (Figure 4.53).

Once inside the host cells, the recombinant viral DNA is replicated. All the genes needed for normal lytic growth are still present in the phage DNA, and so multiplication of the virus takes place by cycles of cell lysis and infection of surrounding cells, giving rise to plaques of lysed cells on a background, or lawn, of bacterial cells. The viral DNA, including the cloned foreign DNA, can be recovered from the viruses in these plaques and analysed further by restriction mapping and agarose gel electrophoresis.

In general, two types of λ phage vectors have been developed, λ **insertion vectors** and λ **replacement vectors** (Figure 4.54). The λ insertion vectors accept less DNA than the replacement type since the foreign DNA is merely inserted into a region of the phage genome with appropriate restriction sites; common examples are λgt10 and

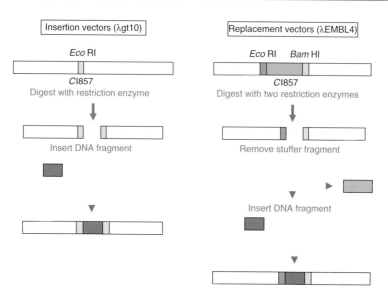

Figure 4.54 General schemes used for cloning into λ insertion and λ replacement vectors. CI857 is a temperature-sensitive mutation that promotes lysis at 42 °C after incubation at 37 °C.

λcharon 16A. With a replacement vector, a central region of DNA not essential for lytic growth is removed (a stuffer fragment) by a double digestion with, for example, *Eco* RI and *Bam* HI. This leaves two DNA fragments termed right and left arms. The central stuffer fragment is replaced by inserting foreign DNA between the arms to form a functional recombinant λ phage. The most notable examples of λ replacement vectors are λEMBL and λZap.

λZap is a commercially produced cloning vector that includes unique cloning sites clustered into a multiple cloning site (MCS) (Figure 4.55). Furthermore, the MCS is located within a *lacZ* region providing a blue/white screening system based on insertional inactivation. It is also possible to express foreign cloned DNA from this vector. This is a very useful feature of some λ vectors since it is then possible to screen for protein product rather than the DNA inserted into the vector. The screening is therefore undertaken with antibody probes directed against the protein of interest. Other features that make this a useful cloning vector are the ability to produce RNA transcripts termed cRNA or **riboprobes**. This is possible because two promoters for RNA polymerase enzymes exist in the vector, a T7 and a T3 promoter, which flank the MCS.

One of the most useful features of λZap is that it has been designed to allow automatic excision in vivo of a small 2.9 kb colony-producing vector termed pBluescript SK, a **phagemid** (see next section). This technique is sometimes termed **single-stranded DNA rescue** and occurs as the result of a process termed **superinfection**, where helper phage are added to the cells, which are then grown for an additional period of approximately 4 hours (Figure 4.56).

The helper phage displaces a strand within the λZap that contains the foreign DNA insert. This is circularised and packaged as a filamentous phage similar to M13. The packaged phagemid is secreted from the *E. coli* cell and may be recovered from the supernatant. The λZap vector therefore allows a number of diverse manipulations to

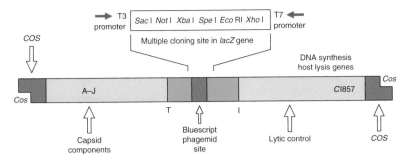

Figure 4.55 General map of λZap cloning vector, indicating important areas of the vector. The multiple cloning site is based on the *lacZ* gene, providing blue/white selection based on the β-galactosidase gene. In between the initiator (I) site and the terminator (T) site lie sequences encoding the phagemid Bluescript.

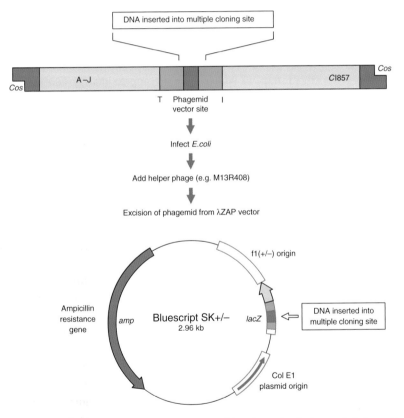

Figure 4.56 Single-stranded DNA rescue of phagemid from λZap. The single-stranded phagemid pBluescript SK may be excised from λZap by addition of helper phage. This provides the necessary proteins and factors for transcription between the I and T sites in the parent phage to produce the phagemid with the DNA cloned into the parent vector.

be undertaken without the necessity of recloning or subcloning foreign DNA fragments. The process of subcloning is sometimes necessary when the manipulation of a gene fragment cloned in a general purpose vector needs to be inserted into a more specialised vector for the application of techniques such as in vitro mutagenesis or protein production.

4.12.3 Phagemid-Based Vectors

Much use has been made of single-stranded bacteriophage vectors such as fd, M13 and vectors that have the combined properties of phage and plasmids, termed phagemids. M13 is a filamentous coliphage with a single-stranded circular DNA genome (Figure 4.57). Upon infection of *E. coli*, the DNA replicates initially as a double-stranded molecule, but subsequently produces single-stranded virions for infection of further bacterial cells (lytic growth). The nature of these vectors makes them ideal for techniques such as chain termination sequencing and in vitro mutagenesis, since both require single-stranded DNA.

M13 or phagemids such as pBluescript SK infect *E. coli* harbouring a male-specific structure termed the F-pilus (Figure 4.58). They enter the cell by adsorption to this structure and once inside, the phage DNA is converted to a double-stranded replicative form or RF DNA. Replication then proceeds rapidly until some 100 RF molecules are produced within the *E. coli* cell. DNA synthesis then switches to the production of single strands and the DNA is assembled and packaged into the capsid at the bacterial periplasm. The bacteriophage DNA is then encapsulated by the major coat protein, gene VIII protein, of which there are approximately 2800 copies with three to six copies of the gene III protein at one end of the particle. The extrusion of the bacteriophage through the bacterial periplasm results in a decreased growth rate of the *E. coli* cell rather than host cell lysis and is visible on a bacterial lawn as an area of clearing. Approximately 1000 packaged phage particles may be released into the medium in one cell division.

In addition to producing single-stranded DNA, the coliphage vectors have a number of other features that make them attractive as cloning vectors. Since the bacteriophage DNA is replicated as a double-stranded RF DNA intermediate, a number of regular DNA manipulations may be performed, such as restriction digestion, mapping and DNA ligation. RF DNA is prepared by lysing infected *E. coli* cells and purifying

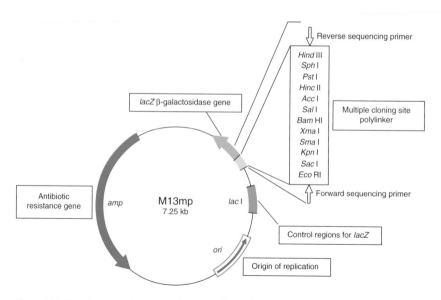

Figure 4.57 Genetic map and important features of bacteriophage vector M13.

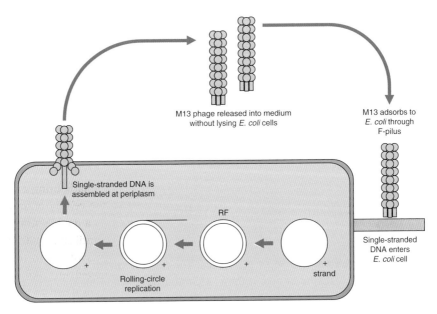

M13 phage released into medium
without lysing *E. coli* cells

M13 adsorbs to
E. coli through
F-pilus

Single-stranded DNA is
assembled at periplasm

RF

Rolling-circle
replication

strand

Single-stranded
DNA enters
E. coli cell

Figure 4.58 Life cycle of bacteriophage M13. The bacteriophage virus enters the *E. coli* cell through the F-pilus. It then enters a stage where the circular single strands are converted to double strands. Rolling-circle replication then produces single strands, which are packaged and extruded through the *E. coli* cell membrane.

the supercoiled circular phage DNA with the same methods used for plasmid isolation. Intact single-stranded DNA packaged in the phage protein coat located in the supernatant may be precipitated with reagents such as polyethylene glycol, and the DNA purified with phenol/chloroform. Thus the bacteriophage may act as a plasmid under certain circumstances and at other times produce DNA in the fashion of a virus. A family of vectors derived from M13 are currently widely used, including M13mp8/9, mp18/19, etc., all of which have a number of highly useful features. They all contain a synthetic MCS, which is located in the *lacZ* gene without disruption of the reading frame of the gene. This allows efficient selection to be undertaken based on the technique of blue/white screening. As the series of vectors were developed, the number of restriction sites was increased in an asymmetric fashion. Thus M13mp8, mp12, mp18 and sister vectors that have the same MCS but in reverse orientation, M13mp9, mp13 and mp19, respectively, have more restriction sites in the MCS making the vector more useful since a greater choice of restriction enzymes is available (Figure 4.59). However, one problem frequently encountered with M13 is the instability and spontaneous loss of inserts that are greater than 6 kb.

Phagemids are very similar to M13 and replicate in a similar fashion. One of the first phagemid vectors, pEMBL, was constructed by inserting a fragment of another phage termed f1, containing a phage origin of replication and elements for its morphogenesis, into a pUC8 plasmid. Following superinfection with helper phage, the f1 origin is activated allowing single-stranded DNA to be produced. The phage is assembled into a phage coat extruded through the periplasm and secreted into the culture medium in a similar way to M13. Without superinfection, the phagemid replicates as a pUC-type plasmid and in the replicative form (RF) the DNA isolated is double-stranded. This allows further manipulations such as restriction digestion, ligation

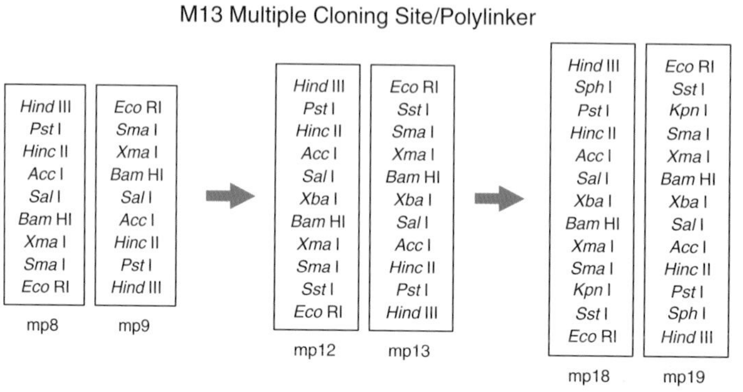

M13 Multiple Cloning Site/Polylinker

Hind III	Eco RI
Pst I	Sma I
Hinc II	Xma I
Acc I	Bam HI
Sal I	Sal I
Bam HI	Acc I
Xma I	Hinc II
Sma I	Pst I
Eco RI	Hind III

mp8   mp9

Hind III	Eco RI
Pst I	Sst I
Hinc II	Sma I
Acc I	Xma I
Sal I	Bam HI
Xba I	Xba I
Bam HI	Sal I
Xma I	Acc I
Sma I	Hinc II
Sst I	Pst I
Eco RI	Hind III

mp12   mp13

Hind III	Eco RI
Sph I	Sst I
Pst I	Kpn I
Hinc II	Sma I
Acc I	Xma I
Sal I	Bam HI
Xba I	Xba I
Bam HI	Sal I
Xma I	Acc I
Sma I	Hinc II
Kpn I	Pst I
Sst I	Sph I
Eco RI	Hind III

mp18   mp19

Figure 4.59 Design and orientation of polylinkers in M13 series. Only the main restriction enzymes are indicated.

and mapping analysis to be performed. The pBluescript SK vector is also a phagemid and can be used in its own right as a cloning vector and manipulated as if it were a plasmid. It may, like M13, be used in nucleotide sequencing and site-directed mutagenesis, and it is also possible to produce RNA transcripts that may be used in the production of labelled cRNA probes or riboprobes.

4.12.4 Cosmid-Based Vectors

The way in which the phage λ DNA is replicated is of particular interest in the development of larger insert cloning vectors termed cosmids (Figure 4.60). These are especially useful for the analysis of highly complex genomes and are an important part of various genome mapping projects. The upper limit of the insert capacity of phage λ is approximately 21 kb. This is because of the requirement for essential genes and the fact that the maximum length between the *cos* sites is 52 kb. Consequently, cosmid vectors have been constructed that incorporate the *cos* sites from phage λ and also the essential features of a plasmid, such as the plasmid origin of replication, a gene for drug resistance, and several unique restriction sites for insertion of the DNA to be cloned. When a cosmid preparation is linearised by restriction digestion, and ligated to DNA for cloning, the products will include concatemers of alternating cosmid vector and insert. Thus, the only requirement for the length of DNA to be packaged into viral heads is that it should contain *cos* sites spaced apart at the correct distance; in practice, this spacing can range between 37 and 52 kb. Such DNA can be packaged in vitro if phage head precursors, tails and packaging proteins are provided. Since the cosmid is very small, inserts of about 40 kb in length will be most readily packaged. Once inside the cell, the DNA recircularises through its *cos* sites, and from then onwards behaves exactly like a plasmid.

4.12.5 Large Insert-Capacity Vectors

The advantage of vectors that accept larger fragments of DNA than phage λ or cosmids is that fewer clones need to be screened when searching for the foreign DNA of interest. They have also had an enormous impact in the mapping of the genomes of

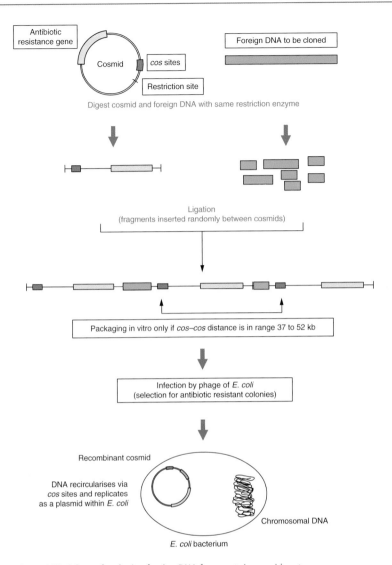

Figure 4.60 Scheme for cloning foreign DNA fragments in cosmid vectors.

organisms such as the mouse and are used extensively in the human genome mapping project. Recent developments have allowed the production of large insert-capac-
ity vectors based on human artificial chromosomes, bacterial artificial chromosomes
(BACs), mammalian artificial chromosomes (MACs) and on the virus P1 artificial
chromosomes (PACs). However, perhaps the most significant development are vectors
based on yeast artificial chromosomes (YACs).

4.12.6 Yeast Artificial Chromosome (YAC) Vectors

Yeast artificial chromosomes (YACs) are linear molecules composed of a centromere,
telomeres and a replication origin termed an ARS element (autonomous replicating sequence). The YAC is digested with restriction enzymes at the SUP4 site (a

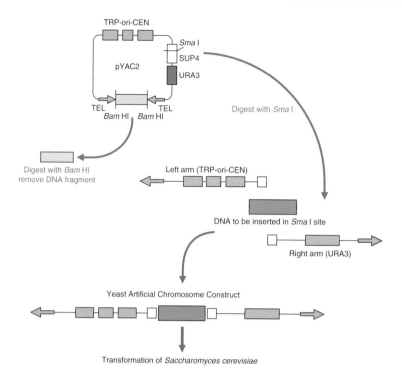

Figure 4.61 Scheme for cloning large fragments of DNA into YAC vectors.

suppressor tRNA gene marker) and *Bam* HI sites separating the telomere sequences (Figure 4.61). This produces two arms and the foreign genomic DNA is ligated to produce a functional YAC construct. YACs are replicated in yeast cells; however, the external cell wall of the yeast needs to be removed to leave a spheroplast. These are osmotically unstable and also need to be embedded in a solid matrix such as agar. Once the yeast cells are transformed, only correctly constructed YACs with associated selectable markers are replicated in the yeast strains. DNA fragments with repeat sequences are sometimes difficult to clone in bacteria-based vectors, but may be successfully cloned in YAC systems. The main advantage of YAC-based vectors, however, is the ability to clone very large fragments of DNA. The stable maintenance and replication of foreign DNA fragments of up to 2000 kb have been carried out in YAC vectors and they are the main vector of choice in the various genome mapping and sequencing projects.

4.12.7 Vectors Used in Eukaryotes

The use of *E. coli* for general cloning and manipulation of DNA is well established; however, numerous developments have been made for cloning in eukaryotic cells. Plasmids used for cloning DNA in eukaryotic cells require a eukaryotic origin of replication and marker genes that will be expressed by eukaryotic cells. At present, the two most important applications of plasmids to eukaryotic cells are for cloning in yeast and in plants.

Although yeast has a natural plasmid, called the 2μ circle, this is too large for use in cloning. Plasmids such as the **yeast episomal plasmid (YEp)** have been created by genetic manipulation using replication origins from the 2μ circle, and by incorporating a gene that will complement a defective gene in the host yeast cell. If, for example, a strain of yeast is used that has a defective gene for the biosynthesis of an amino acid, an active copy of that gene on a yeast plasmid can be used as a selectable marker for the presence of that plasmid. Yeast, like bacteria, can be grown rapidly, and it is therefore well suited for use in cloning. Of particular use has been the creation of **shuttle vectors**, which have origins of replication for yeast and bacteria such as *E. coli*. This means that constructs may be prepared rapidly in the bacteria and delivered into yeast for expression studies.

The bacterium *Agrobacterium tumefaciens* infects plants that have been damaged near soil level, and this infection is often followed by the formation of plant tumours in the vicinity of the infected region. It is now known that *A. tumefaciens* contains a plasmid called the **Ti plasmid**, part of which is transferred into the nuclei of plant cells that are infected by the bacterium. Once in the nucleus, this DNA is maintained through integration with the chromosomal DNA. The integrated DNA carries genes for the synthesis of opines (which are metabolised by the bacteria but not by the plants) and for tumour induction (hence 'Ti'). DNA inserted into the correct region of the Ti plasmid will be transferred to infected plant cells, and in this way it has been possible to clone and express foreign genes in plants (Figure 4.62). This is an essential prerequisite for the genetic engineering of crops.

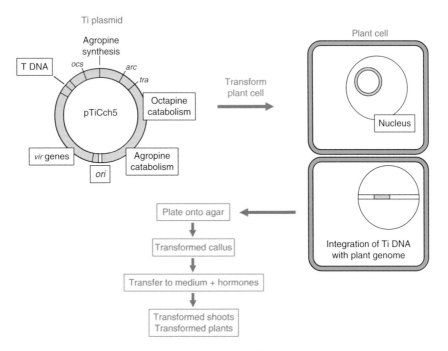

Figure 4.62 Scheme for cloning in plant cells using the Ti plasmid.

4.12.8 Delivery of Vectors into Eukaryotes

Following the production of a recombinant molecule, the so-called constructs are subsequently introduced into cells to enable their replication as the cells replicate. Initial recombinant DNA experiments were performed in bacterial cells, because of their ease of growth and short doubling time. Gram-negative bacteria such as *E. coli* can be made competent for the introduction of extraneous plasmid DNA into cells. The natural ability of bacteriophage to introduce DNA into *E. coli* has also been well exploited and results in 10–100-fold higher efficiency for the introduction of recombinant DNA compared to transformation of competent bacteria with plasmids. These well-established and traditional approaches are the reason why so many cloning vectors have been developed for *E. coli*. The delivery of cloning vectors into eukaryotic cells is, however, not as straightforward as that for the bacterium *E. coli*.

It is possible to deliver recombinant molecules into animal cells by transfection. The efficiency of this process can be increased by first precipitating the DNA with Ca^{2+} or making the membrane permeable with divalent cations. Polymers possessing a high molecular mass, such as DEAE-dextran or polyethylene glycol (PEG), may also be used to maximise the uptake of DNA. The technique is rather inefficient but a selectable marker that provides resistance to a toxic compound such as neomycin can be used to monitor the success. Alternatively, DNA can be introduced into animal cells by electroporation. In this process, the cells are subjected to pulses of a high-voltage gradient, causing many of them to take up DNA from the surrounding solution. This technique has proved to be useful with cells from a range of animal, plant and microbial sources. More recently, the technique of lipofection has been used as another delivery method. The recombinant DNA is encapsulated by lipid-coated particles that fuse with the lipid membrane of cells and thus release the DNA into the cell. Lastly, microinjection of DNA into cell nuclei of eggs or embryos has also been performed successfully in many mammalian cells.

The ability to deliver recombinant molecules into plant cells is not without its problems. Generally, the outer cell wall of the plant must be stripped, usually by enzymatic digestion, to leave a protoplast. The cells are then able to take up recombinants from the supernatant. The cell wall can be regenerated by providing appropriate media. In cases where protoplasts have been generated, transformation may also be achieved by electroporation. An even more dramatic transformation procedure involves propelling microscopically small titanium or gold pellet microprojectiles coated with the recombinant DNA molecule, into plant cells in intact tissues. This biolistic transformation involves the detonation of an explosive charge that is used to propel the microprojectiles into the cells at high velocity; hence it is sometimes referred to as a gene gun. The cells then appear to reseal themselves after the delivery of the recombinant molecule. This is a particularly promising technique for use with plants whose protoplasts will not regenerate whole plants.

4.13 HYBRIDISATION AND GENE PROBES

4.13.1 Cloned cDNA Probes

The increasing accumulation of DNA sequences in nucleic acid databases coupled with the availability of custom synthesis of oligonucleotides has provided a relatively straightforward means to design and produce gene probes and primers for PCR. Such probes and primers are usually designed with bioinformatics software, using sequence information from nucleic acid databases. Alternatively, gene-family-related sequences may also be successfully employed. However, there are many gene probes that have traditionally been derived from cDNA or from genomic sequences and that have been cloned into plasmid and phage vectors. These require manipulation before they can be labelled and used in hybridisation experiments. Gene probes may vary in length from 100 bp to a number of kilobases, although this is dependent on their origin. Many are short enough to be cloned into plasmid vectors and are useful in that they may be manipulated easily and are relatively stable, both in transit and in the laboratory. The DNA sequences representing the gene probe are usually excised from the cloning vector by digestion with restriction enzymes and purified. In this way, vector sequences that may hybridise non-specifically and cause high background signals in hybridisation experiments are removed.

4.13.2 RNA Gene Probes

It is also possible to prepare cRNA probes or riboprobes by in vitro transcription of gene probes cloned into a suitable vector. A good example of such a vector is the phagemid pBluescript SK; at each end of the multiple cloning site where the cloned DNA fragment resides are promoters for T3 or T7 RNA polymerase. The vector can be linearised with a restriction enzyme digest, and T3 or T7 RNA polymerase is used to transcribe the cloned DNA fragment. Provided a labelled NTP is added in the reaction, a riboprobe labelled to a high specific activity will be produced (Figure 4.63).

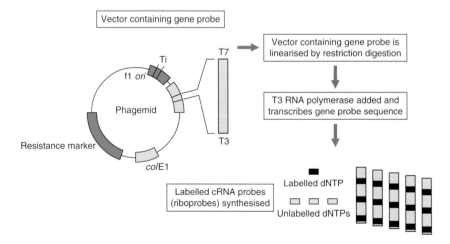

Figure 4.63 Production of cRNA (riboprobes) using T3 RNA polymerase and phagemid vectors.

One advantage of riboprobes is that they are single-stranded and their sensitivity is generally regarded as superior to cloned double-stranded probes. They are used extensively in *in situ* hybridisation (see Section 4.17.2) and for identifying and analysing mRNA.

4.14 SCREENING GENE LIBRARIES

4.14.1 Colony and Plaque Hybridisation

Once a cDNA or genomic library has been prepared, the next task requires the identification of the specific fragment of interest. In many cases, this may be more problematic than the library construction itself, since many hundreds of thousands of clones may be in the library. One clone containing the desired fragment needs to be isolated from the library and therefore a number of techniques, mainly based on hybridisation, have been developed.

Colony hybridisation is one method used to identify a particular DNA fragment from a plasmid gene library (Figure 4.64). A large number of clones are grown up to form colonies on one or more plates, and these are then replica plated (see Section 4.12.1) onto nylon membranes placed on solid agar medium. Nutrients diffuse through the membranes and allow colonies to grow on them. The colonies are then lysed, and liberated DNA is denatured and bound to the membranes, so that the pattern of colonies is replaced by an identical pattern of bound DNA. The membranes are then incubated with a pre-hybridisation mix containing non-labelled non-specific DNA such as salmon sperm DNA to block non-specific sites. Following this denaturation, the labelled gene probe is added. Under hybridising conditions, the probe will bind only to cloned fragments containing at least part of its corresponding gene. The membranes are then washed to remove any unbound probe and the binding detected by autoradiography of the membranes. If non-radioactive labels have been used then alternative methods of detection must be employed. By comparison of the patterns on the autoradiograph with the original plates of colonies, those that contain the desired gene (or part of it) can be identified and isolated for further analysis. A similar procedure is used to identify desired genes cloned into bacteriophage vectors. In this case, the process is termed **plaque hybridisation**. It is the DNA contained in the bacteriophage particles found in each plaque that is immobilised onto the nylon membrane. This is then probed with an appropriately labelled complementary gene probe and detection undertaken as for colony hybridisation.

4.14.2 PCR Screening of Gene Libraries

In many cases it is possible to use the PCR to screen cDNA or genomic libraries constructed in plasmids or bacteriophage vectors. This is usually undertaken with primers that anneal to the vector rather than the foreign DNA insert. The size of an amplified product may be used to characterise the cloned DNA and subsequent restriction mapping is then carried out (Figure 4.65). The main advantage of the PCR over traditional hybridisation-based screening is the rapidity of the technique, as PCR screening may be undertaken in 3–4 h, whereas it may be several days before detection by

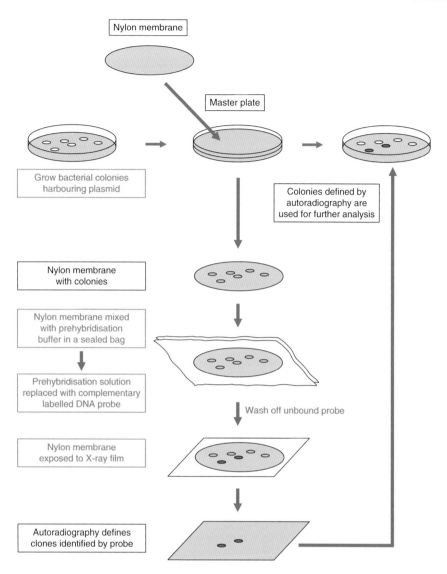

Figure 4.64 Colony hybridisation technique for locating specific bacterial colonies harbouring recombinant plasmid vectors containing desired DNA fragments. This is achieved by hybridisation to a complementary labelled DNA probe and autoradiography.

hybridisation is achieved. The PCR screening technique gives an indication of the size rather than the sequence of the cloned insert; however, PCR primers that are specific for a foreign DNA insert may also be used. This allows a more rigorous characterisation of clones from cDNA and genomic libraries.

4.14.3 Screening Expression cDNA Libraries

In some cases the protein for which the gene sequence is required is partially characterised and in these cases it may be possible to produce antibodies to that protein. This allows immunological screening to be undertaken rather than gene hybridisation.

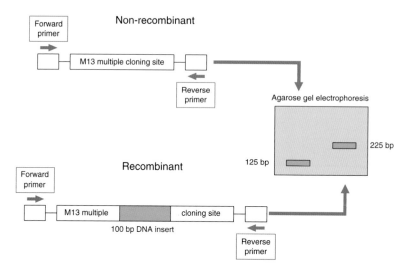

Figure 4.65 PCR screening of recombinant vectors. In this figure, the M13 non-recombinant has no insert and so the PCR undertaken with forward and reverse sequencing primers gives rise to a product 125 bp in length. The M13 recombinant with an insert of 100 bp will give rise to a PCR product of 125 bp + 100 bp = 225 bp and thus may be distinguished from the non-recombinant by analysis on agarose gel electrophoresis.

Such antibodies are useful since they may be used as the probe if little or no gene sequence is available. In these cases it is possible to prepare a cDNA library in a specially adapted vector termed an expression vector, which transcribes and translates any cDNA inserted into it. The protein is usually synthesised as a fusion with another protein such as β-galactosidase. Common examples of expression vectors are those based on bacteriophage such as λgt11 and λZap or plasmids such as pEX. The precise requirements for such vectors are identical to vectors that are dedicated to producing proteins in vitro . In some cases, expression vectors incorporate inducible promoters that may be activated by, for example, increasing the temperature, allowing stringent control of expression of the cloned cDNA molecules (Figure 4.66).

The cDNA library is plated out and nylon membrane filters prepared as for colony/plaque hybridisation. A solution containing the antibody to the desired protein is then added to the membrane. The membrane is washed to remove any unbound protein and a further labelled antibody that is directed to the first antibody is applied. This allows visualisation of the plaque or colony that contains the cloned cDNA for that protein and this may then be picked from the agar plate and pure preparations grown for further analysis.

4.15 APPLICATIONS OF GENE CLONING

4.15.1 Protein Engineering

One of the most powerful developments in molecular biology has been the ability to artificially create defined mutations in a gene and analyse the resulting protein following in vitro expression. Numerous methods are now available for producing site-directed mutations, many of which now involve the PCR. Commonly termed

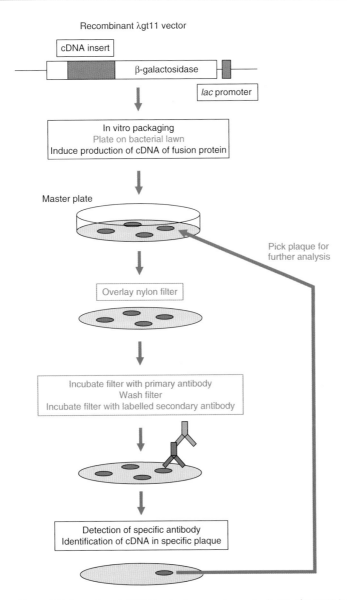

Recombinant λgt11 vector

cDNA insert

β-galactosidase

lac promoter

In vitro packaging
Plate on bacterial lawn
Induce production of cDNA of fusion protein

Master plate

Pick plaque for
further analysis

Overlay nylon filter

Incubate filter with primary antibody
Wash filter
Incubate filter with labelled secondary antibody

Detection of specific antibody
Identification of cDNA in specific plaque

Figure 4.66 Screening of cDNA libraries in expression vector λgt11. The cDNA inserted upstream of the gene for λβ-galactosidase will give rise to a fusion protein under induction (e.g. with IPTG). The plaques are then blotted onto a nylon membrane filter and probed with an antibody specific for the protein coded by the cDNA. A secondary labelled antibody directed to the specific antibody can then be used to identify the location (plaque) of the cDNA.

protein engineering, this process involves a logical sequence of analytical and computational techniques centred around a design cycle. This includes the biochemical preparation and analysis of proteins, the subsequent identification of the gene encoding the protein and its modification. The production of the modified protein and its further biochemical analysis completes the concept of rational redesign to improve or probe a protein's structure and function (Figure 4.67).

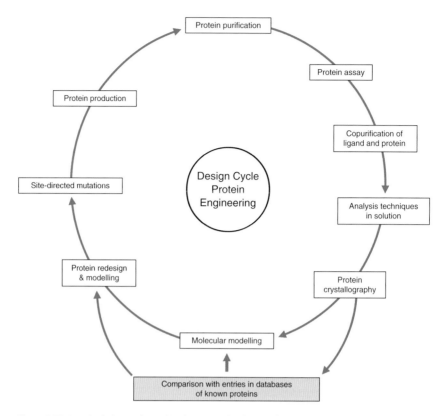

Figure 4.67 Protein design cycle used in the rational redesign of proteins and enzymes.

The use of design cycles and rational design systems are exemplified by the study and manipulation of subtilisin. This is a serine protease of broad specificity and of considerable industrial importance being used in soap powder and in the food and leather industries. Protein engineering has been used to alter the specificity, pH profile and stability to oxidative, thermal and alkaline inactivation. Analysis of homologous thermophiles and their resistance to oxidation has also been improved. Engineered subtilisins of improved bleach resistance and wash performance are now used in many brands of washing powders. Furthermore, mutagenesis has played an important role in the re-engineering of important therapeutic proteins such as the Herceptin® antibody, which has been used to successfully treat certain types of breast cancer.

4.15.2 Kunkel Oligonucleotide-Directed Mutagenesis

This is a traditional method of site-directed mutagenesis and demands that the gene is already cloned. Complete sequencing of the gene is essential to identify a potential region for mutation. Once the precise base change has been identified, an oligonucleotide is designed that is complementary to part of the gene, but has one base difference. This difference is designed to alter a particular codon, which, following translation, gives rise to a different amino acid and hence may alter the properties of the protein.

A single-stranded uracil-containing vector such as M13 is grown in a selectable host deficient of the enzymes dUTPase (*dut⁻*) and uracil *N*-deglycosidase (*ung⁻*). The mutagenic oligonucleotide and the single-stranded vector are annealed and DNA polymerase is added together with the dNTPs. The primer for the reaction is the 3′ end of the oligonucleotide. The DNA polymerase produces a new DNA strand complementary to the existing one, but which incorporates the oligonucleotide with the base mutation. The subsequent transformation of a *dut⁺ ung⁺ E.coli* strain with the recombinant produces multiple copies, one strand containing the sequence with the mutation and the other the parent uracil-containing strand which is preferentially degraded.

Plaque hybridisation using the oligonucleotide as the probe is then used at a stringency that allows only those plaques containing a mutated sequence to be identified (Figure 4.68). Further methods have also been developed that simplify the process of detecting the strands with the mutations.

4.15.3 PCR-Based Mutagenesis

The PCR has been adapted to allow mutagenesis to be undertaken and this relies on single bases mismatched between one of the PCR primers and the target DNA becoming incorporated into the amplified product following thermal cycling.

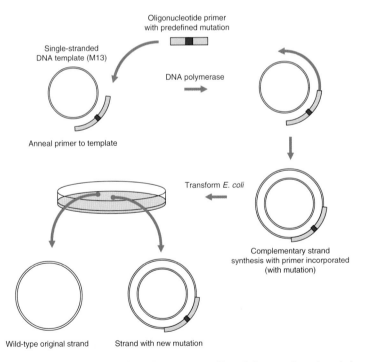

Figure 4.68 Oligonucleotide-directed mutagenesis. This technique requires a knowledge of nucleotide sequence, since an oligonucleotide may then be synthesised with the base mutation. Annealing of the oligonucleotide to complementary (except for the mutation) single-stranded DNA provides a primer for DNA polymerase to produce a new strand and thus incorporates the primer with the mutation.

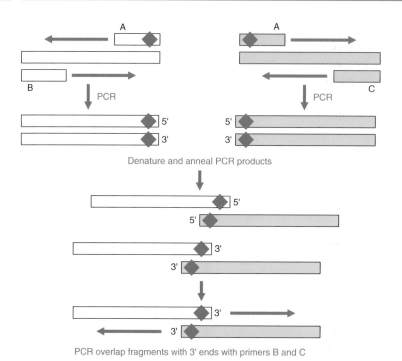

Figure 4.69 Construction of a synthetic DNA fragment with a predefined mutation using overlap PCR mutagenesis.

The basic **PCR mutagenesis** system involves the use of two primary PCR reactions to produce two overlapping DNA fragments, both bearing the same mutation in the overlap region; this technique is thus termed **overlap extension PCR**. The two separate PCR products are made single-stranded and the overlap in sequence allows the products from each reaction to hybridise. Subsequently, one of the two hybrids bearing a free 3′-hydroxyl group is extended to produce a new duplex fragment. The other hybrid with a 5′-hydroxyl group cannot act as substrate in the reaction. Thus, the overlapped and extended product will now contain the directed mutation (Figure 4.69). Deletions and insertions may also be created with this method, although the requirements of four primers and three PCR reactions limits the general applicability of the technique. A modification of overlap extension PCR may also be used to construct directed mutations; this is termed **megaprimer PCR**. This latter method utilises three oligonucleotide primers to perform two rounds of PCR. A complete PCR product, the megaprimer is made single-stranded and this is used as a large primer in a further PCR reaction with an additional primer.

The above are all methods for creating rational defined mutations as part of a design cycle system. However, it is also possible to introduce random mutations into a gene and select for enhanced or new activities of the protein or enzyme it encodes. This accelerated form of artificial molecular evolution may be undertaken using **error-prone PCR**, where deliberate and random mutations are introduced by a low-fidelity PCR amplification reaction. The resulting amplified gene is then translated and its activity assayed. This has already provided novel evolved enzymes such as a *p*-nitrobenzyl esterase, which exhibits an unusual and surprising affinity for organic solvents. This accelerated evolutionary approach to protein engineering has been useful in the production of novel antibodies produced by phage display (see Section 4.16.3) and in the development of antibodies with enzymatic activities (catalytic antibodies).

4.16 EXPRESSION OF FOREIGN GENES

One of the most useful applications of recombinant DNA technology is the ability to artificially synthesise large quantities of natural or modified proteins in a host cell such as bacteria or yeast. The benefits of these techniques have been enjoyed for many years since the first insulin molecules were cloned and expressed in 1982 (Table 4.8). Contamination of proteins purified from native sources, such as in the case of the blood product factor VIII which was often contaminated with infectious agents, has also increased the need to develop effective vectors for production of foreign genes. In general, the expression of foreign genes is carried out in specialised cloning vectors in a host such as *E.coli* (Figure 4.70). It is possible to use cell-free transcription and translation systems that direct the synthesis of proteins without the need to grow and maintain cells. Cell-free in vitro protein expression is carried out with the appropriate amino acids, ribosomes, tRNA molecules, cofactors and isolated template mRNA or DNA. Wheat germ extracts or rabbit reticulocyte lysates can provide the necessary components and are usually the systems of choice for eukaryotic protein production. The resulting proteins may be detected by polyacrylamide

Table 4.8 **A number of recombinant DNA-derived human therapeutic reagents**

Therapeutic area	Recombinant product
Drugs	Erythropoietin
	Insulin
	Growth hormone
	Coagulation factors (e.g. factor VIII)
	Plasminogen activator
Vaccines	Hepatitis B
Cytokines/growth factors	GM-CSF
	G-CSF
	Interleukins
	Interferons

Notes: GM-CSF, granulocyte–macrophage colony-stimulating factor; G-CSF, granulocyte colony-stimulating factor.

gel electrophoresis (see Section 6.3) or by immunological detection using Western blotting (see Section 7.7). These systems are ideal for the rapid production of proteins on a smaller scale than in vivo systems and are particularly useful in the study of protein functional analysis, such as post-translational modification, or folding and stability studies. Furthermore, it is possible to produce proteins that would be toxic to expression hosts in vivo.

4.16.1 Prokaryotic Expression Vectors

For a foreign gene to be expressed in a bacterial cell, it must have particular **promoter** sequences upstream of the coding region, to which the RNA polymerase will bind prior to transcription of the gene. The choice of promoter is vital for correct and efficient transcription, since the sequence and position of promoters are specific to a particular host such as *E. coli*. It must also contain a **ribosome-binding** site, placed just before the coding region. Unless a cloned gene contains both of these sequences, it will not be expressed in a bacterial host cell. If the gene has been produced via cDNA from a eukaryotic cell, then it will certainly not have any such sequences. Consequently, expression vectors have been developed that contain promoter and ribosome-binding sites positioned just before one or more restriction sites for the insertion of foreign DNA (Figure 4.70). These **regulatory sequences**, such as that from the *lac* operon of *E. coli*, are usually derived from genes that, when induced, are strongly expressed in bacteria. Since the mRNA produced from the gene is read as triplet codons, the inserted sequence must be placed so that its reading frame is in phase with the regulatory sequence. This can be ensured by the use of three vectors that differ only in the number of bases between promoter and insertion site, the second and third vectors being respectively one and two bases longer than the first. If an insert is cloned in all three vectors then in general it will subsequently be in the correct reading frame in one of them. The resulting clones can be screened for the production of a functional foreign protein. The final clone should then be checked by Sanger sequencing (see Section 20.2.1).

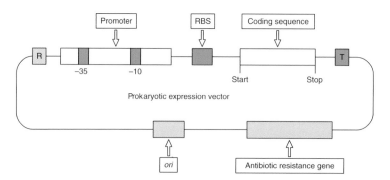

Figure 4.70 Components of a typical prokaryotic expression vector. To produce a transcript (coding sequence) and translate it, a number of sequences in the vector are required. These include the promoter and ribosome-binding site (RBS). The activity of the promoter may be modulated by a regulatory gene (R), which acts in a way similar to that of the regulatory gene in the *lac* operon. T indicates a transcription terminator.

4.16.2 Expression of Eukaryotic Genes

It is not only possible, but usually essential, to use cDNA instead of a eukaryotic genomic DNA to direct the production of a functional protein by bacteria. This is because bacteria are not capable of processing RNA to remove introns, and so any foreign genes must be pre-processed as cDNA if they contain introns. A further problem arises if the protein must be glycosylated, by the conjugation with oligosaccharides at specific sites, in order to become functional. Although the use of bacterial expression systems is somewhat limited for eukaryotic systems, there are a number of eukaryotic expression systems based on plant, mammalian, insect and yeast cells. These types of cell can perform such post-translational modifications, producing a correct glycosylation pattern or phosphorylation. It is also possible to include a signal or address sequence at the 5′ end of the mRNA that directs the protein to a particular cellular compartment or even out of the cell altogether into the supernatant. This makes the recovery of expressed recombinant proteins much easier, since the supernatant may be drawn off while the cells are still producing protein.

One useful eukaryotic expression system is based on the monkey COS cell line. These cells each contain a region derived from a mammalian monkey virus termed simian virus 40 (SV40). A defective region of the SV40 genome has been stably integrated into the COS cell genome. This allows the expression of a protein termed the large T antigen, which is required for viral replication. When a recombinant vector having the SV40 origin of replication and carrying foreign DNA is inserted into the COS cells, viral replication takes place. This results in high-level expression of foreign proteins. The disadvantage of this system is the ultimate lysis of the COS cells and limited insert capacity of the vector. Much interest has thus been focussed on other modified viruses, vaccinia virus and baculovirus. These have been developed for high-level expression in mammalian cells and insect cells, respectively. The vaccinia virus in particular has been used to correct defective ion transport by introducing a wild-type cystic fibrosis gene into cells bearing a mutated cystic fibrosis (CFTR) gene. There is no doubt that the further development of these vector systems will enhance eukaryotic protein expression in the future.

4.16.3 Phage Display Techniques

As a result of the production of phagemid vectors and as a means of overcoming the problems of screening large numbers of clones generated from genomic libraries of antibody genes, a method for linking the phenotype or expressed protein with the genotype has been devised. This is termed phage display, since a functional protein is linked to a major coat protein of a coliphage, whilst the single-stranded gene encoding the protein is packaged within the virion. The initial steps of the method rely on the PCR to amplify gene fragments that represent functional domains or subunits of a protein such as an antibody. These are then cloned into a phage display vector, which is an adapted phagemid vector and used to transform *E. coli*. A helper phage is then added to provide accessory proteins for new phage molecules to be constructed. The DNA fragments representing the protein or polypeptide of interest are also transcribed

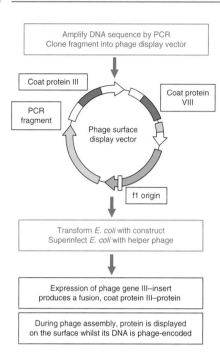

Figure 4.71 Flow diagram indicating the main steps in the phage display technique.

and translated, but linked to the gene for major coat protein III (gIII). Thus when the phage is assembled, the protein or polypeptide of interest is incorporated into the coat of the phage and displayed, whilst the corresponding DNA is encapsulated (Figure 4.71).

There are numerous applications for the display of proteins on the surface of bacteriophage viruses, bacteria and other organisms, and commercial organisations have been quick to exploit this technology. One major application is the analysis and production of engineered antibodies from which the technology was mainly developed. In general, phage-based systems have a number of novel applications in terms of ease of selection rather than screening of antibody fragments, allowing analysis by methods such as affinity chromatography. In this way, it is possible to generate large numbers of antibody heavy and light chain genes by PCR amplification and mix them in a random fashion. This **recombinatorial library** approach may allow new or novel partners to be formed, as well as naturally existing ones. This strategy is not restricted to antibodies and vast libraries of peptides may be used in this combinatorial chemistry approach to identify novel compounds of use in biotechnology and medicine.

Phage-based cloning methods also offer the advantage of allowing mutagenesis to be performed with relative ease. This may allow the production of antibodies with affinities approaching that derived from the human or mouse immune system. This may be brought about by using an error-prone DNA polymerase in the initial steps of constructing a **phage display library**. It is possible that these types of libraries may provide a route to high-affinity recombinant antibody fragments that are difficult to produce by more conventional hybridoma fusion techniques (see Section 7.1.2). Surface display libraries have also been prepared for the selection of ligands, hormones and other polypeptides in addition to allowing studies on protein–protein or protein–DNA interactions or determining the precise binding domains in these receptor–ligand interactions.

4.16.4 Alternative Display Systems

A number of display systems have been developed based on the original phage display technique. One interesting method is ribosome display, where a sequence or even a library of sequences are transcribed and translated in vitro; however, in the DNA library, the sequences are fused to spacer sequences lacking a stop codon. During translation at the ribosome, the protein protrudes from the ribosome and is locked in with the mRNA. The complex can be stabilised by adding salt. In this way, it is possible to select the appropriate protein through binding to its ligand. Thus a high-affinity protein–ligand can be isolated that has the mRNA that originally encoded it. The mRNA may then be reverse transcribed into cDNA and amplified by PCR to allow further methods, such as mutagenesis, to be undertaken. In the similar technique of mRNA display, the association between the protein and mRNA is through a more stable covalent puromycin link rather than the salt-induced link as in ribosome display. Further display systems, based on yeast or bacteria, have also been developed and provide powerful in vitro selection methods.

4.17 ANALYSING GENES AND GENE EXPRESSION

4.17.1 Identifying and Analysing mRNA

The levels and expression patterns of mRNA dictate many cellular processes and therefore there is much interest in the ability to analyse and determine levels of a particular mRNA. Technologies such as real-time or **quantitative PCR**, microarray and now RNA sequencing (RNA-Seq) are employed to perform high-throughput analysis (see also Sections 4.18.1, 20.6.2). A number of other informative techniques have been developed that allow the fine structure of a particular mRNA to be analysed, and the relative amounts of an RNA quantified by non-PCR-based methods. This is important, not only for gene regulation studies, but may also be used as a marker for certain clinical disorders. Traditionally, the **Northern blot** has been used for detection of particular RNA transcripts by blotting extracted mRNA and immobilising it on a nylon membrane. Subsequent hybridisation with labelled gene probes allows precise determination of the size and abundance of a transcript. However, much use has been made of a number of nucleases that digest only single-stranded nucleic acids and not double-stranded molecules. In particular, the **ribonuclease protection assay** (RPA) has allowed much information to be gained regarding the nature of mRNA transcripts (Figure 4.72). In the RPA, single-stranded mRNA is hybridised in solution to a labelled single-stranded RNA probe that is in excess. The hybridised part of the complex becomes protected, whereas the non-hybridised part of the probe made from RNA is digested with RNase A and RNase T1. The protected fragment may then be analysed on a high-resolution polyacrylamide gel. This method may give valuable information

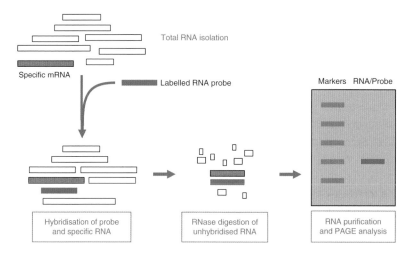

Figure 4.72 Steps involved in the ribonuclease protection assay (RPA). PAGE, polyacrylamide gel electrophoresis.

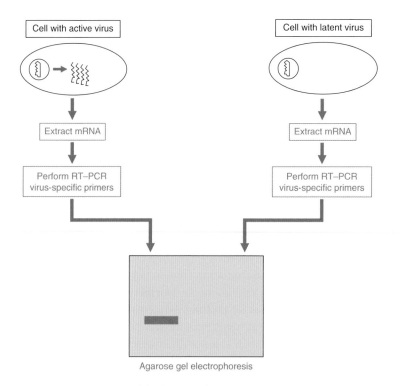

Figure 4.73 Representation of the detection of active viruses using RT-PCR.

regarding the mRNA in terms of the precise structure of the transcript (transcription start site, intron/exon junctions, etc.). It is also quantitative and requires less RNA than a Northern blot. A related technique, S1 nuclease mapping, is similar, although the non-hybridised part of a DNA probe, rather than an RNA probe, is digested, this time with the enzyme S1 nuclease.

The PCR has also had an impact on the analysis of RNA via the development of a technique known as **reverse transcriptase-PCR (RT-PCR)**. Here, the RNA is isolated and a first-strand cDNA synthesis undertaken with reverse transcriptase; the cDNA is then used in a conventional PCR. Under certain circumstances, a number of thermo-stable DNA polymerases have reverse transcriptase activity, which obviates the need to separate the two reactions and allows the RT-PCR to be carried out in one tube. One of the main benefits of RT-PCR is the ability to identify rare or low levels of mRNA transcripts with great sensitivity. This is especially useful when detecting, for example, viral gene expression and furthermore provides the means of differentiating between latent and active virus (Figure 4.73).

In many cases, the analysis of tissue-specific gene expression is required, and again the PCR has been adapted to provide a solution. This technique, termed **differential display**, is also an RT-PCR-based system requiring that isolated mRNA be first converted into cDNA. In a subsequent step, one of the PCR primers, designed to anneal to a general mRNA element such as the poly(A) tail in eukaryotic cells, is used in conjunction with a combination of arbitrary 6–7 bp primers that bind to the 5′ end of the transcripts. Consequently, this results in the generation of multiple PCR products with reproducible patterns (Figure 4.74). Comparative analysis by gel electrophoresis of PCR products generated from different cell types therefore allows the identification and isolation of those transcripts that are differentially expressed. As with many PCR-based techniques, the time to identify such genes is dramatically reduced to a few days compared to several weeks that are required to construct and screen cDNA libraries using traditional approaches.

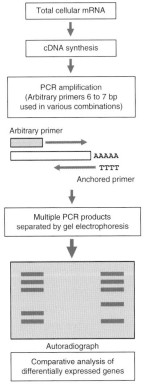

Figure 4.74 Analysis of gene expression using differential display PCR.

4.17.2 Analysing Genes *In Situ*

Gross chromosomal changes are often detectable by microscopic examination of the chromosomes within a karyotype. Single or restricted numbers of base substitutions, deletions, rearrangements or insertions are far less easily detectable, but may induce similarly profound effects on normal cellular biochemistry. *In situ* hybridisation makes it possible to determine the chromosomal location of a particular gene fragment or gene mutation. This is carried out by preparing a radiolabelled DNA or RNA probe and applying this to a tissue or chromosomal preparation fixed to a microscope slide. Any probe that does not hybridise to complementary sequences is washed off and an image of the distribution or location of the bound probe is viewed by autoradiography (Figure 4.75). Using tissue or cells fixed to slides it is also possible to carry out *in situ* PCR and qPCR. This is a highly sensitive technique, where PCR is carried out directly on the tissue slide with the standard PCR reagents. Specially adapted thermal cycling machines are required to hold the slide preparations and allow the PCR to proceed. This allows the localisation and identification of, for example, single copies of intracellular viruses and, in the case of qPCR, the determination of initial concentrations of nucleic acid.

An alternative labelling strategy used in karyotyping and gene localisation is fluorescence *in situ* hybridisation (FISH). This method (sometimes also termed chromosome painting) is based on *in situ* hybridisation, but different gene probes are labelled with different fluorophores, each specific for a particular chromosome. The advantage of this method is that separate gene regions may be identified and comparisons made within the same chromosome preparation. The technique is also of great interest in genome mapping for ordering DNA probes along a chromosomal segment.

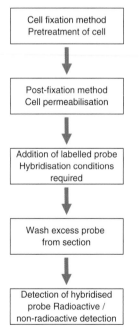

Figure 4.75 General scheme for *in situ* hybridisation.

4.17.3 Analysing Protein–DNA Interactions

To determine potential transcriptional regulatory sequences, genomic DNA fragments may be cloned into specially devised promoter probe vectors. These contain sites for insertion of foreign DNA that lie upstream of a reporter gene. A number of reporter genes are currently used, including the *lacZ* gene encoding β-galactosidase, the *cat* gene encoding chloramphenicol acetyl transferase (CAT) and the *lux* gene, which produces luciferase and is determined in a bioluminescent assay (see Section 13.6). Fragments of DNA potentially containing a promoter region are cloned into the vector and the constructs transfected into eukaryotic cells. Any expression of the reporter gene will be driven by the foreign DNA, which must therefore contain promoter sequences (Figure 4.76). These plasmids and other reporter genes, such as those using green fluorescent protein (GFP) or the firefly luciferase gene, allow quantification of gene transcription in response to transcriptional activators.

The binding of a regulatory protein or transcription factor to a specific DNA site results in a complex that may be analysed by the technique termed **electrophoretic mobility shift assay** (EMSA) or **gel retardation assay**. It may also be adapted to study protein–RNA interactions. In gel electrophoresis, the migration of a DNA fragment bound to a protein of a relatively large mass will be retarded in comparison to the DNA fragment alone. For gel retardation to be useful, the region containing the promoter DNA element must be digested or mapped with a restriction endonuclease before it is complexed with the protein. The location of the promoter may then be defined by finding the position on the restriction map of the fragment that binds to the regulatory protein and therefore retards it during electrophoresis. One potential problem with gel retardation is finding the precise nucleotide binding region of the protein, since this depends on the accuracy and detail of the restriction map and the convenience of the restriction sites. However, it is a useful first step in determining the interaction of a regulatory protein with a

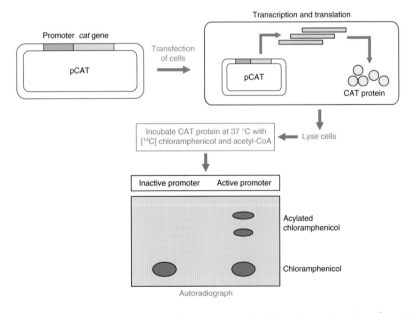

Figure 4.76 Assay for promoters using the reporter gene for chloramphenicol acetyl transferase (CAT).

DNA binding site. Large-scale methods now employ massively parallel DNA sequencing (see Section 20.2.2) after initial immunoprecipitation of the chromatin with an anti-body. This method, termed **ChIP-Seq** (chromatin immunoprecipitation sequencing), has enabled the genome-wide identification of transcription factor binding sites in DNA.

Most methods used in identifying regulatory sites in DNA rely on the fact that the interaction of a DNA-binding protein with a regulatory DNA sequence will protect that DNA sequence from degradation by an enzyme such as DNase I in a **footprint-ing assay**. In its basic form, the DNA regulatory sequence is first labelled at one end and then mixed with the DNA-binding protein (Figure 4.77). DNase I is added under conditions favouring a partial digestion. This limited digestion ensures that a num-ber of fragments are produced where the DNA is not protected by the DNA-binding protein; the region protected by the DNA-binding protein will remain undigested. All the fragments are then separated on a high-resolution polyacrylamide gel alongside a control digestion where no DNA-binding protein is present. The resulting gel will

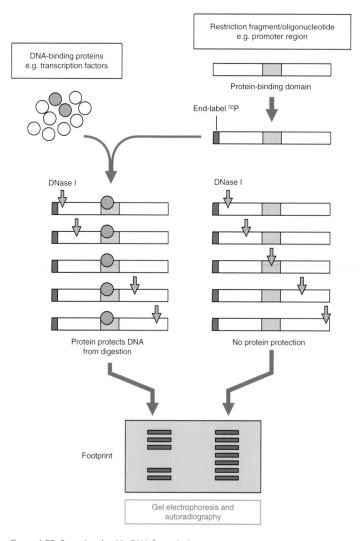

Figure 4.77 Steps involved in DNA footprinting.

contain a ladder of bands representing the partially digested fragments. Where DNA has been protected, no bands appear; this region or hole is termed the DNA footprint. The position of the protein-binding sequence within the DNA may be elucidated from the size of the fragments either side of the footprint region.

Footprinting is a more precise method of locating a DNA–protein interaction than gel retardation; however, it is unable to give any information as to the precise interaction or the contribution of individual nucleotides. Identification of regulatory regions in DNA may alternatively be accomplished on a genome-wide scale by sequencing using techniques such as DNase-Seq. Owing to the large scale, this approach requires bioinformatic analysis in order to provide genome-wide footprints.

In addition to the detection of DNA sequences that contribute to the regulation of gene expression, an ingenious way of detecting the protein transcription factors has been developed. This is termed the **yeast two-hybrid system**. Transcription factors have two domains, one for DNA binding and the other to allow binding to further proteins (**activation domain**). These occur as part of the same molecule in natural transcription factors, for example TFIID. However, they may also be formed from two separate domains. Thus a recombinant molecule is formed, encoding the protein under study as a fusion with the DNA-binding domain. The fusion construct cannot, however, activate transcription. Genes from a cDNA library are expressed as a fusion with the activator domain; these also cannot initiate transcription. However, when the two fractions are mixed together, transcription is initiated if the domains are complementary (Figure 4.78). This is indicated by the transcription of a reporter gene such as

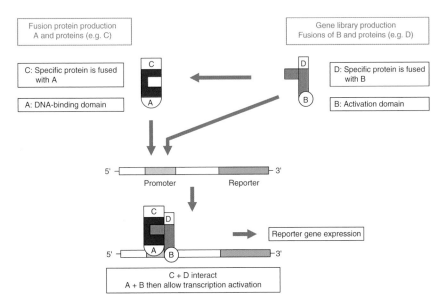

Figure 4.78 Yeast two-hybrid system (interaction trapping technique). Transcription factors have two domains, one for DNA binding (A) and the other to allow binding to further proteins (B). Thus a recombinant molecule is formed from a protein (C) as a fusion with the DNA-binding domain. It cannot, however, activate transcription alone. Genes from a cDNA library (D) are expressed as a fusion with the activator domain (B) but also cannot initiate transcription alone. When the two fractions are mixed together, transcription is initiated if the domains are complementary and expression of a reporter gene takes place.

the *cat* gene. The technique is not just confined to transcription factors and may be applied to any protein system where interaction occurs.

4.17.4 Transgenics and Gene Targeting

In many cases it is desirable to analyse the effect of certain genes and proteins in an organism rather than in the laboratory. Furthermore, the production of pharmaceutical products and therapeutic proteins is also desirable in a whole organism. This also has important consequences for the biotechnology industry (Table 4.9). The introduction of foreign genes into germ-line cells and the production of an altered organism is termed transgenics. There are two broad strategies for transgenesis. The first is direct transgenesis in mammals, whereby recombinant DNA is injected directly into the male pronucleus of a recently fertilised egg. This is then raised in a foster mother animal resulting in an offspring that is all transgenic. Selective transgenesis is where the recombinant DNA is transferred into embryo stem (ES) cells. The cells are then cultured in the laboratory and those expressing the desired protein selected and incorporated into the inner cell mass of an early embryo. The resulting transgenic animal is raised in a foster mother, but in this case the transgenic animal is a mosaic or chimeric since only a small proportion of the cells will be expressing the protein. The initial problem with both approaches is the random nature of the integration of the recombinant DNA into the genome of the egg or embryo stem cells. This may produce proteins in cells where it is not required or disrupt genes necessary for correct growth and development.

A refinement of this approach is gene targeting which involves the production of an altered gene in an intact cell, a form of in vivo mutagenesis as opposed to in vitro mutagenesis. The gene is inserted into the genome of, for example, an ES cell by specialised viral-based vectors. The insertion is non-random, however, since homologous sequences exist on the vector to the gene and on the gene to be targeted. Thus, homologous recombination may introduce a new genetic property to the cell, or inactivate an already existing one, termed gene knockout. Perhaps the most important aspect

Table 4.9 Use of transgenic mice for investigation of selected human disorders

Gene/protein	Genetic lesion	Disorder in humans
Tyrosine kinase (TK)	Constitutive expression of gene	Cardiac hypertrophy
HIV transactivator	Expression of HIV *tat* gene	Kaposis sarcoma
Angiotensinogen	Expression of rat angiotensinogen gene	Hypertension
Cholesterol ester transfer (CET) protein	Expression of *CET* gene	Atherosclerosis
Hypoxanthine-guanine phosphoribosyl transferase (HPRT)	Inactivation of *HPRT* gene	HPRT deficiency

of these techniques is that they allow animal models of human diseases to be created. This is useful since the physiological and biochemical consequences of a disease are often complex and difficult to study, impeding the development of diagnostic and therapeutic strategies.

4.17.5 Modulating Gene Expression by RNAi

There are a number of ways of experimentally changing the expression of genes. Traditionally, methods have focussed on altering the levels of mRNA by manipulation of promoter sequences or levels of accessory proteins involved in control of expression. In addition, post-mRNA production methods have also been employed, such as antisense RNA, where a nucleic acid sequence complementary to an expressed mRNA is delivered into the cell. This antisense sequence binds to the mRNA and prevents its translation. A development of this theme and a process that is found in a variety of normal cellular processes is termed RNA interference (RNAi; see also Section 4.5.8). This process of RNAi is now a well-established and powerful technique that has been widely applied to identify the function of genes and the resulting proteins either in cells grown in culture or even in vivo using model organisms. Essentially the technique may be used to decrease the expression of a gene termed gene knockdown and differs from the method of gene knockout, where the expression of a gene is abolished (see Figure 4.19).

One important feature of the method is the design and synthesis of the dsRNA (double-strand RNA) to the gene of interest. This is introduced into the appropriate cell line after which the RNAi pathway in the cell is activated. A potential problem in the dsRNA design process is the so-called off-target effect, where the expression of multiple genes is inadvertently reduced, as well as the gene under study. This can be addressed in part by employing a number of computational methods that assist in the design of the dsRNA, thus minimising off-target effects. Indeed, improvements of the technique have also included the synthesis and delivery of short interfering RNA (siRNA) directly, rather than using the longer dsRNA that is cleaved by dicer to produce the siRNAs. The precise delivery of siRNA may also prove troublesome, depending on the cell or tissue used, and RNA instability is still a major issue to be fully addressed.

The production of short hairpin RNA (shRNA) from sequences cloned and expressed from plasmid vectors introduced into the cell line or organism has also proved beneficial. The choice of which dsRNA to use largely depends on the organism, for example mammalian cells have a process whereby the introduction of long dsRNA evokes an unwanted interferon-based immune response, whereas this is decreased when using synthesised siRNA. RNAi technology holds enormous promise in the biotechnology industry and also medical fields where, for example, viral infections may be addressed by gene knockdown of specific mRNA targets in HIV-1, hepatitis B and C. Indeed, RNAi-based therapy may have the potential to treat certain types of cancers where aberrant levels of oncogene mRNA such as MYC are found, leading to a highly specific form of treatment.

4.17.6 CRISPR/cas9-Based Genome Editing

One major gene editing system is CRISPR (clustered regularly interspaced short palindromic repeats) involving RNA-guided engineered nucleases (Figure 4.79). The system was first identified as an adaptive immunity pathway in prokaryotes providing resistance to bacterial phage and analogous to the RNA interference process found in eukaryotes. It has since been modified and adapted for the engineering of genomes. Essentially, CRISPR may be thought of as a programmable restriction enzyme system. It utilises double-stranded sequence-specific breaks and repair using HDR (homology directed repair) via homologous recombination. The availability of a vast amount of sequence information has allowed the development of the system into an essential molecular biology technique.

The system consists of two parts, a short synthetic guide RNA (gRNA) and a nonspecific double-stranded endonuclease termed Cas9 (Cas: CRISPR-associated). The guide RNA has a specific 20 nucleotide spacer/targeting sequence (crRNA) that identifies the genomic target to be identified and modified, linked to a trans-activating (tracrRNA) sequence necessary for the recruitment and stability of Cas9 nuclease. The gRNA sequence and Cas9 nuclease elements are constructed into a plasmid expression vector that can be used to transfect cells under study. Transcription of the gRNA and Cas9 results in crRNA that binds to the DNA to be altered, which is held in the Cas9 nuclease with the aid of the tracrRNA. Indeed, it is the fact that the target sequence can be reprogrammed simply by changing the 20 nucleotides in the crRNA that makes the system so elegant. DNA is then digested by the Cas9 nuclease at specific points on both DNA strands. A donor DNA sequence may be included with a desired feature, such as a base change, insertion or deletion. The cell then employs homologous recombination to repair the break and incorporate the new DNA sequence. It is also possible to program Cas9 with multiple guide RNAs to allow multiplex site-specific editing

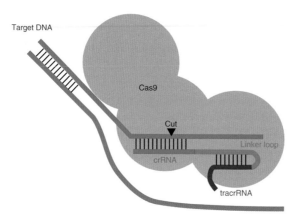

Figure 4.79 The CRISPR/Cas9 gene editing system. The nuclease Cas9 is targeted to DNA by a guide RNA consisting of a specific 20 nucleotide spacer/targeting sequence (crRNA) that identifies the DNA to modified, linked to a trans-activating (tracrRNA) sequence necessary for the recruitment and stability of Cas9 nuclease. DNA is then digested by the Cas9 nuclease at specific points on both DNA strands. Homologous recombination to repair the break and incorporate the new DNA sequence then follows.

for designing large deletions, inversions and translocations. Further refinements of the editing system have allowed the targeting of protein domains for transcriptional regulation and epigenetic modification.

4.18 ANALYSING GENETIC MUTATIONS AND POLYMORPHISMS

There are several types of mutation that can occur in nucleic acids, either transiently or those that are stably incorporated into the genome. During evolution, mutations may be inherited in one or both copies of a chromosome, resulting in polymorphisms within the population. Mutations may potentially occur at any site within the genome; however, there are several instances whereby mutations occur in limited regions. This is particularly obvious in prokaryotes, where elements of the genome (termed hypervariable regions) undergo extensive mutations to generate large numbers of variants, by virtue of the high rate of replication of the organisms. Similar hypervariable sequences are generated in the normal antibody immune response in eukaryotes. Mutations may have several effects upon the structure and function of the genome. Some mutations may lead to undetectable effects upon normal cellular functions, termed conservative mutations. An example of these are mutations that occur in intron sequences and therefore play no part in the final structure and function of the protein or its regulation. Alternatively, mutations may result in profound effects upon normal cell function, such as altered transcription rates or on the sequence of mRNAs necessary for normal cellular processes.

Mutations occurring within exons may alter the amino acid composition of the encoded protein by causing amino acid substitution or by changing the reading frame used during translation. These point mutations were traditionally detected by Southern blotting or, if a convenient restriction site was available, by restriction fragment length polymorphism (RFLP). However, the PCR has been used to great effect in mutation detection since one can amplify the desired region of DNA with a common sequence (e.g. M13) embedded within one of the primers and perform Sanger sequencing. A more high-throughput approach is the allele-specific oligo-nucleotide PCR (ASO-PCR) where two competing primers and one general primer are used in the reaction (Figure 4.80). One of the primers is directly complementary to the known point mutation, whereas the other is a wild-type primer; that is, the primers are identical except for the terminal 3′ end base. Thus, if the DNA contains the point mutation, only the primer with the complementary sequence will bind and be incorporated into the amplified DNA, whereas if the DNA is normal, the wild-type primer is incorporated. The results of the PCR are analysed by agarose gel electrophoresis. A further modification of ASO-PCR has been developed where the primers are each labelled with a different fluorophore. Since the primers are labelled differently, a positive or negative result is produced directly without the need to examine the PCRs by gel electrophoresis.

Various modifications now allow more than one PCR to be carried out at a time (multiplex PCR), and hence the detection of more than one mutation is possible at the same time. Where the mutation is unknown it is also possible to use a PCR system with a gel-based detection method termed denaturing gradient gel electrophoresis

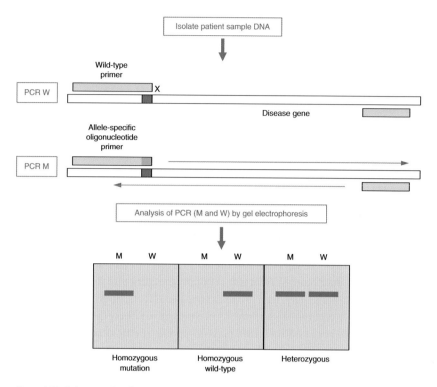

Figure 4.80 Point mutation detection using allele-specific oligonucleotide PCR (ASO–PCR).

(DGGE). In this technique, a sample DNA heteroduplex containing a mutation is amplified by the PCR, which is also used to attach a GC-rich sequence to one end of the heteroduplex. The mutated heteroduplex is identified by its altered melting properties through a polyacrylamide gel that contains a gradient of denaturant such as urea. At a certain point in the gradient the heteroduplex will denature relative to a perfectly matched homoduplex and thus may be identified. The GC clamp maintains the integrity of the end of the duplex on passage through the gel (Figure 4.81). The sensitivity of this and other mutation detection methods has been substantially increased by use of the PCR, and further mutation techniques used to detect known or unknown mutations are indicated in Table 4.10. An extension of this principle is used in a number of detection methods employing denaturing high-performance liquid chromatography (dHPLC). Commonly known as wave technology, the rapid detection of denatured single strands containing mismatches allows a high-throughput analysis of samples to be achieved.

Polymorphisms are particularly interesting elements of the human genome and as such may be used as the basis for differentiating between individuals. All humans carry repeats of sequences known as minisatellite DNA, of which the number of repeats varies between unrelated individuals. Hybridisation of probes that anneal to these sequences using Southern blotting is one method to type and identify those individuals; there are also PCR methods for mini/microsatellites.

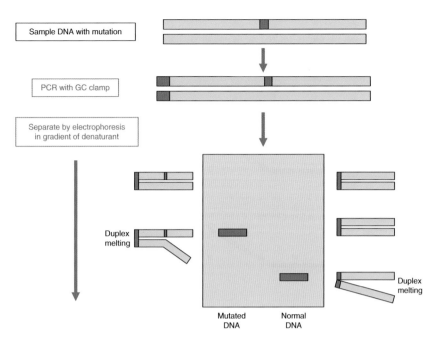

Figure 4.81 Detection of mutations using denaturing gradient gel electrophoresis (DGGE).

Table 4.10 **Main methods of detecting mutations in DNA samples**

Technique	Basis of method	Main characteristics of detection
Southern blotting	Gel/Blot	Labelled probe hybridisation to DNA
Dot/slot blotting	Blot	Labelled probe hybridisation to DNA
Allele-specific oligo-PCR (ASO–PCR)	PCR	Oligonucleotide matching to DNA sample
Denaturing gradient gel electrophoresis (DGGE)	Gel/PCR	Melting temperature of DNA strands
Single-stranded conformation polymorphism (SSCP)	Gel/PCR	Conformation difference of DNA strands
Multiplex ligation-dependent probe amplification (MLPA)	Gel/PCR	Oligonucleotide matching to DNA sample/ligation
DNA sequencing	Capillary/Sanger	Nucleotide sequence analysis of DNA
DNA microchips	Glass chip	Sample DNA hybridisation to oligo arrays

Multiplex ligation–dependent probe amplification (MLPA) is a PCR-based multiplex assay that allows a number of target sites, such as deletions, duplications, mutations or SNPs, to be amplified with a pair of primer-containing probes (Figure 4.82). The process involves a denaturation and hybridisation stage, a ligation stage and a

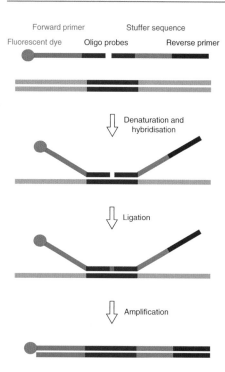

Figure 4.82 Multiplex ligation-dependent probe amplification (MLPA). In this method, two hybridisation probes are used, one containing a forward PCR primer sequence labelled with a fluorescent dye, the other with a stuffer sequence and reverse primer sequence. Denaturation and hybridisation with the target sequence takes place, after which a ligation step is employed. If ligation takes place then PCR amplification can proceed; however, if no ligation occurs PCR does not take place and thus no target DNA is present.

PCR amplification stage, followed by analysis of the products. The technique involves designing two adjacent hybridisation probes that contain the fluorescence-labelled forward sequence and the reverse primer sequence used in the amplification stage. The latter is designed with a stuffer sequence that can be varied to suit the target. The probes are first hybridised to the denatured sample DNA. In the next stage, the two oligonucleotide probes are joined by ligation – only ligated probes can be used in the amplification stage. Oligonucleotide probes that are not ligated will only contain one primer sequence, and as a consequence cannot be amplified and generate a signal.

DNA fingerprinting is the collective term for two distinct genetic testing systems that use either 'multi-locus' probes or 'single-locus' probes. Initially described DNA fingerprinting probes were multi-locus probes and so termed because they detect hypervariable minisatellites throughout the genome, i.e. at multiple locations within the genome. In contrast, several single-locus probes were discovered that under specific conditions only detect the two alleles at a single locus and generate what have been termed DNA profiles because, unlike multi-locus probes, the two-band pattern result is in itself insufficient to uniquely identify an individual.

Techniques based on the PCR have been coupled with the detection of minisatellite loci. The inherent larger size of such DNA regions is not best suited to PCR amplification; however, new PCR developments are beginning to allow this to take place. The discovery of polymorphisms within the repeating sequences of minisatellites has led to the development of a PCR-based method that distinguishes an individual on the basis of the random distribution of repeat types along the length of a person's two alleles for one such minisatellite. Known as **minisatellite variant repeat (MVR)** analysis or digital DNA typing, this technique can lead to a simple numerical coding of the repeat variation detected. Potentially, this combines the advantages of PCR (sensitive and rapid) with the discriminating power of minisatellite alleles. Thus for the future, there are a number of interesting identification systems under development and evaluation. Techniques for genetic detection of polymorphisms have been used in many cases of paternity testing and immigration control, and are becoming central factors in many criminal investigations. They are also valuable tools in plant biotechnology for cereal typing and in the field of pedigree analysis and animal breeding.

4.18.1 Microarrays (DNA Chips)

The use of microarrays or DNA chips allow large-scale analysis and quantification of genes and gene expression. A microarray consists of an ordered arrangement of potentially hundreds of thousands of DNA sequences, such as oligonucleotides or cDNAs deposited onto a solid surface. The solid support may be either glass or silicon, and currently the arrays are synthesised on or off the chip. They require complex fabrication methods similar to those used in producing computer microchips. Most commercial production employs robotic ultrafine microarray deposition instruments that dispense volumes in the picolitre range. Alternatively, on-chip fabrication, as used by Affymetrix, builds up layers of nucleotides using a process borrowed from the computer industry termed photolithography. Here, wafer-thin masks with holes allow photoactivation of specific dNTPs, which are linked together at specific regions on the chip. The whole process allows layers of oligonucleotides to be built up with each nucleotide at each position being defined by computer control.

The arrays themselves may represent a variety of nucleic acid material. This may be mRNA produced in a particular cell type, termed cDNA expression arrays, or may alternatively represent coding and regulatory regions of a particular gene or group of genes. A number of arrays are now available that may be used for determination of mutations in DNA, mRNA transcript levels or other polymorphisms such as SNPs. Sample DNA is placed on the array, and any non-hybridised DNA washed off. The array is then analysed and scanned for patterns of hybridisation by detection of fluorescence signals. Any mutations or genetic polymorphisms in relevant genes may be rapidly analysed by computer-controlled interpretation of the resulting hybridisation pattern and mutation, transcript level or polymorphism defined. Indeed, the collation and manipulation of data from microarrays presents as big a problem as fabricating the chips in the first place. The potential of microarrays appears to be limitless and a number of arrays have been developed for the detection of various genetic mutations, including the cystic fibrosis *CFTR* gene (cystic fibrosis transmembrane regulator) and the breast cancer gene *BRCA*1, and for the study of the human immunodeficiency virus (HIV).

One current application of microarray technology is the generation of a catalogue of SNPs across the human genome. Estimates indicate that there are approximately 85 million SNPs and, importantly, some may point to the development of certain diseases. For example, a SNP in the F5 gene is implicated in a thrombophilia, factor V

Leiden, a genetic blood clotting disorder. SNP analysis is therefore clearly a candidate for microarray analysis and developments such as the Affymetrix Genome-Wide SNP array enable the simultaneous analysis of nearly 1 million SNPs on one gene chip. In order to simplify the problem of the vast numbers of SNPs that need to be analysed, the HapMap project (see also Section 4.3.2) currently analyses SNPs that are inherited as a block, and in theory as few as 500 000 SNPs will be required to genotype an individual.

An extension of microarray technology may also be used to analyse tissue sections. This process, termed tissue microarrays (TMAs), uses tissue cores or biopsies from conventional paraffin-embedded tissues. Thousands of tissue cores are sliced and placed on a solid support such as glass where they may all be subjected to the same immunohistochemical staining process or analysis with gene probes using *in situ* hybridisation. As with DNA microarrays, many samples may be analysed simultaneously, less tissue is required and greater standardisation is possible.

4.18.2 Gene Discovery and Localisation

A number of disease loci have been identified and located to certain chromosomes. This has been facilitated by the use of *in situ* mapping techniques such as FISH. In fact, a number of genes have been identified and the proteins of interest determined where little was initially known about the gene except for its location. This method of gene discovery is known as positional cloning and was instrumental in the isolation of the *cftr* gene responsible for the disorder cystic fibrosis.

The genes that are actively expressed in a cell at any one time are estimated to be as little as 10% of the total. The remaining DNA is packaged and serves an as yet unknown function. Investigations have found that certain active genes may be identified by the presence of so-called HTF islands (*Hpa* II tiny fragments islands) often found at the 5′ end of genes. These are CpG-rich sequences that are not methylated and form tiny fragments on digestion with the restriction enzyme *Hpa* II. A further gene discovery method that has been used extensively in the past few years is a PCR-based technique giving rise to a product termed an expressed sequence tag (EST). This represents part of a putative gene for which a function has yet to be assigned. It is carried out on cDNA by using primers that bind to an anchor sequence such as a poly(A) tail and primers that bind to sequences at the 5′ end of the gene. Such PCRs may subsequently be used to map the putative gene to a chromosomal region or be used itself as a probe to search a genomic DNA library for the remaining parts of the gene. This type of information can be visualised using bioinformatics and useful information determined in a process termed data mining (see also Section 21.8.1). Much interest currently lies in ESTs since they may represent a shortcut to gene discovery.

A further gene isolation system that uses adapted vectors, termed exon trapping or exon amplification, may be used to identify exon sequences. Exon trapping requires the use of a specialised expression vector that will accept fragments of genomic DNA containing sequences for splicing reactions to take place. Following transfection of a eukaryotic cell line, a transcript is produced that may be detected by using specific primers in a RT-PCR. This indicates the nature of the foreign DNA by virtue of the

Table 4.11 **Techniques used to determine putative gene-encoding sequences**

Identification	Key elements of method
Zoo blotting (cross-hybridisation)	Evolutionary conservation of DNA sequences that suggest functional significance
Homology searching	Gene database searching for gene-family-related sequences
Identification of CpG islands	Regions of hypomethylated CpG frequently found 5′ to genes in vertebrate animals
Identification of open reading frame (ORF) promoters/splice sites/RBS	DNA sequences scanned for consensus sequences *in silico*
Northern blot hybridisation	mRNA detection by binding to labelled gene probes
Exon trapping technique	Artificial RNA splicing assay for exon identification
Expressed sequence tags (ESTs)	cDNAs amplified by PCR that represent part of a gene

Notes: RBS, ribosome binding site; cDNA, complementary DNA.

splicing sequences present. A list of further techniques that aid in the identification of a potential gene-encoding sequence is indicated in Table 4.11.

4.19 MOLECULAR BIOTECHNOLOGY AND APPLICATIONS

It is a relatively short period of time since the early 1970s when the first recombinant DNA experiments were carried out. However, huge strides have been made, not only in the development of molecular biology techniques, but also in their practical application. The molecular basis of disease and the new areas of genetic analysis and gene therapy hold great promise. In the past, medical science relied on the measurement of protein and enzyme markers that reflected disease states. It is now possible not only to detect such abnormalities at an earlier stage using mRNA techniques, but also in some cases to predict such states using genome analysis. The complete mapping and sequencing of the human genome and the development of techniques such as DNA microchips will certainly accelerate such events. Perhaps even more difficult is the elucidation of diseases that are multifactorial and involve a significant contribution from environmental factors. One of the best-studied examples of this type of disease is cancer. Molecular genetic analysis has allowed a discrete set of cellular genes, termed oncogenes, to be defined that play key roles in such events. These genes and their proteins are also major points in the cell cycle and are intimately involved in cell regulation (see examples in Table 4.12). In a number of cancers, well-defined molecular events have been correlated with mutations in these oncogenes and therefore in the

Table 4.12 **General classification of oncogenes and their cellular and biochemical functions**

Oncogene	Example	Main details
G-proteins	H-, K- and N-*ras*	GTP-binding protein/GTPase
Growth factors	*sis, nt-2, hst*	β-chain of platelet-derived growth factor (PDGF)
Growth factor receptors	*erbB*	Epidermal growth factor receptor (EGFR)
	fms	Colony-stimulating factor-1 receptor
Protein kinases	*abl, src*	Protein tyrosine kinases
	mos, ras	Protein serine kinases
Nucleus-located transcription factors	*mye*	DNA-binding protein

Table 4.13 **A number of selected examples of targets for gene therapy**

Disorder	Defect	Gene target	Target cell
Emphysema	Deficiency (α1-AT)	α1-Antitrypsin (α1-AT)	Liver cells
Gaucher disease (storage disorder)	GC deficiency	Glucocerebrosidase	GC fibroblasts
Haemoglobinopathies	Thalassaemia	β-Globin	Fibroblasts
Lesch–Nyhan syndrome	Metabolic deficiency	Hypoxanthine guanine phosphoribosyl transferase (HPRT)	HPRT cells
Immune system disorder	Adenosine deaminase deficiency	Adenosine deaminase (ADA)	T and B cells

corresponding protein. It is already possible to screen and predict the fate of some disease processes at an early stage, a point that itself raises significant ethical dilemmas (see also Section 1.4). In addition to understanding cellular processes, both in normal and disease states, great promise is also held in drug discovery and molecular gene therapy. A number of genetically engineered therapeutic proteins and enzymes have been developed and are already having an impact on disease management. In addition, the correction of disorders at the gene level (gene therapy) is also underway. Perhaps one of the most startling applications of molecular biology to date is gene editing and the development of gene modifications methods such as the CRISPR/Cas9 system, which may have a profound impact on treating genetic-based diseases. A number of these developments are indicated in Table 4.13.

Table 4.14 **Current selected plant/crops modified by genetic manipulation**

Crop or plant	Genetic modification
Canola (oil seed rape)	Insect resistance, seed oil modification
Maize	Herbicide tolerance, resistance to insects
Rice	Modified seed storage protein, insect resistance
Soya bean	Tolerance to herbicide, modified seed storage protein
Tomato	Modified ripening, resistance to insects and viruses
Sunflower	Modified seed storage protein

The production of modified crops and animals for farming and as producers of important therapeutic proteins is also one of the most exciting developments in molecular biology. This has allowed the production of modified crops, improving their resistance to environmental factors and their stability (Table 4.14). The production of transgenic animals also holds great promise for improved livestock quality, low-cost production of pharmaceuticals and disease-free or disease-resistant strains. In addition, RNAi has also been shown to be a useful addition in many areas of crop improvement. In the future this may overcome such factors as contamination with agents such as BSE. There is no doubt that improved methods of producing livestock by whole-animal cloning will also be a major benefit. All of these developments do, however, require debate and the many ethical considerations that arise from them require careful consideration.

4.20 PHARMACOGENOMICS

As a result of the developments in genomics, new methods of providing targeted drug treatment are beginning to be developed. This area is linked to the proposal that it is possible to identify those patients who react in a specific way to drug treatment by identifying their genetic make-up. In particular, SNPs may provide a key marker of potential disease development and reaction to an individual treatment. A simple example that has been known for some time is the reaction to a drug used to treat a particular type of childhood leukaemia. Successful treatment of the majority of patients may be achieved with 6-mercaptopurine. A number of patients do not respond well, and in some cases it may be fatal to administer this drug. This is now known to be due to a mutation in the gene encoding the enzyme that metabolises the drug. Thus, it is possible to analyse patient DNA prior to administration of a drug to determine what the likely response will be. The technology to deduce a patient's genotype is available and it is now possible to analyse SNPs that may also correlate with certain disease processes in a microarray-type format. This opens up the possibility of assigning a **pharmacogenetic** profile at birth, in much the same way as blood typing for later treatment. A further possibility is the determination of likely susceptibility to a disease based on genetic information. A number of companies, including the US-based 23andMe and Icelandic genetics company deCode, are able to provide personal genetic information based on modelling and analysis of disease genes in large population studies for certain conditions such as diabetes and others. However, the unprecedented scope of such medical tests has revealed several ethical and legal conundrums (ownership of and access to the data) as well as the requirement to ensure appropriate scientific evidence for interpreting the data by means of risk calculations.

4.21 SUGGESTIONS FOR FURTHER READING

4.21.1 Experimental Protocols

Green M.R. and Sambrook J. (2012) *Molecular Cloning: A Laboratory Manual*, 4th Edn., Cold Spring Harbor Press, Cold Spring Harbor, NY, USA.

Walker J.M and Rapley R. (2008) *Molecular Biomethods Handbook: Cells, Proteins, Nucleic Acids*, Humana Press, Totowa, NJ, USA.

4.21.2 General Texts

Brown T.A. (2016) *Gene Cloning and DNA Analysis*, 7th Edn., Wiley-Blackwell, Oxford, UK.

Divan A. and Royds J. (2013) *Tools and Techniques in Biomolecular Science*, Oxford University Press, Oxford, UK.

McLennan A., Bates A., Turner P. and White M. (2013) *BIOS Instant Notes in Molecular Biology*, Garland Science, New York, USA.

Rapley R. and Whitehouse D. (2015) *Molecular Biology and Biotechnology*, 6th Edn., Royal Society of Chemistry Press, London, UK.

Sidhu S.S. and Geyer C.R. (2015) *Phage Display In Biotechnology and Drug Discovery*, 2nd Edn., CRC Press, Boca Raton, FL, USA.

Strachan T. and Read A.P. (2010) *Human Molecular Genetics*, 4th Edn., Bios, Oxford, UK.

4.21.3 Review Articles

Annas G.J. and Elias S. (2014) 23andMe and the FDA. *New England Journal of Medicine* 370, 985–988.

Cohen S.N. (2013) DNA cloning: A personal view after 40 years. *Proceedings of the National Academy of Sciences* 110, 15521–15529.

Gibson D.G., Young L., Chuang R.-Y., *et al.* (2009) Enzymatic assembly of DNA molecules up to several hundred kilobases. *Nature Methods* 6, 343–345.

Hoseini S.S. and Sauer M.G. (2015) Molecular cloning using polymerase chain reaction, an educational guide for cellular engineering. *Journal of Biological Engineering* 9, 2.

Katsanis S.H. and Katsanis N. (2013) Molecular genetic testing and the future of clinical genomics. *Nature Reviews Genetics* 14, 415–426.

Marcheschi R.J., Gronenberg L.S. and Liao J.C. (2013) Protein engineering for metabolic engineering: current and next-generation tools. *Biotechnology Journal* 8, 545–555.

Pray L. (2008) Discovery of DNA structure and function: Watson and Crick. *Nature Education* 1, 100.

Pray L. (2008) Recombinant DNA technology and transgenic animals. *Nature Education* 1, 51.

Sampson T.R. and Weiss D.S. (2014) Exploiting CRISPR/Cas systems for biotechnology. *BioEssays* 36, 34–38.

Sioud M. (2015) RNA interference: mechanisms, technical challenges, and therapeutic opportunities. *Methods in Molecular Biology* 1218, 1–15.

Wu C.H., Liu I., Lu R. and Wu H. (2016) Advancement and applications of peptide phage display technology in biomedical science. *Journal of Biomedical Science* 23, 8.

4.21.4 Websites

General molecular biology and cloning
www.qiagen.com/gb/resources/molecular-biology-methods (accessed April 2017)
biotechlearn.org.nz/themes/dna_lab/dna_cloning (accessed April 2017)
www.neb.com/tools-and-resources/feature-articles/foundations-of-molecular-cloning-past-present-and-future (accessed April 2017)

DNA microarrays
www.genome.gov/10000533/dna-microarray-technology/ (accessed April 2017)

Molecular oncology
www.mycancergenome.org/content/molecular-medicine/detecting-gene-alterations-in-cancers/ (accessed April 2017)

PCR amplification
www.thermofisher.com/uk/en/home/references/protocols/nucleic-acid-amplification-and-expression-profiling/pcr-protocol.html (accessed April 2017)
www.sigmaaldrich.com/technical-documents/protocols/biology/standard-pcr.html (accessed April 2017)

Plasmids and cloning
www.addgene.org/plasmid-protocols/pcr-cloning/ (accessed April 2017)

Primer design for PCR
www.ncbi.nlm.nih.gov/guide/howto/design-pcr-primers/ (accessed April 2017)
primer3.sourceforge.net/ (accessed April 2017)
www.ncbi.nlm.nih.gov/tools/primer-blast/ (accessed April 2017)
www.genomics.agilent.com/primerDesignProgram.jsp (accessed April 2017)

Protein engineering
www.creative-biolabs.com/Protein-Engineering-Services.html (accessed April 2017)

Restriction enzymes
www.neb.com/products/restriction-endonucleases/restriction-endonucleases
(accessed April 2017)

RNAi
www.ncbi.nlm.nih.gov/probe/docs/techrnai/ (accessed April 2017)
www.rnaiweb.com/RNAi/siRNA_Design/ (accessed April 2017)

UGENE (a bioinformatics software package with many useful tools for cloning work)
ugene.net/ (accessed April 2017)

5 Preparative Protein Biochemistry

SAMUEL CLOKIE

5.1 INTRODUCTION

The term preparative biochemistry describes the purification of biological molecules such as proteins, viruses and nucleic acids to a sufficiently high concentration so as to be useful for a range of processes. Preparative biochemistry can be contrasted with analytical biochemistry that can be summarised as the application of small quantities of analyte to determine the composition or content of the biochemical sample.

At first sight, the purification of *one* protein from a cell or tissue homogenate that will typically contain 10 000–20 000 different proteins may seem a daunting task. However, in practice, on average, only four different fractionation steps are needed to purify a given protein.

The reason for purifying a protein can be to provide material for structural or functional studies; the final degree of purity required depends on the purpose for which the protein will be used. For some applications, protein samples are not required to be of highest purity, but for pharmaceutical use, the desired purity is typically close to 100%.

Theoretically, a protein is pure when a sample contains only a single protein species, although in practice it is almost impossible to achieve this stage. Fortunately, many studies on proteins can be carried out on samples that contain as much as 5–10% or higher contamination with other proteins. This is an important point, since each purification step necessarily involves loss of some of the protein that is the target of purification. An extra (and potentially unnecessary) purification step that increases the purity of the sample from, say, 90% to 98% may result in a protein sample of higher purity, but insufficient sample material for the planned studies.

For example, a protein sample of 90% purity is sufficient for amino-acid sequence determination studies as long as the sequence is analysed quantitatively to ensure that the deduced sequence does not arise from a contaminant protein. Similarly, immunisation of a rodent to provide spleen cells for monoclonal antibody production (Section 7.2.2) can be carried out with a sample that is considerably less than 50% pure. As long as the protein of interest raises an immune response, the fact that antibodies may also be produced against the contaminating proteins is typically ignored. For kinetic studies on an enzyme, a relatively impure sample can be used, provided it does not contain any competing activities. On the other hand, if the goal is to raise a monospecific polyclonal antibody in an animal (see Section 7.2.1), it is necessary to have a highly purified protein as antigen, otherwise immunogenic contaminating proteins will give rise to additional antibodies. Equally, proteins that are to have a therapeutic use must be extremely pure to satisfy regulatory (safety) requirements. Clearly, therefore, the degree of purity required depends on the purpose for which the protein is needed.

Protein purification can be broadly divided into three steps after the initial extraction from the cellular source: capture, intermediate stage to remove bulk impurities and polishing, to achieve highly pure proteins. These stages are discussed below and can be completed using different types of chromatography.

5.2 DETERMINATION OF PROTEIN CONCENTRATIONS

The need to determine protein concentration in solution is a routine requirement during protein purification. The only truly accurate method for determining protein concentration involves acidic hydrolysis of a portion of the sample and subsequent amino-acid analysis of the hydrolysate. However, this is very time-consuming, particularly if multiple samples are to be analysed. Fortunately, there are quicker methods that give a reasonably accurate assessment of protein concentrations.

Most of these (Table 5.1) are colorimetric methods, where a portion of the protein solution is reacted with a reagent that produces a coloured product. This coloured product is then measured spectrophotometrically (see Section 13.2) and the observed absorbance related to the amount of protein present by appropriate calibration. However, these methods are not absolute, since the development of colour is often at least partly dependent on the amino-acid composition of the protein(s). The presence of prosthetic groups (e.g. carbohydrate) also influences colorimetric assays. A standard calibration curve using bovine serum albumin (BSA), is commonly used due to its low cost, high purity and ready availability. However, since the amino-acid composition of BSA will differ from the composition of the sample being tested, any concentration values deduced from the calibration graph can only be approximate.

Table 5.1 **Summary of protein concentration assays**

Method	Applications and notes	Tolerance to contaminants	Sensitivity
Ultraviolet absorption (280 nm)	- General use - Chromatography in-line	Sensitive to compounds that absorb strongly – obscuring the absorption maxima at 280 nm	$10~\mu g~cm^{-3}$
Lowry assay (660 nm)	- General assay	Sensitive to TRIS, PIPES, HEPES, EDTA	$10-1000~\mu g~cm^{-3}$
Bicinchoninic acid (BCA) (562 nm)	- General assay	Tolerant to a broad range of compounds	$0.5~\mu g~cm^{-3}$
Bradford assay (Coomassie Brilliant Blue) (595 nm)	- General assay - Simple - Binds only Arg and Lys		$20~\mu g~cm^{-3}$
Kjeldahl analysis	- Determines N content of any compound - Used for complex samples	Any N-containing compound will add to the result (DNA, Melamine)	$1~mg~cm^{-3}$ in the original method

5.2.1 Ultraviolet (UV) Absorption

The aromatic amino-acid residues tyrosine and tryptophan in a protein exhibit an absorption maximum at a wavelength of 280 nm (see also Section 13.2.4). Since the proportions of these aromatic amino acids in proteins vary, the extinction coefficients are characteristic parameters of individual proteins that can be estimated based on their sequence (see Example 13.2).

The direct determination of protein concentration is relatively sensitive, as it is possible to measure protein concentrations as low as $10~\mu g~cm^{-3}$. Unlike colorimetric assays, this method is non-destructive, i.e. having made the measurement, the sample in the cuvette can be recovered and used further. This is particularly useful when one is working with small amounts of protein and cannot afford to waste any. However, the method is subject to interference by the presence of other compounds that absorb at 280 nm. Nucleic acids fall into this category, having an absorbance as much as 10 times that of proteins at this wavelength. Hence the presence of only a small percentage of nucleic acid (absorption maximum at ~260 nm) can greatly influence the observed absorbance of aromatic amino-acid side chains. However, if the absorbances at both wavelengths 260 and 280 nm are measured, it is possible to apply a correction factor (see Example 13.3).

A further great advantage of the UV absorption method is its usefulness for continuous non-destructive monitoring of protein contents, for example in chromatographic column effluents.

All contemporary column chromatography systems have in-line UV spectrometer units (typically using multiple fixed wavelengths, for example 230, 260 and 280 nm) that monitor protein elution from columns (for examples, see Figure 5.7, Figure 5.8).

5.2.2 Lowry (Folin–Ciocalteau) Assay

Historically, the Lowry assay has been the most commonly used method for determining protein concentration, but has nowadays been replaced by the more sensitive methods described below. The Lowry method is reasonably sensitive, detecting down to 10 µg cm^{-3} of protein, and the sensitivity is moderately constant from one protein to another. When the Folin–Ciocalteau reagent (a mixture of sodium tungstate, molybdate and phosphate), together with a copper sulfate solution, is mixed with a protein solution, a blue-purple colour is produced that can be quantified by its absorbance at 660 nm. As with most colorimetric assays, care must be taken such that other compounds that interfere with the assay are not present. For the Lowry method, this includes TRIS and zwitterionic buffers such as PIPES and HEPES, as well as EDTA. The method is based on both the Biuret and the Folin–Ciocalteau reaction. In the Biuret reaction, the peptide bonds of proteins react with Cu^{2+} under alkaline conditions to produce Cu^+. The Folin–Ciocalteau reaction is poorly understood, but essentially involves the reduction of phosphomolybdotungstate to hetero-polymolybdenum blue by the copper-catalysed oxidation of aromatic amino acids. The resultant strong blue colour is therefore partly dependent on the tyrosine and tryptophan content of the protein sample.

5.2.3 The Bicinchoninic Acid Method

This method is similar to the Lowry method in that it also depends on the conversion of Cu^{2+} to Cu^+ under alkaline conditions. Complexation of Cu^+ by bicinchoninic acid (BCA) results in an intense purple colour with an absorbance maximum at 562 nm. The method is more sensitive than the Lowry method, being able to detect down to 0.5 µg protein cm^{-3}. Perhaps more importantly it is generally more tolerant of the presence of compounds that interfere with the Lowry assay; hence it enjoys higher popularity than the original Lowry method.

5.2.4 The Bradford Method

This method relies on the binding of the dye Coomassie Brilliant Blue to protein. At low pH, the free dye has absorption maxima at 470 and 650 nm; however, when bound to protein, the absorption maximum is observed at 595 nm. The practical advantages of the method are that the reagent is simple to prepare and that the colour develops rapidly and is stable. Although it is sensitive down to 20 µg protein cm^{-3}, it is only a relative method, as the amount of dye binding appears to vary with the content of the basic amino acids arginine and lysine in the protein. This makes the choice of a standard rather difficult. In addition, many proteins will not dissolve properly in the acidic reaction medium.

5.2.5 Kjeldahl Analysis

This is a general chemical method for determining the nitrogen content of any compound. It is not normally used for the analysis of purified proteins or for monitoring column fractions, but is frequently used for analysing complex solid samples and microbiological samples for protein content. The sample is digested by boiling with concentrated sulfuric acid in the presence of sodium sulfate (to raise the boiling point) and a copper and/or selenium catalyst. The digestion converts all the organic nitrogen to ammonia, which is trapped as ammonium sulfate. Completion of the digestion stage is generally recognised by the formation of a clear solution. The ammonia is released by the addition of excess sodium hydroxide and removed by steam distillation in a Markham still. It is collected in boric acid and titrated with standard hydrochloric acid using methyl-red–methylene-blue as indicator. It is possible to carry out the analysis automatically in an autokjeldahl apparatus. Alternatively, a selective ammonium ion electrode may be used to directly determine the content of ammonium ion in the digest. Although Kjeldahl analysis is a precise and reproducible method for the determination of nitrogen, the determination of the protein content of the original sample is complicated by the variation in the nitrogen content of individual proteins and by the presence of nitrogen in contaminants such as DNA. In practice, the nitrogen content of proteins is generally assumed to be 16% by weight. This method has the major limitation that it can be fooled by non-protein sources of nitrogen; this was revealed during the 2008 Chinese milk scandal. Use of the Kjeldahl method by regulatory agencies allowed unscrupulous manufacturers to add melamine to infant milk formula, artificially increasing the apparent protein content.

Example 5.1   **BRADFORD ASSAY TO DETERMINE PROTEIN CONCENTRATION**

Question   A series of dilutions of bovine serum albumin (BSA) was prepared and 0.1 cm^3 of each solution subjected to a Bradford assay. The increase in absorbance at 595 nm relative to an appropriate blank was determined in each case, and the results are shown below. A sample (0.1 cm^3) of a protein extract from *E. coli* gave an absorption reading of $A_{595} = 0.84$ in the same assay. What was the concentration of protein in the *E. coli* extract?

Concentration of BSA (mg cm^{-3})	A_{595}
1.5	1.40
1.0	0.97
0.8	0.79
0.6	0.59
0.4	0.37
0.2	0.17

Answer   If a graph of BSA concentration against A_{595} is plotted it is seen to be linear. From the graph, at an A_{595} of 0.84 it can be seen that the protein concentration of the *E. coli* extracted is 0.85 mg cm^{-3}.

5.3 ENGINEERING PROTEINS FOR PURIFICATION

Historically, proteins have been isolated directly from organisms, tissues or fluids in a non-recombinant manner. The ability to clone and over-express genes for protein synthesis using genetic engineering methodology has changed this dramatically and nowadays the vast majority of protein isolation is carried out using recombinant methods with over-expression in bacterial or eukaryotic cells. Therefore, the manipulation of the gene of interest to engineer particular protein constructs is the first important step in the design of a protein purification procedure.

Some typical manipulations that are carried out to aid purification, improve folding or ensure secretion of recombinant proteins are discussed below.

5.3.1 Fusion Constructs to Aid Protein Purification

This approach requires the introduction of a genetic sequence that codes for a small peptide, or a well-characterised protein, to be placed in frame with the protein of interest, such that the protein is produced as a fusion protein. The peptide or protein (known as a 'tag') can be placed at the N- or C-terminal ends and provides a means for the fusion protein to be selectively purified from the cell extract by affinity chromatography (see Section 5.8.4). A protease site is typically engineered between the tag and protein that is then cleaved to release the protein of interest from the fusion construct. Clearly, the amino-acid sequence of the peptide linkage between tag and protein has to be carefully designed to allow chemical or enzymatic cleavage of this sequence. There are more than 20 published protein and peptide tags and each has specific characteristics beneficial to a particular experiment; the following sections describe some of the most commonly used tags.

Glutathione-S-transferase (GST)

The protein of interest is expressed as a fusion protein with the enzyme glutathione-S-transferase (GST) that was originally cloned from the blood fluke *Schistosoma japonicum*. The cell extract is typically passed through a column packed with glutathione-linked agarose beads, whereupon the GST binds to the glutathione with high affinity. Wash buffer is applied to the beads and once the non-tagged protein is washed from the column, the GST-fusion protein is eluted by adding an excess of reduced glutathione (a tripeptide with a γ-peptide linkage, γ-Glu-Cys-Gly), displacing the bound GST moiety. A typical further step would be removal of the fusion protein. In the pGEX series of vectors, this is achieved using human thrombin, which cleaves a specific amino-acid sequence contained within the linker region (Table 5.2). GST is typically fused via the N-terminal to the target gene and the fusion constructs often possess better solubility than the native proteins, albeit this effect is not as pronounced as with maltose-binding protein. However, GST does contain four cysteine residues that can become oxidised and cause aggregation, which can be a problem for expression of oligomeric proteins.

Maltose-Binding Protein (MBP)

Maltose-binding protein has the desirable property of significantly enhancing the solubility of the target protein. MBP is itself a well-ordered protein and is believed to aid the folding of the target protein, if placed on the N-terminal side. As such, it is often employed when attempting to express 'difficult' proteins that fail to be expressed when attempted with a small tag such as poly-histidine or GST. One-step purification can be achieved using amylose resin (that binds MBP), although care has to be taken to avoid amylose degradation by galactases. This is achieved by supplementing the media with 2 mM glucose to the growth media, as this represses galactase expression.

Poly-Histidine Tag (His-tag)

The essential amino acid histidine contains an imidazole ring that provides a stable structure, enabling co-ordinated binding with metal ions in a pH-dependent manner. The His-tag typically contains six sequential histidine residues that confer binding to immobilised bivalent nickel or cobalt ions with a dissociation constant in the μM range. The captured protein is efficiently eluted with either free imidazole (typically at 50–300 mM) or at acidic pH. Low concentrations of imidazole (10–20 mM) may be included in the wash buffer to increase purity of the final eluate. An extension to this method is the introduction of two hexa-histidine tags in series; this increases the affinity to the capture media to allow more aggressive washing. Correspondingly, the concentration of free imidazole required to achieve elution needs to be increased, typically to 300 mM.

FLAG®

This is a short hydrophilic amino acid sequence (Asp-Tyr-Lys-Asp-Asp-Asp-Asp-Lys) that may be attached to the N- or C-terminal end of the protein, and is designed for purification by immunoaffinity chromatography. A monoclonal antibody against the FLAG® sequence is available on an immobilised support for use in affinity chromatography. The cell extract, which includes the FLAG®-labelled protein, is passed through the column where the antibody binds to the FLAG®-labelled protein, allowing all other proteins to pass through. This is carried out in the presence of calcium ions, since the binding of the FLAG® sequence to the monoclonal antibody is calcium-dependent. Once all unbound protein has been washed from the column, the FLAG®-linked protein is released by passing EDTA through the column, which chelates the calcium ions. If the FLAG® sequence is used as an N-terminal fusion, it can conveniently be removed by the enzyme enterokinase, which recognises the poly-aspartate sequence and cleaves on the C-terminal side of the following lysine residue (see Table 5.2). In some cases, it can be advantageous to leave the tag attached, such as to enable convenient tracking of the protein in downstream experiments.

Dual Tags

Combinations of tags are sometimes used to allow tracing of protein using a convenient method, such as Western blotting, or improved protein production. A commonly used combination is the fusion of the hexa-histidine tag and maltose-binding protein (MBP) used in tandem, which brings together two advantageous properties. The His-tag allows economical and rapid purification and the MBP increases the solubility of the protein. Frequently, a TEV protease site is engineered between the His–MBP fusion and the target protein to allow efficient removal of the tags with minimal cost.

AviTag™ (Acceptor Peptide)

This 15 amino-acid sequence was optimised by phage display and the sequence can be efficiently biotinylated in vitro using the *Escherichia coli* enzyme BirA, which conjugates *D*-(+)-biotin to the lysine residue indicated:

$N_{terminus}$ -GLNDIFEAQKEWHE-$C_{terminus}$

Biotinylation is also possible in transfected mammalian cells. The biotin group is recognised by streptavidin; the complex of both proteins associates with a dissociation constant (K_d) of 4×10^{-14}. As such, this method can be used to efficiently purify in vivo tagged proteins that are natively folded and post-translationally modified. The biotinylated protein may also be useful for a number of applications such as binding assays, or detection of biomolecules.

5.3.2 Proteases Used in the Purification Process of Engineered Proteins

Some commonly used proteases for the removal of fusion peptides and proteins are summarised in Table 5.2. A very popular protease is the tobacco etch virus (TEV) endopeptidase, which is patent-free and economically produced in laboratories in deference to commercially available enzyme systems such as Factor X, etc.

In order to overcome the relatively poor solubility of TEV protease, it is common to express it in bacterial culture as an MBP fusion construct with a TEV protease recognition site. During expression, the produced protein construct cleaves the MBP fusion protein, resulting in the target protease. As TEV protease is prone to autolysis, a single amino-acid change (S219V) is typically engineered and sufficient to remove this property; the mutant is also slightly more efficient than the wild-type enzyme.

Purification pipeline considerations also led to addition of a His-tag to the TEV protease constructs. This not only aids in convenient purification of the enzyme in the laboratory, but also provides an elegant means for the final clean-up after proteolysis of a His-tagged target protein. With immobilised metal ion chromatography, the

Table 5.2 Summary of commonly used proteases in protein production techniques

Name	Sequence	Notes
Thrombin	LVPR^GS	Typically incubated with substrate overnight
Factor Xa	I(E/D)GR^	Less specific than thrombin
PreScission™ Protease (Human Rhinovirus (HRV) 3 C)	L(E/Q)VLF(E/Q)^GP	High activity at 4 °C
Tobacco Etch Virus (TEV)	ENLYFQ^G	Highly specific at a range of temperatures
Enterokinase	DDDDK^(P)	The site will not be cleaved if followed by a proline
Carboxypeptidase B	N(R/K)^	Can be used for microsequencing

Notes: The cleavage site is indicated by ^.

cleaved poly-histidine tag and His-tagged TEV protease, as well as any non-cleaved His-fusion target, are retained on the affinity column.

5.3.3 Intein-Mediated Purification

As discussed above, it is possible to create fusion proteins by engineering the DNA to contain the required sequence. However, it may be desirable to introduce post-translational modifications that cannot be achieved through molecular cloning or reliably achieved through chemical or enzymatic modification. In a process known as intein-mediated protein ligation, or expressed protein ligation, two peptides or proteins are covalently linked to create a fusion protein. In this method, a protein is produced with a thioester at the C-terminus and ligated to a peptide or protein containing an N-terminal cysteine. This methodology utilises the inducible self-cleavage activity of protein splicing elements (termed inteins) to separate the target protein from the affinity tag (see Figure 5.1); there is thus no requirement for a protease.

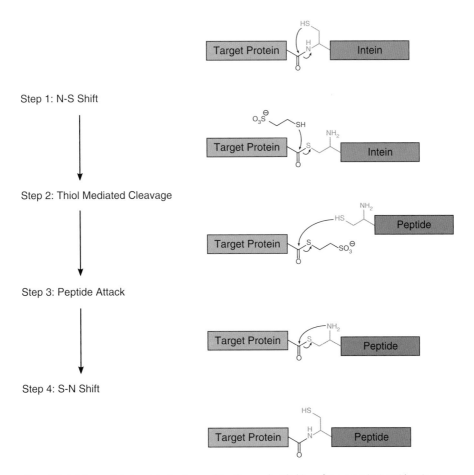

Figure 5.1 Intein-mediated protein ligation enables the covalent linking of two proteins/peptides via an N-terminal cysteine residue on the conjugation partner.

An example application of this technique is the study of protein phosphorylation, where, for the purpose of binding studies, it is essential to know the stoichiometry of phosphorylation. Intein-mediated fusion allows the generation of a protein of interest and its fusion with a separately synthesised phosphorylated peptide (where the degree of phosphorylation is known). This is achieved by separating the incomplete intein-mediated ligation protein from the successfully ligated protein, for example by size-exclusion chromatography (see Section 5.8.3).

5.3.4 Secreted Proteins

For cloned genes that are being expressed in microbial or eukaryotic cells, there are a number of advantages in manipulating the gene to ensure that the protein product is secreted from the cell:

- To facilitate purification: Clearly if the protein is secreted into the growth medium, there will be far fewer contaminating proteins present than if the cells had to be ruptured to release the protein, when all the other intracellular proteins would also be present.
- Prevention of intracellular degradation of the cloned protein: Many cloned proteins are recognised as 'foreign' by the cell in which they are produced and are therefore degraded by intracellular proteases. Secretion of the protein into the culture medium should minimise this degradation.
- Reduction of the intracellular concentration of toxic proteins: Some cloned proteins are toxic to the cell in which they are produced and there is therefore a limit to the amount of protein the cell will produce before it dies. Protein secretion should prevent cell death and result in continued production of protein.
- To allow post-translational modification of proteins: Most post-translational modifications of proteins occur as part of the secretory pathway, and these modifications, for example glycosylation, are a necessary process in producing the final protein structure. Since prokaryotic cells do not glycosylate their proteins, this explains why many proteins have to be expressed in eukaryotic cells (e.g. yeast) rather than in bacteria. The entry of a protein into a secretory pathway and its ultimate destination is determined by a short amino-acid sequence (signal sequence) that is usually at the N-terminus of the protein. For proteins targeted to the membrane or outside the cell, the route is via the endoplasmic reticulum and Golgi apparatus, the signal sequence being cleaved off by a protease prior to secretion. In some cases, secretion can be achieved by using the protein's native signal sequence (example: human γ-interferon expression in *Pichia pastoris*). In addition, there are a number of well-characterised yeast signal sequences (e.g. the α-factor signal sequence) that can be used to ensure secretion of proteins cloned into yeast.

5.4 PRODUCING RECOMBINANT PROTEIN

There exist a number of expression systems and the decision on which to use depends on a range of factors. Ease of use is one important consideration as well as cost. Generally, *Escherichia coli* and yeast systems are explored at first, followed by insect and mammalian systems. Cultures are grown in vessels ranging from cell culture dishes, to shaker flasks and bioreactors (see also Chapter 3).

5.4.1 Bacteria

Shortly after the discovery of the *lac* operon, non-metabolisable analogues of lactose such as isopropyl-β-*D*-thiogalactoside (IPTG), that relieve repression of the *lac* repressor, were exploited to drive transcription and translation of the desired gene in *Escherichia coli*. The main advantage is the relative ease to manipulate, transform and induce bacteria to produce a protein of interest.

Bacteria have cell diameters of the order of 1 to 4 μm and generally have extremely rigid cell walls. Bacteria can be classified as either Gram-positive or Gram-negative, depending on whether or not they are stained by the Gram stain (crystal violet and iodine). In Gram-positive bacteria (Figure 5.2), the plasma membrane is surrounded by a thick shell of peptidoglycan (20–50 nm), which stains with the Gram stain. In Gram-negative bacteria (e.g. *Escherichia coli*), the plasma membrane is surrounded by a thin (2–3 nm) layer of peptidoglycan, but this is compensated for by the presence of a second outer membrane of lipopolysaccharide. The negatively charged lipopolysaccharide polymers interact laterally, being linked by divalent cations such as magnesium. Gram-negative bacteria can secrete proteins into the periplasmic space as well as the cytoplasm.

Transformation and Culturing of Bacteria

When the plasmid containing the gene of interest has been cloned into the appropriate expression vector, it is introduced (transformed) into the bacterial cells. This can be performed by applying a heat shock or a short pulse of electricity (electroporation). In either method, the brief stress to the bacteria momentarily renders the membrane permeable and allows the plasmid entry into the cell. To improve cell viability, the initial culturing conditions are gentle, by use of enriched culture media. This procedure is conducted using a small amount of medium before plating out onto an Agar plate. The plasmid will typically contain an antibiotic resistance cassette that allows selection using an appropriate antibiotic infused into the agar plate. The purpose of this method is to select a clone that has one identical copy of the plasmid that can then be expanded to a larger volume of medium, suitable for induction and to yield a substantial amount of protein.

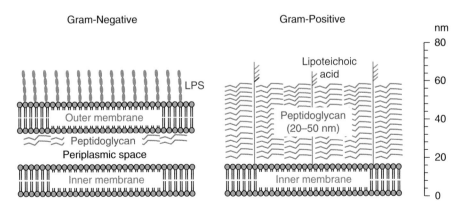

Figure 5.2 The structure of the cell wall of Gram-negative and Gram-positive bacteria. LPS, lipopolysaccharide.

Optimising Expression

The successful over-expression of recombinant genes in bacteria is not guaranteed. The lack of success is often due to the sequence of amino acids containing significant secondary structure that inhibit correct **protein folding**. Also, the gene of interest may utilise rare codons that exist in the source organism, but do not occur in the bacterial expression organism. Bioinformatics analysis coupled with high-throughput DNA synthesis techniques (see also Section 4.9.3) can be used to optimise the codon usage of the gene of interest to use the codons available in *E. coli*, thereby increasing the chances of successful expression.

A successful protein purification is more likely to be achieved if the starting material contains a high percentage of the over-expressed protein in relation to the total protein mixture. It is therefore prudent to design trial induction experiments to ascertain the optimal expression conditions. Small-scale trials would cover a range of different temperatures, different strains of bacteria and concentrations of induction agent (for example, IPTG).

The media used for growing the bacterial culture contains the appropriate antibiotic for selection and the cells are typically grown in a shaker flask until an optical density of $OD_{600} = 0.5–1.0$ is achieved, in order to avoid cellular stress that may be caused by over-expression of the target gene. However, T7 RNA polymerase (required for the most commonly used T7 promoter system in biotechnologically used plasmids) is expressed at a basal level in most strains of *E. coli* and even small amounts of expressed protein may induce cellular stress. Therefore, the choice of bacterial strain has a large impact on protein expression. Several commercially available optimised bacterial strains exist that contain plasmids encoding rare codons, strong repressor function to avoid basal expression and are highly competent (i.e. easy to transform). Also, oxygenation of the culture media during growth increases the yield, hence cultures are typically grown in shaking flasks. Flasks that include baffles will agitate the medium more to increase oxygenation.

One may encounter robust expression of a protein, when assessing entire bacterial cells by SDS-PAGE, but the protein could have been directed towards inclusion bodies, rendering the protein insoluble. This solubility problem can be addressed by cotransformation of **helper plasmids** that prevent the over-expressed protein becoming incorporated into inclusion bodies; examples include plasmids coding for GroEL and GroES. These molecular chaperones aid solubility of proteins in the cytoplasm by assisting in protein folding.

Media

Luria–Bertani (LB) also known as Lysogeny broth is the most commonly used medium to culture bacteria for protein over-expression. It contains a broad range of nutrients that are derived from the enzymatic digestion of casein with trypsin (known as tryptone), yeast extract and sodium chloride in a 2:1:1 ratio (by weight). Before one can grow large cultures, the transformed bacteria are grown in an enriched medium called SOC that allows viable bacteria to grow before plating out. SOC is similar to LB, but has different ratios of nutrients and is supplemented with KCl, $MgCl_2$, $MgSO_4$ and glucose.

There is a deliberate absence of activating sugars in LB in order to prevent basal expression of the plasmid, so that IPTG can be added to cause protein expression. However, auto-induction media can be employed that contain a carefully defined amount of glucose, that is preferentially used by the bacteria until becoming exhausted at the point just before saturation. At this moment, other sugars will induce their operons and cause expression of the target gene without the need to add IPTG.

5.4.2 Yeast and Fungi

Filamentous fungi and yeasts have a rigid cell wall that is composed mainly of polysaccharide (80–90%). In lower fungi and yeast, the polysaccharides are mannan and glucan; in filamentous fungi, it is chitin cross-linked with glucans. Yeasts also have a small percentage of glycoprotein in the cell wall, and there is a periplasmic space between the cell wall and the cell membrane. If the cell wall is removed, the cell content, surrounded by a membrane, is referred to as a spheroplast.

5.4.3 Mammalian Cells

Mammalian cells are of the order of 10 μm in diameter and are enclosed by a plasma membrane, weakly supported by a cytoskeleton. These cells therefore lack any substantial rigidity and are easy to disrupt by shear forces. The major advantage of using mammalian cells is the increased chances of achieving correct folding and the post-translational modifications that may be required. The disadvantages are the high costs of media and the potentially more complex chromatography protocols required.

5.4.4 Plant Cells

Plant cells are of the order of 100 μm in diameter and have a fairly rigid cell wall, comprising carbohydrate complexes and lignin or wax, that surrounds the plasma membrane. Although the plasma membrane is protected by this outer layer, the large size of the cell still makes it susceptible to shear forces.

5.5 CELL-DISRUPTION METHODS

5.5.1 Cell Disruption and Production of Initial Crude Extract

The ability to clone and over-express genes to obtain proteins from any source organism has so many advantages that the starting material will therefore most likely be from one kind of cell culture, as discussed above. However, post-translational modifications, poor folding or poor induction characteristics can still make it necessary in some instances to purify native proteins of interest from an abundant source. Protein purification procedures necessarily start with the disruption of cells or tissue to release the protein content of the cells into an appropriate buffer, unless the protein is secreted. This initial extract is therefore the starting point for protein purification, but thought must first be given to the composition of the buffer used to extract the proteins.

Extraction Buffer

Normally, extraction buffers are at an ionic strength of 0.1–0.2 M of a monovalent salt and a pH between 7 and 8, which is considered to be compatible with the conditions found inside the cell. TRIS or phosphate buffers are most commonly used. Additionally, a range of other reagents may be added to the buffer for specific purposes. These include:

- An antioxidant: Within the cell, proteins are in a fairly reducing environment, but when released into the buffer they are exposed to a more oxidising environment. Since most proteins contain a number of free thiol groups (from the amino acid cysteine) these can undergo oxidation to give inter- and intramolecular disulfide bridges. To prevent this, reducing agents such as dithiothreitol, β-mercaptoethanol, cysteine or reduced glutathione are often included in the buffer.

- Enzyme inhibitors: Once the cell is disrupted, the organisational integrity of the cell is lost, and proteolytic enzymes that were carefully packaged and controlled within the intact cells are released, for example from lysosomes. Such enzymes may degrade proteins in the extract, including the protein of interest. To slow down unwanted proteolysis, all extraction and purification steps are carried out at 4 °C, and, in addition, a range of protease inhibitors is included in the buffer. Each inhibitor is specific for a particular type of protease, for example serine proteases, thiol proteases, aspartic proteases and metalloproteases. Common examples of inhibitors include: di-isopropylphosphofluoridate (DFP), phenylmethyl-sulfonylfluoride (PMSF) and tosylphenylalanyl-chloromethylketone (TPCK) (all serine protease inhibitors); iodoacetate and cystatin (thiol protease inhibitors); pepstatin (aspartic protease inhibitor); EDTA and 1,10-phenanthroline (metalloprotease inhibitors); and amastatin and bestatin (exopeptidase inhibitors).

- Enzyme substrate and cofactors: Low levels of substrate are often included in extraction buffers when an enzyme is purified, since binding of substrate to the enzyme active site can stabilise the enzyme during purification processes. Where relevant, cofactors that otherwise might be lost during purification are also included to maintain enzyme activity so that it can be detected when column fractions are screened.

- Phosphatase inhibitors: Sodium orthovanadate (Na_3VO_4) is a general inhibitor of phosphotyrosyl phosphatases, commonly used in buffers for protein kinase assays. Being a competitive inhibitor, it can be conveniently removed by dilution, inactivation with EDTA or dialysis. Na_3VO_4 is often combined with sodium fluoride, which inhibits phosphoseryl and phosphothreonyl phosphatases. β-glycerol-phosphate is a general serine/threonine phosphatase inhibitor that acts as a pseudo-substrate by providing a ready source of organic phosphate.

- EDTA: This is used to chelate divalent metal cations; each molecule of EDTA will bind one cation. The binding is very efficient due to the coordination provided by four oxygen and two nitrogen atoms. EDTA is added to remove divalent metal ions (M^{2+}) that can react with thiol groups in proteins giving thiolates as per:

$$R\text{-SH} + Me^{2+} \longrightarrow R\text{-S-}Me^+ + H^+ \qquad\qquad (Eq\ 5.1)$$

- The sequestration of free magnesium ions by EDTA decreases the activity of any enzyme that uses ATP as a cofactor. Cation chelation also reduces the activity of metal-loproteases. However, EDTA would obviously interfere with immobilised metal affinity chromatography (IMAC) used to purify His-tagged proteins. Additionally, EDTA interferes with nucleotide binding.
- Polyvinylpyrrolidone (PVP): This is often added to extraction buffers for plant tissue. Plant tissue contains considerable amounts of phenolic compounds (both monomeric, such as p-hydroxybenzoic acid, and polymeric, such as tannins) that can bind to enzymes and other proteins by non-covalent forces, including hydrophobic, ionic and hydrogen bonds, causing protein precipitation. These phenolic compounds are also easily oxidised, predominantly by endogenous phenol oxidases, to form quinones, which are highly reactive and can combine with reactive groups in proteins causing cross-linking, and further aggregation and precipitation. Insoluble PVP (which mimics the polypeptide backbone) is therefore added to adsorb the phenolic compounds, which can then be removed by centrifugation. Thiol compounds (reducing agents) are also added to minimise the activity of phenol oxidases, and thus prevent the formation of quinones.
- Sodium azide: For buffers that are going to be stored for long periods of time, anti-bacterial and/or antifungal agents are sometimes added at low concentrations. Sodium azide is frequently used as a bacteriostatic agent.

Membrane Proteins

Membrane-bound proteins require special conditions for extraction as they are not released by simple cell-disruption procedures alone. Two classes of membrane proteins are identified. Extrinsic (or peripheral) membrane proteins are bound only to the surface of the cell, normally via electrostatic and hydrogen bonds. These proteins are predominantly hydrophilic in nature and are relatively easily extracted, either by raising the ionic concentration of the extraction buffer (e.g. to 1 M NaCl) or by either acidic or basic conditions (pH = 3–5 or pH = 9–12). Once extracted, the peripheral membrane proteins can be purified by conventional chromatographic procedures. Intrinsic membrane proteins are those that are embedded in the membrane. These invariably have significant regions of hydrophobic amino acids (those regions of the protein that are embedded in the membrane, and associated with lipids) and have low solubility in aqueous buffer systems. Hence, once extracted into an aqueous polar environment, appropriate conditions must be used to retain their solubility. Intrinsic proteins are usually extracted with buffer containing detergents. The choice of detergent is mainly one of trial and error, but can include ionic detergents such as sodium dodecyl sulfate (SDS), sodium deoxycholate, cetyl trimethylammonium bromide (CTAB) and CHAPS, and non-ionic detergents such as Triton X-100 and Nonidet P-40.

Once extracted, intrinsic membrane proteins can be purified using conventional chromatographic techniques such as size-exclusion, ion-exchange or affinity chromatography (using lectins). Importantly, in each case, it is necessary to include detergent in all buffers to maintain protein solubility. The level of detergent used is normally 10- to 100-fold less than that used to extract the protein, in order to minimise any interference of the detergent with the chromatographic process.

Sonication

This method is ideal for a suspension of cultured cells or microbial cells. A sonicator probe is lowered into the suspension of cells and high frequency sound waves (<20 kHz, i.e. ultrasound) generated for 30–60 s. These sound waves cause disruption of cells by shear force and cavitation. Cavitation refers to areas where there is alternate compression and rarefaction, which rapidly interchange. The gas bubbles in the buffer are initially under pressure but, as they decompress, shock waves are released and disrupt the cells. This method is suitable for relatively small volumes (50–100 cm^3). Since considerable heat is generated by this method, samples must be kept on ice during treatment, typically in a so-called rosette cell. The efficiency of cell lysis by sonication is thought to be around 50–60%. The possible danger of damaging the target protein is a disadvantage of this method.

Repeated Freeze–Thaw

A more gentle and readily available method for lysis of bacterial cells is to subject the aqueous cell suspension to repeated freezing and thawing conditions. Resuspended cells are frozen (either at −20 °C, −80 °C or in liquid nitrogen) and then thawed in warm water. This procedure is repeated three times. The shear forces due to ice formation cause disruption of the enclosed cells. Frequently, a short sonication of the resulting homogenate (lysate) is carried out to break down the bacterial genomic DNA, which can make the suspension stringy and viscous (see also Section 5.6.2).

Blenders

These are commercially available, although a typical domestic kitchen blender will suffice. This method is ideal for disrupting mammalian or plant tissue by shear force. Tissue is cut into small pieces and blended, in the presence of buffer, for about 1 min to disrupt the tissue. This method is inappropriate for bacteria and yeast, but a blender can be used for these microorganisms if small glass beads are introduced to produce a bead mill. Cells are trapped between colliding beads and physically disrupted by shear forces.

Grinding With Abrasives

Grinding with a pestle in a mortar, in the presence of sand or alumina and a small amount of buffer, is a useful method for disrupting bacterial or plant cells; cell walls are physically ripped off by the abrasive. However, the method is appropriate for handling only relatively small samples. The Dyno®-mill is a large-scale mechanical version of this approach; it comprises a chamber containing glass beads and a number of rotating impeller discs. Cells are ruptured when caught between colliding beads. A 600 cm^3 laboratory scale model can process 5 kg of bacteria per hour.

Presses

The use of homogenisers such as a French press, or the Manton–Gaulin press, which is a larger-scale version, is an excellent means for disrupting microbial cells. A cell suspension (~50 cm³) is forced by a piston-type pump under high pressure (10 000 psi ≈ 1450 kPa) through a small orifice. Breakage occurs due to shear forces as the cells are forced through the small orifice, and also by the rapid drop in pressure as the cells emerge from the orifice, which allows the previously compressed cells to expand rapidly and effectively burst. Multiple passes are usually needed to lyse all the cells, but under carefully controlled conditions it is possible to selectively release proteins from the periplasmic space. The Hughes press is a variation on this method; the cells are forced through the orifice as a frozen paste, often mixed with an abrasive. Both the ice crystals and the abrasive aid in disrupting the cell walls.

Enzymatic Methods

The enzyme lysozyme, isolated from hen egg whites, cleaves peptidoglycan. The peptidoglycan cell wall can therefore be removed from Gram-positive bacteria (see Figure 5.2) by treatment with lysozyme, and if carried out in a suitable buffer, the cell membrane will rupture, owing to the osmotic effect of the suspending buffer.

Gram-negative bacteria can similarly be disrupted by lysozyme, but treatment with EDTA (to remove divalent metal ions, thus destabilising the outer lipopolysaccharide layer) and the inclusion of a non-ionic detergent to solubilise the cell membrane are also needed. This effectively permeabilises the outer membrane, allowing access of the lysozyme to the peptidoglycan layer. If carried out in an isotonic medium so that the cell membrane is not ruptured, it is possible to selectively release proteins from the periplasmic space.

Yeast can be similarly disrupted using enzymes to degrade the cell wall, followed by either osmotic shock or mild physical force to disrupt the cell membrane. Enzyme digestion alone allows the selective release of proteins from the periplasmic space. The two most commonly used enzyme preparations for yeast are zymolyase or lyticase, both of which have β-1,3-glucanase activity, together with a proteolytic activity specific for the yeast cell wall. Chitinase is commonly used to disrupt filamentous fungi. Enzymatic methods tend to be used for laboratory-scale work, since for large-scale work their use is limited by cost.

5.6 PRELIMINARY PURIFICATION STEPS

5.6.1 Removal of Cell Debris

The initial extract, produced by the disruption of cells and tissue, and referred to at this stage as a **homogenate** (or lysate), will invariably contain insoluble matter. For example, for mammalian tissue there will be incompletely homogenised connective and/or vascular tissue, and small fragments of non-homogenised tissue. This is most easily removed by filtering through a double layer of cheesecloth or by low speed (5000×g) centrifugation. Any fat floating on the surface can be removed by coarse filtration through glass wool or cheesecloth. However, the solution will still be cloudy with organelles and membrane fragments that are too small to be conveniently removed by filtration or low speed centrifugation. If necessary, they can be removed first by precipitation using materials such as celite (a diatomaceous earth that provides a large surface area to trap the particles), cell debris remover (a cellulose-based absorber), or any type of flocculants such as starch, gums, tannins or polyamines; the resultant precipitate being removed by centrifugation (typically about 20 000×g for 45 min; or 40 000×g for 30 min in an ultracentrifuge) or filtration.

5.6.2 Non-Protein Components

Even the cleared homogenate contains not only proteins, but also other molecules such as DNA, RNA, carbohydrate and lipid, as well as any number of small-molecular-weight metabolites. Small molecules tend to be removed later on during dialysis steps or steps that involve fractionation based on size (e.g. size-exclusion chromatography) and therefore are of little concern. However, specific attention has to be paid at this stage to macromolecules such as nucleic acids and polysaccharides. This is particularly true for bacterial extracts, which are particularly viscous owing to the presence of chromosomal DNA. Indeed microbial extracts can be extremely difficult to centrifuge to produce a supernatant extract. Some workers include DNase I in the extraction buffer to reduce viscosity, the small DNA fragments generated being removed at later dialysis/size-exclusion steps. Likewise, RNA can be removed by treatment with RNase. DNA and RNA can also be removed by precipitation with protamines – a mixture of small, highly basic (i.e. positively charged) proteins, whose natural role is to bind to DNA in the sperm head. Protamines are usually extracted from fish organs, which are obtained as a waste product at canning factories. These positively charged proteins bind to negatively charged phosphate groups on nucleic acids, thus masking them and rendering them insoluble. The addition of a solution of protamines to the homogenate therefore precipitates most of the DNA and RNA, which can subsequently be removed by centrifugation. An alternative is to use polyethyleneimine, a synthetic long-chain cationic polymer with molecular mass of about 24 kDa. This polymer also binds to the phosphate groups in nucleic acids, and is very effective, precipitating DNA and RNA almost instantly.

For bacterial extracts, carbohydrate capsular gum can also be a problem as this can interfere with protein precipitation methods. This is best removed by the use of lysozyme, which can digest capsular gum and is more efficient in the presence of detergent used to lyse the cells.

5.7 PRINCIPLES OF LIQUID CHROMATOGRAPHY

5.7.1 The Partition Coefficient

Chromatography is a core technique of biochemical investigations and is used extensively to purify a protein of interest from a complex mixture such as obtained with the methods described above. Chromatography can be analytical or preparative, but the basis of all forms of chromatography is the distribution or partition coefficient (P), which describes the way in which a compound (the analyte) distributes between two immiscible phases. For two such phases, A and B, the value for this coefficient is a constant at a given temperature and is given by the expression:

$$P = \frac{c(\text{solute})_{\text{phase A}}}{c(\text{solute})_{\text{phase B}}} \tag{Eq 5.2}$$

The term effective distribution coefficient is defined as the total amount, as distinct from the concentration, of analyte present in one phase divided by the total amount present in the other phase. It is in fact the distribution coefficient multiplied by the ratio of the volumes of the two phases present.

5.7.2 Column Chromatography

Chromatographic systems used for protein purification consist of a stationary phase, which is typically an immobilised solid, and a liquid mobile phase, which is passed through the stationary phase after the mixture of analytes to be separated has been applied to the column. The mobile phase, commonly referred to as the eluent, is passed through the column either by use of a pumping system or gravity (atmospheric pressure). The stationary phase is either coated onto discrete small particles (the matrix) and packed into the column or applied as a thin film to the inside wall of the column. During the chromatographic separation, the analytes continuously pass back and forth between the two phases, exploiting differences in their distribution coefficients, and emerge individually in the eluate as it leaves the column.

5.7.3 Chromatography Components for Protein Purification

A typical chromatographic system suitable for protein purification consists of the following components, ideally be situated in a cold room or refrigerator to maintain protein stability:

- A stationary phase: Chosen to be appropriate for the analytes to be separated; typically an aqueous buffer with sufficient ioinic strength to maintain a soluble protein. Certain additives discussed in Section 5.5.1 are sometimes required.
- A column: Research-lab-sized liquid chromatography columns range in length from 5–100 cm, with internal diameters from 4 mm to 6 cm and made of stainless steel, plastic or glass, depending on the application and hence the system pressure involved. The column has to be carefully packed to generate reliable separations. Often, columns are obtained pre-packed from commercial suppliers.
- A mobile phase and delivery system: Chosen to complement the stationary phase and hence to discriminate between the sample analytes and to deliver a constant rate of flow into the column.
- An injector system: To deliver test samples to the top of the column in a reproducible manner.
- A detector with data acquisition: To give a continuous record of the presence of the analytes in the eluate as it emerges from the column. Detection is usually based on the measurement of a physical parameter such as visible or ultraviolet absorption or fluorescence. A typical chromatography system has a computer to control the system

and acquire readouts of the monitored physical parameter. The plot of the monitored parameter versus elution time is called a **chromatogram**.

- A **fraction collector**: For collecting the separated analytes for further biochemical studies.

Depending on the particle size of the stationary phase (and the desired resolution), the chromatography may be carried out as gravity-driven, low-pressure or high-pressure liquid chromatography. Gravity-driven chromatography is often used with cartridges for a quick single-step purification of small-scale samples and without inline detection systems. Larger samples are processed in pumped systems. In **low-pressure liquid chromatography**, the flow of the eluent through the column is achieved by a low pressure pump, frequently a **peristaltic pump**.

The resolution of a mixture of analytes increases as the particle size of the stationary phases decreases, but such a decrease leads to a high back-pressure from the eluent flow. This is addressed by high-pressure liquid chromatography (HPLC). This technique is therefore most frequently applied for preparative protein biochemistry and employed in commercially available purification systems that are fully computer-controlled. The much higher resolution of HPLC as compared to low-pressure systems is further surpassed by so-called ultra-performance liquid chromatography (UPLC) systems, which employ particle sizes of down to 1.7 µm in the stationary phase and back-pressures of up to 150 Mpa.

The application of samples onto HPLC columns in the correct way is a particularly important factor in achieving successful separations. The most common method of sample introduction is by use of a **loop injector** (Figure 5.3). This consists of a metal loop, of fixed small volume, that can be filled with the sample. The eluent from the pump is then channelled through the loop by means of a valve switching system and the sample flushed onto the column via the loop outlet without interruption of the flow of eluent to the column.

Analyte Development and Elution

Analyte development and elution relates to the separation of the mixture of analytes applied to the stationary phase by the mobile phase and their elution from the

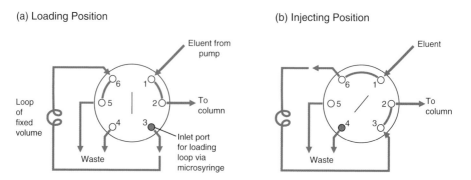

Figure 5.3 HPLC loop injector. (a) The loop is loaded via port 3 with excess sample going to waste via port 5. In this position the eluent from the pump passes to the column via ports 1 and 2. (b) In the injecting position eluent flow is directed through the loop via ports 1 and 6 and then onto the column.

column. Column chromatographic techniques can be subdivided on the basis of the development and elution modes.

- In zonal development, the analytes in the sample are separated on the basis of their distribution coefficients between the stationary and mobile phases. The sample is dissolved in a suitable solvent and applied to the stationary phase as a narrow, discrete band. The mobile phase is then allowed to flow continuously over the stationary phase, resulting in the progressive separation and elution of the sample analytes. If the composition of the mobile phase is constant (as is the case in gas chromatography and some forms of liquid chromatography), the process is called isocratic elution. However, to facilitate separation, the composition of the mobile phase may be gradually changed, for example with respect to pH, salt concentration or polarity. This is referred to as gradient elution. The composition of the mobile phase may be changed continuously or in a stepwise manner.
- In displacement or affinity development (practically only possible in liquid chromatography), the analytes in the sample are separated on the basis of their affinity for the stationary phase. The sample of analytes dissolved in a suitable solvent is applied to the stationary phase as a discrete band. The analytes bind to the stationary phase with a strength determined by their affinity constant for the phase. The analytes are then selectively eluted by using a mobile phase containing a specific solute that has a higher affinity for the stationary phase than have the analytes in the sample. Thus, as the mobile phase is added, this agent displaces the analytes from the stationary phase in a competitive fashion, resulting in their repetitive binding and displacement along the stationary phase and eventual elution from the column in the order of their affinity for the stationary phase, the one with the lowest affinity being eluted first.

5.7.4 Chromatographic Performance Parameters

The successful chromatographic separation of analytes in a mixture depends upon the selection of the most appropriate process of chromatography, followed by the optimisation of the experimental conditions associated with the separation. Optimisation requires an understanding of the processes that occur during the development and elution, and of the calculation of a number of experimental parameters characterising the behaviour of each analyte in the mixture.

   In any chromatographic separation, two processes occur concurrently and affect the behaviour of each analyte and hence the success of the separation of the analytes from each other. The first involves the basic mechanisms defining the chromatographic process such as adsorption, partition, ion exchange, ion pairing and molecular exclusion. These mechanisms involve the unique kinetic and thermodynamic processes that characterise the interaction of each analyte with the stationary phase. The second general process comprises diffusion and non-specific interactions (see also Section 5.7.8), which tend to oppose the separation and result in non-ideal behaviour of each analyte. These processes manifest themselves as a broadening and tailing of each analyte peak. The challenge is to minimise these secondary processes.

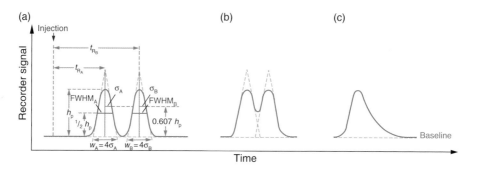

Figure 5.4 (a) Chromatogram of two analytes showing complete resolution and the calculation of retention times; (b) chromatogram of two analytes showing incomplete resolution (fused peaks); (c) chromatogram of an analyte showing excessive tailing, For explanation of parameters shown see main text.

5.7.5 Retention Time

A chromatogram is a pictorial record of the detector response as a function of elution volume or retention time. It consists of a series of peaks, ideally symmetrical in shape, representing the elution of individual analytes, as shown in Figure 5.4. The retention time t_R for each analyte has two components. The first is the time it takes the analyte molecules to pass through the free spaces between the particles of the matrix coated with the stationary phase. This time is referred to as the dead time, t_M. The volume of the free space is referred to as the column void volume, V_0. The value of t_M will be the same for all analytes and can be measured by using an analyte that does not interact with the stationary phase, but simply spends all of the elution time in the mobile phase travelling through the void volume. The second component is the time the stationary phase retains the analyte, referred to as the **adjusted retention time**, t'_R. This time is characteristic of the analyte and is the difference between the observed retention time and the dead time:

$$t'_R = t_R - t_M$$
(Eq 5.3)

5.7.6 Retention Factor

One of the most important parameters in column chromatography is the retention factor, k (previously called capacity factor). It is simply the additional time that the analyte takes to elute from the column relative to an unretained or excluded analyte that does not interact with the stationary phase and which, by definition, has a k value of 0. Thus:

$$k = \frac{t_R - t_M}{t_M} = \frac{t'_R}{t_M}$$
(Eq 5.4)

Note that k has no units.

It is apparent from this equation that if the analyte spends an equal time in the stationary and mobile phases, its t_R would equal $2 \times t_M$ and its k would thus be 1.

Similarly, if it spent four times as long in the stationary phase as the mobile phase, t_R would equal $5 \times t_M$ and so the retention factor would be:

$$k = \frac{5 \times t_M - t_M}{t_M} = 4.$$

If an analyte has a k of 4, it follows that there will be four times the amount of analyte in the stationary phase than in the mobile phase at any point in the column at any time. It is evident, therefore, that k is related to the distribution coefficient of the analyte (Equation 5.2), which was defined as the relative concentrations of the analyte between the two phases. Since amount and concentration are related by volume, we can write:

$$k = \frac{t_R'}{t_M} = \frac{m_S}{m_M} = P \times \frac{V_S}{V_M} \qquad \text{(Eq 5.5)}$$

where m_S is the mass of analyte in the stationary phase, m_M is the mass of analyte in the mobile phase, V_S is the volume of stationary phase and V_M is the volume of mobile phase. The ratio V_S/V_M is referred to as the volumetric phase ratio, β. Hence:

$$k = P \times \beta \qquad \text{(Eq 5.6)}$$

A larger retention factor results in the distribution ratio favouring the stationary phase, which increases retention time. Therefore, the retention factor for an analyte will increase with both the distribution coefficient between the two phases and the volume of the stationary phase. Values of k normally range from 1 to 10. Retention factors are important, because they are independent of the physical dimensions of the column and the rate of flow of mobile phase through it. They can therefore be used to compare the behaviour of an analyte in different chromatographic systems. They are also a reflection of the selectivity of the system that, in turn, is a measure of its inherent ability to discriminate between two analytes. Such selectivity is expressed by the selectivity or separation factor, α, which can also be viewed as simply the relative retention ratio for the two analytes:

$$\alpha = \frac{k_A}{k_B} = \frac{P_A}{P_B} = \frac{t_{R_A}'}{t_{R_B}'} \qquad \text{(Eq 5.7)}$$

The selectivity factor is influenced by the chemical nature of the stationary and mobile phases. Some chromatographic mechanisms are inherently highly selective, for example affinity chromatography (see Section 5.8.1).

5.7.7 Plate Height

Chromatography columns are considered to consist of a number of adjacent zones in each of which there is sufficient space for an analyte to completely equilibrate between the two phases. Each zone is called a theoretical plate (of which there are N in total in the column). The length of column containing one theoretical plate is referred to as the plate height, H, which has units of length (normally micrometres).

The smaller the value of H and the greater the value of N, the more efficient the column in separating a mixture of analytes. The numerical value of both N and H for a particular column is expressed by reference to a particular analyte. Plate height is simply related to the width of the analyte peak, expressed in terms of its standard deviation σ (Figure 5.4), and the distance it travelled within the column, x. Specifically:

$$H = \frac{\sigma^2}{x} \qquad \text{(Eq 5.8)}$$

For symmetrical Gaussian peaks, the base width is equal to $4 \times \sigma$ and the peak width at the point of inflection is equal to $2 \times \sigma$. Hence, the value of H can be calculated from the chromatogram by measuring the peak width. The number of theoretical plates in the whole column of length L is equal to L divided by the plate height:

$$N = \frac{L}{H} = \frac{L \times x}{\sigma^2} \qquad \text{(Eq 5.9)}$$

If the position of a peak emerging from the column is such that $x = L$, Equation 5.9 can be converted to

$$N = \frac{L^2}{\sigma^2} = 16 \times \frac{L^2}{w^2} \qquad \text{(Eq 5.10)}$$

knowing that the width of the peak at its base, w, is equal to $4 \times \sigma$ and hence $\sigma = \frac{1}{4} \times w$.

If both L and w are measured in units of time rather than length, then Equation 5.10 becomes:

$$N = 16 \times \left(\frac{t_R}{w}\right)^2 \qquad \text{(Eq 5.11)}$$

Rather than expressing N in terms of the peak base width, it is possible to express it in terms of the peak width at half height (full width at half maximum, FWHM) and this has the practical advantage that this is more easily measured:

$$N = 5.54 \times \left(\frac{t_R}{\text{FWHM}}\right)^2 \qquad \text{(Eq 5.12)}$$

Equations 5.11 and 5.12 represent alternative ways to calculate the column efficiency in theoretical plates. The value of N, which has no units, can be as high as 50 000 to 100 000 per metre for efficient columns and the corresponding value of H can be as little as a few micrometres. The smaller the plate height (the larger the value of N), the narrower the analyte peak (Figure 5.5).

5.7.8 Peak Broadening

A number of processes oppose the formation of a narrow analyte peak, thereby increasing the plate height:

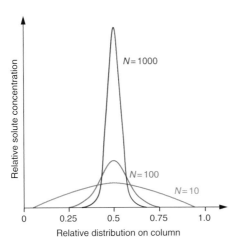

Figure 5.5 Relationship between the number of theoretical plates (*N*) and the shape of the analyte peak.

- **Application of the sample to the column:** It takes a finite time to apply the analyte mixture to the column, so that the part of the sample applied first will already be moving along the column by the time the final part is applied. The part of the sample applied first will elute at the front of the peak.
- **Longitudinal diffusion:** Fick's law of diffusion states that an analyte will diffuse from a region of high concentration to one of low concentration at a rate determined by the concentration gradient between the two regions and the diffusion coefficient of the analyte. Thus, the analyte within a narrow band will tend to diffuse outwards from the centre of the band, resulting in band broadening.
- **Multiple pathways:** The random packing of the particles in the column results in the availability of many routes between the particles for both mobile phase and analytes. These pathways will vary slightly in length and hence elution time; the smaller the particle size the less serious this problem.
- **Equilibration time between the two phases:** It takes a finite time for each analyte in the sample to equilibrate between the stationary and mobile phases as it passes down the column. As a direct consequence of the distribution coefficient (*P*), some of each analyte is retained by the stationary phase, whilst the remainder stays in the mobile phase and continues its passage down the column. This partitioning automatically results in some spreading of each analyte band. Equilibration time, and hence band broadening, is also influenced by the particle size of the stationary phase. The smaller the size, the less time it takes to establish equilibration.

Two of these four factors promoting the broadening of the analyte band are influenced by the flow rate of the eluent through the column. Longitudinal diffusion, defined by Fick's law, is inversely proportional to flow rate, whilst equilibration time due to the partitioning of the analyte is directly proportional to flow rate. These two factors together with the problem of multiple pathways affect the value of the plate height (*H*) for a particular column and, as previously stated, plate height determines the width of the analyte peak. The precise relationship between the three factors and plate height is expressed by the van Deemter equation (Equation 5.13), which is illustrated graphically in Figure 5.6:

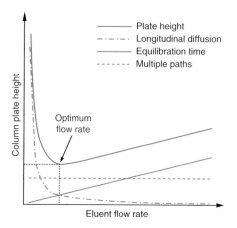

Figure 5.6 Illustration of the van Deemter equation. The plot shows that the optimum flow rate for a given column is the net result of the influence of flow rate on longitudinal diffusion, equilibration time and multiple pathways.

$$H = A + \frac{B}{u} + C \times u \qquad\qquad \text{(Eq 5.13)}$$

where u is the flow rate of the eluent (in units of m s^{-1}) and A, B and C are constants for a particular column and stationary phase relating to multiple paths (A), longitudinal diffusion (B) and equilibration time (C).

Figure 5.6 gives a clear demonstration of the importance of establishing the optimum flow rate for a particular column. Longitudinal diffusion is much faster in a gas than in a liquid and, as a consequence, flow rates are higher in gas chromatography than in liquid chromatography.

As previously stated, the width of an analyte peak is expressed in terms of the standard deviation σ, which is half the peak width at the point of inflection ($0.607 \times h_p$, where h_p is the peak height; see Figure 5.4). It can be shown that

$$\sigma = \sqrt{2 \times D \times t_R} \qquad\qquad \text{(Eq 5.14)}$$

where D is the diffusion coefficient of the analyte. The diffusion coefficient is a measure of the rate at which the analyte moves randomly in the mobile phase from a region of high concentration to one of lower concentration; it has units of m^2 s^{-1}.

Since the value of σ is proportional to the square root of t_R, it follows that if the elution time increases by a factor of four, the width of the peak will double. Thus, the longer it takes a given analyte to elute, the wider will be its peak. This phenomenon counteracts the improvement in resolution achieved by increasing the column length (see Section 5.7.10, Equation 5.16).

5.7.9 Asymmetric Peaks

In some chromatographic separations, the ideal Gaussian-shaped peaks are not obtained, but rather asymmetrical peaks are produced. In cases where there is a gradual rise at the front of the peak and a sharp fall after the peak, the phenomenon is known as fronting. The most common cause of fronting is overloading the column, so that reducing the amount of mixture applied to the column often resolves the problem. In cases where the rise in the peak is normal, but the tail is protracted, the phenomenon is known as tailing (see Figure 5.4c). The probable explanation for tailing is the retention of analyte by a few sites (frequently hydroxyl groups) on the stationary phase, commonly on the inert support matrix. Such sites strongly adsorb molecules of the analyte and only slowly release them. This problem can be overcome by chemically removing the sites, for example by treating the matrix with a silanising reagent such as hexamethyldisilazine. This process is sometimes referred to as cap-ping. Peak asymmetry is usually expressed as the ratio between the width of the peak at the centre (i.e. at full peak height h_p) and the width of the peak at $0.1 \times h_p$.

5.7.10 Resolution

The success of a chromatographic separation is judged by the ability of the system to resolve one analyte peak from another. Resolution (R_S) is defined as the ratio of the difference in retention time (Δt_R) between the two peaks to the mean of their base widths (w_A and w_B), w_{av}:

$$R_S = \frac{\Delta t_R}{w_{av}} = \frac{2 \times \left(t_{R_A} - t_{R_B} \right)}{w_A + w_B} \qquad \text{(Eq 5.15)}$$

When $R_S = 1.0$, the separation of the two peaks is 97.7% complete (thus the overlap is 2.3%). When $R_S = 1.5$, the overlap is reduced to 0.2%. Unresolved peaks are referred to as fused peaks (Figure 5.4b). Provided the overlap is not excessive, the analysis of the individual peaks can be made on the assumption that their characteristics are not affected by the incomplete resolution.

Resolution is influenced by column efficiency, selectivity factors (see Equation 5.7) and retention factors according to:

$$R_S = \frac{\sqrt{N}}{4} \times \frac{\alpha - 1}{\alpha} \times \frac{k_2}{1 + k_{av}} \qquad \text{(Eq 5.16)}$$

where k_2 is the retention factor for the longest retained peak and k_{av} is the mean retention factor for the two analytes. Equation 5.16 is one of the most important in chromatography as it enables a rational approach to the improvement of the resolu-tion between analytes. For example, it can be seen that resolution increases with the square root of N. Since N is linked to the length of the column, doubling the length of the column will increase resolution by $\sqrt{2} \approx 1.4$ and in general is not the preferred way to improve resolution. Since both retention factors and selectivity factors are linked to retention times and retention volumes, altering the nature of the two phases or their relative volumes will impact on resolution. Retention factors are also dependent upon distribution coefficients, which in turn are temperature dependent; hence altering the column temperature may improve resolution.

The capacity of a particular chromatographic separation is a measure of the amount of material that can be resolved into its components without causing peak overlap or fronting. Ion-exchange chromatography resins (Section 5.8.2) have a high capacity, which is why it this type of chromatography is often used in the earlier stages of a purification process.

5.8 CHROMATOGRAPHIC METHODS FOR PROTEIN PURIFICATION

5.8.1 Affinity Chromatography

Certain proteins bind strongly to specific small molecules. One can take advantage of this by developing an affinity chromatography system where the small molecule (ligand) is bound to an insoluble support. When a crude mixture of proteins containing the protein of interest is passed through the column, the ligand binds the protein to the matrix, whilst all other proteins pass through the column. The bound protein can then be eluted from the column by changing the pH, increasing salt strength or passing through a high concentration of unbound free ligand. For example, glutathione-S-transferase (GST) is a protein with high affinity for the

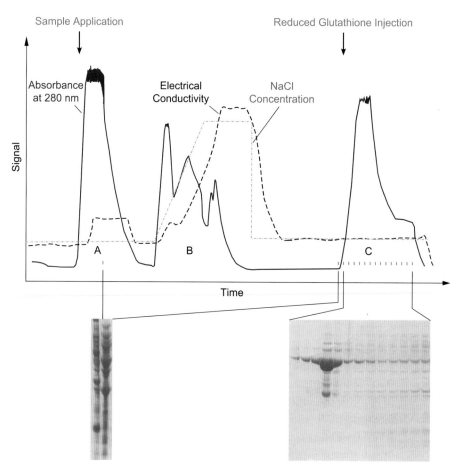

Figure 5.7 Example of an affinity chromatography purification procedure. Absorbance at 280 nm is indicated by the blue line, conductivity by the dashed line and salt concentration (NaCl) by the grey dotted line. The red ticks indicate different fractions isolated, the contents of which are analysed on SDS-PAGE (shown at the bottom). A: Sample injection; due to the high protein concentration, the absorbance reaches the maximum recordable value; most protein passes through the column, unbound. B: Contaminating proteins eluting by way of a salt gradient. C: Elution of the fusion protein by injection of reduced glutathione.

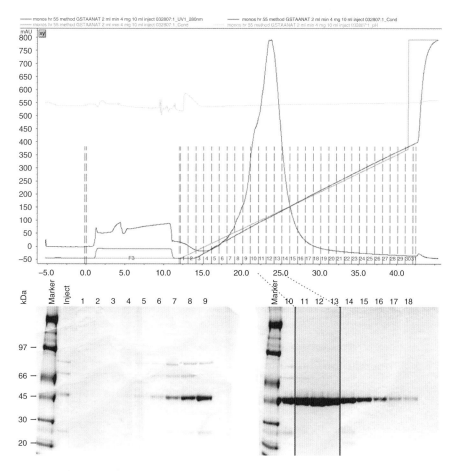

Figure 5.8 Illustration of ion exchange chromatography. The top panel shows the chromatogram from a purification of a GST-fusion protein. Although the initial affinity purification step with glutathione was relatively pure, there are contaminating bands at ~28, ~66 and ~75 kDa. A column packed with a strong cation exchanger (MonoS®) was used to bind the target protein and a gradient of 0–100% 1 M NaCl in HEPES (pH 8.0) was applied using an ÄKTA HPLC system. Analysis of the isolated fractions show highly pure target protein in fractions 11–13, as judged by Coomassie Brilliant Blue staining.

tripeptide glutathione (Figure 5.7). The technique is most frequently used with pro-tein fusion constructs (see Section 5.3.1), but also has applications for a wide variety of other targets, including nucleotides, nucleic acids, immunoglobulins, membrane receptors, and even whole cells and cell fragments. Beaded agarose resins with reac-tive groups (activated agarose) for conjugation allow immobilisation of suitable small and macro-molecules. Most frequently used groups for conjugation include N-hydroxysuccinimide (NHS) or cyanogen bromide (CNBr) for amino groups, and carbodiimides for carboxyl groups.

5.8.2 Ion Exchange Chromatography

Proteins differ from one another in the proportions of the charged amino acids (aspartic and glutamic acids, lysine, arginine and histidine) that they contain.

Hence, different proteins possess different net charges at a particular pH. This difference is exploited in ion exchange chromatography, where the protein of interest is adsorbed to a solid support material bearing charged groups of the opposite sign (ion exchange resin). Proteins with the same charge as the resin pass through the column to waste; subsequently, bound proteins are selectively released from the column by gradually increasing the concentration of salt ions in the buffer passing through the column. These ions compete with the protein for binding to the resin, the more weakly charged protein being eluted at the lower salt concentration and the more strongly charged protein being eluted at higher salt concentrations. Alternatively, one may keep the salt concentration constant, but gradually change the pH of the eluting buffer.

Ion exchange chromatography can be used as an initial purification step, or as a 'polishing' step after a certain purity has been achieved because it is a relatively gentle purification method.

Frequently, fairly high purity of the target protein can be achieved with one simple gradient elution of an ion exchange column (Figure 5.8), but often it may also be necessary to perform scouting experiments to determine the optimal run conditions involving buffer composition, length of gradient and type of media (anionic or cationic exchanger).

5.8.3 Size-Exclusion Chromatography (SEC)

Size differences between proteins can be exploited in molecular exclusion chromatography (also known as gel filtration). The size-exclusion medium consists of a range of beads with slightly differing amounts of cross-linking and therefore slightly different pore sizes. The separation process depends on the different abilities of the various proteins to adsorb to some, all or none of the cavities on the beads, which relates the retention time on the resin to the size of the protein. The method has limited resolving power, but can be used to obtain a separation between large and small protein molecules and therefore be useful when the protein of interest is either particularly large or particularly small. A major advantage is that this method is very gentle on proteins and is often used as a final stage in preparations destined for protein crystallography and other applications that require functional protein. Size-exclusion chromatography can also be used to determine the relative molecular mass of a protein. Furthermore, owing to the large size difference between inorganic ions and proteins, it is frequently used for desalting of protein solutions.

An important parameter to consider for size-exclusion chromatography is the concentration of the analyte; it should be as high as practicably possible. The higher the concentration the better the resolution due to reduced diffusion (see Section 5.7.8). Large proteins or protein complexes will pass through the medium and elute first, therefore it is very important to thoroughly equilibrate the column before use. Some amount of material will be lost in a column through dilution and surface interactions, therefore it is important to select the appropriate column size.

5.8.4 Hydrophobicity Interaction Chromatography (HIC)

Proteins differ in the amount of hydrophobic amino acids that are present on their surfaces. This difference can be exploited in salt fractionation (see Section 5.9.2), but can also be used in a higher-resolution method using hydrophobic interaction chromatography (HIC). Typical resins include Butyl- and Phenyl-Sepharose, where butyl or phenyl groups are bonded to the agarose support matrix. The protein mixture is loaded on the column typically under high salt conditions ('salting out'), where due to the hydrophobic effect (see Section 2.2) interactions will occur between the immobilised hydrophobic groups on the resin and hydrophobic regions on the proteins. Proteins are then eluted by applying a decreasing salt gradient to the column and should emerge from the column in order of increasing hydrophobicity. However, some highly hydrophobic proteins may not be eluted, even in the total absence of salt. In this case, it is necessary to add a small amount of water-miscible organic solvent, such as propanol or ethylene glycol to the column buffer solution. This will compete with the proteins for binding to the hydrophobic matrix and will elute any remaining proteins. With individual proteins, it may also be possible to employ a triggered

---

Example 5.2 **ESTIMATION OF RELATIVE MOLECULAR MASS**

Question The molecular mass of a protein was investigated by exclusion chromatography using a Sephacryl S300 column and using aldolase, catalase, ferritin, thyroglobulin and Blue Dextran as standards. The following elution data were obtained:

Protein	Molecular mass M (kDa)	Retention volume V (cm³)
Aldolase	158	22.5
Catalase	210	21.4
Ferritin	444	18.2
Thyroglobulin	669	16.4
Blue Dextran	2000	13.6
Unknown		19.5

What is the approximate molecular mass of the unknown protein?

Answer A plot of the logarithm of the molecular mass of individual proteins versus their retention volume has a linear section from which it can be estimated that the unknown protein with a retention volume of 19.5 cm³ has a molecular mass of 330 kDa.

conformational switch to change between two different states of hydrophobicity (for example, the calcium-myristoyl switch observed in Visinin-like proteins), thus eliminating the requirement for high salt concentrations.

5.9 OTHER METHODS OF PROTEIN PURIFICATION

5.9.1 Stability-Based Purification

Denaturation fractionation exploits differences in the heat sensitivity of proteins. The three-dimensional (tertiary) structure of proteins is maintained by a number of forces, mainly hydrophobic interactions, hydrogen bonds and sometimes disulfide bridges. The denatured state of a protein is characterised by a disruption of some or all of these bonds, thus yielding an unfolded protein chain that is, in most cases, insoluble. One of the easiest ways to denature proteins in solution is to heat them (thermal denaturation). However, different proteins will denature at different temperatures, depending on their different thermal stabilities; this, in turn, is a measure of the number of bonds holding the tertiary structure together. If the protein of interest is particularly heat stable, then heating the extract to a temperature at which the protein is stable yet other proteins denature can be a very useful preliminary step. The temperature at which the protein being purified is denatured is first determined by a small-scale experiment. Once this temperature is known, it is possible to remove more thermo-labile proteins by heating the mixture to a temperature 5–10 °C below this critical temperature for a period of 15–30 min. The denatured, unwanted proteins are then removed by centrifugation. The presence of the substrate, product or a competitive inhibitor of an enzyme often stabilises it (see also differential scanning fluorimetry, Section 14.4.4) and allows an even higher heat denaturation temperature to be employed.

In a similar way, proteins differ in the ease with which they are denatured by extremes of pH (< 3 and > 10). The sensitivity of the protein under investigation to extreme pH is determined by a small-scale trial. The protein mixture is then adjusted to a pH not less than 1 pH unit from that at which the test protein is precipitated. More sensitive proteins will precipitate and are removed by centrifugation.

5.9.2 Solubility-Based Purification

Proteins differ in the balance of charged, polar and hydrophobic amino acids that they display on their surfaces. Charged and polar groups on the surface are solvated by water molecules, thus making the protein molecule soluble, whereas hydrophobic residues are masked by water molecules that are necessarily found adjacent to these regions. Since solubility is a consequence of solvation of charged and polar groups on the surfaces of the protein, it follows that, under particular fixed conditions, proteins will differ in their solubilities. In particular, it is possible to exploit the fact that proteins precipitate differentially from solution upon the addition of species such as neutral salts or organic solvents ('salting out', solubility fractionation). It should be stressed here that these methods precipitate native (i.e. active) protein by aggregation; the insoluble protein typically does not denature during this process.

Salt Fractionation

Salt fractionation is frequently carried out using ammonium sulfate. With increasing salt concentration, freely available water molecules that can solvate the salt ions become scarce. Water molecules that have been forced into contact with hydrophobic groups on the surface of a protein are increasingly deployed (rather than those involved in solvating polar groups on the protein surface, which are bound by electrostatic interactions and are far less easily given up). Therefore, more and more water molecules are removed from the hydrophobic surface areas of the protein, thus exposing the hydrophobic patches. The exposed hydrophobic patches cause proteins to aggregate by hydrophobic interaction, resulting in precipitation. The first proteins to aggregate are therefore those with the most hydrophobic residues on the surface, followed by those with fewer hydrophobic residues. Clearly, in protein mixtures, the aggregates formed are made of mixtures of more than one protein. Individual identical molecules do not seek out each other, but simply bind to another adjacent molecule with an exposed hydrophobic patch. However, many proteins are precipitated from solution over a narrow range of salt concentrations, making this a suitably simple procedure for enriching the proteins of interest.

As an example, Table 5.3 illustrates the purification of a protein using ammonium sulfate precipitation. As increasing amounts of ammonium sulfate are dissolved in a protein solution, certain proteins start to aggregate and precipitate out of solution. By carrying out a controlled pilot experiment where the percentage of ammonium sulfate is increased stepwise from, say, 10% to 20% to 30% etc., the resultant precipitate at each step being recovered by centrifugation, redissolved in buffer and analysed for the protein of interest, it is possible to determine a fractionation procedure that will give a significantly purified sample. In the example shown in Table 5.3, the original homogenate was made in 45% ammonium sulfate and the precipitate recovered and discarded. The supernatant was then made in 70% ammonium sulfate, the precipitate collected, redissolved in buffer, and kept, with the supernatant being discarded. This produced a purification factor of 2.7. As can be seen, a significant amount of protein has been removed at this step (237 g of protein) while 81% of the total enzyme present was recovered, i.e. the yield was good. This step has clearly produced an enrichment of the protein of interest from a large volume of extract and at the same time has concentrated the sample.

Organic Solvent Fractionation

Organic solvent fractionation is based on differences in the solubility of proteins in aqueous solutions containing water-miscible organic solvents such as ethanol, acetone and butanol. The addition of organic solvent effectively 'dilutes out' the water present and reduces the dielectric constant. At the same time, water molecules are deployed to hydrate the organic solvent molecules, thereby removing water molecules from the charged and polar groups on the surface of proteins. This process gradually exposes more and more charged surface groups and leads to aggregation of proteins due to charge (ionic) interactions between molecules. Proteins consequently precipitate in decreasing order of the number of charged groups on their surface as the organic solvent concentration is increased.

Table 5.3 **Example of a protein purification schedule**

Fraction	Volume (cm³)	Protein concentration (mg cm⁻³)	Total protein (mg)	Total activity (10^{-6} katal)	Activity[a] (10^{-9} katal cm⁻³)	Specific activity (katal mg⁻¹)	Purification factor[b]	Overall yield[c] (%)
Homogenate	8500	40	340 000	260	30.6	$7.65 \cdot 10^{-10}$	1	100
45%–70%(NH$_4$)$_2$SO$_4$	530	194	103 000	210	396	$2.04 \cdot 10^{-9}$	2.7	81
CM-cellulose	420	19.5	8190	179	425	$2.19 \cdot 10^{-8}$	28.6	69
Affinity chromatography	48	2.2	105.6	162	3360	$1.53 \cdot 10^{-6}$	2000	62
DEAE-Sepharose	12	2.3	27.6	129	10 761	$4.67 \cdot 10^{-6}$	6105	50

Notes: [a]The unit of enzyme activity (katal) is defined as the amount that produces 1 mol of product per second under standard assay conditions. [b]Defined as: purification factor = (specific activity of fraction)/(specific activity of homogenate). [c]Defined as: overall yield = (total activity of fraction)/(total activity of homogenate). Adapted with permission from Doonan S., Ed. (1996) *Protein Purification Protocols*, Methods in Molecular Biology (Vol 59), Humana Press Inc., Totowa, NJ.

Organic polymers can also be used for the fractional precipitation of proteins. This method resembles organic solvent fractionation in its mechanism of action, but requires lower concentrations to cause protein precipitation and is thus less likely to cause protein denaturation. The most commonly used polymer is polyethylene glycol (PEG), with a relative molecular mass in the range 6000–20 000.

Isoelectric Precipitation Fractionation

Isoelectric precipitation fractionation is based upon the observations that proteins have a solubility minimum at their isoelectric point (pI). At this pH, there are equal numbers of positive and negative charges on the protein molecule; intermolecular repulsions are therefore minimised and protein molecules can approach each other. This therefore allows opposite charges on different molecules to interact, resulting in the formation of insoluble aggregates. The principle can be exploited to remove unwanted protein, by adjusting the pH of the protein extract so as to cause the precipitation of these proteins, but not that of the target protein. Alternatively, the target protein can be removed by adjusting the pH of the extract to the pI of the target protein. In practice, the former alternative is preferable, since some denaturation of the precipitating protein inevitably occurs.

Inclusion Body Purification

Finally, an unusual solubility phenomenon can be utilised in some cases for protein purification from bacteria. Early workers who were over-expressing heterologous proteins in *Escherichia coli* at high levels were alarmed to discover that, although their protein was expressed in high yield (up to 40% of the total cell protein), the protein aggregated to form insoluble particles that became known as inclusion bodies. Initially this was seen as a major impediment to the production of proteins in bacteria, since the inclusion bodies effectively carry a mixture of monomeric and polymeric denatured proteins formed by partial or incorrect folding, probably due to the reducing environment of the cytoplasm. However, it was soon realised that this phenomenon could be used to advantage in protein purification. The inclusion bodies can be separated from a large proportion of the bacterial cytoplasmic protein by centrifugation, giving an effective purification step. The recovered inclusion bodies must then be solubilised and denatured, typically by addition of 6 M guanidinium hydrochloride (as a chaotropic agent) in the presence of a reducing agent to disrupt any disulfide bridges. For refolding, the denatured protein is then either diluted into or dialysed against a suitable buffer to attain the active, native conformation. The efficiency of the refolding step varies widely with individual proteins. In some cases, reasonably high yields of pure, refolded protein are obtained. However, there are also other cases, where the refolding of denatured protein proves challenging and/or rather inefficient.

5.10 MONITORING PROTEIN PURIFICATION

The purification of a protein invariably involves the application of several individual steps, each of which generates a relatively large number of fractions containing buffer and protein, either eluted from chromatography columns (Section 5.8) or obtained from other fractionation methods (Section 5.9). It is thus necessary to determine how much protein is present in each individual fraction so that an elution or separation profile (a plot of protein concentration versus fraction number) can be produced. Appropriate methods for detecting and quantifying protein in solution are summarised in Table 5.1. However, the estimation of protein content only informs about the total amount of protein present and does not allow conclusions to be drawn as to the identity of individual proteins in a mixture or indeed the presence of the desired target protein.

Therefore, methods are required to determine which fractions contain the protein of interest so that their contents can be pooled and progressed to the next purification step. The most easily accessible method is to analyse aliquots of each fraction by gel electrophoresis (SDS-PAGE; see Section 6.3.1 and the examples in Figures 5.7 and 5.8). If an antibody to the protein of interest is available, then samples from each fraction can be dried onto nitrocellulose and the antibody used to detect the protein-containing fractions using the dot blot method (Section 4.9.2). Alternatively, an immunoassay such as ELISA or radioimmunoassay (Section 7.3.1) may be employed.

For recombinant proteins that are expressed as a fusion protein, i.e. linked to a tag that aids purification by an affinity step (Section 5.3.1), the presence of the target protein can be assessed by evaluating binding and release to/from the affinity resin.

An alternative approach that can be used for cloned genes that are expressed in cells is to express the protein as a fusion protein with a second protein that can be easily assayed, such as, for example, using a simple colorimetric assay, as is the case with β-galactosidase. The presence of the protein of interest can be detected by the presence of the linked β-galactosidase activity.

If the target protein is an enzyme, each fraction can be analysed by evaluating the presence or absence of enzymatic activity. A successful fractionation step is recognised by an increase in the specific activity of the sample, where the specific activity of the enzyme is given by the enzyme activity (see Equation 23.1) in relation to the total mass of protein present in the preparation:

$$\text{Specific activity} = \frac{\text{Enzyme activity}}{m(\text{total protein in fraction})} = \frac{\frac{\Delta n(\text{product formed})}{\Delta t}}{m(\text{total protein in fraction})} \quad \text{(Eq 5.17)}$$

The measurement of units of an enzyme relies on an appreciation of certain basic kinetic concepts and upon the availability of a suitable analytical procedure (discussed in more detail in Chapter 23).

For a purification step to be successful, the specific activity of the sample must be greater after the purification step than it was before. This increase is best represented as the fold purification:

$$\text{Fold purification} = \frac{\text{specific activity after purification}}{\text{specific activity before purification}} \quad \text{(Eq 5.18)}$$

Additionally, another important factor is the yield of the step; obviously, it is not very useful to have an increased specific activity, while losing a substantial amount of the target protein. The yield is defined as follows:

$$\text{Yield} = \frac{\text{units of enzyme after purification}}{\text{units of enzyme in original preparation}} \qquad \text{(Eq 5.19)}$$

A yield of 70% or more in any purification step would normally be considered to be acceptable. As an example, Table 5.3 shows how yield and specific activity vary during a purification schedule.

Example 5.3 **ENZYME FRACTIONATION**

Question A tissue homogenate was prepared from pig heart tissue as the first step in the preparation of the enzyme aspartate aminotransferase (AAT). Cell debris was removed by filtration and nucleic acids removed by treatment with polyethyleneimine, leaving a total extract (solution A) of 2 dm³. A sample of this extract (50 mm³) was added to 3 cm³ of buffer in a 1 cm pathlength cuvette and the absorbance at 280 nm observed as 1.7.

(i) Determine the approximate protein concentration in the extract, assuming that an absorbance of 1.0 at 280 nm in a cuvette of 1 cm pathlength indicates a concentration of 1 mg cm⁻³ protein. Determine the total protein content of the extract.

(ii) One unit of AAT enzyme activity is defined as the amount of enzyme in 3 cm³ of substrate solution that causes an absorbance change at 260 nm of 0.1 min⁻¹. To determine enzyme activity, 100 mm³ of extract was added to 3 cm³ of substrate solution and an absorbance change of 0.08 min⁻¹ was recorded. Determine the number of units of AAT actively present per cm³ of extract A, and hence the total number of enzyme units in the extract.

(iii) The initial extract (solution A) was then subjected to ammonium sulfate fractionation. The fraction precipitating betweeen 50% and 70% saturation was collected and redissolved in 120 cm³ of buffer (solution B). A 5 mm³ aliquot of solution B was added to 3 cm³ of buffer and the absorbance at 280 nm determined to be 0.89 using a 1 cm pathlength cuvette. Determine the protein concentration, and hence total protein content, of solution B.

(iv) An aliquot of 20 mm³ of solution B was used to assay for AAT activity and an absorbance change of 0.21 per minute at 260 nm was recorded. Determine the number of AAT units per cm⁻³ in solution B and hence the total number of enzyme units in solution B.

(v) From your answers to (i) to (iv), determine the specific activity of AAT in both solutions A and B.

(vi) From your answers to question (v), determine the fold purification achieved by the ammonium sulfate fractionation step.

(vii) Finally, determine the yield of AAT following the ammonium sulfate fractionation step.

Answer

(i) Assuming the approximation that a protein solution of the concentration 1 mg cm^{-3} possesses an absorbance of 1.0 at 280 nm using a 1 cm path-length cell, we can deduce that the protein concentration in the cuvette as per 1.0 mg cm^{-3} × 1.7 = 1.7 mg cm^{-3}.

Since 50 mm^3 = 0.05 cm^3 of solution A was added to 3.0 cm^3, the sample had been diluted by a factor of 3.05/0.05 = 61. Therefore, the protein concentration of solution A is 61 × 1.7 mg cm^{-3} = 104 mg cm^{-3}.

Since the total volume of solution A is 2 dm^3 = 2000 cm^3, the *total* amount of protein in solution A is
2000 cm^3 × 104 mg cm^{-3} = 208 000 mg = 208 g.

(ii) Since one enzyme unit causes an absorbance change at 260 nm of 0.1 per minute (in a total volume of 3 cm^3), there were 0.08 min^{-1} / (0.1 min^{-1}) = 0.8 enzyme units in the cuvette. These 0.8 enzyme units came from an aliquot of 100 mm^3 of solution A that was added to the cuvette. Normalisation per volume therefore yields

$$\frac{0.8 \text{ units}}{100 \text{ mm}^3} = \frac{0.8 \text{ units}}{0.1 \text{ cm}^3} = 8.0 \frac{\text{units}}{\text{cm}^3}$$

Since the total volume of solution A is 2 dm^3 = 2000 cm^3, there is a total of 2000 × 8.0 units = 16 000 units in solution A.

(iii) Using the same approach as in (i), the protein concentration of solution B is 3.005 cm^3 / (0.005 cm^3) × 0.89 × 1 mg cm^{-3} = 535 mg cm^{-3}.

Since the total volume of solution B is 120 cm^3, the total mass of protein present in solution B is 120 cm^3 × 535 mg cm^{-3} = 64.2 g.

(iv) Using the same approach as in (ii), there are 0.21 min^{-1}/(0.1 min^{-1}) = 2.1 units of enzyme activity in the cuvette. Since these units came from an aliquot of 20 mm^3 = 0.02 cm^3 volume, and we are interested in the

number of units per 1 cm^3, we need to normalise the result as per
2.1 units / (0.02 cm^{-3}) = 105 units cm^{-3}.

Since the total volume of solution B is 120 cm^3, the total units of enzyme in solution B is
120 cm^3 × 105 units cm^{-3} = 12 600 units.

(v)   For solution A, the specific activity is
16 000 units / (208 000 mg) = 0.077 units mg^{-1}.

For solution B, the specific activity is
12 600 units / (64 200 mg) = 0.196 units mg^{-1}.

(vi)   The fold purification of the ammonium sulfate fractionation step is
0.197 units mg^{-1} / (0.077 units mg^{-1}) = 2.6.

(vii)   The yield of AAT in the ammonium sulfate fractionation step is
12 600 units / (16 000 units) × 100% = 79%.

5.11 STORAGE

Care should be taken when storing proteins; the final use needs to be considered and will guide the most appropriate storage plan. Most preparations can be stored up to 24 h or longer, if kept at 4 °C in appropriate buffers that typically contain about 100 mM of a monovalent salt (see ionic strength; Section 2.3.2). For longer periods of time, low protein concentrations (< 5 mg cm^{-3}) are preferred and antibacterial agents can be used (e.g. sodium azide). Frequently, proteins are stored longer term at temperatures below 0 °C. In these cases, protein samples should be dialysed into the buffer required for the final use and are best snap-frozen in aliquots (to avoid freeze/thaw cycles) by placing into liquid nitrogen or onto dry ice. Glycerol is typically added to the purified protein to prevent the formation of ice crystals. The stability of frozen protein samples is different for every protein, but, in general, many can be stored at −20 °C for months; longer periods of time require −80 °C. For the end use, frozen samples should be thawed quickly to avoid pH gradients forming when buffers are differentially warmed up, as many buffers change pH depending on the temperature.

Alternatively, one can freeze-dry proteins to produce a lyophilised pellet. This generally involves snap-freezing the protein sample and applying a strong vacuum until all water is removed. However, not all proteins will retain their structure and/or activity throughout this procedure.

5.12 SUGGESTIONS FOR FURTHER READING

5.12.1 Experimental Protocols

Chong S., Mersha F.B., Comb D.G., *et al.* (1997) Single-column purification of free recombinant proteins using a self-cleavable affinity tag derived from a protein splicing element. *Gene* 192, 271–281.

Diaz A.A., Tomba E., Lennarson R., *et al.* (2009) Prediction of protein solubility in *Escherichia coli* using logistic regression. *Biotechnology and Bioengineering* 105, 374–383.

Structural Genomics Consortium *et al.* (2008) Protein production and purification. *Nature Methods* 5, 135–146.

Studier F.W. (2014) *Stable expression of clones and auto-induction for protein production in E. coli.* Methods in Molecular Biology (Vol 1091), Springer, New York, USA, pp 17–32.

5.12.2 General Texts

Cutler P. (2004) *Protein Purification Protocols*, Humana Press, Totowa, NJ, USA.

Nedelkov D. (2006) *New and Emerging Proteomics Techniques*, Humana Press, New York, USA.

Simpson R.J., Adams P.D. and Golemis E.A. (2008) *Basic Methods in Protein Purification and Analysis: A Laboratory Manual*, CSH Press, New York, USA.

Thompson J.D. (2008) *Functional Proteomics*, Humana Press, New York, USA.

Walker J.M. (2005) *Proteomics Protocols Handbook*, Humana Press, Totowa, NJ, USA.

5.12.3 Review Articles

Stevens R.C. (2000) Design of high-throughput methods of protein production for structural biology. *Structure* 8, R177–R185.

5.12.4 Websites

FLAG® system (Sigma-Aldrich)
www.sigmaaldrich.com/life-science/proteomics/recombinant-protein-expression/purification-detection/flag-system.html (accessed April 2017)

Intein Mediated Purification (IMPACT™; New England BioLabs)
www.neb.com/products/e6901-impact-kit (accessed April 2017)

Protein purification handbooks (GE Healthcare Life Sciences)
www.gelifesciences.com/handbooks (accessed April 2017)

6 Electrophoretic Techniques

RALPH RAPLEY

6.1 GENERAL PRINCIPLES

The term electrophoresis describes the migration of a charged particle under the influence of an electric field. Many important biological molecules, such as amino acids, peptides, proteins, nucleotides and nucleic acids, possess ionisable groups and, therefore, at any given pH, exist in solution as electrically charged species, either cations (positively charged) or anions (negatively charged). Under the influence of an electric field these charged particles will migrate either to the cathode or to the anode, depending on the nature of their net charge.

The equipment required for electrophoresis consists basically of two items, a power pack and an electrophoresis unit. Electrophoresis units are available for running either vertical or horizontal gel systems. Vertical slab gel units are commercially available and routinely used to separate proteins in acrylamide gels (Section 6.2). The gel is formed between two glass plates that are clamped together, but held apart by plastic spacers. A commonly used equipment in this context is the so-called minigel apparatus (Figure 6.1). Gel dimensions are typically 8.5 cm wide × 5 cm high, with a thickness of 0.5–1 mm. A plastic comb is placed in the gel solution and is removed after polymerisation to provide loading wells for up to 10 samples. When the apparatus is assembled, the lower electrophoresis tank buffer surrounds the gel plates and affords some cooling of the gel plates. A typical horizontal gel system is shown in Figure 6.2. The gel is cast on a glass or plastic sheet and placed on a cooling plate (an insulated surface through which cooling water is passed to conduct away generated heat). Connection between the gel and electrode buffer is made using a thick wad of wetted filter paper (Figure 6.2); note, however, that agarose gels for DNA electrophoresis are run submerged in the buffer

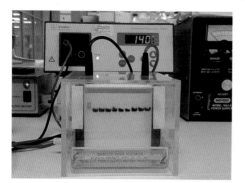

Figure 6.1 Photograph showing an assembled SDS-PAGE minigel. Nine wells that have been loaded can be identified by the blue dye (bromophenol blue) that is incorporated into the loading buffer. The outermost left lane was loaded with a protein marker.

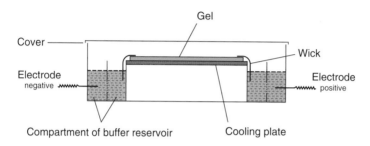

Figure 6.2 A typical horizontal apparatus, such as that used for immunoelectrophoresis, isoelectric focussing and the electrophoresis of DNA and RNA in agarose gels.

(Section 6.4.1). The power pack supplies a direct current between the electrodes in the electrophoresis unit. All electrophoresis is carried out in an appropriate buffer, which is essential to maintain a constant state of ionisation of the molecules being separated. Any variation in pH would alter the overall charge and hence the mobilities (rate of migration in the applied field) of the molecules being separated.

In order to understand fully how charged species separate, it is necessary to look at some simple equations relating to electrophoresis. When a **potential difference** ΔV (voltage) is applied across the electrodes, it generates a potential gradient, $\bar{E}$, which is the applied voltage, divided by the distance, d, between the electrodes. When this potential gradient $\bar{E}$ is applied, the force on a molecule bearing a charge of Q (measured in coulombs) is

$$\bar{F} = \bar{E} \times Q \tag{Eq 6.1}$$

It is this force (measured in newtons) that drives a charged molecule towards an electrode. However, there is also a frictional resistance that retards the movement of this charged molecule. This frictional force depends on the hydrodynamic size, as well as the shape of the molecule, the pore size of the medium in which electrophoresis is taking place and the viscosity of the buffer. The velocity, v, of a charged molecule in an electric field is therefore given by the equation:

$$v = \frac{E \times Q}{f} = \frac{\Delta V \times Q}{d \times f} \tag{Eq 6.2}$$

where f is the frictional coefficient.

More commonly, the term electrophoretic mobility (u; see also Section 6.5) of an ion is used, which is the ratio of the velocity of the ion to field strength (v/E). When a potential difference is applied, therefore, molecules with different overall charges will begin to separate owing to their different electrophoretic mobilities. Even molecules with similar charges will begin to separate if they have different molecular sizes, since they will experience different frictional forces. As will be seen below, some forms of electrophoresis rely almost totally on the different charges on molecules to effect separation, whilst other methods exploit differences in molecular size and therefore encourage frictional effects to bring about separation.

Provided the electric field is removed before the molecules in the sample reach the electrodes, the components will have been separated according to their electrophoretic mobilities. Electrophoresis is thus an incomplete form of electrolysis. The separated samples are then located by staining with an appropriate dye or by autoradiography (Section 9.3.3) if the sample is radiolabelled.

The current in the solution between the electrodes is conducted mainly by the buffer ions, a small proportion being conducted by the sample ions. Ohm's law expresses the relationship between current (I), voltage (ΔV) and resistance (R):

$$\frac{\Delta V}{I} = R \qquad \text{(Eq 6.3)}$$

It therefore appears that it is possible to accelerate an electrophoretic separation by increasing the applied voltage, which would result in a corresponding increase in the current flowing. The distance migrated by the ions will be proportional to both current and time. However, this would ignore one of the major problems for most forms of electrophoresis, namely the generation of heat.

During electrophoresis, the power P (measured in watts) generated in the supporting medium is given by

$$P = I^2 \times R \qquad \text{(Eq 6.4)}$$

most of which is dissipated as heat. Heating of the electrophoretic medium has the following effects:

- An increased rate of diffusion of sample and buffer ions, leading to broadening of the separated samples.
- The formation of convection currents, which leads to mixing of separated samples.
- Thermal instability of samples that are rather sensitive to heat. This may include denaturation of proteins (and thus the loss of enzyme activity).
- A decrease of buffer viscosity, and hence a reduction in the resistance of the medium.

If a constant voltage is applied, the current increases during electrophoresis owing to the decrease in resistance (see Ohm's law, Equation 6.3) and the rise in current increases the heat output still further. For this reason, workers often use a stabilised power supply, which provides constant power and thus eliminates fluctuations in heating.

Constant heat generation is, however, a problem. The answer might appear to be to run the electrophoresis at very low power (low current) to overcome any heating problem, but this can lead to poor separation of sample components as a result of the increased amount of diffusion resulting from long separation times. Compromise conditions, therefore, have to be found with reasonable power settings, to give acceptable separation times, and an appropriate cooling system, to remove liberated heat. While such systems work fairly well, the effects of heating are not always totally eliminated. For example, for electrophoresis carried out in cylindrical tubes or in slab gels, although heat is generated uniformly through the medium, heat is removed only from the edges, resulting in a temperature gradient within the gel, the temperature at the centre of the gel being higher than that at the edges. Since the warmer fluid at the centre is less viscous, electrophoretic mobilities are therefore greater in the central region (electrophoretic mobilities increase by about 2% for each 1 °C rise in temperature), and electrophoretic zones develop a bowed shape, with the zone centre migrating faster than the edges.

A final factor that can affect electrophoretic separation is the phenomenon of electroendosmosis (also known as electro-osmotic flow), which is due to the presence of charged groups on the surface of the support medium. For example, paper has some carboxyl groups present, agarose (depending on the purity grade) contains sulfate groups and the surface of glass walls used in capillary electrophoresis (Section 6.5) contains silanol (Si–OH) groups. Figure 6.3 demonstrates how electroendosmosis occurs in a capillary tube, although the principle is the same for any support medium that has charged groups on it. In a fused-silica capillary tube, above a pH value of about 3, silanol groups on the silica capillary wall will ionise, generating negatively charged sites. It is these charges that generate electroendosmosis. The ionised silanol

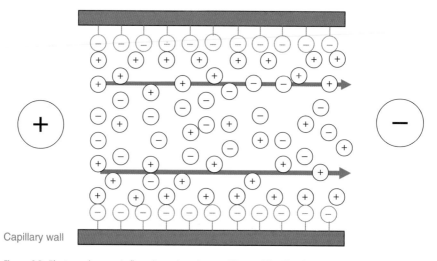

Figure 6.3 Electroendosomotic flow through a glass capillary. Acidic silanol groups impart a negative charge on the capillary walls. Electrolyte cations are thus attracted to the capillary walls, forming an electrical double layer. When a voltage is applied, the cations migrate toward the cathode, dragging solvent along. The resultant net movement of electrolyte solution towards the cathode is known as electroendosmotic flow.

groups create an electrical double layer, or region of charge separation, at the capillary wall/electrolyte interface. When a voltage is applied, cations in the electrolyte near the capillary wall migrate towards the cathode, pulling electrolyte solution with them. This creates a net electroendosmotic flow towards the cathode.

6.2 SUPPORT MEDIA AND BUFFERS

The pioneering work on protein electrophoresis by Arne Tiselius (for which he received the Nobel Prize in Chemistry in 1948) was performed in free solution. However, it was soon realised that many of the problems associated with this approach, particularly the adverse effects of diffusion and convection currents, could be minimised by stabilising the medium. This was achieved by carrying out electrophoresis on a porous mechanical support, which was wetted in electrophoresis buffer and in which electrophoresis of buffer ions and samples could occur. The support medium cuts down convection currents and diffusion so that the separated components remain as sharp zones. The earliest supports used were filter paper or cellulose acetate strips, wetted in electrophoresis buffer. Nowadays, these media are infrequently used, although cellulose acetate still has its uses (see Section 6.3.6). In particular, for many years small molecules such as amino acids, peptides and carbohydrates were routinely separated and analysed by electrophoresis on supports such as paper or thin-layer plates of cellulose, silica or alumina. Although occasionally still used nowadays, such molecules are now more likely to be analysed by more modern and sensitive techniques such as high-performance liquid chromatography (HPLC; see Section 5.7). While paper or thin-layer supports are fine for resolving small molecules, the separation of macromolecules such as proteins and nucleic acids on such supports is poor.

However, the introduction of the use of gels as a support medium led to a rapid improvement in methods for analysing macromolecules. The earliest gel system to be used was the starch gel and, although this still has some uses, the vast majority of electrophoretic techniques used nowadays involve either agarose gels or polyacrylamide gels.

6.2.1 Agarose Gels

Agarose is a linear polysaccharide (average relative molecular mass about 12 000) made up of the basic repeat unit agarobiose, which comprises alternating units of galactose and 3,6-anhydrogalactose (Figure 6.4). Agarose is one of the components of agar, which is a mixture of polysaccharides isolated from certain seaweeds. Agarose is usually used at concentrations of between 1% and 3%. Agarose gels are formed by suspending dry agarose in aqueous buffer, then boiling the mixture until a clear solution forms. This is poured and allowed to cool to room temperature to form a rigid gel. The gelling properties are attributed to both inter- and intramolecular hydrogen bonding within and between the long agarose chains. This cross-linked structure gives the gel good anticonvectional properties. The pore size in the gel is controlled by the initial concentration of agarose; large pore sizes are formed from low concentrations and smaller pore sizes are formed from the higher concentrations. Although essentially free from charge, substitution of the alternating sugar residues with carboxyl, methoxyl, pyruvate and especially sulfate groups occurs to varying degrees.

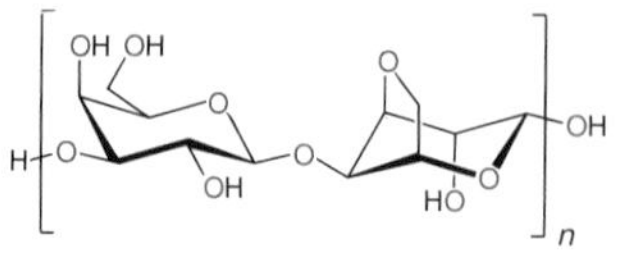

Figure 6.4 Agarobiose, the repeating unit of agarose.

This substitution can result in electroendosomosis during electrophoresis and ionic interactions between the gel and sample in all uses, both unwanted effects. Agarose is therefore sold in different purity grades, based on the sulfate concentration – the lower the sulfate content, the higher the purity.

Agarose gels are used for the electrophoresis of both proteins and nucleic acids. For proteins, the pore sizes of a 1% agarose gel are large relative to the sizes of proteins. Agarose gels are therefore used in techniques such as flat-bed isoelectric focussing (Section 6.3.4), where the proteins are required to move unhindered in the gel matrix according to their native charge. Such large pore gels are also used to separate much larger molecules such as DNA or RNA, because the pore sizes in the gel are still large enough for DNA or RNA molecules to pass through the gel. In these cases, however, the pore size and molecule size are more comparable and fractional effects begin to play a role in the separation of these molecules (Section 6.4). A further advantage of using agarose is the availability of low melting temperature agarose (62–65 °C). As the name suggests, these gels can be reliquified by heating to 65 °C and thus, for example, DNA samples separated in a gel can be cut out of the gel, returned to solution and recovered.

Owing to the poor elasticity of agarose gels and the consequent problems of removing them from small tubes, the gel rod system sometimes employed for acrylamide gels is not used. Horizontal slab gels are invariably used for isoelectric focussing or immunoelectrophoresis in agarose. Horizontal gels are also used routinely for DNA and RNA gels (Section 6.4), although vertical systems have been used by some workers.

Different buffers can be used and are selected depending on the samples to be analysed. The most commonly used buffer for separating DNA is TRIS-acetate containing EDTA (TAE) with a typical pH of 8.3 and used to dissolve the matrix (agarose) as well as the running buffer. This buffer suits the separation of dsDNA and allows rapid run times, even though it is a weak buffer and can warm considerably. For separation of RNA, TRIS-borate with EDTA (TBE) is typically used due to superior buffering capacity (beneficial for long run times). However, borate acts as an inhibitor for many enzymes, which is problematic for downstream enzymatic steps such as ligation reactions in molecular biology. TBE gives superior separation of smaller fragments (< 2 kb) in comparison to TAE and is often used for separation of small RNA molecules such as microRNA.

6.2.2 Polyacrylamide Gels

Electrophoresis in acrylamide gels is frequently referred to as PAGE, being an abbreviation for PolyAcrylamide Gel Electrophoresis. Cross-linked polyacrylamide gels are formed from the polymerisation of acrylamide monomer in the presence of smaller amounts of N,N´-methylene-bis-acrylamide (normally referred to as 'bis-acrylamide') (Figure 6.5). Note that bis-acrylamide is essentially two acrylamide molecules linked by a methylene group, and is used as a cross-linking agent. Acrylamide monomer is polymerised in a head-to-tail fashion into long chains and occasionally a bis-acrylamide molecule is built into the growing chain, thus introducing a second site for chain extension. Proceeding in this way, a cross-linked matrix of fairly well-defined structure is formed (Figure 6.5). The polymerisation of acrylamide is an example of free-radical catalysis, and is initiated by the addition of ammonium persulfate and the base N,N,N',N'-tetramethylenediamine (TEMED). TEMED catalyses the decomposition of the persulfate ion to give a free radical (i.e. a molecule with an unpaired electron):

$$S_2O_8^{2-} + e^- \longrightarrow SO_4^{2-} + SO_4^{-\bullet} \qquad \text{(Eq 6.5)}$$

If this free radical is represented as $R^\bullet$ (where the dot represents an unpaired electron) and M as an acrylamide monomer molecule, then the polymerisation proceeds as follows:

$$R^\bullet \quad + \quad M \longrightarrow RM^\bullet$$

$$RM^\bullet \quad + \quad M \longrightarrow RMM^\bullet \qquad \text{(Eq 6.6)}$$

$$RMM^\bullet + \quad M \longrightarrow RMMM^\bullet \text{ etc.}$$

Free radicals are highly reactive species due to the presence of an unpaired electron that needs to be paired with another electron to stabilise the molecule. $R^\bullet$ therefore reacts with M, forming a single bond by sharing its unpaired electron with one from the outer shell of the monomer molecule. This therefore produces a new free radical molecule $R{-}M^\bullet$, which is equally reactive and will attack a further monomer molecule. In this way long chains of acrylamide are built up, being cross-linked by the introduction of the occasional bis-acrylamide molecule into the growing chain. Oxygen reacts with remaining free radicals and therefore all gel solutions are normally degassed (the solutions are briefly placed under vacuum to remove loosely dissolved air) prior to

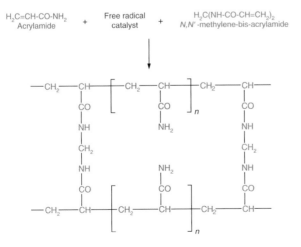

Figure 6.5 The formation of a polyacrylamide gel network from acrylamide and bis-acrylamide.

use. The degassing of the gel solution also serves a second purpose. The polymerisation of acrylamide is an exothermic reaction (i.e. heat is liberated) and the warming up of the gel solution as it sets can liberate air bubbles that become trapped in the polymerised gel. The degassing step prevents this possibility.

Photopolymerisation is an alternative method that can be used to polymerise acrylamide gels. Ammonium persulfate and TEMED are replaced by riboflavin and when the gel is poured it is placed in front of a bright light for 2–3 h. Photodecomposition of riboflavin generates a free radical that initiates polymerisation.

Acrylamide gels are defined in terms of the total percentage of acrylamide present, and the pore size in the gel can be varied by changing the concentrations of both acrylamide and bis-acrylamide. Acrylamide gels can be made with a content of between 3% and 30% acrylamide; low-percentage gels (e.g. 4%) have large pore sizes and are used, for example, in the electrophoresis of proteins, where free movement of the proteins by electrophoresis is required without any noticeable frictional effect, for example in flat-bed isoelectric focussing (Section 6.3.4) or the stacking gel system of an SDS–polyacrylamide gel (Section 6.3.1). Low-percentage acrylamide gels are also used to separate DNA (Section 6.2). Gels of between 10 and 20% acrylamide are used in techniques such as SDS–gel electrophoresis, where the smaller pore size now introduces a sieving effect that contributes to the separation of proteins according to their size (Section 6.3.1).

Proteins were originally separated on polyacrylamide gels that were polymerised in glass tubes, approximately 7 mm in diameter and about 10 cm in length. The tubes were easy to load and run, with minimum apparatus requirements. However, only one sample could be run per tube and, because conditions of separation could vary from tube to tube, comparison between different samples was not always accurate. The later introduction of vertical gel slabs allowed up to 20 samples to be run under identical conditions at the same time. Vertical slabs are used routinely both for the analysis of proteins (Section 6.3) and for the separation of DNA fragments during DNA sequence analysis (Section 6.2). Although some workers prepare their own acrylamide gels, others purchase commercially available ready-made gels for techniques such as SDS–PAGE, native gels and isoelectric focussing (IEF) (see below).

6.3 ELECTROPHORESIS OF PROTEINS

6.3.1 Sodium Dodecyl Sulfate (SDS)-Polyacrylamide Gel Electrophoresis

SDS–polyacrylamide gel electrophoresis (SDS–PAGE) is the most widely used method for analysing protein mixtures qualitatively. It is particularly useful for monitoring protein purification and, because the method is based on the separation of proteins according to size, it can also be used to determine the relative molecular mass of proteins. SDS ($H_3C-(CH_2)_{10}-CH_2-OSO_3^- Na^+$) is an anionic detergent. For denaturing-reducing SDS-PAGE, samples to be run on SDS–PAGE are firstly boiled for 5 min in sample buffer containing β-mercaptoethanol or dithiothreitol (DTT) and SDS. Mercaptoethanol or DTT reduce any disulfide bridges present that are holding together the protein tertiary structure, and the SDS binds strongly to the thermally denatured protein (albeit SDS on its own also causes protein denaturation to some extent). Each protein in the mixture is therefore fully denatured by this treatment and opens up into a rod-shaped structure with a series of negatively charged SDS molecules along the polypeptide chain. On average, one SDS molecule binds for every two amino-acid residues. The original native charge on the molecule is therefore completely swamped by the negatively charged SDS molecules. The rod-like structure remains, as any rotation that tends to fold up the protein chain would result in repulsion between negative charges on different parts of the protein chain, returning the conformation back to the rod shape. The sample buffer also contains an ionisable tracking dye, usually bromophenol blue, that allows the electrophoretic run to be monitored, and sucrose or glycerol, which gives the sample solution density thus allowing the sample to settle easily through the electrophoresis buffer to the bottom when injected into the loading well (see Figure 6.1). Once the samples are all loaded, a current is passed through the gel. In conventional **discontinuous gel electrophoresis**, the samples to be separated are not in fact loaded directly into the main separating gel. When the main separating gel (normally about 5 cm long) has been poured between the glass plates and allowed to set, a shorter (approximately 0.8 cm) stacking gel is poured on top of the separating gel and it is into this gel that the wells are formed and the proteins loaded. The purpose of this stacking gel is to concentrate the protein sample into a sharp band before it enters the main separating gel. This is achieved by utilising differences in ionic strength and pH between the electrophoresis buffer (pH 8.8) and the stacking gel buffer (pH 6.8) and involves a phenomenon known as **isotachophoresis**. The stacking gel has a very large pore size (4% acrylamide), which allows the proteins to move freely and concentrate, or stack, under the effect of the electric field. The band-sharpening effect relies on the fact that negatively charged glycinate ions (in the electrophoresis buffer) have a lower electrophoretic mobility than do the protein–SDS complexes, which, in turn, have lower mobility than the chloride ions (Cl^-) of the loading buffer and the stacking gel buffer. When the current is switched on, all the ionic species have to migrate at the same speed, otherwise there would be a break in the electrical circuit. The glycinate ions can move at the same speed as Cl^- ions only if they are in a region of higher field strength. Field strength is inversely proportional to conductivity, which is proportional to concentration. The result is that the three species of interest adjust their concentrations so that $[Cl^-] > [protein–SDS] > [glycinate]$. There is only a small quantity of protein–SDS complexes, so they concentrate in a very tight band between glycinate and Cl^- boundaries. Once the glycinate reaches the separating gel it becomes more ionised in the higher pH environment and its mobility increases. Thus, the interface between glycinate and Cl^- leaves behind the protein–SDS complexes, which are left to electrophorese at their own rates. The negatively charged protein–SDS complexes now continue to move towards the anode, and, because they have the same charge per unit length, they travel into the separating gel under the applied electric field with the same mobility. However, as they pass through the separating gel the proteins separate, owing to the molecular sieving properties of the gel. Quite simply, the smaller the protein the more easily it can pass through the pores of the gel, whereas large proteins are successively retarded by frictional resistance due to the sieving effect of the gels. Being a small molecule, the dye bromophenol blue migrates in a non-retarded fashion and therefore indicates the electrophoresis front. When the dye reaches the bottom of the gel, the current is turned off, and the gel is removed from between the glass plates and shaken in an appropriate stain solution (usually Coomassie Brilliant Blue, see Section 6.3.7) and then washed in destain solution. The destain solution removes unbound background dye from the gel, leaving stained proteins visible as blue bands on a clear background. A typical minigel would take about 1 h to prepare and set, 40 min to run at 200 V and have a 1 h staining time with Coomassie Brilliant Blue. Upon destaining, strong protein bands would be seen in the gel within 10–20 min, but overnight destaining is needed to completely remove all background stain. Vertical slab gels are invariably run, since this allows up to 10 different samples to be loaded onto a single gel. A typical SDS–polyacrylamide gel is shown in Figure 6.6.

Typically, the separating gel contains between 5 and 15% polyacrylamide. This gives a gel of a certain pore size in which proteins of relative molecular mass (M_r) 10 000 move through the gel relatively unhindered, whereas proteins of M_r = 100 000 can only just enter the pores of this gel. Gels of 15% polyacrylamide are therefore useful for separating proteins in the range M_r = 100 000 to 10 000. However, a protein of M_r = 150 000, for example, would be unable to enter a 15% gel. In this case a larger-pored gel (e.g. a 10% or even 7.5% gel) would be used so that the protein could now enter the gel and be stained and identified. It is obvious, therefore, that the choice of gel to be used depends on the size of the protein being studied. The fractionation range of different percentage acrylamide gels is shown in Table 6.1, illustrating, for example, that in a 10% polyacrylamide gel, proteins greater than 200 kDa in mass cannot enter the gel, whereas proteins with relative molecular mass in the range 200 000 to 15 000 will separate. Proteins of M_r = 15 000 or less are too small to experience the sieving effect of the gel matrix, and all run together as a single band at the electrophoresis front.

More recently, a system comprising only a single gel, as opposed to the conventional stacking and separation gels, has been introduced. These single gels are simpler and more convenient to prepare, as the need for two different gel layers has been eliminated. The performance of these gels in PAGE, staining/destaining, and electroblotting is virtually indistinguishable from that of discontinuous gel systems. Different from conventional gels, the matrix of single gels contains three amino acids (serine, glycine and asparagine), yet no SDS. This provides a further advantage of the single gel system, because a prepared gel can be used either as SDS, SDS/denaturing or native gel (see Section 6.3.2), simply depending on the choice of sample preparation and running buffer.

The M_r of a protein can be determined by comparing its mobility with those of a number of standard proteins of known M_r that are run on the same gel. By plotting a

Table 6.1 **The relationship between acrylamide gel concentration and protein fractionation range**

Acrylamide concentration (%)	Protein fractionation range ($M_r \times 10^{-3}$)
5	60–350
10	15–200
15	10–100

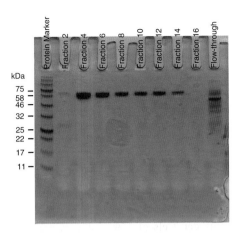

Figure 6.6 A typical Coomassie-stained SDS–polyacrylamide single gel. The first track shows a protein marker, whereas tracks 2–10 were loaded with elution fractions and flow-through (unbound sample) from a protein chromatography experiment.

graph of distance moved against log M_r for each of the standard proteins, a calibration curve can be constructed. The distance moved by the protein of unknown M_r is then measured, and then log M_r and hence M_r can be determined from the calibration curve (see Example 6.1).

SDS–gel electrophoresis is often used after each step of a purification protocol to assess the purity or otherwise of the sample. A pure protein should give a single band on an SDS–polyacrylamide gel, unless the molecule is made up of two unequal subunits. In the latter case, two bands, corresponding to the two subunits, will be seen. Since only submicrogram amounts of protein are needed for the gel, very little material is used in this form of purity assessment and at the same time a value for the relative molecular mass of the protein can be determined on the same gel run (as described above), with no more material being used.

6.3.2 Native (Buffer) Gels

While SDS–PAGE is the most frequently used gel system for studying proteins, the method is of no use if one is aiming to detect a particular protein (often an enzyme) on the basis of its biological activity, because the protein (enzyme) is denatured by the SDS–PAGE procedure. In this case, it is necessary to use non-denaturing conditions. In native or buffer gels, polyacrylamide is again used as a support matrix (normally at 7.5%), but the SDS is absent and the proteins are *not* denatured prior to loading. Since all the proteins in the sample being analysed carry their native charge at the pH of the gel (normally pH 8.7), proteins separate according to their different electrophoretic mobilities *and* the sieving effects of the gel. It is therefore not possible to predict the behaviour of a given protein in a buffer gel, but, because of the range of different charges and sizes of proteins in a given protein mixture, good resolution is achieved. The enzyme of interest can be identified by incubating the gel in an appropriate substrate solution such that a coloured product is produced at the site of the enzyme. An alternative method for enzyme detection is to include the substrate in an agarose gel that is poured over the acrylamide gel and allowed to set. Diffusion and interaction of enzyme and substrate between the two gels results in colour formation at the site of the enzyme. Often, duplicate samples will be run on a gel, the gel cut in half and

Example 6.1   **MOLECULAR MASS DETERMINATION BY ELECTROPHORESIS**

Question   The following table shows the distance moved in an SDS–polyacrylamide gel by a series of marker proteins of known relative molecular mass (M_r). A newly purified protein (X) run on the same gel showed a single band that had moved a distance of 45 mm. What is the M_r of protein X?

Protein	M_r	Distance moved (mm)
Transferrin	78 000	6.0
Bovine serum albumin	66 000	12.5
Ovalbumin (egg albumin)	45 000	32.0
Glyceraldehyde-3-phosphate dehydrogenase	36 000	38.0
Carbonic anhydrase	29 000	50.0
Trypsinogen	24 000	54.0
Soyabean trypsin inhibitor	20 100	61.0
β-Lactoglobulin	18 400[a]	69.0
Myoglobin	17 800	69.0
Lysozyme	14 300	79.0
Cytochrome c	12 400	86.5

Note: [a]β-lactoglobulin has a relative molecular mass of 36 800, but is a dimer of two identical subunits of 18 400 relative molecular mass. Under the reducing conditions of the sample buffer the disulfide bridges linking the subunits are reduced and thus the monomer chains are seen on the gel.

Answer   Construct a calibration graph by plotting log M_r versus distance moved for each of the marker proteins. From this graph you can determine a relative molecular mass for protein X of approximately 31 000. Note that this method is accurate to ±10%, so your answer is 31 000 ± 31.

one half stained for activity, the other for total protein. In this way, the total protein content of the sample can be analysed and the particular band corresponding to the enzyme identified by reference to the activity stain gel.

6.3.3 Gradient Gels

This is again a polyacrylamide gel system, but instead of running a slab gel of uniform pore size throughout (e.g. a 15% gel) a gradient gel is formed, where the acrylamide concentration varies uniformly from, typically, 5% at the top of the gel to 25% acrylamide at the bottom of the gel. The gradient is formed via a gradient mixer and run down between the glass plates of a slab gel. The higher percentage acrylamide (e.g. 25%) is poured between the glass plates first and a continuous gradient of decreasing acrylamide concentration follows. Therefore, at the top of the gel there is a large pore size (5% acrylamide), but as the sample moves down through the gel the acrylamide concentration slowly increases and the pore size correspondingly decreases. Gradient gels are normally run as SDS gels with a stacking gel.

There are two advantages to running gradient gels. First, a much greater range of protein M_r values can be separated than on a fixed-percentage gel. In a complex mixture, very low-molecular-weight proteins travel freely through the gel to begin with, and start to resolve when they reach the smaller pore sizes towards the lower part of the gel. Much larger proteins, on the other hand, can still enter the gel, but start to separate immediately due to the sieving effect of the gel. The second advantage of gradient gels is that proteins with very similar M_r values may be resolved, although they cannot otherwise be resolved in fixed percentage gels. As each protein moves through the gel, the pore sizes become smaller until the protein reaches its pore size limit. The pore size in the gel is now too small to allow passage of the protein, and the protein sample stacks up at this point as a sharp band. A similar-sized protein, but with slightly lower M_r will be able to travel a little further through the gel before reaching its pore size limit, at which point it will form a sharp band. These two proteins, of slightly different M_r values, therefore separate as two, close, sharp bands.

6.3.4 Isoelectric Focussing Gels

This method is ideal for the separation of amphoteric substances (which behave as acids and as bases) such as proteins because it is based on the separation of molecules according to their different isoelectric points (pIs; see Section 2.6). The method has high resolution, being able to separate proteins that differ in their isoelectric points by as little as 0.01 of a pH unit. The most widely used system for isoelectric focussing (IEF) utilises horizontal gels on glass plates or plastic sheets. Separation is achieved by applying a potential difference across a gel that contains a pH gradient. The pH gradient is formed by the introduction into the gel of compounds known as ampholytes, which are complex mixtures of synthetic polyamino-polycarboxylic acids (Figure 6.7). Ampholytes can be purchased in different pH ranges covering either a wide band (e.g. pH 3–10) or various narrow bands (e.g. pH 7–8), and a pH range is chosen such that the samples being separated will have their isoelectric points within this range. Commercially available ampholytes include BioLyte® and Pharmalyte®.

Traditionally, 1–2 mm thick IEF gels have been used by research workers, but the relatively high cost of ampholytes makes this a fairly expensive procedure if a number of gels are to be run. However, the introduction of thin-layer IEF gels, which are only 0.15 mm thick and which are prepared using a layer of electrical insulation tape as the spacer between the gel plates, has considerably reduced the cost of preparing IEF gels, and such gels are now commonly used. Since this method requires the proteins to move freely according to their charge under the electric field, IEF is carried out in

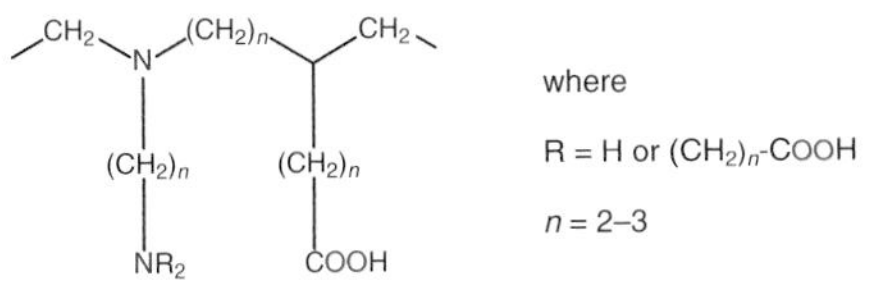

where

R = H or $(CH_2)_n$-COOH

$n = 2$–3

Figure 6.7 The general formula for ampholytes.

low-percentage gels to avoid any sieving effect within the gel. Polyacrylamide gels (4%) are commonly used, but agarose is also used, especially for the study of high M_r proteins that may undergo some sieving, even in a low-percentage acrylamide gel.

To prepare a thin-layer IEF gel, carrier ampholytes, covering a suitable pH range, and riboflavin are mixed with the acrylamide solution, and the mixture is then poured over a glass plate (typically 25 cm × 10 cm), which contains the spacer. The second glass plate is then placed on top of the first to form the gel cassette, and the gel photopolymerised by placing it in front of a bright light. The photodecomposition of the riboflavin generates a free radical, which initiates polymerisation (Section 6.2.2); this takes 2–3 h. Once the gel has set, the glass plates are prised apart to reveal the gel stuck to one of the glass sheets. Electrode wicks, which are thick (3 mm) strips of wetted filter paper (the anode chamber contains phosphoric acid, the cathode chamber sodium hydroxide) are laid along the long length of each side of the gel and a potential difference applied. Under the effect of this potential difference, the ampholytes form a pH gradient between the anode and cathode. The power is then turned off and samples applied by laying on the gel small squares of filter paper soaked in the sample. A voltage is again applied for about 30 min to allow the sample to electrophorese off the paper and into the gel, at which time the paper squares can be removed. Depending on which point on the pH gradient the sample has been loaded, proteins that are initially at a pH region below their isoelectric point will be positively charged and will initially migrate towards the cathode. As they proceed, however, the surrounding pH will be steadily increasing, and therefore the positive charge on the protein will decrease correspondingly until eventually the protein arrives at a point where the pH is equal to its isoelectric point. The protein will now be in the zwitterion form with no net charge, so further movement will cease. Likewise, substances that are initially at pH regions above their isoelectric points will be negatively charged and will migrate towards the anode until they reach their isoelectric points and become stationary. It can be seen that as the samples will always move towards their isoelectric points it is not critical where on the gel they are applied. To achieve rapid separations (2–3 h) relatively high voltages (up to 2500 V) are used. As considerable heat is produced, gels are run on cooling plates (10 °C) and power packs used to stabilise the power output to minimise thermal fluctuations. Following electrophoresis, the gel must be stained to detect the proteins. However, this cannot be done directly, because the ampholytes will stain too, giving a totally blue gel. The gel is therefore first washed with fixing solution (e.g. 10% (v/v) trichloroacetic acid). This precipitates the proteins in the gel and allows the much smaller ampholytes to be washed out. The gel is stained with Coomassie Brilliant Blue and then destained (Section 6.3.7). A typical IEF gel is shown in Figure 6.8.

The pI of a particular protein may be determined conveniently by running a mixture of proteins of known isoelectric points on the same gel. A number of mixtures of proteins with differing pI values are commercially available, covering the pH range 3.5–10. After staining, the distance of each band from one electrode is measured and a graph of distance for each protein against its pI (effectively the pH at that point) plotted. By means of this calibration line, the pI of an unknown protein can be determined from its position on the gel.

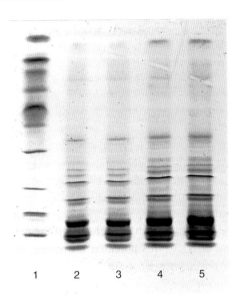

Figure 6.8 A typical isoelectric focussing gel. Track 1 contains a mixture of standard proteins of known isoelectric points. Tracks 2–5 show increasing loadings of venom from the Japanese water moccasin snake. (Courtesy of Bio-Rad Laboratories Ltd.)

IEF is a highly sensitive analytical technique and is particularly useful for studying microheterogeneity in a protein. For example, a protein may show a single band on an SDS gel, but may show three bands on an IEF gel. This may occur, for example, when a protein exists in mono-, di- and tri-phosphorylated forms. The difference of a couple of phosphate groups has no significant effect on the overall relative molecular mass of the protein, hence a single band on SDS gels, but the small charge difference introduced on each molecule can be detected by IEF.

The method is particularly useful for separating isoenzymes (Section 23.1.2), which are different forms of the same enzyme, often differing by only one or two amino-acid residues. Since the proteins are in their native forms, enzymes can be detected in the gel either by washing the unfixed and unstained gel in an appropriate substrate or by overlayering with agarose containing the substrate. The approach has found particular use in forensic science, where traces of blood or other biological fluids can be analysed and compared according to the composition of certain isoenzymes.

Although IEF is used mainly for analytical separations, it can also be used for preparative purposes. In vertical column IEF, a water-cooled vertical glass column is used, filled with a mixture of ampholytes dissolved in a sucrose solution containing a density gradient to prevent diffusion. When the separation is complete, the current is switched off and the sample components run out through a valve in the base of the column. Alternatively, preparative IEF can be carried out in beds of granulated gel, such as Sephadex® G-75.

6.3.5 Two-Dimensional Polyacrylamide Gel Electrophoresis

This technique combines the technique of IEF (first dimension), which separates proteins in a mixture according to charge (pI), with the size separation technique of SDS–PAGE (second dimension). The combination of these two techniques to give two-dimensional (2D) PAGE provides a highly sophisticated analytical method for analysing protein mixtures. To maximise separation, most workers use large format 2D gels (20 cm × 20 cm), although the minigel system can be used to provide useful separation in some cases. For large-format gels, the first dimension (isoelectric focussing) is carried out in an acrylamide gel that has been cast on a plastic strip (18 cm × 3 mm wide). The gel contains ampholytes (for forming the pH gradient) together with 8 M urea and a non-ionic detergent, both of which denature and maintain the solubility of the proteins being analysed. The denatured proteins therefore separate in this gel according to their isoelectric points. The IEF strip is then incubated in a sample buffer containing SDS (which associates with the denatured proteins) and then placed between the glass plates of, and on top of, a previously prepared 10% SDS–PAGE gel. Electrophoresis is commenced and the SDS-bound proteins run into the gel and separate according to size, as described in Section 6.3.1. The IEF gels are provided as dried strips and need rehydrating overnight. The first-dimension IEF run takes 6–8 h, the equilibration step with SDS sample buffer about 15 min, and then the SDS–PAGE step takes about 5 h. A typical 2D gel is shown in Figure 6.9. Using this method, one can routinely resolve between 1000 and 3000 proteins from a cell or tissue extract, and in some cases workers have reported the separation of between 5000 and 10 000 proteins.

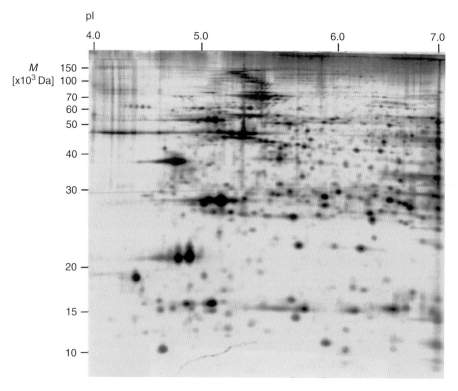

Figure 6.9 A typical two-dimensional gel. The sample applied was 100 μg of total protein extracted from a normal dog heart ventricle. The first dimension was carried out using a pH 4–7 isoelectric focussing gel. The second dimension was a 12% SDS–PAGE vertical slab gel. The pattern was visualised by silver staining. (Courtesy of Monique Heinke and Dr Mike Dunn, Division of Cardiothoracic Surgery, Imperial College School of Medicine, Heart Science Centre, Harefield, UK.)

6.3.6 Cellulose Acetate Electrophoresis

Although one of the older methods, cellulose acetate electrophoresis still has a number of applications. In particular, it has retained a use in the clinical analysis of serum samples (see Section 10.2.1), but is also used in laminar flow devices. Cellulose acetate has the advantage over paper in that it is a much more homogeneous medium, with uniform pore size, and does not adsorb proteins in the way that paper does. There is therefore much less trailing of protein bands and resolution is better, although nothing like as good as that achieved with polyacrylamide gels. The method is, however, far simpler to set up and run. Single samples are normally run on cellulose acetate strips (2.5 cm × 12 cm), although multiple samples are frequently run on wider sheets. The cellulose acetate is first wetted in electrophoresis buffer (pH 8.6 for serum samples) and the sample (1–2 mm³) loaded as a 1 cm wide strip about one-third of the way along the strip. The ends of the strip make contact with the electrophoresis buffer tanks via a filter paper wick that overlaps the end of the cellulose acetate strip, and electrophoresis is conducted at 6–8 V cm⁻¹ for about 3 h. Following electrophoresis, the strip is stained for protein (see Section 6.3.7), destained, and the bands visualised. A typical serum protein separation shows about six major bands. However, in many disease states, this serum protein profile changes and a clinician can obtain information concerning the disease state of a patient from the altered pattern. Although still frequently used for serum analysis, electrophoresis on cellulose acetate is being replaced by the use of agarose gels, which give similar, but somewhat better, resolution. A typical example of the analysis of serum on an agarose gel is shown in Figure 6.10. Similar patterns are obtained when cellulose acetate is used.

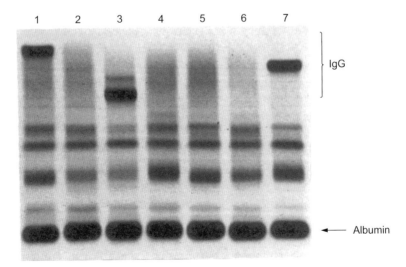

Figure 6.10 Electrophoresis of human serum samples on an agarose gel. Tracks 2, 4, 5 and 6 show normal serum protein profiles. Tracks 1, 3 and 7 show myeloma patients, who are identified by the excessive production of a particular monoclonal antibody seen in the IgG fraction. (Courtesy of Charles Andrews and Nicholas Cundy, Edgware General Hospital, London.)

Enzymes can easily be detected in samples subjected to electrophoresis on cellulose acetate, by using the zymogram technique. The cellulose strip is laid on a strip of filter paper soaked in buffer and substrate. After an appropriate incubation period, the strips are peeled apart and the paper zymogram treated accordingly to detect enzyme product; hence, it is possible to identify the position of the enzyme activity on the original strip. An alternative approach to detecting and semi-quantifying *any* particular protein on a strip is to treat the strip as the equivalent of a protein blot and to probe for the given protein using primary antibody and then enzyme-linked secondary antibody (Section 6.3.8). Substrate colour development indicates the presence of the particular protein and the amount of colour developed in a given time is a semiquantitative measure of the amount of protein. Thus, for example, large numbers of serum samples can be run on a wide sheet, the sheet probed using antibodies, and elevated levels of a particular protein identified in certain samples by increased levels of colour development.

6.3.7 Detection, Estimation and Recovery of Proteins in Gels

The most commonly used general protein stain for detecting protein on gels is the sulfated trimethylamine dye Coomassie Brilliant Blue R-250 (CBB). Staining is usually carried out using 0.1% (w/v) CBB in methanol:water:glacial acetic acid (45:45:10, by volume). This acid–methanol mixture acts as a denaturant to precipitate or fix the protein in the gel, which prevents the protein from being washed out whilst it is being stained. Staining of most gels is accomplished in about 2 h and destaining, usually overnight, is achieved by gentle agitation in the same acid–methanol solution, but in the absence of the dye. The Coomassie stain is highly sensitive; a very weakly staining band on a polyacrylamide gel would correspond to about 0.1 μg (100 ng) of protein. The CBB stain is not used for staining cellulose acetate (or indeed protein blots) because it binds quite strongly to the paper. In this case, proteins are first denatured by brief immersion of the strip in 10% (v/v) trichloroacetic acid, and then immersed in a solution of a dye that does not stain the support material, for example Procion Blue, Amido Black or Procion S.

Although the Coomassie stain is highly sensitive, many workers require greater sensitivity, such as that provided by silver staining. Silver stains are based either on techniques developed for histology or on methods based on the photographic process. In either case, silver ions (Ag^+) are reduced to metallic silver on the protein, where the silver is deposited to give a black or brown band. Silver stains may commence immediately after electrophoresis, or, alternatively, after staining with CBB. With the latter approach, the major bands on the gel can be identified with CBB and then minor bands, not detected with CBB, resolved using the silver stain. The silver stain is at least 100 times more sensitive than CBB, detecting proteins down to 1 ng amounts. Other stains with similar sensitivity include the fluorescent stains SYPRO® Orange (30 ng) and SYPRO® Ruby (10 ng). These dyes have the advantage of binding to proteins in a reversible manner; a useful property for downstream analysis by, for example, mass spectroscopy.

Glycoproteins have traditionally been detected on protein gels by use of the periodic acid–Schiff (PAS) stain. This allows components of a mixture of glycoproteins to be distinguished. However, the PAS stain is not very sensitive and often gives very weak, red-pink bands, difficult to observe on a gel. A far more sensitive method used nowadays is to blot the gel (Section 6.3.8) and use lectins to detect the glyco-proteins. Lectins are protein molecules that bind carbohydrates, and different lectins have been found that have different specificities for different types of carbohydrate. For example, certain lectins recognise mannose, fucose or terminal glucosamine of the carbohydrate side-chains of glycoproteins. The sample to be analysed is run on a number of tracks of an SDS–polyacrylamide gel. Coloured bands appear at the point where the lectins bind if each blotted track is incubated with a different lectin, washed, incubated with a horseradish-peroxidase-linked antibody to the lectin, and then peroxidase substrate added. In this way, by testing a protein sample against a series of lectins, it is possible to determine not only that a protein is a glycoprotein, but to obtain information about the type of glycosylation.

Quantitative analysis (i.e. measurements of the relative amounts of different pro-teins in a sample) can be achieved by scanning densitometry. A number of commer-cial scanning densitometers are available, and work by passing the stained gel track over a beam of light and measuring the transmitted light; standard office desktop scanners can also be used for this purpose. A graphics presentation of protein zones (peaks of absorbance) against migration distance is produced, and peak areas can be calculated to obtain quantitative data (e.g. using the software ImageJ). However, such data must be interpreted with caution because there is only a limited range of pro-tein concentrations over which there is a linear relationship between absorbance and concentration. Also, equal amounts of different proteins do not always stain equally with a given stain, so any data comparing the relative amounts of protein can only be semi-quantitative. An alternative way of obtaining such data is to cut out the stained bands of interest, elute the dye by shaking overnight in a known volume of 50% pyri-dine, and then to measure spectrophotometrically the amount of colour released. More recently gel documentation systems have been developed, which are replacing scan-ning densitometers. Such benchtop systems comprise a video imaging unit (computer linked) attached to a small 'darkroom' unit that is fitted with a choice of white or ultra-violet light (transilluminator). Gel images can be stored on the computer, enhanced accordingly and printed as required on a printer, thus eliminating the need for wet developing in a purpose-built darkroom, as is the case for traditional photography.

Although gel electrophoresis is used generally as an analytical tool, it can be uti-lised to separate proteins in a gel to achieve protein purification. Protein bands can be cut out of protein blots and sequence data obtained by subjecting the blot to mass spectrometric analysis (see Section 21.3). Stained protein bands can be cut out of protein gels and the protein recovered by electrophoresis of the protein out of the gel piece (electroelution). A number of different designs of electroelution cells are com-mercially available, but perhaps the easiest method is to seal the gel piece in buffer in a dialysis sac and place the sac in buffer between two electrodes. Protein will migrate out of the gel piece towards the appropriate electrode, but will be retained by the dial-ysis sac. After electroelution, the current is reversed for a few seconds to drive off any protein that has adsorbed to the wall of the dialysis sac and then the protein solution within the sac is recovered.

6.3.8 Protein (Western) Blotting

Although essentially an analytical technique, PAGE does of course achieve fractionation of a protein mixture during the electrophoresis process. It is possible to make use of this fractionation to examine further individual separated proteins. The first step is to transfer or blot the pattern of separated proteins from the gel onto a sheet of nitrocellulose paper or polyvinylidene fluoride (PVDF) membrane, hereafter referred to as membrane. The method is known as protein blotting, or Western blotting by analogy with Southern blotting (Section 4.9.2), the equivalent method used to recover DNA samples from an agarose gel. Transfer of the proteins from the gel to nitrocellulose is achieved by a technique known as electroblotting. In this method, a sandwich of gel and nitrocellulose is compressed in a cassette and immersed, in buffer, between two parallel electrodes (Figure 6.11).

A current is passed at right angles to the gel, which causes the separated proteins to migrate out of the gel and into the nitrocellulose sheet. The nitrocellulose with its transferred protein is referred to as a blot. Once transferred onto nitrocellulose, the separated proteins can be examined further. This involves probing the blot, usually using an antibody to detect a specific protein. The blot is first incubated in a protein solution, for example 10% (w/v) bovine serum albumin, or 5% (w/v) non-fat dried milk (known as the so-called blotto technique), which will block all remaining hydrophobic binding sites on the nitrocellulose sheet. The blot is then incubated in a dilution of an antiserum (primary antibody) directed against the protein of interest. This immunoglobulin G (IgG) molecule will bind to the blot if it detects its antigen, thus identifying the protein of interest. In order to visualise this interaction, the blot is incubated further in a solution of a secondary antibody, which is directed against the IgG of the species that provided the primary antibody. For example, if the primary antibody was raised in a rabbit, then the secondary antibody would be anti-rabbit IgG. This secondary antibody is appropriately labelled so that the interaction of the secondary antibody with the primary antibody can be visualised on the blot. Anti-species IgG molecules are readily available commercially, with a choice of different labels attached. One of the most common detection methods is to use an

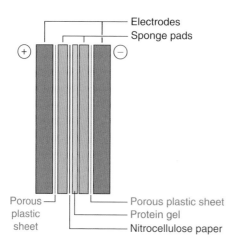

Figure 6.11 Diagrammatic representation of electroblotting. The gel to be blotted is placed on top of a sponge pad saturated in buffer. The nitrocellulose sheet is then placed on top of the gel, followed by a second sponge pad. This sandwich is supported between two rigid porous plastic sheets and held together with two elastic bands. The sandwich is then placed between parallel electrodes in a buffer reservoir and an electric current passed. The sandwich must be placed such that the immobilising medium is between the gel and the anode for SDS–polyacrylamide gels, because all the proteins carry a negative charge.

enzyme-linked secondary antibody (Figure 6.12; see also Section 7.3.2); the principle behind the use of enzyme-linked antibodies to detect antigens in blots is highly analogous to that used in enzyme-linked immunosorbent assays (ELISAs). In this case, following treatment with enzyme-labelled secondary antibody, the blot is incubated in enzyme–substrate solution, when the enzyme converts the substrate into an insoluble coloured product that is precipitated onto the nitrocellulose. The presence of a coloured band therefore indicates the position of the protein of interest. By careful comparisons of the blot with a stained gel of the same sample, the protein of interest can be identified. The enzyme used in enzyme-linked antibodies is usually either alkaline phosphatase, which converts colourless 5-bromo-4-chloro-indolylphosphate (BCIP) substrate into a blue product, or horseradish peroxidase, which, with H_2O_2 as a substrate, oxidises either 3-amino-9-ethylcarbazole into an insoluble brown product, or 4-chloro-1-naphthol into an insoluble blue product. An alternative approach to the detection of horseradish peroxidase is to use the method of **enhanced chemiluminescence** (ECL). In the presence of hydrogen peroxide and the chemiluminescent substrate luminol (Figure 6.13), horseradish peroxidase oxidises the luminol with concomitant production of light, the intensity of which is increased 1000-fold by the presence of a chemical enhancer. The light emission can be detected by exposing the blot to a photographic film. Corresponding ECL substrates are available for use with alkaline-phosphatase-labelled antibodies.

Although enzymes are commonly used as markers for second antibodies, other markers can also be used. These include:

- ^{125}I-labelled secondary antibody: Binding to the blot is detected by autoradiography (Section 9.3.3).
- **Fluorescein-labelled secondary antibody**: The fluorescent label is detected by exposing the blot to ultraviolet light.

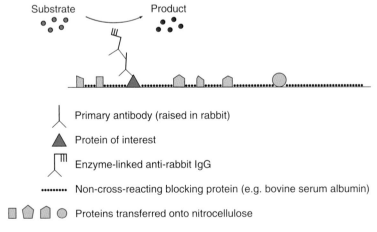

Figure 6.12 The use of enzyme-linked second antibodies in immunodetection of protein blots. First, the primary antibody (e.g. raised in a rabbit) detects the protein of interest on the blot. Second, enzyme-linked anti-rabbit IgG detects the primary antibody. Third, addition of enzyme substrate results in coloured product deposited at the site of the protein of interest on the blot.

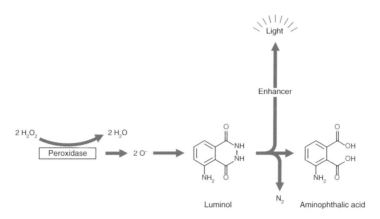

Figure 6.13 The use of enhanced chemiluminescence to detect horseradish peroxidase.

- ^{125}I-labelled protein A: Protein A is purified from *Staphylococcus aureus* and specifically binds to the Fc region of IgG molecules. ^{125}I-labelled protein A is therefore used instead of a second antibody, and binding to the blot is detected by autoradiography.
- Biotinylated secondary antibodies: Biotin is a small-molecular-weight vitamin that binds strongly to the egg protein avidin ($K_d = 10^{-15}$). The blot is incubated with biotinylated secondary antibody, then incubated further with enzyme-conjugated avidin. Since multiple biotin molecules can be linked to a single antibody molecule, many enzyme-linked avidin molecules can bind to a single biotinylated antibody molecule, thus providing an enhancement of the signal. The enzyme used is usually alkaline phosphatase or horseradish peroxidase.
- Quantum dots: These are engineered semiconductor nanoparticles, with diameters of the order of 2–10 nm, which fluoresce when exposed to UV light. Quantum dot nanocrystals comprise a semiconductor core of CdSe surrounded by a shell of ZnS. This crystal is then coated with an organic molecular layer that provides water solubility, and conjugation sites for biomolecules. Typically, therefore, secondary antibodies will be bound to a quantum dot, and the position of binding of the secondary antibody on the blot identified by exposure to UV light.

In addition to the use of labelled antibodies or proteins, other probes are sometimes used. For example, radioactively labelled DNA can be used to detect DNA-binding proteins on a blot. The blot is first incubated in a solution of radiolabelled DNA, then washed, and an autoradiograph of the blot made. The presence of radioactive bands, detected on the autoradiograph, identifies the positions of the DNA-binding proteins on the blot.

6.4 ELECTROPHORESIS OF NUCLEIC ACIDS

6.4.1 Agarose Gel Electrophoresis of DNA

For the majority of DNA samples, electrophoretic separation is carried out in agarose gels. This is because most DNA molecules and their fragments that are analysed routinely are considerably larger than proteins and since such fragments would be unable to enter a polyacrylamide gel, the larger pore size of an agarose gel is required. For example, the commonly used plasmid pBR322 has an M_r of 2.4×10^6. However, rather than using such large numbers, it is more convenient to refer to DNA size in terms of the number of base-pairs. Although, originally, DNA size was referred to in terms of base-pairs (bp) or kilobase-pairs (kbp), it has now become the accepted nomenclature to abbreviate kbp to simply kb when referring to double-stranded DNA. In that nomenclature, pBR322 has a size of 4.36 kb; note that even a small restriction fragment of 1 kb has an M_r of 620 000. When talking about single-stranded DNA it is common to refer to size in terms of nucleotides (nt). Since the charge per unit length (owing to the phosphate groups) in any given fragment of DNA is the same, all DNA samples should move towards the anode with the same mobility under an applied electrical field. However, separation in agarose gels is achieved because of resistance to their movement caused by the gel matrix. The largest molecules will have the most difficulty passing through the gel pores (very large molecules may even be blocked completely), whereas the smallest molecules will be relatively unhindered. Consequently, the mobility of DNA molecules during gel electrophoresis will depend on size, the smallest molecules moving fastest. This is analogous to the separation of proteins in SDS–polyacrylamide gels (Section 6.3.1), although the analogy is not perfect, as double-stranded DNA molecules form relatively stiff rods and while it is not completely understood how they pass through the gel, it is probable that long DNA molecules pass through the gel pores end-on. While passing through the pores, a DNA molecule will experience drag; so the longer the molecule, the more it will be retarded by each pore. Sideways movement may become more important for very small double-stranded DNA and for the more flexible single-stranded DNA. It will be obvious from the above that gel concentrations must be chosen to suit the size range of the molecules to be separated. Gels containing 0.3% agarose will separate double-stranded DNA molecules of between 5 and 60 kb in size, whereas 2% gels are used for samples of between 0.1 and 3 kb. Many laboratories routinely use 0.8% gels, which are suitable for separating DNA molecules in the range 0.5–10 kb. Since agarose gels separate DNA according to size, the M_r of a DNA fragment may be determined from its electrophoretic mobility by running a number of standard DNA markers of known M_r on the same gel. This is most conveniently achieved by running a sample of bacteriophage λ DNA (49 kb) that has been cleaved with a restriction enzyme such as *Eco* RI. Since the base sequence of λ DNA is known, and the cleavage sites for *Eco* RI are known, this generates fragments of accurately known size (Figure 6.14).

DNA gels are invariably run as horizontal, submarine or submerged gels; so named because such a gel is totally immersed in buffer. Agarose, dissolved in gel buffer by boiling, is poured onto a glass or plastic plate, surrounded by a wall of adhesive tape or a plastic frame to provide a gel about 3 mm in depth. Loading wells are formed by placing a plastic well-forming template or comb in the poured gel solution, and removing this comb once the gel has set. The gel is placed in the electrophoresis tank, covered with buffer, and samples loaded by directly injecting the sample into the wells. Samples are prepared by dissolving them in a buffer solution that contains sucrose, glycerol or Ficoll®, which makes the solution dense and allows it to sink to the bottom of the well. A dye such as bromophenol blue is also included in the sample

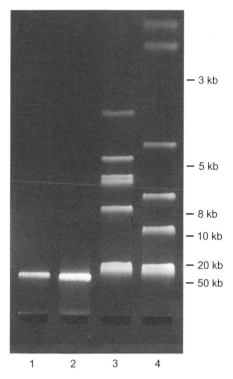

Figure 6.14 Photograph showing four tracks from a 0.8% agarose submarine gel. The gel was run at 40 V in TRIS/borate/EDTA buffer for 16 h, stained with ethidium bromide and viewed under ultraviolet light. Sample loadings were about 0.5 µg of DNA per track. Tracks 1 and 2, λ DNA (49 kb). Track 3, λ DNA cleaved with the enzyme *Eco* RI to generate fragments of the following size (in order from the origin): 21.80 kb, 7.52 kb, 5.93 kb, 5.54 kb, 4.80 kb, 3.41 kb. Track 4, λ DNA cleaved with the enzyme *Hind* III to generate fragments of the following size (in order from the origin): 23.70 kb, 9.46 kb, 6.75 kb, 4.26 kb, 2.26 kb, 1.98 kb. (Courtesy of Stephen Boffey, University of Hertfordshire.)

solvent; it makes it easier to see the sample that is being loaded and also acts as a marker of the electrophoresis front. No stacking gel (Section 6.3.1) is needed for the electrophoresis of DNA, because the mobilities of DNA molecules are much greater in the well than in the gel, and therefore all the molecules in the well pile up against the gel within a few minutes of the current being turned on, forming a tight band at the start of the run. General purpose gels are approximately 25 cm long and 12 cm wide, and are run at a voltage gradient of about 1.5 V cm^{-1} overnight. A higher voltage would cause excessive heating. For rapid analyses that do not need extensive separation of DNA molecules, it is common to use minigels that are less than 10 cm long. In this way information can be obtained in 2–3 h.

Once the system has been run, the DNA in the gel needs to be stained and visualised. The reagent most widely used include the fluorescent dyes ethidium bromide and SYBR® Green. The gel is rinsed gently in a solution of the fluorescent dye and then viewed under ultraviolet light (300 nm wavelength); alternatively, the dye can be dissolved in liquefied agarose just before pouring the gel. Dyes such as ethidium bromide and SYBR® Green are molecules that bind between the stacked base-pairs of DNA (i.e. they intercalate) (Section 4.7.4). The dye concentration therefore builds up at the site of the DNA bands and under ultraviolet light, the DNA bands fluoresce orange-red (ethidium bromide) or green (SYBR® Green). As little as 10 ng of DNA can be visualised as a 1 cm wide band. It should be noted that extensive viewing of the DNA with ultraviolet light can result in damage of the DNA by nicking and base-pair dimerisation. This is of no consequence if a gel is only to be viewed, but obviously viewing of the gel should be kept to a minimum if the DNA is to be recovered (see below). It is essential to protect one's eyes by wearing goggles when ultraviolet light is used. If viewing of gels under ultraviolet is carried out for long periods, a plastic mask that covers the whole face should be used to avoid 'sunburn'.

6.4.2 DNA Sequencing Gels

DNA sequencing gels have been a 'workhorse' technique for the molecular biologist, but have now been replaced by automated methods such as dideoxy sequencing for routine applications (see Section 20.2.1). However, for some particular applications, such as DNA footprinting (see Section 4.17.3), sequencing gels are still used.

Whereas agarose gel electrophoresis of DNA is highly suitable for relatively short DNA molecules, a different form of electrophoresis has to be used when DNA sequences are to be determined. Whichever DNA sequencing method is used (Section 20.2), the final analysis usually involves separating single-stranded DNA molecules shorter than about 1000 nt and differing in size by only 1 nt. To achieve this, it is necessary to have a small-pored gel and so acrylamide gels are used instead of agarose. For example, 3.5% polyacrylamide gels are used to separate DNA in the range 80–1000 nt and 12% gels to resolve fragments of between 20 and 100 nt. If a wide range of sizes needs to be analysed, it is often convenient to run a gradient gel, for example from 3.5% to 7.5%. Sequencing gels are run in the presence of denaturing agents, urea and formamide. Since it is necessary to separate DNA molecules that are very similar in size, DNA sequencing gels tend to be very long (100 cm) to maximise the separation achieved. A typical DNA sequencing gel is shown in Figure 6.15.

As mentioned above, electrophoresis in agarose can be used as a preparative method for DNA. The DNA bands of interest can be cut out of the gel and the DNA recovered by: (a) electroelution, (b) macerating the gel piece in buffer, centrifuging and collecting the supernatant or (c), if low melting point agarose is used, melting the gel piece and diluting with buffer. In each case, the DNA is finally recovered by precipitation of the supernatant with ethanol.

6.4.3 Pulsed-Field Gel Electrophoresis

The agarose gel methods for DNA described above can fractionate DNA of 60 kb or less. The introduction of pulsed-field gel electrophoresis (PFGE) and the further development of variations on the basic technique means that nowadays DNA fragments up to 2×10^3 kb can be separated. Essentially, this allows the separation of whole chromosomes by electrophoresis. The method basically involves electrophoresis in agarose, where two electric fields are applied alternately at different angles for defined time periods (e.g. 60 s). Activation of the first electric field causes the coiled molecules to be stretched in the horizontal plane and start to move through the gel. Interruption of this field and application of the second field force the molecule to move in the new direction. Owing to a length-dependent relaxation behaviour when a long-chain molecule undergoes conformational change in an electric field, the smaller a molecule, the quicker it realigns itself with the new field and is able

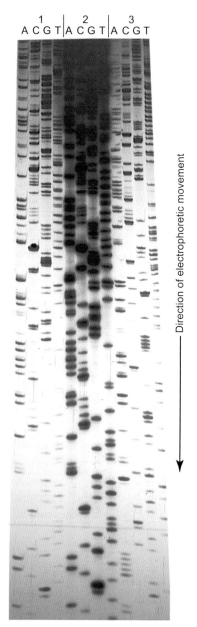

1   2   3
A C G T A C G T A C G T

Direction of electrophoretic movement →

Figure 6.15 Autoradiograph of a DNA sequencing gel. Samples were prepared using the Sanger dideoxy method of DNA sequencing (Section 20.2.1). Each set of four samples was loaded into adjacent tracks, indicated by A,C, G and T, depending on the identity of the dideoxyribonucleotide used for that sample. Two sets of samples were labelled with [35]S (1 and 3) and one was labelled with [32]P (2). It is evident that ^{32}P generates darker, but more diffuse bands than does ^{35}S, making the bands nearer the bottom of the autoradiograph easy to see. However, the broad bands produced by ^{32}P cannot be resolved near the top of the autoradiograph, making it impossible to read a sequence from this region. The much sharper bands produced by ^{35}S allow sequences to be read with confidence along most of the autoradiograph and so a longer sequence of DNA can be obtained from a single gel.

to continue moving through the gel. Larger molecules take longer to realign. In this way, with continual reversing of the field, smaller molecules draw ahead of larger molecules and separate according to size. PFGE has proved particularly useful in identifying the course of outbreaks of bacterial food-borne illness (e.g. *Salmonella* infections). Having isolated the bacterial pathogen responsible for the illness from an individual, the DNA is isolated and cleaved into large fragments, which are separated by PFGE. For example, DNA from *Salmonella* species, when digested with the restriction enzyme *Xba* I, gives around 15 fragments ranging from 25 kb to 680

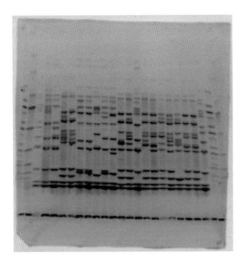

Figure 6.16 PFGE separation of the digestion pattern produced with the restriction enzyme *Nhe* I, of 21 strains of *Neisseria meningitidis*. There are two molecular-weight marker tracks at either end of the gel. (Courtesy of Dr Giovanna Morelli, Max-Planck Institute for Molecular Genetics, Berlin, Germany.)

kb. This pattern of fragments, or 'fingerprint', is unique to that strain. If the same fingerprint is found from bacteria from other infected people, then it can be assumed that they were all infected from a common source. Thus, by comparing their eating habits, food sources, etc. the source of the infection can be traced to a restaurant, food item, etc. Figure 6.16 shows the restriction patterns from different strains of *Neisseria meningitidis*.

6.4.4 Electrophoresis of RNA

Like that of DNA, electrophoresis of RNA is usually carried out in agarose gels, and the principle of the separation, based on size, is the same. Often, one requires a rapid method for checking the integrity of RNA immediately following extraction and before deciding whether to process it further. This can be achieved easily by electrophoresis in a 2% agarose gel in about 1 h. Ribosomal RNAs (18 S and 28 S) are clearly resolved and any degradation (seen as a smear) or DNA contamination is seen easily. This can be achieved on a 2.5–5% acrylamide gradient gel with an overnight run. Both these methods involve running native RNA. There will almost certainly be some secondary structure within the RNA molecule owing to intramolecular hydrogen bonding (see, for example, the clover leaf structure of tRNA, Figure 4.6). For this reason, native RNA run on gels can be stained and visualised with ethidium bromide. However, if the study objective is to determine RNA size by gel electrophoresis, then full denaturation of the RNA is needed to prevent hydrogen bond formation within or even between polynucleotides that will otherwise affect the electrophoretic mobility. There are three denaturing agents (formaldehyde, glyoxal and methylmercuric hydroxide) that are compatible with both RNA and agarose. Either one of these may be incorporated into the agarose gel and electrophoresis buffer, and the sample is heat denatured in the presence of the denaturant prior to electrophoresis. After heat denaturation, each of these agents forms adducts with the amino groups of guanine and uracil, thereby preventing hydrogen bond reformation at room temperature

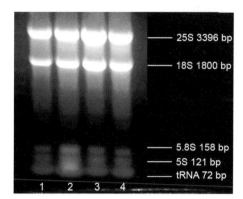

Figure 6.17 Separation of yeast RNA on a 1.5% agarose gel. (Courtesy of Dr Tomas Masek, Department of Genetics and Microbiology, Charles University, Prague, Czech Republic.)

during electrophoresis. It is also necessary to run denaturing gels if the RNA is to be blotted (Northern blots, Section 4.9.2) and probed, to ensure that the base sequence is available to the probe. Denatured RNA stains only very weakly with ethidium bromide, so **acridine orange** is commonly used to visualise RNA on denaturing gels. However, it should be noted that many workers will be using radiolabelled RNA and accordingly identify bands by autoradiography. An example of the electrophoresis of RNA is shown in Figure 6.17.

6.5 CAPILLARY ELECTROPHORESIS

The technique has variously been referred to as high-performance capillary electrophoresis (HPCE), capillary zone electrophoresis (CZE), free-solution capillary electrophoresis (FSCE) and capillary electrophoresis (CE), but the term CE is the most common nowadays. The microscale nature of the capillary used, where only microlitres of reagent are consumed by analysis and only nanolitres of sample needed for analysis, together with the ability for on-line detection down to femtomole (10^{-15} mol) sensitivity in some cases has for many years made capillary electrophoresis the method of choice for many biomedical and clinical analyses. Capillary electrophoresis can be used to separate a wide spectrum of biological molecules, including amino acids, peptides, proteins, DNA fragments (e.g. synthetic oligonucleotides) and nucleic acids, as well as any number of small organic molecules such as drugs or even metal ions (see below). The method has also been applied successfully to the problem of chiral separations.

As the name suggests, capillary electrophoresis involves electrophoresis of samples in very narrow-bore tubes (typically 50 μm internal diameter, 300 μm external diameter). One advantage of using capillaries is that they reduce problems resulting from heating effects. Because of the small diameter of the tubing, there is a large surface-to-volume ratio, which gives enhanced heat dissipation. This helps to eliminate both convection currents and zone broadening, owing to increased diffusion caused by heating. It is therefore not necessary to include a stabilising medium in the tube, allowing free-flow electrophoresis.

Theoretical considerations of CE generate two important equations:

$$t = \frac{l^2}{u \times \Delta V}$$

(Eq 6.7)

where t is the migration time for a solute, l is the tube length, u is the electrophoretic mobility of the solute, and ΔV is the applied voltage.

The separation efficiency, in terms of the total number of theoretical plates, N, is given by

$$N = \frac{u \times \Delta V}{2 \times D}$$

(Eq 6.8)

where D is the diffusion coefficient of the solute.

From these equations, it can be seen, first, that the column length plays no role in separation efficiency (Equation 6.8), but that it has an important influence on migration time and hence analysis time (Equation 6.7) and, second, high separation efficiencies are best achieved through the use of high voltages (u and D are dictated by the solute and are not easily manipulated).

It therefore appears that the ideal situation is to apply as high a voltage as possible to as short a capillary as possible. However, there are practical limits to this approach. As the capillary length is reduced, the amount of heat that must be dissipated increases, owing to the decreasing electrical resistance of the capillary. At the same time, the surface area available for heat dissipation is decreasing. Therefore at some point a significant thermal effect will occur, placing a practical limit on how short a tube can be used. Also, the higher the voltage that is applied, the greater the current, and therefore the heat generated. In practical terms, a compromise between voltage used and capillary length is required. Voltages of 10–50 kV with capillaries of 50–100 cm are commonly used.

The basic apparatus for CE is shown diagrammatically in Figure 6.18. A small plug of sample solution (typically 5–30 µm^3) is introduced into the anode end of a fused

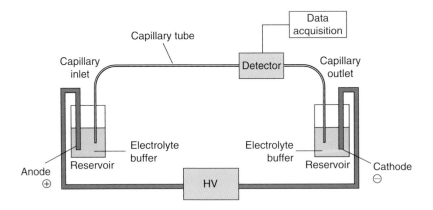

Figure 6.18 Diagrammatic representation of a typical capillary electrophoresis apparatus.

silica capillary tube containing an appropriate buffer. Sample application is carried out in one of two ways: by high-voltage injection or by pressure injection.

6.5.1 High-Voltage Injection

With the high voltage switched off, the buffer reservoir at the positive electrode is replaced by a reservoir containing the sample, and a plug of sample (e.g. 5–30 μm^3 of a 1 mg cm^{-3} solution) is introduced into the capillary by briefly applying high voltage. The sample reservoir is then removed, the buffer reservoir replaced, voltage again applied and the separation is then commenced.

6.5.2 Pressure Injection

The capillary is removed from the anodic buffer reservoir and inserted through an air-tight seal into the sample solution. A second tube provides pressure to the sample solution, which forces the sample into the capillary. The capillary is then removed, replaced in the anodic buffer and a voltage applied to initiate electrophoresis.

A high voltage (up to 50 kV) is then put across the capillary tube and component molecules in the injected sample migrate at different rates along the length of the capillary tube. Electrophoretic migration causes the movement of charged molecules in solution towards an electrode of opposite charge. Consequently, positive and negative sample molecules migrate at different rates. However, although analytes are separated by electrophoretic migration, they are all drawn towards the cathode by electroendosmosis. Since this flow is quite strong, the rate of electroendosmotic flow usually being much greater than the electrophoretic velocity of the analytes, all ions, regardless of charge sign, and neutral species are carried towards the cathode. Positively charged molecules reach the cathode first because the combination of electrophoretic migration and electro-osmotic flow causes them to move fastest. As the separated molecules approach the cathode, they pass through a viewing window where they are detected by an ultraviolet monitor that transmits a signal to a recorder, integrator or computer. Typical run times are between 10 and 30 min. An example of a capillary electrophoretograph is shown in Figure 6.19.

This free solution method is the simplest and most widely practised mode of capillary electrophoresis. However, while the generation of ionised groups on the capillary wall is advantageous via the introduction of electroendosmotic flow, it can also sometimes be a disadvantage. For example, protein adsorption to the capillary wall can occur with cationic groups on protein surfaces binding to the ionised silanols. This can lead to smearing of the protein as it passes through the capillary (recognised as peak broadening) or, worse, complete loss of protein due to total adsorption on the walls. Some workers therefore use coated tubes, where a neutral coating group has been used to block the silanol groups. This of course eliminates electroendosmotic flow. Therefore, during electrophoresis in coated capillaries, neutral species are immobile, while acid species migrate to the anode and basic species to the cathode. Since detection normally takes place at only one end of the capillary, only one class of species can be detected at a time in an analysis using a coated capillary.

Peptide	
1	Lys-Arg-Pro-Pro-Gly-Phe-Ser-Pro-Phe-Arg
2	Met-Lys-Arg-Pro-Pro-Gly-Phe-Ser-Pro-Phe-Arg
3	Arg-Pro-Pro-Gly-Phe-Ser-Pro-Phe-Arg
4	Ser-Arg-Pro-Pro-Gly-Phe-Ser-Pro-Phe-Arg
5	Ile-Ser-Arg-Pro-Pro-Gly-Phe-Ser-Pro-Phe-Arg

Figure 6.19 Capillary electrophoresis of five structurally related peptides. Column length was 100 cm and the separation voltage 50 kV. Peptides were detected by their ultraviolet absorbance at 200 nm. (Courtesy of Patrick Camilleri and George Okafo, GSK Ltd.)

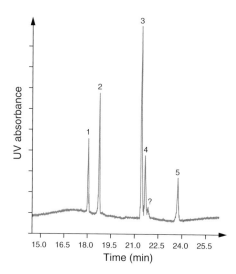

A range of variations on this basic technique also exist. For example, as seen above, in normal CE, neutral molecules do not separate, but rather travel as a single band. However, separation of neutral molecules can be achieved by including a surfactant such as SDS with the buffer. Above a certain concentration, some surfactant molecules agglomerate and form micelles, which, under the influence of an applied electric field, will migrate towards the appropriate electrode. Solutes will interact and partition with the moving micelles. If a solute interacts strongly, it will reach the detector later than one which partitions to a lesser degree. This method is known as micellular electrokinetic capillary electrophoresis (MECC). Since ionic solutes will also migrate under the applied field, separation by MECC is due to a combination of both electrophoresis and chromatography.

Original developments in CE concentrated on the separation of peptides and proteins, but in recent years, CE has been successfully applied to the separation of a range of other biological molecules. The following provides a few examples.

- In the past, peptide analysis has been performed routinely using reversed-phase HPLC, achieving separation based on hydrophobicity differences between peptides. Peptide separation by CE is now also routinely carried out, and is particularly useful, for example as a means of quality (purity) control for peptides and proteins produced by preparative HPLC. Figure 6.19 shows the impressive separation that can be achieved for peptides with very similar structures.
- High purity synthetic oligodeoxyribonucleotides are necessary for a range of applications, including use as hybridisation probes in diagnostic and gene-cloning

experiments, use as primers for DNA sequencing and the polymerase chain reaction (PCR), use in site-directed mutagenesis and use as antisense therapeutics. CE can provide a rapid method for analysing the purity of such samples. For example, analysis of an 18-mer antisense oligonucleotide containing contaminant fragments (8-mer to 17-mer) can be achieved in only 5 min.

- Point mutations in DNA, such as occur in a range of human diseases, can be identified by CE.
- CE can be used to quantify DNA. For example, CE analysis of PCR products from HIV-I allowed the identification of between 200 000 and 500 000 viral particles per cubic centimetre of serum.
- Chiral compounds can be resolved using CE. Most work has been carried out in free solution using cyclodextrins as chiral selectors.
- A range of small molecules, drugs and metabolites can be measured in physiological solutions such as urine and serum. These include amino acids (over 50 are found in urine), nucleotides, nucleosides, bases, anions such as chloride and sulfate (and can be separated in human plasma) and cations such as Ca^{2+} and Fe^{3+}.

6.6 MICROCHIP ELECTROPHORESIS

The further miniaturisation of electrophoretic systems has led to the development of microchip electrophoresis (MCE), which has many advantages over conventional electrophoresis methods, allowing very high-speed analyses at very low sample sizes. For example, microchip analysis can often be completed in tens of seconds, whereas capillary electrophoresis (CE) can take 20 min and conventional gel electrophoresis at least 2 h. Using new detection systems, such as laser-induced fluorescence, picomole (10^{-12} mol) to attomole (10^{-18} mol) sensitivity can be achieved, which is at least two orders of magnitude greater than for conventional CE. Detection systems for molecules that do not fluoresce include electrochemical detectors, pulsed amperometric detection (PAD) and sinusoidal voltammetry. All these detection techniques offer high sensitivity, are ideally suited to miniaturisation, are very low cost and are highly compatible with advanced micromachining and microfabrication technologies (see below). Finally, a potential gradient of only a few volts is required, which eliminates the need for the high voltage power supplies used by CE.

The manufacturing process that produces microchips is called microfabrication (see also Section 14.6.2). The process etches precise and reproducible capillary-like channels (typically, 50 µm wide and 10 µm deep; slightly smaller than a strand of human hair) on the surface of sheets of quartz, glass or plastic. A second sheet is then fused on top of the first sheet, turning the etched channels into closed microfluidic channels. The end of each channel connects to a reservoir through which fluids are introduced/removed. Typically, the size of chips can be as small as 2 cm². Basically, the microchip provides an electrophoretic system similar to CE, but with more flexibility.

Current developments of this technology are based on integrating functions other than just separation into the chip. For example, sample extraction, pre-concentration of samples prior to separation, PCR amplification of DNA samples

Assay Class: DNA 7500
Data Path: C:\...–08\2100 expert_DNA 7500_DE54704329_2
Gel Image

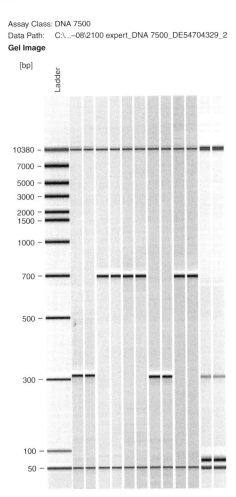

Figure 6.20 Example of a computer-generated image from an Agilent 2100 Bioanalyzer. Here, data from the sample being analysed are transformed into gel-like images similar to a stained agarose gel (electropherogram).

using infrared-mediated thermocycling for rapid on-chip amplification and the extraction of separated molecules using microchamber-bound solid phases are all examples of where further functions have been built into a microchip electrophoresis system. An interface has also been developed for microchip electrophoresis mass spectrometry (MCE-MS) where drugs have been separated by MCE and then identified by MS.

One example of a microfluidic circuit system is the Agilent Bioanalyzer. The sample is applied in one area of a small cassette and driven through microchannels (that are pre-filled with dye-laden agarose immediately before use) under computer-controlled electrophoresis. The channels lead to reservoirs allowing, for example, incubation with other reagents, such as dyes, for a specified time. Electrophoretic separation is thus carried out in a microscale format. The small sample size minimises sample and reagent consumption and the units, being computer controlled, allow data to be captured within a very short timescale. The resulting data are presented as an **electropherogram** with a digital representation of an agarose gel in an easy to interpret format (Figure 6.20).

6.7 SUGGESTIONS FOR FURTHER READING

6.7.1 Experimental Protocols

Ahn T., Yim S., Choi H. and Yun C. (2001) Polyacrylamide gel electrophoresis without a stacking gel: use of amino acids as electrolytes. *Analytical Biochemistry* **291**, 300–303.

Makovets S. (2013) Basic DNA electrophoresis in molecular cloning: a comprehensive guide for beginners. *Methods in Molecular Biology* **1054**, 11–43.

Yim S., Ahn T., Kim J. and Yun C. (2002) Polyacrylamide gel electrophoresis without a stacking gel: application for separation of peptides. *Analytical Biochemistry* **305**, 277–279.

6.7.2 General Texts

Hames B.D. and Rickwood D. (2002) *Gel Electrophoresis of Proteins: A Practical Approach,* 3rd Edn., Oxford University Press, Oxford, UK.

Magdeldin S. (2012) *Gel Electrophoresis: Principles and Basics,* InTech Publishers, Rijeka, Croatia.

Walker J.M. (2009) *The Protein Protocols Handbook,* 3rd Edn., Humana Press, New York, USA.

6.7.3 Review Articles

Lee P.Y., Costumbrado J., Hsu C.Y. and Kim Y.H. (2012) Agarose gel electrophoresis for the separation of DNA fragments. *Journal of Visualized Experiments* **20**, e3923.

Stellwagen N.C. (2009) Electrophoresis of DNA in agarose gels, polyacrylamide gels and in free solution. *Electrophoresis* **30**, S188–S195.

6.7.4 Websites

Image Processing and Analysis in Java (ImageJ)
imagej.nih.gov/ij/ (accessed April 2017)

Protein biology tools (ThermoFisher Scientific)
www.thermofisher.com/uk/en/home/life-science/protein-biology (accessed April 2017)

Protein electrophoresis (Bio-Rad)
www.bio-rad.com/en-uk/applications-technologies/introduction-protein-electrophoresis (accessed April 2017)

7 Immunochemical Techniques

KATJA FISCHER

7.1 INTRODUCTION

The immune system of mammals has evolved over millions of years and provides an incredibly elegant protection system that is capable of responding to infective challenges as they arise. The cells of immunity and their products are transported throughout the body, primarily in the blood and also through fluid within the tissues and organs themselves, thereby providing immunity to all areas of the body. There are several cell types involved in immune responses, each with a role to play and each controlled by chemical mediators known as cytokines. As the immune system is such a powerful tool, it needs careful management to ensure its effective operation. Both over- and under-activity could have fatal consequences.

There are two main layers of defence: the innate immune systems, providing an immediate, but non-specific, response is found in all plants and animals. Vertebrates possess a second layer of protection, the adaptive immune system, which is activated by the innate response and adapts its reaction during an infection to provide an improved response if the pathogen is encountered again. The immune system is broadly additive, i.e. more complex animals have elements analogous to those found in primitive species, but have extra features as well.

Immunity is monitored, delivered and controlled by specialised cells derived from stem cells in the bone marrow. There are phagocytes, including many types of white blood cell (neutrophils, monocytes, macrophages, mast cells and dendritic cells), which move around the body removing debris and foreign materials. There are also other assorted cells (eosinophils, basophils and natural killer cells) whose function is to rapidly gather in areas of the body where a breach of security has occurred and to deliver potent chemicals capable of dealing with any foreign bodies. The adaptive immune system includes two lineages of lymphocytes: B cells (bone marrow- or bursa-derived) are primarily involved in humoral immunity and produce antibodies that bind to foreign proteins; T cells (matured in thymus) are involved in cell-mediated immunity and produce either cytokines to direct the immune response (T helper cells) or enzymes that induce the death of pathogen-infected cells (cytotoxic T cells).

In mammals, the skin forms a barrier against foreign attack. Any breaches are countered by cells of the immune system, even though foreign material might not be present. This response is mediated by cytokines, which can be released from damaged tissue or from cells of the immune system near the site of injury (macrophages, B lymphocytes, T lymphocytes and mast cells). As part of the innate immune response, mammals have a pre-programmed ability to recognise and immediately act against substances derived from fungal and bacterial microorganisms. This is mediated through a sequential biochemical cascade involving over 30 different proteins (the complement system), which, as a result of signal amplification, triggers a rapid killing response. The complement cascade produces peptides that attract immune cells, increases vascular permeability, coats (opsonises) the surface of a pathogen, marking it for destruction, and directly disrupts foreign plasma membranes.

The antibody response in mammals creates immunological memory after an initial response to a specific pathogen, achieving protection against that pathogen. Importantly, this process forms the basis of vaccination. This chapter focusses on such amazing immunological phenomena and how they can be leveraged for medical research and therapy.

7.1.1 Development of the Immune System

Mammalian embryos develop an immune system before birth to provide the newborn with immediate protection. Additional defences are acquired from maternal milk to cover the period during which the juvenile immune system matures. The cells of the mammalian immune system are descended from distinct lineages derived initially from stem cells. B cells (B lymphocytes) produce antibodies that recognise specific epitopes (molecular shapes) of foreign substances (called antigens). An antigen is defined as a substance capable of causing an immune response and leading to the production of antibodies specific to it. An antibody (Ab), also known as an immunoglobulin (Ig), is a protein produced by the immune system to specifically bind a unique antigen. Consequently, antigens are the specific targets of those antibodies whose production they initiated. Dendritic cells, macrophages and other antigen-presenting cells (APCs) recognise antigens within the body and will endocytose (engulf) them when encountered. The majority of antigens found by the body are from invading viruses, bacteria, parasites and fungi. These pathogens display many proteins and other structures that are **antigenic** (i.e. they behave as an antigen) when encountered by the immune system. Upon uptake of antigens, the APCs process these into small fragments and present them to the B cells. The fragments contain epitopes that are typically about 15 amino-acid residues in size. This corresponds to the size that the antibody binding site can bind to. After ingesting the antigen fragments, the B cells recruit 'help' in the form of cytokines from T cells, which stimulate cell division, differentiation and secretion of antibody. Each B cell that was capable of binding an antigen fragment and has ingested it will then model an antibody on the shape of the epitope and secrete it into the blood.

During embryonic development, the immune system has to learn what constitutes self and what is foreign. Failure to do this would lead to self-destruction or to an inability to mount an immune response to foreign substances. This learning process is achieved by selective **clonal deletion** of self-recognising B cells. Early in the development of the immune system, B cell lineages randomly rearrange the antibody-creating genes to produce a 'starter pack' of B cells that respond to a huge number of molecular shapes. These B cells have randomly produced antibodies expressed on their surface, and are ready to engage should an antigen fit an antibody-binding site. This process provides crude but instant protection to a large number of foreign substances immediately after birth, and forms the basis for the B cells that eventually provide protection for the rest of the life of the animal. However, within the population of randomly produced B cells are some that will be responsive to self-antigens. These are extremely dangerous as they could lead to the destruction of parts of the body. Embryos are derived exclusively from cells originating from the fusion of egg and sperm. Immunologically, the entire foetus can thus be regarded as being part of self. Any B cells that start to divide within the embryo prior to birth are responding to self-antigens and are destroyed as they are potentially dangerous. This selective clonal deletion is fundamental to the development of the immune system and without it the organism could not continue to develop. The remainder of the B cells that have not undergone cell division only recognise non-self-antigens and are retained within the bone marrow of the animal as a quiescent cell population, waiting for stimulation from passing stimulated antigen-presenting cells. Stimulation of B cells requires the presence of both macrophages and T lymphocytes. T cells are responsible for 'helping' and 'suppressing' the immune response. T cells also undergo clonal selection during development to ensure they do not recognise self-antigens. In addition, they are positively selected to ensure that they do recognise proteins of the **major histocompatibility complex** (MHC) found on cell surfaces. The balance of appropriate immune response is governed by the interplay of T and B lymphocytes along with macrophages and other antigen-presenting cells to ensure that an individual is protected, but not endangered by inappropriate responses.

After birth, exposure to foreign materials causes an immediate response resulting in antibody production and secretion by B cells. The antibody binds to the target, marks it as foreign and it is then removed. Dendritic cells and macrophages are responsible for much of the removal of foreign material that they ingest by receptor-mediated endocytosis (uptake of membrane-bound particles by cells) and phagocytosis (uptake of solid particles from outside the surface membrane into the cell body), respectively. The material is digested, exported to the cell surface as small antigen fragments and presented to B and T cells; lymphocytes with T cell receptors play a central role in cell-mediated immunity. Should a B cell carry an antibody that binds the presented antigen fragment, it will internalise it, causing a number of intracellular changes, known as B cell activation. This process can involve the recruitment of T cells, stimulating cell growth and metabolism. B cell activation may also occur without the presence of antigen-presenting cells, when the lymphocyte is directly exposed to antigens. B cell activation leads to two further major changes, apart from cell division and antibody secretion. It leads to a larger population of cells being retained in the bone marrow, termed memory B cells, which will recognise and rapidly respond should the antigen be encountered again. Binding of antigens also causes the B cells to refine the quality of the antibody they produce. Avidity is the strength with which the antibody binds to the antigen and affinity describes the fit of the antibody shape to the target. Both of these can be improved after the B cells have been stimulated, but require more than one exposure to the antigen. This process is known as affinity maturation and is characterised by a change in antibody type from the predominantly low-affinity pentameric (five molecules linked together) immunoglobulin M (IgM) to the high-affinity immunoglobulin G (IgG). Other antibody types may be produced in specific tissues and in response to particular antigens. For example, parasites in the intestines often induce high levels of immunoglobulin E (IgE) in the gut mucosa – the innermost layer of the gut, which secretes large amounts of mucus. After several encounters with an antigen, a background level of specific antibody will be found in the blood, along with a population of memory cells capable of rapidly responding to the antigen presence by initiating high levels of antibody secretion. This status is known as being immune and is the basis of both artificial immunisations for protection against disease (vaccination) and also for the production of antibodies for both diagnostic and therapeutic use.

7.1.2 Harnessing the Immune System for Antibody Production

Three main types of antibodies can be produced: polyclonal, monoclonal and recombinant. Each of these antibody types has advantages, but also limitations, and they should thus be viewed as complementary to each other, as each is particularly useful in specific areas (see comparison in Table 7.1).

Polyclonal antibodies are produced in animals by injecting them with antigen and biochemically recovering the antibodies from the animal blood serum. Polyclonal antibodies are essentially a population of antibody molecules contributed by many B cell clones. The multiple B cell lineages involved produce a range of antibodies to many different epitopes, all binding the same original antigen. Mammals produce antibodies to practically any foreign protein, providing it has a molecular mass greater than 5000 Da and the antigen is not closely related to substances found in the animal itself. Many mammalian proteins and other biochemical substances are highly conserved and thus elicit a very similar immune response in many species. This can lead to problems in producing antibodies for diagnostic and therapeutic use. As discussed, the immune system does not generally mount a response to self and because of this, if the antigen donor and antibody-producing host are closely related species, some antigens may not raise an antibody response. Historically, the first antibodies produced artificially for diagnostic purposes were polyclonal. They are generated in a number of animal species by immunising the host with the substance of interest, usually three or four times. Blood is collected and serum tested at regular intervals; when the antibody titre is high enough, the antibody fraction is purified from the blood serum. Generally, larger animals are used, since bigger volumes can be obtained from larger species. The use of polyclonal antibodies in therapeutics is limited, as they themselves can be antigenic when injected into other animals. Polyclonal antibodies are cheap to produce, robust, but often less specific than other antibodies and therefore have variable qualities depending on the batch, as well as the specific donor animal.

Monoclonal antibodies (monoclonals) are produced in tissue culture by cells originating from a single hybridoma, a cell line stemming from the fusion of a single immortal mammalian cancer cell with a single antibody-producing B cell. B lymphocytes have a limited lifespan in tissue culture, but the cloned hybridoma has immortality conferred by the tumour parent and continues to produce antibody. In contrast to whole-antigen-specific polyclonal antibodies produced by multiple immune cells, monoclonals are specific to single epitopes and are produced by clonal cells. This difference (summarised in Table 7.1) is fundamental to the way in which monoclonal antibodies can be used for both diagnostics and therapeutics. Once cloned, the hybridoma cell lines are reasonably stable and can be cultured to produce large quantities of antibody, which they secrete into the culture medium. The antibody they produce has the same properties as the parent lymphocyte and it is this uniqueness that makes monoclonal antibodies so useful. As monoclonal antibodies are epitope specific and there are thousands of potential epitopes on an average antigen, a cell fusion process generates many hundreds of hybridoma clones, each making an individual antibody. Consequently, to ensure that the antibodies produced have the correct qualities needed for the final intended use, the screening process that is used to select the clones of value is the most important part of making hybridomas. Monoclonal antibodies can be used for human and veterinary therapeutics; however, they can be themselves antigenic, i.e. while structurally similar, the differences between murine and human antibody proteins can invoke an immune response against mouse-specific epitopes when murine monoclonal antibodies are injected into humans. This can result in their rapid clearance from the recipients' blood, systemic inflammation and the production of human anti-mouse antibodies. Chimeric or humanised hybridomas can be engineered by joining mouse DNA encoding the binding portion of a monoclonal antibody with human antibody-encoding DNA. This construct can be cloned into an immortal cell to produce an antibody that will recognise human epitopes, but not be immunogenic when injected into humans. Transgenic mouse technology is the most successful approach to generate human monoclonal antibody therapeutics. Humanised antibodies have been used very successfully for treating a range of human conditions, including breast cancer, lymphoma and rejection symptoms after organ transplantation.

Table 7.1 **Comparison of polyclonal and monoclonal antibodies**

Polyclonal antibodies (pAbs)	Monoclonal antibodies (mAbs)
Recognise multiple epitopes on any one antigen	Recognise only one epitope on an antigen
Multiple epitopes generally provide more robust detection. pAbs can help amplify signal from target protein with low expression level, as the target protein will bind more than one antibody molecule on the multiple epitopes	As mAbs detect one target epitope they are highly specific, their use usually results in less background and they are less likely to cross-react with other proteins
More tolerant of minor changes in the antigen (e.g. polymorphism, heterogeneity of glycosylation or slight denaturation) than homogeneous mAbs	May be too specific (e.g. less likely to detect across a range of species). Detection techniques using mAbs are vulnerable to the loss of epitope through chemical treatment of the antigen. Pooling two or more monoclonal antibodies to the same antigen may be necessary
Production affordable for average laboratories, requires basic skills, short production time	Expensive production, technology requires expertise and training, long production time
Sera usually contain large amounts of non-specific antibodies. This can give background signal in some applications	Technology can produce large amounts of homogeneous, specific antibodies
Batch-to-batch variability (between individual hosts and within hosts over time)	No or low batch-to-batch variability
The resource is limited to the host individual and is not renewable	Once a hybridoma is made, it is a constant and renewable source

Recombinant antibodies are commonly generated by molecular methodologies in either prokaryotic or eukaryotic expression systems. The idea of producing antibodies in crop plants is attractive, as the costs of growing are negligible and the amounts of antibody produced could be very large. Expression in plants has had some initial success.

Two methods are most commonly used to produce recombinant antibodies. **DNA libraries** can be used to produce a **bacteriophage** expressing antibody fragments on the surface. Useful antibodies can be identified by assay and the bacteriophage producing it is then used to transfect the antibody DNA into bacteria for expression in culture by the recombinant cells. The antibodies produced are monoclonal, but do not have the full structure of those expressed by animals or cell lines derived from animals. They are less robust and as they are much smaller than native antibodies it may not be possible to modify them without losing binding function. However, the great advantage of using this system is the speed with which antibodies can be generated, generally in a matter of weeks. In contrast, the typical timescale for producing monoclonal antibodies from cell fusions is about 6 months.

Antibodies can also be generated from donor lymphocyte (B cell) DNA. The highest concentration of B cells is found in the spleen after immunisation and so this is the tissue usually used for DNA extraction. The antibody-coding genes are then selectively amplified by polymerase chain reaction (PCR) and transfected into the genome of a eukaryotic cell line, whereupon a small percentage of cells will incorporate the PCR product into the genomic DNA. Usually, a resistance gene is cotransfected so that only recombinant cells containing antibody genes will grow in culture. The cells chosen for this work are often those most easily grown in culture and may be rodent or other mammalian lines. Typically, Chinese hamster ovary (CHO) cells are used for recombinant antibody production and have in fact become the industry standard amongst biotechnology companies. Yeasts, filamentous algae and insect cells have also been used as recipients for antibody genes, albeit with varying degrees of success.

7.1.3 Antibody Structure and Function

Mammalian antibodies are all based on a Y-shaped molecule consisting of four polypeptide chains held together by disulfide bonds (see Figure 7.1). There are two pairs of chains, known as heavy (H) and light (L).The base of the Y is known as the constant region (C_H1, C_H2 and C_H3) and the tips of the arms are the variable region (V_L, V_H). The amino-acid sequence in the constant region is conserved within a species. The variable region is composed of between 110 and 130 amino acids and varies greatly among different antibodies produced by an individual. Variations in these regions of both the heavy and the light chains form the antigen binding site of the antibody and define its specificity for a particular epitope on the corresponding antigen. Enzymatic digestion of antibodies is performed to generate useful biochemical reagents. Treatment with the enzyme papain gives rise to three fragments: two antigen-binding fragments (Fab) and one constant fragment (Fc). The protease digests the molecule at the hinge region, and the resulting Fab fragments retain their antigen-binding capability. The small-sized Fab fragments can be useful for some immunochemical applications where the larger native antibody molecule would have difficulty binding.

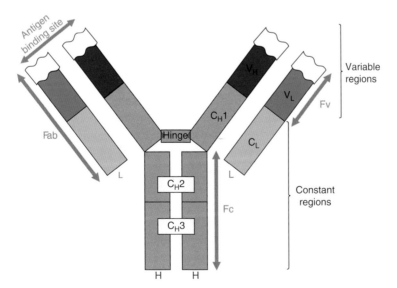

Figure 7.1 Immunoglobulin G. See also Figure 12.11 in Section 12.5.2 for low-resolution structural information by hydrodynamic methods.

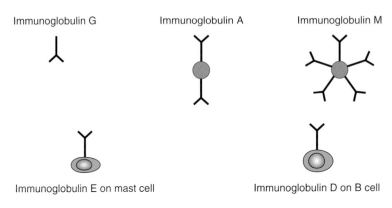

Immunoglobulin G   Immunoglobulin A   Immunoglobulin M

Immunoglobulin E on mast cell   Immunoglobulin D on B cell

Figure 7.2 Immunoglobulin classes.

There are five major classes of antibody molecules, also known as immunoglobulins (Igs). Immunoglobulin G (IgG) is the most common and it is characterised by its Y-shaped structure. The other classes of antibody are immunoglobulins M, A, D and E (Figure 7.2).

Immunoglobulin M (IgM) is produced by immature and newly activated B lymphocytes in response to initial exposure to an antigen (primary immune response). Structurally it is formed from a circular arrangement of five or six immunoglobulin G molecules. Being a polymer with many binding sites, IgM possesses high avidity. IgM is generally only found in serum as its large size prevents it from crossing tissue boundaries. The pentameric form is particularly useful for complexing antigens such as bacteria into aggregates, either for disposal or for further processing by the immune system through complement activation. Cells secreting IgM can progress to IgG production, if the animal is challenged again by the antigen. This progression to IgG production is known as affinity maturation and requires maturation of the cells to memory cell status. After several encounters with an antigen, a background level of specific antibody will be found in the blood along with a population of B memory cells capable of rapidly responding to the antigen by initiating high levels of antibody secretion – and thus reaching the status of being immune. Extreme levels of antibody (hyperimmunity) may be reached after repeated exposure to an antigen. In this status, additional exposure to antigen can lead to anaphylactic shock due to the overwhelmingly large immune response. Conversely, repeated immunisation can lead to a total loss of immune response to an antigen. This acquired immune tolerance as a response to antigen over-stimulation is characterised by a loss of circulating B cells reactive to the antigen and also by a loss of T cell response to the antigen. This can be used therapeutically to protect individuals against allergic responses.

Immunoglobulin A (IgA) is a dimeric form of immunoglobulin, essentially two IgG molecules arranged end to end with the binding sites facing outwards. This structural topology makes it resistant to enzyme degradation. It is produced by B cells of the mucosal surface of the mouth, nose, eyes, digestive tract and genitourinary system, and is directly secreted into the fluids associated with these tissues, playing a critical role in mucosal immunity.

Immunoglobulin D (IgD) is an antibody resembling IgG and is found on the surface of immature B cells along with IgM. It is a cellular marker that indicates that an immature B cell is ready to mount an immune response and may be responsible for the migration of the cells from the spleen into the blood. It is used by macrophages to identify the cells to which they can present antigen fragments.

Immunoglobulin E (IgE) also resembles IgG and is produced in response to allergens and parasites. It is secreted by B lymphocytes and attaches itself to the surface of mast cells and basophils. Binding of allergen to IgE molecules on the cell surface cause the antibodies to cross-link and move together in the cell membrane. This cross-linking triggers the release of histamine. Histamine is responsible for the symptoms suffered by individuals as a result of exposure to allergens.

Immunoglobulin G and, to a lesser extent, IgM are the only two antibodies that are of practical use in immunochemistry. IgG is the antibody of choice for assay development as it is easily purified from serum and tissue culture medium. It is very robust and can be modified by labelling with marker molecules without losing function (see, for example, immuno microscopy; Section 7.4 and Chapter 11). It can be stored for extended periods of time at 4 °C or lower. Occasionally, antigens will not generate IgG responses in vivo and IgM is produced instead. This is caused by the antigen being unable to fully activate the B cells and, as a result, no memory cells are produced. Such antigens are often highly glycosylated and it is the large number of sugar residues that block the full activation of the B cells. IgM can be used for assay development, but is more difficult to work with since it is rather unstable. The molecules are difficult to label without losing their function as the presence of labels in adjacent binding sites can easily interfere with antigen recognition due to steric hindrance. IgM can be used directly from cell tissue culture supernatant in assays with an appropriate secondary anti-IgM enzyme conjugate.

7.2 ANTIBODY PREPARATION

All methods used in immunochemistry rely on the antibody molecule or derivatives of it. Antibodies are incredibly useful molecules that can be designed to detect an almost limitless number of antigens. They are adaptable and will operate in many conditions. They can be used in both diagnostic and therapeutic scenarios. Antibodies can be made in various ways and the choice of which method to use is very much dependent on the final assay format. For an antibody to be of use, it has to have a defined specificity, affinity and avidity, as these are the qualities that determine its usefulness in the method to be used.

7.2.1 Polyclonal Antibody Production

Polyclonal antibodies are raised in appropriate donor animals, generally mice, rats or rabbits for smaller amounts and sheep or goats for larger quantities. It is important that animals are free of infection, as this would raise unwanted antibodies.

Usually, antigens are mixed with an appropriate adjuvant prior to immunising the animals. Adjuvants increase the immunogenicity of the antigen, thereby reducing the amount of antigen required, as well as stimulating specific immunity to it. Adjuvants may be detergents, oils or complex proprietary products containing bacterial cell walls or preparations of them. Prior to immunising an animal, pre-immune blood samples are taken such that the baseline IgG level can be appraised (Figure 7.3). Immunisations are spaced at intervals between 2 and 4 weeks to maximise antibody production. Blood samples are taken throughout the immunisation program and the serum is tested for specific activity to the antigen by **enzyme-linked immunosorbent assay** (ELISA) or other methods (see Section 7.3). Typically, a range of doubling serum dilutions are made (1/100–1/12 800) and tested against the antigen. Detection of antigen at 1/6400 dilution indicates high levels of circulating antibody. **Western blot analysis** of the serum can confirm the specificity of the antibodies produced (see Section 7.7). Once a high level of the desired antibody is detected in test bleeds, larger samples can be taken. Animal welfare legislation governs permissible amounts and frequency of bleeds. Blood donations are allowed to clot and the serum collected. Serum can be stored at 4 °C or lower for longer periods.

Occasionally, antigens that give a poor response in mammals can give much higher yields in chickens. Chickens secrete avian **immunoglobulin Y** (IgY) into their eggs to provide protection for the developing embryo. This can be utilised for effective polyclonal antibody production. The chickens are immunised three or four times with the antigen and the immune status is monitored by test bleeds. Eggs are collected and can yield up to 50 mg of antibody per yolk. The antibody has to be purified from the egg yolks prior to use and a number of proprietary kits can be used to do this.

Very pure polyclonal antibodies can be produced in rats and mice in **ascitic fluid**, which is the intra-peritoneal fluid extracted from mice with a peritoneal tumour. Non-secretory myeloma cells are injected into the peritoneal cavity and the growing

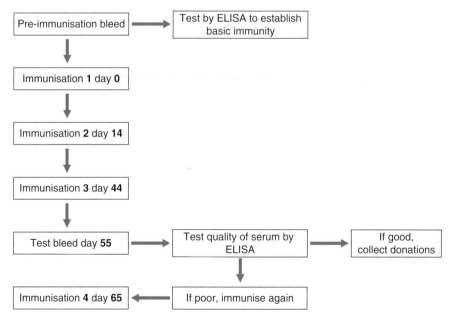

Figure 7.3 Immunisation schedule.

tumour cells cause the animal to produce ascitic fluid which contains very high levels of the antibodies that the animal is currently secreting in its blood. Animals that are immunised with a particular antigen of interest will produce high titers of the respective polyclonal antibodies in the ascitic fluid, which can be removed by aspiration with a syringe and needle.

7.2.2 Monoclonal Antibody Production

Mice, rats or other rodents are the donor animal of choice for monoclonal antibody production, as they are low cost and easy to manage and handle. Balb/C is the mouse strain typically used for monoclonal antibody production and most of the tumour cell lines used for fusions are derived from this mouse.

Mice are immunised, usually three or four times over the course of 3–4 months, by the intraperitoneal route using antigen mixed with an appropriate adjuvant (Figure 7.4). Samples from test bleeds can be taken to monitor the immune status of the animals. Once the mice are sufficiently immune, they are left for 2–3 months to rest. This is important as the cells that will be used for the hybridoma production are memory B cells and require the rest period to become quiescent.

Mice are sacrificed and the spleens removed; a single spleen typically provides sufficient cells for two or three cell fusions. Three days prior to cell fusion, the partner cell line NS-0 is cultured to provide a log phase culture. For cell fusions, a number of methods are established, but one of the most commonly used is fusion by centrifugation in the presence of polyethylene glycol (PEG). Cells from spleen and fusion partner are mixed in a centrifuge tube, PEG is added to solubilise the cell membranes and the fusion is carried out by gentle centrifugation. In order to suppress functional interference, PEG levels are lowered by dilution with culture medium and clones are grown from single recombinant parent cells on microtitre wells. Cell

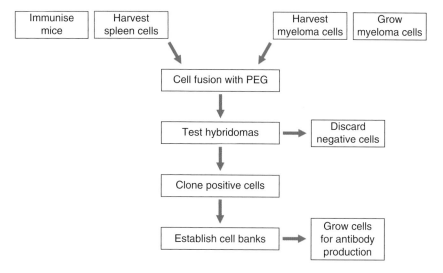

Figure 7.4 Monoclonal antibody production.

lines used for fusion partners have a defective enzyme pathway and thus allow for selection after cell fusion. NS-0 cells lack the enzyme hypoxanthine-guanine phosphoribosyl transferase (HGPRT), which prevents them from using a nucleoside salvage pathway when the primary pathway is disabled due to the presence of the antibiotic aminopterin. Accordingly, the tissue culture additive HAT, which contains hypoxanthine, aminopterin and thymidine is used to select for hybridomas after cell fusion. They inherit an intact nucleoside salvage pathway from the spleen cell parent, which allows them to grow in the presence of aminopterin. Unfused NS-0 cells are unable to assimilate nucleosides and die after a few days. Unfused spleen cells are unable to divide more than a few times in tissue culture and will die as well. Two weeks after the cell fusion, the only cells surviving under the above conditions are hybridomas. The immunisation process ensures that many of the spleen cells that have fused will be secreting antibody targeting the antigen used for immunisation; however, this cannot be relied upon and rigorous screening is required to ensure that the hybridomas selected are secreting an antibody of interest. Screening is often carried out by ELISA, but other antibody assessment methods may be used. It is important that hybridomas are assessed repeatedly as they can lose the ability to secrete antibody after a few cell divisions. This occurs as chromosomes are lost during division to return the hybridoma to its normal chromosome quota.

Once hybridomas have been selected they have to be cloned to ensure that they are truly monoclonal. Cloning involves the derivation of cell colonies from individual cells grown isolated from each other. In limiting dilution cloning, a cell count is carried out and cells are diluted into media, ensuring that only one cell is present in each well of the tissue culture plate. The plates are incubated for 7 days and cell growth assessed after this time. Colonies derived from single cells are then tested for antibody production by ELISA. It is desirable that a cell line should exhibit 100% cloning efficiency in terms of antibody secretion, but some cell lines are inherently unstable and will always produce a small number of non-secretory clones. Provided such cell lines are not subcultured excessively, this problem may not be substantial, though. Recloning these lines regularly ensures that cultures are never too far from an authenticated clone.

It is very important to know the antibody isotype of the hybridomas and a number of commercial kits are available to determine this. Most are based on lateral flow technology, which will be discussed in Section 7.5. Once the isotype of the antibody is established and it is clonally stable, cultures can be grown to provide both cell banks and antibody for use in testing or for reagent development. It is absolutely essential to record the pedigree of every cell line and to be vigilant in handling and labelling flasks to prevent cross-contamination of cell lines.

7.2.3 Cell Banking

Cell banks are established from known positive clones and are produced in a way that maximises reproducibility between frozen cell stocks and minimises the risk of cellular change (Figure 7.5). A positive clone derived from a known positive clone is rapidly expanded in tissue culture until enough cells are present to produce 12 vials of frozen cells simultaneously (see Section 3.5.10 for cryopreservation of

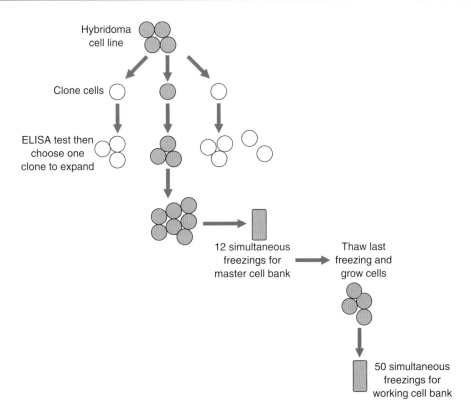

Figure 7.5 Cell banking.

cells). This master cell bank is stored at −196 °C under liquid nitrogen vapour. The working cell bank is derived from the master cell bank. One of the frozen vials from the master cell bank is thawed and rapidly grown until there are enough cells to produce 50 vials of frozen cells simultaneously. This is the working cell bank and it is also stored at −196 °C under liquid nitrogen vapour. This strategy ensures that all of the vials of the working cell bank are identical. Each of the 12 identical vials of the master cell bank will provide a new working bank, providing 550 further identical working vials before the process of deriving a new master cell bank by selecting a positive clone for expansion from the last master cell bank vial is required.

7.2.4 Growing Hybridomas for Antibody Production

All monoclonal antibodies are secreted by growing hybridomas into the tissue culture media and a number of methodologies exist for optimising antibody production with respect to yield, ease of purification and cost. The simplest method for antibody production relies on static bulk cultures of cells growing in T flasks. T flasks are designed for tissue culture and have various media capacities and cell culture surface areas. For most applications, a production run requires between 250 ml and 1000 ml of medium. Most cell lines produce between 4 and 40 mg of antibody per litre. The cells from a working cell bank vial are thawed rapidly into 15 ml medium containing

10% fetal bovine serum and placed in an incubator at 37 °C supplemented with 5% CO_2. Once cell division has started, the culture sizes are increased using medium supplemented with 5% fetal bovine serum until the desired volume is reached. Once the working volume has been reached, the cells are left to divide until all nutrients are utilised and cell death occurs; this takes about 10 days. The cell debris can then be removed by centrifugation and the antibody harvested from the tissue culture medium. For some applications, the antibody can be used in this form without further processing.

Monoclonal antibodies can also be produced in ascitic fluid in mice by injecting hybridoma cells into the peritoneum. The hybridoma cells grow to high densities and secrete high levels of the monoclonal antibody of interest into the ascitic fluid, which is harvested by aspiration with a syringe and needle. Nude mice have no T cells and thus possess a poor immune system. They are often used for ascitic fluid production as they do not mount an immune response to the implanted cells. The mice should be naïve, i.e. not be immunised prior to use, as it is important that the only antibody present in the ascitic fluid is derived from the implanted cells.

A number of in vitro bioreactor systems have been developed to produce high yields of monoclonal antibody in small volumes of fluid, which mimics ascitic fluid production (Figure 7.6). All of them rely on physically separating the cells from the culture medium by a semi-permeable membrane that allows nutrient transfer, but prevents monoclonal antibodies from crossing. The culture medium can be changed to maximise cell growth and health, and fluid can be removed from around the cells

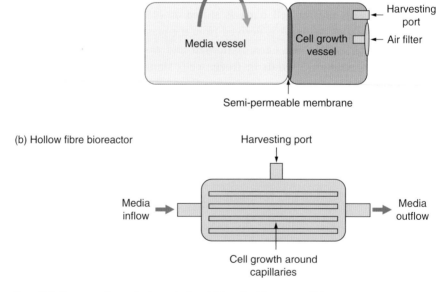

(a) Rotating bioreactor

Media vessel

Cell growth vessel

Harvesting port

Air filter

Semi-permeable membrane

(b) Hollow fibre bioreactor

Harvesting port

Media inflow

Media outflow

Cell growth around capillaries

Figure 7.6 Bioreactors for antibody production. (a) Rotating bioreactor utilising two compartments to separate growth medium and cells from antibodies. (b) Hollow fibre bioreactor utilising a capillary network to separate growth medium from cells and antibodies.

to harvest antibody. Some are based on a rotating cylinder with a cell-growing compartment at one end and separated from the media container by a membrane. Others have capillary systems formed from membrane running through the cell culture compartment and in these, the media is pumped through the cartridge to facilitate nutrient and gas exchange. These systems do produce high yields of antibody, but their set up and maintenance requires a significant effort. Nevertheless, they are ideal where large quantities of monoclonal antibody are needed and space is at a premium. They are, however, prone to contamination by yeasts and great care must be exercised when handling them. Cells are grown in bioreactors for up to 6 weeks, so the clone used must be stable and it is advisable to monitor its long-term growth prior to embarking on this form of culture. The major advantage of bioreactor cultures is that the antibody is produced in high concentration without the presence of media components, hence making it easy to purify. Total quantities per bioreactor run may be several hundred milligrams to gram quantities.

7.2.5 Antibodies Recognising Small Molecules

The immune system will recognise foreign proteins if they have a molecular mass above 2000–5000 Da. The magnitude of the response increases proportionally to the molecular mass. If an antibody is to be generated against a smaller molecule, it has to be conjugated to a carrier in order to effectively increase its size above the threshold for immune surveillance. A small molecule that is not antigenic by itself, but is eliciting an immune response when attached to a large carrier molecule is called a hapten. Carriers are macromolecules that bind these small haptens and enable them to induce an immune response. Most carriers are secretory or cell surface proteins that are naturally exposed to the immune system. Additionally, some polymers are known to act as carriers. If a polyclonal antibody is generated, it is advisable to change the carrier protein at least once in the immunisation procedure as this favours more antibody being made to the hapten and less to each of the carrier proteins. If a monoclonal antibody is generated, then the carrier protein can be the same throughout the immunisations. When screening hybridomas for monoclonal antibody production, it is necessary to screen against the hapten and carrier separately and to select antibodies responding to the hapten only.

7.2.6 Anti-Idiotypic Antibodies

An idiotope is the unique set of epitopes of the variable portion of an antibody. An antibody can have multiple idiotopes, which together form the idiotype of the antibody. If a new antibody is generated that specifically binds to an idiotope of a previously described antibody, it is termed an anti-idiotypic antibody. The binding site of an anti-idiotypic antibody is a copy of an original epitope. Some human cancers have unique cell surface tumour antigens that can be targeted for antibody therapy. Anti-idiotypic vaccines contain mouse monoclonal antibodies with designated idiotopes, mimicking these antigen sequences and thus triggering an immune response against the tumour antigen. By way of example, the recently developed anti-idiotypic vaccine racotumomab acts as a therapeutic vaccine for the treatment of solid tumours.

7.2.7   Phage Display for Development of Antibody Fragments

Bacteriophage or, in short, phage, are viruses that infect and replicate within bacteria. They can be engineered by molecular methods to express proteins, and if the nucleotide sequence corresponding to the foreign protein is fused to the coat protein gene, the foreign protein will be expressed on the virus surface. It is possible to isolate the variable (V) antibody coding genes from various sources and insert these into the phage, resulting in single-chain antigen-binding (scFv) fragments. Whereas whole antibodies are too large and complex to be expressed by this system, the scFv fragments can be expressed and used for diagnostic purposes. The DNA used in this process may come from immunised mouse B cells. The V genes are cloned into the phage producing a library, which is then screened to isolate clones that specifically bind antigen immobilised onto a solid surface. During the library screen, the desirable clones bind to the antigen and those that do not bind (no recognition) are washed off. Bound clones are eluted and put through further screens. The repeated cycling of these steps results in a phage mixture that is enriched with relevant (i.e. binding) phage. This process is called panning, referring to the gold-panning technique used by nineteenth-century prospectors. Selected phages can be multiplied in their host bacterium, sequenced and the 'best' scFv fragments can subsequently be used to develop ELISA and other immunoassays.

7.2.8   Antibody Purification

The choice of method used for the purification of antibodies depends very much on the fluid that they are in. Antibody can be purified from serum by the addition of chaotropic ions, typically ammonium sulfate. At around 60% saturation with ammonium sulfate, a precipitate is obtained that mainly contains antibody, thus providing a rapid method for IgG purification. This method does not work well in tissue culture supernatant, since media components such as ferritin are coprecipitated. However, ammonium sulfate precipitation may be used as a preparatory method prior to further chromatographic purification.

Prior to purification by affinity chromatography, the tissue culture supernatant is often concentrated, as a reduced volume simplifies the process. Tangential flow devices and centrifugal concentrators may be used to reduce the volume to 10% of the starting amount (Figure 7.7). Antibodies from both polyclonal and monoclonal sources can be purified by similar means. In both cases, the antibody type is IgG, which allows purification by protein A/G affinity chromatography. Protein A/G is a recombinant fusion protein that combines the IgG binding domains of both protein A and protein G. These proteins are derivatives of bacterial cells and the ability to reversibly bind IgG molecules forms the pivotal step in this affinity purification. Binding to the column occurs at neutral pH and the pure antibody fraction can be eluted at pH 2.0. Fractions are collected and neutralised to yield pH 7.0. Antibody-containing fractions are identified by spectrophotometry using absorbance at 280 nm (specific wavelength for protein absorbance; see Section 13.2.4) and are pooled. A solution of IgG at 1 mg cm^{-3} has an absorbance reading of approximately 1.4 at 280 nm. This can be used to calculate the amount of antibody in specific aliquots after purification.

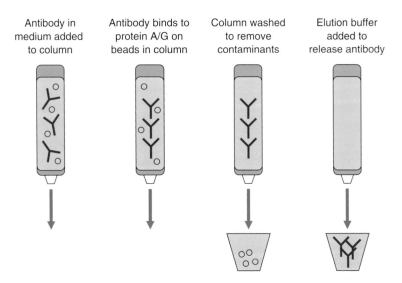

Figure 7.7 Affinity chromatography of antibodies using immobilised protein A/G.

Purified antibody should be adjusted to 1 mg cm^{-3} and kept at 4 °C, or −20 °C for long-term storage. Typically, 0.02% sodium azide is added to the antibody solution in order to increase shelf-life by suppressing the growth of adventitious microorganisms. Antibodies can be stored for several years at 4 °C, and for decades if kept below −20 °C, without losing activity.

7.2.9 Antibody Modification

For most of the common immunoassays, Western blotting, ELISA, immunofluorescence, immunohistochemistry and immunocytochemistry, as well as flow cytometry (see Chapter 8), there are numerous antibody labelling protocols for the detection and quantification of antigens. The label can be attached by one of two approaches:

- direct: the label is attached via a covalent bond to the primary antibody
- indirect: unlabelled primary antibody is bound to the antigen and a secondary labelled reagent is used for quantification.

Antibodies can be labelled by the addition of a marker enzyme such as horseradish peroxidase (HRP) or alkaline phosphatase (AP). Linkage is achieved by cross-linking chemistry (e.g. glutaraldehyde) to provide stable antibody–enzyme conjugates. Conjugation of an antibody to HRP is carried out in two stages. First, glutaraldehyde is coupled to reactive amino groups on the enzyme. The glutaraldehyde-tagged HRP is then purified by size-exclusion chromatography (Section 5.8.3) to remove unreacted cross-linker. In a second step, glutaraldehyde-tagged HRP is added to antibody solution, and the cross-linker reacts with amino groups on the antibody, thus forming a covalent link between the antibody and HRP. Alkaline phosphatase can be linked to the antibody by glutaraldehyde using a one-step conjugation.

A number of proprietary labelling reagents are also available for preparing antibody–enzyme conjugates. Fluorescent labels are required in immunofluorescent assays; here, fluorescein is usually the molecule of choice. Often, the derivative used for labelling is **fluorescein isothiocyanate (FITC)**. In FITC, an isothiocyanate group (–N=C=S) is incorporated as a reactive group that readily couples to primary amines and thus yields fluorescent conjugates of antibodies.

The interaction of biotin and avidin or streptavidin is one of the strongest known non-covalent interactions between two proteins and has been exploited for use in many applications, including detection of proteins, nucleic acids and lipids, as well as protein purification. Biotin is a small molecule that when introduced to biologically active macromolecules, such as antibodies, in most cases doesn't change activity. Streptavidin is a tetrameric protein, with each subunit binding one molecule of biotin. The bond is highly specific and very strong. Due to these properties, the **streptavidin–biotin bridge** as a very high affinity protein–ligand interaction and often utilised to link proteins in immunoassays. In typical applications, the antibody is conjugated to biotin, and streptavidin is conjugated to a fluorophore (see Section 13.5.1) or enzyme, thereby providing signal amplification and increased sensitivity so that antigens that are expressed at low levels are more likely to be detected (Figure 7.8).

It is possible to link antibodies to gold particles for use in **immunosorbent electron microscopy** (ISEM; see Section 7.4.2) and **lateral flow devices** (Section 7.5). Gold particles are prepared by citrate reduction of auric acid. The size of particle is predictable and can be controlled by pH manipulation. The gold particles are reactive and will bind antibodies to their surface, thus forming **immunogold**. The immunogold particles are stable and can be stored at 4 °C until required.

Similarly, rare earth elements (**lanthanides**) can be used as labels and because each lanthanide fluoresces at a different wavelength, a single assay can detect the presence of two or three different antibodies that can be individually visualised at the same time (multiplexing). The lanthanides are attached to the antibodies as a chelate, with the most common one being diethylenetriamine pentaacetate (DTPA).

Antibodies attached to **latex particles** can be used either as the solid phase for an immunoassay or as markers in lateral flow devices. Magnetised latex particles

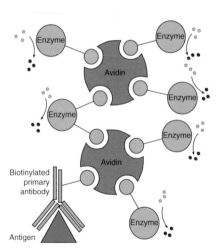

Figure 7.8 Streptavidin–biotin complex.

are available, allowing the easy separation of the latex particle–antibody–antigen complex from a liquid phase. There are also variants of latex and magnetic particles that have protein A covalently attached to their surface. Since protein A binds the Fc portion of an antibody, the latter is orientated such that the antigen binding site is facing outwards, thus maximising the chance of successful encounter.

7.3 IMMUNOASSAY FORMATS

The first immunoassay formats described were methods based on the agglutination reaction (Figure 7.9), which is characterised either by gel formation in a liquid phase or as an opaque band in an agar plate assay. Agglutination only occurs when the right amounts of antibody and antigen are present. It relies on the fact that the two binding sites of an antibody can bind to different antigen particles, thereby forming a bridge between two antigen molecules. The resulting lattice forms a stable structure where antigen and antibody particles are suspended in solution by their attachments to each other. For this to take place, equivalent amounts of antigen and antibody have to be present. If too much antibody is present, each antigen molecule will bind multiple antibodies and the meshwork will not develop. In contrast, if too much antigen is present, each antibody will bind only one antigen particle and no lattice will form. To find the condition of equivalence, a dilution series of antibody is often made; the determination of antigen concentration is then possible from the end point at which agglutination occurs.

The Ouchterlony double diffusion agar plate method is the most common gel-based assay system used (Figure 7.10). Wells are cut in an agar plate and samples are loaded containing either an antigen or an antibody solution. The antigen and the antibody diffuse through the agar and if the antibody recognises the antigen within the gel a precipitin band is formed due to agglutination. Depending on how many surface epitopes of the antigen match the antibody tested, the precipitin bands will be more or less pronounced. A diffusion gradient is formed through the agar as the reagents progress; hence there is no requirement for dilutions to ensure that agglutination occurs.

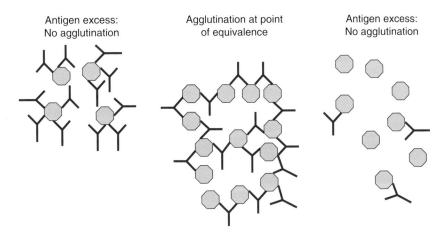

Antigen excess:
No agglutination

Agglutination at point
of equivalence

Antigen excess:
No agglutination

Figure 7.9 Agglutination reaction.

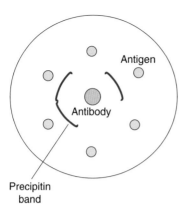

Figure 7.10 Ouchterlony double diffusion plate. Bands occur where antigen and antibody form an immunoprecipitate.

Antigen

Antibody

Precipitin band

Antibodies raised against an antigen from one organism will react to a greater or lesser degree with similar antigens from a related organism, depending on how many surface epitopes they share. This has given rise to a systematic method of identification of microorganisms known as serotyping. The system works well with closely related organisms, but is not definitive, as it only assesses surface markers. The pattern of precipitin bands obtained for reference antibodies is specific and can be used to classify samples into serotypes. By way of example, the method was used until very recently to characterise *Salmonella* strains. Modifications of the agglutination reaction involve the use of antibody-bound latex particles, which allow the reaction to be observed more easily in the liquid phase, typically carried out in round-bottomed microtitre plates. As discussed before, the agglutination reaction only occurs at the point of equivalence. A positive agglutination test appears cloudy to the eye as latex particles are suspended in solution. A negative result is characterised by a 'button' at the bottom of the reaction vessel that is formed by non-reacted particles. A negative result may indicate the mismatch of antibody to antigen or it may be obtained from either excess antigen or excess antibody. Agglutination immunoassays are commonly used for the detection of microbial pathogen antigens in blood serum as they provide rapid results with a minimum of equipment.

7.3.1 Enzyme Immunosorbent Assays

The vast majority of immunoassays carried out today fall into the category of enzyme-linked immunosorbent assays (ELISAs). These are routinely used as diagnostic methods that detect antigens of infectious agents or antibodies against them in bodily fluid. In this technique, the antigen or antibody from the sample tested is immobilised from the liquid phase (plasma or serum) onto a solid phase, typically wells of a microtitre plate. Immobilisation is achieved either directly, or indirectly by the use of a coating antibody (also called capture antibody), which actively traps antigen in the solid phase. The immobilised antigen will bind specific antibodies; detection of specifically bound antibody or of the antigen itself is thus achieved with another antibody, which is labelled with a reporter enzyme producing a signal (usually a colour change) when the enzyme substrate is added to the antibody–antigen–enzyme complex (Figure 7.11).

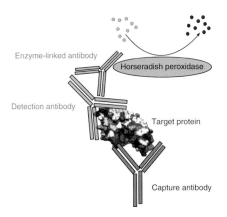

Figure 7.11 Schematic diagram of the sandwich ELISA.

There are many variations of this method, but all of them rely on the formation of the antibody–antigen complex and its detection by the reaction of the reporter enzyme. Since these assays rely on a stepwise addition of layers, with each one being linked to the one before, they are collectively termed sandwich ELISA. In the first layer, the coating antibody irreversibly binds the antigen; in this process, the available antigen is concentrated until saturation has been reached. This is particularly useful when testing for low levels of antigens in fluids such as blood serum. After the antigen has been applied and prior to addition of the detecting antibody, plate blocking may be done by addition of a protein that does not specifically interfere with the ELISA (e.g. casein from milk powder or bovine serum albumin). Depending on how many detection layers are employed, enzyme immunosorbent assays can be classified into direct (e.g. double antibody sandwich) and indirect (e.g. triple antibody sandwich) ELISAs (see also Table 7.2).

Table 7.2 **Comparison of direct and indirect ELISA**

Direct (double antibody sandwich) ELISA	Indirect (triple antibody sandwich) ELISA
Quick – only one antibody and few experimental steps	Additional incubation and wash steps required
Cross-reactivity of secondary antibody is eliminated	Cross-reactivity might occur with the secondary antibody, resulting in a non-specific signal
Immunoreactivity of the primary antibody might be adversely affected by labelling	Maximum immunoreactivity of the unlabelled primary antibody is retained
Labelling primary antibodies is time-consuming and expensive. No flexibility in choice of primary antibody label from one experiment to another	High flexibility – many primary antibodies can be combined with the same labelled secondary antibody. Different visualisation systems can be trialled with the same primary antibody
Minimal signal amplification	Sensitivity may be increased because each primary antibody contains several epitopes that can be bound by the labelled secondary antibody, allowing for signal amplification

7.3.2 Double Antibody Sandwich ELISA (DAS ELISA)

Double antibody sandwich (DAS) ELISA (also called direct ELISA) is probably the most widely used immunochemical technique in diagnostics. The principle is that the antigen is immobilised on a solid phase by a primary coating antibody and detected with a second antibody that has been labelled with a marker enzyme (Figure 7.12). The antigen creates a bridge between the two antibodies and the presence of the enzyme causes a colour change in the chromogenic (colour-producing) substrate (see also Figure 6.12). The marker enzyme used is usually either horseradish peroxidase or alkaline phosphatase. Other enzymes have been reported to give higher sensitivity, but this is at the expense of more complex substrates and buffers. In some systems, the enzyme is replaced with a radioactive label and this format is known as the **immunoradiometric assay** (IRMA). DAS ELISA is used extensively in horticulture and agriculture to ensure that plant material is free of virus. Potato tubers, for example, that are to be used as seed for growing new crops have to be free of potato viruses, and screening in particular for potato leafroll virus (PLRV) is carried out by DAS ELISA. PLRV antibodies are coated onto the wells of ELISA plates and then the sap to be tested is added. After incubation, the plates are washed and PLRV antibody conjugated to alkaline phosphatase is added. After further incubation and washing, the substrate is used to identify the positive wells, caused by the presence of the antigen (PLRV) in the sandwich of antibodies.

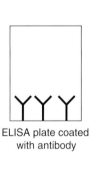
ELISA plate coated with antibody

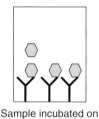
Sample incubated on plate

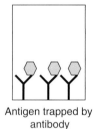
Antigen trapped by antibody

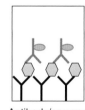
Antibody/enzyme conjugate incubated on plate

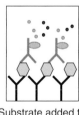
Substrate added to plate causes a colour change in positive wells

Figure 7.12 Schematic diagram of the double antibody sandwich (DAS) ELISA.

7.3.3 Triple Antibody Sandwich ELISA (TAS ELISA)

The triple antibody sandwich (TAS) ELISA, also known as indirect ELISA, is a method often used to identify antibodies against pathogens in patient blood to diagnose infection. Again, layers of reagents are built up, each dependent on the binding of the

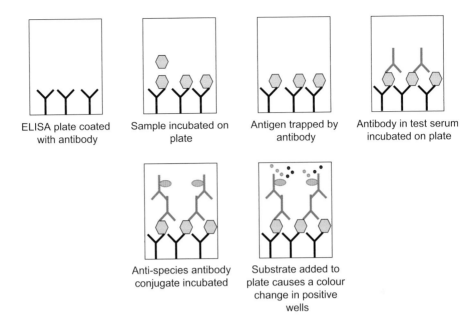

ELISA plate coated with antibody	Sample incubated on plate	Antigen trapped by antibody	Antibody in test serum incubated on plate

Anti-species antibody conjugate incubated	Substrate added to plate causes a colour change in positive wells

Figure 7.13 Schematic diagram of a triple antibody sandwich (TAS) ELISA.

previous one (Figure 7.13). As an example of a diagnostic test, hepatitis B virus (HBV) capture antibody is bound to the wells of a microtitre plate and coat protein from the virus (the antigen) is added. After incubation and washing, patient serum is added, which, if it contains antibodies, reacts with the antigen. Anti-human antibody conjugated to an enzyme marker is added, which will bind to the patient antibodies. Finally, substrate is added to identify samples which were positive for HBV antigen. The test works well for the diagnosis of HBV infection and is also used to ensure that blood donations given for transfusion are free from this virus.

7.3.4 Enhanced ELISA Systems

The maximum sensitivity of an ELISA is in the picomole range and there have been many attempts to increase the detection threshold beyond this. The physical limitations are based on the dynamics of the double binding event and the subsequent generation of signal above the background value. In some assays, antibody binding can be marginally improved by temperature optimisation. There are two basic ways to amplify the signal in ELISA:

- More enzyme can be bound by using **multi-valent** attachment molecules; e.g. the properties of the streptavidin–biotin binding systems allow amplification through this route. The small biotin molecule combined with the tetravalent structure of streptavidin (i.e. having four binding sites) is the property that produces the amplification (Figure 7.8). The detection antibody is labelled with biotin and the reporter enzyme with streptavidin. The high affinity and multi-valency of the reagents allows larger complexes of enzyme to be linked to the detection antibody, producing an increase in substrate conversion and improved colour development in positive samples.

- Alternatively, the substrate reaction can be enhanced by using a two-enzyme system. In this case, the primary enzyme bound to the antigen catalyses a change in a second enzyme, which then generates a signal. Both of these methods will increase the signal generated, but may also increase the background reaction.

7.3.5 Competitive ELISA

Competitive ELISA is used when the antigen is small and has only one epitope (Figure 7.14). The principle is based on the competition between the natural unlabelled antigen to be tested for a labelled (e.g. enzyme-conjugated) form of the antigen which is the detection reagent. The test sample and a defined amount of enzyme-conjugated antigen are mixed together and placed into the antibody-coated wells of a microtitre plate. The antigen and the conjugated derivative compete for the available spaces on the coating antibody layer. The more natural unlabelled antigen present the more it will displace (compete out) the labelled form, leading to a reduction in enzyme bound to the plate. A decrease in signal from labelled antigen indicates the presence of the antigen in the test samples when compared to assay wells with labelled antigen alone. This form of ELISA is routinely used for testing blood samples for thyroxine, a hormone that is responsible for regulating metabolic rate; a deficiency (hypothyroidism) or excess (hyperthyroidism) results in retarded or accelerated metabolism, respectively. Patients suffering from hypothyroidism can be given additional thyroxine if required, but it is important to establish the baseline level before treating the condition. Competitive ELISA provides an accurate measure of the circulating level of the hormone compared to a standard curve of known dilutions.

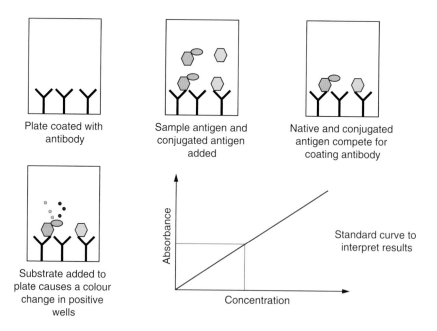

Figure 7.14 Competitive ELISA.

Most frequently, competitive ELISA involves enzyme-labelled probes, but radioactive labelling is done as well, and this form of competitive ELISA is known as radioimmunoassay (RIA).

Competitive ELISA using conjugated antibody can also be used to quantify levels of circulating antibody in test serum. Here, the solid phase holds the antigen (coated onto the microtiter plate wells) and the antibody is delivered in solution. The test serum and the labelled form of the antibody are mixed together and added to the reaction wells to compete for binding to the antigen. As before, the level of antibody is determined by the reduction in signal observed by addition of the substrate.

7.3.6 Modifications of the Traditional Sandwich ELISA

Dissociation-Enhanced Lanthanide Fluorescence Immunoassay (DELFIA)

DELFIA is a time-resolved fluorimetric assay that relies on the unique properties of lanthanide chelate antibody labels (see Section 7.2.9). Lanthanides emit a fluorescent signal when stimulated with light of a specific wavelength. The fluorescence emission (see Section 13.5) has a long decay period, which allows time for autofluorescence to diminish, thereby increasing sensitivity due to a reduced background signal. DELFIA offers a signal enhancement greater than that possible from conventional enzyme-linked assays. The lanthanide chelates are conjugated to the secondary antibody and as there are a number of lanthanides available, each with a unique signal, multiplexing (more than one test carried out in the same reaction vessel) is possible. The assay is carried out in a similar way to standard ELISA and may be competitive or non-competitive.

Enzyme-Linked Immunospot Assay (ELISPOT)

ELISPOT involves an ELISA-like capture and measurement of proteins secreted by cells that are plated on membrane-coated microtitre plate wells. Using the sandwich principle, the proteins are captured locally as they are secreted by the plated cells, and detection is achieved with a precipitating substrate. The spots can be counted manually with a dissecting microscope or using an automated reader. In recent years, ELISPOT in vitro tests for the diagnosis of tuberculosis and Lyme disease became commercially available.

7.4 IMMUNO MICROSCOPY

7.4.1 Immunofluorescence Microscopy

Immunofluorescence (IF) microscopy uses antibodies conjugated to fluorescent markers in order to locate specific structures in specimens and allows them to be visualised by illuminating them with ultraviolet light. Very frequently used labels include fluorescein (producing green light) and rhodamine (emitting red light). Microscopes equipped for IF have dual light sources, allowing the operator to view the specimen under white light before illuminating with ultraviolet (fluorescence excitation) to look for specific fluorescence emission. The technique is particularly useful for the whole-cell staining technique in bacteriology, studies of membranes and surface markers on eukaryotic cells, as well as real-time studies of migration, endocytosis and the fate of labelled receptors in living cells (see also Section 11.2.5).

7.4.2 Immunosorbent Electron Microscopy

Immunosorbent electron microscopy (ISEM) is a diagnostic technique used primarily in virology. Virus-specific antibodies conjugated to gold particles are used to visualise the virus particles. The gold is electron-dense and is seen as a dark shadow against the light background of the specimen field. The technique can be used for both transmission or scanning systems. If gold-labelled primary antibodies are not available, anti-IgG–gold conjugated antibodies can be used, together with a primary antibody, in a double antibody system. Both monoclonal and polyclonal antibodies can be used for ISEM, depending on the required specificity.

7.5 LATERAL FLOW DEVICES

Lateral flow devices (LFD; see also Section 14.6.2) are used as rapid diagnostic platforms allowing for almost instantaneous results from fluid samples (Figure 7.15). They are usually supplied as a plastic cassette with a port for applying the sample and an observation window for viewing the result. The technology is based on a solid phase consisting of a nitrocellulose or polycarbonate membrane, which has a detection zone coated with a capture antibody. The detection antibody is conjugated to a solid coloured marker, usually latex or colloidal gold, and is stored in a fibre pad, which acts as a reservoir. The solid phase has a layer of transparent plastic overlaying it,

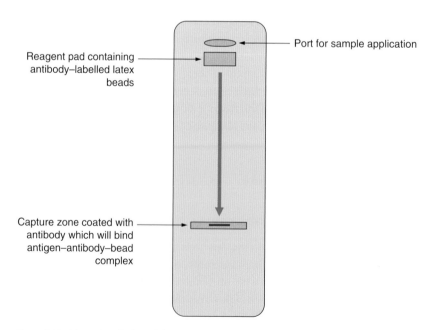

Figure 7.15 Schematics of a lateral flow device. The sample migrates from the application port to the capture zone by capillarity. If antigen is present in the sample, a coloured line is formed. See also Section 14.6.2 for the working schematics of the device.

leaving a very narrow gap, which will draw liquid by capillarity. The sample is applied to the reservoir pad through the sample port, where it can react with the conjugated antibody if the specific antigen is present. The liquid then leaves the reservoir and travels up the solid phase pad to the location of the capture antibody. If the sample contains the specific antigen it will bind to both the conjugated and capture antibodies. This results in a coloured line if the sample is positive. If the sample is negative, then no coloured line develops. The system lends itself to multiplexing and up to three different antigens can be tested for simultaneously with appropriate capture antibodies and different coloured marker particles. The technology has been applied to home pregnancy testing and various other self-diagnostic kits. Lateral flow devices are also used by police forces and regulatory authorities for the rapid identification of illicit drugs.

7.6 EPITOPE MAPPING

Epitope mapping is carried out to establish where on the target protein the antibody binds. For example, when testing novel monoclonal antibodies it is often necessary to know the precise epitope to which binding occurs. This, however, can only be performed where the epitope is linear. A linear epitope is formed by a contiguous sequence of amino acids in the protein sequence of the antigen and the antibody binds to the structure that they form. Non-linear epitopes are formed by non-contiguous amino acids that are nevertheless positioned close to each other in space due to the fold adopted by the protein (see Section 2.2.2), as is the case, for instance, in helical or hairpin structures. To carry out epitope mapping, the amino-acid sequence of the target antigen must be known. The sequence is then used to design and generate synthetic peptides, each around 15 amino-acid residues in length and overlapping with the previous one by about five residues. The synthetic peptides are then coated onto the wells of microtitre plates or onto nitrocellulose membranes and mixed with the antibody of interest. The eventual engagement is visualised by using a secondary antibody–enzyme conjugate and substrate. Based on the results obtained for the individual peptides and the position of the sequence in the native protein, it is possible to predict where the epitope lies and also what its sequence is.

7.7 IMMUNOBLOTTING

This technique is also known as **Western blot analysis** and is used to identify proteins from biological samples after electrophoresis. The sample may be subjected to electrophoresis under reducing or non-reducing conditions until separation of individual proteins by size is achieved. The separated proteins are usually visualised by staining with a general protein stain (see Sections 5.2 and 6.3). They can be transferred onto a nitrocellulose or polyvinyl membrane by using an electroblotter. After protein transfer, the membrane is treated with a protein-blocking solution to prevent non-specific binding of antibody to the membrane itself. Popular blocking compounds are casein (milk powder) or bovine serum albumin (BSA). Often, indirect methods are used for reasons of cost. For indirect labelling, the membrane is incubated in antigen-specific primary antibody solution and, after washing, treated with a solution of a secondary enzyme-linked antibody that specifically recognises the species of the primary antibody. Both peroxidase and alkaline phosphatase have substrates that produce a visible colour on the blot where the antibodies have bound. However, sensitivity can be increased by the use of luminol as a substrate for peroxidase; the emitted light is detected by exposure to photographic film or a charge-coupled device (CCD). The substrate reaction can be stopped after optimum colour development, and the blot dried and stored for reference. Further details can be found in Section 6.3.8.

7.8 FLUORESCENCE-ACTIVATED CELL SORTING (FACS)

Fluorescence-activated cell sorting (FACS; see also Sections 14.6.3 and 8.3) machines are devices that are capable of separating populations of cells into groups of cells with similar characteristics based on antibody binding (Figure 7.16). The technique is used on fixed or live cells, the latter allowing recovery and subsequent culture of the cells after separation. Many cell markers identify subsets of cell types and specific antibodies for these markers are available. The method can be used quite successfully to separate normal from abnormal cells in bone marrow samples from patients with leukaemia. This can be used as a method of cleaning the marrow prior to autologous (i.e. derived from the patient themselves) marrow transplantation.

The technique is also used for diagnostic tests where the numbers of cell subtypes need to be known. This is of particular use when looking at blood-borne cells, such as lymphocytes, where the ratios of cell types can be of diagnostic significance. For example, in HIV infection, the numbers of specific T cell subtypes are of great diagnostic significance in the progress of the infection to AIDS. The cells are labelled with fluorescent-tagged antibodies to specific cell markers and passed through a narrow

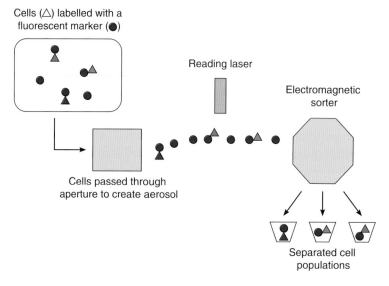

Figure 7.16 Fluorescence-activated cell sorting.

gauge needle to produce an aerosol. The droplet size is adjusted so that each one should contain only one cell. The aerosol is passed through a scanning laser, which allows detection of the fluorescent label. The droplets have a surface charge and can be deflected by an electromagnet based on their fluorescent label status. The system relies on computer control to effectively sort the cells into labelled and non-labelled populations. The desired cell population can subsequently be recovered and counted or cultured. More than one label can be used simultaneously, allowing multiple sorting.

7.9 CELL AND TISSUE STAINING TECHNIQUES

There are many antibodies available that recognise receptors on and structural proteins in cells and tissues, and these can be of diagnostic use. Generally, immunostaining is carried out on fixed tissues, but staining of living tissues may be required in some cases, since fixation may destroy the structure of epitopes or dynamic events need to be analysed.

Fixed tissues are prepared by standard histological methods. The tissue is fixed with a preservative that kills the cells, but maintains structure and makes the cell membranes permeable. The sample is embedded in wax or epoxy resin and fine slices are taken using a microtome; the slices are then mounted onto microscope slides. The antibodies used for immunopathology may carry enzyme, fluorescent markers or labels such as gold particles. They may also be unconjugated, and in this case require a secondary antibody conjugate and substrate for visualisation. It is important to remember that enzymes such as alkaline phosphatase may be endogenous (found naturally) in the tissue sample and their activity is not easily blocked; therefore, horseradish peroxidase is often used as an alternative. Any endogenous peroxidase activity in the sample can be blocked by treating the sample with hydrogen peroxide. Antibodies may recognise structural proteins within the cells and can access them in fixed tissues through the permeabilised membranes. More than one antibody can be used to produce a composite stain if different coloured markers are used. Combinations of fluorescent and enzyme staining may also be used, but this has to be carried out sequentially. Fluorescent stains can be used in conjunction with standard histological stains viewed with a microscope equipped with both white and ultraviolet light. As fluorescence will decay over time, photographs can be taken through the microscope and kept as a permanent record.

7.10 IMMUNOCAPTURE POLYMERASE CHAIN REACTION (PCR)

Immunocapture PCR is a hybrid method that uses the specificity of antibodies to capture antigen from the sample and the diagnostic power of PCR to provide a result. The method is particularly useful in diagnostic virology, where the technique allows the capture of virus from test samples and subsequent diagnosis by PCR. It is valuable where levels of virus are low, such as in water samples and other non-biological sources. The technique can be carried out in standard PCR microtitre plates or in PCR tubes. The antibody is bound passively to the plastic of the plate or tube, and subsequently incubated with the sample containing the antigen (see Figure 7.17).

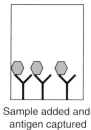

Plate coated with antibody   Sample added and antigen captured   Sample lysed to release nucleic acid, followed by PCR with specific primers

Figure 7.17 Immunocapture PCR.

After washing to remove excess sample material, the PCR reagents can be added and thermocycling carried out on the bound viral nucleic acid. RNA viruses will require an additional reverse transcription step prior to PCR (see also Section 4.10.4).

7.11   IMMUNOAFFINITY CHROMATOGRAPHY

Immunoaffinity chromatography (IAC) can be used for a number of applications. The principle is based on the immobilisation of antibody on a matrix (resin), normally beads, which are then placed in a chromatography column. Antibody may be permanently linked to the beads by covalent linkage to reactive sites on a resin or bound using proteins A or G. Since direct binding of the antibody to the beads is more stable, such columns can be repeatedly used following appropriate regeneration. IAC may be used as a clean-up method in analytical chemistry to extract small quantities of biological compounds, such as hormones, from patient samples. The columns are made by cross-linking highly purified antibody (monoclonal or polyclonal) to functional groups on the chromatography beads to form an affinity matrix. Harsh conditions must be avoided, since denaturation of the antibody molecules could occur.

A number of proprietary resins are available that have reactive sites suitable for antibody immobilisation. The affinity matrix is loaded into the chromatography column prior to use. Antibody binding of antigen generally occurs best at around pH 7.4, but individual monoclonal antibodies may vary considerably from this pH. Once the sample has been loaded onto the column, it should be washed to remove contaminating material from the sample fluid. Conditions for elution differ for individual antibodies and antigens, but pH 2.0 buffer, methanol or 10% acetonitrile have all been used successfully. The column can be regenerated after elution by incubating with pH 7.4 buffer. Samples eluted from IAC columns may be tested further by high-performance liquid chromatography, ELISA or other analytical techniques.

7.12   ANTIBODY-BASED BIOSENSORS

A biosensor (see also Section 14.6) is a device that is composed of a biological element and an electronic transducer that converts signal reception by the biological entity into an electrical impulse. The biorecognition elements that can be used are enzymes, whole cell receptors, DNA and antibodies. The reasoning in developing effective biosensors is that biosensors can cost less than traditional analytical techniques, require fewer reagents, provide faster turnaround times and higher sample throughput. Antibodies have the potential to be excellent biological molecules to use for this technology as they can be developed to detect virtually any molecule. The main problem with developing this technology with antibodies has been the lack of adequate physico-chemical transduction systems. Three methods have been developed that will provide a signal from antibody binding. Antibodies may be bound onto thin layers of gold, which in turn are coated onto refractive glass slides. If the slides are illuminated at a precise angle with fixed-wavelength laser light then electron waves are produced on the surface of the gold. This is known as surface plasmon resonance and only occurs if the incident angle and wavelength of light are precisely right. If the antibody binds antigen then the surface plasmon resonance pattern is changed and a measurable change in emitted energy is observed.

Fibre optic sensors have also been developed which rely on the natural ability of biological materials to fluoresce with light at defined frequency. The reaction vessel is coated with antibody and the fibre optic sensor used to illuminate and read light scatter from the vessel. The sample is then applied and the sample vessel washed. The fibre optic sensor is again used to illuminate and read backscatter from the vessel. Changes in the fluorescence will give a change in the observed returned light.

A third approach relies on changes in crystals as a result of surface molecules bound to them. Piezoelectric crystals generate a characteristic signature resonance when stimulated with an alternating current. The crystals are elastic and changes to their surface will produce a change in the signature resonance. The binding of antigen to antibody located on the surface of the crystal can be sufficient to alter the signature and therefore induce a signal indicating that antigen has been detected by antibody.

7.13 LUMINEX® TECHNOLOGY

Luminex® assays combine bead-based assay technology, flow cytometry and fluorescence detection into a fast-performing technique that delivers data for significant statistical analysis. The technology is based on polystyrene and paramagnetic microspheres (beads) that contain red and infrared fluorophores of differing intensities, allowing them to be differentiated. Individual bead sets are coated with a capture antibody specific for an analyte. Multiple analyte-specific beads can be combined in a single well of a microplate to detect and quantify up to 100 targets simultaneously using a Luminex® instrument for analysis. This allows high throughput analysis of many samples at one time in the homogeneous format of a multi-well plate. The required sample volumes are low. At the core of this technology are the particular colour-coded beads, on the surface of which the assay of choice is performed. In combination with a high-speed multi-well plate reader these assays can be automated, allowing faster throughput.

7.14 THERAPEUTIC ANTIBODIES

Therapeutic antibodies fall into a number of different classes, but are all designed to bind to specific structures or molecules with the aim of altering cellular or systemic responses in vivo . The simplest of these are the **inhibitory systemic** (found throughout the body) **antibodies** that will bind to substances to render them ineffective. In their crudest form, they consist of hyperimmune serum used to alleviate the symptoms of bites and stings from a number of poisonous animals. Antivenom produced in horses for treatment of snake bite is a good example of this. Hyperimmune serum derived from human patients who have had the disease has also been used prophylactically (i.e. to reduce the risk of disease) after exposure to pathogenic viruses. Hyperimmune serum is available to help treat a number of pathogenic viral conditions such as West Nile fever, AIDS and hepatitis B. The appropriate serum is used after exposure to the pathogen, for example by needle-stick injury, and help to reduce the risk of infection occurring.

The next class of therapeutic antibodies comprises those that bind bioactive molecules and reduce their effects in vivo . They are all monoclonal and have a number of targets that help to alleviate the symptoms of a number of human diseases. For example, targeting systemic cytokines that have been implicated in the progression of arthritis has shown encouraging results. Monoclonal antibodies can also be used to reduce the numbers of specific cell types in vivo by binding to their surface markers. The binding of the antibody to the cells alerts the immune system and causes the cells to be cleared from circulation. Chimeric (formed from two sources) monoclonal antibodies consisting of mouse variable regions and human constant regions that are specific to the B-cell marker CD20 have been used successfully for the treatment of systemic *lupus erythematosus*. This disease is characterised by the development of aberrant B cells secreting auto-antibodies that cause a number of autoimmune phenomena. The decrease in circulating B cells reduces the production of auto-antibodies and thus alleviates some of the symptoms.

Agonistic antibodies are monoclonal therapeutic antibodies which cause up-regulation of cellular processes and have the ability to influence living cells in vivo. They up-regulate cellular systems by binding to surface receptor molecules. Normally, cell receptors are stimulated briefly by their native ligand, which is cleared by a cellular process shortly after the binding event, thus resulting in a brief up-regulation. Agonistic monoclonal antibodies bind to the receptor molecule and mimic their native ligand, but have the capacity to remain in place for much longer since they cannot be cleared as quickly as the native ligand. The action of agonistic antibodies is incredibly powerful and the internal system cascades that can be triggered are potentially catastrophic for both the cell and the organism. Their use has been mainly restricted to induction of apoptosis (programmed cell death) in cancer cells and only where a known unique cellular receptor is being stimulated.

There are also a number of therapeutic inhibitory antibodies available and all of them down-regulate cellular systems by blocking the binding of antigen to receptor. They behave as competitive analogues to the inhibitor and remain bound to the receptor for a very long time (dwell time), thus increasing their potency. They may block the binding of hormones, cytokines and other cellular messengers. Such antibodies have been used successfully for the management of some hormone-dependent tumours (e.g. breast cancer) and also for the down-regulation of the immune system to help prevent rejection after organ transplantation.

Therapeutic antibodies need to be carefully engineered to make them effective as treatment agents. The avidity and affinity of the antibodies is critical to their therapeutic efficacy as their specific binding ability is critical to the length and specificity of action. Additionally, they must not appear as foreign to the immune system or they will be rapidly cleared by the body. Often, the original monoclonal antibody developed in the research phase has been derived using a mouse system and thus constitutes a murine antibody. These antibodies can be humanised by retaining the murine binding site and replacing the constant region genes with human ones. The resulting antibody escapes immune surveillance, but retains the effective binding capacity. Each engineered, therapeutic antibody has a different half-life in vivo, and this factor is of great importance when baseline dosage is being established. All of the currently used therapeutic antibodies may cause side effects in patients. Hence this line of therapy is only exploited when the benefits outweigh the potential problems that may be encountered. Great success has been achieved by using humanised monoclonal antibodies in the treatment of prostate cancer in men and breast cancer in women. In the future, there will be a rise in the availability of therapeutic antibodies both for the up- and down-regulation of cellular and systemic responses. Cancer therapy and immune modulation of autoimmune phenomena are probably the two areas where the greatest developments will take place.

7.15 SUGGESTIONS FOR FURTHER READING

7.15.1 Experimental Protocols

Burns R. (2005) *Immunochemical Protocols*, 3rd Edn., Humana Press, Totowa, NJ, USA.

Coligan J. (2005) *Short Protocols in Immunology*, John Wiley, New York, USA.

Howard G. and Kaser M. (2007) *Making and Using Antibodies: A Practical Handbook*, CRC Press, Boca Raton, FL, USA.

Subramanian G. (2004) *Antibodies: Volume 1, Production and Purification*, Kluwer Academic, Dordrecht, the Netherlands.

Wild D. (2005) *Immunoassay Handbook*, 3rd Edn., Elsevier, New York, USA.

7.15.2 General Texts

Cruse J. and Lewis R. (2002) *Illustrated Dictionary of Immunology*, 2nd Edn., CRC Press, Boca Raton, FL, USA.

Murphy K. and Weaver C. (2016) *Janeway's Immunobiology*, 9th Edn., Garland Science, New York, USA.

Parham P. (2014) *The Immune System*, 4th Edn., Garland Science, New York, USA.

7.15.3 Review Articles

Chattopadhyay P.K., Gierahn T.M., Roederer M. and Love J.C. (2014) Single-cell technologies for monitoring immune systems. *Nature Immunology* 15, 128–135.

Geering B. and Fussenegger M. (2015) Synthetic immunology: modulating the human immune system. *Trends in Biotechnology* 33, 65–79.

Shukla A.A. and Thömmes J. (2010) Recent advances in large-scale production of monoclonal antibodies and related proteins. *Trends in Biotechnology* 28, 253–261.

Tomar N. and De R.K. (2014) Immunoinformatics: a brief review. *Methods in Molecular Biology* 1184, 23–55.

7.15.4 Websites

Current Protocols in Immunology
dx.doi.org/10.1002/0471142735 (accessed May 2017)

8 Flow Cytometry

JOHN GRAINGER AND JOANNE KONKEL

8.1 INTRODUCTION

Flow cytometry is a technique that allows for the rapid measurement of optical and fluorescence properties of large numbers of individual particles. This is achieved by creating a stream of particles that move past one or multiple light beams and collect information about the light that is given off in all directions as single particles intersect the beam. By employing flow cytometry, multiple features of each particle can be investigated simultaneously, allowing for detailed characterisation of distinct populations within heterogeneous mixtures of particles. Most typically, the particles in question are eukaryotic cells, although they could be anything from bacteria to latex beads. Given that cells are by far the most common sample type analysed by flow cytometry, throughout this chapter, cell and particle will to some degree be used interchangeably.

The first flow cytometers, as we know them today, were developed in the late 1960s. These early machines were capable of detecting two fluorescence parameters, as well as using scattering of light to resolve certain cell populations based on their size and internal complexity. Over the subsequent 50 years, technical developments of flow cytometers themselves as well as fluorescence dyes enhanced the usage of this technology, which now allows the routine investigation of up to 20 fluorescence parameters. Given the vast array of parameters that can be measured, from specific proteins on cell membranes or inside cells to cellular components such as DNA and RNA, flow cytometry is used for a diverse range of purposes outside of its original research setting from genetic disease screening to detection of microbial contaminants in drinking water (Section 8.6).

An additional feature of some flow cytometers is the capability to sort specific particles based on the fluorescence properties interrogated. These cytometers, termed fluorescence–activated cell sorting (FACS) machines, allow for particles with desired parameters to be isolated, which can then be subjected to further biochemical or molecular analysis. One particularly exciting use of FACS in recent years has been in concert with high-throughput sequencing technologies (Chapter 20) to explore the molecular heterogeneity of human cell populations on a genome-wide scale. Thus, over the past 50 years, flow cytometry has found a role in numerous research and clinical settings and new uses continue to be established for flow cytometry in many cutting-edge applications.

Figure 8.1 The photo shows a typical flow cytometry instrumentation with the cytometer itself and attached computer used for data acquisition. Underneath the flow cytometer are the sheath fluid tank and the waste tank.

8.2 INSTRUMENTATION

In terms of construction, flow cytometers (Figure 8.1) comprise three main components:

- Fluidics system – this is critical to draw particles, such as blood cells, into the machine and channel them into single file
- Optical system – this system consists of lasers and the lenses that are used to focus the laser beam
- Signal detection and processing system – this converts the light signals to voltages that are then recorded.

Basic knowledge of how these components function aids the user to ensure not only optimal, but appropriate, set-up of the flow cytometer for different applications.

8.2.1 Fluidics System

Perhaps the key benefit of flow cytometry over many other widely used techniques is the capability to simultaneously assess multiple characteristics of individual cells or particles. Initially, when particles enter the flow cytometer, they pass through the sample injection port (SIP; Figure 8.2) and subsequently become distributed randomly in the space provided by the sample line, since the diameter of the latter is larger than that of most particles that are being analysed. In order to characterise each particle individually, the particles are first organised into a single stream that is then allowed to pass through the 'interrogation point' where it intersects the focussed light from the laser beams.

The fluidics system has two components: a central core, through which the sample fluid is injected, and an outer sheath fluid. These two components are pushed through

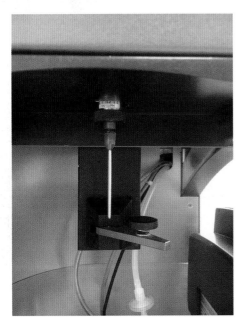

Figure 8.2 Sample injection port (SIP). This is the port through which the cells are brought into the flow cytometer and subsequently channelled into a single stream of cells.

the system at slightly different pressures. Of note, the sheath fluid is pushed through at higher pressure than the sample under laminar flow. A drag effect is created, causing the central core of sample to become narrower, aligning the particles in single file and allowing for uniform illumination. This is termed **hydrodynamic focussing** (Figure 8.3).

It is possible to manipulate the flow rate of cells passing the laser beam, depending on the purpose for which the flow cytometer is being used. For example, high flow rates are appropriate for immunophenotyping cells (as outlined in Section 8.5); in contrast, if the DNA content of the cell is assessed (as outlined in Section 8.6.4) then a slower injection rate may be required to ensure sufficiently high resolution. A slow flow rate decreases the diameter of the sample stream and increases the uniformity and accuracy of illumination. It is important to note that proper functioning of the fluidics system is critical to allow for appropriate interception between the particle and the laser beam. Importantly, the user needs to ensure that the fluidics system is free of air bubbles and debris, and is properly pressurised at all times.

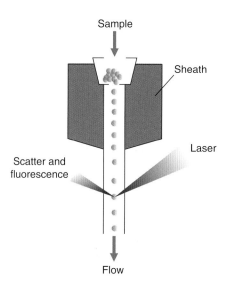

Figure 8.3 Hydrodynamic focussing to generate a stream of single particles. Following entry into the flow cytometer via the SIP, cells are focussed into a single stream.

8.2.2 Optical System

Following alignment of the particles into a single file they pass through the interrogation point where they are intercepted by a laser beam, providing the incident light. Together with the lenses and collection optics, these components form the optical system of the flow cytometer.

Lenses are used to focus the laser light that hits the particle resulting in **light scattering** (see also Section 14.3.5) around the edges of the particle (Figure 8.4). Two types of scattered light are analysed:

- **Forward scatter** (FSC) is the light collected along the same direction as the incident laser beam (and up to about 20° offset from that direction). The intensity of

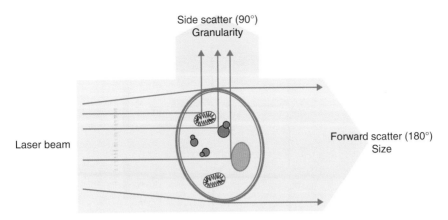

Figure 8.4 Light scattering by a cell flowing past the interrogation point. As the cell passes through the laser beam, scattered light is acquired along the same direction as the laser beam (forward scatter, FSC; information about the cell size) and at a perpendicular direction (side scatter, SSC; information about granularity or internal complexity).

forward-scattered light is proportional to the cell surface area and thus provides an estimate of cell size. Larger cells tend to diffract and refract more light than smaller cells. Diffracted light arises due to bending of the incident light when it hits the edge of the cell. In contrast, refracted light occurs due to bending when it enters the translucent cell and then enters the air again. For either phenomenon, a larger cell results in more scattered photons.

- Side scatter (SSC) is a measure of reflected light (i.e. light bouncing off the surface) that is collected at a direction perpendicular to the incident laser beam. SSC is related to the cell granularity or internal complexity (more granules result in more pronounced reflection of light). For example, eosinophils that are highly granular will have a higher SSC than lymphocytes.

FSC and SSC properties can be used to differentiate specific cell types within a heterogeneous cell population such as the blood (Section 8.5, see also Figure 8.10).

If fluorescent dyes or fluorescently conjugated antibodies are used to detect specific attributes of the particle, then, depending on the dyes employed, different wavelengths of incident light and laser types (gas lasers, diode lasers) can be used to excite the fluorescent moiety by either red, yellow-green, blue, violet or ultraviolet light. Collection of the emitted light from the fluorophore uses a system of optical filters and mirrors that separate and point light of defined wavelengths towards the appropriate detectors (Figure 8.5). Depending on the required bandwidth and the fluorescence signal to be detected, different types of filters are used (Figure 8.6):

- Band pass – these allow a narrow range of wavelengths, close to the emission peak of the fluorescent dye to reach detector
- Short pass – transmits wavelengths of light equal to or less than a specific wavelength
- Long pass – filters transmit light equal to or longer than the specified wavelength.

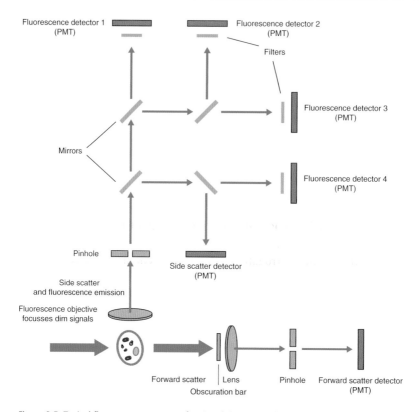

Figure 8.5 Typical flow cytometer set-up for signal detection. The light scattered by particles passing the interrogation point is reoriented and filtered to be ultimately detected by photodiodes or photomultiplier tubes (PMT).

(a) 575 nm Short Pass Filter

λ < 575 nm transmitted

Light source ➡

(b) 520 nm Long Pass Filter

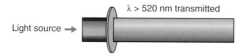

λ > 520 nm transmitted

Light source ➡

(c) 630/20 nm Band Pass Filter

λ = 620–640 nm transmitted

Light source ➡

Figure 8.6 Different types of optical filter are employed, depending on the fluorescence signal to be detected.

8.2.3 Signal Detection and Processing

The light emitted and refracted/reflected from the single stream of particles is sensed by photodetectors. These photodetectors convert the light signal into a stream of electrons (current). Two different types of detectors are commonly used in flow cytometers: silicon photodiodes (PDs) and photomultiplier tubes (PMTs). PDs convert light into photoelectrons with greater efficiency than PMTs, but PMTs have a greater sensitivity. Historically, the stronger FSC light signals have been detected by PDs, while the weaker SSC and fluorescence signals have been detected by PMTs. In contemporary instruments, PMTs are frequently used, even for the FSC channel.

In response to incoming photons, the photodetector will generate a current, the extent of which is proportional to the number of photons detected, and thus the scatter or fluorescence signal emitted by the particle. When the particle enters the cross-section of the interrogation point, the current produced by the detector will start to increase. Ultimately, when the cell is at the centre of the incident laser beam, the current will reach maximum. As the particle leaves the cross-section of the interrogation point, the electrical current/signal will return to baseline levels. Such a pulse of signal is termed an event (Figure 8.7).

The initial signal measured is an analogue signal, but this is converted to a digital signal by an analogue-to-digital converter (ADC) to allow for subsequent computer analysis. Additionally, prior to conversion, the electrical current can be amplified by a linear or logarithmic amplifier. The choice of linear or logarithmic amplification is dependent on the application for which the flow cytometer is being used and the dynamic range required. For example, log amplification is typically used for fluorescence studies of immune cell populations as it expands weak signals and reduces strong signals allowing them to be easily displayed on a histogram. In contrast, something like DNA analysis typically employs linear scaling as here it is important to be able to detect very small differences.

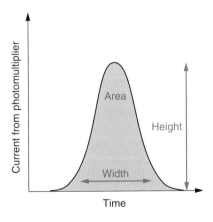

Area: The integral of the pulse.

Height: The maximum current output by the photomultiplier.

Width: The time interval during which the pulse occurs.

The signal intensity equals the peak area, but is often estimated by the height.

The width parameter measures the time the cell spends in the laser beam.

Figure 8.7 Parameters used to quantify a signal pulse (event) as a cell moves through the interrogation point.

8.3 FLUORESCENCE-ACTIVATED CELL SORTING (FACS)

As well as being employed to qualitatively and quantitatively assess cell phenotype and function, flow cytometry can be used to separate particles or cells based on fluorescence labelling. Following isolation, these cells can then be used for further functional, molecular or biochemical studies (some examples are discussed in Section 8.6.3). To allow sorting, a special flow cytometer is needed that is slightly more complex than one required for normal flow cytometric analysis. However, this method will allow cells expressing different markers (and therefore bound by different fluorescently conjugated antibodies) to be separated from one another.

A key benefit of fluorescence-activated cell sorting (FACS; see also Section 7.8), over other cell-sorting methods such as panning and bead enrichment, is that cells can be sorted to a high degree of purity. Populations can be isolated up to 95–98% purity, and there are ample downstream applications of such a pure population of cells, both experimentally and clinically. Cells isolated and purified via FACS can be expanded in vitro, transferred in vivo, used in cell assays or subjected to RNA or DNA analysis.

The general principles of FACS are the same as those for analytical flow cytometry in that cells are assessed based on fluorescence characteristics of the mixed cell population, either by binding of fluorescently conjugated antibodies or genetic encoding/expression of a fluorescent protein. Cells are then gated to identify the cell population of interest to be sorted. For separation, the cell suspension is forced into a single stream of cells, but broken into droplets in such a way that each fluid droplet is likely to only contain one cell. The basic principle of this droplet generation is the same as that used in ink-jet printing technology. Fluorescence characteristics of each droplet, and therefore cell, are then measured. If the observed parameters match the required characteristics defined by the gating strategy, a charge will be applied to the droplet. The droplet then flows past an electrostatic deflection system that allows droplets to be deflected, or guided, into specific collection tubes based on their electrostatic charge (Figure 8.8).

At present, commercially available research-grade FACS instruments are able to sort and collect up to six separate populations at one time. However, to successfully separate cell populations, good markers are required in order to fluorescently mark populations of interest. Obviously, the brighter the marker the easier it is to separate the population of interest. The success of the sorting also depends upon the machine being appropriately set up and calibrated. Most frequently, FACS will be performed by a trained worker who is familiar with the instrument and the methodological background. This requires specialist training, often provided by the manufacturer of the instrument.

8.4 FLUORESCENCE LABELS

Intrinsically fluorescent compounds in the cell are limited; therefore, in order to visualise different aspects of the cell anatomy, cells are stained with fluorescent dyes (probes) to allow us to see biological components that would otherwise not be visible. These probes are used to identify a wide variety of components from cell surface

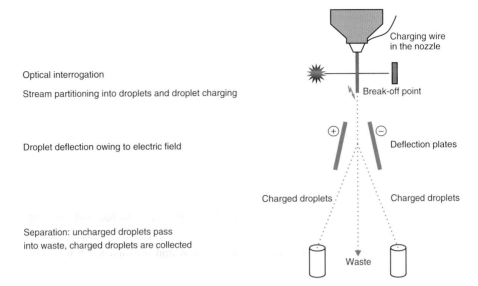

Optical interrogation

Stream partitioning into droplets and droplet charging

Droplet deflection owing to electric field

Separation: uncharged droplets pass
into waste, charged droplets are collected

Charging wire
in the nozzle

Break-off point

Deflection plates

Charged droplets

Charged droplets

Waste

Figure 8.8 Basic principle of electrostatic cell sorting. A stream of droplets is generated and then, by means of electrostatic deflection, either sorted into specific compartments or moved into the waste flow, depending on their surface marker expression.

receptors or intracellular organelles to nucleic acid components and markers of apoptosis.

Following staining and uptake into the flow cytometer, the labelled cells pass through the interrogation point and the fluorophores are excited by incident light of appropriate wavelength (see Section 13.5.1). Upon return to their electronic ground state, the fluorophore releases photons that are detected by the photo detectors. By employing several fluorescent dyes with differing excitation and emission wavelengths, simultaneous measurement of multiple cell properties can be achieved (Figure 8.9).

8.4.1 Fluorescent Dyes

Perhaps the most commonly used fluorescent dye, in part because it was one of the first employed for flow cytometry, is **fluorescein isothiocyanate** (FITC). FITC absorbs light within the range 400–550 nm, with a maximum absorbance at 490 nm; it is often excited by an argon-ion laser delivering blue light with a wavelength of 488 nm. As this is very close to the absorption (=excitation) maximum, the fluorescence emission intensity will, consequently, be maximised. The FITC emission spectrum shows maximum intensity at a wavelength of 518 nm. The difference between the maxima of the absorption and emission spectra is known as the Stokes shift (see also Section 13.5.1). FITC is particularly sensitive to **photobleaching** and a number of FITC derivatives have been developed, including Alexa Fluor® 488, to overcome this issue.

A set of very versatile fluorescent dyes can be used to stain nucleic acids, as well as to assess membrane integrity or analyse the cell cyle. Commonly used dyes in this context include propidium iodide, 7-aminoactinomycin D (7-AAD) and 4′,6-diamidino-2-phenylindole (DAPI). These dyes are excited at different wavelengths and the

Laser Wavelength	Laser Type	Popular Fluorophores
~405 nm	Diode	DAPI, Hoechst, Alexa Fluor® 405, BFP
~440 nm	Diode	CFP
473 nm	Doubled DPSS	
488 nm	Ar-ion gas	GFP, FITC, Alexa Fluor® 488
~488 nm	Doubled OPS	
515 nm	Ar-ion gas, Doubled DPSS	YFP, Rhodamine
561 nm	Doubled DPSS	TRITC, Cy3®, RFP
568 nm	Kr-ion gas	
594 nm	Doubled DPSS	Texas Red, mCherry (mRFP)
633 nm	HeNe gas	
~635 nm	Diode	Cy5®, Alexa Fluor® 647
647 nm	Kr-ion gas	

Figure 8.9 Commonly used fluorescent dyes and laser sources used for excitation (see also Table 11.3).

choice of which dye to use is often dependent on the surface markers being stained in tandem with them. For example 7-AAD can be excited by blue light, whereas DAPI is excited by UV light.

Tandem dyes comprise another tool that has added to the versatility of fluorescence labels; examples include phycoerythrin-Cy7 and allophycocyanin-Cy7. The tandem dye concept employs the process of fluorescence resonance energy transfer (FRET; see Section 13.5.3), whereby the fluorophore with the lower excitation wavelength is excited by the incident light and the fluorophore with the higher excitation wavelength emits the fluorescence photons. Therefore, the wavelength difference between the excitation and emission wavelength is increased as compared to a single fluorophore, resulting in a larger Stokes shift.

Most recently, a new family of organic polymers (including Brilliant Violet™ (BV) 605, BV650, BV711 and others) that are excited by violet and ultraviolet light has enabled a rapid expansion of available fluorescence labels, allowing for 14–20 colours to be regularly employed on flow cytometers for research and clinical purposes.

8.4.2 Toxins

Additionally, fluorescently labelled toxins can also be used to look at specific features of the cell. One of the best examples of this approach is the fluorescent conjugate of phalloidin, a phallotoxin from the death cap mushroom that binds and stabilises filamentous actin (F-actin) and allows for assessment of total F-actin by flow cytometry. Toxins typically have a dissociation constant (K_d) in the low nanomolar range and thus make exquisitely sensitive markers when coupled to fluorescent dyes.

8.4.3 Fluorescent Proteins

Another type of molecule that can also be used in flow cytometry is fluorescent proteins. One of the most commonly used is green fluorescent protein (GFP), which, unlike the other proteins discussed below, will be expressed by the cell population of interest. GFP is excited at 489 nm and can be used in multiple applications, including sorting of GFP transfected cells and fate mapping of specific cell populations in trans-genic animals.

Another two regularly employed fluorescent labels are phycoerythrin (PE) and allophycocyanin (APC). These are phycobiliproteins, water-soluble proteins that are important in capturing light energy and are present in cyanobacteria and certain types of algae. Compared to FITC labelling, use of either of the two phycobiliproteins results in a five- to tenfold greater fluorescence emission intensity. Like FITC, an argon laser can be used to excite PE. Owing to its higher absorption maximum (λ_{exc} = 650 nm), APC is usually excited by a helium-neon laser producing red light.

8.4.4 Antibodies

Monoclonal or polyclonal antibodies are typically conjugated to fluorescence dyes to enable specific labelling of cell populations. Different dyes use different conjugation chemistry; for example, FITC reacts with the amino group of lysine residues. Given that immunophenotyping is one of the most common uses of flow cytometry, a vast array of antibodies is available that are directed towards cell surface markers defining immune cell populations and their activation states, as well as intracellular factors such as cyto-kines and transcription factors. Surface markers in immunology are often referred to as clusters of differentiation (CD; currently 371 for humans) while cytokines are typically referred to as interleukins (Ils). Similar to antibodies, another type of molecule that can be used to identify specific receptors are affibodies. Affibodies imitate monoclonal anti-bodies but are designed to be expressed in a bacterial expression system.

8.4.5 Label Selection and Experiment Design

Emission spectra of fluorescent dyes and reporter molecules extend over a wide wave-length range and hence there can be substantial overlap (spectral overlap) when sev-eral dyes are used; for example, a fraction of the FITC signal overlaps with that of PE and vice versa. Such overlap can be corrected in a process called colour compensation (see Section 8.5.3), whereby a fraction of the emission by one fluorophore (acquired in one channel measured for a particular sample) is subtracted from the emission by another (acquired in a different channel). When only a small number of fluorophores are involved, this process is relatively simple and can be easily performed manually. For large numbers of fluorophores to be analysed, automated methods have been established (Section 8.5.3). The assessment and correction of this spectral overlap involves running single stains of each colour into the flow cytometer to establish the emission of each fluorophore individually.

When compiling panels of fluorescent dyes for an experiment to investigate multiple aspects of a cell's characteristics, it is important to appraise the parameters of different flu-orophores and their spectral overlap profile. The reduction of spectral overlap helps to min-imise the risk of incorrect analysis. For example, some fluorescent dyes are brighter than others and should therefore be employed for surface markers that are expressed at lower levels on cells. Additionally, some fluorophores have a fairly high molecular mass, such as PE and APC, and are thus less useful for intracellular staining. PE and APC also tend to generate much higher background staining than some of the other fluorescent labels.

Example 8.1   **IDENTIFYING CELL POPULATIONS**

Question   You are required to analyse basic differences between cells in two individuals. Apart from using antibodies, what other techniques could you use to characterise cell populations?

Answer   On a very simple level, samples can be examined based on forward and side scatter. Becoming more sophisticated and similar to antibodies, another type of molecule that can be used to identify specific receptors are fluorescently labelled affibodies. Affibodies imitate monoclonal antibodies, but are designed to be expressed in a bacterial expression system. These can be generated against a plethora of cell surface markers. To examine specific aspects of the cell populations the following approaches can be used:

1. Fluorescently labelled toxins can be employed to specific cellular features; for example fluorescently labelled phalloidin can be used to examine total F-actin by flow cytometry.

2. DNA dyes, such as propidium iodide, can be used to assess cell viability and proliferation.

3. Calcium binding dyes can be employed to assess intracellular calcium influx.

8.5 PRACTICAL CONSIDERATIONS

Perhaps the most common, and indeed most famous, use for flow cytometry is in immunophenotyping. This is used by research scientists to define populations of cells, but also used by clinical scientists to identify immunodeficiencies and genetic disorders (see Section 8.6.4). One of the first uses of flow cytometry was during the 1980s when the number of cases of acquired immunodeficiency syndrome (AIDS) was rapidly increasing. In this context, flow cytometry was used to assess lymphocyte depletion in AIDS patients.

In this section, we will use an example of simple immunophenotyping of one of the most commonly investigated tissues – human blood – to understand the basic protocols involved in flow cytometry, as well as the practicalities and limitations of this technique. Sample preparation for flow cytometric analysis can be very straightforward, but in some cases may also involve highly specialised tissue preparation protocols. Examining cell populations in human blood is considered a simple protocol and the basic outline is shown in Table 8.1.

Table 8.1 **General protocol for preparation of human blood for flow cytometry**

1.	Generate a single cell suspension of tissue of interest.
2.	Centrifuge sample, 400×g for 5 min.
3.	Add Fc block reagents to sample; incubate at 4 °C for 15–20 min.
4.	Wash sample in PBS containing 0.5% bovine serum albumin (here called FACS buffer); top up with buffer, then centrifuge as above.
5.	Add antibody staining cocktail; incubate at 4 °C for 15–20 min.
6.	Wash sample in PBS; top up with buffer, then centrifuge as above.
7.	Add live/dead discriminator stain (e.g. DAPI); incubate at 4 °C for 15–20 min.
8.	Wash sample in FACS buffer; top up with buffer, then centrifuge as above.

Note: For buffer description see main text. PBS – phosphate-buffered saline.

8.5.1 Sample Processing

One of the most important aspects of acquiring good flow cytometry data for immuno-phenotyping is the quality of the sample. Typically, it is best to process the sample as rapidly as possible following isolation of cells. Extracted cells should be kept on ice at all times during the staining procedure.

The tissue of interest then needs to be prepared so as to generate a suspension containing only a single cell type. This can be straightforward and involve simply lysing red blood cells in human blood or specialised protocols where physical disruption or enzymatic digestion is employed to break down the structure of a tissue, generating a single-cell suspension. This sample preparation step is important, not only for cells to be analysed in isolation by flow cytometry, but also to ensure the sample is homogeneous and the absence of interference with the normal fluidics flow of the flow cytometer. Human blood is usually taken in tubes containing EDTA or heparin to prevent clotting. Typically, the red blood cells are lysed (using water or ammonium-chloride–potassium buffer) as only white blood cells are of interest for phenotyping. If one is interested in examining a specific type of cell, then extracted cells can be separated by density-gradient centrifugation (see Section 12.4.2) prior to investigation by flow cytometry. In that case, there is no need for red blood cell lysis as these will be removed by the preceding separation step. Frequently, such a separation step is used to separate mononuclear cells (termed peripheral blood mononuclear cells or PBMCs) from granular cells such as neutrophils and eosinophils.

Once a single-cell suspension has been achieved, cell samples are first incubated with blocking agents, such as Fc block (also known as anti-CD16/32), mouse and/or human serum. These blocking agents are incubated with the sample to prevent non-specific binding of antibodies to cells that express surface receptors capable of binding to antibodies, even if they are not expressing the surface marker of interest, as is the case in macrophages and neutrophils. Another way of thinking about this is that these cells are sticky and without Fc block or serum are likely to have high levels of background binding. Following this incubation, cells are stained for specific surface markers alongside a marker such as 4',6-diamidino-2-phenylindole (DAPI) that will stain only dead cells in the preparation and hence allow for discrimination between live and dead cells. This is particularly important when examining immune cells isolated from tissue preparations, where cell viability can be influenced by the tissue processing protocol.

8.5.2 Gating

In flow cytometry, the discrimination of a particular cell population from all cells present is called gating. A gated population has been purposefully selected by the person acquiring or analysing the data. The expression of a specific marker is then analysed only for a certain cell population rather than all cell types present in a sample. This selection not only ensures efficient analysis, but also increases the chances of identifying potential variations in marker expression.

When acquiring data on a sample, there are a few important points to bear in mind to ensure the validity of the data:

- The sample should not be processed through the cytometer too fast. This can be determined by the flow rate or events per second that the instrument is able to detect. Typically, one should not exceed 10 000 events per second, as otherwise the instrument may not be able to read all the events and discard some of them.
- The number of events per second should be stable and remain reasonably constant throughout data acquisition. If this is not the case, either the fluidics of the instrument are blocked or the sample is not of sufficient homogeneity.

First, it is common to use the ratio of forward- and side-scatter parameters to either gate on cell populations of interest or to exclude very small cells, such as remaining red blood cells or lysed cellular debris. Subsequently, the ratio of forward-scatter height to area is employed to identify and remove any events that do not indicate the presence of single cells. This gate is sometimes called the singlet gate and will remove any cells that are stuck together. Examining cells based on forward and side scatter can inform about the cell populations within the sample (Figure 8.10a). Certain populations can be distinguished based upon this analysis, which provides an approximation of cellular morphology based on size and internal structures of the cell. Although this should never be used as a definitive way to identify cell populations, it is a helpful indicator.

In a second discrimination, any dead cells are gated out based on the dead cell marker to ensure that only viable cells will be analysed (Figure 8.10b).

After identifying the viable cell populations by these preliminary analyses, the population of interest can be gated on, for example the CD14+CD16+ blood monocytes (a specific population of white blood cells), as illustrated in Figure 8.11.

8.5.3 Problems of Spectral Overlap and Colour Compensation

As discussed in Section 8.4, a major problem faced when undertaking flow cytometry experiments is the degree of spectral overlap between the emission spectra of different fluorescent dyes (e.g. FITC and PE). In some cases, this can be taken care of at the

(a)   (b)

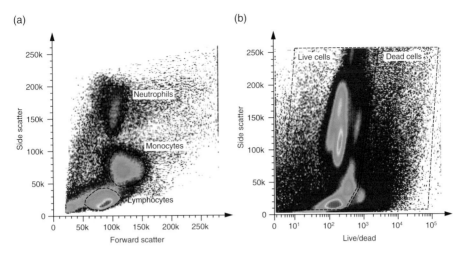

Figure 8.10 Assessment of forward- and side-scatter profile, as well as live cell discrimination via flow cytometry. (a) The correlation between forward and side scatter can be used to provide an estimation of cellular morphology and size. The plot shows forward- against side-scatter intensity on whole human blood after red blood cell lysis. Multiple cell populations can be distinguished based on these simple parameters. (b) The plot shows side-scatter intensity against a dead cell marker, in this case a fixable UV dye. Live and dead cell populations are indicated.

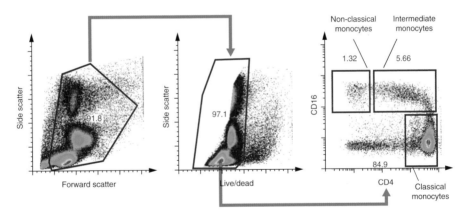

Figure 8.11 Gating strategy for human blood monocytes. The plots show a sequential gating strategy to identify human blood monocytes by flow cytometry.

stage of designing the experiment by carefully choosing the antibody panel employed to stain the sample such that the spectral overlap is minimised. However, in order to take advantage of the ability of flow cytometry to analyse multiple labels at once, it may not be possible to avoid spectral overlap. Indeed, panels employed to stain samples for flow cytometry can utilise up to 18 different fluorescent dyes conjugated to antibodies. The more fluorophores are included in a panel, the more complicated spectral overlap can become. Figure 8.12 highlights the overlaps between the most commonly employed fluorescent colours in flow cytometry.

It is possible to correct for spectral overlap in a process called **colour compensation** when processing the data. The correction to reduce 'spill over' from one fluorescence channel to another may be done manually or computationally using compensation tools included with the flow cytometry software. One common way to do this involves a control experiment with compensation beads where each fluorophore is bound to an individual aliquot of beads and their fluorescence data are acquired in separate runs so that the spectrum of each fluorescence label can be established in isolation. Examples of colour compensation are illustrated in Figure 8.13, showing a correlation of fluorescence intensity from two fluorescence labels (FITC and PE) that are under- and over-compensated (too much of the emission from one channel is bleeding into another), as well as correctly adjusted. Clearly, correct colour compensation improves the accuracy of any results obtained.

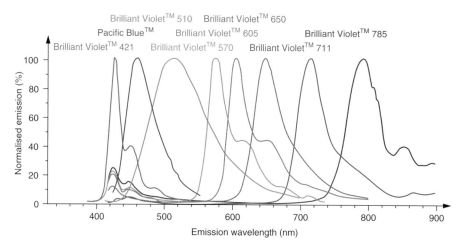

Figure 8.12 Comparison of the emission spectra for commonly used fluorescent dyes illustrates the spectral overlaps. The emission profiles represent the probability a photon will be emitted following excitation of the dye by a specific wavelength of light.

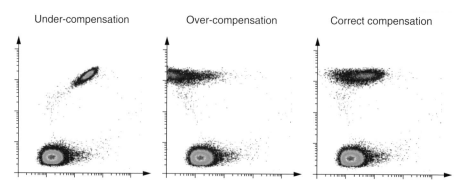

Figure 8.13 Illustration of colour compensation due to spectral overlap of FITC and PE emission spectra. The plots show PE under-compensated (left), over-compensated (centre) and correctly compensated (right) in the FITC channel.

8.5.4 Fluorescence Minus One (FMO) Controls

In order to address uncertainties in multi-colour flow cytometry as to whether a particular stain results from a true positive stain or spill over from another channel (i.e. inappropriate colour compensation), a strategy of control experiments is typically applied. Given that completely unstained cells have a very different fluorescence profile to cells that have been stained with multiple reagents, the current best control is to stain with all of the labels employed except for the one of interest. This control is termed a fluorescence minus one, or FMO, control. For example, in a five-colour staining panel, five FMO controls should be generated in each of which four labels are deployed, but a different one omitted each time (see Figure 8.14). This allows for a rigorous determination of truly positive staining on a sample.

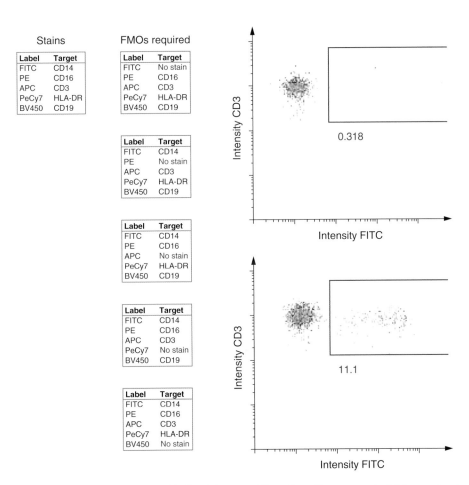

Figure 8.14 Fluorescence minus one (FMO) staining control. The tables outline an example staining panel and the FMOs that should be employed. The comparison of two flow cytometry plots illustrates the results of one such FMO control. The top plot shows staining on a sample for all colours except FITC. The fluorescence in the FITC channel (plotted on the x-axis) in the absence of an FITC-labelled antibody indicates the background staining and therefore where the positive gate should be drawn. The bottom plot shows a sample in which all colours have been stained, including FITC. The numbers indicate the frequency of cells staining positive for FITC.

8.5.5 Back Gating

Back gating is a useful technique to validate whether a specific marker can be employed to look at a defined cell population. Practically, the sample is gated on the marker of interest and the resulting forward/side-scatter profile is assessed in order to find the location of the immune cell population on the plot. This approach ascertains whether the stained cell population is homogeneous. As such, back gating is also a good quality control measure.

8.5.6 Analysis and Quantification

Depending on the number of labels used for an experiment, it is possible to get a lot of high-resolution information about different cell populations. The analysis of these data can be time-consuming, but is aided by commercially available special-ised software that requires the operator to gate on populations of interest and allow them to query not only percentages of cell types of interest, but the extent to which a population is positive for a specific marker. For example, one could determine the frequency of monocytes within a particular cell preparation and also assess the staining intensity of a particular marker on monocytes. A major area of current software development is automated flow cytometry analysis where gates will be determined by the software rather than the user, thereby streamlining analysis with less user input.

Example 8.2   **FLUOROCHROME PANELS**

Question   You are designing a panel of antibodies using four fluorochromes to identify a specific subset of immune cells. What are some of the key factors you need to consider when designing this panel?

Answer   Key factors to be considered include:

1. Possible compensation issues, due to spectral overlap, between the fluoro-phores to be employed.

2. Autofluorescence of cells to be examined. For example, macrophages pos-sess a considerable autofluorescence.

3. Brightness of fluorophores to be employed. For example, are you using a dim fluorophore conjugated to an antibody with low affinity?

4. Number of FMO controls to use – have you got the right number of controls?

8.6 APPLICATIONS

In the previous section, we focussed on the use of flow cytometry specifically for immunophenotyping. However, flow cytometry can be used for a vast array of purposes by research scientists and clinicians. In this section, we will look more broadly at some of these possible applications of flow cytometry, new technologies that utilise flow cytometry and look to the future to discuss emerging applications.

8.6.1 Assessment of Apoptosis Markers

Flow cytometry can be used to rapidly quantify markers of cell death. Measurement of cell viability is important for almost all flow cytometry experiments, as previously discussed in Section 8.5; dead cells can confound analysis of antibody staining. Viability of living cells can easily be assessed by adding DNA binding compounds. These probes intercalate with DNA, but will only have access to DNA in cells with damaged membranes, in other words during death. Discrimination of live and dead cells is mostly frequently undertaken using 7-AAD, propidium iodide or DAPI. Furthermore, there are now a number of commercially available fixable live/dead stains, which allow discrimination of live and dead cells in tandem with, for example, intracellular cytokine staining. Antibody stains are also available to assess cell survival factors and other hallmarks of apoptosis, such as active caspase 3 or Bcl-2.

Detection of Changes to Surface Membrane

An early change during apoptosis is translocation of phosphatidylserine from the inner to the outer leaflet of the plasma membrane. Ultimately, the integrity of the plasma membrane is compromised, but early in the process, only phosphatidylserine is exposed. The calcium-dependent membrane binding protein annexin A5 possesses substantial specificity for phosphatidylserine; fluorescently labelled annexin A5 is thus able to stain phosphatidylserine on the surface of apoptotic cells. In tandem with a cell viability dye, such as 7-AAD, it is possible to detect annexin A5 positive and 7-AAD negative cells, which represent early apoptotic cells with an altered plasma membrane. In contrast, a simultaneous positive staining with annexin A5 and 7-AAD indicates dead cells.

8.6.2 Functional Assays

There are a number of assays that can be undertaken with flow cytometry to assess cellular functions. These can be combined with detailed cellular phenotyping (outlined in Section 8.5) to provide comprehensive datasets on cell populations present in blood and tissues. A few of the most commonly employed assays will be discussed in the following.

Calcium Flux Analysis

A key cellular function that can be assessed by flow cytometry is calcium signalling, a downstream signal in multiple activation pathways. Through use of calcium-binding dyes, intracellular calcium influx can be assessed in cell populations following stimulation by means of establishing baseline intracellular calcium levels, as well as the kinetics (magnitude and duration) of the calcium flux.

Indo-1 is the most commonly employed dye in this context. Its excitation wavelength is in the ultraviolet region (λ_{exc} = 346 nm) and it possesses distinct emission spectra depending on whether it is in its free form (λ_{em} = 475 nm; green) or calcium-bound (λ_{em} = 401 nm; violet). Typically, it is the ratio of intensities at both wavelengths that is monitored over time (so-called **ratiometric technique**). Prior to activation, the signal will be predominately green; upon cell stimulation, and release of intracellular calcium, the dye molecules will bind Ca^{2+} ions and emit at the violet wavelength. Indo-1 can be employed to assess calcium-signalling kinetics in a wide variety of cell types; all that is required is an agonist that will illicit a calcium flux in the cell type of interest. For example, activation of the T cell receptor on lymphocytes can be employed to assess T cell calcium flux. Key control experiments include positive and negative stains to ensure Indo-1 loading and emission. Most frequently, the calcium ionophore **ionomycin** is used as a positive control, generating the maximum calcium release into the cytoplasm and therefore greatest emission of Indo-1 in the calcium-bound state. In contrast, calcium chelators, which will blunt calcium signals, are employed as negative controls.

Practically, Indo-1 is used to stain a sample and the loading of cell populations is assessed on the flow cytometer in order to allow a baseline of Indo-1 fluorescence to be established. Without halting data acquisition, the sample tube is briefly removed from the SIP and the cell stimulant is added to the tube. The tube is then returned to the SIP and ratiometric Indo-1 emission is followed against time, as shown in Figure 8.15.

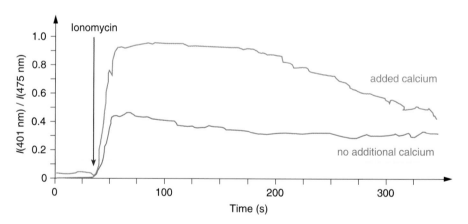

Figure 8.15 Calcium flux analysis via flow cytometry. Flow cytometry plots (in the absence and presence of additional calcium in the media) show ratiometric change in Indo-1 fluorescence following activation of calcium signals. By monitoring the change in fluorescence ratio for Indo-1 against time, a calcium baseline, calcium signalling kinetics and magnitude can be determined, as indicated.

Phosflow

Another method that can be employed to query cellular function by flow cytometry is phospho-flow cytometry, or 'Phosflow'. Here, intracellular signalling pathways are examined by assessing phosphorylation of protein sites on key signalling proteins. By using antibodies against specific phosphorylated proteins, Phosflow can be employed to measure signalling cascades in individual cells. This method is therefore not merely an alternative to Western blotting, but a complementary technique to this staple method of assessing cellular signalling pathways, as it offers a reliable way to measure signalling events in individual cells. Western blotting, by contrast, measures the average level of phosphorylation within the entire population. A disadvantage of Phosflow in this context is that it is not possible to determine the relative molecular mass of the targeted protein – information that can be useful to confirm the specificity of the antibody.

Practically, cells are stimulated and then fixed using paraformaldehyde-based buffers allowing the induced states of protein phosphorylation to be preserved. Next, cells are permeabilised and then stained with fluorescently conjugated antibodies against the phosphorylated proteins of interest. This can occur simultaneously with antibodies against surface markers, allowing cellular signalling events to be assessed in multiple populations of cells simultaneously. For example, Phosflow can be used to assess mitogen-activated protein kinase (MAPK) signalling events or mechanistic target of rapamycin (mTOR) signalling. The bottleneck in assessment of signalling pathways in this context remains the generation of phospho-specific antibodies that work in cells that have been fixed with paraformaldehyde.

Cell Proliferation

There are a number of ways to assess cellular proliferation by flow cytometry. Here, we will discuss the most commonly employed, which allow assessment of cell proliferation in different ways. Alongside the applications outlined below, straightforward flow cytometry staining for the intracellular proliferation marker Ki-67 can also be conducted to assess cell proliferation. Ki-67 is an antigen that is only exposed during interphase, and therefore positive staining for Ki-67 marks proliferating cells.

- Cell cycle analysis: DNA-binding dyes such as propidium iodide are stoichiometric; as more DNA is present in the cell it will bind more dye, resulting in greater fluorescence emitted by propidium iodide. This can be exploited to determine the current phase of the cell cycle of the examined cells. Specifically, cells in S phase will have more DNA than cells in G1 phase; furthermore, cells in G2 phase should be twice as bright as those in G1 phase, as they contain double the amount of DNA (Figure 8.16a).
- Carboxyfluorescein succinimidyl ester (CFSE) dilution: CFSE is a fluorescent dye that covalently binds free amine groups present on intracellular molecules via its succinimidyl group. This conjugation leads to incorporation of CFSE within the cell and prohibits transfer of the dye to other cells. With this approach, the proliferation of the CFSE-labelled cells can be monitored. Each time the labelled cells divide, the CFSE fluorescence emission will be halved, with the daughter cells receiving half the amount of CFSE-labelled proteins. In this way, the number of cell divisions over time can be determined from the CFSE dilution within a population of cells (Figure 8.16b). This method can typically trace up to eight generations of cell division.

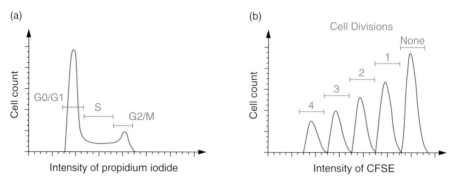

Figure 8.16 Examination of cell proliferation by flow cytometry. Two common assays employed to assess cell cycle progression and cellular proliferation are outlined. (a) The plot shows staining for propidium iodide on gated cells to assess cell cycle progression. Intensity of propidium iodide staining indicates whether the cells are in G0/G1, S or G2/M phase. (b) This plot shows cell proliferation assessment based on staining for CFSE. The intensity of CFSE staining decreases as cells proliferate, since half the amount of CFSE is passed on to the daughter cells. As such, the number of times a cell has divided can be monitored by examining how much the CFSE fluorescence has been diluted.

8.6.3 Flow Cytometry Combined With Other Technologies

Flow cytometric technology has been coupled with other techniques to develop novel applications for use in research and also the clinic. A few of these will be outlined in this section and demonstrate the versatile use of flow cytometry throughout the scientific world.

ImageStream®

Imaging with flow cytometry combines flow cytometric analysis with microscopy, allowing specific cell populations to be identified by conventional flow cytometry, but, at the same time, imaging each cell. These specialised flow cytometers take high-resolution images of each cell. The images can be bright-field images (see Section 11.2.5) to assess morphology or enhanced with fluorescence information. Experimentally, this method allows assessment of protein localisation within a cell alongside examination of cell morphology and the usual benefits of flow cytometry. Although generating large data files, employing an ImageStream® allows for a substantial number of cells to be examined, providing statistical assessment of cellular phenotypes, as well as generating good-quality fluorescence images of the cells.

Multiple Analyte Detection Systems

Often referred to as cytometric bead arrays, this application of flow cytometry allows an array of soluble proteins to be analysed simultaneously. This technology employs fluorescently labelled beads that are coated with a capture antibody targeting a specific analyte, most commonly a cytokine. Multiple analytes can be analysed simultaneously, as beads coated with different capture antibodies can be distinguished from each other based on fluorescence. By employing cytokine standards, similar to those used in ELISAs (see also Section 7.3), concentrations of analytes in serum, tissues or cell culture can be determined. The key advantage of this application over more traditional methods such as ELISAs is that multiple analytes can be simultaneously quantified using as little as 25–50 µl of sample.

Uses of FACS

As well as sorting highly pure populations of cells for further analysis, FACS is also a key experimental tool with a broad array of applications. The ability to separate cell populations based on specific markers is invaluable for research and a few of the FACS applications are discussed below:

- Enrichment of transfected cells: Importantly, FACS is the only method for cellular purification that is able to separate cell populations based on internal staining. For example, FACS can be employed to sort GFP-expressing cells. The applications of this advantage would allow, for example, transfected (and therefore GFP-positive) cells to be separated from non-transfected cells.

- Genomic technologies and single cell examination: The ability to isolate highly pure populations of cells has heralded the onset of in-depth genomic assessment of these cells. Microarray analysis and RNA-sequencing (RNA-Seq; see Section 4.17.1) are frequently coupled with FACS to examine the transcriptome of the sorted cells. In the immunology field, the ImmGen consortium aims to probe the transcriptome of all immune cells in this way. Through the use of established as well as novel surface markers for cell populations, very specific populations are isolated and examined, providing detailed information on gene expression and gene networks in immune cells. Building upon this application, more recently, groups have begun to isolate individual cells and subject these to next-generation sequencing technologies (Section 20.2.2). Importantly, FACS can be employed to isolate single cells in a high-throughput manner, allowing many single cells to be obtained for further genomic examination. This approach allows the heterogeneity of cell populations to be resolved at a transcriptional level, providing a better understanding of the cellular response of the isolated cell population.

- Ig-sequencing; 'Bug FACS': As well as being employed to assess cell populations, phenotype and functional attributes, flow cytometry can also be utilised to query interactions between the immune system and the commensal bacterial communities that cover our bodies. These microbial communities, termed the microbiome, are key for full maturation of the immune system, but full elucidation of the extensive and intricate interactions between these microbes and the host immune system is outstanding. One way to assess these interactions is to determine which microbes have been encountered by the immune system and subsequently initiated the production of microbe-specific immunoglobulins (Igs; see also Section 7.1.1). By employing flow cytometry, bacterial communities, from any site in the body (including the gastrointestinal tract, saliva or oral biofilm) can be examined to see which bacteria are recognised, and therefore bound, by Igs. Antibodies against different Igs, for example immunoglobulin A (IgA), immunoglobulin M (IgM) and immunoglobulin G (IgG), can be used to ascertain the type of immune recognition driven by the Ig-bound bacteria. Through use of cell, or in this case bacteria, sorting, the Ig-bound bacteria can be separated out from the non-bound bacteria and subjected to deep-sequencing techniques such as 16S-sequencing (see Section 16.6), providing an assessment of bacteria capable of driving antibody responses at specific sites.

8.6.4 Translational Applications in Clinical Practice

Flow cytometry is a routine tool used by clinical pathologists and immunologists to diagnose patients with certain types of cancer or immunodeficiencies. There are novel patterns of surface-marker expression on certain cancers and unique pathophysiologies that, when analysed via flow cytometry, lead to the determination of specific treatment strategies. In addition to being employed as a diagnostic tool, by utilising FACS it is possible to isolate tumour-free populations of stem cells for stem cell transplantation. Flow cytometry also provides a rapid method for looking at cytokine production and HLA genotypes, and as such the translational applications of flow cytometry are considerable.

Cancer Diagnosis and Prognosis

Flow cytometry can be used to help identify the distinct immune phenotypes of malignancies and also cell DNA content, both of which can be used to help guide selection of therapies. One of the best examples is in CD20+ B cell malignancies that are now commonly treated with anti-CD20 antibody (rituximab). Flow cytometry can be used to distinguish between B cell malignancy and more aggressive lymphomas, such as mantle cell lymphoma. This is done by looking for a reproducible expression pattern of multiple antigens or antigen density on malignant cells. Additionally, because flow cytometry can detect individual changes in cells, even in a heterogeneous population, it can also be employed to identify residual malignant cells following therapy. This is termed minimal residual disease monitoring as, frequently, malignancies in which there is disease persistence eventually result in worse disease outcome.

DNA Content Analysis

Relative cell DNA content can easily be measured by flow cytometry and this can be used to help guide treatment in acute lymphoblastic leukaemia. As described, DNA content can be measured by employing a fluorescent dye such as propidium iodide that intercalates with DNA. The fluorescence emission is directly proportional to the amount of DNA in the nucleus and so it can be used to establish gross gains or losses in DNA. This can be used in a tumour cell population to establish abnormal DNA content or 'DNA content aneuploidy'. This is typically associated with malignancy; however, it can also occur in certain benign conditions. DNA aneuploidy can be associated with worse or better outcomes, depending on the type of cancer.

Immunologic Diseases

A major use of flow cytometry is to determine CD4+ T cell counts following HIV infection. It can also be used to classify and assess prognosis of various primary immune deficiencies, including severe combined immune deficiency (SCID), autoimmune lymphoproliferative syndrome and antibody deficiencies. Additionally, flow cytometry can be used in allogeneic stem cell grafts to look at specific immune cell populations and also assess immune reconstitution following allogeneic bone marrow transplant to predict survival post-transplant. Flow cytometry is similarly employed to look for rare CD34+ endothelial cell progenitors that can help to predict clinical outcome in patients that have peripheral artery and cardiovascular disease.

Cellular Transplantation

FACS can be used to enrich for specific cell populations that are required for cell-based therapy, such as, for example, CD34+ human hematopoietic stem and progenitor cells. In essence, autologous CD34+ stem cells can be sorted to high purity with the possibility of achieving a clinically significant depletion of contaminating tumour cells. Studies with patients suffering from non-Hodgkin's lymphoma, breast cancer and multiple myeloma provided the proof of concept that durable hematopoietic reconstitution can occur and robust remission achieved when these highly purified cells are transferred.

Flow cytometry is also frequently employed in the clinic to measure antibodies to human leukocyte antigen (HLA) molecules. Individuals generate antibodies to these molecules, for example following exposure during pregnancy or blood transfusion. The presence of donor-directed anti-HLA antibodies poses a significant risk of allograft rejection. Flow cytometry provides a high-throughput method to determine the presence of antibodies to HLA in patients due to undergo transplantation. HLA-genotyping can also be performed by flow cytometry by employing a multiplexing platform. Here, high throughput examination of HLA-genotypes can be determined in samples derived from patients and donors undergoing transplantation.

8.6.5 The Future of Flow Cytometry

Since its advent, multi-colour flow cytometry has advanced significantly, marked in part by the expansion of the historic two-colour flow cytometry to the 20+ colours that are detected in some settings today. Moreover, the technology has advanced in such a way as to allow not only more detailed and complex flow cytometry panels to be assessed, but also to make instruments available that are smaller, simpler to use and less expensive. Further features include more automated packages and, in some cases, even portable instruments, thus widening the user base and applications for this technology.

The continuing evolution of this technology will involve the ability to visualise and differentiate between even more tagged antibodies. Although there may be a maximum number of fluorescent labels that can feasibly be used, there are other ways to tag antibodies. Mass cytometry, or CyTOF, allows visualisation of antibodies labelled with heavy metal ions instead of fluorescent labels. Whether a cell is labelled with an antibody or not is determined by time-of-flight analysis, similar to mass spectrometry (see Section 15.3.3). This new methodology overcomes the issues of spectral overlap between fluorescent dyes, allowing many more differentially labelled antibodies to be employed.

In line with the expansion of the technology, analysis packages have also developed. Moving forward, a clear advancement on most of the currently employed analysis packages will be the development of automated systems for data analysis. Whereas automation exists to a certain degree, the ability to assess multiple experimental variables in continuously changing staining panels is certainly a requirement for further advancements. Similarly, increased abilities of automated data interpretation will enhance the high-throughput applicability of this technology. Clearly, flow cytometry promises to continue to be used in a vast number of different arenas in the years to come, as it is today.

8.7 SUGGESTIONS FOR FURTHER READING

8.7.1 Experimental Protocols

Baumgarth N. and Roederer M. (2000) A practical approach to multicolor flow cytometry for immunophenotyping. *Journal of Immunological Methods* **243**, 77–97.

Biancotto A., Fuchs J. C., Williams A., Dagur P. K. and McCoy Jr. J. P. (2011) High dimensional flow cytometry for comprehensive leukocyte immunophenotyping (CLIP) in translational research. *Journal of Immunological Methods* **363**, 245–261.

Hawley T.S. and Hawley R.G. (2016) *Flow Cytometry Protocols*. Methods in Molecular Biology, Vol. 699, Humana Press, Totowa, NJ, USA.

Leach M., Drummond M., Doig A., *et al.* (2015) *Practical Flow Cytometry in Haematology: 100 Worked Examples*, Wiley-Blackwell, New York, USA.

8.7.2 General Texts

Watson J.V. (2004) *Introduction to Flow Cytometry*, Cambridge University Press, Cambridge, UK.

8.7.3 Review Articles

Brown M. and Wittwer C. (2000) Flow cytometry: principles and clinical applications in hematology. *Clinical Chemistry* **46**, 1221–1229.

Jaye D. L., Bray R. A., Gebel H. M., Harris W.A.C. and Waller E. K. (2012) Translational applications for flow cytometry in clinical practice. *Journal of Immunology* **188**, 4715–4719.

Virgo P. F. and Gibbs G. J. (2012) Flow cytometry in clinical pathology. *Annals of Clinical Biochemistry* **49**, 17–28.

8.7.4 Websites

Fluorescence tutorials (Chromocyte)
www.chromocyte.com/educate/ (accessed May 2017)

ImmGen consortium
www.ImmGen.org/ (accessed May 2017)

Purdue University Cytometry Laboratories
cyto.purdue.edu/ (accessed May 2017)

Tutorial collection by The George Washington University
flowcytometry.gwu.edu/tutorial (accessed May 2017)

9 Radioisotope Techniques

ROBERT J. SLATER

9.1 WHY USE A RADIOISOTOPE?

When researchers contemplate using a radioactive compound there are several things they have to consider. First and foremost, they must ask the question: is a radioisotope necessary or is there another way to achieve the objectives? The reason for this is that use of radioisotopes is governed by very strict legislation. The rules are based on the premise that radioactivity is potentially unsafe (if handled incorrectly) and should therefore only be used if there are no alternatives. Then, once it is decided that there is no alternative, the safest way of carrying out the work needs to be planned. Essentially this means using the safest isotope and the smallest amount possible. But why do we use radioisotopes in the first place? First, it is possible to detect radioactivity with exquisite sensitivity. This means that, for example, the progress of an element through a metabolic pathway or in the body of a plant or animal can be followed relatively easily. Very small amounts of a radioactive molecule are needed, and detection methods are well established. Second, it is possible to follow what happens in time. Imagine a metabolic pathway such as carbon dioxide fixation (the Calvin cycle). All the metabolites in the cycle are present simultaneously; so a good way to establish the order of the metabolism is to add a radioactive molecule (in this example, ^{14}C-labelled carbon dioxide in the form of sodium bicarbonate) and see what happens to it by extracting the metabolites from the plant and identifying the radioactive ones. Third, it is possible to trace what happens to individual atoms in a pathway. This is done, for example, by creating compounds with a particular isotope in specific locations in the molecule. Fourth, we can identify a part or end of a molecule, and follow reactions very precisely. This has been very useful in molecular biology, where it is often necessary to label one end of a DNA molecule (e.g. for techniques such as gel mobility shift assay or DNA footprinting, methods for investigating sequence-specific DNA protein binding), or immunochemical diagnostics; see Section 7.2.9). Fifth, γ-ray emitters (^{60}Co or ^{137}Cs) are employed in irradiators that may be used for a variety of purposes in research and industry, including sterilisation of medical and pharmaceutical supplies (see also Section 3.6.2), preservation of foodstuffs, eradication of insects through sterile male release programs and calibration of thermoluminescent dosimeters. Finally, there is a use that sometimes seems too obvious for it to be considered. In chemistry and biochemistry, we are used to chemical reactions where one compound is turned into another. We can identify and measure ('assay') the reactants and products and learn something about the atomic rearrangement during the reaction. A prominent example in this context is DNA replication where the use of radioisotopes provides a method for detecting particular products of the reaction.

The use of radioactivity in biochemical research has made a very significant contribution to knowledge, for example with the above-mentioned use of carbon-14 in the 1940s to study photosynthesis, through to modern drug discovery methods and DNA sequencing. It is fair to say, however, that many techniques that began with the use of a radioisotope have been replaced by other forms of detection system such as fluorescence (see Section 13.5), mainly due to the special safety provisions required when working with radioisotopes. Despite these developments, there are still some recent research approaches that rely on radioactivity (e.g. the scintillation proximity assay as used in the pharmaceutical industry).

Having understood the principles behind why radioisotopes are useful as described above, we now need to understand what radioactivity is and how to use it.

9.2 THE NATURE OF RADIOACTIVITY

9.2.1 Atomic Structure

An atom is composed of a positively charged central nucleus inside a much larger cloud of negatively charged electrons. The nucleus accounts for only a small fraction of the total size of the atom, but it has more than 99.9% of the atomic mass. Atomic nuclei are composed of two major particles, protons and neutrons (collectively called nucleons) that are attracted to each other by a nuclear force. Protons have a positive electrical charge. The electrons are attracted to the nucleus by an electromagnetic force. If the number of protons and electrons is the same, the atom is electrically neutral, but if there are more protons than electrons then the atom is a positively charged ion; similarly, if the electron number exceeds the proton number then the atom is a negatively charged ion.

The number of protons present in the nucleus is known as the atomic number (Z), and it determines the identity of the element; for example, atoms with six protons are carbon. Neutrons are uncharged particles with a mass approximately equal to that of a proton. The sum of protons and neutrons in a given nucleus is the mass number (A). Thus

$$A = Z + N \tag{Eq 9.1}$$

where N is the number of neutrons.

Since the number of neutrons in a nucleus is not related to the atomic number, it does not affect the chemical properties of the atom. Atoms of a given element may not necessarily contain the same number of neutrons. In that case, their mass numbers differ and they are called isotopes. Symbolically, an isotope is represented by a subscript number for the atomic number, and a superscript number for the mass number, followed by the symbol of the element (Figure 9.1).

$$^A_Z \text{Element} \quad ^{12}_6C \quad ^{14}_6C \quad ^{16}_8O \quad ^{18}_8O$$

Figure 9.1 Denotation of isotopes.

However, in practice it is more conventional to just cite the mass number (e.g. ^{14}C), because the atomic number and element symbol are equivalent in meaning. The number of isotopes of a given element varies: for example, there are three isotopes of hydrogen (^1H, ^2H and ^3H), seven of carbon (^{10}C to ^{16}C inclusive) and 20 or more of some of the elements of high atomic number.

Very importantly, the number of neutrons N does not affect the 'chemistry' of an atom. The chemical behaviour of an element is determined by the electrons, since electrons from one atom interact with other atoms to form molecules. As we have seen above, in a neutral atom, the number of electrons equals the number of protons Z, which is synonymous with an individual element. To sum up this section: the proton number Z defines the element, the neutron number N defines the isotope and the electron number determines the chemistry.

9.2.2 Atomic Stability and Radiation

In general, the ratio of neutrons to protons will determine whether an isotope of an element is stable enough to exist in nature. Stable isotopes for elements with low atomic numbers tend to have an equal number of neutrons and protons, whereas stability for elements of higher atomic numbers requires more neutrons. Unstable isotopes are called radioisotopes. They become stable isotopes by the process of radioactive decay: changes occur in the atomic nucleus, and particles and/or electromagnetic radiation are emitted.

9.2.3 Types of Radioactive Decay

There are several types of radioactive decay; only those most relevant to biochemists are considered below. A summary of properties is given in Table 9.1.

Table 9.1 Properties of different types of radiation

Alpha	Beta	Gamma, X-rays and Bremsstrahlung
Heavy charged particle	Light charged particle	Electromagnetic radiation
More toxic than other forms of radiation	Toxicity same as electromagnetic radiation per unit of energy	Toxicity same as beta radiation per unit of energy
Not penetrating	Penetration varies with source	Highly penetrating

Decay by Negatron Emission

In this case a neutron is converted to a proton by the ejection of a negatively charged beta (β)-particle called a negatron (β^-):

$$\text{neutron} \longrightarrow \text{proton} + \text{negatron} \tag{Eq 9.2}$$

To all intents and purposes, a negatron is an electron, but the term negatron is preferred, although not always used, since it serves to emphasise the nuclear origin of the particle. As a result of negatron emission, the nucleus loses a neutron, but gains a proton. The mass number A remains constant. An isotope frequently used in biological work that decays by negatron emission is carbon-14, which decays to nitrogen.

$$^{14}_{6}\text{C} \longrightarrow {^{14}_{7}}\text{N} + \beta^- \tag{Eq 9.3}$$

Negatron emission is very important in the biological sciences, because many of the radionuclides that decay by this mechanism can be incorporated into the molecules of life. Examples are: ^{3}H and ^{14}C, which can be used to label any organic compound; ^{35}S is used to label methionine, for example to study protein synthesis and ^{33}P or ^{32}P are powerful tools in molecular biology when used as nucleic acid labels.

Decay by Positron Emission

Some isotopes decay by emitting positively charged β-particles, referred to as positrons (β^+). This occurs when a proton is converted to a neutron:

$$\text{proton} \longrightarrow \text{neutron} + \text{positron} \tag{Eq 9.4}$$

Positrons are extremely unstable and have only a transient existence. They interact with electrons and are annihilated, whereby the mass and energy of these two particles are converted to two γ-rays emitted at 180° to each other (back–to–back emission). This straight line of emissions in opposite directions is the basis of three-dimensional imaging of biological samples labelled with positron emitters.

   As a result of positron emission, the nucleus loses a proton and gains a neutron, the mass number A stays the same. An example of an isotope decaying by positron emission is sodium-22, which decays to neon:

$$^{22}_{11}\text{Na} \longrightarrow {^{22}_{10}}\text{Ne} + \beta^+ \tag{Eq 9.5}$$

Positron emitters are detected by the same instruments used to detect γ-radiation. They are used in biological sciences to spectacular effect in brain scanning with the technique positron emission tomography (PET scanning) used, for example, to identify active and inactive areas of the brain (see Section 9.3.5).

Decay by α-Particle Emission

Isotopes of elements with high atomic numbers frequently decay by emitting alpha (α)-particles. An α-particle is a helium nucleus; it consists of two protons and two neutrons (^{4}He^{2+}). Emission of α-particles results in a considerable lightening of the nucleus, a decrease in atomic number (Z) of 2 and a decrease in the mass number (A)

of 4. Isotopes that decay by α-emission are not frequently encountered in biological work, although they can be found in instruments such as scintillation counters and smoke alarms. Radium-226, for example, decays by α-emission to radon-222, which is itself radioactive. This begins a complex decay series, which culminates in the formation of lead-206:

$$^{226}_{88}\text{Ra} \longrightarrow {}^{4}_{2}\text{He}^{2+} + {}^{222}_{86}\text{Rn} \longrightarrow \longrightarrow \longrightarrow {}^{206}_{82}\text{Pb} \quad \text{(Eq 9.6)}$$

Alpha emitters are extremely toxic if ingested, due to the large mass and the ionising power of the α-particle.

Electron Capture

In this form of decay, a proton captures an electron near the nucleus:

$$\text{proton} + \text{electron} \longrightarrow \text{neutron} + \gamma\text{-ray} \quad \text{(Eq 9.7)}$$

The proton becomes a neutron and electromagnetic radiation (γ-rays) is released. For example, iodine-125 decays to tellurium-125 by an electron capture mechanism:

$$^{125}_{53}\text{I} \longrightarrow {}^{125}_{52}\text{Te} + \gamma\text{-ray} \quad \text{(Eq 9.8)}$$

γ-Rays constitute electromagnetic radiation at a very short wavelength, and therefore possess the same sort of energy as X-rays. The latter describe electromagnetic radiation that is emitted when matter is bombarded with fast electrons. The difference between γ-rays and X-rays is thus that γ-rays are produced continuously in a naturally occurring process, but X-rays are produced by a machine that can be switched off. There has also been a historic distinction that classified X-rays as having slightly lower energy than γ-rays; however, this has become obsolete since modern synchrotron facilities (see Section 14.3.4) can produce highly energetic X-rays.

Decay by Emission of γ-Rays

In some cases, α- and β-particle emission also gives rise to γ-rays. The γ-radiation has low ionising power, but high penetration. For example, the radiation from cobalt-60 will penetrate 15 cm of steel. Being the same type of electromagnetic radiation, the toxicity of γ-radiation is the same as that of X-rays. An example of a γ-ray emitter is iodine-131, often used in diagnosis and treatment of thyroid disorders (thyroid hormone contains iodine); it decays to xenon-131:

$$^{131}_{53}\text{I} \longrightarrow {}^{131}_{54}\text{Xe} + \gamma\text{-ray} \quad \text{(Eq 9.9)}$$

Another γ-emitter that is used extensively in diagnosis but also fundamental research is the metastable technetium-99. Metastable means it has an unstable nucleus that first of all loses some energy by emission of γ-rays without change of atomic number. The γ emission has a very short half-life (6 hours) and the metastable technetium-99 (extracted from molybdenum-99 which is obtained from nuclear reactors) thus has to be used immediately. Of importance for clinical applications is the fact that the isotope has a very short **biological half–life** (the time taken for an animal to lose half of the isotope from the body) and so it results in very low doses, for example when used in diagnostic procedures or studying labelled pharmaceuticals.

$$^{99m}_{43}\text{Tc} \longrightarrow {}^{99}_{43}\text{Tc} + \gamma\text{-ray} \longrightarrow {}^{99}_{44}\text{Ru} + \beta \qquad\qquad (\text{Eq } 9.10)$$

9.2.4 Radioactive Decay Energy

The usual unit used in expressing energy levels associated with radioactive decay is the electron volt (eV). One electron volt is the energy acquired by one electron in accelerating through a potential difference of 1 V and is equivalent to 1.6×10^{-19} J. For the majority of isotopes, the term million or mega electron volts (MeV) is more applicable. Isotopes emitting α-particles are normally the most energetic, falling in the range 4.0 to 8.0 MeV, whereas β- and γ-emitters generally have decay energies of less than 3.0 MeV. The higher the energy of radiation the more it can penetrate matter and the more hazardous it becomes.

9.2.5 Rate of Radioactive Decay

Radioactive decay is a spontaneous process and measured as disintegrations per second (d.p.s.) or per minute (d.p.m.). It occurs at a rate that is characteristic for the radioisotope, and is thus defined by the rate constant λ, i.e. the fraction of an isotope decaying per time interval.

Importantly, radioactive decay is a nuclear event, not a chemical reaction; so the rate of decay is not affected by temperature or pressure. Clearly, the number of atoms N is always falling (as atoms decay) and so the rate of decay decreases with time. Therefore, the slope of a graph showing the number of atoms N (or the relative amount) at time t is continuously decreasing (see Figure 9.2). The radioactive decay can be described with the same mathematical form as first-order kinetics and a plot of radioactivity against time thus yields a so-called **exponential decay curve**.

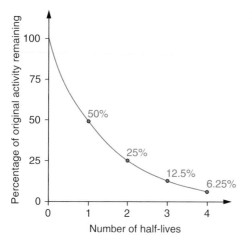

Figure 9.2 Demonstration of the exponential nature of radioactive decay.

The mathematical equation that underpins the graph shown is as follows:

$$\ln \frac{N_t}{N_0} = -\lambda \times t \qquad \text{(Eq 9.11)}$$

where λ is the decay constant for an isotope, N_t is the number of radioactive atoms present at time t, and N_0 is the number of radioactive atoms originally present. There is a natural logarithm ($\ln = \log_e$) in the equation; this means if we were to plot the logarithm of d.p.s. values against time, we would get a graph with a straight line and a negative slope (λ).

In practice it is often convenient to express the decay constant in terms of half-life ($t_{1/2}$). This is defined as the time taken for the activity to fall from any value to half that value (see Figure 9.2). When N_t in Equation 9.11 is equal to $0.5 \times N_0$, then t will equal the half-life of the isotope. Thus:

$$\ln \frac{1}{2} = -\lambda \times t_{1/2} \quad \text{or} \quad t_{1/2} = \frac{0.693}{\lambda} \qquad \text{(Eq 9.12)}$$

The values of $t_{1/2}$ vary widely from over 10^{19} years for lead-204 to 3×10^{-7} seconds for polonium-212. The half-lives of some isotopes frequently used in biological work are given in Table 9.2.

9.2.6 Units of Radioactivity

The Système International d'Unités (SI system) uses the becquerel (Bq) as the unit of radioactivity. This is defined as one **disintegration per second** (1 d.p.s.; see Section 9.2.5). However, an older unit, not in the SI system and still frequently used, is the curie (Ci). This is defined as the quantity of radioactive material in which the number of d.p.s. is the same as that in 1 g of radium, that is 3.7×10^{10} s^{-1} (or 37 GBq). For biological purposes, this unit is too large to be practical and so the microcurie (μCi) and millicurie (mCi) are used. It is important to realise that the units Bq and Ci refer to the number of disintegrations actually occurring in a sample not to the disintegrations detected, which generally will be only a proportion of the disintegrations occurring. Detected decays are referred to as counts and the corresponding rate is thus reported as counts per second (c.p.s.). For quick reference, a list of units and definitions frequently used in radioisotope work is provided in Table 9.5 at the end of the chapter.

9.2.7 Interaction of Radioactivity With Matter

α-Particles
These particles are the same as a helium nucleus and possess very high energy (4–8 MeV), with α-particles emitted by a particular isotope all having the same energy. They are relatively slow (about one-twentieth of the speed of light) and carry a double positive charge. α-Particles interact with matter in two ways:

Table 9.2 Properties of radioisotopes commonly used in the biological sciences

Property	³H	¹⁴C	³⁵S	³²P	³³P	¹²⁵I	¹³¹I
$t_{1/2}$	12.3 years	5730 years	87.4 days	14.3 days	25.4 days	59.5 days	8.02 days
Mode of decay	β	β	β	β	β	X-ray (by electron capture) and Auger electrons	γ and β
Max. β energy (MeV)	0.019	0.156	0.167	1.709	0.249	Auger electrons 0.035	0.806
ALI[a]	480 (MBq)[b]	34 (MBq)	15 (MBq)	6.3 (MBq)	14 (MBq)	1.3 (MBq)[c]	0.9 (MBq)[c]
Maximum range in air	6mm	24 cm	26 cm	790 cm	49 cm	>10m	>10 cm
Shielding required	None	1 cm acrylic	1 cm acrylic	1 cm acrylic	1 cm acrylic	Lead 0.25 m or lead-impregnated acrylic	Lead 13 mm
γ dose rate (μSv h⁻¹ from 1 GBq at 1 m)	–	–	–	(β dose rate 760 μSv, 10 cm from 1 MBq)	–	41	51
Čerenkov counting	–	–	–	Yes	–	–	–

Notes: [a]Annual limit on intake, based on a dose limit of 20 mSv using the most restrictive dose coefficients for inhalation or ingestion. For explanation of the unit Sv see Section 9.5. [b]Bound ³H. [c]Based on dose equivalent limit of 500 mSv to thyroid.

Example 9.1   **THE EFFECT OF HALF-LIFE**

Question   The half-life of ^{32}P is 14.3 days. How long would it take a solution containing
42 000 d.p.m. to decay to 500 d.p.m.?

Answer   Use Equation 9.12 to calculate the value of λ:

$$\lambda = \frac{0.693}{t_{1/2}} = \frac{0.693}{14.3 \text{ days}} = 0.0485 \text{ day}^{-1}$$

Then use Equation 9.11 to calculate the time taken for the counts to decrease. In
this equation, $N_0 = 42\ 000$ d.p.m. and $N_t = 500$ d.p.m.:

$$t = \frac{\ln\dfrac{N_t}{N_0}}{-\lambda} = \frac{1}{\lambda} \times \ln\frac{N_0}{N_t} = \frac{1}{0.0485 \text{ day}^{-1}} \times \ln\frac{42000 \text{ d.p.m.}}{500 \text{ d.p.m.}} = 91.4 \text{ days}$$

You can check this by a rough estimation: the calculated time is approximately
6 half-lives:

$$6 \times 13.2 \text{ days} = 79.2 \text{ days} \approx 91.4 \text{ days}$$

$$\frac{42000 \text{ d.p.m.}}{2^6} = 656 \text{ d.p.m.} \approx 500 \text{ d.p.m.}$$

- they cause excitation, where energy is transferred from the α-particle to electrons of neighbouring atoms. During this process, these electrons are elevated to higher orbitals, but they eventually fall back and simultaneously emit energy as photons
- they ionise atoms in their path by removing electrons from a target atom, which becomes ionised and forms an ion pair, consisting of a positively charged ion and an electron.

Despite their initial high energy, α-particles frequently collide with atoms in their path and so the radiation is not very penetrating (a few centimetres through air) and the energy is lost quickly. The particles cannot penetrate paper or the outer layers of skin. The safety consequences of this are that on the whole, α-particle emitters are only dangerous when ingested or inhaled, as for example with inhalation of polonium-210 in tobacco smoke.

Negatrons
Negatrons (β^-) are small, rapidly moving particles that carry a single negative charge. They interact with matter to cause ionisation and excitation. Their high speed and small size means they are less likely to interact with matter than α-particles and therefore are less ionising and more penetrating. Another difference between α-particles

and negatrons is that the latter are emitted over a range of energies from a single source. Negatron emitters have a characteristic energy spectrum (see Figure 9.5b below). The **maximum energy level** (E_{max}) varies from one isotope to another, ranging from 0.018 MeV for ^{3}H to 4.81 MeV for ^{38}Cl. The difference in E_{max} affects the penetration of the radiation and therefore the safety measures that are required: β-particles from ^{3}H can travel only a few millimetres in air and do not penetrate skin, whereas those from ^{32}P can penetrate over 1 m of air and easily enter human tissue. Therefore, radiation shields are needed when working with ^{32}P.

γ-Rays and X-rays

These rays (henceforth collectively referred to as γ-rays for simplicity) are electromagnetic radiation and therefore have no charge or mass. They are very penetrating (just think of diagnostic X-rays in hospital). They cause excitation and ionisation, and interact with matter to create secondary electrons that behave as per negatron emission.

Bremsstrahlung Radiation

When high-atomic-number materials absorb high-energy β-particles, the absorber emits a secondary radiation, a γ-ray, called Bremsstrahlung radiation. For this reason, shields for ^{32}P use low-atomic-number materials such as acrylic ('plastic glass'), made of carbon, oxygen and hydrogen.

9.3 DETECTION AND MEASUREMENT OF RADIOACTIVITY

There are three commonly used methods of detecting and quantifying radioactivity. These are based on the ionisation of gases, on the excitation of solids or solutions, and imaging techniques for example autoradiography (the use of photographic emulsions and films) or phosphor imaging.

9.3.1 Methods Based Upon Gas Ionisation

If a charged particle passes through a gas, its electrostatic field dislodges electrons from atoms sufficiently close to its path and causes ionisation (Figure 9.3a). The ability to induce ionisation decreases in the order

α > β > γ with a ratio of about 10000:100:1.

If ionisation occurs between a pair of electrodes enclosed in a suitable chamber, a current flows. Ionisation counters like those shown in Figure 9.3a are sometimes called proportional counters ('proportional' because small voltage changes can affect the count rate; see Figure 9.3c). The Geiger–Müller counter (Figure 9.3b,c, Figure 9.4a) has a cylindrical-shaped gas chamber and operates at a high voltage, thus eliminating problems with voltage stablity; therefore, the counter is cheaper and lighter than proportional counters.

In ionisation counters, the ions have to travel to their respective electrodes; other ionising particles entering the tube during this time (the so-called dead time) are not detected and this reduces the counting efficiency.

Example 9.2   **RADIATION MONITORING WITH IONISATION COUNTERS**

Question   Using the information described in Sections 9.2.3 and 9.2.7, what types of radiation can be detected by an ionisation counter?

Answer   The counters will detect α-particles and medium- to high-energy β-particles; they will detect γ-radiation, but with lower efficiency. They will not detect tritium (^{3}H) because the β-particles have very low energy and will not pass into the gas chamber.

(a)

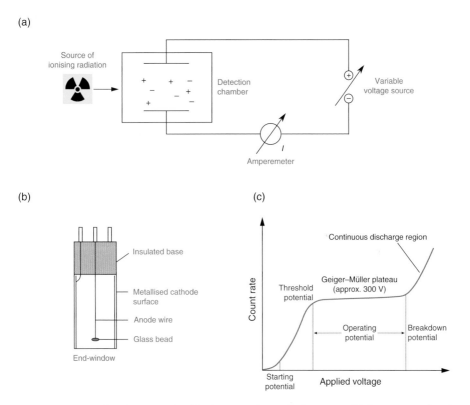

(b)   (c)

Figure 9.3 Detection based on ionisation. (a) Schematics of an ionisation counter. (b) The Geiger–Müller tube. (c) The effect of applied voltage on count rate.

Ionisation counters are used for routine monitoring of the laboratory to check for contamination (but not ^{3}H). They are also useful in experimental situations where the presence or absence of radioactivity needs to be known, rather than the absolute quantity; for example, quick screening of radioactive gels prior to autoradiography, checking that a labelled DNA probe is where you think it is (and not down the sink!) or checking chromatographic fractions for labelled components.

(a)   (b)

(c)

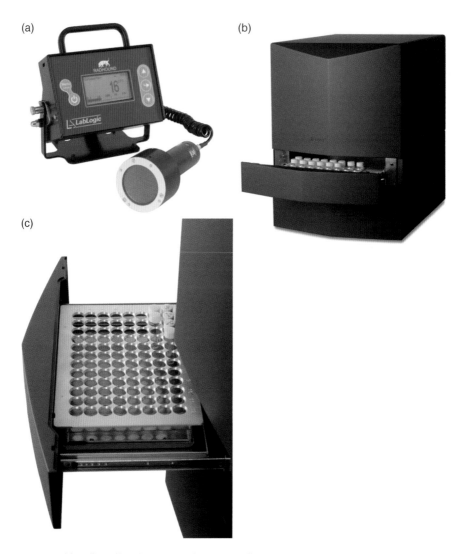

Figure 9.4 (a) Radhound bench monitor with Geiger–Müller counter; (b) Hidex 300 SL liquid scintillation counter; (c) sample rack for liquid scintillation counter. (Reproduced with permission from LabLogic and Hidex.)

9.3.2 Methods Based Upon Excitation

Radioactive isotopes interact with matter in two ways, ionisation and excitation. The latter effect leads an excited atom or compound to emit light photons, similar to the phenomenon of fluorescence (see Section 13.5). The species emitting photons after excitation is thus often called a fluor. In the context of radioisotopes, this process is known as scintillation. When the emitted light is detected by a photomultiplier, it forms the basis of scintillation counting. Essentially, a photomultiplier converts the energy of radiation into an electrical signal, and the magnitude of the current that results is directly proportional to the energy of the radiation. This means that two, or even more, isotopes can be separately detected and measured in the same sample,

Example 9.3   **THE EFFECT OF DEAD TIME**

Question   What will happen to the counting efficiency of a Geiger–Müller counter as the count rate rises?

Answer   The efficiency will fall since there will be an increased likelihood that two or more β-particles will enter the tube during the dead time.

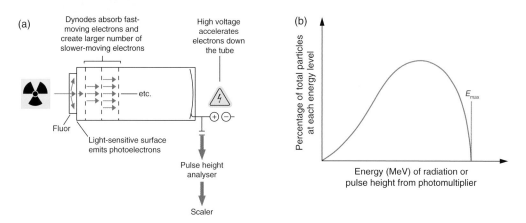

Figure 9.5 (a) The mode of action of a photomultiplier and (b) the energy spectrum of a typical β-emitter.

provided they have sufficiently different energy spectra. The mode of action of a photomultiplier is shown in Figure 9.5a, and the energy spectrum of a β-particle emitter in Figure 9.5b.

Types of Scintillation Counting

There are two types of scintillation counting, which are illustrated diagrammatically in Figure 9.6. In solid scintillation counting, the sample is placed adjacent to a solid fluor (e.g. sodium iodide). Solid scintillation counting is particularly useful for γ-emitting isotopes, mainly because γ-rays can penetrate the fluor. The counters can be small hand-held devices with the fluor attached to the photomultiplier tube (Figure 9.5a), or larger bench-top machines with a well-shaped fluor designed to automatically count many samples (Figure 9.6a).

In liquid scintillation counting (Figure 9.6b; see also Figure 9.4b, c), the sample is mixed with a scintillation fluid containing a solvent and one or more dissolved fluors. This method is particularly useful in quantifying weak β-emitters such as ^{3}H, ^{14}C and ^{35}S, which are frequently used in biological work. Scintillation fluids are called cocktails because there are different formulations, made of a solvent (such as toluene or diisopropylnaphthalene) plus fluors such as 2,5-diphenyloxazole (PPO), 1,4-bis(5-phenyloxazol-2-yl)benzene (nicknamed POPOP, pronounced as it reads: 'pop op') or 2-(4′-t-butylphenyl)-5-(4″-bi-phenyl)-1,3,4-oxydiazole (butyl-PBD).

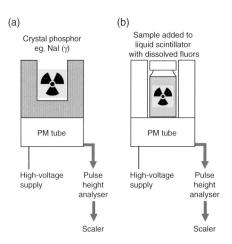

Figure 9.6 Diagrammatic illustration of (a) solid and (b) liquid scintillation counting methods.

Cocktails can be designed for counting organic samples, or may contain detergent to facilitate counting of aqueous samples.

There are various specialised types of scintillation counters, other than the hand-held monitors or vial readers already discussed. Many are made specifically for certain applications, for example when large numbers of a particular type of sample are being monitored. There are counters designed for counting single- or dual-labelled radio-isotopic samples separated by liquid chromatography and others that are microtitre plate readers, so they combine liquid scintillation counting with, for example, high-throughput screening (see Section 24.5.3).

Advantages of Scintillation Counting

Scintillation counting is widely used in biological work and it has several advantages over gas ionisation counting:

- like fluorescence, scintillation is very fast so there is effectively no dead time
- counting efficiencies are high (from about 50% for low-energy β-emitters to 90% for high-energy emitters)
- the ability to count samples of many types, including liquids, solids, suspensions and gels
- the general ease of sample preparation
- the ability to separately count different isotopes in the same sample (used in dual-labelling experiments)
- highly automated (hundreds of samples can be counted automatically and built-in computer facilities carry out many forms of data analysis, such as efficiency correction, graph plotting, radio-immunoassay calculations, etc.).

Disadvantages of Scintillation Counting

No technique is without disadvantages, so the following have to be considered or overcome in the design of the instruments:

- cost of the instrument and cost per sample (for scintillation fluid, the counting vials and disposal of the organic waste)

- **quenching**: this is the name for reduction in counting efficiency caused by coloured compounds that absorb the scintillated light, or chemicals that interfere with the transfer of energy from the radiation to the photomultiplier (correcting for quenching contributes significantly to the cost of scintillation counting)
- **chemiluminescence**: this happens when chemical reactions between components of the samples to be counted and the scintillation cocktail produce scintillations that are unrelated to the radioactivity; modern instruments can detect chemiluminescence and subtract it from the results automatically
- **phospholuminescence**: this results from pigments in the sample absorbing light and re-emitting it (the samples should thus be kept in the dark prior to counting).

Using Scintillation Counting for Dual-Labelled Samples

Different β-particle emitters have different energy spectra, so it is possible to quantify two isotopes separately in a single sample, provided their energy spectra can be distinguished from each other. Examples of pairs of isotopes that can be counted together are:

- ^{3}H and ^{14}C
- ^{3}H and ^{35}S
- ^{3}H and ^{32}P
- ^{14}C and ^{32}P
- ^{35}S and ^{32}P.

The principle of the method is illustrated in Figure 9.7, where it can be seen that the spectra of two isotopes (referred to as S and T) overlap only slightly. By setting a **pulse height analyser** to reject all pulses of energy below a threshold X and to reject all pulses of energy above a window XY and also to reject below a threshold of A and a window of AB, it is possible to separately count the two isotopes. A pulse height analyser set with a threshold and window for a particular isotope is known as a **channel**. Modern counters operate with a so-called multi-channel analyser that records the entire energy spectrum simultaneously. This greatly facilitates multi-isotope counting and in particular allows the effect of quenching on dual-label counting to be assessed adequately. Dual-label counting has proven useful in many aspects of molecular biology (e.g. nucleic acid hybridisation and transcription), metabolism (e.g. steroid synthesis) and drug development.

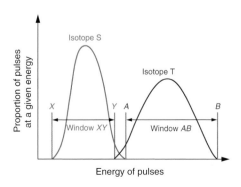

Figure 9.7 Diagram to illustrate the principle of counting dual-labelled samples.

Determination of Counting Efficiency

When detecting radioactivity, it is often the *actual* rate of decay (as opposed to the *recorded* rate) that is of particular interest. This is because we want to record and publish data that are independent of the types of equipment used to do the detection. To calculate the actual rates (in units of d.p.m. or d.p.s.; see Section 9.2.5) we need to know the efficiency of counting. Furthermore, in liquid scintillation counting, one has to contend with quenching. Samples may vary in nature, so the levels of quenching may vary from one to the next. Therefore, the efficiency of counting needs to be determined for every sample.

One way to do this is to use an internal standard called a spike. The sample is counted and gives a reading of A counts per minute (c.p.m.). Then, a small amount of standard material of known disintegrations per minute with B d.p.m. is added. The sample is measured again, yielding C c.p.m. This allows calculation of the counting efficiency of the sample:

$$\text{Counting efficiency} = \frac{C - A}{B} \times 100\% \qquad \text{(Eq 9.13)}$$

Carefully carried out, this is the most accurate way of correcting for quenching. Admittedly, it is a tedious process, since this has to be done for every sample. Automated methods have thus been devised, but all of these use the internal standard as the basis for establishing the parameters.

As a sample is quenched, the efficiency of light production falls, and it therefore creates an illusion that the radiation has a lower energy. The principle is straightforward: observe how the energy spectrum shifts as the efficiency falls, store this relationship in a computer and then analyse the energy spectrum of the sample to determine the efficiency. Different isotopes have different energies, so this system works best when there is only one, and the same, radioisotope in all the samples.

An alternative way is to use an external standard source of radioactivity built into the counter. This is called an external standard because the standard is not mixed with the sample. The source is usually a γ-emitter such as barium-133. The quenching in the sample is observed by counting this standard, the external source is then removed from the vicinity of the sample by the instrument, and the sample is counted. This is done for every sample in turn and the data are then reported as corrected values.

Yet another way of determining the counting efficiency does not require any standard, but addresses the issue by way of statistics. Some instruments (e.g. the Hidex 300 SL, shown in Figure 9.4 and Figure 9.8) use three photomultiplier tubes that count the sample simultaneously. The chances of either three or two tubes detecting a signal is affected by the extent of quenching in the sample, and the ratio of triple to double coincidence is therefore related to the counting efficiency.

Sample Preparation

For solid scintillation counting, sample preparation is easy and only involves transferring the sample to a glass or plastic vial (or tube) compatible with the counter. In liquid scintillation counting, sample preparation is more complex and starts with

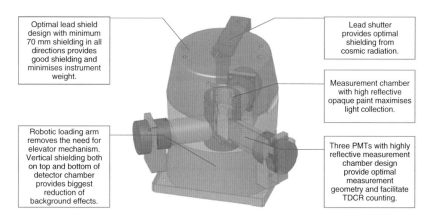

Optimal lead shield design with minimum 70 mm shielding in all directions provides good shielding and minimises instrument weight.

Lead shutter provides optimal shielding from cosmic radiation.

Measurement chamber with high reflective opaque paint maximises light collection.

Robotic loading arm removes the need for elevator mechanism. Vertical shielding both on top and bottom of detector chamber provides biggest reduction of background effects.

Three PMTs with highly reflective measurement chamber design provide optimal measurement geometry and facilitate TDCR counting.

Figure 9.8 The arrangement of photomultiplier tubes in a liquid scintillation counter; shown here is the Hidex 300SL. (Reproduced with permission from LabLogic.)

consideration of the types of sample and the particular formulation of scintillation fluid to use. The types of sample vials include low-potassium glass vials (with low levels of potassium-40 so as to reduce background count), or polyethylene tubes (cheaper but not reusable). Vials need to be chemically resistant, have good light transmission and give low background counts. The trend has been towards mini-vials, which use smaller volumes of scintillation fluid. Some counters are designed to accept very small samples in special polythene bags split into an array of many compartments; these are particularly useful to the pharmaceutical industry where there are laboratories that routinely run large numbers of receptor binding assays. Accurate counting requires that the sample is in the same phase as the counting cocktail. As described above, the scintillation fluid should be chosen as appropriate to aqueous or organic samples.

Čerenkov Counting

The Čerenkov effect occurs when a particle passes through a substance with a speed higher than that of light passing through the same substance. If a β-emitter has a decay energy value in excess of 0.5 MeV, then this causes water to emit a bluish white light usually referred to as Čerenkov light. It is possible to detect this light using a conventional liquid scintillation counter. Since there is no requirement for organic solvents and fluors, this technique is relatively cheap, sample preparation is very easy and there is no problem with chemical quenching. Table 9.2 indicates which isotopes can be counted this way.

Scintillation Proximity Assay

A scintillation proximity assay (SPA) is an application of scintillation counting that facilitates automation and rapid throughput of experiments. It is therefore highly suited to work such as screening for biological activity in new drugs. The principle of SPA is illustrated in Figure 9.9. The beads for SPA are often constructed from polystyrene and combine a binding site for a molecule of interest with a scintillant. This

Example 9.4   **CALCULATION OF THE ACTUAL RATE OF DECAY**

Question A scintillation counter analyses the energy spectrum of an external standard, it records a point on the spectrum (the quench parameter) and assesses the shift in the spectrum as the efficiency falls. The efficiency of detecting ^{14}C in a scintillation counter is determined by counting a standard sample containing 105 071 d.p.m. at different degrees of quenching (by adding increasing amounts of a quenching chemical such as chloroform). The results look like this:

Counts (c.p.m.)	Quench parameter
87 451	0.90
62 361	0.64
45 220	0.46
21 014	0.21

An experimental sample gives 2026 c.p.m. at a quench parameter of 0.52. What is the true count rate?

Answer First, the counting efficiency of the quenched standards needs to calculated; the efficiency in this example is defined as:

$$counting\ efficiency = \frac{detected\ counts\ in\ c.p.m.}{actual\ rate\ of\ decay\ in\ d.p.m.}.$$

This can be calculated for the four data points as shown below:

Counts (c.p.m.)	Counting efficiency	Quench parameter
87 451	$\dfrac{87\ 451}{105\ 071} = 0.83$	0.90
62 361	$\dfrac{62\ 361}{105\ 071} = 0.59$	0.64
45 220	$\dfrac{45\ 220}{105\ 071} = 0.43$	0.46
21 014	$\dfrac{21\ 014}{105\ 071} = 0.20$	0.21

Then, the efficiency is plotted against the quench parameter. Using the value of the quench parameter for the experimental sample (0.52), an efficiency can be extrapolated from the plot; in this case 0.48.

The actual rate of decay is then calculated using the formula:

$$\text{actual rate of decay in d.p.m.} = \frac{\text{detected counts in c.p.m.}}{\text{counting efficiency}} = \frac{2026 \text{ c.p.m.}}{0.48} = 4221 \text{ d.p.m.}$$

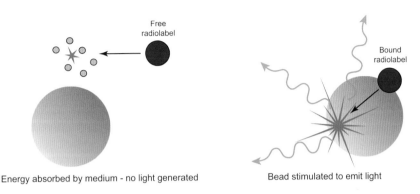

Free
radiolabel

Bound
radiolabel

Energy absorbed by medium - no light generated

Bead stimulated to emit light

Figure 9.9 The concept behind the scintillation proximity assay (SPA).

is particularly useful for those types of radiation that do not travel far, e.g. β-particle emission from weak energy emitters such as tritium (^{3}H) and carbon-14. If molecules containing such radioisotopes are in solution with a suspension of SPA beads, the radiation does not stimulate the scintillant in the beads and cannot be detected efficiently by a scintillation counter. This is because the radiation is absorbed by the solution; it does not reach the scintillant. However, if the radioisotope binds to the bead, it is close enough to stimulate the scintillant in the bead, so light is emitted and the isotope is detected.

There are many applications of this technology, such as enzyme and binding assays, indeed any situation where the interaction between two molecules needs to be investigated, for example, a receptor binding assay. This assay involves attachment of the receptor to the SPA beads; the radiolabelled ligand (a drug or a hormone) is then mixed with the beads to allow potential binding to occur. Any binding event of a radiolabelled ligand will stimulate the scintillant and thus be counted. Interference by chemicals such as drugs can be investigated (which is very often the mode of action of therapeutics); they can be added at increasing concentration to study the effect and quantitatively analysed.

A summary of the advantages of SPA technology:

- Versatile: use with enzyme assays, receptors and any molecular interaction
- Works with a range of appropriate isotopes such as tritium, carbon-14, phosphorus-33 and sulfur-35
- No need for separation step (e.g. isolation of the bound ligand)
- Less manipulation, therefore reduced toxicity
- Amenable to automation.

9.3.3 Methods Based Upon Exposure of Photographic Emulsions

Ionising radiation acts upon a photographic emulsion or film to produce a latent image very much like visible light; this is called autoradiography. The emulsion or film contains silver halide crystals. As energy from the radioactive material is dissipated, the silver halide becomes negatively charged and is reduced to elementary silver, thus forming a particulate latent image. Photographic developers show these silver grains as a blackening of the film, then a fixation step is used to remove any remaining silver halide and obtain a permanent image. This is a very sensitive technique and has been used in a wide variety of biological experiments. A good example is autoradiography of nucleic acids separated by gel electrophoresis (see Figure 9.10).

Here, an autoradiography emulsion or X-ray film is placed as close as possible to the sample and exposed at any convenient temperature. Quantitative images are produced until saturation is reached. The shades of grey in the image are related to a combination of levels of radiation and length of exposure, until a black image results. Isotopes with an energy of radiation equal to, or higher than, carbon-14 (E_{max} = 0.156 MeV) are required. The higher the energy the quicker the results will be obtained.

Suitable Isotopes

In general, weak β-emitting isotopes (e.g. tritium, carbon-14 and sulfur-35) are most suitable for autoradiography, particularly for cell and tissue localisation experiments.

Since the sample is placed close to the film, the low energy of the radiation has a comparatively low spread and a clear image results. Emitters with higher energy (e.g. phosphorus-32) yield faster results, but poorer resolution, since the higher-energy negatrons produce much longer track lengths, therefore affecting a greater surface area of the film, and so result in less discrete images. This is illustrated in Figure 9.10, showing autoradiography with three different isotopes.

Fluorography

If low-energy β-emitters are used, it is possible to enhance the sensitivity by several orders of magnitude using fluorography. A fluor (e.g. 2,5-diphenyloxazole (PPO) or

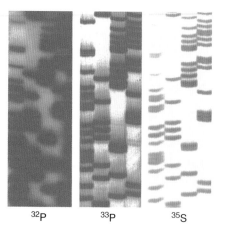

^{32}P ^{33}P ^{35}S

Figure 9.10 Three autoradiographs showing the use of different radioisotopes in DNA sequencing. The isotope with the highest energy (phosphorus-32) leads to the poorest resolution, because the radiation spreads out further, making the DNA bands appear thicker. The lowest-energy radiation (from sulfur-35) yields images with the best resolution. (Reproduced with permission from Cunningham M.W., Patel A., Simmonds A.C. and Williams D (2002) In vitro labelling of nucleic acids and proteins. In *Radioisotopes in Biology*, 2nd Edn., ed. Slater R.J., Oxford University Press, Oxford, UK.)

sodium silicate) can be used to enhance the image. The β-particles emitted from the isotope will excite the fluor, giving rise to emission of light and thus registration on the film. This has been used, for example, for detecting radioactive nucleic acids in gels. The fluor is infiltrated into the gel following electrophoresis and subsequently the gel is dried and placed in contact with a pre-flashed film (see below).

Intensifying screens

Intensifying screens are used when obtaining a fast result is more important than high resolution. This is frequently the case when analysing electrophoresis gels or membrane filters where high-energy β-emitters (e.g. ^{32}P-labelled DNA) or γ-emitting isotopes (e.g. ^{125}I-labelled protein) are used. The intensifying screen consists of a solid mixture called phosphor, which is placed on the side of the film that is opposite the sample. High-energy radiation passes through the film, causes the phosphor to fluoresce and emit light, which in turn registers on the film. The reduction in resolution is due to the spread of light emanating from the screen.

Low-Temperature Exposure

Ideally, when intensifying screens or fluorography are used, the exposure should be conducted at low temperature. The light registered in fluorography or by using intensifying screens is of low intensity and the resulting latent image on the film may suffer from a back reaction. Exposure at low temperature (–70 °C) slows the back reaction and therefore provides higher sensitivity. Direct autoradiography, in contrast, does not require low temperature as the kinetics of light registration on the film are different.

Pre-Flashing

The response of a photographic emulsion to radiation is not linear and usually involves a slow initial phase (called lag), followed by a linear phase. The sensitivity of films may be increased by pre-flashing. This involves a millisecond light flash prior to the sample being brought into juxtaposition with the film and is often used where high sensitivity is required or if results are to be quantified.

Quantification

Autoradiography is usually used for qualitative rather than quantitative analysis of radioactivity. However, it is possible to obtain quantitative data directly from autoradiographs by using digital image analysis. Quantification is not reliable at low or high levels of exposure, because of the lag phase (see pre-flashing above) or saturation, respectively. Pre-flashing combined with fluorography or intensifying screens create the best conditions for quantitative working.

9.3.4 Phosphor Imaging

Phosphor imaging is a technique similar to autoradiography, but with images being developed onto reusable phosphor screens (also called image plates). Since results can be obtained in a fraction of the time and with similar resolution when compared to autoradiography, this has become a popular method. The detection process is based on scintillation counting, whereby light emitted from excited electrons in the fluor of the image plate is read by a laser scanner. It can be used for example for tritium, carbon-11, carbon-14, iodine-125, fluorine-18, phosphorus-32, phosphorus-33, sulfur-35, metastable technetium-99 and other sources of ionising radiation such as X-rays (image plate detectors are often found in diffractometers; see Section 14.3.4). There are instruments that look very much like traditional desk scanners. The technique can be used to quantify and visualise the location of labelled molecules such as nucleic acids, proteins and bound ligands. Samples can be any thin solid object such as tissue slices, gels, blots etc.

9.3.5 Positron Emission Tomography (PET) Scanning

This powerful technique is used to observe biochemical processes in living organisms, including humans, and is thus of clinical importance (nuclear medicine). Frequently, fluoro-deoxyglucose (FDG) is used as the labelled molecule as it includes the β-emitter fluorine-18. This glucose analogue concentrates in tissues with high metabolic activity, but it cannot be metabolised through glycolysis, so it is chemically relatively stable in the cells. It is detected by γ-emission using a special scanner that detects the 180° γ-emissions (see Section 9.2.3). It is an expensive technique, but has the great advantage that living cells, tissues and organisms can be scanned. The results are compiled as computer-generated three-dimensional images of high resolution (called tomograms; see Figure 9.11). It is used in fundamental research (e.g. brain scanning), disease diagnosis (e.g. tumour growth) and applied research, such as location of drug target molecules. A related technique called immuno-PET uses radioactive monoclonal antibodies to detect target proteins with great sensitivity. For example, fluorine-18 conjugated affibodies (antibody mimetics in the form of small proteins engineered to bind to a large number of target proteins) directed toward human epidermal growth factor (HER2) have been used to image tumour sites in living organisms.

9.4 OTHER PRACTICAL ASPECTS OF COUNTING RADIOACTIVITY AND ANALYSIS OF DATA

9.4.1 Self-Absorption

Self-absorption is primarily a problem with low-energy β-emitters: radiation is absorbed by the sample itself. Self-absorption can be a serious problem in the counting of low-energy radioactivity by scintillation counting if the sample is, for instance, a membrane filter. In these cases, automated methods for calculating counting efficiency in a scintillation counter do not work.

9.4.2 Specific Activity

The specific activity of a radioisotope is defined as its radioactivity in relation to the amount of the element or the molecule it is attached to, expressed as Bq mmol^{-1}, Ci mmol^{-1} or d.p.m. mmol^{-1}. It is a very important aspect of the use of radioisotopes in

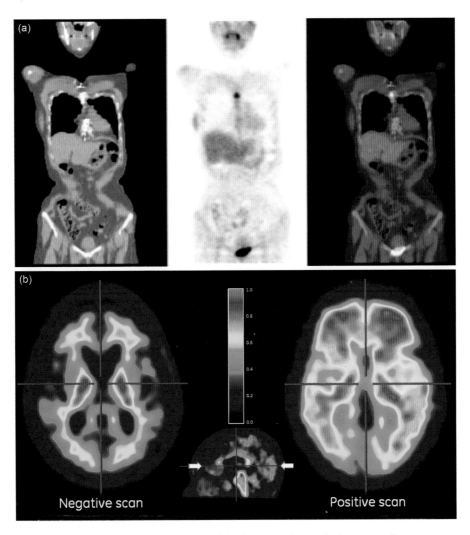

Figure 9.11 (a) PET scanning of the central part of the chest cavity (the mediastinum) using fluoro-deoxyglucose (FDG) combined with computed tomography (i.e. imaging by sections). A higher than normal uptake of FDG suggests the presence of a tumour. (b) Normal (left) and abnormal (right) brain scans of uptake of 2-[3-(^{18}F)fluoro-4-(methylamino)phenyl]-1,3-benzothiazol-6-ol (flutemetamol, tradename Vizamyl™, GE Healthcare), a PET scanning radiopharmaceutical containing the radionuclide fluorine-18, used as a diagnostic tool for Alzheimer's disease. (Images reproduced with permission of GE Heathcare.)

biological work, because the higher the specific activity the smaller the quantities of labelled substance that can be detected and therefore the more sensitive the experiment. Specific activity is related to the rate of decay; therefore, the shorter the half-life the higher the specific activity.

Sometimes, it is not necessary to purchase the highest specific activity available. For example, in vitro enzyme assays often require a relatively high substrate concentration and thus the specific activity may need to be lowered. Consider the example below (for definitions of units, see Table 9.5 at the end of the chapter):

[³H]Leucine is purchased with a specific activity of 5.55 TBq mmol⁻¹ (150 Ci mmol⁻¹) and a concentration of 9.25 MBq 250 mm⁻³ (250 mCi 250 mm⁻³). A 10 cm³ solution of 250 mM and 3.7 kBq cm⁻³ (0.1 mCi cm⁻³) is required. This solution is prepared as follows:

- 10 cm³ at 3.7 kBq cm⁻³ is 37 kBq (1 mCi), therefore 1 mm³ of stock radioisotope is required. This is formulated by first preparing a 100× dilution of stock radioisotope in water and pipetting 100 mm³ of the dilution into a suitable vessel.
- Add 2.5 cm³ of a 1 M stock solution of unlabelled leucine (also called cold carrier), and make up to 10 cm³ final volume with distilled water.
- Note that there is no need to take into account the amount of the leucine carrier in the [³H]leucine preparation; it is a negligible quantity due to the high specific activity.

In order to manipulate solutions of relatively low specific activity, the following formula is often used to calculate the mass of cold carrier:

$$m = M \times A \times \left[\frac{1}{a'} - \frac{1}{a} \right] \qquad \text{(Eq 9.14)}$$

where m is the mass of cold carrier required (mg), M is the molecular mass of the compound, A is the amount of radioactivity present (MBq), a is the original specific activity (MBq mmol⁻¹) and a' is the required specific activity (MBq mmol⁻¹).

9.4.3 Statistics

The emission of radioactivity is a stochastic process and the spread of counts forms a normal distribution. The standard deviation can be calculated very simply (using mathematics devised by Poisson) by taking a square root of the counts:

$$\sigma = \sqrt{\text{total counts taken}} \qquad \text{(Eq 9.15)}$$

Essentially, this means that the more counts that are registered during a measurement, the smaller the standard deviation is with respect to the mean count rate. It follows that the more counts measured, the more accurate the data (see Table 9.3). It is thus common practice to count to 10 000 counts or for a period of 10 min, whichever is the quicker, although for very low count rates longer counting times are required. Another common practice is to quote mean results plus or minus 2 standard deviations, since 95.5% of results lie within this range, the so-called confidence interval (see Section 19.3.4).

Table 9.3 Accuracy increases with the number of counts, illustrated for a series of 1 min counts

Counts	σ	
100	10	σ is 10% of the mean
1000	33	σ is 3% of the mean
10 000	100	σ is 1% of the mean

Example 9.5   **PREPARATION OF A SOLUTION OF KNOWN ACTIVITY**

Question   One litre of [³H]uridine with a concentration of 100 mmol dm⁻³ and 50 000 c.p.m. cm⁻³ is required. If all measurements are made on a scintillation counter with an efficiency of 40%, how would you prepare this solution using non-labelled uridine as well as the purchased supply of [³H]uridine that has a radioactive concentration of 1 mCi cm⁻³ and a specific activity of 20 Ci mol⁻¹ (0.75 TBq mol⁻¹)?

[NB: M(uridine) = 244 g mol⁻¹; 1 Ci = 22.2 × 10¹¹ d.p.m.]

Answer   This problem is similar to the leucine example given in the main text.

First, one needs to correct for the 40% counting efficiency

$$\text{actual rate of decay} = \frac{\text{detected counts}}{\text{counting efficiency}} =$$

$$\frac{50\,000 \text{ c.p.m. cm}^{-3}}{0.40} = 125\,000 \text{ d.p.m. cm}^{-3}$$

Since this rate is with respect to 1 cm⁻³ = 1 ml, it needs to be multiplied by 10³ so it refers to 1 litre:

$$\text{actual rate of decay} = 125\,000 \text{ d.p.m. cm}^{-3} = 125 \times 10^6 \text{d.p.m. l}^{-1}$$

which is equivalent to:

$$\text{actual rate of decay} = \frac{125 \times 10^6}{22.2 \times 10^{11}} \text{Ci l}^{-1} = 56.3 \, \mu \text{ Ci l}^{-1}.$$

Given a specific activity of 20 Ci mol⁻¹ of the purchased sample of [³H]uridine, we next need to work out the molar amount required to achieve 56.3 µCi l⁻¹:

$$n = \frac{56.3 \, \mu \text{ Ci l}^{-1}}{20 \text{ Ci mol}^{-1}} \times 1\,1 = 2.815 \, \mu \text{mol}$$

Since a concentration of 100 mmol dm⁻³ of [³H]uridine is required, we can now calculate the mass of unlabelled uridine (cold carrier) to be added to the radio-isotope-labelled uridine:

$$m\,(\text{uridine}) = M \times n = M \times c \times V_L = 244 \text{g mol}^{-1} \times 100 \text{ mmol l}^{-1} \times 1\,1 = 24.4 \text{ g}$$

The 2.815 µmol of [³H]uridine correspond to 0.687 mg uridine. Compared to the 24.4 g of cold uridine, this is a negligible contribution to the molar amount of uridine in the final sample and can be ignored.

Therefore, we require 56.3 mm³ (56.3 µCi, 2.08 MBq) of [³H]uridine plus 24.4 g of uridine.

Example 9.6   **ACCURACY OF COUNTING**

Question   A radioactive sample was recorded with 564 c.p.m. over 10 min. What is the accuracy of the measurement for 95.5% confidence?

Answer   The rate of decay was determined to be 564 c.p.m.; knowing the time period of the measurement, we can calculate the number of counts:

$$\text{Rate of decay} = \frac{\text{counts}}{\text{time}}$$

$$\Rightarrow \text{counts} = \text{rate of decay} \times \text{time} = 564 \text{ c.p.m.} \times 10 \text{ min} = 5640$$

Using Equation 9.15, the standard deviation of the measurement is calculated as the square root of the counts:

$$\sigma = \sqrt{5640} = 75$$

In order to achieve a 95.5% confidence interval, we need to report a range of $\pm 2\sigma$, i.e. 5640 ± 150 counts (95.5%). Expressed as a rate of decay, this yields 564 ± 15 c.p.m. (95.5%).

9.4.4 The Choice of Radionuclide

This is a complex question depending on the precise requirements of the experiment. A summary of some of the key features of radioisotopes commonly used in biological work is shown in Table 9.4. The key factors are:

- safety
- the type of detection to be used
- the sensitivity required (see Section 9.4.2)
- the cost.

For example, phosphorus-33 may be chosen for work with DNA, because it has high enough energy to be detected easily; it is safer than phosphorus-32 and its half-life is short enough to give high specific activity, but long enough to be convenient to use. On the other end of the scale are the isotopes used for PET scanning. They deliver a penetrating radiation, necessary for detection, they have a very short half-life giving huge specific activity and therefore great sensitivity, but are expensive. Additionally, as the half life of fluorine-18 is so short (~110 minutes), it is imperative to correct for decay during image acquisition to obtain reliable results.

Although different isotopes undergo the same reactions, they may do so at different rates. This is known as the isotope effect. The different rates are approximately proportional to the differences in mass between the isotopes. This can be a problem in the case of hydrogen and tritium, but the effect is small for carbon-12 and carbon-14 and almost insignificant for phosphorus-32 and phosphorus-33. The isotope effect may be taken into account when choosing which part of a molecule to label with tritium.

Table 9.4 **The relative merits of commonly used radioisotopes**

Isotope	Advantages	Disadvantages
^{3}H	Relative safety	Low efficiency of detection
	High resolution in imaging techniques	Isotope exchange with environment
	Wide choice of positions in organic compounds	Isotope effect
^{14}C	Relative safety	Low specific activity
	Wide choice of labelling position in organic compounds	
	Good resolution in imaging techniques	
^{35}S	High specific activity	Short half-life
	Good resolution in imaging techniques	Relatively long biological half-life
^{33}P	High specific activity	Lower specific activity than ^{32}P
	Good resolution in imaging techniques	Less sensitive than ^{32}P
	Less hazardous than ^{32}P	Cost
^{32}P	Ease of detection	Short half-life affects costs and experimental design
	High specific activity	
	Short half-life simplifies disposal	High β energy, so external radiation hazard
	Čerenkov counting	Poor resolution in autoradiography
^{125}I	Ease of detection	High penetration of radiation
	High specific activity	
	Good for labelling proteins	
^{131}I	Ease of detection	High penetration of radiation
	High specific activity	Short half-life
^{18}F	Very high specific activity	Cost (due to very short half-life)
	Exquisite sensitivity	
	Ease of detection	
	Disposal (due to very short half-life)	

9.5 SAFETY ASPECTS

The greatest practical disadvantage of using radioisotopes is the toxicity. When absorbed, radiation causes ionisation and free radicals form that interact with macro-molecules in the cell, causing mutation of DNA and hydrolysis of proteins. The toxicity of radiation is dependent not just on the amount present, but rather the amount absorbed by the body, the energy of the absorbed radiation and its biological effect. The absorbed dose of radiation is used to describe this and measured in grays (Gy), an SI unit; 1 Gy is an absorption of 1 J per kg of absorber.

A key aspect of determining toxicity is to know how much energy might be absorbed; this is a similar risk assessment as determining how much exposure to sunlight gives you sunburn and potentially skin cancer. The higher the energy of the radiation the greater the potential hazard. Different types of radiation of the same energy are associated with differing degrees of biological hazard. It is therefore necessary to introduce a dose equivalent, whereby the biological effects of any type of radiation is weighted by the biological hazard (obtained by way of comparison with the hazards presented by X-rays of the same energy). The dose equivalent is measured in units of sievert (Sv). For β-radiation, the values of absorbed dose and dose equivalent are the same, but for α-radiation, 1 Gy corresponds to 20 Sv. In other words, α-radiation is 20 times as toxic to humans as X-rays for the same energy absorbed. Clearly, from the point of view of safety, it is advisable to use radioisotopes with low energy wherever possible.

Absorbed dose from known sources can be calculated from knowledge of the rate of decay of the source, the energy of radiation, the penetrating power of the radiation and the distance between the source and the laboratory worker. As the radiation is emitted from a source in all directions, the level of irradiation is related to the area of a sphere. Thus, the absorbed dose is inversely related to the square of the distance from the source; or, put another way, if the distance is doubled the dose is quartered. This is also known as the inverse-square law, and a useful formula is:

$$\text{Dose}_1 \times \left(\text{Distance}_1\right)^2 = \text{Dose}_2 \times \left(\text{Distance}_2\right)^2 \qquad \text{(Eq 9.16)}$$

The relationship between radioactive source and absorbed dose is illustrated in Figure 9.12. The rate at which a dose is delivered is referred to as the dose rate, expressed in Sv h^{-1}. It can be used to calculate the total dose received by a person. For example, a source may be delivering 10 mSv h^{-1}. If you worked with the source for 6 h, your total dose would be 60 mSv. Dose rates for isotopes are provided in Table 9.2.

Currently, the dose limit for workers exposed to radiation is 20 mSv in a year to the whole body, but this is rarely ever approached by biologists because the levels of radiation used are so low. Dose limits are set for individual organs. The most important of these to know are for hands (500 mSv per year) and for lens of the eye (150 mSv per year). For health and safety reasons, work that involves radioisotopes (or other forms of high-energy radiation) must be carried out in

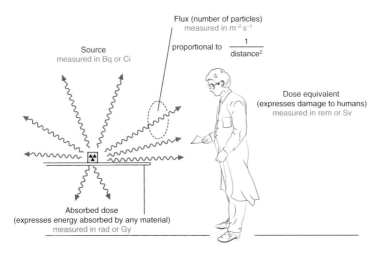

Figure 9.12 The relationship between radioactivity of source and absorbed dose.

dedicated areas (supervised areas for low-dose work and controlled areas for high-dose work).

In general, the **ALARA principle** is applied to work with radiation; it stands for: As Low As Reasonably Achievable. In practice, work in the biosciences rarely involves a worker receiving a measurable dose. Supervised areas are common, but not always required (e.g. for tritium or carbon-14 experiments). Controlled areas are required only in certain circumstances, for example for isotope stores or radioiodination work. A potential problem in biosciences, however, is the internal radiation hazard. This is caused by radiation entering the body, for example by inhalation, ingestion, absorption or puncture. This is a likely source of hazard where work involves open sources (i.e. liquids and gases), as is the case in most work in biology. Control of contamination is assisted by:

- complying with applicable state laws and regulations
- complying with local rules, written by an employer
- conscientious personal conduct in the laboratory
- regular monitoring
- carrying out work in some kind of containment.

A useful guide for assessing internal risks is the **annual limit on intake** (ALI). The ingestion of one ALI results in a person receiving a dose limit to the whole body or to a particular organ. Some ALIs are shown in Table 9.2. Management of radiation protection is similar in most countries. In the USA, there is a Code of Federal Regulations. In the UK, the Ionising Radiations Regulations (1999). Every institution requires certification (monitored by the Environmental Protection Agency in the USA or the Environment Agency in the UK) and employs a Radiation Protection Advisor.

When planning to use a radioisotope, consider the following (see also Figure 9.13):

- Is a radioisotope absolutely necessary? If the answer is no, then a non-radioactive method should be used.
- Which isotope to use? Ideally the one with the lowest energy that can deliver your needs.

When handling radioisotopes the rules are:

- Wear protective clothing, gloves and glasses
- Use the smallest amount possible (the ALARA principle)
- Keep radioactive materials safe, secure and well labelled
- Work in defined areas and use a spill tray
- Monitor the working area frequently (by liquid scintillation counting for low energy isotopes)
- Have no foods or drinks in the laboratory
- Wash and monitor hands after the work is done

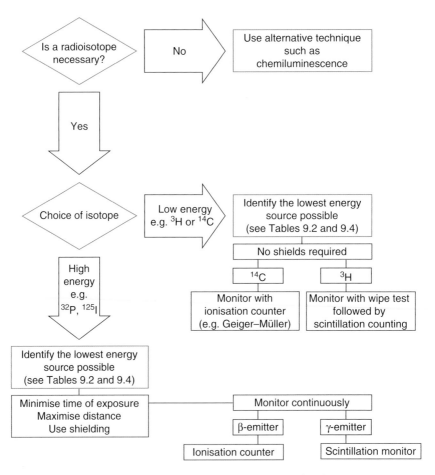

Figure 9.13 Flowchart of considerations when planning experiments with radioisotopes.

- Follow all local rules such as for the dispensing of stock and the disposal of waste
- Do not create radioactive aerosols or dust.

In particular for penetrating radiation (such as, for example, phosphorus-32 and λ-emitters):

- Maximise the distance between yourself and the source
- Minimise the time of exposure
- Maintain shielding at at all times (not necessary for tritium).

Table 9.5 Units commonly used to describe radioactivity

Unit	Abbreviation	Definition
counts per minute or second	c.p.m. c.p.s.	The *recorded* rate of decay
disintegrations per minute or second	d.p.m. d.p.s.	The *actual* rate of decay
curie	Ci	The number of d.p.s. equivalent to 1 g of radium (3.7×10^{10} d.p.s.)
millicurie	mCi	10^{-3} Ci or 2.22×10^9 d.p.m.
microcurie	μCi	10^{-6} Ci or 2.22×10^6 d.p.m.
becquerel (SI unit)	Bq	1 d.p.s.
megabecquerel (SI unit)	MBq	10^6 Bq or 27.027 μCi
gigabecquerel (SI unit)	GBq	10^9 Bq or 27.027 mCi
terabecquerel (SI unit)	TBq	10^{12} Bq or 27.027 Ci
electron volt	eV	The energy attained by an electron accelerated through a potential difference of 1 volt. Equivalent to 1.6×10^{-19} J
roentgen	R	The amount of radiation that produces 1.61×10^{15} ion-pairs kg^{-1}
rad	rad	The dose that gives an energy absorption of 0.01 J kg^{-1}
gray	Gy	The dose that gives an energy absorption of 1 J kg^{-1}. Thus 1 Gy = 100 rad
rem	rem	The amount of radiation that gives a dose in humans equivalent to 1 rad of X-rays
sievert	Sv	The amount of radiation that gives a dose in humans equivalent to 1 Gy of X-rays. Thus 1 Sv=100 rem

9.6 SUGGESTIONS FOR FURTHER READING

9.6.1 Experimental Protocols

Slater R.J. (2002) *Radioisotopes in Biology: A Practical Approach, 2nd Edn.*, Oxford University Press, Oxford, UK.

9.6.2 General Texts

Billington D., Jayson G.G. and Maltby P.J. (1992) *Radioisotopes*, Bios Scientific, Oxford, UK.

Creager A.N.H. (2013) *Radioactivity Traced Life Atomic: A History of Radioisotopes in Science and Medicine*, University of Chicago Press, Chicago, IL, USA.

L'Annunziata M. (2016) *Radioactivity: Introduction and History, from the Quantum to Quarks*, Elsevier, Amsterdam, Netherlands.

Martin A., Harbison S., Beach K. and Cole P. (2012) *An Introduction to Radiation Protection*, 6th Edn. Chapman and Hall, London, UK.

Slater R.J. (1996) *Radioisotopes in molecular biology.* In *Molecular Biology and Molecular Medicine*, Myers R.A. Ed., Wiley-VCH, New York, USA, pp. 209–219.

The Health and Safety Executive (2000) *Work with Ionising Radiation: Approved Codes of Practice*, HSE Books, London, UK.

Wolfe R.R. and Chinkes D.L. (2004) *Isotope Tracers in Metabolic Research: Principles and Practice of Kinetic Analysis*, 2nd Edn., John Wiley & Sons, Inc., New York, USA.

9.6.3 Review Articles

Deakin T. (2015) Radisotopic characterization as an analytical tool: current status, limitations and future challenges. *Bioanalysis* **7**, 541–555.

Elmore C.S. and Bragg R.A. (2015) Isotope chemistry: a useful tool in the drug discovery arsenal. *Bioorganic and Medicinal Chemistry Letters* **25**, 167–171.

Sing, N. Ed. (2011) *Radioisotopes: Applications in Bio-Medical Science*, InTech, Rijeka, Croatia. A series of reviews at www.intechopen.com/books/radioiso topes-applications-in-bio-medical-science (accessed May 2017).

9.6.4 Websites

Radiometric detectors (*PerkinElmer*)
www.perkinelmer.com/Catalog/Family/ID/Radiomatic%20Flow%20Scintillation%20 Analyzers (accessed May 2017)

Radiation information by the British Health and Safety Executive
www.hse.gov.uk/radiation/ (accessed May 2017)

Radiation protection information by the US National Regulatory Commission and Environmental Protection Agency
www.nrc.gov/about-nrc/radiation.html (accessed May 2017)
www.epa.gov/radiation (accessed May 2017)

Radiation protection information by the Australian Radiation Protection and Nuclear Safety Agency
www.arpansa.gov.au/RadiationProtection/index.cfm (accessed May 2017)

Positron emission tomography tutorial (University of Ghent)
www.analchem.ugent.be/radiochemie/funct_beeldvorming/Let%27s_Play_PET_static/
laxmi.nuc.ucla.edu_8000/lpp/lpphome.html (accessed May 2017)

10 Principles of Clinical Biochemistry

GILL RUMSBY

10.1 PRINCIPLES OF CLINICAL BIOCHEMICAL ANALYSIS

10.1.1 Basis of Analysis of Body Fluids for Diagnostic, Prognostic and Monitoring Purposes

Underlying most human diseases is a change in the amount or function of one or more proteins that in turn triggers changes in cellular, tissue or organ function. The dysfunction is commonly characterised by a significant change in the biochemical profile of body fluids. The application of quantitative analytical biochemical tests to a large range of biological analytes in body fluids and tissues is a valuable aid to the diagnosis and management of the prevailing disease state. In this section, the general biological and analytical principles underlying these tests will be discussed and related to the general principles of quantitative biochemical analysis discussed in Section 2.7.

Body fluids such as blood, cerebrospinal fluid and urine, in both healthy and diseased states, contain a large number of inorganic ions and organic molecules. While some of these chemical species exert their normal biological function within that fluid, the majority of these take a passive role and are only transported by the fluid. The presence of this latter group of chemical species within the fluid is due to the fact that normal cellular secretory mechanisms and the temporal synthesis and turnover of individual cells and their organelles within the major organs of the body both result in the release of cell components into the surrounding extra-cellular fluid and eventually into the blood circulatory system. This, in turn, transports them to the main excretory organs, namely the liver, kidneys and lungs, so that these cell components and/or their degradation products are eventually excreted in faeces, urine, sweat and expired air. Examples of cell components in this category include enzymes, hormones, intermediary metabolites, small organic molecules and inorganic ions.

The concentration, amount or enzyme activity of a given cell component that can be detected in these fluids of a healthy individual at any point in time depends on many factors that can be classified into one of three categories, namely chemical characteristics of the component, endogenous factors characteristic of the individual and exogenous factors that are imposed on the individual.

- Chemical characteristics: Some molecules are inherently unstable outside their normal cellular environment. For example, some enzymes are reliant on the presence of their substrate and/or coenzyme for stability and these may be absent or present only in low concentrations in the extra-cellular fluid. Molecules that can act as substrates of catabolic enzymes found in extra-cellular fluids, in particular blood, will also be quickly metabolised. Cell components that fall into these two categories therefore have a short half-life outside the cell and are normally present in low concentrations in fluids such as blood.
- Endogenous factors: These include age, gender, body mass and pregnancy. For example:
  (a) Serum cholesterol concentrations are higher in men than pre-menopausal women, but the differences decrease post-menopause
  (b) Serum alkaline phosphatase activity is higher in children than in adults and is raised in women during pregnancy
  (c) Serum insulin and triglyceride concentrations are higher in obese individuals than in the lean
  (d) Serum creatinine, a metabolic product of creatine important in muscle metabolism, is higher in individuals with a large muscle mass
  (e) Serum sex hormone concentrations differ between males and females and change with age.
- Exogenous factors: These include time, exercise, food intake and stress. Several hormones are secreted in a time-related fashion. For example, cortisol and to a lesser extent thyroid-stimulating hormone (TSH) and prolactin all show a diurnal rhythm in their secretion. In the case of cortisol, its secretion peaks around 9.00 am and declines during the day, reaching a trough between 11.00 pm and 5.00 am. The secretion of female sex hormones varies during a menstrual cycle and that of 25-hydroxycholecalciferol (vitamin D_3) varies with the seasons, peaking during the late summer months. The concentrations of glucose, triglycerides and insulin in blood rise shortly after the intake of a meal. Stress, including that imposed by the process of taking a blood sample by puncturing a vein (venipuncture), can stimulate the secretion of a number of hormones and neurotransmitters, including prolactin, cortisol, adrenocorticotropic hormone (ACTH) and adrenaline.

The influence of these various factors on the extent of release of cell components into extra-cellular fluids inevitably means that even in healthy individuals there is a considerable intra-individual variation (i.e. variation from one occasion to another) in the value of any chosen test analyte of diagnostic importance and an even larger inter-individual variation (i.e. variation between individuals). More importantly, the superimposition of a disease state onto these causes of intra- and inter-individual variation will result in an even greater variability between test occasions.

Many clinical conditions compromise the integrity of cells located in the organs affected by the condition. This may result in the cells becoming more 'leaky' or, in more severe cases, actually dying (necrosis) and releasing their contents into the surrounding extra-cellular fluid. In the vast majority of cases, the extent of release of specific cell components into the extra-cellular fluid, relative to the healthy reference range, will reflect the extent of organ damage and this relationship forms the basis of diagnostic clinical biochemistry. If the cause of the organ damage continues for a prolonged time and is essentially irreversible (i.e. the organ does not undergo self-repair), as is the case in cirrhosis of the liver, for example, then the mass of cells remaining to undergo necrosis will progressively decline so that eventually the release of cell components into the surrounding extra-cellular fluid will decrease, even though organ cells are continuing to be damaged. In such a case, the measured amounts will not reflect the extent of organ damage.

Clinical biochemical tests have been developed to complement in four main ways a provisional clinical diagnosis based on the patient's medical history and clinical examination:

- To support or reject a provisional diagnosis by detecting and quantifying abnormal amounts of test analytes consistent with the diagnosis. For example, troponin T (a component of the contractile apparatus of cardiac muscle), creatine kinase (CK, specifically the CK-MB isoform) and aspartate transaminase all rise following a myocardial infarction (MI, heart attack) that results in cell death in some heart tissue. The pattern and degree of response can be used to support a diagnosis of MI. Tests can also help a differential diagnosis, for example in distinguishing the various forms of jaundice (yellowing of the skin due to the presence of the yellow pigment bilirubin, a metabolite of haem) by the measurement of alanine transaminase (ALT) and aspartate transaminase (AST) activities and by determining whether or not the bilirubin is conjugated with β-glucuronic acid.

- To monitor recovery following treatment by repeating the tests on a regular basis and monitoring the return of the test values to those within the reference range. Following hepatitis, for example, raised serum ALT returns to reference range values within 10–12 days. Similarly, the measurement of serum tumour markers such as CA125 can be used to follow recovery or recurrence after treatment for ovarian cancer.

- To screen for latent disease in apparently healthy individuals by testing for raised levels of key analytes. For example, measuring plasma glucose for diabetes mellitus and immunoreactive trypsinogen for newborn screening of cystic fibrosis. It is now common for serum cholesterol and other factors, for example weight and blood pressure, to be used as a measure of the risk of an individual developing heart disease. This approach is particularly important for individuals with a family history of the disease. A serum cholesterol <=5 mM is recommended and individuals with values greater than this should be counselled on the importance of a healthy diet and regular exercise and, if necessary, on the requirement for cholesterol-lowering 'statin' drugs to be prescribed.

- To detect toxic side effects of treatment, for example in patients receiving hepatotoxic drugs, by undertaking regular liver function tests. An extension of this is therapeutic drug monitoring in which patients receiving drugs such as phenytoin and carbamazepine (both of which are used in the treatment of epilepsy) that have a low therapeutic index (ratio of the dose required to produce a toxic effect relative to the dose required to produce a therapeutic effect) are regularly monitored for drug levels and liver function to ensure that they are receiving effective and safe therapy.

10.1.2 Reference Ranges

In order for a biochemical test for a specific analyte to be routinely used as an aid to clinical diagnosis, it is essential that the test has the required performance indicators (Section 10.2), especially sensitivity and specificity. Sensitivity expresses the proportion of patients with the disease who are correctly identified by the test:

$$\text{Sensitivity} = \frac{\text{Number of true positive tests}}{\text{Number of total patients with the disease}} \times 100\% \qquad \text{(Eq 10.1)}$$

Specificity expresses the proportion of patients without the disease who are correctly identified by the test:

$$\text{Specificity} = \frac{\text{Number of true negative tests}}{\text{Number of total patients without the disease}} \times 100\% \qquad \text{(Eq 10.2)}$$

Ideally, both of these indicators for a particular test should be 100%, but this is not always the case. Such a discrepancy most likely occurs in cases where the change in the amount of the test analyte in the clinical sample is small compared with the reference range values found in healthy individuals.

Whereas both of the above indicators express the performance of the test, it is equally important to be able to quantify the probability that the patient with a positive test has the disease in question. This is best achieved by the predictive power of the test. This expresses the proportion of patients with a positive test that are correctly diagnosed as disease positive:

$$\text{Positive predictive value} = \frac{\text{Number of true positive patients}}{\text{Number of total positive tests}} \qquad \text{(Eq 10.3)}$$

$$\text{Negative predictive value} = \frac{\text{Number of true negative patients}}{\text{Number of total negative tests}} \qquad \text{(Eq 10.4)}$$

A low positive predictive value means that there will be many more false positives that will need to be followed up with additional testing. A test with a high negative predictive value can be useful as it means that a negative result makes the disease unlikely.

The concept of predictive power can be illustrated by reference to fetal screening for Down's syndrome and neural tube defects. Preliminary tests for these conditions in unborn children are based on the measurement of α-fetoprotein (AFP), human chorionic gonadotropin (hCG) and unconjugated oestriol (uE3) in the mother's blood and an ultra-sound assessment of nuchal translucency thickness – this is a space at the back of the baby's neck. The presence of Down's syndrome results in an increased hCG, inhibin and nuchal translucency (NT) as well as decreased AFP and uE3 relative to the average in healthy pregnancies, whereas open neural tube defects tend to have elevated AFP. The results of the tests are used in conjunction with the gestational and maternal ages to calculate the risk of the baby suffering from these conditions. If the risk is high, further tests are undertaken, including the recovery of some fetal cells for genetic screening from the amniotic fluid surrounding the foetus in the womb by inserting a hollow needle into the womb (amniocentesis). The tests have a detection rate of 90% for Down's syndrome and 85% for open neural tube defects. Accordingly, the performance indicators of the tests are not 100%, but they are sufficiently high to justify their routine use.

The correct interpretation of all biochemical test data is heavily dependent on the use of the correct reference range against which the test data are to be judged. As previously pointed out, the majority of biological analytes of diagnostic importance are subject to considerable inter- and intra-individual variation in healthy adults, and the analytical method chosen for a particular analyte assay will have its own precision, accuracy and selectivity (see Section 2.7) that will influence the analytical results. In view of these biological and analytical factors, individual laboratories should ideally establish their own reference range for each test analyte using their chosen methodology and a statistically determined number (usually hundreds) of 'healthy' individuals. The recruitment of individuals to be included in reference range studies presents a considerable practical and ethical problem due to the difficulty of defining 'normal' and of using invasive procedures, such as venipuncture, to obtain the necessary biological samples. The establishment of reference ranges for children, especially neonates, is a particular problem. It is therefore usual to accept the manufacturer's quoted ranges where available.

Reference ranges are most commonly expressed as the range that covers the mean ±1.96 standard deviations of the mean of the experimental population. This range covers 95% of the population. The majority of reference ranges are based on a normal distribution of individual values, but in some cases the experimental data are asymmetric and then often skewed to the upper limits. In such cases it is normal to use logarithmic data to establish the reference range; nevertheless, the range may overlap with values found in patients with the test disease state. Typical reference ranges are shown in Table 10.1.

10.2 CLINICAL MEASUREMENTS AND QUALITY CONTROL

10.2.1 The Operation of Clinical Biochemistry Laboratories

The clinical biochemistry laboratory in a typical general hospital in the UK serves a population of about 400 000 containing approximately 60 general practitioner (GP) groups, depending upon the location in the UK. This population generates approximately 1200 requests from GPs and hospital doctors each weekday for clinical biochemical tests on their patients. Each patient request requires the laboratory to undertake an average of seven specific analyte tests. The result is that a typical general hospital laboratory carries out between 2.5 and 3 million analyte tests each year. The majority of clinical biochemistry laboratories offers the local medical community as many as 200 different clinical biochemical tests that can be divided into eight categories, as shown in Table 10.2.

Most of the requests for biochemical tests arise on a routine daily basis, but some will arise from emergency medical situations at any time of the day. The large number of daily test samples coupled with the need for a 24-hour 7-days a week service dictates that the laboratory must rely heavily on automated analysis to carry out the tests and on information technology to process the data.

To achieve an effective service, a clinical biochemistry laboratory has three main functions:

Table 10.1 **Typical reference ranges for biochemical analytes (IU: international unit)**

Analyte	Reference range	Comment
Sodium	133–146 mM	
Potassium	3.5–5.3 mM	Values increased by haemolysis or prolonged contact with cells.
Urea	2.5–7.8 mM	Range varies with sex and age.
Creatinine	75–115 mM (males) 58–93 mM (females)	Creatinine (a metabolite of creatine) production relates to muscle mass and is also a reflection of renal function. Values for both sexes increase by 5–20% in the elderly.
Aspartate transaminase (AST)	< 40 IU l^{-1}	Perinatal levels are < 80 IU l^{-1} and fall to adult values by the age of 18. Some slightly increased values up to 60 IU l^{-1} may be found in females over the age of 50. Results are increased by haemolysis.
Alanine transaminase (ALT)	< 40 IU l^{-1}	Higher values are found in males up to the age of 60.
Alkaline phosphatase (ALP)	30–130 IU l^{-1} (adults)	Significantly raised results of up to two- or threefold occur during growth spurts through teenage years. Slightly raised levels are seen in the elderly and in women during the third trimester of pregnancy.
Cholesterol	No reference range, but recommended value of ≤5.0 mM	The measurement of cholesterol in an adult 'well' population does not show a Gaussian distribution, but a very tailed distribution with relatively few low results. The majority are < 10 mM but there is a long tail up to over 20 mM. There is a tendency for males to have higher cholesterol than females of the same age, but after the menopause female values revert to those of males. Generally values increase with age.

Note: IU: international unit.

- to advise the requesting GP or hospital doctor on the appropriate tests for a particular medical condition and on the collection, storage and transport of the patient samples for analysis
- to provide a quality analytical service for the measurement of biological analytes in an appropriate and timely way
- to provide the requesting doctor with a data interpretation and advice service on the outcome of the biochemical tests and possible further tests.

The advice given to the clinician is generally supported by a user handbook, prepared by senior laboratory personnel, that includes a description of each test offered, instructions on sample collection and storage, normal laboratory working hours and the approximate time it will take the laboratory to undertake each test. This turnaround time varies from less than one hour to several weeks, depending upon the speciality of the test. The vast majority of biochemical tests are carried out on serum

Table 10.2 **Examples of biochemical analytes used to support clinical diagnosis**

Type of analyte	Examples
Foodstuffs entering the body	Cholesterol, glucose, fatty acids, triglycerides, vitamin D
Waste products	Bilirubin, creatinine, urea
Tissue-specific messengers	Adrenocorticotropic hormone (ACTH), follicle-stimulating hormone (FSH), luteinising hormone (LH), thyroid-stimulating hormone (TSH)
General messengers	Cortisol, insulin, thyroxine
Response to messengers	Glucose tolerance test assessing the appropriate secretion of insulin; tests of pituitary function
Organ function	Adrenal function – cortisol, adrenocorticotropic hormone (ACTH); renal function – K^+, Na^+, urea, creatinine; thyroid function – free thyroxine (FT_4), free tri-iodothyronine (FT_3), thyroid-stimulating hormone (TSH)
Organ disease markers	Heart – troponins, creatine kinase (CKMB), AST, lactate dehydrogenase (LD); liver – ALT, ALP, γ-glutamyl transferase (GGT), bilirubin, albumin
Disease-specific markers	Specific proteins ('tumour markers') secreted from specific organs – prostate-specific antigen (PSA), CA125 (ovary), calcitonin (thyroid), α-fetoprotein (liver)

or plasma derived from a blood sample. Serum is the preferred matrix for biochemical tests, but the concentrations of most test analytes are almost the same in the two fluids. Serum is obtained by allowing the blood to clot and removing the clot by centrifugation. To obtain plasma, it is necessary to add an anticoagulant to the blood sample and remove red cells by centrifugation. The two most common anticoagulants are heparin and EDTA, the choice depending on the particular biochemical test required. For example, EDTA complexes calcium ions, therefore calcium in 'EDTA plasma' would be undetectable. For the measurement of glucose, fluoride-oxalate is added to the sample; the oxalate as an anticoagulant, the fluoride to inhibit glycolysis during the transport and storage of the sample. Special vacuum collection tubes containing specific anticoagulants or other additives are available for the storage of blood samples. Collection tubes are also available containing clot enhancers to speed the clotting process for serum preparation. Many containers incorporate a gel with a specific density designed to float between cells and serum providing a barrier between the two for up to 4 days. During these 4 days, the cellular component will experience lysis, so any subsequent contamination of the serum will include intracellular components. Some biochemical tests may also require whole blood, urine, cerebrospinal fluid (the fluid surrounding the spinal cord and brain), faeces, sweat, saliva and amniotic fluid. It is essential that the samples are collected in the appropriate container at the correct time (particularly important if the test is for the measurement of hormones such as cortisol that are subject to diurnal release) and labelled with appropriate patient details. Samples submitted to the laboratory for biochemical tests are accompanied by a request form, signed by the requesting clinician, which gives details of the tests required and brief details of the reasons for the request to aid data interpretation and to help identify other appropriate tests.

Laboratory Reception

On receipt in the laboratory, both the sample and the request form are assigned an acquisition number, usually in an optically readable form that includes a bar code. A check is made of the validity of the sample details on both the request form and sample container to ensure that the correct container for the tests required has been used. Samples may be rejected at this stage, if details are not in accordance with the set protocol. Correct samples are then split from the request form and prepared for analysis, typically by centrifugation to prepare serum or plasma. The request form is processed into the computer system, which identifies the patient against the sample acquisition number, and the tests requested by the clinician typed into the database. It is vital at this reception phase that the sample and patient data match and that the correct details are placed in the database. These details must be adequate to uniquely identify the patient, bearing in mind the number of potential patients in the catchment area, and will include name, address, date of birth, NHS number (a unique identifier for each individual in the UK for health purposes), hospital or accident and emergency number and acquisition number.

10.2.2 Analytical Organisation

The analytical organisation of the majority of clinical biochemical laboratories is divided into two major work areas:

- automated section
- manual section.

Automated Section

Auto-analysers dedicated to clinical biochemical analysis are available from many commercial manufacturers (see an example in Figure 10.1). These are fully automated and have carousels for holding the test samples in racks each carrying up to 15 samples, one or two carousels each for up to 60 different reagents that are identified by a unique bar code, carousels for sample washing/preparation and a reaction carousel containing up to 200 cuvettes for initiating and monitoring individual test reactions. Analysers use three methodologies: spectrophotometry (see Section 13.2), immunoturbidimetry (based on the agglutination reaction, see also Section 7.3) and potentiometry. Spectrophotometry systems are capable of measuring at 16 wavelengths simultaneously, whilst the potentiometric system, based on the use of ion-selective electrodes, which work in a similar fashion to the pH electrode (Section 2.4), that are combined into a single unit, is used to measure electrolyte concentrations of Na^+, K^+ and Cl^- simultaneously, with an analytical time of less than 4 minutes on a sample size of only 25 µl.

Figure 10.1 A Cobas® 8000 series chemistry analyser. The figure shows the ISE module (centre) and the c702 module for photometric analysis (right). (Reproduced with permission of Hoffmann-La Roche Ltd.)

Table 10.3 **Examples of analytical techniques used to quantify analytes by auto-analysers**

Analytical technique	Examples of analytes
Ion-selective electrodes	K^+, Na^+, Li^+, Cl^-
Visible and UV spectrophotometry	Urea, creatinine, calcium, urate
Turbidimetry	IgG, IgA, IgM
Reaction rate	Enzymes – aspartate transaminase, alanine transaminase, γ-glutamyl transferase, alkaline phosphatase, creatine kinase, lactate dehydrogenase
Kinetic interaction of microparticles in solution (KIMS)	Therapeutic drug monitoring – phenytoin, carbamazepine
Immunoassay	Cortisol, luteinising hormone, thyroid-stimulating hormone

The reaction carousel contains cuvettes that are automatically cleaned and dried before each assay. Supplementary washes are available to reduce carry-over when necessary. Reagents for up to 62 assays based on different analytical techniques (Table 10.3) may be loaded onto the analyser. In order to maximise throughput, testing is organised in two parallel lines, each one of which uses separate resources and is individually controlled. The analyser uses a fixed cycle time, and the total time for a run is dependent upon the assay being performed and determined by the assay parameters. The majority of tests are complete in less than 10 minutes, but some, e.g. troponin T, may take up to 20 minutes. Individual assay results are reported immediately and complete results for a particular patient sample are assembled as soon as the final assay is complete. The theoretical throughput for the Cobas® c702 analyser is 2000 tests per hour. Each laboratory will have at least two analysers, each offering a similar analytical repertoire so that one can back up the other in case of failure. The analyser reads the bar code acquisition number for each sample and on the basis of the reading interrogates the host computer database to identify the tests to be carried out on the sample. The identified tests are then automatically prioritised into the most efficient order and the analyser programmed to take the appropriate volume of sample by means of a sampler that may also be capable of detecting microclots in the sample, add the appropriate volume of reagents in a specified order and monitor the progress of the reaction. Internal quality control samples are also analysed at regular intervals. The analyser automatically monitors the use of all reagents so that it can identify when each will need replenishing. When the test results are calculated, the operator can review them either directly at the analyser or at a remote computer, both of which are connected to the main database. When appropriate, the results can also be checked against previous results for the same patient.

Potentiometry (microsensor analysers)

Recent advances in microsensor technology have stimulated the development of miniaturised multi-analyte sensor blocks that are widely used in the clinical biochemistry laboratory for the routine measurement of pH, pCO_2, Na^+, K^+, Ca^{2+}, Cl^-, glucose and lactate (see also Section 14.6). In addition, such sensors are used in equipment sited closer to the patient (i.e. point of care), e.g. intensive care units, where rapid readouts are needed. The basic component of the block is a 'cartridge' that contains the analyte sensors, a reference electrode, a flow system, wash solution, waste receptacle and a process controller. The block is thermostatically controlled at 37 °C and the sensors are embedded in three layers of plastic, the size and shape of a credit card (Figure 10.2a). Each card may contain up to 24 sensors. A metallic contact under each sensor forms the electrical interface with the cartridge. As the test sample passes over the sensors, a current is generated by mechanisms specific to the individual analyte and recorded. The size of the current is proportional to the concentration of analyte in the fluid in the sample path. Calibration of the sensors with standard solutions of the analyte allows the concentration of the test sample to be evaluated. The sensor card and the sample path are automatically washed after each test sample and can be used for the analysis of up to 750 whole blood samples before being discarded. The microsensor instruments contain an active quality control process controller that monitors the operation of the system, validates the integrity of the cartridge and monitors the electrode response to detect microclots in the test sample that may invalidate the analytical results.

The glucose and lactate sensors have a platinum amperometric electrode with a positive potential relative to the reference electrode. In the case of the measurement of glucose, the enzyme glucose oxidase catalyses a reaction generating hydrogen peroxide, which then diffuses through a controlling layer (Figure 10.2b) and is oxidised by the platinum electrode to release electrons and create a current flow, the size of which is proportional to the rate of hydrogen peroxide diffusion.

Spectrophotometry

Reactions monitored by spectrophotometry may either assess end-point results or rates. An end-point assay describes an experiment where measurements are taken after reaction completion and the intensity of the coloured (or turbidity) product is an indicator of the product concentration (see also Section 23.3.5). For rate determination, measurements are taken as the reaction proceeds, and the rate of reaction

(a)

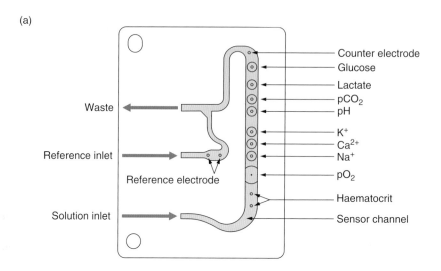

(b)

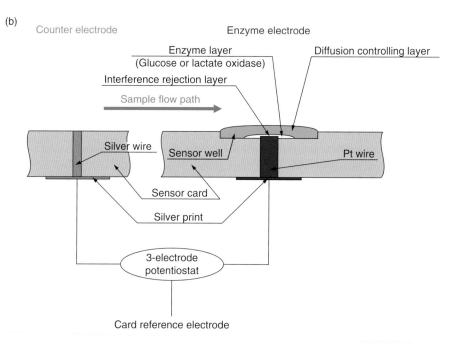

Figure 10.2 Microsensor analyte detectors. (a) The GEM 3000 Sensor Card; (b) the amperometric glucose/lactate analyte sensors. (Reproduced by permission of IL Critical Care, Lexington, USA.)

is proportional to the concentration or enzyme activity of the analyte in the sample (see also Section 23.2). A number of absorbance measurements are made during the analysis, in some cases up to 70 times in 10 minutes. Typically, a reaction is monitored at two wavelengths (bichromatic analysis) – one of these is at or near the peak absorbance of the chromophore produced by the reaction; the second wavelength is where the chromophore has little or no absorbance. This strategy can help to reduce interference.

Immunoassay

Automated immunoassays may be one of three types: competitive, sandwich and bridging (see also Section 7.3). Competitive assays are used for small analytes, e.g. cortisol, and the signal produced is inversely proportional to the analyte concentration. The sandwich principle is used for larger analytes, e.g. thyroid-stimulating hormone (TSH), and utilises two labelled antibodies binding to different sites on the analyte. In this case, the signal is directly proportional to the analyte concentration. Bridging assays are used for the detection of antibodies, e.g. immunoglobulin M (IgM). Immunoassay procedures undertaken by modern auto-analysers are mostly based on light-emitting techniques such as electrochemiluminescence (ECL), chemiluminescence, fluorescence or polarised fluorescence techniques. The range of analytes varies for different manufacturers, but usually includes important factors in endocrinology (e.g. thyroid function tests), therapeutic drugs (theophylline, digoxin) and drugs of abuse (opiates, cannabis). The operation of auto-analysers in immunoassay mode is similar to that described for photometric tests and the results are generally reported on the same day.

Manual Assays Section

This approach to biochemical tests is generally more labour intensive than the automated section, often requiring a higher degree of skill in both analysis and interpretation. It covers a range of analytical techniques such as gel and capillary electrophoresis (see Section 6.5), immuno-electrophoresis (see Section 7.7), liquid chromatography-mass spectrometry (LC-MS; see Section 15.1.1), selected ion monitoring by MS (Section 21.7.2) and gas chromatography. Examples include the assays for plasma proteins (see Section 10.3.8), metanephrines (for the diagnosis of phaeochromocytoma, a neuroendocrine tumour), or haemoglobin A_{1c} (for the diagnosis and monitoring of diabetes).

Result Reporting

The instrument operator or the section leader initially validates analytical results. This validation process will, in part, be based on the use of internal quality control procedures for individual analytes. Quality control samples are analysed at least twice daily or are included in each batch of test analytes. The analytical results are then subject to an automatic process that identifies results that are either significantly abnormal or require clinical comment or interpretation against rules set by senior laboratory staff.

Neonatal Screening

Neonatal or newborn screening is the process of testing newborn babies for certain potentially dangerous disorders. If these conditions are detected early, preventative measures can be adopted that help to protect the child from the disorders. However, such testing is not easy due to the difficulty of obtaining adequate samples of biological fluids for the tests. The development of tandem MS techniques (see Section 21.3) has significantly alleviated this problem. It is possible to screen for a range of metabolic diseases using a single dried blood spot of just 3 mm in diameter. There is no need for prior separation of components, since online liquid chromatography is used to deliver the sample to the mass spectrometer (LC-MS). A large number of inherited metabolic diseases can be screened by the technique, including:

- aminoacidopathies such as phenylketonuria (PKU) caused by a deficiency of the enzyme phenylalanine hydroxylase
- fatty acid oxidation defects such as medium chain acyl-CoA dehydrogenase deficiency (MCADD)
- organic acidaemias such as maple syrup urine disease, a disorder affecting the metabolism of the branched chain amino acids.

10.2.3 Quality Assessment Procedures

In order to validate the analytical precision and accuracy of the biochemical tests conducted by a clinical biochemistry department, the laboratory participates in external quality assessment schemes in addition to routinely carrying out internal quality control procedures that involve the repeated analysis of reference samples covering the full analytical range for the test analyte. In the UK, there are two main national clinical biochemistry external quality assessment schemes: the UK National External Quality Assessment Scheme (UK NEQAS) and the Wales External Quality Assessment Scheme (WEQAS). The majority of UK hospital clinical biochemistry departments subscribe to both schemes. UK NEQAS and WEQAS distribute test samples on a fortnightly basis, the samples being human serum based. In the case of UK NEQAS, the samples contain multiple analytes each at an undeclared concentration within the analytical range. The concentration of each analyte is varied from one distribution to the next. In contrast, WEQAS distributes four or five test samples containing the test analytes at a range of concentrations within the analytical range. Both UK NEQAS and WEQAS offer a number of quality assessment schemes in which the distributed test samples contain groups of related analytes, such as general chemistry analytes, peptide hormones, steroid hormones and therapeutic drug-monitoring analytes. Participating laboratories elect to subscribe to schemes relevant to their analytical services.

The participating laboratories are required to analyse the external quality assessment samples alongside routine clinical samples and to report the results to the organising centre. Each centre undertakes a full statistical analysis of all the submitted results and reports them back to the individual laboratories on a confidential basis. The statistical data record the individual laboratory's data in comparison with all the submitted data and with the compiled data broken down into individual methods (e.g. the glucose oxidase and hexokinase methods for glucose) and for specific manufacturers' systems. Results are presented in tabular, histogram and graphical form and are compared with the results from recent previously submitted samples. This comparison with previous performance data allows longer-term trends in analytical performance for each analyte to be monitored. Laboratory data regarded as unsatisfactory are identified and followed up. Selected data from typical UK NEQAS reports are presented in Table 10.4 and Figure 10.3 and from a WEQAS report in Table 10.5.

These data are for a laboratory that used the hexokinase method for glucose using an Aeroset instrument. Accordingly, the WEQAS report includes the results submitted by all laboratories using the hexokinase method and all results for the method using an Aeroset. The overall results refer to all methods irrespective of instrument. All

Table 10.4 **Selected UK NEQAS quality assessment data for serum glucose**

Analytical method	n	Mean (mM)	Standard deviation (mM)	CV (%)
All methods (ALTM)	485	10.90	0.22	2.0
Dry slide	33	10.59	0.16	1.6
OCD (J&J) slides	33	10.59	0.16	1.6
Glucose oxidase electrode	16	10.72	0.20	1.9
Beckman	12	10.74	0.22	2.1
Hexokinase + G6PDH	426	10.91	0.21	1.9
Abbott	129	10.89	0.19	1.7
Beckman (Olympus)	56	11.05	0.28	2.5
Roche Cobas/Modular	201	10.90	0.20	1.8
Siemens (Bayer)	19	10.74	0.23	2.2
Glucose oxidase/dehydrogenase	39	10.90	0.28	2.6
Siemens (Bayer)	25	10.81	0.17	1.6

Notes: These data are for a laboratory that used the hexokinase method and reported a result of 10.90 mM. UK NEQAS calculates a method laboratory trimmed mean (MLTM) as a target value. It is the mean value of all the results returned by all laboratories using the same method principle with results ± 2 SD outside the mean omitted. Its value was 10.90 mM. On the basis of the difference between the MLTM and the laboratory's result, UK NEQAS also calculates a score of the specimen accuracy and bias together with a measurement of the laboratory's consistency of bias. This involves aggregating the bias data from all specimens of that analyte submitted by the laboratory within the previous 6 months representing the 12 most recent distributions. This score is an assessment of the tendency of the laboratory to give an over- (positive) or under- (negative) estimate of the target MLTM values. The score indicated that the laboratory was performing consistently in agreement with the MLTM. CV: coefficient of variation (see also Section 19.3.3). The results embodied in this table are shown in histogram form in Figure 10.3. The data are reproduced with permission of UK NEQAS, Wolfson EQA Laboratory, Birmingham.

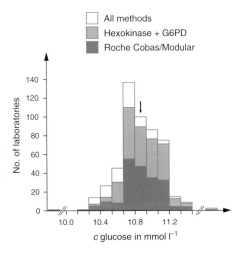

Figure 10.3 Histogram of UK NEQAS quality assessment data for serum glucose based on data in Table 10.4. The arrow denotes the performance of the individual laboratory.
(Reproduced by permission of UK NEQAS, Wolfson EQA Laboratory, Birmingham, UK.)

Table 10.5 **WEQAS quality assessment data for serum glucose**

Analytical method		Sample number			
		1	2	3	4
Reported result	(mM)	7.2	3.7	17.2	8.7
Hexokinase	Mean (mM)	6.9	3.6	17.0	8.4
	SD (mM)	0.2	0.1	0.5	0.3
	Number	220	221	219	219
Aeroset	Mean (mM)	7.2	3.8	17.3	8.5
	SD (mM)	0.15	0.08	0.045	0.2
	Number	8	8	9	8
Overall	Mean (mM)	7.0	3.7	17.2	8.6
	SD (mM)	0.28	0.20	0.54	0.33
	Number	388	392	388	389
WEQAS	SD (mM)	0.26	0.16	0.6	0.3
SDI		1.15	0.63	0.33	1.0

Notes: SD, standard deviation; SDI, standard deviation index. The data are reproduced with permission of WEQAS, Directorate of Laboratory Medicine, University Hospital of Wales, Cardiff.

the data are 'trimmed' in that results outside ± 2 SD of the mean are rejected, which explains why the total number for each test sample varies slightly. WEQAS SD is calculated from the precision profiles for each analyte. The SDI is defined as:

$$\text{SDI} = \frac{\text{Laboratory result} - \text{Method mean result}}{\text{WEQAS SD}} \qquad \text{(Eq 10.5)}$$

SDI is a measure of total error and includes components of inaccuracy and imprecision. The four SDIs for the laboratory are used to calculate an overall analyte SDI, in this case 0.78. A value of less than 1 indicates that all estimates were within ±1 SD and is regarded as a good performance. An unacceptable performance would be indicated by a value greater than 2.

10.2.4 Clinical Audit and Accreditation

In addition to participating in external quality assessment schemes, clinical laboratories are also subject to clinical audit. This is a systematic and critical assessment of the general performance of the laboratory against its own declared standards and procedures and against nationally agreed standards. In the context of analytical procedures, the audit evaluates the laboratory performance in terms of the appropriateness of the use of the tests offered by the laboratory, the clinical interpretation of the results and the procedures that operate for the receipt, analysis and reporting of the test samples. Thus, whilst it includes the evaluation of analytical data, the audit is primarily concerned with processes leading to the test data with a view to implementing change and improvement. The ultimate objective of the audit is to ensure that the patient receives the best possible care and support in a cost-effective way. The audit is normally undertaken by junior doctors, laboratory staff or external assessors, lasts for several days, and involves interaction with all laboratory personnel.

Closely allied to the process of clinical audit is that of accreditation. Whereas clinical audit is carried out primarily for the local benefit of the laboratory and its staff, and ultimately for the patient, accreditation is a public and national recognition of the professional quality and status of the laboratory and its personnel. The accreditation process and assessment is the responsibility of either a recognised public professional body or a government department or agency. Different models operate in different countries. In the UK, accreditation of clinical biochemistry laboratories is required by government bodies and is carried out by the United Kingdom Accreditation Service (UKAS). In the USA, accreditation is mandatory and may be carried out by one of a number of 'deemed authorities', such as the College of American Pathologists. Accreditation organisations also exist for non-clinical analytical laboratories. Examples in the UK include the National Measurement Accreditation Service (NAMAS) and the British Standards Institution (BSI). The International Accreditation Cooperation (ILAC), the European Cooperation for Accreditation (EA) and the Asia-Pacific Laboratory Accreditation Cooperation (APLAC) are three of many international fora for the harmonisation of national standards of accreditation for analytical laboratories.

Assessors appointed by the accreditation body assess the compliance by the laboratory with standards set by the accreditation body. The standards cover a wide range of issues, such as those of accuracy and precision, timeliness of results, clinical relevance of the tests performed, competence to carry out the tests, as judged by the training and qualifications of the laboratory staff, health and safety, the quality of administrative and technical support systems, and the quality of the laboratory management systems and document control. The successful outcome of an assessment is the national recognition that the laboratory is in compliance with the standards and hence provides quality healthcare. The accreditation normally lasts for 4 years with intermittent surveillance visits.

10.3 EXAMPLES OF BIOCHEMICAL AIDS TO CLINICAL DIAGNOSIS

10.3.1 Principles of Diagnostic Enzymology

The measurement of the activities or concentrations of selected enzymes in serum is a long-established aid to clinical diagnosis and prognosis. The enzymes found in serum can be divided into three categories based on the location of their normal physiological function:

- Serum-specific enzymes: The normal physiological function of these enzymes is based in serum. Examples include the enzymes associated with lipoprotein metabolism and with the coagulation of blood.

- Secreted enzymes: These are closely related to the serum-specific enzymes. Examples include pancreatic lipase and salivary amylase.
- Non-serum-specific enzymes: These enzymes have no physiological role in serum. They are released into the extra-cellular fluid and consequently appear in serum as a result of normal cell turnover or more abundantly as a result of cell membrane damage, cell death or morphological changes to cells such as those in cases of malignancy. Their normal substrates and/or cofactors may be absent or in low concentrations in serum.

Serum enzymes in this third category are of the greatest diagnostic value. When a cell is damaged, the contents of the cell are released over a period of several hours with enzymes of the cytoplasm appearing first, since their release is dependent only on the impairment of the integrity of the plasma membrane. The release of these enzymes following cell membrane damage is facilitated by their large concentration gradient, in excess of a thousandfold, across the membrane. The integrity of the cell membrane is particularly sensitive to events that impair energy production, for example by the restriction of supply of oxygen. It is also sensitive to toxic chemicals, including some drugs, microorganisms, certain immunological conditions and genetic defects. Enzymes released from cells by such events may not necessarily be found in serum in the same relative amounts as were originally present in the cell. Such variations reflect differences in the rate of their metabolism and excretion from the body and hence of differences in their serum half-lives. This may be as short as a few hours (intestinal alkaline phosphatase, glutathione S-transferase, creatine kinase) or as long as several days (liver alkaline phosphatase, alanine aminotransferase, lactate dehydrogenase).

The clinical exploitation of non-serum-specific enzyme activities is influenced by several factors:

- Organ specificity: Few enzymes are unique to one particular organ, but some enzymes are present in much larger amounts in some tissues than in others. As a consequence, the relative proportions (pattern) of a number of enzymes found in serum are often characteristic of the organ of origin.
- Isoenzymes: Some clinically important enzymes exist in isoenzyme forms (see Section 23.1.2) and in many cases the relative proportion of the isoenzymes varies considerably between tissues so that measurement of the serum isoenzymes allows their organ of origin to be deduced.
- Reference ranges: The activities of enzymes present in the serum of healthy individuals are invariably smaller than those in the serum of individuals with a diagnosed clinical condition such as liver disease. In many cases, the extent to which the activity of a particular enzyme is raised by the disease state is a direct indicator of the extent of cellular damage to the organ of origin.
- Variable rate of increase in serum activity: The rate of increase in the activity of released enzymes in serum following cell damage in a particular organ is a characteristic of each enzyme. Moreover, the rate at which the activity of each enzyme decreases towards the reference range following the event that caused cell damage and the subsequent treatment of the patient is a valuable indicator of the patient's recovery from the condition.

The practical implications of these various points to the applications of diagnostic enzymology are illustrated by its use in the management of heart disease and liver disease.

10.3.2 Ischaemic Heart Disease and Myocardial Infarction

The healthy functioning of the heart is dependent upon the availability of oxygen. This oxygen availability may be compromised by the slow deposition of cholesterol-rich atheromatous plaques in the coronary arteries. As these deposits increase, a point is reached at which the oxygen supply cannot be met at times of peak demand, for example at times of strenuous exercise. As a consequence, the heart becomes temporarily ischaemic ('lacking in oxygen') and the person experiences severe chest pain, a condition known as angina pectoris ('angina of effort'). Although the pain may be severe during such events, the cardiac cells temporarily deprived of oxygen are not damaged and do not release their cellular contents. However, if the arteries become completely blocked either by the plaque or by a small thrombus (clot) that is prevented from flowing through the artery by the plaque, the patient experiences a myocardial infarction (MI, 'heart attack') characterised by the same severe chest pain, but in this case the pain is accompanied by irreversible damage to the cardiac cells and the release of their cellular contents. This release is not immediate, but occurs over a period of many hours. From the point of view of the clinical management of the patient, it is important for the clinician to establish whether or not the chest pain was accompanied by a myocardial infarction. In about one-fifth of the cases of a myocardial event, the patient does not experience the characteristic chest pain ('silent myocardial infarction'), but again it is important for the clinician to be aware that the event has occurred. Electrocardiogram (ECG) patterns are a primary indicator of these events, but in atypical presentations ECG changes may be ambiguous and additional evidence is sought in the form of changes in specific serum proteins and enzymes. The time course of release of these MI marker proteins is illustrated in Figure 10.4. In current practice, usually only creatinine kinase (CK) and one of the cardiac muscle proteins, troponin T or troponin I, are used:

- Creatine kinase (CK): This enzyme converts phosphocreatine (important in muscle metabolism) to creatine. CK is a dimeric protein composed of two monomers, one denoted as M (muscle), the other as B (brain), so that three isoforms exist: CK-MM, CK-MB and CK-BB. The tissue distribution of these isoenzymes is significantly different such that heart muscle consists of 80–85% MM and 15–20% MB, skeletal muscle 99% MM and 1% MB and brain, stomach, intestine and bladder predominantly BB. CK enzyme activity is raised in a number of clinical conditions, but since the CK-MB form is almost unique to the heart, its raised activity in serum gives unambiguous support for a myocardial infarction, even in cases in which the total CK activity remains within the reference range. A rise in total serum CK activity is detectable within 6 hours of a myocardial infarction and the serum activity reaches a peak after 24–36 hours (Figure 10.4). However, a rise in CK-MB is detectable within 3–4 hours, has 100% sensitivity within 8–12 hours and reaches a peak within 10–24 hours. It remains raised for 2–4 days.
- Troponin: The troponins I and T are protein components of a complex that regulates the contractility of the myocardial cells. The concentration in serum increases at the

same rate as CK-MB after a myocardial infarction, has a similar time for 100% sensitivity and for peak time, but it remains raised for several days after the onset of symptoms. Troponin T is assayed by a 'sandwich' immunological assay in which troponin T in the sample reacts with a biotinylated monoclonal cardiac troponin T-specific antibody and a monoclonal troponin T antibody labelled with a ruthenium complex (see also Section 7.3.6). The measurement of serum troponin I or T is widely used to rule out cardiac damage in patients with chest pain – if there is no significant change when measured at presentation and again at 3–6 hours, then MI can be ruled out and the patient discharged or monitored under less intensive conditions. Newer troponin assays with high sensitivity allow this time interval to be reduced. Since troponin levels remain raised for several days following a myocardial infarction they can be helpful in identifying delayed presentation of MI (Figure 10.4).

- Total CK activity: This is assessed by coupled reactions (Section 23.3.2) with hexokinase and glucose-6-phosphate dehydrogenase in the presence of N-acetylcysteine as activator, and the measurement of increase in absorbance at 340 nm or by fluorescence polarisation (primary wavelength 340 nm, reference wavelength 378 nm):

$$\text{creatine phosphate} + \text{ADP} \rightleftharpoons \text{creatine} + \text{ATP}$$

$$D\text{-glucose} + \text{ATP} \rightleftharpoons D\text{-glucose-6-phosphate} + \text{ADP} \qquad \text{(Eq 10.6)}$$

$$D\text{-glucose-6-phosphate} + \text{NAD(P)}^+ \rightleftharpoons D\text{-6-phosphogluconate} + \text{NAD(P)H} + \text{H}^+$$

- CK-MB activity: This is assessed by the inhibition of the activity of the M monomer by the addition to the serum sample of an antibody to the M monomer. This inhibits CK-MM and the M unit of CK-MB. The activity of CK-BB is unaffected, but is normally undetectable in serum, hence the remaining enzyme activity in serum is due to the B unit of CK-MB. It is assayed by the above coupled assay procedure and the activity doubled to give an estimate of the CK-MB activity. An alternative assay uses a double antibody technique: CK-MB is bound to anti-CK-MB coated on microparticles, the resulting complex washed to remove non-bound forms of CK and anti-CK-MM conjugated to alkaline phosphatase added. It binds to the antibody–antigen complex, is washed to remove unbound materials and assayed using 4-methylumbelliferone phosphate as substrate, the released 4-methylumbelliferone being measured by its fluorescence and expressed as a concentration (μg l^{-1}) rather than as enzyme activity.

The measurement of CK and troponin concentrations, together with plasma potassium, glucose and arterial blood gases, is routinely used to monitor the recovery of patients following a myocardial infarction. A patient may experience a second myocardial infarction within a few days of the first. In such cases, the pattern of serum enzymes shown in Figure 10.4 is repeated, with the new pattern being superimposed on the remnants of the first profile. CK-MB is the best initial indicator of a second infarction since the levels of troponin may not reflect a secondary event.

The sensitivity and specificity (Section 10.1.2) of an electrocardiogram (ECG) and diagnostic enzymology in the management of heart disease are complementary. The specificity of an ECG is 100%, whilst that of enzyme measurements is 90%; however, the sensitivity of ECG is 70%, whilst that of troponin measurement can be up to 94% depending on the cut-off value used.

10.3.3 Liver Disease

Diagnostic enzymology is routinely used to discriminate between several forms of liver disease, including:

- Hepatitis: General inflammation of the liver most commonly caused by viral infection, but which may also be a consequence of blood poisoning (septicaemia) or glandular fever. Acute hepatitis can show a marked increase (60-fold) in aminotransferases over a few days before returning to normal.
- Cirrhosis: A general destruction of the liver cells and their replacement by fibrous tissue. It is most commonly caused by excess alcohol intake, but is also a result of prolonged hepatitis, various autoimmune diseases and genetic conditions. Enzyme activity may only be mildly elevated when chronic liver damage is present.
- Malignancy: Primary and secondary tumours.
- Cholestasis: The prevention of bile from reaching the gut either due to blockage of the bile duct by gallstones or tumours, or to liver cell destruction as a result of cirrhosis or prolonged hepatitis. This gives rise to obstructive jaundice (presence of bilirubin, a yellow metabolite of haem, in the skin).
- Obesity: Serum aminotransferases can be mildly but persistently elevated in obese subjects reflecting non-alcoholic fatty liver disease.

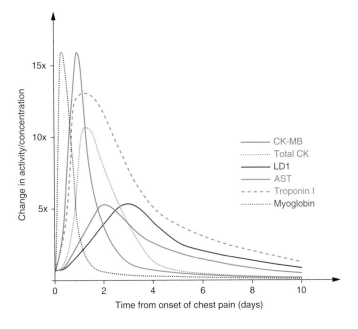

Figure 10.4 Serum enzyme activity as well as myoglobin and troponin I concentration changes following a myocardial infarction. Changes are expressed as a multiple of the upper limit of the reference range. Values vary according to the severity of the event, but the time course of each profile is characteristic of all events. In current practice in the UK, only troponin and CK or CK-MB are used.

Patients with these various liver diseases often present to their doctor with similar symptoms and a differential diagnosis needs to be made on the basis of a range of investigations, including imaging techniques, especially ultrasonography (ultra-sound), magnetic resonance imaging (MRI; see Section 14.3.3), computerised tomog-raphy (CT) scanning, microscopic examination of biopsy samples and liver function tests. Four enzymes are routinely assayed to aid differential diagnosis:

- Aspartate aminotransferase (AST) and alanine aminotransferase (ALT): As previ-ously stated, these enzymes are widely distributed, but their ratios in serum are characteristic of the specific cause of liver cell damage. For example, an AST/ALT ratio of less than 1 is found in non-alcoholic fatty liver disease, a ratio of about 1 in obstructive jaundice caused by viral hepatitis and a ratio of greater than 1 in alcohol abuse.

- Aspartate aminotransferase activity: This is assessed by a coupled assay with malate dehydrogenase and the measurement of the decrease in absorbance at 340 nm:

$$L\text{-aspartate} + 2\text{-oxoglutarate} \rightleftharpoons \text{oxaloacetate} + L\text{-glutamate}$$

$$\text{oxaloacetate} + NADH + H^+ \rightleftharpoons \text{malate} + NAD^+$$

(Eq 10.7)

- Alanine aminotransferase activity: This is also assessed by a coupled assay, in this case with lactate dehydrogenase and the measurement of the decrease in absorbance at 340 nm:

$$L\text{-alanine} + 2\text{-oxoglutarate} \rightleftharpoons \text{pyruvate} + L\text{-glutamate}$$

$$\text{pyruvate} + NADH + H^+ \rightleftharpoons L\text{-lactate} + NAD^+$$

(Eq 10.8)

- γ-Glutamyl transferase (GGT): This enzyme transfers a γ-glutamyl group between substrates and may be assayed by the use of γ-glutamyl-3-carboxy-4-nitroanilide as substrate and glycylglycine as acceptor. The amount of 5-amino-2-nitrobenzo-ate released is proportional to GGT activity. GGT is widely distributed and is abun-dant in liver, especially bile canaliculi, kidney, pancreas and prostate. Raised enzyme activities are found in cirrhosis, secondary hepatic tumours and cholestasis, and tend to parallel increases in the enzyme activity of alkaline phosphatase, especially in cholestasis. GGT enzyme activity in the serum is raised by alcohol and some drugs; for this reason, it is not a particularly useful marker.

- Alkaline phosphatase (ALP): This enzyme is found in most tissues, but is especially abundant in the bile canaliculi, kidney, bone and placenta. It may be assayed using 4-nitrophenyl phosphate as substrate and monitoring the release of 4-nitrophenol at 400 nm. Its enzyme activity is raised in obstructive jaundice and, when measured in conjunction with ALT, can be used to distinguish between obstructive jaundice and hepatitis since ALP enzyme activity is raised more than that of ALT in obstruc-tive jaundice. Decreasing serum activity of ALP is valuable in confirming an end of cholestasis. Raised serum ALP levels can also be present in various bone diseases and during growth and pregnancy.

10.3.4 Kidney Disease

The kidneys, together with the liver, are the major organs responsible for the removal of waste material from the body. The kidneys also have other specific functions, including the control of electrolyte and water homeostasis, as well as the synthesis of erythropoietin. Each of the two kidneys contains approximately 1 million nephrons that receive the blood flowing to the kidneys. Blood flowing to the kidneys is first presented to the glomerulus of each nephron, which filters the plasma fluid to produce the ultrafiltrate or primary urine by removing all the contents of the plasma except proteins. Each nephron produces approximately 100 mm^3 of primary urine per day giving a total production of primary urine by the two kidneys of approximately 100–140 ml per minute or 200 litres per day in a healthy adult person. This is referred to as the glomerular filtration rate (GFR). The primary urine then encounters the tubule of the nephron that is the site of the reabsorption of water, the active and passive reabsorption of lipophilic compounds and cellular nutrients such as sugars and amino acids, and the active secretion of others. These two processes in combination result in the production of approximately 2 litres of urine per day, which is collected in the urinary bladder.

The glomerular filtration rate is the accepted best indicator of kidney function. Any pathology of the kidneys is reflected in a decreased GFR and this in turn has serious physiological consequences, including anaemia and severe cardiovascular disease. Kidney disease is progressive and proceeds through subacute or intrinsic renal disease, such as glomerular nephritis, into chronic kidney disease (CKD). Complete kidney failure leads to the need for kidney dialysis and organ transplantation. There is evidence that the incidence of CKD is increasing in developed countries and is associated with increasing risk of diabetes and an increasingly elderly population. There is thus a great clinical demand for accurate measurements of GFR in order to detect the onset of kidney disease, to assess its severity and to monitor its subsequent progression.

Measurement of Glomerular Filtration Rate

The measurement of GFR is based on the concept of renal clearance, which is defined as the volume of serum cleared of a given substance by glomerular filtration in unit time. It therefore has units of ml min^{-1}. In principle, any endogenous or exogenous substance that is subject to glomerular filtration and is not reabsorbed could form the basis of the measurement. The polysaccharide inulin meets these criteria and is subject to few variables or interferences. However, because it is not naturally occurring in the body, it is inconvenient for routine clinical use; it is commonly used as a standard for alternative methods. In practice, serum creatinine is the most commonly used marker. It is the end product of creatine metabolism in skeletal muscle and meets the excretion criteria, so that its serum concentration is inversely related to GFR. However, it is subject to a number of non-renal variables, including:

- Muscle mass: Serum values are influenced by extremes of muscle mass as in athletes and in individuals with muscle-wasting disease or malnourished patients.
- Gender: Serum creatinine is higher in males than females for a given GFR.

- Age: Children under 18 years have a reduced serum creatinine and the elderly have an increased value.
- Ethnicity: African–Caribbeans have a higher serum creatinine for a given GFR than Caucasians.
- Drugs: Some commonly used drugs such as cimetidine, trimethoprim and cephalosporins interfere with creatinine excretion and hence give elevated GFR values.
- Diet: Recent intake of red meat and oily fish can raise serum creatinine levels.

Routine laboratory estimations of GFR (referred to as eGFR) are based on the measurement of serum creatinine concentration and then the calculation of eGFR using an equation that makes corrections for four of the above variables (see Equation 10.10). Serum creatinine is routinely measured by one of two ways:

- Spectrophotometric method based on the Jaffe reaction: This involves the use of alkaline picric acid reagent, which produces a red-coloured product that is measured at 510 nm. A limitation is that the reagent also reacts with other substances such as ketones, ascorbic acid and cephalosporins and, as a result, gives high values.
- Coupled enzyme assay: One method uses creatininase, creatinase and sarcosine oxidase to produce hydrogen peroxide. The hydrogen peroxide, in the presence of 4-aminophenazone, tri-iodo-3-hydroxybenzoic acid (HTIB) and peroxidase, yields a quinone imine chromophore, whose colour intensity is directly proportional to the creatinine concentration:

$$\text{creatinine} + H_2O \underset{}{\overset{\text{creatininase}}{\rightleftharpoons}} \text{creatine}$$

$$\text{creatine} + H_2O \underset{}{\overset{\text{creatinase}}{\rightleftharpoons}} \text{sarcosine} + \text{urea}$$

$$\text{sarcosine} + O_2 + H_2O \underset{}{\overset{\text{sarcosine oxidase}}{\rightleftharpoons}} \text{glycine} + HCHO + H_2O_2$$

(Eq 10.9)

$$H_2O_2 + \text{4-aminophenazone} + HTIB \underset{}{\overset{\text{peroxidase}}{\rightleftharpoons}} \text{quinone imine chromophore} + H_2O + HI$$

Two other methods are commonly used for research:

- HPLC or GC-MS: HPLC uses a C18 column and water/acetonitrile (95:5 v/v) eluent containing 1-octanesulfonic acid as a cation-pairing agent. GC-MS is based on the formation of the *tert*-butyldimethylsilyl derivative of creatinine.
- Isotopic dilution coupled with mass spectrometry (ID-MS). This involves the addition of ^{13}C- or ^{15}N-labelled creatinine to the serum sample, isolation of creatinine by ion-exchange chromatography and quantification by mass spectrometry using selected ion monitoring (SIM; see also Section 21.7.2). The lower limit of detection is about 0.5 ng.

The lack of an internationally or even nationally agreed standard assay for creatinine leads to significant inter-laboratory differences in both bias and imprecision (see Section

2.7) so that national external quality assurance schemes, such as UK NEQAS and WEQAS (Section 10.2.3) have important roles in alerting laboratories to assays that stray outside national control values. In the UK, biological reference materials with ascertained concentrations are often provided by the National Institute for Biological Standards and Control (NIBSC). UK NEQAS provides clinical laboratories that participate in the eGFR scheme with an assay-specific adjustment factor to correct for methodological variations in estimations of serum creatinine. The factor is obtained using calibration against a GC-MS creatinine assay. It is updated at 6-monthly intervals. A number of equations have been derived to calculate eGFR from serum creatinine values, but the one currently used throughout the UK is the four-variable isotope dilution mass spectrometry (ID-MS) traceable Modification of Diet in Renal Disease (MDRD) Study equation:

$$eGFR = 175 \times \left(\frac{c\,(\text{serum creatinine})}{88.4\,\mu M} \right)^{-1.154} \times \left(\frac{age}{1\ year} \right)^{-0.203}$$
$$\times 0.742\,[\text{if female}] \times 1.210\,[\text{if black}] \times 1\,\frac{ml}{min}$$

(Eq 10.10)

Serum creatinine concentrations are expressed in μM to the nearest whole number and are adjusted for variations in body size by normalising using a factor for body surface area (1.73 m^2). The units of eGFR are therefore ml min^{-1} (1.73 m^2)$^{-1}$ and values are reported in whole numbers. The equation has been validated in a large-scale study against the most accurate method to measure GFR based on the use of ^{125}I-iothalamate. This equation should not be used in individuals with extremes of muscle mass, e.g. amputees, or those with unstable renal function or pregnant. Alternative equations exist for use with children.

Reference values for eGFR are 130 ml min^{-1} (1.73 m^2)$^{-1}$ for males in the age range 20–30 and 125 ml min^{-1} (1.73 m^2)$^{-1}$ for females of the same age range, although in practice, values greater than 90 are simply reported as > 90 ml min^{-1} (1.73 m^2)$^{-1}$. Values decline with increasing age.

Clinical Assessment of Renal Disease

Acute kidney injury is the failure of renal function over a period of hours or days and is defined by increasing serum creatinine and urea. It is a life-threatening disorder caused by the retention of nitrogenous waste products and salts such as sodium and potassium. The rise in potassium may be visible by changes in the electrocardiogram and pose a risk of cardiac arrest. Acute kidney injury may be classified into pre-renal, renal and post-renal. Prompt identification of pre- or post-renal factors and appropriate treatment action may allow correction before damage to the kidneys occurs. Pre-renal failure occurs due to a lack of renal perfusion. This situation can occur as a result of volume loss in haemorrhage, gastrointestinal fluid loss and burns, or because of a decrease in cardiac output caused by cardiogenic shock, massive pulmonary embolus or cardiac tamponade (application of pressure) or other causes of hypertension such as sepsis. Post-renal causes include bilateral uretic obstruction because of calculi or tumours or by decreased bladder outflow/urethral obstruction, e.g. urethral stricture or prostate enlargement through hypertrophy of carcinoma. Correction of the underlying problem can avoid any kidney damage. Renal causes of acute renal failure include glomerular nephritis, vascular disease, severe hypertension, hypercalcaemia, invasive disorders such as sarcoidosis or lymphoma and nephrotoxins including animal and plant toxins, heavy metals, aminoglycosides, antibiotics and non-steroidal anti-inflammatory drugs.

Chronic kidney disease (CKD) is a progressive condition affecting both glomerular and tubular function and is characterised by a declining eGFR (Table 10.6). All CKD patients are subject to regular clinical and laboratory assessment and, once Stage 3 has been reached, to additional clinical management. This monitoring is aimed at attempting to reverse or arrest the disease by drug therapy, e.g. to treat hypertension, saving the patient the inconvenience and the paying authority the cost of dialysis or transplantation.

10.3.5 Endocrine Disorders

Endocrine hormones are synthesised in the brain, adrenal gland, pancreas, testes and ovary, and perhaps most importantly in the hypothalamus and pituitary, but they act elsewhere in the body since they are released into the circulatory system. The result is a hormonal cascade that incorporates an amplification of the amount of successive hormone released into the circulatory system, increasing from micrograms to milligrams, as well as a negative feedback that operates to control the cascade when the level of the 'action' hormone has reached its optimum value. Most signals originate in the central nervous system either as a result of an environmental (external) signal, such as trauma or temperature, or an internal signal. The response is a signal to the hypothalamus and the release of a hormone such as corticotropin-releasing hormone (CRH). This travels in the bloodstream to the anterior pituitary gland, where it acts on its receptor and results in the release of a second hormone, adrenocorticotropic hormone (ACTH). ACTH, in turn, circulates in the blood to reach its target gland, the adrenal cortex, where it acts to release the 'action' hormone, cortisol, known as the stress hormone. The released cortisol raises blood pressure and blood glucose and is subject to a natural diurnal variation, peaking in early morning and being lowest around midnight. It has a negative feedback effect on the pituitary and adrenal

Table 10.6 **Stages of chronic kidney disease (CKD)**

CKD stage	eGFR (ml min^{-1} 1.73^{-1} m^{-2})	Clinical relevance
1	> 90	Regard as normal unless other symptoms present[a]
2	60–89	Regard as normal unless other symptoms present[a]
3A	45–59	Mild to moderate renal impairment
3B	30–44	Moderate to severe impairment
4	15–29	Severe renal impairment
5	< 15	Advanced renal failure

Note: [a]Symptoms include persistent proteinuria, haematuria, weight loss, hypertension.

cortex. Glands linked by the action of successive hormones are referred to as an axis, e.g. the hypothalamus–pituitary–adrenal axis. These coordinated cascades regulate the growth and function of many types of cells (Table 10.7). The hormones released act at specific receptors, commonly G-protein coupled receptors (GPCRs), which trigger the release of second messengers such as cAMP, cGMP, inositol triphosphate, Ca^{2+} and protein kinases. Diseases of the endocrine system result in dysregulated hormone release, inappropriate signalling response or, in extreme cases, the destruction of the gland. Examples include diabetes mellitus, Addison's disease, Cushing's syndrome, hyper- and hypothyroidism and obesity. Such medical conditions are characterised by their long-term nature. Laboratory tests are commonly employed to measure hormone levels in order to assist in the diagnosis of the condition and the subsequent care of the patient.

Thyroid Function Tests

Approximately 1% of the population suffers from some form of thyroid disease, although in many cases the symptoms may be non-specific. Even so, over 1 million thyroid function tests are conducted annually in the UK. As shown in Table 10.7, the hypothalamus releases thyrotropin-releasing hormone (TRH) which acts directly on the pituitary to produce thyroid-stimulating hormone (TSH); the latter, in turn, stimulates the thyroid gland to produce two thyroid hormones, thyroxine (T4) and tri-iodothyronine (T3; Figure 10.5). The gland produces approximately 10% of the circulating T3, the remainder being produced by the metabolism of T4 mainly in the liver and kidney. The majority of T4 and T3 are bound to thyroxine-binding globulin (TBG), but only the free unbound forms (fT4, fT3) are biologically active. Although the concentration of T3 is approximately one-tenth of that of T4, T3 is ten times more active. Both hormones act on nuclear receptors to increase cell metabolism and both have a negative feedback effect on the hypothalamus to switch off the secretion of

Table 10.7 **Examples of hormones of the hypothalamus–pituitary axis**

Secreted hormone	Pituitary effect	Gland effect
Thyrotropin-releasing hormone (TRH)	Release of thyroid-stimulating hormone (TSH)	Release of thyroxine (T4) and tri-iodothyronine (T3) by thyroid gland
Growth-hormone-releasing hormone (GHRH)	Release of growth hormone (GH)	Stimulates cell and bone growth
Corticotropin-releasing hormone (CRH)	Release of adrenocorticotropic hormone (ACTH)	Stimulates production and secretion of cortisol in adrenal cortex
Gonadotropin-releasing hormone (GnRH)	Release of follicle-stimulating hormone (FSH) and luteinising hormone (LH)	FSH – maturation of follicles/spermatogenesis; LH – ovulation/production of testosterone

TRH and on the pituitary to switch off TSH secretion. Hyperthyroidism is a consequence of the over-production of the two hormones and common causes are thyroiditis, Grave's disease and TSH-producing pituitary tumours. Hypothyroidism, characterised by weakness, fatigue, weight gain and joint or muscle pain, may be primary due to the under-secretion of T4 and T3, possibly due to irradiation or drugs such as lithium, autoimmune disease or secondary due to damage to the hypothalamus or pituitary. Normal laboratory tests for these conditions are based on the measurement of TSH and usually fT4 and fT3, occasionally total (bound and unbound) T4, all by immunoassay. See the case study in Example 10.1 for the importance of continuous fT3/fT4 monitoring, when prescribing supplemental thyroxine.

10.3.6 Hypothalamus–Pituitary–Gonad Axis

In both sexes, the hypothalamus produces gonadotropin-releasing hormone (GRH) that stimulates the pituitary to release luteinising hormone (LH) and follicle-stimulating hormone (FSH). In males, the release of LH and FSH is fairly constant, whereas in females the release is cyclical. In males, LH stimulates Leydig cells in the testes to produce testosterone, which together with FSH stimulates the production of sperm. The testosterone has a negative feedback effect on both the hypothalamus and the pituitary, thereby controlling the release of GRH. The testosterone acts on various body tissues to give male characteristics. In females, FSH acts on the ovaries to produce both oestradiol and the development of the follicle. The oestradiol and LH then act to stimulate ovulation. Oestradiol has a negative feedback effect on the hypothalamus and the pituitary and acts on body tissues to produce female characteristics.

10.3.7 Diabetes Mellitus

Diabetes is the most common metabolic disorder of carbohydrate, fat and protein metabolism, and is primarily due to either a deficiency or complete lack of the secretion of insulin by the β-cells of the islets of Langerhans in the pancreas. It affects

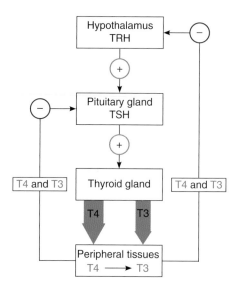

Figure 10.5 Hypothalamic–pituitary–thyroid axis, an example of a negative feedback system. In this case, the products of the thyroid gland, thyroxine (T4) and tri-iodothyronine (T3) feedback on the hypothalamus and pituitary to stop secretion of TRH and TSH, respectively.

1–2% of Western populations and 5–10% of the population over the age of 40. The disease is characterised by hyperglycaemia (elevated blood glucose level) leading to long-term complications. Diabetes can be classified into a number of types:

- Insulin-dependent diabetes (Type 1) (also called juvenile diabetes and brittle diabetes) is due to the autoimmune destruction of β-cells in the pancreas. Generally it has a rapid onset with a strong genetic link.
- Non-insulin-dependent diabetes (Type 2) (also called adult-onset diabetes and maturity-onset diabetes), is a complex progressive metabolic disorder characterised by β-cell failure and variable insulin resistance. A subtype is maturity-onset diabetes of the young (MODY) which usually occurs before the age of 25 years. It is the first form of diabetes for which a genetic cause and molecular consequence have been established. Mutations of the genes for hepatocyte nuclear factor 4α (MODY1), glucokinase (MODY2), HNF1α (MODY3), insulin promotor factor 1 (MODY4), HNF1β (MODY5) and neurogenic differentiation factor 1 (MODY6) have all been described.
- Impaired glucose tolerance where there is an inability to metabolise glucose in the 'normal' way, but not so impaired as to be defined as diabetes.
- Gestational diabetes that is any degree of glucose intolerance developed during pregnancy. It is characterised by a decrease in insulin sensitivity and an inability to compensate by increased insulin secretion. The condition is generally reversible after the termination of pregnancy, but up to 50% of women who develop it are prone to develop Type 2 diabetes later in life.
- Other types, which include certain genetic syndromes, pancreatic disease, endocrine disease and drug- or chemical-induced diabetes.

Insulin-Dependent Diabetes (Type 1)

Between 5% and 10% of all diabetics have the insulin-dependent form of diabetes requiring regular treatment with insulin. Type 1 develops in young people with a peak incidence of around 12 years of age. In this type of diabetes the degree of insulin deficiency is so severe that only insulin replacement can avoid the complications of diabetes that are discussed later. Dietary control or oral drugs are not sufficient. The disease is caused by the autoimmune destruction of β-cells in the pancreas, thus reducing the ability of the body to produce insulin. Islet cell antibodies (ICA), IA-2 antibodies to the transmembrane protein tyrosine phosphatase-like molecule in islet cells, auto-antibodies to glutamic acid decarboxylase (GAD) found in β-cells and insulin auto-antibodies (IAA) are all used as diagnostic markers of the disease.

Non-Insulin-Dependent Diabetes (Type 2)

Type 2 diabetes accounts for 90% of all cases and develops later in life and can be exacerbated by obesity. MODY versions account for 1–5% of all cases and are not associated with obesity. From population screening studies it is thought that only half of those individuals with Type 2 have been diagnosed. Control of blood glucose levels in this group is normally by a combination of diet and oral drug therapy, but occasionally it may require insulin injection.

Example 10.1   **CASE STUDY OF A HYPOTHYROID CONDITION**

Description   A 59-year-old woman presented with a history of lethargy, cold intolerance
and weight gain. On examination, the doctor noticed that the patient's hair
appeared thin and her skin dry. Several tests were requested, including thyroid
function tests the results of which were:

TSH : 46.9 mU l^{-1} (normal range 0.4–4.5 mU l^{-1}); mU: milliunits

c(fT4) = 5.6 pM (normal range 9.0–25 pM).

These results indicate overt primary hypothyroidism. As the patient suffered
from cardiovascular disease, the doctor commenced thyroxine replacement
therapy at an initial dose of 25 µg daily. After 2 weeks, the tests were repeated:

TSH : 37.6 mU l^{-1}

c(fT4) = 8.2 pM.

These results remain abnormal, so it was agreed that the tests should be
repeated in 6 weeks' time. At this stage the results were:

TSH : 19.1 mU l^{-1}

c(fT4) = 11.8 pM.

These results indicate that either the thyroxine replacement dosage was inadequate or that compliance was poor. As the patient confirmed that she had been
taking the therapy as prescribed, the doctor increased the dose to 50 µg per day.
After a further 8 weeks the tests were repeated:

TSH : 1.5 mU l^{-1}

c(fT4) = 14.8 pM.

The patient reported feeling much better and had improved clinically.

Comment   The above results are typical of patients with hypothyroidism, also referred to
as myxoedema. Primary hypothyroidism due to thyroid gland dysfunction is
by far the most common cause of the condition but secondary (pituitary) and
tertiary (hypothalamic) causes also exist. In these latter two cases the main
biochemical abnormality is a low fT4 concentration. The TSH level may be low
or within the reference range in secondary and tertiary hypothyroidism, i.e.
it does not respond to low fT4. Patients with primary hypothyroidism require
lifelong therapy. Patients with cardiovascular disease, as in this case, must be
initiated at a lower dose than normal as over-treatment can lead to angina, cardiac arrhythmia and myocardial infarction. Elderly patients are also started at
a lower dose for the same reason. Once therapy has been commenced, thyroid
function tests should be carried out after 2–3 months to check for steady-state
conditions and thereafter repeated on an annual basis.

Example 10.2 **CASE STUDY OF PREGNANCY**

Description A 28-year-old female PE teacher presented to her GP with non-specific symptoms of increased tiredness, nausea, stomach cramps and amenorrhoea with a
last menstrual period 3 months previously. She was a previously fit, healthy lady
who had a normal menstrual history. Having recently moved house, she thought
that stress might be the cause of her symptoms. Her GP requested routine biochemistry tests including thyroid function tests, all of which were within normal
reference range. A urine pregnancy test was also performed and to the patient's
surprise was positive and confirmed by laboratory serum β-human chorionic
gonadotropin (β-hCG) of 150 640 IU l^{-1} (IU: international units).

Comment These results confirm that this lady was approximately 10 weeks pregnant. The
serum β-hCG levels during pregnancy are shown in the figure below. A level
> 25 IU l^{-1} is indicative of pregnancy. Implantation of the developing embryo
into the endometrial lining of the uterus results in the secretion of β-hCG and
as pregnancy continues its synthesis increases at an exponential rate, doubling
every 2 days, and reaching a peak of 100 000–200 000 IU l^{-1} at 60–90 days
(first trimester). Levels then decline to approximately 1000 IU l^{-1} at around
20 weeks pregnancy (during second trimester) to a stable plateau for the
remainder of the pregnancy. Oestradiol, oestrone, oestriol and progesterone all
increase in the early stages of pregnancy as a result of the action of β-hCG on
the corpus luteum of the ovaries. Unlike β-hCG levels, the levels of the three
oestrogens and progesterone continue to rise during pregnancy, playing a vital
role in the sustenance and maintenance of the foetus. At the end of pregnancy,
the placental production of progesterone falls, stimulating contractions leading
to birth.

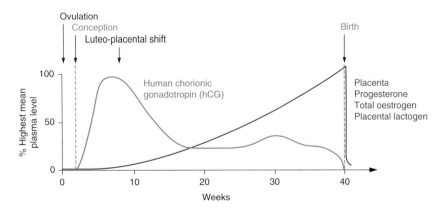

Figure: Hormonal profile during pregnancy. (Adapted with permission from Professor Alan S. McNeilly,
MRC Human Reproductive Science Unit, Edinburgh, UK.)

Diagnosis and Monitoring Control of Diabetes

Diabetes is frequently recognised by the symptoms it causes, but can be confirmed by clinical biochemical measurements based on World Health Organization (WHO) recommendations in which symptoms are accompanied by:

- a fasting (12 hours) plasma glucose level ≥ 7 mM;
- a random plasma glucose level ≥ 11.1 mM;
- application of an oral glucose tolerance test in which a 75 g dose of glucose is administered and the plasma glucose level measured after 2 hours. Diabetes is characterised by a value ≥ 11.1 mM.

The diagnostic cut-off values of 7.0 and 11.1 mM are based on the level at which retinopathy begins to appear in a population. The clinical aim in the treatment of Type 1 diabetes is to maintain plasma glucose levels in the healthy range of 4–6 mM. This is typically monitored by patients themselves by measuring their blood glucose at predetermined times that are inter-related to their mealtimes during the day. For example, the lowest blood glucose of the day is likely to be after the longest fast before breakfast and the highest blood glucose of the day is likely to be 1 hour after the main meal. By manipulating treatment around these highs and lows, good glycaemic control is generally maintained. The patients measure their blood glucose using hand-held, portable blood glucose meters based on glucose oxidase using dry stick technology to measure finger-prick blood samples.

Another measure of glycaemic control is by using haemoglobin A_{1c} (HbA_{1c}) measurements. This testing strategy works on the basis that most proteins (in this case haemoglobin A) will bind glucose dependent on the length of time they are in contact with glucose, the temperature and the concentration of glucose. Hence haemoglobin, having a typical half-life of 120 days, will bind the appropriate amount of glucose depending on the concentration of glucose. HbA_{1c} is typically measured in the clinic using HPLC to separate the different haemoglobins and was initially expressed as a percentage of total haemoglobin. Following introduction of standardisation that allowed traceability to the International Federation of Clinical Chemistry and Laboratory Medicine (IFCC) standard and reference method, the units were changed from % to mmol per mol haemoglobin. This change was introduced in 2009 in the UK although some laboratories may still report results in both units for education purposes. The lower the result the better the control and a target of 48 mmol per mol haemoglobin has been proposed for good diabetic control. This test is extremely useful in measuring long-term control of diabetes, but is not without its pitfalls. For example, if the patient has very brittle diabetes with equal numbers of hypoglycaemic and hyperglycaemic periods (see below), then the hypo- will cancel out the hyperglycaemic periods and the HbA_{1c} levels will falsely indicate good glycaemic control.

The test has also been adopted for the diagnosis of Type 2 diabetes as it does not require fasting or a glucose tolerance test (Table 10.8). There are a number of caveats to its use in this regard, including pregnancy, children and anyone with a genetic or haematological abnormality that can influence HbA_{1c} or its measurement. In addition, a value less than 48 mmol per mol haemoglobin cannot exclude a diagnosis of diabetes made on blood glucose results.

Table 10.8 **Diagnostic criteria for diabetes using HbA$_{1c}$**

HbA$_{1c}$ indication	HbA$_{1c}$ concentration	
	mmol mol^{-1} haemoglobin	%
Normal	< 42	< 6
Pre-diabetic	42–47	6–6.4
Diabetic	≥ 48	≥ 6.5

10.3.8 Plasma Proteins

Plasma contains a very large number of proteins, many of which are present only in trace amounts. The ones that have their main physiological role in plasma have three main functions:

- osmotic regulation
- transport of ligands such as hormones, metal ions, bilirubin, fatty acids, vitamins and drugs
- response to infection or foreign bodies entering the body.

All plasma proteins are synthesised in the liver, with the exception of the immuno-globulins, which are synthesised in the bone marrow. Plasma proteins are readily separated by electrophoresis (see Section 6.3) and this technique forms the basis of several clinical diagnostic tests. The tests are normally recorded subjectively, but a densitometer may be used to obtain a semi-quantitative result.

Albumin
Albumin is the most common plasma protein, constituting some 50% of all plasma protein. Its half-life in plasma is about 20 days, and in a good nutritional state the liver produces about 15 g albumin per day to replace this loss. Albumin is the main regulator of the osmotic pressure of plasma, but also acts as a transporter of haem, bilirubin (a metabolite of haem), biliverdin (a metabolite of bilirubin), free fatty acids, steroids and metal ions (e.g. Cu^{2+}, Fe^{3+}). Some drugs also bind to albumin. Other specific transport proteins found in plasma include steroid-binding proteins such as cortisol-binding globulin, sex-hormone-binding globulin (androgens and oestrogens) and metal-binding proteins, e.g. caeruloplasmin (Cu^{2+}) and transferrin (Fe^{3+}). Other transport plasma proteins include thyroid-binding globulin (thyroxine T4 and triiodo-thyronine T3) and haptoglobin (haemoglobin).

Immunoglobulins
Immunoglobulins (see Chapter 7) are synthesised in bone marrow in response to the exposure to a specific foreign body. Immunoglobulins share a common Y-shaped structure of two heavy and two light chains, the light chains (κ, λ) forming the upper arms of the Y. The class of immunoglobulin is determined by the heavy chain that gives rise to five types – IgG, IgA, IgM, IgD and IgE.

IgG accounts for approximately 75% of the immunoglobulins present in the plasma of adults and has a half-life of approximately 22 days. It is present in extra-cellular fluids and appears to eliminate small proteins through aggregation and the reticuloendothelial system. IgA is the secretory immunoglobulin protecting the mucosal surfaces. IgA is synthesised by mucosal cells and represents approximately 10% of plasma immunoglobulins and has a half-life of 6 days. It is found in bronchial and intestinal secretions and is a major component of colostrum (the form of milk produced by the mammary gland immediately after giving birth). IgA is the primary immunological barrier against pathogenic invasion of the mucosal membranes. IgM is found in the intravascular space and its role is to eliminate circulating microorganisms and antigens. IgM accounts for about 8% of plasma immunoglobulins and has a half-life of 5 days. IgM is the first antibody to be synthesised after an antigenic challenge. IgD and IgE are minor immunoglobulins whose roles are not clear since a deficiency of either seems to be associated with no obvious pathology. IgE plays a major part in allergy and may be significantly raised in situations of allergic response, for example in hay fever and atopic eczema.

Myeloma

Myeloma, also called multiple myeloma, is a malignant pathology of plasma cells in which there is an unregulated replication of a single β-cell clone in the bone marrow that proliferates and effectively behaves as a tumour. The cells produce large quantities of a single protein, which migrates as a single dense band in the γ-globulin region during electrophoresis of a serum sample. The protein is called the **paraprotein** and thus belongs to the class of immunoglobulins with two light chains and two heavy chains. Some myelomas produce an excess of light chains that appear in the serum, and, due to their small size, also in the urine. They are detected by electrophoresis and are referred to as **Bence Jones proteins**. Their presence is typically a cause of great concern as it indicates that the cell line may be more aggressive and replicating faster than in the case of exclusive appearance of the paraprotein. In rare cases of myeloma, the marrow cells only produce light chains.

Acute Phase Response

Following a stimulus of tissue injury or infection, the body will respond by producing an **acute phase response** characterised by the release of a number of acute phase proteins from the liver, which is detectable by a change in the pattern of plasma protein electrophoresis. The acute phase response registers as an increased synthesis of some proteins such as α-1-antitrypsin, a proteinase inhibitor that down-regulates inflammation, fibrinogen and C-reactive protein (CRP). These are referred to as **positive acute phase proteins**. There will also be a decrease in the production of other proteins such as albumin and transferrin. These are known as **negative acute phase proteins**. The clinical measurement of acute phase proteins, particularly CRP, by immuno-turbidimetry is widely used as a marker of inflammation in a variety of clinical conditions.

Example 10.3   **CASE STUDY OF MYELOMA**

Description   A 72-year-old woman presented to her GP with a 3-week history of painful hips, chest and shoulders, and with shortness of breath on exertion. She was constipated and had lost 12.5 kg over the last 6 months. She complained that she was very thirsty and had to get up to urinate during the night, something that she had previously never had to do.

Initial laboratory investigations found her to be hypercalcaemic, dehydrated and anaemic. Her biochemistry results showed marked renal impairment, with raised urea and creatinine. Her ALP level was within the reference range. Her serum protein concentration was raised, despite having a low albumin level. Serum protein electrophoresis showed a large monoclonal band in the γ-globulin region. By comparison of the area under the peak of the γ-protein with the combined areas (total protein), the band was quantified as 61 g l^{-1}. The band was typed as IgAκ by immunofixation, using antisera specific against individual immunoglobulin subclasses to bind the monoclonal protein to the electrophoresis before staining. An early morning urine sample was requested for Bence Jones protein analysis by electrophoresis. This detected a large band of free κ light chains in the urine. The results indicated that her other immunoglobulins were suppressed, leaving her susceptible to infection. An isotope bone scan using diphosphonates labelled with ^{99}Te demonstrated osteolytic lesions (bone loss) that are characteristic of multiple myeloma.

Comment   The two most common causes of hypercalcaemia are primary hyperparathyroidism and malignancy. The signs and symptoms in a person of this age are typical of multiple myeloma, especially in the context of hypercalcaemia with a normal ALP, which is raised in primary hyperparathyroidism. Hypercalcaemia results from stimulation of osteoclasts (a type of bone cell) by factors from myeloma cells and can cause polyuria, polydipsia (need to drink excessive fluid) and dehydration. The impaired renal function in this lady may be a result of hypercalcaemia and Bence Jones protein, as both are nephrotoxic. As the malignant plasma cells proliferate throughout the bone marrow, the bone marrow has a reduced capacity to produce normal cells, causing anaemia and immunosuppression. The difference between the concentration of serum total protein and albumin is attributed to the immunoglobulins and consequently myeloma patients may have a high total protein concentration in the presence of normal or low albumin.

The patient was rehydrated with intravenous saline and was given bisphosphonate to lower her calcium levels. Her renal failure resolved over time (N.B. some patients with advanced myeloma will require haemodialysis). A bone marrow aspiration was performed, which showed >80% infiltration of plasma cells, confirming the diagnosis of multiple myeloma. A quantification of serum β2-microglobulin was requested as a prognostic indicator. Once she had stabilised, she was commenced on a course of dexamethasone, a synthetic glucocorticoid that binds immunoglobulins and hence relieves some of the symptoms of the malignancy. Her response to treatment was monitored by regular quantification of her monoclonal band.

10.4 SUGGESTIONS FOR FURTHER READING

10.4.1 Experimental Protocols

Chace D.H. and Kalas T.A. (2005) A biochemical perspective on the use of tandem mass spectrometry for newborn screening and clinical testing. *Clinical Biochemistry*, 38, 296–309.

Wild, D. (2013) *The Immunoassay Handbook*, 4th Edn., Elsevier Science, London, UK.

10.4.2 General Texts

Blau N., Duran M., Blaskovics M.E. and Gibson K.M. (2003) *Physician's Guide to the Laboratory Diagnosis of Metabolic Diseases*, 2nd Edn., Springer Verlag, Berlin, Germany.

Bruns D. E. and Ashwood E. R. (2007) *Tietz Fundamentals of Clinical Chemistry*, 6th Edn., W.B. Saunders, Philadelphia, PA, USA.

Jones R. and Payne B. (1997) *Clinical Investigation and Statistics in Laboratory Medicine*, ACB Ventures, London, UK.

Saunders G.C. and Parkes H.C. (1999) *Analytical Molecular Biology: Quality and Validation*, LGC, Teddington, UK.

Walker S.W, Beckett G.I., Rae P. and Ashby P. (2013) *Lecture Notes on Clinical Biochemistry*, 9th Edn., Blackwell Science, Oxford, UK.

10.4.3 Review Articles

Campbell C., Caldwell G., Coates P., *et al.* (2015) Consensus statement for the management and communication of high risk laboratory results. *The Clinical Biochemist Reviews* 36, 97–105.

McNeil A.R. and Stanford P.E. (2015) Reporting thyroid function tests in pregnancy. *The Clinical Biochemist Reviews* 36, 109–126.

10.4.4 Websites

Laboratory quality management system: Handbook (World Health Organization) www.who.int/ihr/publications/lqms/en/ (accessed May 2017)

UK National External Quality Assessment Scheme www.ukneqas.org.uk (accessed May 2017)

UK National Institute for Biological Standards and Control www.nibsc.org (accessed May 2017)

Wales External Quality Assessment Scheme www.weqas.com (accessed May 2017)

11 Microscopy

STEPHEN W. PADDOCK

11.1 INTRODUCTION

Biochemical and molecular analyses are frequently complemented by microscopic examination of tissue, cell or organelle preparations. Examples range from relatively simple observations by eye to more complex experimental manipulations and measurements. For example, the evaluation of the integrity of samples during an experiment; mapping the fine details of the spatial distribution of macromolecules within cells; direct measurements of biochemical events within living tissues or the elucidation of the macromolecular structure of organelles. In the recent past, the measurement of cellular dynamics has entered a new realm of sophistication and accuracy with the introduction of super-resolution techniques that have enabled the direct viewing of macromolecular dynamics in living cells (Section 11.4.7).

There are two fundamentally different types of microscope: the light microscope and the electron microscope (Figure 11.1). Light microscopes use a series of glass lenses to focus light as it interacts with a specimen in order to form an image, whereas electron microscopes use electromagnetic lenses to focus a beam of electrons. Light microscopes are able to magnify to a maximum of approximately 1500 times, whereas electron microscopes are capable of magnifying to a maximum of approximately 200 000 times.

Magnification is not the best measure of a microscope, however. Rather, resolution, the ability to distinguish between two closely spaced points in a specimen, is a much more reliable estimate of a microscope's utility. Standard light microscopes have a

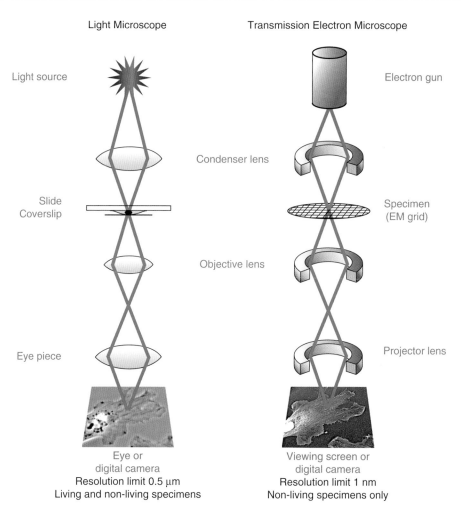

Figure 11.1 Light and electron microscopy. Schematic that compares the path of light through a compound light microscope (LM) with the path of electrons through a transmission electron microscope (TEM). Light from a lamp (LM) or a beam of electrons from an electron gun (TEM) is focussed at the specimen by glass (LM) or electromagnetic condenser lenses (TEM). For the LM, the specimen is mounted on a glass slide with a coverslip placed on top, and for the TEM the specimen is placed on a copper electron microscope grid. The image is magnified with an objective lens (glass in the LM and electromagnetic lens in the TEM), and projected onto a detector with the eyepiece lens in the LM or the projector lens in the TEM. The detector can be the eye or a digital camera in the LM or a phosphorescent viewing screen or digital camera in the TEM.

lateral resolution limit of about 0.5 micrometres (μm) for routine analysis. In contrast, electron microscopes have a lateral resolution of up to 1 nanometre (nm). Computer enhancement methods have improved upon the theoretical maximum 0.5 μm resolution limit of the light microscope down to 0.1 μm resolution in some specialised instruments called **super-resolution microscopes** (Section 11.4.7).

Applications of the microscope in biomedical research can be relatively simple and routine; for example, a quick check of the status of a preparation during an experiment or of the health of cells growing on a plastic dish in tissue culture. Here, a simple bench-top light microscope is perfectly adequate. Other applications may be more involved, for example, measuring the concentration of calcium in a living embryo over a millisecond timescale. In this case, a more advanced light microscope (often called an imaging system) is required. Or, if determining the macromolecular details of a cellular organelle is of interest, then a high-power electron microscope and associated computer software would be the instrument of choice.

Some microscopes are more suited to specific applications than others. Both living and non-living specimens can be viewed with a light microscope, and often in real colour, whereas only non-living ones can be viewed with an electron microscope, and never in real colour.

Images may be required from specimens of vastly different sizes and at different magnifications (Figure 11.2). For imaging whole animals (metres), through tissues and embryos (micrometres), and down to cells, proteins and DNA (nm). In addition, the study of living cells may require time resolution from days, for example when imaging neuronal development or disease processes, to milliseconds, for example when imaging cell signalling events.

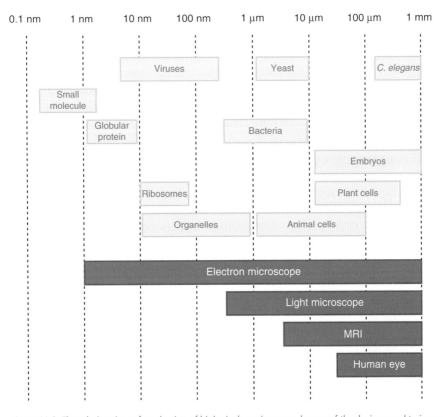

Figure 11.2 The relative sizes of a selection of biological specimens and some of the devices used to image them. The range of resolution for each instrument is included in the dark bars at the base of the figure.

11.2 THE LIGHT MICROSCOPE

The simplest form of light microscope consists of a single glass lens mounted in a metal frame – a magnifying glass. Here, the specimen requires very little preparation, and is usually held close to the eye in the hand. Focussing of the region of interest is achieved by moving the lens and the specimen relative to one another. The source of light is usually the sun or ambient indoor light, and the detector is the human eye. The recording device is a hand drawing or an anecdote. This is how the early light microscopists, Anton Van Leeuwenhoek (1632–1723) and Robert Hooke (1635–1703) collected and recorded their images. Examples of Hooke's hand-drawn images can be found in his classic text *Micrographia*, first published in 1665 (a 'best-seller' of its time).

11.2.1 Compound Microscopes

All modern light microscopes are made up of more than one glass lens in combination. The major components are the condenser lens, the objective lens and the eyepiece lens, and, such instruments are therefore called compound microscopes (Figure 11.1). Each of these components is in turn made up of combinations of lenses, which are necessary to produce magnified images with reduced artefacts and aberrations. For example, chromatic aberration occurs in a lens when different wavelengths of light are separated and pass through a lens at different angles. This results in rainbow colours around the edges of objects in the image. All modern lenses feature corrections in order to avoid this problem as far as possible.

The main components of the compound light microscope include a light source that is focussed at the specimen by a condenser lens. Light that either passes through the specimen (transmitted light) or is reflected back from the specimen (reflected light) is focussed by the objective lens into the eyepiece lens. The image is either viewed directly by eye through the eyepiece or it is most often projected onto a detector, for example a photographic film or, more likely, a digital camera. Digitally captured images are displayed on the screen of a computer imaging system, stored in a digital format and reproduced using digital methods.

The part of the microscope that holds all of the components firmly in position is called the stand. There are two basic types of compound light microscope stands – an upright or an inverted microscope stand (Figure 11.3). The light source is below the condenser lens in the upright microscope and the objectives are above the specimen stage. This is the most commonly used format for viewing specimens. The inverted microscope is engineered so that the light source and the condenser lens are above the specimen stage, and the objective lenses are beneath it. Moreover, the condenser and light source can often be swung out of the light path. This allows additional room for manipulating the specimen directly on the stage, for example, for the microinjection of macromolecules into tissue culture cells, for in vitro fertilisation of eggs, or for viewing and manipulating developing embryos over time.

The correct illumination of the specimen is critical for achieving high-quality images and photomicrographs. The light sources typically used for this purpose are mercury or xenon lamps, lasers or light-emitting diodes (LEDs). Light from the light

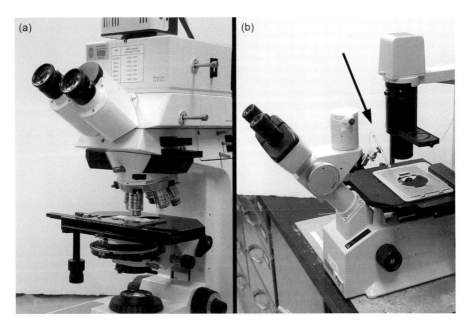

Figure 11.3 Two basic types of compound light microscope. An upright light microscope (a) and an inverted light microscope (b). Note how there is more room available on the stage of the inverted microscope (b). This instrument is set up for microinjection with a needle holder to the left of the stage (arrow).

source passes into the condenser lens, which is mounted beneath the microscope stage in an upright microscope (and above the stage in an inverted microscope) in a bracket that can be raised and lowered for focussing (Figure 11.3). The condenser focusses the light and illuminates the specimen with parallel beams of light. A correctly positioned condenser lens produces illumination that is uniformly bright and free from glare across the viewing area of the specimen (Köhler illumination). Condenser misalignment and an improperly adjusted condenser aperture diaphragm are major sources of poor images in the light microscope.

The specimen stage is a mechanical device that is finely engineered to hold the specimen firmly in place (Figure 11.4). Any movement or vibration will be detrimental to the final image. The stage enables the specimen to be moved and positioned in fine and smooth increments, both horizontally and transversely, in the x- and the y-directions, for locating a region of interest. The stage is moved vertically in the z-direction for focussing the specimen or for inverted microscopes, the objectives themselves are moved and the stage remains fixed. There are usually coarse and fine focussing controls for low magnification and high magnification viewing, respectively. The fine focus control can be moved in increments of 1 μm or better in the best research microscopes. The specimen stage can either be moved by hand or by a stepper motor attached to the fine focus control of the microscope, and controlled by a computer.

The objective lens is responsible for producing the magnified image, and can be the most expensive component of the light microscope (Figure 11.4). Objectives are available in many different varieties, and there is a wealth of information inscribed

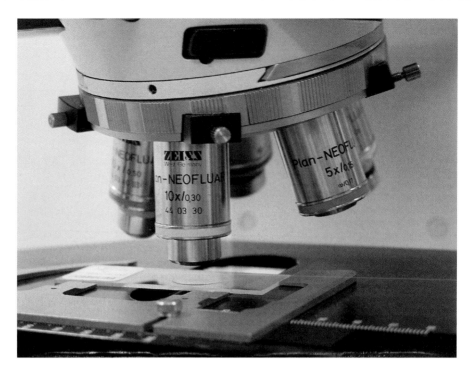

Figure 11.4 The objective lens. A selection of objective lenses mounted on an upright research-grade compound light microscope. From the inscription on the two lenses in focus, they provide relatively low magnification of 10× and 5× that of the respective numerical aperture (NA), here 0.30 and 0.16. Both lenses are Plan-NEOFLUAR, which means they are relatively well corrected to avoid chromatic aberration. The 10× lens is directly above a specimen mounted on a slide and coverslip, and held in place on the specimen stage.

on each one. This may include the manufacturer, magnification (4×, 10×, 20×, 40×, 60×, 100×), immersion requirements (air, oil or water), coverslip thickness (usually 0.17 mm) and often with more-specialised optical properties of the lens (Section 11.2.5). In addition, lens corrections for optical artefacts such as chromatic aberration and flatness of field may also be included in the lens description. For example, words such as fluorite, the least corrected (often shortened to 'fluo'), or plan apochromat, the most highly corrected (often shortened to 'plan' or 'plan apo'), may appear somewhere on the lens.

The **numerical aperture** (NA) is always marked on the lens. This is a number usually between 0.025 and 1.4. The NA is a measure of the ability of a lens to collect light from the specimen. Lenses with a low NA collect less light than those with a high NA. The resolution varies inversely with NA, which infers that higher NA objectives yield the best resolution. Generally speaking, the higher-power objectives have a higher NA and better resolution than the lower-power lenses with lower numerical apertures. For example, 0.2 μm resolution can only be achieved using a 100× plan-apochromat oil immersion lens with an NA of 1.4. Should there be a choice between two lenses of the same magnification, then it is usually best to choose the one of higher NA.

Lower-power lenses have the advantage of imaging a greater area of the specimen than higher-power ones, albeit at a lower resolution. When larger areas of specimens need to be imaged at high resolution, several images are collected and subsequently 'stitched' together into a single image using a computer algorithm.

The resolution achieved by a lens is a measure of its ability to distinguish between two objects in the specimen. The shorter the wavelengths of illuminating light, the higher the resolving power of the microscope (Figure 11.5). The limit of resolution for a microscope that uses visible light is about 300 nm with a dry lens (in air) and 200 nm with an oil immersion lens. By using ultraviolet light (UV) as a light source, the resolution can be improved to 100 nm because of the shorter wavelength of the light (200–300 nm). These limits of resolution are often difficult to achieve practically, because of aberrations in the lenses and the poor optical properties of many biological specimens. The lateral resolution is usually better than the axial resolution for any given objective lens (Table 11.1).

11.2.2 Recording Images

In addition to the human eye and photographic film, there are two types of electronic detectors employed on modern light microscopes. These are area detectors that actually form an image directly, for example video cameras and charge coupled devices (CCDs). Alternatively, point detectors can be used to measure intensities in the image; for example photomultiplier tubes (PMTs) and photodiodes. Point detectors are capable of producing images in scanning microscopy (see Section 11.3).

Table 11.1 **Resolution in optical imaging**

	x-y (lateral)	z (axial)
Standard microscope	0.5 µm	1.6 µm
Confocal/multiple-photon	0.25 µm	0.7 µm
Super-resolution	0.1 µm	0.4 µm

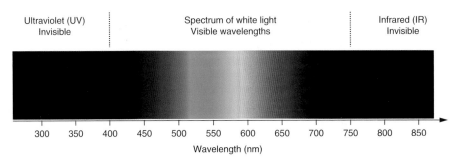

Figure 11.5 The visible spectrum – the spectrum of white light visible to the human eye. Our eyes are able to detect colour in the visible wavelengths of the spectrum; usually in the region between 400 nm (violet) and 750 nm (red). Most modern electronic detectors are sensitive beyond the visible spectrum of the human eye.

11.2.3 Stereomicroscopes

A second type of light microscope, the stereomicroscope is used for the observation of the surfaces of large specimens (Figure 11.6). The stereomicroscope is used when three-dimensional information is required, such as the routine observation of whole organisms, for example for screening through vials of fruit flies or zebrafish embryos. Stereomicroscopes are useful for micromanipulation and dissection, where the wide field of view and the ability to change the magnification (that is, to zoom in and out) is invaluable. A wide range of objectives and eyepieces are available for different applications. The light sources can be positioned from above or below the specimen; encircling the specimen using a ring light or by positioning the source on the side yields a dark–field effect (see Section 11.2.5). These different light angles serve to add contrast or shadow relief to the images.

11.2.4 The Specimen

The specimen (sometimes called the sample) can be an entire organism or a dissected organ (whole mount), an aliquot collected during a biochemical protocol for a quick check of the preparation, a small part of an organism (biopsy) or smear of blood or spermatozoa. In order to collect images from it, the specimen must be in a form that is compatible with the microscope. This is achieved using a preparation protocol, many of which are available in the literature. The goal of a specimen preparation protocol is to render the tissue of interest into a form for optimal study with the microscope.

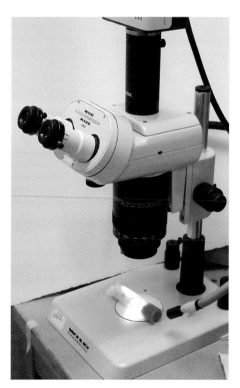

Figure 11.6 A research-grade stereomicroscope. Note the light source is projecting from the side, which can give a shadow effect to the specimen, in this example a vial of fruit flies. The large objective lens above the specimen can be rotated to zoom the image.

This usually involves placing the specimen in a suitable medium (water, tissue culture medium or glycerol) on a glass slide with a coverslip over it. The end product of such a preparation is a relatively thin and somewhat transparent piece of tissue. Preparation protocols can be relatively simple (see Table 11.2 for an example) or they may involve a lengthy series of many steps that take several days to complete.

Coverslips are graded by their thickness. The thinnest ones are labelled #1, which corresponds to a thickness of approximately 0.17 mm. The coverslip side of the specimen is always placed closest to the objective lens. It is essential to use a coverslip that is matched to the objective lens in order to achieve optimal resolution. This is critical for high-magnification imaging; if the coverslip is too thick, it will be impossible to achieve an image using a high-magnification objective lens.

An example of a simple protocol would be taking an aliquot of a biological preparation (for example, living spermatozoa isolated into a balanced salt solution), placing an aliquot of it onto a slide and gently placing a clean coverslip onto the top. The entire protocol would take less than a minute. Shear forces from the movement of the coverslip over the glass slide can cause damage to the specimen or the objective lens. Therefore, the coverslip is sealed to the glass slide in some way, for example, using nail polish for dead cells or perhaps a mixture of beeswax and Vaseline for living cells. In order to keep cells alive on the stage of the microscope, they are usually mounted in some form of chamber, and if necessary heated.

Many specimens are too thick to be mounted directly onto a slide, and these are cut into thin sections using a device called a microtome. The tissue is usually mounted in a block of wax and cut with the knife of the microtome into thin sections (between 10 μm and 500 μm in thickness). The sections are then placed onto a glass slide, stained, immersed in mounting medium and sealed with a coverslip. Some samples are frozen, and cut on a cryostat, which is basically a microtome that can keep a specimen in the frozen state, and produce frozen sections more suitable for immunolabelling (see also Section 11.6.2).

Prior to sectioning, the tissue is usually treated with a chemical agent called a fixative to preserve it. Popular fixatives include formaldehyde and glutaraldehyde, which

Table 11.2 Generalised indirect immunofluorescence protocol

1.	Fix in 1% formaldehyde for 30 minutes
2.	Rinse in cold buffer
3.	Block buffer
4.	Incubate with primary antibody, e.g. mouse antitubulin
5.	Wash 4× in buffer
6.	Incubate with secondary antibody, e.g. fluorescein-labelled rabbit antimouse
7.	Wash 4× in buffer
8.	Incubate with antifade reagent, e.g. Vectashield
9.	Mount on slide with a coverslip
10.	View using epifluorescence microscopy

act by cross-linking proteins, or alcohols – which act by precipitation. All of these fixatives are designed to maintain the structural integrity of the cell. After fixation, the specimen is usually permeabilised in order to allow a stain to infiltrate the entire tissue. The amount of permeabilisation (time and severity) depends upon several factors, including the size of the stain or the density of the tissue. These parameters are found by trial and error for a new specimen, but are usually available for many tissues in published protocols. The goal is to infiltrate the entire tissue with a uniform staining.

Preparation protocols continue to be developed to improve the visibility of biological structures within specimens. Examples include a method called 'CLARITY', which improves the transparency of the specimen to give improved depth penetration, and another is called 'expansion microscopy' where the specimen is expanded during the protocol to give improved optical properties by producing more room between organelles.

11.2.5 Contrast in the Light Microscope

Most cells and tissues are colourless and almost transparent, and lack contrast when viewed in a light microscope. Therefore, to visualise any details of cellular components, it is necessary to introduce contrast into the specimen. This is achieved either by optical means using a specific configuration of microscope components, or by staining the specimen with a dye or, more usually, using a combination of optical and staining methods. Different regions of the cell can be stained selectively with different stains.

Optical Contrast

Contrast is achieved optically by introducing various elements into the light path of the microscope and using lenses and filters that change the pattern of light passing through the specimen and the optical system. This can be as simple as adding a piece of coloured glass or a neutral density filter into the illuminating light path, by changing the light intensity or by adjusting the diameter of a condenser aperture. Usually, all of these operations are adjusted until an acceptable level of contrast is achieved to collect a useful image.

The most basic mode of the light microscope is called bright-field (bright background), which can be achieved with the minimum of optical elements. Contrast in bright-field images is usually produced by the colour of the specimen itself. Bright-field is therefore used most often to collect images from pigmented tissues, or histological sections or tissue culture cells that have been stained with colourful dyes (Figures 11.7a and 11.8b).

Several configurations of the light microscope have been introduced over the years specifically to add contrast to the final image. Dark-field illumination produces images of brightly illuminated objects on a black background (Figures 11.7b and 11.8a). This technique has traditionally been used for viewing the outlines of objects in liquid media such as living spermatozoa, microorganisms, cells growing in tissue culture or for a quick check of the status of a biochemical preparation. For lower magnifications, a simple dark-field setting on the condenser will be sufficient. For more critical dark-field imaging at a higher magnification, a dark-field condenser with a dark-field objective lens will be required.

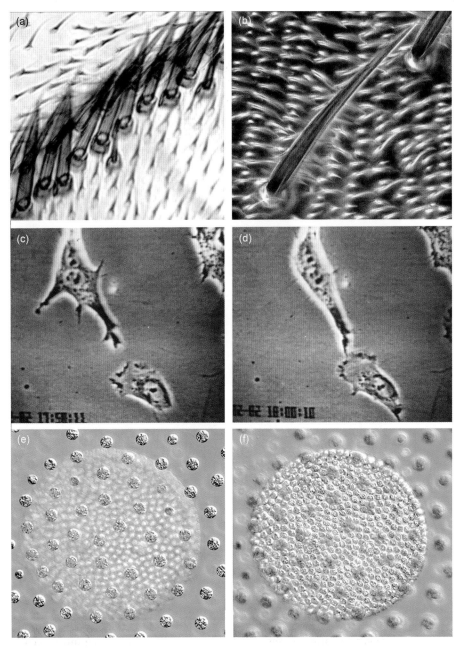

Figure 11.7 Contrast methods in the light microscope. (a) and (b) A comparison of bright-field (a) and dark-field images (b). Here, the bristles on the surface of the fly appear dark on a white background in the bright-field image (a) and white on a black background in a dark-field image (b). The dark colour in the larger sensory bristles in (a) and (b) is produced. (c) and (d) Phase contrast view of cells growing in tissue culture (c) and (d). Two images extracted from a time-lapse video sequence (time between each frame is five minutes). The sequence shows the movement of a mouse 3T3 fibrosarcoma cell (top) and a chick heart fibroblast. Note the bright 'phase halo' around the cells. (e) and (f) Differential interference contrast (DIC) image of two focal planes of the multicellular alga Volvox (e) and (f). (Images (e) and (f) courtesy of the late Michael Davidson, Florida State University, USA.)

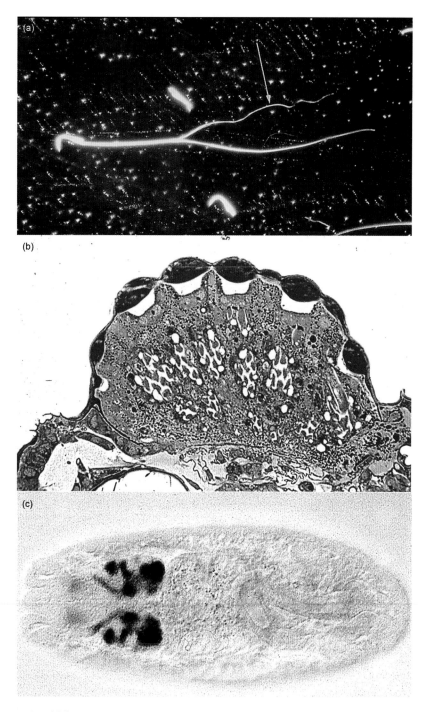

Figure 11.8 Examples of different preparations in the light microscope. (a) Dark-field image of rat sperm preparation. An aliquot was collected from an experimental protocol in order to assess the amount of damage incurred during sonication of a population of spermatozoa. Many sperm heads can be seen in the preparation, and the fibres of the tail are starting to fray (arrowed). (b) A bright-field image of total protein staining on a section of a fly eye cut on a microtome, and stained with Coomassie Blue. (c) DIC image of a stained *Drosophila* embryo – the image shows the outline of the embryo with darker regions of neuronal staining. The DIC image of the whole embryo provides structural landmarks for placing the specific neuronal staining in context of the anatomy.

Phase contrast is used for viewing unstained cells growing in tissue culture and for testing cell and organelle preparations for lysis (Figure 11.7c,d). This method images differences in the refractive index of cellular structures. Light that passes through thicker parts of the cell is held up relative to the light that passes through thinner parts of the cytoplasm. It requires a specialised phase condenser and phase objective lenses (both labelled 'ph'). Each phase setting of the condenser lens is matched with the phase setting of the objective lens. These are usually numbered as Phase 1, Phase 2 and Phase 3, and are found on both the condenser and the objective lens.

Differential interference contrast (DIC) is a form of interference microscopy that produces images with a shadow relief (Figure 11.7e,f). It is used for viewing unstained cells in tissue culture, eggs and embryos, and in combination with some stains. Here, the overall shape and relief of the structure is viewed using DIC and a subset of the structure is stained with a coloured dye (Figure 11.8c).

Fluorescence microscopy is currently the most widely used contrast technique, since it gives superior signal-to-noise ratios (typically white on a black background) for many applications. The most commonly used fluorescence technique is called epifluorescence light microscopy, where 'epi' simply means 'from above'. Here, the light source comes from above the sample, and the objective lens acts as both condenser and objective lens (Figure 11.9). Fluorescence is popular because of the ability to achieve highly specific labelling of cellular compartments. The images usually

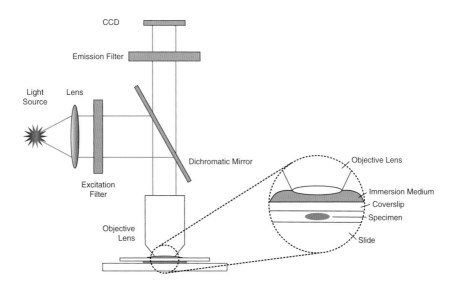

Figure 11.9 Epifluorescence microscopy. Light from a xenon or mercury arc lamp passes through a lens and the excitation filter and reflects off the dichromatic mirror into the objective lens. The objective lens focusses the light at the specimen via the immersion medium (usually immersion oil) and the glass coverslip (see insert). Any light resulting from the fluorescence excitation in the specimen passes back through the objective lens, and since it is of longer wavelength than the excitation light, it passes through the dichromatic mirror. The emission filter only allows light of the specific emission wavelength of the fluorophore of interest to pass through to the CCD array, where an image is formed.

consist of distinct regions of fluorescence (white) over large regions of no fluorescence (black), which gives excellent signal-to-noise ratios (Figure 11.10).

The light source is usually a high-pressure mercury or xenon vapour lamp, and more recently lasers and LED sources, which emit from the UV into the red wavelengths (Figure 11.5). A specific wavelength of light is used to excite a fluorescent molecule or fluorophore in the specimen (Figure 11.9; see also Section 13.5). Light of longer wavelength from the excitation of the fluorophore is then imaged. This is achieved in the fluorescence microscope using combinations of filters that are specific for the excitation and emission characteristics of the fluorophore of interest. There are usually three main filters: an excitation, a dichromatic mirror (often called a dichroic) and a barrier filter, mounted in a single housing above the objective lens. For example, the commonly used fluorophore, fluorescein is optimally excited at a wavelength of 488 nm, and emits maximally at 518 nm (Table 11.3).

A set of glass filters for viewing a particular fluorophore requires that all wavelengths of light from the lamp be blocked except for the excitation wavelength. Such filters are called excitation filters and, for fluorescein, allow a maximum amount of light with the wavelength of 488 nm to pass through it. This light is then directed to the specimen via the dichromatic mirror. Any fluorescein label in the specimen is excited by excitation light, and, as a result, emits light of the wavelength 518 nm that returns from the specimen, and passes through both the dichromatic mirror and the barrier filter to the detector. The emission filters only allow light of 518 nm to pass through to the detector, and thus ensure that only the signal emitted from the fluorophore of interest is registered.

Chromatic mirrors and filters can be designed to filter two or three specific wavelengths for imaging specimens labelled with two or more fluorophores (multiple

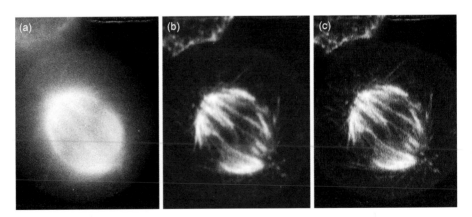

Figure 11.10 Fluorescence microscopy. Comparison of epifluorescence and confocal fluorescence imaging of a mitotic spindle labelled using indirect immunofluorescence labelling with antitubulin (primary antibody) and a fluorescently labelled secondary antibody. The specimen was imaged using (a) conventional epifluorescence light microscopy or (b) and (c) using laser-scanning confocal microscopy. Note the improved resolution of microtubules in the two confocal images (b) and (c) as compared with the conventional image (a). (b) and (c) represent two different resolution settings of the confocal microscope. Image (b) was collected with the pinhole set to a wider aperture than (c). (Image kindly provided by Brad Amos, University of Cambridge, UK.)

Table 11.3 **Commonly used dyes in fluorescence microscopy**

Dye	Excitation maximum (nm)	Emission maximum (nm)
Commonly used fluorophores		
Fluorescein (FITC)	488	518
Cy3	554	568
Tetramethylrhodamine	554	576
Lissamine rhodamine	572	590
Texas Red	592	610
Cy5	652	672
Nuclear dyes		
Hoechst 33342	346	460
DAPI	359	461
Acridine Orange	502	526
Feulgen	570	625
TOTO3	642	661
Calcium indicators		
Fluo-3	506	526
Calcium Green	506	533
Reporter molecules		
CFP	430/490	475/503
GFP	395/489	509
YFP	490	520
DsRed	558	583

Note: See also Figure 8.9 for popular fluorescence dyes.

labelling). The fluorescence emitted from the specimen is often too low to be detected by the human eye or it may be out of the wavelength range of detection by eye, for example in the far-red wavelengths (Figure 11.5). A sensitive digital camera (e.g. a CCD or PMT) easily detects such signals.

Specimen Stains

Contrast can be introduced into the specimen using one or more coloured dyes or stains. These can be non-specific stains, for example, a general protein stain such as Coomassie Blue (Figure 11.8b) or a stain that specifically labels an organelle for example, the nucleus, mitochondria etc. Combinations of such dyes may be used to stain different organelles in contrasting colours. Many of these histological stains are usually observed using bright-field imaging. Other light-microscopy techniques may also be employed in order to view the entire tissue along with the stained tissue. For example, differential interference contrast can be used to view the entire morphology of an embryo and a coloured stain to image the spatial distribution of the protein of interest within the embryo (Figure 11.8c).

More specific dyes are usually used in conjunction with fluorescence microscopy. Immunofluorescence microscopy is used to map the spatial distribution of macromolecules in cells and tissues (see also Section 7.4). The method takes advantage of the highly specific binding of antibodies to proteins. Antibodies are raised to the protein of interest and labelled with a fluorescent probe. This probe is then used to label the protein of interest in the cell and can be imaged using fluorescence microscopy. In practice, cells are usually labelled using indirect immunofluorescence. Here the antibody to the protein of interest (primary antibody) is further labelled with a second antibody carrying the fluorescent tag (secondary antibody). Such a protocol gives a higher fluorescence signal than one that uses a single fluorescently labelled antibody (Table 11.2). Additional methods are available for amplifying the fluorescence signal in the specimen, for example using the tyramide amplification method. Alternatively, at the microscope, a more sensitive detector can improve the resultant image.

A related technique, fluorescence *in situ* hybridisation (FISH) employs the specificity of fluorescently labelled DNA or RNA sequences. The nucleic acid probes are hybridised to chromosomes, nuclei or cellular preparations. Regions that bind the probe are then imaged using fluorescence microscopy. Many different probes can be labelled with different fluorophores in the same preparation. Multiple colour FISH is used extensively for clinical diagnoses of inherited genetic diseases. This technique has been applied to rapid screening of chromosomal and nuclear abnormalities in genetic diseases, such as the Down syndrome.

There are many different types of fluorescent molecules that can be attached to antibodies, DNA or RNA probes for fluorescence analysis (Table 11.3). All of these reagents, including primary antibodies, are available commercially or often from the laboratories that produced them. An active area of development is the production of the brightest fluorescent probes that are excited by the narrowest wavelength band and that are not damaged by light excitation (photobleaching). Traditional examples of such fluorescent probes include fluorescein, rhodamine, the Alexa range of dyes and the Cyanine dyes. A recent addition to the extensive list of probes for imaging are quantum dots, which have the advantage that they do not suffer from photobleaching. Quantum dots are nanocrystals of different sizes that do not fluoresce per se, but glow in different colours when exposed to laser light. The colours depend on the size of the dots.

11.3 OPTICAL SECTIONING

Many images collected from relatively thick specimens using epifluorescence microscopy are not very clear. The reason for this is that the image is made up of the optical plane of interest together with contributions from fluorescence above and below the focal plane of interest. Since the conventional epifluorescence microscope collects all of the information from the specimen, it is often referred to as a wide-field microscope. However, the 'out-of-focus fluorescence' can be removed using a variety of optical electronic techniques to produce optical sections (Figures 11.10 and 11.13).

The term 'optical section' refers to the ability of a microscope to produce sharper images of specimens than those produced using a standard wide-field epifluorescence microscope by removing the contribution from out-of-focus light to the image, without resorting to physically sectioning the tissue. Such methods have revolutionised the ability to collect images from thick and fluorescently labelled specimens such as eggs, embryos and tissues. Optical sections can also be produced using high-resolution differential interference contrast optics (Figure 11.7e,f), micro-computed tomography (CT) scanning or optical-projection tomography. However, currently by far the most prevalent method is to use some form of confocal or associated microscopical approach.

11.3.1 Laser-Scanning Confocal Microscopes

Optical sections are produced in the laser-scanning confocal microscope by scanning the specimen point by point with a laser beam focussed on the specimen, and using a spatial filter, usually a pinhole or a slit, to remove unwanted fluorescence from above and below the focal plane of interest (Figure 11.11). The power of this confocal approach lies in the ability to image structures at discrete levels within an intact biological specimen.

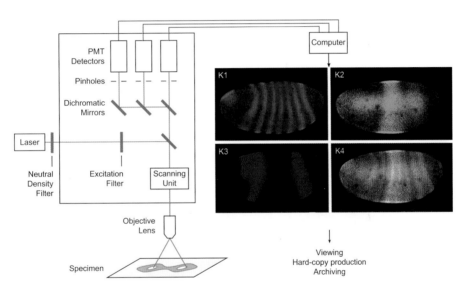

Figure 11.11 Information flow in a generic laser-scanning confocal microscope. Light from the laser passes through a neutral density filter and an excitation filter on its way to the scanning unit. The scanning unit produces a scanned beam at the back focal plane of the objective lens, which focusses the light at the specimen. The specimen is scanned in the x- and the y-direction in a raster pattern, and in the z-direction by fine focussing. Any fluorescence from the specimen passes back through the objective lens and the scanning unit and is directed via dichromatic mirrors to three pinholes. The pinholes act as spatial filters to block any light from above or below the plane of focus in the specimen. The point of light in the specimen is confocal with the pinhole aperture. This means that only distinct regions of the specimen are sampled. Light that passes through the pinholes strikes the photomultiplier tube (PMT) detectors and the signal from the PMTs is built into an image in the computer. The image is displayed on the computer screen often as three monochrome images (K1 – K3) together with a merged colour image of the three monochrome images (K4) and (Figure 11.13g). The computer synchronises the scanning mirrors with the build-up of the image in the computer frame store. The computer also controls a variety of peripheral devices; for example, it correlates movement of a stepper motor connected to the fine focus of the microscope with image acquisition in order to produce a Z–stack. Furthermore, the computer controls the area of the specimen to be scanned by the scanning unit so that zooming is easily achieved by scanning a smaller region of the specimen. In this way, a range of magnifications is imparted to a single objective lens so that the specimen does not have to be moved when changing magnification. Images are written to a hard disk and exported to various devices for viewing, hard-copy production or archiving.

There are two major advantages of using laser-scanning confocal microscopy in preference to conventional epifluorescence light microscopy. First, glare from out-of-focus structures in the specimen is reduced and resolution is increased, both laterally in the x- and the y-directions (0.14 µm) and axially in the z-direction (0.23 µm). Second, the image quality of relatively thick specimens such as fluorescently labelled multi-cellular embryos is substantially improved. Note that for thinner specimens, though, such as for example chromosome spreads and the leading lamellipodium of cells growing in tissue culture (<0.2 µm thick), there is no significant improvement of image quality when using laser-scanning confocal microscopy.

For successful confocal imaging, a minimum number of photons should be used to efficiently excite each fluorescent probe in the specimen, and as many of the emitted photons from the fluorophores as possible should make it through the light path of the instrument to the detector.

Laser-scanning confocal microscopy has found many different applications in biomedical imaging. Some of these applications have been made possible by the ability of the instrument to produce a series of optical sections at discrete steps through the specimen (Figure 11.12). This Z-series (Z-stack) of optical sections collected with a confocal microscope are all in register with each other, and can be merged together to form a single projection of the image (Z-projection) or a three-dimensional representation of the image (3D reconstruction).

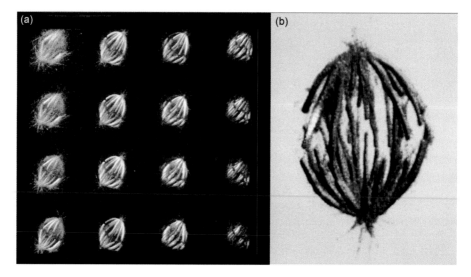

Figure 11.12 Computer 3D reconstruction of confocal images. (a) Sixteen serial optical sections collected at 0.3 µm intervals through a mitotic spindle of a PtK1 cell stained with antitubulin and a second rhodamine-labelled antibody. Using the Z-series macro program, a preset number of frames can be summed, and the images transferred into a file on the hard disk. The stepper motor moves the fine focus control of the microscope by a preset increment. (b) Three-dimensional reconstruction of the dataset produced using computer 3D reconstruction software. Such software can be used to view the dataset from any specified angle or to produce movies of the structure rotating in three dimensions.

Multiple label images can be collected from a specimen labelled with more than one fluorescent probe using multiple laser light sources for excitation (Figure 11.13). Since all of the images collected at different excitation wavelengths are in register, it is relatively easy to combine them into a single multi-coloured image. Here, any overlap of staining is viewed as an additive colour change. Most confocal microscopes are able to routinely image three or four different wavelengths simultaneously.

The scanning speed of most laser-scanning systems is around one full frame per second. This is designed for collecting images from fixed and brightly labelled fluorescent specimens. Such scan speeds are not optimal for living specimens, and laser-scanning instruments are available that scan at faster rates for more optimal live-cell imaging. In addition to point scanning, swept field scanning rapidly moves a micrometre-thin beam of light horizontally and vertically through the specimen.

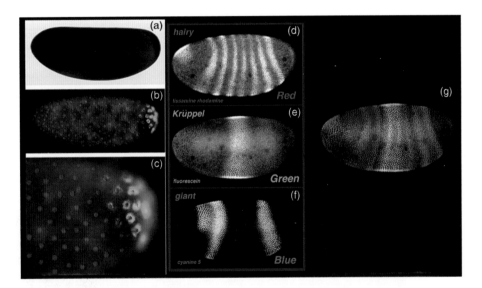

Figure 11.13 Optical sections produced using laser-scanning confocal microscopy. Comparison of alkaline phosphatase (a) and tyramide-amplified detection of mRNAs (b,c). Staining patterns obtained using DIG-labelled antisense probes directed against the CG14217 mRNAs, through conventional AP-based detection (a) or tyramide signal amplification (b), using tyramide-Alexa Fluor 488 (green fluorescence). Close-up images of tyramide-amplified samples are also shown (c). In (b) and (c), nuclei were labelled in red with propidium iodide. Triple-labelled *Drosophila* embryo at the cellular blastoderm stage (d,e,f,g) . The images were produced using an air-cooled 25 mW krypton argon laser, which has three major excitation wavelengths at 488 nm (blue), 568 nm (yellow) and 647 nm (red). The three fluorophores used were fluorescein (λ_{exc} = 496 nm; λ_{em} = 518 nm), lissamine rhodamine (λ_{exc} = 572 nm; λ_{em} = 590 nm) and Cyanine 5 (λ_{exc} = 649 nm; λ_{em} = 666 nm). The images were collected simultaneously as single optical sections into the red, the green and the blue channels, respectively, and merged as a three colour (red/green/blue) image (Figure 11.11). The image shows the expression of three genes; hairy (in red), Krüppel (in green) and giant (in blue). Regions of overlap of gene expression appear as an additive colour in the image, for example, the two yellow stripes of hairy expression in the Krüppel domain (f,g) . (Images a,b and c were kindly provided by Henry Krause, University of Toronto, Canada.)

11.3.2 Spinning-Disc Confocal Microscopes

The spinning-disc confocal microscope employs a different scanning system from the conventional laser-scanning confocal microscope. Rather than scanning the specimen with a single beam, multiple beams scan the specimen simultaneously and optical sections are viewed in real time. Modern spinning-disc microscopes have been improved significantly by the addition of laser light sources and high quality CCD detectors to the instrument. Spinning-disc systems are generally used in experiments where high-resolution images are collected at a fast rate (high spatial and temporal resolution), and are used to follow the dynamics of fluorescently labelled proteins in living cells (Figure 11.14).

11.3.3 Multiple-Photon Microscopes

Many instruments have been further advanced, but still use the same scanning system as the laser-scanning confocal microscope or the spinning-disc systems. In the

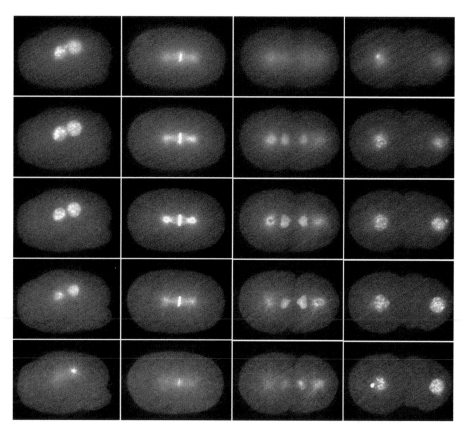

Figure 11.14 Time-lapse imaging of *Caenorhabditis elegans* development. Z-series were collected every 90 seconds of a developing *C. elegans* embryo genetically labelled with GFP-histone (nuclear material) and GFP-α tubulin (microtubules – cytoskeleton), and imaged with a spinning-disc confocal microscope. Each column consists of six optical sections collected 2 μm apart, and the columns are separated by 90 second increments of time. (Image kindly provided by Kevin O'Connell, National Institutes of Health, USA.)

multiple-photon microscope, the light source is a high-energy pulsed laser with tunable wavelengths, and the fluorophores are excited by multiple rather than single photons. Optical sections are produced simply by focussing the laser beam onto the specimen, since multiple-photon excitation of a fluorophore only occurs where energy levels are high enough – statistically confined to the point of focus of the objective lens (Figure 11.15).

Since red light is used in multiple-photon microscopes, optical sections can be collected from deeper within the specimen than those collected with the laser-scanning confocal microscope. Multiple-photon imaging is generally chosen for imaging fluorescently labelled living cells, because red light (= longer wavelengths, lower energy) is less damaging to living cells than light with shorter wavelengths usually employed by confocal microscopes. In addition, since the excitation of the fluorophore is restricted to the point of focus in the specimen, there is less chance of photobleaching the fluorescent probe and causing damage to the specimen itself (Figure 11.15).

11.3.4 Deconvolution

Optical sections can be produced using an image-processing method called deconvolution to remove the out-of-focus information from the digital image. Such images are computed from conventional wide-field microscope images. There are two basic types of deconvolution algorithm: deblurring and restoration. The general approach relies upon knowledge of the **point spread function** of the imaging system. This is usually measured by imaging a point source, for example, a small subresolution fluorescent bead (0.1 µm), and evaluating how the point is spread out in the micrograph. Since it is assumed that the real image of the bead should be a point, it is possible to calculate the amount of distortion in the image of the bead imposed by the imaging system. The actual image of the point can then be restored using a mathematical function, which can be applied to any subsequent images collected under identical settings of the microscope.

Early versions of the deconvolution method were relatively slow; for example, it could take some algorithms in the order of hours to compute a single optical section. Deconvolution is now much faster using today's fast computers and improved

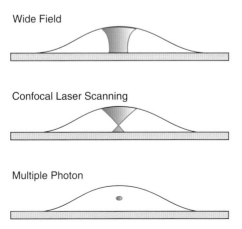

Figure 11.15 Illumination in wide field, confocal and multiple-photon microscopes. The diagram shows a schematic of a side view of a fluorescently-labelled cell on a coverslip. The shaded areas in each cell represents the volume of fluorescent excitation produced by each of the different microscopes in the cell. Conventional epifluorescence microscopy illuminates throughout the cell. In the laser-scanning confocal microscope, fluorescence illumination is throughout the cell but the pinhole in front of the detector excludes the out-of-focus light from the image. In the multiple-photon microscope, excitation only occurs at the point of focus where the light flux is high enough.

software, and the method compares favourably with the confocal approach for producing optical sections. Deconvolution is practical for multiple-label imaging of both fixed and living cells, and excels over the scanning methods for imaging relatively dim and thin specimens, for example yeast cells. The method can also be used to remove additional background from images that were collected with the laser-scanning confocal microscope, the spinning-disc microscope or a multiple-photon microscope.

11.3.5 Total Internal Reflection Microscopy

Another area of active research is in the development of single molecule detection techniques. For example, total internal reflection (TIR) microscopy uses the properties of an evanescent wave (see also Section 14.6.1) close to the interface of two media (Figure 11.16), for example the region between the specimen and the glass coverslip. The technique relies on the fact that the intensity of the evanescent field falls off rapidly so that the excitation of any fluorophore is confined to a region of just 100 nm above the glass interface. This is thinner than the optical-section thickness achieved using confocal methods and allows the imaging of single molecules at the interface.

11.3.6 Optical-Projection Tomography

Optical-projection tomography is useful for imaging specimens that are too big to be imaged using other microscope-based imaging methods, e.g. vertebrate embryos. Here, the resolution is better than that achieved using magnetic resonance imaging (MRI; see Section 14.3.3), but not as good as confocal microscopy. Optical-projection tomography can take advantage of some of the dyes used in confocal microscopy.

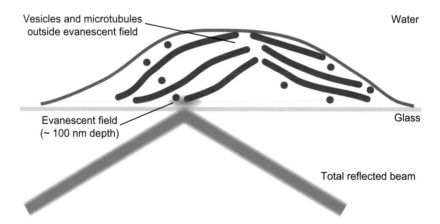

Figure 11.16 Total internal reflection microscopy. A 100 nm thick region of excitation is produced at the glass–water interface when illumination conditions are right for internal reflection. In this example, only those vesicles and microtubules within the evanescent field will contribute to the fluorescence image at 100 nm z-resolution.

11.4 IMAGING LIVE CELLS AND TISSUES

There are two fundamentally different approaches to imaging biochemical events over time. One strategy is to collect images from a series of fixed and stained tissues at different developmental ages. Each animal represents a single time point in the experiment. Alternatively, the same cells and tissues can be imaged in the living state during development. Here, the events of interest are captured directly. The second approach, imaging living cells and tissues, is technically more challenging than the first because great care must be taken to maintain cells in a healthy state during the time course of an experiment, with checks at the end of the experiment for cell damage.

11.4.1 Avoidance of Artefacts

The only way to eliminate artefacts from specimen preparation is to view the specimen in the living state. Many living specimens are sensitive to light, especially those labelled with fluorescent dyes. This is because the excitation of fluorophores can release cytotoxic free radicals into the cell. Moreover, some wavelengths are more damaging than others. Generally, the shorter wavelengths (higher-energy light) are more harmful than the longer ones (lower-energy light) and, therefore, near-infrared light rather than ultraviolet light is preferred for imaging (Figure 11.5).

The intensity of light used for imaging must not compromise the cells. This is achieved using extremely low intensities of light, using relatively bright fluorescent dyes and extremely sensitive photodetectors. The viability of cells may also depend upon the cellular compartment that has been labelled with the fluorophore. For example, imaging the nucleus with a dye that is excited with a short wavelength will cause more cellular damage than imaging in the cytoplasm with a dye that is excited in the far red. For these reasons, it is necessary, when possible, to control for artefacts introduced by the imaging system itself.

Great care has to be taken in order to maintain the tissue in the living state on the microscope stage. A live cell chamber is usually required for mounting the specimen on the microscope stage. This is basically a modified slide and coverslip arrangement that allows access to the specimen by the objective and condenser lenses. It also supports the cells in a constant environment, and, depending on the cell type of interest, the chamber may have to provide a constant temperature, humidity, pH, carbon dioxide and/or oxygen levels. Many chambers have the facility for introducing fluids or perfusing the preparation with drugs for experimental treatments.

11.4.2 Time-Lapse Imaging

Time-lapse imaging continues to be used for the study of cellular dynamics. Here, images are collected at pre-determined time intervals (Figure 11.14). Usually, a shutter is placed in the light path so that the the specimen is only exposed to light when an image is collected and the amount of energy impacting the cells is reduced to an absolute minimum. When the images are played back in real time, a movie of the process of interest is produced.

Time-lapse is used to study cell behaviour in tissues and embryos, and the dynamics of macromolecules within single cells (Figure 11.7c,d). The event of interest and also the amount of light energy absorbed and tolerated by the cells govern the time interval used. For example, a cell in tissue culture moves relatively slowly and a time interval of 30 seconds between images might be used. Stability of the specimen and of the microscope is extremely important for successful time-lapse imaging; importantly, the focus should not drift during the experiment.

Whereas phase contrast is the traditional choice for imaging cell movement and behaviour of cells growing in tissue culture, differential interference contrast (DIC) or fluorescence microscopy is generally chosen for imaging the development of eggs and embryos. Computer imaging methods can be used in conjunction with DIC to improve resolution. Such enhancement is based on subtraction of a background image from each time-lapse frame in order to improve the contrast electronically. Such an approach has enabled the motility assays for motor proteins (for example, kinesin and dynein) whereby microtubules have been visualised on glass, assembled in vitro from tubulin in the presence of microtubule-associated proteins.

11.4.3 Photo Animation

The problems of presenting time-lapse series in a publication have been largely solved by the ability to publish QuickTime (a multimedia framework by Apple Inc.) movie files on the web pages of various journals. The software Photoshop (Adobe) also provides a bridge to additional image processing. For example, sequences of confocal images of different stages of development have been manipulated using Photoshop, and subsequently transferred to a commercially available animation program such as Morpheus (Morpheus Development LLC), and processed into short animated sequences of development. These sequences can be further edited and compiled using video editing software such as Final Cut Pro (now by Apple Inc.), and viewed as a digital movie using a video player directly on the computer or exported to DVD for presentation purposes.

11.4.4 Fluorescent Stains of Living Cells

Relatively few cells possess any inherent fluorescence (autofluorescence) although some endogenous molecules are fluorescent and can be used for imaging, for example, NAD(P)H. However, small fluorescent molecules can be loaded into living cells using many different methods, including diffusion, microinjection, bead loading, virus entry and electroporation. Larger molecules such as fluorescently labelled proteins are usually injected into cells.

Many reporter molecules are available for recording the expression of specific genes in living cells using fluorescence microscopy, as well as viewing whole transgenic animals using fluorescence stereomicroscopes (Table 11.3). In particular, the green fluorescent protein (GFP) is a very convenient and valuable reporter of gene expression, because it is directly visible in the living cell using epifluorescence light microscopy with standard filter sets.

The GFP gene can be linked to another gene of interest so that its expression is accompanied by GFP fluorescence in the living cell. No fixation, substrates, coenzymes or cell-loading techniques are required. The fluorescence of GFP is extremely bright, and variants with enhanced brightness continue to be isolated or designed/bio-engineered. Moreover, spectral variants of GFP and additional reporters such as DsRed are available for multiple labelling of living cells. These probes have revolutionised the ability to image living cells and tissues using light microscopy (Figures 11.14 and 11.17).

11.4.5 Multi-Dimensional Imaging

The collection of Z-series over time is called four-dimensional (4D) imaging, where individual optical sections (x- and y-dimensions) are collected at different depths in the specimen (z-dimension) at different times (the fourth dimension), thus resulting in one time and three space dimensions (Figure 11.18). Moreover, multiple wavelength images can also be collected over time. This approach has been called 5D imaging. Software is available for the analysis and display of such 4D and 5D datasets. For example, the movement of a structure through the consecutive stacks of images can be traced, changes in volume of a structure can be measured, and the 4D datasets can be displayed as series of Z-projections or stereo movies. Multi-dimensional experiments can present problems with respect to handling large amounts of data, since it is not unusual to acquire gigabytes of information for a single 4D imaging experiment.

11.4.6 Scanned-Light-Sheet Microscopy

This method uses a thin sheet of laser light for optical sectioning with an objective lens and CCD camera detector system oriented perpendicular to it. The technique was developed to improve the penetration of living specimens and enables the imaging of live samples from many different angles at a cellular resolution. This approach is used by selective plane illumination microscopy (SPIM) where the specimen itself is rotated in the beam. Advantages of the technique include extremely low photo damage and high acquisition speed. It has been used to image every nucleus in zebrafish embryos over 24 hours of development at stunning resolution.

11.4.7 Super-Resolution Methods

Several advanced methods are now breaking the theoretical resolution limit of the light microscope, as first proposed by Abbe in the early 1800s. Up until recently, the dogma was that the limit of resolution of the light microscope was dependent on the wavelength of light used, and was fixed at around 0.5 µm. Better resolution could only be achieved using electron microscopy, and, consequently, only fixed/non-living specimens could be imaged at higher resolution.

New super-resolution light microscopes are able to achieve resolutions down to 0.1 µm in the lateral dimension and 0.6 µm in the axial direction, and in living cells. Such instruments include fluorescence photoactivation localisation microscopes

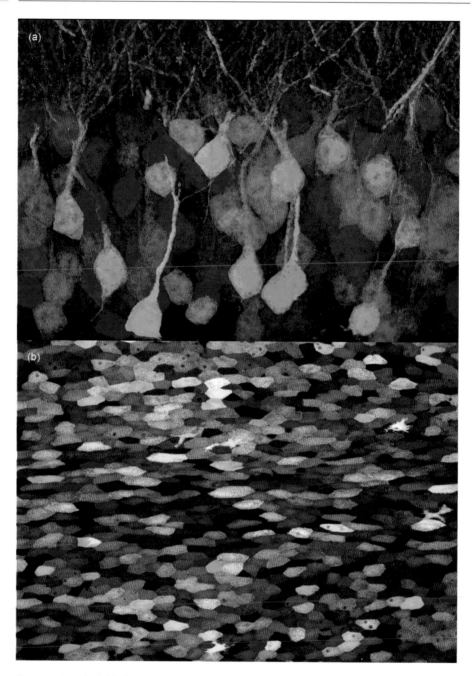

Figure 11.17 Multiple labelling in the living brain using the **Brainbow** technique. Unique colour combinations in individual neurons are achieved by the relative levels of three or more fluorescent proteins (XFPs – spectral variants of GFP). The images are collected using a multi-channel laser-scanning confocal microscope. Up to 90 different colours (neurons) can be distinguished using this technique.

(a) Hippocampus. (Image courtesy of Jeff Lichtman, Harvard University, USA.)

(b) Multiple labelling in living zebrafish skin using the **Fishbow** technique – developed from the Brainbow technique. Every cell on the surface of a zebrafish glows in a slightly different colour hue when viewed in a fluorescence light microscope. This method effectively gives each cell a unique identity, like a living bar code, that allows tracing over the eight-day lifespan. It is thus possible to simultaneously track hundreds of cells during the development and regeneration of fish tissues. (Chen-Hui Chen and Kenneth Poss, Department of Cell Biology, Duke University Medical Center, Durham, NC.)

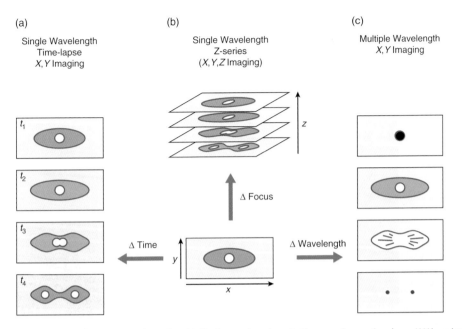

Figure 11.18 Multi-dimensional imaging. (a) Single wavelength excitation over time or time-lapse X,Y imaging; (b) Z-series or X,Y,Z imaging. The combination of (a) and (b) results in 4D imaging. (c) Multiple wavelength imaging. The combination of (a), (b) and (c) comprises a 5D imaging experiment.

(FPALMs), stimulated emission depletion microscopes (STEDs), stochastic optical reconstruction microscopes (STORMs) and 3D structured illumination microscopes (SIMs). All of these methodologies make use of the excitation and emission properties of fluorescent probes, together with imaginative instrumentation and computer programs to break the diffraction limit. They provide valuable information on macromolecular dynamics in living cells that are not available from conventional microscopy techniques (see, for example, Figure 11.19).

11.4.8 Whole-Animal Methods

Various instruments have been designed over the years for imaging cells in living animals. There are two main approaches; mini microscopes that can be mounted on an animal for long-term observations or hand-held probes that can be pressed against an animal for immediate diagnostic imaging. This continues to be an area of ongoing efforts, mainly resting on the development of new lenses for efficient light capturing and fibre-based endoscopes that can capture the signal in vivo .

11.5 MEASURING CELLULAR DYNAMICS

Understanding the function of proteins within the context of the intact living cell is one of the main aims of contemporary biological research. The visualisation of specific cellular events has been greatly enhanced by modern microscopy. In addition to qualitatively viewing the images collected with a microscope, quantitative

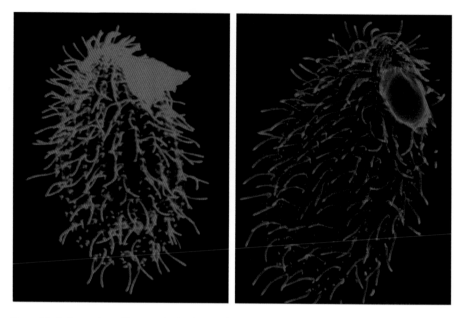

Figure 11.19 Comparison of images obtained by confocal and super-resolution microscopy. *Tetrahymena* cells labelled with antitubulin polyglycylation antibody and rhodamine secondary antibody. The confocal microscopy image (left) shows a uniform signal throughout the cilia, whereas the super-resolution microscopy image (right) shows that poly-glycylated tubulin is present in the outer doublets of the cilia only and not in the central pair microtubules. The cilia show two lines of poly-glycylated tubulin, which are the doublet microtubules. The central space occupied by the central singlet microtubule lacks the red poly-glycylated tubulin signal. (Image kindly provided by Mayukh Guha and Jacek Gaertig, University of Georgia, USA).

information can be gleaned from the images. The collection of meaningful measurements has been greatly facilitated by the advent of digital image processing. Subtle changes in intensity of probes of biochemical events can be detected with sensitive digital detectors. These technological advancements have allowed insight into the spatial aspects of molecular mechanisms.

Relatively simple measurements include counting features within a 2D image or measuring areas and lengths. Measurements of depth and volume can be made in 3D, 4D and 5D datasets. Images can be calibrated by collecting an image of a calibration grid at the same settings of the microscope that was used for collecting the images during the experiment. Many computer image-processing systems allow for a calibration factor to be added into the program, and all subsequent measurements will then be comparable.

The rapid development of fluorescence microscopy, together with digital imaging and, above all, the development of fluorescent probes of biological activity, such as GFP and DsRed, have enabled a new level of sophistication into quantitative imaging. Most of the measurements are based on the ability to accurately measure the brightness of and the wavelength emitted from a fluorescent probe within a sample using a digital imaging system. This is also the basis of flow cytometry (see Chapter 8), which measures the individual brightness of a population of cells as they pass through a laser beam. Cells can be sorted into different populations using a related technique, fluorescence-activated cell sorting (see Sections 7.8, 14.6.3).

The brightness of the fluorescence emitted by the probe can be calibrated to the amount of probe present at any given location in the cell at high resolution. For example, the concentration of calcium is measured in different regions of living embryos using calcium indicator dyes, such as fluo-3, whose fluorescence increases in proportion to the amount of free calcium in the cell (Figure 11.20). Many probes have been developed for making such measurements in living tissues. Controls are a necessary part of such measurements, since photobleaching and various dye arte-facts during the experiment can obscure the true measurements. This can be achieved by staining the sample with two ion-sensitive dyes, and comparing their measured brightness during the experiment. Such measurements are usually expressed as ratios (ratio imaging) and provide a control for dye loading problems, photobleaching and instrument variation.

Fluorescence recovery after photobleaching (FRAP; see also Section 13.5.3) uses the high light flux from a laser to locally destroy fluorophores labelling the macro-molecules to create a bleached zone (photobleaching). The observation and recording

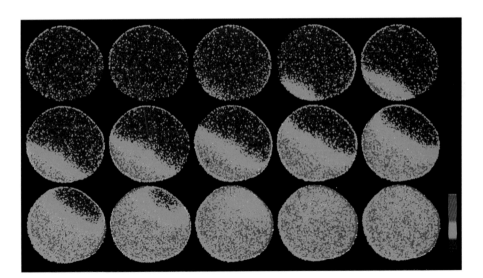

Figure 11.20 Calcium imaging in living cells. A fertilisation-induced calcium wave in the egg of the starfish. The egg was microinjected with the calcium-sensitive fluorescent dye fluo-3 and subsequently fertilised by the addition of sperm during observation using time-lapse confocal microscopy with a 40× water immersion lens and a laser-scanning confocal microscope. An optical section located near the egg equator was collected every 4 seconds using the normal scan mode accumulated for two frames. Afterwards, the images were corrected for offset and subjected to ratio generation by linearly dividing the initial pre-fertilisation image into each successive frame of the time-lapse run. After ratio generation, the images were prepared as a montage and rendered with a pseudo-colour look-up table in which blue regions represent low ratios of free calcium levels, and red areas depict high ratios of free calcium levels. Note that the wave sweeps through the entire ooplasm, rather than being cortically restricted. (Image kindly provided by Steve Stricker, University of New Mexico, USA.)

of the subsequent movement of undamaged fluorophores into the bleached zone gives a measure of molecular mobility. This enables biochemical analysis within the living cell. A second technique related to FRAP, **photoactivation**, uses a probe whose fluorescence can be induced by a flash of short-wavelength light (typically UV). The method depends upon **caged fluorescent probes** that are locally activated (uncaged) by a pulse of UV light. Alternatively, variants of GFP can be expressed in cells and selectively photoactivated. The activated probe is imaged using a longer wavelength of light. Here, the signal-to-noise ratio of the images can be better than that for photobleaching experiments.

A third method, **fluorescence speckle microscopy** was discovered as a chance observation while microinjecting fluorescently labelled proteins into living cells at a very low concentration. Since the fluorescently labelled protein participates in the same functions as the endogenous (and thus non-labelled) protein, the distribution of the fluorophore is 'diluted' and, when viewed in the microscope, structures inside cells that have been labelled in this way have a speckled appearance. The dark regions act as fiduciary marks for the observation of dynamics.

Fluorescence resonance energy transfer (FRET; see also Section 13.5.3), is a fluorescence-based method that can take fluorescence microscopy past the theoretical resolution limit of the light microscope, allowing the observation of protein–protein interactions in vivo (Figure 11.21). FRET occurs between two fluorophores when the emission of the first one (the donor) serves as the excitation source for the second one (the acceptor). This will only occur when two fluorophore molecules are very close to one another, at a distance of 60 Å or less.

As an example, the complex formation between two proteins can be monitored using FRET. One of the two proteins is tagged with cyan fluorescent protein (CFP), the other with yellow fluorescent protein (YFP). The excitation wavelength is chosen

(a)

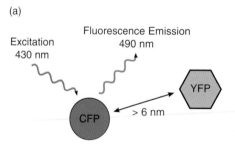

(b)

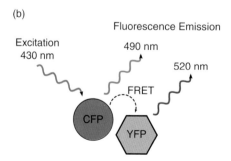

Figure 11.21 Fluorescence resonance energy transfer (FRET). In the situation shown in (a), the donor cyan fluorescent protein (CFP) and the acceptor yellow fluorescent protein (YFP) are not close enough for FRET to occur (more than 6 nm separation). Therefore, excitation with light of a wavelength of 430 nm (blue) results in the emission of light at 490 nm (green) – the fluorescence emission of CFP. In the situation shown in (b), CFP and YFP are close enough for energy transfer to occur (closer than 60 Å). Here, excitation with light at 430 nm (blue) results in fluorescence emission of CFP (490 nm; green) and of YFP (520 nm; red).

Example 11.1 **LOCATING AN UNKNOWN PROTEIN TO A SPECIFIC CELLULAR COMPARTMENT**

Question You have isolated and purified a novel protein from a biochemical preparation. How might you determine its subcellular distribution and possible function in the cell?

Answer Many fluorescent probes are available that label specific cellular compartments. For example, TOTO3 labels the nucleus and fluorescent phalloidins label cell outlines. An antibody to your protein could be raised and used to label cells by immunofluorescence. Using a multiple-labelling approach and perhaps an optical-sectioning technique such as laser-scanning confocal microscopy, the distribution of the protein in the cell relative to known distributions can be ascertained. For higher resolution immuno-electron microscopy or FRET studies could be performed.

---

such that CFP is excited. If the two proteins of interest are spatially close, so are their fusion partners, and energy is transferred from the excited CFP to YFP. Fluorescence emission is then observed at the emission wavelength of YFP.

A more complex technique, fluorescence lifetime imaging (FLIM) measures the amount of time a fluorophore is fluorescent after excitation with a 10 ns pulse of laser light. FLIM is a method used for detecting multiple fluorophores with different fluorescent lifetimes and overlapping emission spectra.

11.6 THE ELECTRON MICROSCOPE

11.6.1 Principles

Electron microscopy is used when the greatest resolution is required, and when the living state can be ignored. The images produced in an electron microscope reveal the so-called ultrastructure of cells. There are two different types of electron microscope – the transmission electron microscope (TEM) and the scanning electron microscope (SEM). In the TEM, electrons that pass through the specimen are imaged. In contrast, in the SEM, electrons that are reflected back from the specimen (secondary electrons) are collected, and the surfaces of specimens are imaged.

The equivalent of the light source in an electron microscope is the electron gun. When a high voltage of between 40 000 and 100 000 volts (the accelerating voltage) is clamped between the cathode and the anode, a tungsten filament emits electrons (Figure 11.1). The negatively charged electrons pass through a hole in the anode, forming an electron beam. The beam of electrons passes through a stack of electro-magnetic lenses (referred to as the column). Focussing of the electron beam is achieved by changing the voltage across the electromagnetic lenses. When the electron beam passes through the specimen, some of them are scattered, while others are focussed by the projector lens onto the detector. The detector can be a phosphorescent screen, photographic film or a digital camera. Since electrons have a limited penetration power, specimens for TEM must be thin (50–100 nm) to allow them to pass through.

Thicker specimens can be viewed by using a higher accelerating voltage, for example in the intermediate-voltage electron microscope (IVEM) or the high-voltage electron microscope (HVEM), which use acceleration voltages of 400 000 V and 1 MV (= 10^6 V), respectively. Stereo images can be collected by acquiring two images at 8–10° tilt angles. Such images are useful in assessing the 3D relationships of organelles within cells when viewed in a stereoscope or with a digital stereo-projection system.

11.6.2 Preparation of Specimens

Contrast in the electron microscope (EM) depends on the atomic number; the higher the atomic number, the greater the scattering and the contrast. Therefore, heavy metals such as uranium, lead and osmium are used to add contrast in the EM. Labelled structures possess high electron density and appear black in the image (Figure 11.22).

The major drawback of EM observation of biological specimens therefore is the non-physiological conditions necessary for their observation. All liquid has to be removed from the specimen before it can be imaged in the EM; for biological specimens, this means that all water needs to be removed. This is because the electron beam can only be produced and focussed in a vacuum, which causes any liquids present to boil. Nevertheless, the improved resolution afforded by the EM has provided much information about biological structures and biochemical events within cells that could otherwise have not been collected using any other microscopical technique. Extensive specimen preparation is required for EM analysis, and, for this reason, there can be issues of interpreting the images because of artefacts from specimen preparation.

Specimens for the TEM are traditionally prepared by fixation in glutaraldehyde to cross-link proteins, followed by osmium tetroxide soaking to fix and stain lipid membranes. Subsequent dehydration is achieved by immersion of the sample in a series of alcohols to remove the water, and then embedding in a plastic such as Epon for thin sectioning (Figure 11.22).

Small pieces of the embedded tissue are sectioned using an ultramicrotome with either a glass or a diamond knife. Ultrathin sections are cut to a thickness of approximately 60 nm. The ribbons of sections are floated onto the surface of water and their interference colours are used to assess their thickness. The desired 60 nm section thickness has a silver/gold-like interference colour on the water surface. The sections are then mounted onto copper or gold EM grids, and are subsequently stained with heavy metals, for example uranyl acetate and lead citrate.

For the SEM, samples are fixed in glutaraldehyde, dehydrated through a series of solvents and dried completely either in air or by critical-point drying. The latter method removes all of the water from the specimen instantly and avoids surface tension in the drying process, thereby avoiding artefacts of drying. The specimens

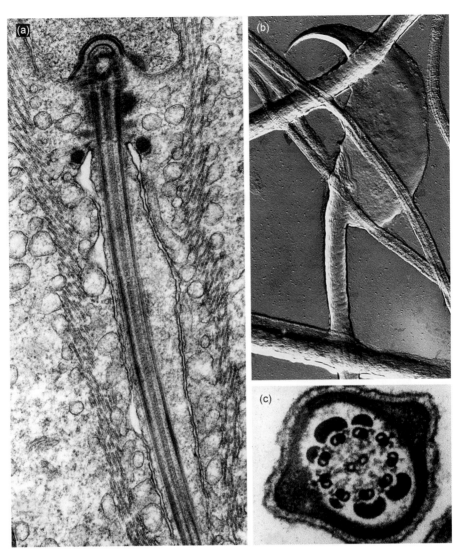

Figure 11.22 Transmission electron microscopy (TEM). (a), (c) Ultrathin Epon sections (60 nm thick) of developing rat sperm cells stained with uranyl acetate and lead citrate. (b) Carbon surface replica of a mouse sperm.

are then mounted onto a special metal holder or stub and coated with a thin layer of gold before viewing in the SEM (Figure 11.23). Surfaces can also be viewed in the TEM using either negative stains or carbon replicas of air-dried specimens (Figure 11.22).

Immuno–electron microscopy methods allow the localisation of molecules within the cellular micro-environment for TEM and on the cell surface for SEM (Figure 11.24). Cells are prepared in a similar way to that for indirect immunofluorescence, with the exception that rather than a fluorescent probe bound to the secondary antibody electron, dense colloidal gold particles are used. Multiple labelling can be achieved using different sizes of gold particles (up to 10 nm) attached to antibodies to the proteins of

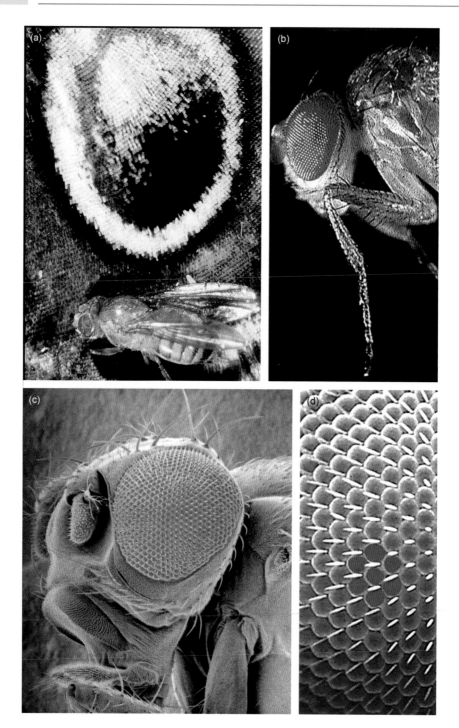

Figure 11.23 Image surfaces using the light microscope (stereomicroscope) (a, b) and the electron microscope (scanning electron microscope) (c, d). (a) A stereomicroscope view of a fly (*Drosophila melanogaster*) on a butterfly wing (*Precis coenia*); (b) zoomed in to view the head region of the red-eyed fly. (c) SEM image of a similar region of the fly's head; (d) zoomed more to view the individual ommatidia of the eye. Note that the stereomicroscope images can be viewed in real colour, whereas those produced using the SEM are in grey scale. Colour can only be added digitally to EM images (d). (Images (b), (c) and (d) kindly provided by Georg Halder, VIB Centre for the Biology of Disease, Leuven, Belgium.)

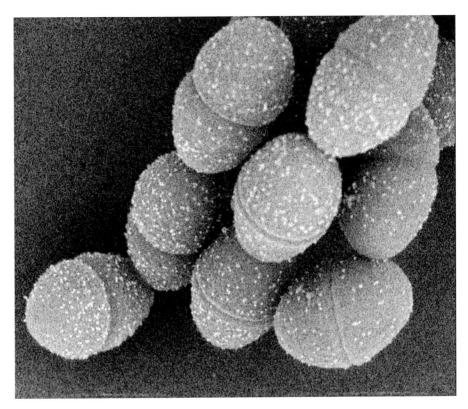

Figure 11.24 Immuno-electron microscopy. SEM of microbes *Enterococcus faecalis* labelled with 10 nm colloidal gold for the surface adhesion protein 'aggregation substance'. This protein facilitates exchange of DNA during conjugation. The gold labels appear as white dots on the surface of the bacteria. (Image kindly provided by the late Stan Erlandsen, University of Minnesota, USA.)

interest. The method depends upon the binding of protein A to the gold particles; protein A, in turn, binds to antibody fragments. Certain resins, for example Lowicryl and LR White, have been formulated to allow antibodies and gold particles to be attached to ultrathin sections for immunolabelling.

11.6.3 Electron Tomography

New methods of fixation continue to be developed in an attempt to avoid the artefacts of specimen preparation and to observe the specimen more closely to its living state. Specimens are rapidly frozen in milliseconds by high-pressure freezing. Under these conditions, the biochemical state of the cell is more likely to be preserved. Many of such frozen hydrated samples can be observed directly in the electron microscope or they can be chemically fixed using freeze substitution methods. For this latter method, fixatives are infused into the preparation at low temperature, after which the specimen is slowly warmed to room temperature.

With **cryo-electron tomography** (Cryo-ET), the 3D structure of cells and macromolecules can be visualised at 5–8 nm resolution. Typically, cells are rapidly frozen, fixed by freeze substitution and embedded in epoxy resin. Thick 200 nm sections are cut and imaged in the TEM equipped with a tilting stage. A typical tilt series of 100 or so images is collected in a digital form and exported to a computer reconstruction program for analysis. The collection of 2D digital EM images is then converted into a high-resolution 3D representation of the specimen (Figure 11.25). The method is especially useful for imaging the fine connections within cells, especially the cytoskeleton and nuclear pores (Figure 11.26) and elucidating the surface structures of viruses.

11.6.4 Integrated Microscopy

When preparing a sample for microscopic studies, it is possible to use the same specimen for viewing in the light microscope and subsequently in the electron microscope. This approach is called integrated microscopy. The correlation of images of

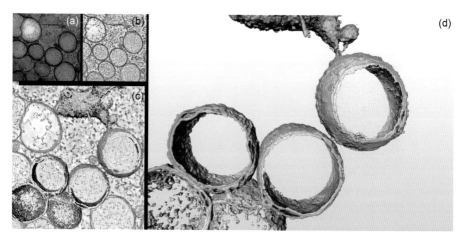

Figure 11.25 Electron tomography revealing the interconnected nature of SARS-Coronavirus–induced double-membrane vesicles. Monkey kidney cells were infected with SARS-Coronavirus in a Biosafety Level-3 laboratory and pre-fixed using 3% paraformaldehyde at 7 hours post-infection. Subsequently, the cells were rapidly frozen by plunge-freezing and freeze substitution was performed at low temperature, using osmium tetraoxide and uranyl acetate in acetone to optimally preserve cellular ultrastructure and gain maximal contrast. After washing with pure acetone at room temperature, the samples were embedded in an epoxy resin and polymerised at 60 °C for 2 days. Using an ultramicrotome, 200 nm thick sections were cut, placed on a 100 mesh EM grid, and used for electron tomography. (a) To facilitate the image alignment that is required for the final 3D reconstruction, a suspension of 10 nm gold particles was layered on top of the sections as fiducial markers. The scale bar represents 100 nm. Images were recorded with an FEI T12 transmission electron microscope operating at an acceleration voltage of 120 kV. A tilt series consisted of 131 images recorded using 1° tilt increments between −65° and 65°. For dual-axis tomography, which improves resolution in the x- and y-directions, the specimen was rotated 90° around the z-axis and a second tilt series was recorded. (b) This image shows a single tomogram slice through the 3D reconstruction with a digital thickness of 1.2 nm. To compute the final electron tomogram, the dual-axis tilt series was aligned by means of the fiducial markers using the IMOD software package. (c) The 3D surface-rendered reconstruction of viral structures and adjacent cellular features were generated by thresholding and subsequent surface rendering using the AMIRA Visualization Package (TGS Europe). (d) The final 3D surface-rendered model shows interconnected double-membrane vesicles (outer membrane, gold; inner membrane, silver) and their connection to an endoplasmic reticulum stack (depicted in bronze). (Images kindly provided by Kevin Knoops and Eric Snijder, Leiden University, The Netherlands.)

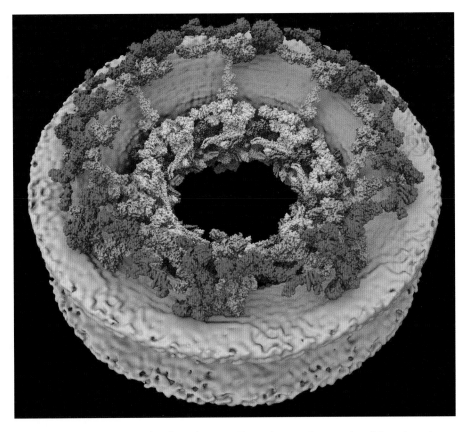

Figure 11.26 Electron tomography of a nuclear pore. The nuclear pore is a complex of three rings of proteins that allows the passage of materials in both directions between the nucleus and the cytoplasm. The image of the nuclear pore was produced using cryo-electron microscopy and computer modelling. (Jan Kosinski and Martin Beck, Structural and Computational Biology Unit, EMBL, Heidelberg, Germany.)

the same cell collected using the high temporal resolution of the light microscope and the high spatial resolution of the electron microscope adds additional information to imaging compared to using the two techniques separately (Figure 11.27). The integrated approach also addresses the problem of artefacts. In this context, there are suitable probes available that are fluorescent in the light microscope and provide electron-dense features in the electron microscope.

11.7 IMAGE MANAGEMENT

Most images produced by any kind of modern microscope are collected in a digital form. In addition to greatly speeding up the collection of the images (and experiment times), the use of digital imaging has allowed the use of digital image databases and the rapid transfer of information between laboratories across the World Wide Web. Moreover there is no loss in resolution or colour balance from the images collected at the microscope as they pass between laboratories and journal web pages.

11.7.1 Image Storage

It is not advisable to store image files on the computer hard disk of the microscope for a long period of time or even on a server, due to the space limitations caused by large image files. Additionally, single hard drives should not be relied upon in case

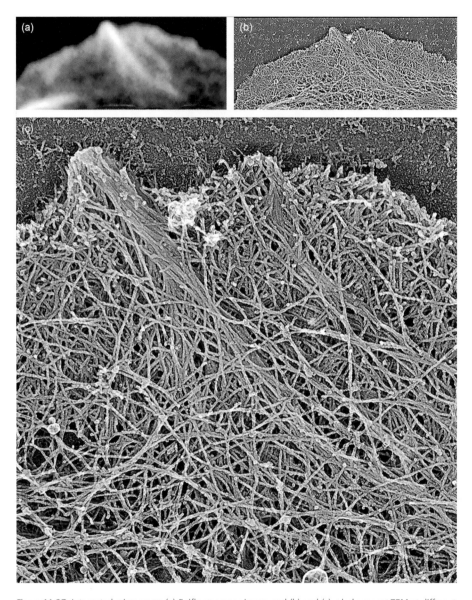

Figure 11.27 Integrated microscopy. (a) Epifluorescence image and (b) and (c) whole-mount TEM at different magnifications of the same cell. The fluorescence image is labelled with rhodamine phalloidin, which stains polymerised actin. A stress fibre at the periphery of the cell appears as a white line in the fluorescence image (a), and when viewed in the TEM, the stress fibres appear as aligned densities of actin filaments. The TEM whole mount was prepared using detergent extraction, chemical fixation, critical-point drying and platinum/carbon coating. (Image kindly provided by Tatyana Svitkina, University of Pennsylvania, USA.)

Example 11.2   **CHOOSING A MICROSCOPE FOR A SPECIFIC APPLICATION**

Question   Which microscope and contrast methodology might you use for the following applications?

1. Fixed and fluorescently labelled zebrafish embryos.
2. Sonicated cell extract to quickly assess the extent of cellular damage.
3. Macromolecular structure of a virus.
4. Screen GFP-labelled gene expression in living fruit flies.
5. Fixed and stained section of liver biopsy to check for cancer cells.
6. Surface structure of yeast cells.
7. Macromolecular interactions in a living cell.
8. Swimming spermatozoa.
9. Colourful surface structure of a butterfly wing.
10. Imaging deep into a living GFP-labelled developing zebrafish embryo over time.

Answer

1. Epifluorescence light microscopy.
2. Light microscopy – dark-field or phase-contrast optics.
3. Electron microscope tomography.
4. Fluorescence stereomicroscopy.
5. Light microscopy – bright-field optics.
6. Scanning electron microscopy.
7. Super-resolution light microscopy.
8. Dark-field, phase-contrast or differential-interference contrast light microscopy.
9. Stereomicroscopy.
10. Multiple-photon or light-sheet microscopy.

of sudden drive failure, causing irreversible loss of data. A professional storage and archive solution should be sought that would involve storing the data on RAID arrays at multiple locations. Current possibilities to store data in the cloud offer convenient, secure and affordable solutions that can allow backup of data as soon as they are generated.

11.7.2 Image Informatics

While much emphasis is usually placed on the development of specimen preparation techniques and microscope instrumentation for collecting images, the ability to cope with large numbers of images and correspondingly large datasets should be kept in mind when designing experiments. With the increasing use of digital image-capture microscopy, it has become a major challenge to locate, view and interpret large numbers of images.

Multi-dimensional images, such as 4D images from multi-focal plane time-lapse recordings, present a challenge in keeping track and archiving. Without careful organisation, important research data can be difficult or impossible to find, much less visualise and analyse effectively. This need has spawned the field of image informatics to develop tools to aid in the management, sharing, visualisation and analysis of datasets collected using many different imaging systems. An example of an image informatics platform is the Open Microscopy Environment (OME), a consortium of companies and academics with the mission of developing open-source tools for biological image data management.

11.7.3 Further Information

More detailed information on any of the microscopes and their applications in biochemistry and molecular biology can found on the World Wide Web. Several websites have been included as starting points for further study (Section 11.8.4) together with a list of video resources (Section 11.8.5).

11.8 SUGGESTIONS FOR FURTHER READING

11.8.1 Experimental Protocols

Abramowitz M. (2003) *Microscope Basics and Beyond*, Olympus of America Inc., Melville, New York, USA.

Chen C.H., Puliafito A., Cox B.D., *et al.* (2016) Multicolor cell barcoding technology for long-term surveillance of epithelial regeneration in zebrafish. *Developmental Cell* 36, 668–680. (Details of the FISHBOW technique).

Chung K. and Deisseroth K. (2013) CLARITY for mapping the nervous system. *Nature Methods* 10, 508–513.

Keller P.J., Schmidt A.D., Wittbrodt J. and Stelzer E.H. (2008) Reconstruction of zebrafish early embryonic development by scanned light sheet microscopy. *Science* 322, 1065–1069. (Application of scanning light microscopy to image living zebrafish embryos – stunning movies of zebrafish embryogenesis available online).

Knoops K., Kikkert M., van den Worm S.H.E., *et al.* (2008) SARS-Coronavirus replication is supported by a reticulovesicular network of modified endoplasmic reticulum. *PloS Biology* 6, 1957–1974. (Cryo-electron tomography in action).

Lichtman J.W. and Fraser S.E. (2001) The neuronal naturalist: watching neurons in their native habitat. *Nature Neuroscience* 4, 1215–1220.

Livet J., Weissman T.A., Kang H., *et al.* (2007) Transgenic strategies for combinatorial expression of fluorescent proteins in the nervous system. *Nature* 450, 56–62. (Imaginative use of reporter gene technology to label multiple neurons in living brains).

Stewart M.P., Sharei A., Ding X., *et al.* (2016) In vitro and *ex vivo* strategies for intracellular delivery. *Nature* **538**, 183–192.

Swedlow J.R., Lewis S.E. and Goldberg I.G. (2006) Modelling data across labs, genomes, space and time. *Nature Cell Biology* **8**, 1190–1194.

Van Roessel P. and Brand A.H. (2002) Imaging into the future: visualizing gene expression and protein interaction with fluorescent proteins. *Nature Cell Biology* **4**, E15–E20.

Volpi E.V. and Bridger J.M. (2008) FISH glossary: an overview of the fluorescence *in situ* hybridization technique. *BioTechniques* **45**, 385–409.

Wallace W., Schaefer L.H. and Swedlow J.R. (2001) Working person's guide to deconvolution in light microscopy. *BioTechniques* **31**, 1076–1097.

11.8.2 General Texts

Alberts B. and Johnson A. (2014) *Molecular Biology of the Cell*, 6th Edn. Walter Garland Science, New York, USA.

Cornea A. and Conn P.M. (2014) *Fluorescence Microscopy: Super-Resolution and other Novel Techniques*, Academic Press, Cambridge, MA, USA.

Cox G.C. (2012) *Optical Imaging Techniques in Cell Biology*, CRC Press, Boca Raton, FL, USA.

Goldman R.D. and Spector D.L. (2004) *Live Cell Imaging: A Laboratory Manual*, Cold Spring Harbor Laboratory Press, Cold Spring Harbor, NY, USA.

Matsumoto B. (2010) *Practical Digital Photomicrography: Photography Through the Microscope for the Life Sciences*, Rocky Nook, Santa Barbara, CA, USA.

Mueller-Reichert T. (2010) *Electron Microscopy of Model Systems*, Methods in Cell Biology, Vol. 96, 1st Edn., Academic Press, Cambridge, MA, USA.

Murphy D.B. and Davidson M.W. (2013) *Fundamentals of Light Microscopy and Electronic Imaging*, 2nd Edn., Wiley-Blackwell, Hoboken, NJ, USA.

Paddock S.W. (2014) *Confocal Microscopy: Methods and Protocols*, Methods in Molecular Biology (Vol 1075), 2nd Edn., Springer, New York, USA.

Sedgewick J. (2008) *Scientific Imaging with Photoshop: Methods, Measurement, and Output*, Pearson Education, Peachpit Press, Berkeley, CA, USA. (Practical manual on the use of PhotoShop for measuring and preparing images for publication).

Spector D.L. and Goldman R.D. (2006) *Basic Methods in Microscopy*, Cold Spring Harbor Laboratory Press, Cold Spring Harbor, NY, USA.

Watkins S. and St. Croix C. (2013) *Methods and Applications in Microscopy and Imaging*, 1st Edn., Current Protocols, John Wiley & Sons Inc., Hoboken, NJ, USA.

11.8.3 Review Articles

Afzelius B.A. and Maunsbach A.B. (2004) Biological ultrastructure research: the first 50 years. *Tissue Cell* **36**, 83–94.

Baker M. (2010) Whole-animal imaging: the whole picture. *Nature* **463**, 977–980.

Baumeister W. (2004) Mapping molecular landscapes inside cells. *Biological Chemistry* **385**, 865–872. (Review of electron tomography).

Chen F., Tillberg P.W. and Boyden E.S. (2015) Expansion microscopy. *Science* 347, 543–548.

Dunn G.A. and Jones G.E. (2004) Cell motility under the microscope: Vorsprung durch Technik. *Nature Reviews Molecular Cell Biology* 5, 667–672. (Review of techniques used to study cell motility).

Eliceiri K.W., Berthold M.R., Goldberg I.G., *et al.* (2012) Biological imaging software tools. *Nature Methods* 9, 697–710.

Giepmans B.N.G., Adams S.R., Ellisman M.H. and Tsien R.Y. (2006) The fluorescent toolbox for assessing protein location and function. *Science* 312, 217–224.

Liu Z. and Keller P. J. (2016) Emerging and genomic tools for developmental systems biology. *Developmental Cell* 36, 597–610.

McIntosh R., Nicastor D. and Mastronarde D. (2005) New views of cells in 3D: an introduction to electron tomography. *Trends in Cell Biology* 15, 43–51.

Pampaloni F., Reynaud E.G. and Stelzer E.H.K. (2007) The third dimension bridges the gap between cell culture and live tissue. *Nature Reviews Molecular Cell Biology* 8, 839–845.

Zhang J., Campbell R.E., Ting A.Y. and Tsien R.Y. (2002) Creating new fluorescent probes for cell biology. *Nature Reviews Molecular Cell Biology* 3, 906–918.

11.8.4   Websites

The Brainbow technique (Harvard University)
cbs.fas.harvard.edu/science/connectome-project/brainbow (accessed May 2017)

CLARITY Resource Center
http://clarityresourcecenter.org/ (accessed May 2017)

Electron microscopy (John Innes Centre)
www.jic.ac.uk/microscopy/intro_EM.html (accessed May 2017)

Expansion microscopy
expansionmicroscopy.org/ (accessed May 2017)

History of the microscope
www.history-of-the-microscope.org/ (accessed May 2017)

Microscopy links by The Wellcome Trust Centre for Human Genetics
www.well.ox.ac.uk/external-website-links (accessed May 2017)

Microscopy Resource Center (Olympus)
www.olympusmicro.com/primer/opticalmicroscopy.html (accessed May 2017)

Microscopy Society of America
www.msa.microscopy.org (accessed May 2017)

MicroscopyU (educational website by Nikon)
www.microscopyu.com/ (accessed May 2017)

Molecular expressions: Images from the Microscope (Florida State University)
micro.magnet.fsu.edu/ (accessed May 2017)

The Open Microscopy environment
www.openmicroscopy.org (accessed May 2017)

Royal Microscopical Society
www.rms.org.uk (accessed May 2017)

Structure and function of large macromolecular assemblies (Beck Group at EMBL)
www.embl.de/research/units/scb/beck/ (accessed May 2017)

Super-resolution microscopy
www.sciencemag.org/custom-publishing/technology-features/superresolution-microscopy (accessed May 2017)
zeiss-campus.magnet.fsu.edu/articles/superresolution/introduction.html (accessed May 2017)

11.8.5 Online Video Resources

The electron microscope (Colin Humphreys)
www.youtube.com/watch?v=Kaz4fqSze60 (accessed May 2017)

Electron tomography of mitochondria
www.youtube.com/watch?v=GXaqrNbf-NU (accessed May 2017)

Fluorescence microscopy introduction (Nico Stuurman)
www.youtube.com/watch?v=AhzhOzgYoqw (accessed May 2017)

Imaging life at high spatiotemporal resolution (Eric Betzig)
www.youtube.com/watch?v=mWdD0fRo4CE (accessed May 2017)

The scanning electron microscope
www.youtube.com/watch?v=GY9lfO-tVfE (accessed May 2017)

Seeing the invisible
www.hhmi.org/biointeractive/seeing-the-invisible (accessed May 2017)

Structured illumination microscopy (David Agard)
www.youtube.com/watch?v=i73HhpLJrqs (accessed May 2017)

12 Centrifugation and Ultracentrifugation

KAY OHLENDIECK AND STEPHEN E. HARDING

12.1 INTRODUCTION

Biological centrifugation is a process that uses centrifugal force to separate and purify mixtures of biological particles in a liquid medium. The smaller the particles, the higher the g-forces (see next section) required for the separation. It is a key technique for isolating and analysing cells, subcellular fractions, supramolecular complexes and, with higher g-force instruments or 'ultra'-centrifuges (up to 60 000 revolutions per minute corresponding to $\sim 200\,000{\times}g$) isolated macromolecules such as proteins or nucleic acids. Such high-speed devices require a vacuum to avoid overheating of samples. The development of the first analytical ultracentrifuge – with a specially designed optical system for monitoring and recording the sedimentation process – by Svedberg in the late 1920s and the technical refinement of the preparative centrifugation technique by Claude and colleagues in the 1940s positioned centrifugation technology at the centre of biological and biomedical research for many decades. Today, centrifugation techniques represent a critical tool for modern biochemistry and are employed in almost all invasive subcellular studies. While analytical ultracentrifugation is mainly concerned with the study of purified macromolecules or isolated supramolecular assemblies, preparative centrifugation methodology is devoted to the actual separation of tissues, cells, subcellular structures, membrane vesicles and other particles of biochemical interest.

Most undergraduate students will be exposed to preparative centrifugation protocols during practical classes and might also experience a demonstration of analytical centrifugation techniques. This chapter is accordingly divided into a short introduction into the theoretical background of sedimentation, an overview of practical aspects of using centrifuges in the biochemical laboratory, an outline of preparative centrifugation and a description of the usefulness of ultracentrifugation techniques in the biochemical characterisation of macromolecules. To aid in the understanding of the basic principles of centrifugation, the general designs of various rotors and separation processes are diagrammatically represented. Often, the learning process of undergraduate students is hampered by the lack of a proper linkage between theoretical knowledge and practical applications. To overcome this problem, the description of preparative centrifugation techniques is accompanied by an explanatory flow chart and the detailed discussion of the subcellular fractionation protocol for a specific tissue preparation. Taking the isolation of fractions from skeletal muscle homogenates as an example, the rationale behind individual preparative steps is explained. Since affinity isolation methods not only represent an extremely powerful tool in purifying biomolecules (see Chapter 5), but can also be utilised to separate intact organelles and membrane vesicles by centrifugation, lectin affinity agglutination of highly purified plasmalemmal vesicles from skeletal muscle is described. Traditionally, marker enzyme activities are used to determine the overall yield and enrichment of particular structures within subcellular fractions following centrifugation. As an example, the distribution of key enzyme activities in mitochondrial subfractions from liver is given. However, most modern fractionation procedures are evaluated by more convenient methods, such as protein gel analysis in conjunction with immunoblot analysis (Chapter 7). Miniature gel and blotting equipment can produce highly reliable results within a few hours, making it an ideal analytical tool for high-throughput testing. Since electrophoretic techniques are introduced in Chapter 6 and are used routinely in biochemical laboratories, the protein gel analysis of the distribution of typical marker proteins in affinity-isolated plasmalemma fractions is graphically represented and discussed.

Although monomeric peptides and proteins are capable of performing complex biochemical reactions, many physiologically important elements do not exist in isolation under native conditions. Therefore, if one considers individual proteins as the basic units of the proteome (see Chapter 21), protein complexes actually form the functional units of cell biology. This gives investigations into the supramolecular structure of protein complexes a central place in biochemical research. To illustrate this point, the sedimentation analysis of a high-molecular-mass membrane assembly, the dystrophin–glycoprotein complex of skeletal muscle, is shown and the use of sucrose gradient centrifugation explained.

Analytical ultracentrifugation – which unlike other analytical separation techniques does not require a separation medium, i.e. it is 'matrix-free' – has become a preferred or 'gold standard' technique for establishing the purity or homogeneity and state of aggregation of macromolecular or nanoparticle solutions, and to illustrate this point, we show how the purity of preparations of monoclonal antibodies can be routinely analysed with the modern ultracentrifuge, and how the inclusion of a density gradient, when appropriate, can enhance the resolution of the method even further.

12.2 BASIC PRINCIPLES OF SEDIMENTATION

From everyday experience, the effect of sedimentation due to the influence of the Earth's gravitational field ($G = g = 9.81$ m s^{-2}) versus the increased rate of sedimentation in a centrifugal field ($G > 9.81$ m s^{-2}) is apparent. To give a simple but illustrative example, crude sand particles added to a bucket of water travel slowly to the bottom of the bucket by gravitation, but sediment much faster when the bucket is swung around in a circle. Similarly, biological structures exhibit a drastic increase in sedimentation when they undergo acceleration in a centrifugal field. The relative centrifugal field is usually expressed as a multiple of the acceleration due to gravity.

Below is a short description of equations used in practical centrifugation classes. When designing a centrifugation protocol, it is important to keep in mind that:

- the more dense a biological structure is, the faster it sediments in a centrifugal field
- the more massive a biological particle is, the faster it moves in a centrifugal field
- the denser the biological buffer system is, the slower the particle will move in a centrifugal field
- the greater the frictional coefficient is, the slower a particle will move
- the greater the centrifugal force is, the faster the particle sediments
- the sedimentation rate of a given particle will be zero when the density of the particle and the surrounding medium are equal.

Biological particles moving through a viscous medium experience a frictional drag, whereby the frictional force acts in the opposite direction to sedimentation and equals the velocity of the particle multiplied by the frictional coefficient. The frictional coefficient depends on the size and shape of the biological particle. As the sample moves towards the bottom of a centrifuge tube in swing-out or fixed-angle rotors (see Section 12.3.2), its velocity will increase due to the increase in radial distance. At the same time, the particles also encounter a frictional drag that is proportional to their velocity. The frictional force of a particle moving through a viscous fluid is the product of its velocity and its frictional coefficient, and acts in the opposite direction to sedimentation.

When the conditions for the centrifugal separation of a biological particle are described, a detailed listing of rotor speed and radial dimensions of centrifugation has to be provided. Essentially, the rate of sedimentation, v, is dependent upon the applied centrifugal field G (measured in cm s^{-2}). G is determined by the radial distance, r, of the particle from the axis of rotation (in cm) and the square of the angular velocity, ω, of the rotor (in radians per second):

$$G = \omega^2 \times r \qquad\qquad\qquad (\text{Eq } 12.1)$$

The average angular velocity of a rigid body that rotates around a fixed axis is defined as the ratio of the angular displacement in a given time interval. One radian, usually abbreviated as 1 rad, represents the angle subtended at the centre of a circle by an arc with length equal to the radius of the circle. Since 360° equals 2π radians (or rad), one revolution of the rotor can be expressed as 2π rad. Accordingly, the angular velocity of the rotor, given in rad s^{-1}. Note that rad is treated as a scalar and is related to the rotor speed in revolutions per minute (rpm $\equiv$ 1 min^{-1}) by

$$\omega = 2\pi \text{ rad} \times rpm \qquad\qquad\qquad (\text{Eq } 12.2)$$

and therefore the centrifugal field can be expressed as:

$$G = 4\pi^2 \text{rad}^2 \times rpm^2 \times r \qquad\qquad\qquad (\text{Eq } 12.3)$$

where the variable rpm is the rotor speed (measured in revolutions per minute, i.e. the non-italicised 'rpm' denotes the unit) and r is the radial distance from the centre of rotation. Note that 60 revolutions per minute is the same speed as one revolution per second, i.e. $rpm = 60$ min$^{-1} = 1$ s^{-1}.

Example 12.1   CALCULATION OF CENTRIFUGAL FIELD

Question   What is the applied centrifugal field at a point equivalent to 5 cm from the centre of rotation and an angular velocity of 3000 rad s^{-1}?

Answer   The centrifugal field, G, at a point 5 cm from the centre of rotation may be calculated using Equation 12.1:

$$G = \omega^2 \times r = \left(3000\,\frac{rad}{s}\right)^2 \times 5\ cm = 4.5 \times 10^7\,\frac{rad^2\ cm}{s^2}$$

Example 12.2   CALCULATION OF ANGULAR VELOCITY

Question   For the pelleting of the microsomal fraction from a liver homogenate, an ultracentrifuge is operated at a speed of 40 000 rpm. Calculate the angular velocity, ω, in radians per second.

Answer   For a rotor speed of 40 000 rpm we obtain:

$$rpm = 40\,000\ rpm = \frac{40\,000}{1\,min} = \frac{40\,000}{60\,s}$$

The angular velocity may be calculated using Equation 12.2:

$$\omega = 2\pi \times s = 2\pi \times 40\,000\ min^{-1} = \frac{2\pi \times 40\,000}{1\,min} = \frac{2\pi \times 40\,000}{60\,s} = 4189\ rad\ s^{-1}$$

The centrifugal field is generally expressed in multiples of the earth's gravitational field, g (9.81 m s^{-2}). The relative centrifugal field RCF (or g-force) is the ratio of the centrifugal acceleration at a specified radius and the speed to the standard acceleration of gravity. The RCF can be calculated from the following equation:

$$RCF = \frac{G}{g} = \frac{4\pi^2\ rad^2 \times rpm^2 \times r}{g} \qquad \text{(Eq 12.4)}$$

RCF units are therefore dimensionless as they denote multiples of the gravitational constant g. Grouping numerical constants together leads to a more convenient form of Equation 12.4:

$$RCF = 1.12 \times 10^{-5} \times \frac{r}{1\,cm} \times \left(\frac{rpm}{1\,min^{-1}}\right)^2 \qquad \text{(Eq 12.5)}$$

with r given in cm. Although the relative centrifugal force can easily be calculated and can often be displayed on modern instruments, many centrifugation manuals contain a nomograph for the convenient conversion between relative centrifugal force and speed of the centrifuge at different radii of the centrifugation spindle to a point along

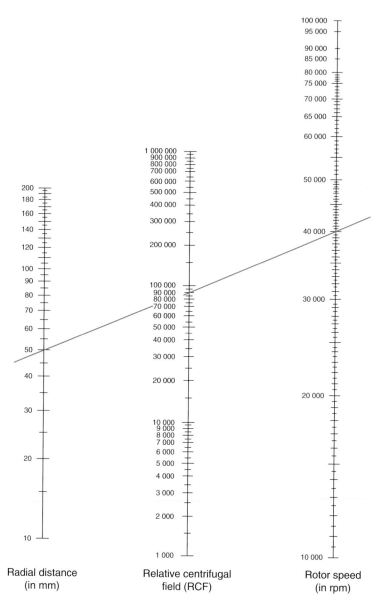

Radial distance   Relative centrifugal   Rotor speed
(in mm)   field (RCF)   (in rpm)

Figure 12.1 Nomograph for the determination of the relative centrifugal field for a given rotor speed and radius. The three columns represent the radial distance (in mm), the relative centrifugal field and the rotor speed *rpm* (in min^{-1}). For the conversion between relative centrifugal force and speed of the centrifuge spindle in revolutions per minute at different radii, draw a straight edge through known values in two columns. The desired figure can then be read where the straight edge intersects the third column. The example shown determines the relative centrifugal force for a 5 cm rotor operating at *rpm* = 40 000 min^{-1}, yielding an applied centrifugal field of about 90 000×g.

the centrifuge tube. A nomograph consists of three columns representing the radial distance (in mm), the relative centrifugal field and the rotor speed (in min⁻¹). For the conversion between relative centrifugal force RCF and speed of the centrifuge spindle (*rpm*, in min⁻¹) at different radii, a straight edge is aligned through known values in two columns, then the desired figure is read where the straight edge intersects the third column. See Figure 12.1 for an illustration of the usage of a nomograph.

In a suspension of biological particles, the rate of sedimentation (v) is dependent not only upon the applied centrifugal field, but also on the nature of the particle, i.e. its density ρ_p, its hydrodynamic radius R_{hydro}, (see Section 2.2.4) and also the density (ρ_m) and viscosity (η) of the surrounding medium. Stokes' law describes this relationship for the sedimentation of a rigid spherical particle:

$$v = \frac{2}{9} \times \frac{R_{hydro}^2 \times \left(\rho_p - \rho_m\right)}{\eta} \times g \qquad\qquad \text{(Eq 12.6)}$$

Example 12.3 CALCULATION OF RELATIVE CENTRIFUGAL FIELD

Question A fixed-angle rotor exhibits a minimum radius, r_{min}, at the top of the centrifuge tube of 3.5 cm, and a maximum radius, r_{max}, at the bottom of the tube of 7.0 cm. See Figure 12.2a for a cross-sectional diagram of a fixed-angle rotor illustrating the position of the minimum and maximum radius. If the rotor is operated at a speed of 20 000 rpm, what is the relative centrifugal field (RCF) at the top and bottom of the centrifuge tube?

Answer The relative centrifugal field may be calculated using Equation 12.5:

$$RCF = 1.12 \times 10^{-5} \times \frac{r}{1\,cm} \times \left(\frac{rpm}{1\,min^{-1}}\right)^2$$

At the top of the centrifuge tube:

$$RCF_{top} = 1.12 \times 10^{-5} \times \frac{3.5\,cm}{1\,cm} \times \left(\frac{20\,000\,min^{-1}}{1\,min^{-1}}\right)^2 = 15\,680$$

At the bottom of the centrifuge tube:

$$RCF_{bottom} = 1.12 \times 10^{-5} \times \frac{7.0\,cm}{1\,cm} \times \left(\frac{20\,000\,min^{-1}}{1\,min^{-1}}\right)^2 = 31\,360$$

This calculation illustrates that with fixed-angle rotors the centrifugal field at the top and bottom of the centrifuge tube might differ considerably, in this case exactly twofold.

where v is the sedimentation rate of the sphere, R_{hydro} is the radius of particle, ρ_p is the density of particle, ρ_m is the density of medium, g is the gravitational acceleration and η is the viscosity of the medium. R_{hydro} will depend on the shape of the particle. Following from Equation 12.6, the sedimentation time can be simply calculated as the ratio of distance sedimented and v.

Accordingly, a mixture of biological particles exhibiting an approximately spherical shape can be separated in a centrifugal field based on their density and/or their size. The time of sedimentation (in seconds) for a spherical particle is:

$$t = \frac{2}{9} \times \frac{\eta}{\omega^2 \times R_{hydro}^2 \times (\rho_p - \rho_m)} \times \ln \frac{r_b}{r_t}$$

(Eq 12.7)

where t is the sedimentation time, η is the viscosity of medium, R_{hydro} is the hydrodynamic radius of the particle, r_b is the radial distance from the centre of rotation to bottom of tube, r_t is the radial distance from the centre of rotation to liquid meniscus, ρ_p is the density of the particle, ρ_m is the density of the medium and ω is the angular velocity of the rotor.

The sedimentation rate or velocity of a biological particle can also be expressed as its sedimentation coefficient (s), whereby:

$$s = \frac{v}{\omega^2 \times r}$$

(Eq 12.8)

measured in the units of time (i.e. seconds).

Since the sedimentation rate per unit centrifugal field can be determined at different temperatures and with various media of different densities and viscosities, experimental values of the sedimentation coefficient are often corrected or 'normalised' for comparison purposes to standard solvent conditions. These standard conditions are the density and viscosity of water at 20.0 °C and the symbol used for this normalised sedimentation coefficient is $s_{20,w}$. Importantly, $s_{20,w}$ will depend on the size and shape or conformation of the macromolecule. The sedimentation coefficients (corrected or non-corrected) of biological macromolecules are relatively small, and are usually expressed as svedberg units denoted as S (see Section 12.5); one svedberg unit equals 10^{-13} s.

12.3 TYPES, CARE AND SAFETY ASPECTS OF CENTRIFUGES

12.3.1 Types of Centrifuges

Centrifugation techniques take a central position in modern biochemical, cellular and molecular biological studies. Depending on the particular application, centrifuges differ in their overall design and size. However, a common feature in all centrifuges is the central motor that spins a rotor containing the samples to be separated. Particles of biochemical interest are usually suspended in a liquid buffer system contained in specific tubes or separation chambers that are located in specialised rotors. The biological medium is chosen for the specific centrifugal application and may differ considerably between preparative and analytical approaches. As outlined below, the optimum pH value, salt concentration, stabilising cofactors and protective ingredients such as protease inhibitors have to be carefully evaluated in order to preserve biological function. The most obvious differences between centrifuges are:

- the maximum speed at which biological specimens are subjected to increased sedimentation
- the presence or absence of a vacuum
- the potential for refrigeration or general manipulation of the temperature during a centrifugation run
- the maximum volume of samples and capacity for individual centrifugation tubes.

Many different types of centrifuges are commercially available including:

- large-capacity low-speed preparative centrifuges
- preparative high-speed ultracentrifuges
- refrigerated preparative centrifuges/ultracentrifuges
- analytical ultracentrifuges
- large-scale clinical centrifuges
- small-scale laboratory microfuges.

Some large-volume centrifuge models are quite demanding on space and also generate considerable amounts of heat and noise, and are therefore often centrally positioned in special instrument rooms in biochemistry departments. However, the development of small-capacity bench-top centrifuges for biochemical applications, even in the case of ultracentrifuges, has led to the introduction of these models in many individual research laboratories.

The main types of centrifuge encountered by undergraduate students during introductory practicals may be divided into microfuges (so called because they centrifuge small volume samples in Eppendorf tubes), large-capacity preparative centrifuges, high-speed refrigerated centrifuges and ultracentrifuges. Simple bench-top centrifuges vary in design and are mainly used to collect small amounts of biological material, such as blood cells. To prevent denaturation of sensitive protein samples, refrigerated centrifuges should be employed. Modern refrigerated microfuges are equipped with adapters to accommodate standardised plastic tubes for the sedimentation of 0.5 to 1.5 cm^3 volumes. They can provide centrifugal fields of approximately 10 000×g and sediment biological samples in minutes, making microfuges an indispensable separation tool for many biochemical methods. Microfuges can also be used to concentrate protein samples. For example, the dilution of protein samples, eluted by column chromatography, can often represent a challenge for subsequent analyses. Accelerated ultrafiltration with the help of plastic tube-associated filter units, spun at low g-forces in a microfuge, can overcome this problem. Depending on the proteins of interest, the biological buffers used and the molecular mass cut-off point of the particular filters, a 10- to 20-fold concentration of samples can be achieved within minutes. Larger preparative bench-top centrifuges develop maximum centrifugal fields of 3000×g to 7000×g and can be used for the spinning of various types of containers. Depending on the range of available adapters, considerable quantities of 5 to 250 cm^3 plastic tubes or 96-well ELISA plates can be accommodated. This gives simple and relatively inexpensive bench centrifuges a central place in many high-throughput biochemical assays, where the quick and efficient separation of coarse precipitates or whole cells is of importance.

High-speed refrigerated centrifuges are absolutely essential for the sedimentation of protein precipitates, large intact organelles, cellular debris derived from tissue homogenisation and microorganisms. As outlined in Section 12.4, the initial bulk separation of cellular elements prior to preparative ultracentrifugation is performed by these kinds of centrifuges. They operate at maximum centrifugal fields of approximately 100 000×g. Such centrifugal force is not sufficient to sediment smaller microsomal vesicles or ribosomes, but can be employed to differentially separate nuclei, mitochondria or chloroplasts. In addition, bulky protein aggregates can be sedimented using high-speed refrigerated centrifuges. An example is the contractile apparatus released from muscle fibres by homogenisation, mostly consisting of myosin and actin macromolecules aggregated in filaments. In order to harvest yeast cells or bacteria from large volumes of culture media, high-speed centrifugation may also be used in a continuous flow mode with zonal rotors. This approach does not therefore use centrifuge tubes, but a continuous flow of medium. As the medium enters the moving rotor, biological particles are sedimented against the rotor periphery and excess liquid removed through a special outlet port.

Ultracentrifugation has decisively advanced the detailed biochemical analysis of subcellular structures and isolated biomolecules. Preparative ultracentrifugation can be operated at relative centrifugal fields of up to 900 000×g. In order to minimise excessive rotor temperatures generated by frictional resistance between the spinning rotor and air, the rotor chamber is sealed, evacuated and refrigerated. Depending on the type, age and condition of a particular ultracentrifuge, cooling to the required running temperature and the generation of a stable vacuum might take a considerable amount of time. To avoid delays during biochemical procedures involving ultracentrifugation, the cooling and evacuation system of older centrifuge models should be switched on at least an hour prior to the centrifugation run. In contrast, modern ultracentrifuges can be started even without a fully established vacuum and will proceed in the evacuation of the rotor chamber during the initial acceleration process. For safety reasons, heavy armour plating encapsulates the ultracentrifuge to prevent injury to the user in case of uncontrolled rotor movements or dangerous vibrations. A centrifugation run cannot be initiated without proper closing of the chamber system. To prevent unfavourable fluctuations in chamber temperature, excessive vibrations or operation of rotors above their maximum rated speed, newer models of ultracentrifuges contain sophisticated temperature regulation systems, flexible drive shafts and an over-speed control device. Although slight rotor imbalances can be absorbed by modern ultracentrifuges, a more severe misbalance of tubes will cause the centrifuge to switch off automatically. This is especially true for swinging-bucket rotors. The many safety features incorporated into modern ultracentrifuges make them a robust piece of equipment that tolerates a certain degree of misuse by an inexperienced operator (see Sections 12.3.2 and 12.3.4 for a more detailed discussion of safety and centrifugation). In contrast to preparative ultracentrifuges, analytical ultracentrifuges contain a solid rotor that incorporates one counterbalancing cell and typically either three or seven analytical cells. A specialised optical system enables the sedimenting material to be observed throughout the duration of a centrifuge run. Using either an absorption optical system (based on ultraviolet/visible light absorption; see also Section 13.2) or a Rayleigh interference optical system (based on light refraction; see also Section 12.5), or a combination of both, concentration distributions of macromolecules in solution can be recorded at any time during ultracentrifugation. From these records, information about the purity/heterogeneity, sedimentation coefficient distribution, average molar mass and molar mass distributions, and ligand interaction information can be obtained.

12.3.2 Types of Rotor

To illustrate the difference in design of fixed–angle rotors, vertical tube rotors and swinging–bucket rotors, Figure 12.2 outlines cross-sectional diagrams of these three main types of rotor. Companies usually name rotors according to their design type, the maximum allowable speed and sometimes the material composition. Depending

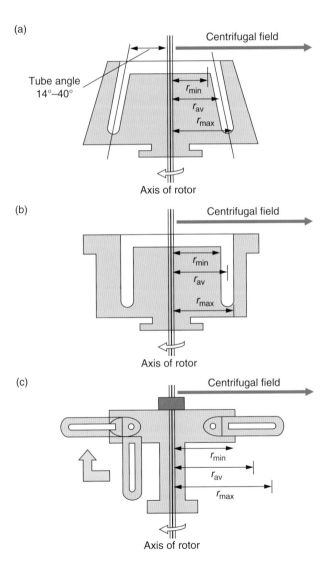

Figure 12.2 Design of the three main types of rotor used in routine biochemical centrifugation techniques. Shown is a cross-sectional diagram of a fixed-angle rotor (a), a vertical tube rotor (b) and a swinging-bucket rotor (c). A fourth type of rotor is represented by the class of near-vertical rotors (not shown).

on the use in a simple low-speed centrifuge, a high-speed centrifuge or an ultra-centrifuge, different centrifugal forces are encountered by a spinning rotor (see also Example 12.3). Accordingly, different types of rotors are made from different materials. Low-speed rotors are usually made of steel or brass, while high-speed rotors consist of aluminium, titanium or fibre-reinforced composites. The exterior of specific rotors might be finished with protective paints. For example, rotors for ultracentrifugation made out of titanium alloy are covered with a polyurethane layer. Aluminium rotors are protected from corrosion by a tough, electrochemically formed layer of aluminium oxide. In order to avoid damaging these protective layers, care should be taken during rotor handling.

Fixed-angle rotors are an ideal tool for pelleting during the differential separation of biological particles where sedimentation rates differ significantly, for example when separating nuclei, mitochondria and microsomes. In addition, isopycnic ('matching density') banding – where the density of the substance matches that of the gradient at that radial position – may also be routinely performed with fixed-angle rotors. For isopycnic separation, centrifugation is continued until the biological particles of interest have reached their isopycnic position in a gradient. This means that the particle has reached a position where the sedimentation rate is zero because the density of the biological particle and the surrounding medium are equal. Centrifugation tubes are held at a fixed angle of between 14° and 40° to the vertical in this class of rotor (Figure 12.2a). Particles move radially outwards and since the centrifugal field is exerted at an angle, they only have to travel a short distance until they reach their isopycnic position in a gradient using an isodensity technique or before colliding with the outer wall of the centrifuge tube using a differential centrifugation method. Vertical rotors (Figure 12.2b) may be divided into true vertical rotors and near-vertical rotors. Sealed centrifuge tubes are held parallel to the axis of rotation in vertical rotors and are restrained in the rotor cavities by screws, special washers and plugs. Since samples are not separated down the length of the centrifuge tube, but across the diameter of the tube, isopycnic separation time is significantly shorter as compared to swinging-bucket rotors. In contrast to fixed-angle rotors, near-vertical rotors exhibit a reduced tube angle of 7° to 10° and also employ quick-seal tubes. The reduced angle results in much shorter run times as compared to fixed-angle rotors. Near-vertical rotors are useful for gradient centrifugation of biological elements that do not properly participate in conventional gradients. Hinge pins or a crossbar is used to attach rotor buckets in swinging-bucket rotors (Figure 12.2c). They are loaded in a vertical position and during the initial acceleration phase, the rotor buckets swing out horizontally and then position themselves at the rotor body for support.

To illustrate the separation of particles in the three main types of rotors, Figure 12.3 outlines the path of biological samples during the initial acceleration stage, the main centrifugal separation phase, de-acceleration and the final harvesting of separated particles in the rotor at rest. In the case of isopycnic centrifugation in a fixed-angle rotor, the centrifuge tubes are gradually filled with a suitable gradient, the sample carefully loaded on top of this solution and then the tubes placed at a specific fixed-angle into the rotor cavities. During rotor acceleration, the sample solution and the gradient undergo reorientation in the centrifugal field, followed by the separation of particles with different sedimentation properties (Figure 12.3a). The gradient returns to its original position during the de-acceleration phase and separated particle bands can be taken from the tubes once the rotor is at rest. In analogy, similar reorientation of gradients and banding of particles occurs in a vertical rotor system (Figure 12.3b). Although run times are reduced and this kind of rotor can usually hold a large number of tubes, resolution of separated bands during isopycnic centrifugation is less when compared with swinging-bucket applications. Since a greater variety of gradients exhibiting different steepness can be used with swinging-bucket rotors, they are the method of choice when maximum resolution of banding zones is required (Figure 12.3c), such as in rate zonal studies based on the separation of biological particles as a function of sedimentation coefficient.

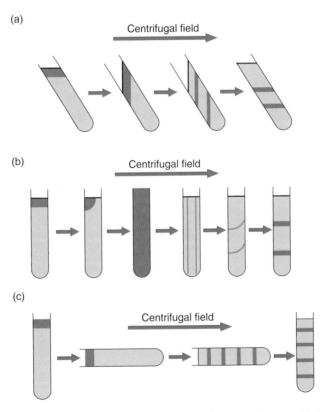

Figure 12.3 Operation of the three main types of rotor used in routine biochemical centrifugation techniques. Shown is a cross-sectional diagram of a centrifuge tube positioned in a fixed-angle rotor (a), a vertical tube rotor (b) and a swinging-bucket rotor (c). The diagrams illustrate the movement of biological samples during the initial acceleration stage, the main centrifugal separation phase, deceleration and the final harvesting of separated particles in the rotor at rest. Using a fixed-angle rotor, the tubes are filled with a gradient, the sample loaded on top of this solution and then the tubes placed at a specific fixed-angle into the rotor cavities. The sample and the gradient undergo reorientation in the centrifugal field during rotor acceleration, resulting in the separation of particles with different sedimentation properties. Similar reorientation of gradients and banding of particles occurs in a vertical rotor system. A great variety of gradients can be used with swinging-bucket rotors, making them the method of choice when maximum resolution of banding zones is required.

12.3.3 Care and Maintenance of Centrifuges

Corrosion and degradation due to biological buffer systems used within rotors or contamination of the interior or exterior of the centrifuge via spillage may seriously affect the lifetime of this equipment. Another important point is the proper balancing of centrifuge tubes. This is not only important with respect to safety, as outlined below, but might also cause vibration-induced damage to the rotor itself and the drive shaft of the centrifuge. Thus, proper handling and care, as well as regular maintenance of both centrifuges and rotors, is an important part of keeping this biochemical method available in the laboratory. In order to avoid damaging the protective layers of rotors, such as polyurethane paint or aluminium oxide, care should be taken in the cleaning of the rotor exterior. Coarse brushes that may scratch the finish should not be used and only non-corrosive detergents employed. Corrosion may be triggered by long-term exposure of rotors to alkaline solutions, acidic buffers, aggressive detergents or salt. Thus, rotors should be thoroughly washed with distilled or deionised water after every run. For overnight storage, rotors should be first left upside down to drain excess liquid and then positioned in a safe and dry place. To avoid damage to the hinge pins of swinging-bucket rotors, they should be dried with tissue paper following removal of biological buffers and washing with water. Centrifuge rotors are often not properly stored in a clean environment; this can quickly lead to the destruction of the protective rotor coating and should thus be avoided. It is advisable to keep rotors in a special clean room, physically separated from the actual centrifugation facility, with dedicated places for individual types of rotors. Some researchers might prefer to pre-cool their rotors prior to centrifugation by transferring them to a cold room. Although this is an acceptable practice and might keep proteolytic degradation to a minimum, rotors should not undergo long-term storage in a wet and cold environment. Regular maintenance of rotors and centrifuges by engineers is important for ensuring the safe operation of a centralised centrifugation facility. In order to judge properly the need for replacement of a rotor or parts of a centrifuge, it is essential that all users of core centrifuge equipment participate in proper book-keeping. Accurate record-keeping of run times and centrifugal speeds is important, since cyclic acceleration and deceleration of rotors may lead to metal fatigue.

12.3.4 Safety and Centrifugation

Modern centrifuges are not only highly sophisticated, but also relatively sturdy pieces of biochemical equipment that incorporate many safety features. Rotor chambers of high-speed and ultracentrifuges are always enclosed in heavy armour plating. Most centrifuges are designed to buffer a certain degree of imbalance and are usually equipped with an automatic switch-off mode. However, even in a well-balanced rotor, tube cracking during a centrifugation run might cause severe imbalance, resulting in dangerous vibrations. When the rotor can only be partially loaded, the order of tubes must be organised according to the manufacturer's instructions, so that the load is correctly distributed. This is important not only for ultracentrifugation with enormous centrifugal fields, but also for both small- and large-capacity bench centrifuges, where the rotors are usually mounted on a more rigid suspension. When using swinging-bucket rotors, it is important always to load all buckets with their caps properly screwed on. Even if only two tubes are loaded with solutions, the empty swinging buckets also have to be assembled since they form an integral part of the overall balance of the rotor system. In some swinging-bucket rotors, individual rotor buckets are numbered and should not be interchanged between their designated positions on similarly numbered hinge pins. Centrifugation runs using swinging-bucket rotors are usually set up with low acceleration and deceleration rates, as to avoid any disturbance of delicate gradients, and reduce the risk of disturbing bucket attachment. This practice also avoids the occurrence of sudden imbalances due to tube deformation or cracking and thus eliminates potentially dangerous vibrations.

Generally, safety and good laboratory practice are important aspects of all research projects and the awareness of the exposure to potentially harmful substances should be a concern for every biochemist. If you use dangerous chemicals, potentially infectious material or radioactive substances during centrifugation protocols, refer to up-to-date safety manuals and the safety statement of your individual department. Perform mock runs of important experiments in order to avoid the loss of precious specimens or expensive chemicals. As with all other biochemical procedures, experiments should never be rushed and protective clothing should be worn at all times. Centrifuge tubes should be handled slowly and carefully so as not to disturb pellets, bands of separated particles or unstable gradients. To help you choose the right kind of centrifuge tube for a particular application, the manufacturers of rotors usually give detailed recommendations of suitable materials. For safety reasons and to guarantee experimental success, it is important to make sure that individual centrifuge tubes are chemically resistant to solvents used, have the right capacity for sample loading, can be used in the designated type of rotor and are able to withstand the maximum centrifugal forces and temperature range of a particular centrifuge. In fixed-angle rotors, large centrifugal forces tend to cause a collapse of centrifuge tubes, making thick-walled tubes the choice for these rotors. The volume of liquid and the sealing mechanisms of these tubes are very important for the integrity of the run and should be done according to manufacturer's instructions. In contrast, swinging-bucket rotor tubes are better protected from deformation and usually thin-walled polyallomer tubes are used. An important safety aspect is the proper handling of separated biological particles following centrifugation. In order to perform post-centrifugation analysis of individual fractions, centrifugation tubes often have to be punctured or sliced. For example, separated vesicle bands can be harvested from the pierced bottom of the centrifuge tube or can be collected by slicing of the tube following quick-freezing. If samples have been pre-incubated with radioactive markers or toxic ligands, contamination of the centrifugation chamber and rotor cavities or buckets should be avoided. If centrifugal separation processes have to be performed routinely with a potentially harmful substance, it makes sense to dedicate a particular centrifuge and accompanying rotors for this work and thereby eliminate the potential of cross-contamination.

12.4 PREPARATIVE CENTRIFUGATION

12.4.1 Differential Centrifugation

Cellular and subcellular fractionation techniques are indispensable methods used in biochemical research. Although the proper separation of many subcellular structures is absolutely dependent on preparative ultracentrifugation, the isolation of large cellular structures, the nuclear fraction, mitochondria, chloroplasts or large protein precipitates can be achieved by conventional high-speed refrigerated centrifugation. Differential centrifugation is based upon the differences in the sedimentation rate of biological particles of different size and density. Crude tissue homogenates containing organelles, membrane vesicles and other structural fragments are divided into different fractions by the stepwise increase of the applied centrifugal field. Following the initial sedimentation of the largest particles of a homogenate (such as cellular debris) by centrifugation, various biological structures or aggregates are separated into pellet and supernatant fractions, depending upon the speed and time of individual centrifugation steps and the density and relative size of the particles. To increase the yield of membrane structures and protein aggregates released, cellular debris pellets are often rehomogenised several times and then subjected to further centrifugation. This is especially important in the case of rigid biological structures such as muscular or connective tissues, or in the case of small tissue samples, as is the case with human biopsy material or primary cell cultures.

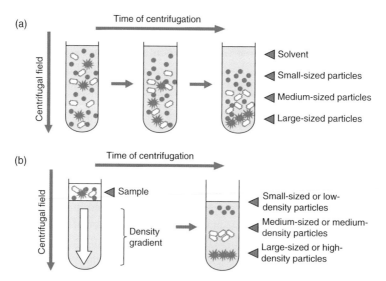

Figure 12.4 Diagram of particle behaviour during differential and isopycnic separation. During differential sedimentation (a) of a particulate suspension in a centrifugal field, the movement of particles is dependent upon their density, shape and size. For separation of biological particles using a density gradient (b), samples are carefully layered on top of a preformed density gradient prior to centrifugation. For isopycnic separation, centrifugation is continued until the desired particles have reached their isopycnic position in the liquid density gradient. In contrast, during rate separation, the required fraction does not reach its isopycnic position during the centrifugation run.

The differential sedimentation of a particulate suspension in a centrifugal field is diagrammatically shown in Figure 12.4a. Initially, all particles of a homogenate are evenly distributed throughout the centrifuge tube and then move down the tube at their respective sedimentation rate during centrifugation. The largest class of particles forms a pellet on the bottom of the centrifuge tube, leaving smaller-sized structures within the supernatant. However, during the initial centrifugation step, smaller particles also become entrapped in the pellet, causing a certain degree of contamination. At the end of each differential centrifugation step, the pellet and supernatant fraction are carefully separated from each other. To minimise cross-contamination, pellets are usually washed several times by resuspension in buffer and subsequent centrifugation under the same conditions. However, repeated washing steps may considerably reduce the yield of the final pellet fraction, and are therefore omitted in preparations with limiting starting material. Resulting supernatant fractions are centrifuged at a higher speed and for a longer time to separate medium-sized and small-sized particles. With respect to the separation of organelles and membrane vesicles, crude differential centrifugation techniques can be conveniently employed to isolate intact mitochondria and microsomes.

12.4.2 Density-Gradient Centrifugation

To further separate biological particles of similar size but differing densities, ultracentrifugation with pre-formed or self-establishing density gradients is the method of choice. Both rate separation or equilibrium methods can be used. In Figure 12.4b, the preparative ultracentrifugation of low- to high-density particles is shown. A mixture of particles, such as is present in a heterogeneous microsomal membrane preparation, is layered on top of a pre-formed liquid density gradient. Depending on the particular biological application, a great variety of gradient materials are available. Caesium chloride is widely used for the banding of DNA and the isolation of plasmids, nucleoproteins and viruses. Sodium bromide and sodium iodide are employed for the fractionation of lipoproteins and the banding of DNA or RNA molecules, respectively. Various companies offer a range of gradient material for the separation of whole cells and subcellular particles, e.g. Percoll®, Ficoll®, dextran, metrizamide and Nycodenz®. For the separation of membrane vesicles derived from tissue homogenates, ultra-pure DNase-, RNase and protease-free sucrose represents a suitable and widely employed medium for the preparation of stable gradients. If one wants to separate all membrane species spanning the whole range of particle densities, the maximum density of the gradient must exceed the density of the most dense vesicle species. Both step-gradient and continuous-gradient systems are employed to achieve this. If automated gradient-makers are not available, which is probably the case in most undergraduate practical classes, the manual pouring of a stepwise gradient with the help of a pipette is not so time-consuming or difficult. In contrast, the formation of a stable continuous gradient is much more challenging and requires a commercially available gradient-maker. Following pouring, gradients are usually kept in a cold room for temperature equilibration and are moved extremely slowly in special holders so as to avoid mixing of different gradient layers. For rate separation (sedimentation velocity;

see Section 12.5.1) of subcellular particles, the required fraction does not reach its isopycnic position within the gradient. For isopycnic separation, density centrifugation is continued until the buoyant density of the particle of interest and the density of the gradient are equal.

12.4.3   Practical Applications of Preparative Centrifugation

To illustrate practical applications of differential centrifugation, density gradient ultracentrifugation and affinity methodology, the isolation of the microsomal fraction from muscle homogenates and subsequent separation of membrane vesicles with a differing density is described (Figure 12.5), the isolation of highly purified sarcolemmal vesicles outlined (Figure 12.6) and the subfractionation of liver mitochondrial membrane systems shown (Figure 12.7). Skeletal muscle fibres are highly specialised structures involved in contraction, and the membrane systems that maintain the regulation of excitation–contraction coupling, energy metabolism and the stabilisation of the cell periphery are diagrammatically shown in Figure 12.5a. The surface membrane consists of the sarcolemma and its invaginations, the transverse tubular membrane system. The transverse tubules may be subdivided into the non-junctional region and the triad part that forms contact zones with the terminal cisternae of the sarcoplasmic reticulum. Motor-neuron-induced depolarisation of the sarcolemma travels into the transverse tubules and activates a voltage-sensing receptor complex that directly initiates the transient opening of a junctional calcium-release channel. The membrane system that provides the luminal ion reservoir for the regulatory calcium cycling process is represented by the specialised endoplasmic reticulum. It forms membranous sheaths around the contractile apparatus whereby the longitudinal tubules are mainly involved in the uptake of calcium ions during muscle relaxation and the terminal cisternae provide the rapid calcium-release mechanism that initiates muscle contraction. Mitochondria are the site of oxidative phosphorylation and exhibit a complex system of inner and outer membranes involved in energy metabolism.

For the optimum homogenisation of tissue specimens, mincing of tissue has to be performed in the presence of a biological buffer system that exhibits the right pH value, salt concentration, stabilising cofactors and chelating agents. The optimum ratio between the wet weight of tissue and buffer volume, as well as the temperature (usually 4 °C) and presence of a protease inhibitor cocktail is also essential to minimise proteolytic degradation. Prior to the 1970s, researchers did not widely use protease inhibitors or chelating agents in their homogenisation buffers. This resulted in the degradation of many high-molecular-mass proteins. Since protective measures against endogenous enzymes have been routinely introduced into subcellular fractionation protocols, extremely large proteins have been isolated in their intact form, such as 427 kDa dystrophin, the 565 kDa ryanodine receptor, 800 kDa nebulin and the longest known polypeptide of 3800 kDa, named titin. Commercially available protease inhibitor cocktails usually exhibit a broad specificity for the inhibition of cysteine proteases, serine proteases, aspartic proteases, metalloproteases and aminopeptidases. They are used in the micromolar concentration range and are best added

(a) Subcellular membrane systems that can be isolated by differential centrifugation

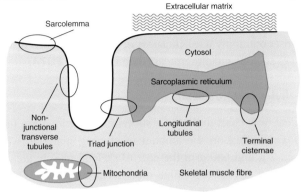

Extracellular matrix

Sarcolemma

Cytosol

Sarcoplasmic reticulum

Non-junctional transverse tubules

Triad junction

Longitudinal tubules

Terminal cisternae

Mitochondria

Skeletal muscle fibre

(b) Scheme of subcellular fractionation of membranes from muscle homogenates

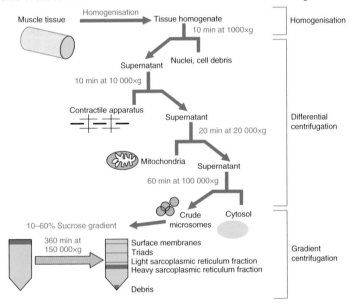

Muscle tissue

Homogenisation

Tissue homogenate
10 min at 1000×g

Homogenisation

Supernatant

Nuclei, cell debris

10 min at 10 000×g

Contractile apparatus

Supernatant
20 min at 20 000×g

Differential centrifugation

Mitochondria

Supernatant
60 min at 100 000×g

10–60% Sucrose gradient

Crude microsomes

Cytosol

360 min at 150 000×g

Surface membranes
Triads
Light sarcoplasmic reticulum fraction
Heavy sarcoplasmic reticulum fraction

Debris

Gradient centrifugation

Figure 12.5 Scheme of the fractionation of skeletal muscle homogenate into various subcellular fractions. Shown is a diagrammatic presentation of the subcellular membrane system from skeletal muscle fibres (a) and a flow chart of the fractionation protocol of these membranes from tissue homogenates using differential centrifugation and density gradient methodology (b).

to buffer systems just prior to the tissue homogenisation process. Depending on the half-life of specific protease inhibitors, the length of a subcellular fractionation protocol and the amount of endogenous enzymes present in individual fractions, tissue suspensions might have to be replenished with a fresh aliquot of a protease inhibitor cocktail. Protease inhibitor kits for the creation of individualised cocktails are also available and consist of substances such as trypsin inhibitor, E-64, aminoethyl-benzenesulfonyl-fluoride, antipain, aprotinin, benzamidine, bestatin, chymostatin, ε-aminocaproic acid, N-ethylmaleimide, leupeptin, phosphoramidon and pepstatin. The most commonly used chelators of divalent cations for the inhibition of degrading enzymes such as metalloproteases are EDTA and EGTA.

12.4.4 Subcellular Fractionation

A typical flow chart outlining a subcellular fractionation protocol is shown in Figure 12.5b. Depending on the amount of starting material, which would usually range between 1 g and 500 g in the case of skeletal muscle preparations, a particular type of rotor and size of centrifuge tube is chosen for individual stages of the isolation procedure. The repeated centrifugation at progressively higher speeds and longer centrifugation periods will divide the muscle homogenate into distinct fractions. Typical values for centrifugation steps are 10 min for 1000×g to pellet nuclei and cellular debris, 10 min for 10 000×g to pellet the contractile apparatus, 20 min at 20 000×g to pellet a fraction enriched in mitochondria, and 1 h at 100 000×g to separate the microsomal and cytosolic fractions. Mild salt washes can be carried out to remove myosin contamination of membrane preparations. Sucrose gradient centrifugation is then used to further separate microsomal subfractions derived from different muscle membranes. Using a vertical rotor or swinging-bucket rotor system at a sufficiently high g-force, the crude surface membrane fraction, triad junctions, longitudinal tubules and terminal cisternae membrane vesicles can be separated. To collect bands of fractions, the careful removal of fractions from the top can be achieved manually with a pipette. Alternatively, in the case of relatively unstable gradients or tight banding patterns, membrane vesicles can be harvested from the bottom by an automated fraction collector. In this case, the centrifuge tube is pierced and fractions collected by gravity or slowly forced out of the tube by a replacing liquid of higher density. Another method for collecting fractions from unstable gradients is the slicing of the centrifuge tube after freezing. Both latter methods destroy the centrifuge tubes and are routinely used in research laboratories.

Cross-contamination of vesicular membrane populations is an inevitable problem during subcellular fractionation procedures. The technical reason for this is the lack of adequate control in the formation of various types of membrane species during tissue homogenisation. Membrane domains originally derived from a similar subcellular location might form a variety of structures, including inside-out vesicles, right-side-out vesicles, sealed structures, leaky vesicles and/or membrane sheets. In addition, smaller vesicles might become entrapped in larger vesicles. Different membrane systems might aggregate non-specifically or bind to or entrap abundant solubilised proteins. Hence, if highly purified membrane preparations are needed for sophisticated cell biological or biochemical studies, affinity separation methodology has to be employed. The flow chart and immunoblotting diagram in Figure 12.6 illustrates both the preparative and analytical principles underlying such a biochemical approach. Modern preparative affinity techniques using centrifugation steps can be performed with various biological or chemical ligands. In the case of immuno-affinity purification, antibodies are used to specifically bind to their respective antigen (see also Section 7.11). For a list of references outlining the use of subcellular fractionation methods in routine biochemical and proteomic applications, please consult the articles listed in Section 12.6.

(a)   Scheme of subcellular fractionation of muscle sarcolemma

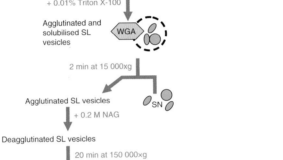

(b)   Diagram of immunoblot analysis of subcellular fractionation procedures

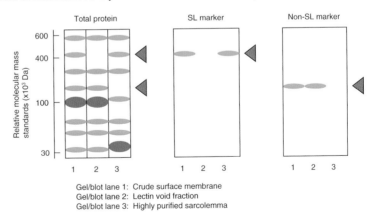

Gel/blot lane 1: Crude surface membrane
Gel/blot lane 2: Lectin void fraction
Gel/blot lane 3: Highly purified sarcolemma

Figure 12.6 Affinity separation method using centrifugation of lectin-agglutinated surface membrane vesicles from skeletal muscle. Shown is a flow chart of the various preparative steps in the isolation of highly purified sarcolemma vesicles (a) (NAG, N-acetylglucosamine; SL, sarcolemma; SN, supernatant; WGA, wheat germ agglutinin) and a diagram of the immunoblot analysis of this subcellular fractionation procedure (b). The sarcolemma and non-SL markers are surface-associated dystrophin of 427 kDa and the transverse-tubular α_{1S}-subunit of the dihydropyridine receptor of 170 kDa, respectively.

12.4.5 Affinity Purification of Membrane Vesicles

In Figure 12.6a is shown a widely employed lectin agglutination method. Lectins are plant proteins that bind tightly to specific carbohydrate structures. The rationale behind using purified wheat germ agglutinin (WGA) lectin for the affinity purification of sarcolemmal vesicles is the fact that the muscle plasmalemma forms mostly right-side-out vesicles following homogenisation. By contrast, vesicles derived from the transverse tubules are mostly inside out and thus do not expose their carbohydrates. Glycoproteins from the abundant sarcoplasmic reticulum do not exhibit carbohydrate moieties that are recognised by this particular lectin species. Therefore, only sarcolemmal vesicles are agglutinated by the wheat germ lectin and the aggregate can be separated from the transverse tubular fraction by centrifugation for 2 min at 15 000×g. The electron microscopical characterisation of agglutinated surface membranes revealed large smooth sarcolemmal vesicles that had electron-dense entrapments. To remove these vesicular contaminants, originally derived from the sarcoplasmic reticulum, immobilised surface vesicles are treated with low concentrations of the non-ionic detergent Triton X-100. This procedure does not solubilise integral membrane proteins, but introduces openings in the sarcolemmal vesicles for the release of the much smaller sarcoplasmic reticulum vesicles. Low g-force centrifugation is then used to separate the agglutinated sarcolemmal vesicles and the contaminants. To remove the lectin from the purified vesicles, the fraction is incubated with the competitive sugar N-acetylglucosamine, which eliminates the bonds between the surface glycoproteins and the lectin. A final centrifugation step for 20 min at 150 000×g results in a pellet of highly purified sarcolemmal vesicles. A quick and convenient analytical method of confirming whether this subcellular fractionation procedure has resulted in the isolation of the muscle plasmalemma is immunoblotting with a mini electrophoresis unit. Figure 12.6b shows a diagram of the protein and antigen banding pattern of crude surface membranes, the lectin void fraction and the highly purified sarcolemmal fraction. Using antibodies to mark the transverse tubules and the sarcolemma, such as the α_{1s}-subunit of the dihydropyridine receptor of 170 kDa and dystrophin of 427 kDa, respectively, the separation of both membrane species can be monitored. This analytical method is especially useful for the characterisation of membrane vesicles, when no simple and fast assay systems for testing marker enzyme activities are available.

In the case of the separation of mitochondrial membranes, the distribution of enzyme activities rather than immunoblotting is routinely used for determining the distribution of the inner membrane, contact zones and the outer membrane in density gradients. Binding assays or enzyme testing represents the more traditional way of characterising subcellular fractions following centrifugation. Figure 12.7a outlines diagrammatically the micro compartments of liver mitochondria and the associated marker enzymes. While the monoamino oxidase (MAO) is enriched in the outer membrane, the enzyme succinate dehydrogenase (SDH) is associated with the inner membrane system and a representative marker of contact sites between both membranes is glutathione transferase (GT). Membrane vesicles from intact mitochondria can be generated by consecutive swelling, shrinking and sonication of the suspended organelles. The vesicular mixture is then separated by sucrose density centrifugation into the three main types of mitochondrial membranes (Figure 12.7b). The distribution of marker enzyme activities in the various fractions demonstrates that the outer membrane has a lower density compared to the inner membrane. The glutathione transferase-containing contact zones are positioned in a band between the inner and outer mitochondrial membrane and contain enzyme activities characteristic for both systems (Figure 12.7c). Routinely used enzymes as subcellular markers would be Na$^+$/K$^+$-ATPase for the plasmalemma, glucose-6-phosphatase for the endoplasmic reticulum, galactosyl transferase for the Golgi apparatus, succinate dehydrogenase for

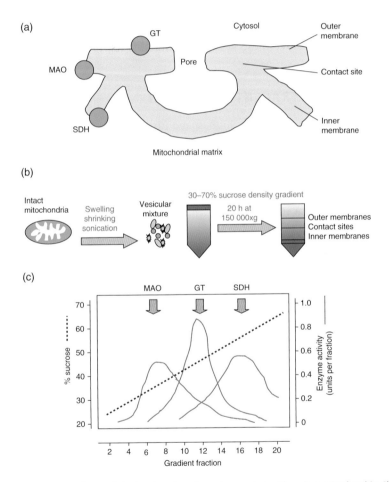

Figure 12.7 Scheme of the fractionation of membranes derived from liver mitochondria. Shown is the distribution of marker enzymes in the micro-compartments of liver mitochondria (MAO, monoamino oxidase; SDH, succinate dehydrogenase; GT, glutathione transferase) (a), the separation method to isolate fractions highly enriched in the inner cristae membrane, contact zones and the outer mitochondrial membrane (b), as well as the distribution of mitochondrial membranes after density-gradient centrifugation (c).

mitochondria, acid phosphatase for lysosomes, catalase for peroxisomes and lactate dehydrogenase for the cytosol.

12.4.6 Separation of DNA Components

A recent application of ultracentrifugation is in the genome-wide identification of gene regulatory regions, i.e. the open regions of DNA not protected by nucleosomes at that point in time. Intact nuclei are incubated with limiting amounts of DNase I, which is able to enter the nuclei and digest accessible DNA from chromatin. The digested DNA is then recovered and applied to a step gradient typically made with 10–40% sucrose (40% at the bottom of the tube, rising to 10% at the top), and subjected to 24 hours at 90 000×g (at 25 °C). Alternatively a 'sucrose cushion' can be used that is simply a fixed concentration of sucrose; typically 9%. The gentle separation of DNA fragments allows efficient molecular cloning of the DNA fragments that are then sequenced using massively parallel sequencing (Chapter 20).

12.5 ANALYTICAL ULTRACENTRIFUGATION

Analytical ultracentrifuges are high-speed ultracentrifuges with optical system(s) for recording the sedimentation process. As biological macromolecules exhibit random thermal motion, their relative uniform distribution in an aqueous environment is not significantly affected by the Earth's gravitational field. Isolated biomolecules in solution only exhibit distinguishable sedimentation when they undergo immense accelerations, e.g. in an ultracentrifugal field. A typical analytical ultracentrifuge can generate a centrifugal field of up to 200 000×g in its analytical cell. Within these extremely high gravitational fields, the ultracentrifuge cell has to allow light passage through the biological particles for proper measurement of the concentration distribution. The schematic diagram in Figure 12.8 outlines one of the two principal optical systems for a modern analytical ultracentrifuge. The availability of high-intensity xenon flash lamps and the advance in instrumental sensitivity and wavelength range has made the measurement of highly dilute protein samples below 230 nm possible. Analytical ultracentrifuges such as the Beckman Optima XL-A allow the use of wavelengths between 190 nm and 800 nm. Sedimentation of isolated proteins or nucleic acids can be useful in the determination of the molecular mass, purity and shape of these biomolecules. A second optical system (not shown) is the Rayleigh interference optical system, which detects the distribution of proteins or nucleic acids – or other macromolecules such as polysaccharides, glycoproteins and synthetic macromolecules – based on their different refractive index compared to the solvent they are dissolved in. The Beckman Optima XL-I, for example, has both optical systems. These two different optical systems can be applied individually, or simultaneously, which can be helpful for looking at interactions involving biomolecules with ligands. Other less common optical systems are based on fluorescence or refractive gradient "Schlieren" optics.

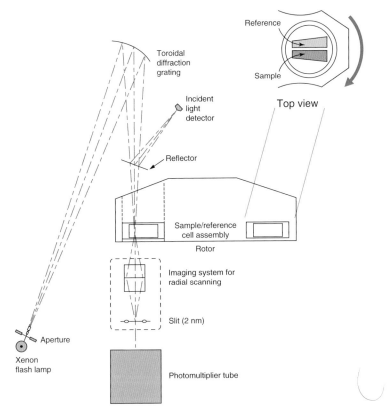

Figure 12.8 Schematic diagram of the absorption optical system of an analytical ultracentrifuge (courtesy of Beckman-Coulter). The high-intensity xenon flash lamp of the Beckman Optima XL-A analytical ultracentrifuge shown here allows the use of wavelengths between 190 nm and 800 nm. The high sensitivity of the absorbance optics allows the measurement of highly dilute protein samples below 230 nm. This is the simplest of the two main optical systems available on a modern analytical ultracentrifuge. An alternative optical system, based on refractive index properties, is known as the Rayleigh interference optical system.

12.5.1 Sedimentation Velocity, Sedimentation Equilibrium and Their Applications

There are two main types of analytical ultracentrifuge experiments: sedimentation velocity and sedimentation equilibrium experiments. In sedimentation velocity experiments, the change in concentration distribution in the ultracentrifuge cell at high rotor speeds is recorded. A boundary is formed between solution and clear solvent left behind and the rate of movement of the boundary per unit centrifugal field yields the sedimentation coefficient (Equation 12.8), which will be a function of the size (molecular mass) and shape (based on friction properties) of the system. For heterogeneous systems the method will give a distribution of sedimentation coefficient.

In contrast, in sedimentation equilibrium experiments an equilibrium or steady-state distribution of concentration is obtained. Here, lower rotor speeds are used and the centrifugal forces are countered by the back forces due to diffusion and no boundary is formed. Instead, after a period of time (usually at least several hours) an equilibrium is reached, and there will be no net movement of macromolecules, hence no friction forces: the pattern only depends on molecular mass. The two methods – sedimentation velocity and sedimentation equilibrium – provide complementary information about a macromolecular system. Sedimentation velocity has a high resolution, which is excellent for monitoring heterogeneity and aggregation and also for evaluating molecular conformation in solution. Sedimentation equilibrium is an 'absolute' method (not requiring standards) for molecular mass and (in hetero-geneous systems) average molecular mass determination. For systems containing non-covalent assemblies, both methods can provide valuable information about the stoichiometry and strength of self-association reactions (i.e. the quaternary struc-ture: protein dimerisation, tetramerisation, etc.) or interactions with other molecules.

Analytical ultracentrifugation is most often employed for:

- the determination of the purity (including the presence of aggregates) and oligomeric state of macromolecules, from recording the distribution of sedimentation coefficients from sedimentation velocity
- the determination of the average molecular mass, or distribution of molecular mass of solutes in their native state, from sedimentation equilibrium
- the examination of changes in the molecular mass of supramolecular complexes, using either sedimentation velocity or sedimentation equilibrium (or both)
- the detection of conformation and conformational changes using sedimentation velocity
- ligand-binding studies (see also Section 23.5.3).

Since the mass of one molecule is extremely small, when researchers refer to the 'molec-ular mass' of a molecule, they really mean the molar mass M which describes the mass of 1 mol (= 6.023×10^{23} molecules) of macromolecules, in units of g mol^{-1}, or, equiva-lently, the relative molecular mass M_r, which is the mass of a macromolecule per one-twelfth of the mass of a carbon-12 atom (see also Section 2.3.1). Molar mass and relative molecular mass are numerically the same, but being a relative measure, M_r has no units.

For a list of references outlining the applicability of ultracentrifugation to the char-acterisation of macromolecular behaviour in solution, please consult the review articles listed in Section 12.6. In addition, manufacturers of analytical ultracentrifuges offer a large range of excellent brochures on the theoretical background of this method and its specific applications available. These introductory texts are usually written by research biochemists and are well worth reading to become familiar with this field.

12.5.2 Sedimentation Coefficient and Sedimentation Coefficient Distribution

The sedimentation coefficient, as defined by Equation 12.8 (after normalisation to stan-dard conditions), will depend on the size (molecular mass) of a macromolecule. For particles of near-globular shape, $s_{20,w}$ is approximately proportional to $M^{2/3}$. The value of $s_{20,w}$ lies for many macromolecules of biochemical interest typically between 1 and 20, and for larger biological particles such as ribosomes, microsomes and mitochon-dria between 80 and several thousand. Figure 12.9 shows the optical records from a sedimentation velocity experiment on a tetanus toxoid protein used in the production of glycoconjugate vaccines against serious disease. The experiment was conducted to determine the sedimentation coefficient of the monomer species and any other com-ponents present, to establish the amount of monomer compared to other species. The presence of dimer can clearly be seen for a whole range of different concentrations used.

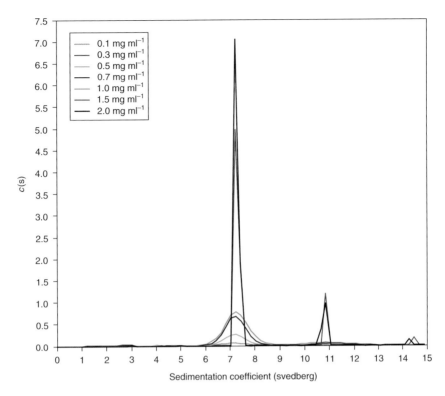

Figure 12.9 Sedimentation coefficient distribution c(s) versus s for tetanus toxoid protein showing ~86% monomer with a sedimentation coefficient of 7.6 S and ~14% of dimer at higher s (11.6 S). A rotor speed of rpm = 45 000 min⁻¹ was used at a temperature of 20.0 °C. Protein solutions were dissolved in a phosphate chloride buffer, pH 6.8, ionic strength 0.1 M. The relative proportions of monomer and dimer do not change with loading concentration, showing the dimerisation process is not reversible. If it was reversible then the proportion of dimer should increase with increasing concentration. Some evidence of a trace amount of a higher-molecular-weight species (with a sedimentation coefficient of approximately 14.5 S) is also seen.

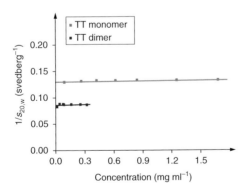

Figure 12.10 Extrapolation of the reciprocal sedimentation coefficients to zero concentration to correct for non-ideality effects of tetanus toxoid monomers and dimers.

If the molecular mass M is known, the sedimentation coefficient can be used to calculate the frictional coefficient f, which is a measure of the conformation of a macromolecule:

$$f = \frac{M \times \left(1 - 0.9982 \text{ g ml}^{-1} \times \bar{v}\right)}{N_A \times s_{20,w}}$$

(Eq 12.9)

0.9982 g ml^{-1} is the density of water at 20.0 °C, N_A is the Avogadro constant and $\bar{v}$ is the partial specific volume (in ml g^{-1}) of the protein. For studies attempting accurate characterisation of conformation, another correction of the sedimentation coefficient is usually necessary. Because of the large size of macromolecules compared to the solvent they are dissolved in, they exhibit non-ideality: (i) they get in the way of each other, and (ii) if they are carrying charge, macromolecules can repel each other. The charge effect can be reduced by increasing the ionic strength (concentration of low-molecular-weight salt ions) of the solvent or, in the case of proteins, working near or at the isoelectric pH. Both effects can be reduced by working at low concentrations of macromolecules, or by taking measurements of $s_{20,w}$ at a series of concentrations, c and extrapolating to zero concentration to yield corrected $(s^0_{20,w})$ values. More accurate extrapolations are generally obtained if the reciprocal $1/s_{20,w}$ is plotted versus c. Figure 12.10 shows such an extrapolation for the tetanus toxoid data of Figure 12.9.

The conformation can be interpreted in terms of ellipsoid models for globular proteins (Figure 12.11a) or bead models (Figure 12.11b) for more complicated structures such as antibodies. It is usually necessary to combine f with other physical data, such as from viscosity or small-angle X-ray scattering measurements (see Section 14.3.5).

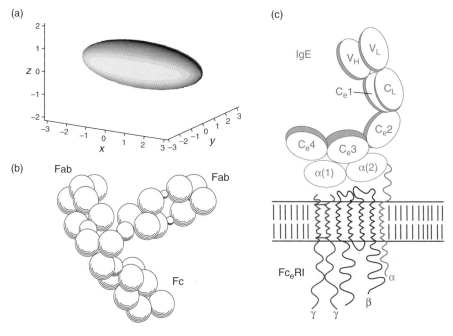

Figure 12.11 Models for the conformations of proteins from the sedimentation coefficient, molecular mass and other information. (a) Ellipsoid model of axial ratio (ratio of long axis to short axis) of 3:1 for tetanus toxoid protein, and was obtained by combining analytical ultracentrifugation data with intrinsic viscosity data. (b) Bead model for antibody IgE (note the individual beads do not represent atoms or domains). This was the first demonstration of the cusp shape of the IgE molecule and was obtained by combining analytical ultracentrifugation data with the radius of gyration from X-ray scattering, later confirmed by spectroscopic and crystallographic data. (c) Schematic representation showing how the cusp shape of IgE facilitates its binding via its $C_\varepsilon 3$ domain to the α-chain of the high affinity membrane IgE receptor known as Fc$_\varepsilon$RI, shown with its constituent α, β, and two γ polypeptide chains.

Density Gradients

For very complex systems, the addition of a density gradient material can assist with the resolution of components. Figure 12.12 illustrates the sedimentation analysis of the dystrophin–glycoprotein complex (DGC) from skeletal muscle fibres. The size of this complex was estimated to be approximately 18 S by comparing its migration to that of the standards β-galactosidase (16 S) and thyroglobulin (19 S). When the membrane cytoskeletal protein dystrophin was first identified, it was shown to bind to a lectin matrix, although the protein does not possess any carbohydrate conjugation; this suggested that dystrophin might exist in a complex with surface glycoproteins. Sedimentation velocity analysis confirmed the existence of such a dystrophin–glycoprotein complex and centrifugation following various biochemical modifications of the protein assembly led to a detailed understanding of the composition of this complex. Alkaline extraction, acid treatment or incubation with different types of detergent causes the differential disassembly of the dystrophin–glycoprotein complex. It is now known that dystrophin is tightly associated with at least 10 different surface proteins that are involved in membrane stabilisation, receptor anchoring and signal transduction processes. The successful characterisation of the dystrophin–glycoprotein complex by sedimentation analysis is an excellent example of how centrifugation methodology can be exploited to quickly gain biochemical knowledge of a newly discovered protein.

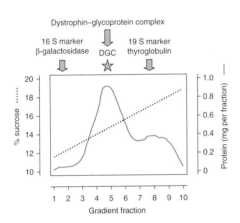

Figure 12.12 Sedimentation analysis of a supramolecular protein complex. Shown is the sedimentation of the dystrophin–glycoprotein complex (DGC). Its size was estimated to be approximately 18 S by comparing its migration to that of the standards β-galactosidase (16 S) and thyroglobulin (19 S).

12.5.3 Molecular Mass Determination

After equilibrium has been obtained in a sedimentation equilibrium experiment, the dataset of concentration in the ultracentrifuge cell $c(r)$ as a function of radial position r can be processed to yield the molar mass M (or the average molecular mass if the system is heterogeneous). For a single solute, a plot of ln $c(r)$ versus r^2 yields a straight line with a slope proportional to M (Figure 12.13). The proportionality constant depends on the rotor speed, temperature, density of the solvent, and the partial specific volume $\bar{v}$. For a heterogeneous system, the plot will be curved upwards, and an average slope is defined, yielding what is known as the weight average molecular mass M_w. In addition, it is possible to see if M_w changes with concentration. If M_w increases with increasing concentration this is symptomatic of an associating or interacting system – further analysis of this can yield the stoichiometry and strength of the interaction.

Molecular (molar) mass determination by analytical ultracentrifugation is applicable to values from a few hundred to several million Da. It is therefore used for the analysis of small carbohydrates, proteins, nucleic acid macromolecules, viruses and subcellular particles such as mitochondria.

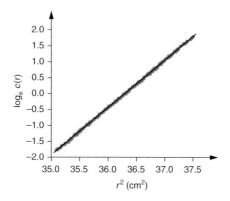

Figure 12.13 Sedimentation equilibrium experiment on IgG1 in phosphate-chloride buffer, pH 6.8, ionic strength 0.1 M, $\theta = 20.0°$; rotor speed $s = 13\,000$ rpm, loading concentration $\rho^* = 1$ mg ml^{-1}. Data points (filled squares), fitted line (black). M_w (from the slope) = (148 ± 2) kDa.

12.6 SUGGESTIONS FOR FURTHER READING

12.6.1 Experimental Protocols

Abdelhameed A.S., Morris G.A., Adams G.G., *et al.* (2012) An asymmetric and slightly dimerized structure for the tetanus toxoid protein used in glycoconjugate vaccines. *Carbohydrate Polymers* **90**, 1831–1835.

Burgess N. K., Stanley A. M. and Fleming K. G. (2008). Determination of membrane protein molecular weights and association equilibrium constants using sedimentation equilibrium and sedimentation velocity. *Methods in Cell Biology* **84**, 181–211.

Carberry S., Zweyer M., Swandulla D. and Ohlendieck K. (2014) Comparative proteomic analysis of the contractile-protein-depleted fraction from normal versus dystrophic skeletal muscle. *Analytical Biochemistry* **446**, 108–115.

Davis K.G., Glennie M., Harding S.E. and Burton, D.R. (1990) A model for the solution conformation of rat IgE. *Biochemical Society Transactions* **18**, 935–936.

Klassen R., Fricke J., Pfeiffer A. and Meinhardt F. (2008) A modified DNA isolation protocol for obtaining pure RT-PCR grade RNA. *Biotechnology Letters* **30**, 1041–1044.

Lee Y.H., Tan H.T. and Chung M.C. (2010) Subcellular fractionation methods and strategies for proteomics. *Proteomics* **10**, 3935–3956.

Sutton B.J. and Gould H.J. (1993) The human IgE network. *Nature* **366**, 421–428.

12.6.2 General Texts

Schuck P., Zhao H., Brautigam C.A. and Ghirlando R. (2015) *Basic Principles of Analytical Ultracentrifugation*, CRC Press, Boca Raton, FL, USA.

Scott D., Harding S.E. and Rowe, A. (2005) *Analytical Ultracentrifugation: Techniques and Methods*, Royal Society of Chemistry, London, UK.

12.6.3 Review Articles

Cole J.L., Lary J.W., Moody T.P. and Laue T. M. (2008) Analytical ultracentrifugation: sedimentation velocity and sedimentation equilibrium, *Methods in Cell Biology* 84, 143–179.

12.6.4 Websites

SEDFIT Tutorials (Peter Schuck)

www.analyticalultracentrifugation.com/tutorials.htm (accessed May 2017)

13 Spectroscopic Techniques

ANNE SIMON AND ANDREAS HOFMANN

13.1 INTRODUCTION

Spectroscopic techniques are based on the interaction of light with matter and probe certain features of a sample to learn about its consistency or structure. Light is electromagnetic radiation, a phenomenon exhibiting different energies. When light interacts with matter, it can emerge with other characteristics and properties, and therefore different molecular features of matter can be probed. Conceptually, matter can either annihilate (absorb) or create (emit) light (discussed in this chapter), or it can scatter light (discussed in the next chapter, Section 14.3.5). An understanding of the properties of electromagnetic radiation and its interaction with matter leads to an appreciation of the variety of types of spectrum and different spectroscopic techniques, and their applications to the solution of biological problems.

In this section, we introduce the basic principles of interaction of electromagnetic radiation with matter and then examine absorption and emission of light by matter and the corresponding molecular spectroscopic techniques of absorption (Sections 13.2, 13.3, 13.4) or emission (Sections 13.5, 13.6). Atomic spectroscopy (Section 13.7) is based on the same principles, but uses light of higher energy.

The applications considered use visible or UV light to probe consistency and conformational structure of biological molecules. Usually, these methods are the first analytical procedures used by a biochemical scientist, and are routinely employed in a variety of experimental approaches (e.g. enzyme kinetics, Section 13.8).

13.1.1 Properties of Electromagnetic Radiation

Electromagnetic radiation (Figure 13.1) is composed of an electric ($\bar{E}$) and a perpendicular magnetic vector ($\bar{M}$), each one oscillating in plane at right angles to the direction of propagation, resulting in a sine-like waveform of each, the $\bar{E}$ and the $\bar{M}$ vectors. The wavelength λ is the spatial distance between two consecutive peaks (one cycle) in the sinusoidal waveform and is measured in submultiples of metre, usually in nanometres (nm). The maximum length of the vector is called the amplitude. The frequency v of the electromagnetic radiation is the number of oscillations made by the wave within the time frame of 1 s. It therefore has the units of $1\ s^{-1} = 1$ Hz. The frequency is related to the wavelength via the speed of light *in vacuo*, $c = 2.998 \times 10^8$ m s^{-1} by

$$v = \frac{c}{\lambda} \qquad \text{(Eq 13.1)}$$

Another parameter in this context is the wavenumber $\tilde{v}$, which describes the number of completed wave cycles per distance and is typically measured in 1 cm^{-1}.

Electromagnetic phenomena are explained in terms of quantum mechanics. The photon is the elementary particle responsible for electromagnetic phenomena. It carries the electromagnetic radiation and has properties of a wave, as well as of a particle, albeit having a mass of zero. As a particle, it interacts with matter by transferring its energy E:

$$E = \frac{h \times c}{\lambda} = h \times v \qquad \text{(Eq 13.2)}$$

where h = 6.63×10^{-34} J s is the Planck constant and v is the frequency of the radiation as introduced above.

13.1.2 Interaction With Matter

The interaction of electromagnetic radiation with matter is a **quantum phenomenon** and dependent upon both the properties of the radiation and the appropriate structural parts of the samples involved. This is not surprising, since the origin of electromagnetic radiation is due to energy changes within matter itself. Energy absorption by a molecule induces different types of transition.

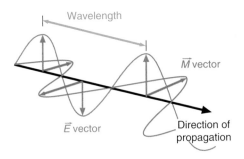

Figure 13.1 Light is electromagnetic radiation and can be described as a wave propagating transversally in space and time. The electric ($\bar{E}$) and magnetic ($\bar{M}$) field vectors are directed perpendicular to each other. For UV/vis, circular dichroism and fluorescence spectroscopy, the electric field vector $\bar{E}$ is of most importance. For nuclear magnetic resonance, the important component is the magnetic field vector $\bar{M}$.

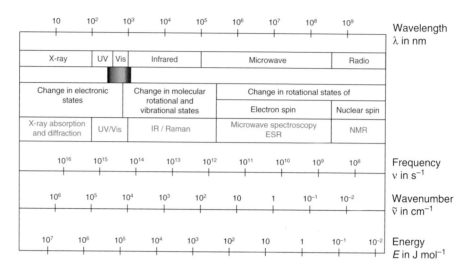

Figure 13.2 The electromagnetic spectrum and its uses for spectroscopic methods.

These transitions are also quantum phenomena and correspond to either the absorption or emission of an energy quantum (a photon). Depending on the order of magnitude of the energy of the electromagnetic radiation, given by the frequency or the wavelength, different states of the molecule are affected. Molecular (sub)structures responsible for interaction with electromagnetic radiation are called **chromophores**. The spectra which arise from such transitions are principally predictable and give access to various types of information about the molecule.

Figure 13.2 shows the spectrum of electromagnetic radiation organised by increasing wavelength, and thus decreasing energy, from left to right. Also annotated are the types of radiation and the various interactions with matter and the resulting spectroscopic applications, as well as the interdependent parameters of frequency ν and wavenumber $\tilde{\nu}$.

When considering a diatomic molecule (see Figure 13.3), energy absorption by the molecule can induce two transition types: **rotational** or **vibrational**. Rotational and vibrational levels possess discrete energies that only merge into a continuum at very high energy. Each electronic state of a molecule possesses its own set of rotational and vibrational levels. Since the kind of schematics shown in Figure 13.3 is rather complex, the **Jablonski diagram** is used instead, where electronic and vibrational states are schematically drawn as horizontal lines, and vertical lines depict possible transitions (see Figure 13.15).

In order for a **transition** to occur in the system, energy must be absorbed. The energy change ΔE needed is defined in quantum terms by the difference in absolute energies between the final and the starting state as $\Delta E = E_{final} - E_{start} = h \times \nu$.

Electrons in either atoms or molecules may be distributed between several energy levels, but principally reside in the lowest levels (**ground state**). In order for an electron to be promoted to a higher level (**excited state**), energy must be put into the

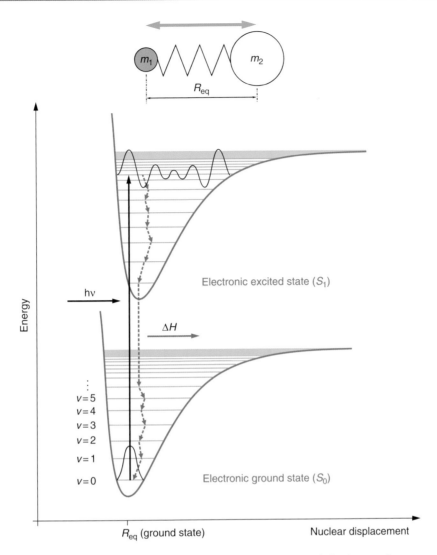

Figure 13.3 Energy diagram for a diatomic molecule exhibiting rotation and vibration, as well as an electronic structure. The distance between two masses m_1 and m_2 (nuclear displacement) is described as a Lennard–Jones potential curve with different equilibrium distances (R_{eq}) for each electronic state. Energetically lower states always have lower equilibrium distances. The vibrational levels (horizontal lines) are superimposed on the electronic levels. Rotational levels are superimposed on the vibrational levels and not shown for reasons of clarity.

system. If this energy $E = h\nu$ is derived from electromagnetic radiation, this gives rise to an absorption spectrum, and an electron is transferred from the electronic ground state (S_0) into the first electronic excited state (S_1). The molecule will also be in an excited vibrational and rotational state. Subsequent relaxation of the molecule into the vibrational ground state of the first electronic excited state will occur. The electron can then revert back to the electronic ground state. For non-fluorescent molecules, this is accompanied by the emission of heat (ΔH).

13.1.3 Polarised Light

A light source usually consists of a collection of randomly oriented emitters. Therefore, the emitted light is a collection of waves with all possible orientations of the electric ($\bar{E}$) field vectors. This light is non-polarised. Linearly or plane-polarised light is obtained by passing light through a polariser that transmits light with only a single plane of polarisation, i.e. it passes only those components of the vector that are parallel to the axis of the polariser (Figure 13.4).

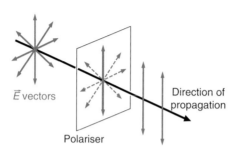

Figure 13.4 Generation of linearly polarised light.

$\bar{E}$ vectors

Polariser

Direction of propagation

The intensity of transmitted light depends on the orientation of the polariser. Maximum transmission is achieved when the plane of polarisation is parallel to the axis of the polariser; the transmission is zero when the orientation is perpendicular. The polarisation P is defined as:

$$P = \frac{I_{\parallel} - I_{\perp}}{I_{\parallel} + I_{\perp}}$$

(Eq 13.3)

(a) Linearly (plane) polarised light

(b) Circularly polarised light

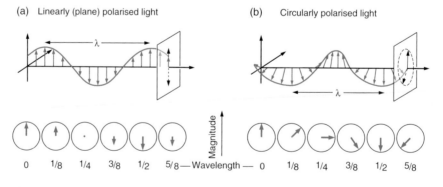

| 0 | 1/8 | 1/4 | 3/8 | 1/2 | 5/8 |

Magnitude

Wavelength

| 0 | 1/8 | 1/4 | 3/8 | 1/2 | 5/8 |

Direction of electric vector is constant
Magnitude of electric vector varies

Direction of electric vector direction varies
Magnitude of electric vector is constant

Figure 13.5 (a) Linearly (plane) and (b) circularly polarised light. While the ($\bar{E}$) vector of circularly polarised light always has the same magnitude, but a varying direction, the direction of the ($\bar{E}$) vector of linearly polarised light is constant; it is its magnitude that varies. With the help of vector algebra, one can now reversely think of linearly polarised light as a composite of two circularly polarised beams with opposite handedness (Figure 13.6a).

$I_\parallel$ and $I_\perp$ are the intensities observed parallel and perpendicular to an arbitrary axis, respectively. The polarisation can vary between -1 and $+1$; it is zero when the light is non-polarised. Light with $0 < |P| < 0.5$ is called partially polarised.

If the $\bar{E}$ vectors of two electromagnetic waves are ¼ wavelength out of phase and perpendicular to each other, the vector that is the sum of the $\bar{E}$ vectors of the two components rotates around the direction of propagation so that its tip follows a helical path. Such light is called **circularly polarised** (Figure 13.5), which itself is a special case of **elliptically polarised** light (Figure 13.6).

Applications of polarised light are, for example, circular dichroism spectroscopy (Section 13.3) and fluorescence polarisation (Section 13.5).

(a)   (b)

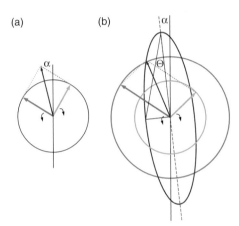

Figure 13.6 (a) Linearly polarised light can be thought of consisting of two circularly polarised components with opposite 'handedness'. The vector sum of the left- and right-handed circularly polarised light yields linearly polarised light. The angle α defines the angle between the plane of the incident light and the plane of the resulting polarised light (a consequence of the phenomenon of optical rotatory dispersion). (b) If the amplitudes of left- and right-handed polarised components differ, the resulting light is elliptically polarised. The composite vector will trace the ellipse shown in grey. The ellipse is characterised by a major and a minor axis. The ratio of minor and major axis yields tan Θ, where Θ is called the ellipticity.

13.1.4 Lasers

Laser is an acronym for light amplification by stimulated emission of radiation. A detailed explanation of the theory of lasers is beyond the scope of this textbook. A simplified description starts with the use of photons of a defined energy to excite an absorbing material. This results in elevation of an electron to a higher-energy level. If, whilst the electron is in the excited state, another photon of precisely that energy arrives, then, instead of the electron being promoted to an even higher level, it can return to the original ground state. However, this transition is accompanied by the emission of two photons with the same wavelength and exactly in phase (coherent photons). Multiplication of this process will produce coherent light with extremely narrow spectral bandwidth. In order to produce an ample supply of suitable photons, the absorbing material is surrounded by a rapidly flashing light of high intensity (pumping).

Lasers are indispensable tools in many areas of science, including biochemistry and biophysics. Several modern spectroscopic techniques utilise laser light sources, due to their high intensity and accurately defined spectral properties. Probably one of the most revolutionising applications in the life sciences is the use of lasers in DNA sequencing with fluorescence labels (see Sections 4.9.7 and 13.5.3), which enabled the breakthrough in whole-genome sequencing.

13.2 ULTRAVIOLET AND VISIBLE LIGHT SPECTROSCOPY

Ultraviolet and visible light spectroscopy gives information on size, form, flexibility of molecules or electronic structure in its fundamental and excited states.

In the spectral domain of ultraviolet and visible light, energy absorption by a molecule can affect three transition types: electronic, rotational or vibrational transitions. These regions of the electromagnetic spectrum and their associated techniques are probably the most widely used for analytical work and research into biological problems. The UV/vis chromophores, i.e. the molecular structures that absorb UV/vis light, typically possess several conjugated double bonds. Many organic molecules with such characteristics are very useful tools in colorimetric applications (see Table 13.1).

Table 13.1 **Common colorimetric and UV/vis absorption assays**

Substance	Reagent	λ_{max} (nm)
Metal ions	Complexation → charge-transfer complex	
Amino acids	(a) Ninhydrin	570 (proline: 420)
	(b) Cupric salts	620
Cysteine residues, thiolates	Ellman reagent (di-sodium-bis-(3-carboxy-4-nitrophenyl)-disulfide)	412
Protein (see also Section 5.2)	(a) Folin (phosphomolybdate, phosphotungstate, cupric salt)	660
	(b) Biuret (reacts with peptide bonds)	540
	(c) BCA reagent (bicinchoninic acid)	562
	(d) Coomassie Brilliant Blue (Bradford assay)	595
	(e) Direct	Tyr, Trp: 278 peptide bond: 190
Coenzymes	Direct	FAD: 438 NADH: 340 NAD$^+$: 260
Carotenoids	Direct	420, 450, 480
Porphyrins	Direct	400 (Soret band)
Carbohydrate	(a) Phenol, H_2SO_4	Glucose: 490 Xylose: 480
	(b) Anthrone (anthrone, H_2SO_4)	620 or 625

Table 13.1 (cont.)

Substance	Reagent	λ_{max} (nm)
Reducing sugars	Dinitrosalicylate, alkaline tartrate buffer	540
Pentoses	(a) Bial (orcinol, ethanol, $FeCl_3$, HCl)	665
	(b) Cysteine, H_2SO_4	380–415
Hexoses	(a) Carbazole, ethanol, H_2SO_4	540 or 440
	(b) Cysteine, H_2SO_4	380–415
	(c) Arsenomolybdate	500–570
Glucose	Glucose oxidase, peroxidase, o-dianisidine, phosphate buffer	420
Ketohexose	(a) Resorcinol, thiourea, ethanoic acid, HCl	520
	(b) Carbazole, ethanol, cysteine, H_2SO_4	560
	(c) Diphenylamine, ethanol, ethanoic acid, HCl	635
Hexosamines	Ehrlich (dimethylaminobenzaldehyde, ethanol, HCl)	530
DNA	(a) Diphenylamine	595
	(b) Direct	260
RNA	Bial (orcinol, ethanol, $FeCl_3$, HCl)	665
Sterols and steroids	Liebermann–Burchard (acetic anhydride, H_2SO_4, chloroform)	425, 625
Cholesterol	Cholesterol oxidase, peroxidase, 4-amino-antipyrine, phenol	500
ATPase assay	Coupled enzyme assay with ATPase, pyruvate kinase, lactate dehydrogenase:	NADH: 340
	ATP → ADP (consumes ATP)	
	phosphoenolpyruvate → pyruvate (consumes ADP)	
	pyruvate → lactate (consumes NADH)	

13.2.1 Chromophores

In proteins, there are three types of chromophores relevant for UV/vis spectroscopy:

- peptide bonds (amide bond)
- certain amino-acid side chains (mainly tryptophan and tyrosine)
- certain prosthetic groups and coenzymes (e.g. porphyrine groups such as in haem).

The electronic transitions of the peptide bond occur in the far UV. The intense peak at 190 nm, and the weaker one at 210–220 nm are relevant transitions. A number of amino acids (Asp, Glu, Asn, Gln, Arg and His) have weak electronic transitions

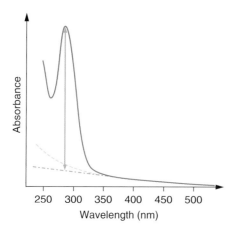

Figure 13.7 The presence of larger aggregates in biological samples gives rise to **Rayleigh scatter** visible by a considerable slope in the region from 500 to 350 nm. The dashed line shows the correction to be applied to spectra with Rayleigh scatter, which increases with λ^{-4}. Practically, linear extrapolation of the region from 500 to 350 nm is often performed to correct for the scatter. The corrected absorbance is indicated by the double arrow.

at around 210 nm. Usually, these cannot be observed in proteins because they are masked by the more intense peptide bond absorption. The most useful range for proteins is above 230 nm, where there are absorptions from aromatic side chains. While a very weak absorption maximum of phenylalanine occurs at 257 nm, tyrosine and tryptophan dominate the typical protein spectrum with their absorption maxima at 274 nm and 280 nm, respectively (Figure 13.7). *In praxi*, the presence of these two aromatic side chains gives rise to a band at ~278 nm. Cystine (Cys_2) possesses a weak absorption maximum of similar strength as phenylalanine at 250 nm. This band can play a role in rare cases in protein optical activity or protein fluorescence.

Proteins that contain prosthetic groups (e.g. haem, flavin, carotenoid) and some metal–protein complexes, may have strong absorption bands in the UV/vis range. These bands are usually sensitive to local environment and can be used for physical studies of enzyme action. Carotenoids, for instance, are a large class of red, yellow and orange plant pigments composed of long carbon chains with many conjugated double bonds. They contain three maxima in the visible region of the electromagnetic spectrum (~420 nm, 450 nm, 480 nm).

Porphyrins are the prosthetic groups of haemoglobin, myoglobin, catalase and cytochromes. Electron delocalisation extends throughout the cyclic tetrapyrrole ring of porphyrins and gives rise to an intense transition at ~400 nm called the Soret band. The spectrum of haemoglobin is very sensitive to changes in the iron-bound ligand. These changes can be used for structure–function studies of haem proteins.

Molecules such as FAD (flavin adenine dinucleotide), NADH and NAD^+ are important coenzymes of proteins involved in electron-transfer reactions (redox reactions). They can be conveniently assayed by using their UV/vis absorption: 438 nm (FAD), 340 nm (NADH) and 260 nm (NAD^+).

In genetic material, the absorption of UV light occurs between 260 nm and 275 nm. The absorption spectra of the purine (adenine, guanine) and pyrimidine (cytosine, thymine, uracil) bases in nucleic acids or in polymers are sensitive to pH and greatly influenced by electronic interactions between bases.

Example 13.1   **ELECTRONIC TRANSITIONS IN UV/VIS CHROMOPHORES**

Question   While interaction with electromagnetic radiation of comparably low energy (e.g. infrared light) causes molecules to undergo vibrational transitions, the higher-energy radiation (i.e. light in the visible and UV region) can cause many molecules to undergo electronic transitions. In such cases, when photons with energy in the UV/vis range are absorbed by a molecule, one of its electrons jumps from the highest occupied molecular orbital (HOMO) to the lowest unoccupied molecular orbital (LUMO). On the energy scale, the HOMO thus ranks lower than the LUMO.

In β-carotene, the molecule responsible for the orange colour of carrots, a peak wavelength of λ_{max} = 450 nm is observed. What is the energy gap between the HOMO and the LUMO?

Answer   As the absorbed photon with a wavelength of λ_{max} provides just enough energy to cause the transition of an electron from the HOMO to the LUMO, the energy gap between the two molecular orbitals can be calculated from the energy of the photon. The latter is given by Equation 13.2:

$$E = \frac{h}{v} = \frac{h \times c}{\lambda} = \frac{6.63 \times 10^{-34} \, \text{J s} \times 2.99 \times 10^{8} \, \text{m s}^{-1}}{450 \times 10^{-9} \, \text{m}} = 0.044 \times 10^{-17} \, \text{J}$$

The molar gap energy is then obtained as per:

$$\Delta E_{gap} = E \times N_{A} = 0.044 \times 10^{-17} \text{J} \times 6.022 \times 10^{23} \, \text{mol}^{-1} = 265 \, \text{kJ mol}^{-1}$$

13.2.2 Principles

Quantification of Light Absorption

If light with wavelength λ and intensity I_0 passes through a sample with appropriate transparency and path length (thickness) d, the intensity I drops along the pathway in an exponential manner (Figure 13.8). I is the number of photons by unit of surface and by unit of time.

The chance of a photon being absorbed by matter is given by an extinction coefficient, which itself is dependent on the wavelength λ of the photon. The characteristic absorption parameter for the sample is the extinction coefficient α, yielding the correlation

$$I = I_0 \times e^{-\alpha \times d} \tag{Eq 13.4}$$

The ratio $T = I/I_0$ is called the transmission.

Biochemical samples usually comprise aqueous solutions, where the substance of interest is present at a molar concentration c. Algebraic transformation of the exponential correlation into an expression based on the decadic logarithm yields the Beer–Lambert law:

Figure 13.8 When light is absorbed by a sample of thickness d, the intensity I decreases in an exponential fashion along the path.

$$\lg \frac{I_0}{I} = \lg \frac{1}{T} = \varepsilon \times c \times d = A \qquad\qquad \text{(Eq 13.5)}$$

where $[d] = 1$ cm, $[c] = 1$ mol dm^{-3} and $[\varepsilon] = 1$ dm^3 mol^{-1} cm^{-1}. ε is the molar absorption coefficient (also molar extinction coefficient). Due to the change of base from e to 10, the extinction coefficient α is related to the molar absorption coefficient ε as $\alpha = 2.303 \times \varepsilon$. A is the absorbance of the sample, which is displayed on the spectrophotometer.

The Lambert–Beer law is valid for low concentrations only. Higher concentrations might lead to association of molecules and therefore cause deviations from ideal behaviour. Absorbance and extinction coefficients are additive parameters, which complicates determination of concentrations in samples with more than one absorbing species. Note that in dispersive samples or suspensions, scattering effects increase the absorbance, since the scattered light is not reaching the detector for readout. The absorbance recorded by the spectrophotometer is thus overestimated and needs to be corrected (Figure 13.7).

Deviations from the Beer–Lambert Law

According to the Beer–Lambert law, absorbance is linearly proportional to the concentration of chromophores. This might not be the case in samples with high absorbance. Every spectrophotometer has a certain amount of stray light, which is light received at the detector, but not anticipated in the spectral band isolated by the monochromator. In order to obtain reasonable signal-to-noise ratios, the intensity of light at the chosen wavelength (I_λ) should be 10 times higher than the intensity of the stray light (I_{stray}). If the stray light gains in intensity, the effects measured at the detector have nothing or little to do with chromophore concentration. Secondly, molecular events might lead to deviations from the Beer–Lambert law. For instance, chromophores might dimerise at high concentrations and, as a result, might possess different spectroscopic parameters.

Absorption or Light Scattering: Optical Density

In some applications, for example measurement of turbidity of cell cultures (determination of biomass concentration), it is not the absorption but the scattering of light (see Section 14.3.5) that is actually measured with a spectrophotometer. Extremely turbid samples like bacterial cultures do not absorb the incoming light. Instead, the light is scattered and thus, the spectrometer will record an apparent absorbance (sometimes also called attenuance). In this case, the observed parameter is called optical density (OD). Instruments specifically designed to measure turbid samples are nephelometers or Klett meters; however, most biochemical laboratories use the general UV/vis spectrometer for determination of optical densities of cell cultures.

Factors Affecting UV/Vis Absorption

Biochemical samples are usually buffered aqueous solutions, which have two major advantages. First, proteins and peptides are comfortable in water as a solvent, which is also the 'native' solvent. Second, in the wavelength interval of UV/vis (700–200 nm) the water spectrum does not show any absorption bands and thus acts as a silent component of the sample.

The absorption spectrum of a chromophore is only partly determined by its chemical structure. The environment also affects the observed spectrum, which can be described by three main parameters:

- protonation/deprotonation (pH, redox)
- solvent polarity (dielectric constant of the solvent)
- orientation effects.

Vice versa, the immediate environment of chromophores can be probed by assessing their absorption, which makes chromophores ideal reporter molecules for environmental factors. Four effects, two each for wavelength and absorption changes, have to be considered:

- a wavelength shift to higher values is called a red shift or bathochromic effect
- similarly, a shift to lower wavelengths is called a blue shift or hypsochromic effect
- an increase in absorption is called hyperchromicity ('more colour')
- a decrease in absorption is called hypochromicity ('less colour').

Protonation/deprotonation arises either from changes in pH or oxidation/reduction reactions, which makes chromophores pH- and redox-sensitive reporters. As a rule of thumb, λ_{max} and ε increase, i.e. the sample displays a batho- and hyperchromic shift, if a titratable group becomes charged.

Furthermore, solvent polarity affects the difference between ground and excited states. Generally, when shifting to a less polar environment one observes a batho- and hyperchromic effect. Conversely, a solvent with higher polarity elicits a hypso- and hypochromic effect.

Lastly, orientation effects, such as an increase in order of nucleic acids from single-stranded to double-stranded DNA, lead to different absorption behaviour (see also Section 4.7.1). A sample of free nucleotides exhibits a higher absorption than a sample with identical amounts of nucleotides, but assembled into a single-stranded polynucleotide. Accordingly, double-stranded polynucleotides exhibit an even smaller absorption than two single-stranded polynucleotides. This phenomenon is known as the **hypochromicity of polynucleotides**. The increased exposure (and thus stronger absorption) of the individual nucleotides in the less ordered states provides a simplified explanation for this behaviour.

13.2.3 Instrumentation

The optical density measurement of a compound in solution is a result of two measurements:

(a) The absorbance of all molecular chromophores present in the buffer (also called control solution):

$$A_{control} = \lg \frac{I_0}{I_{control}}$$

(b) The absorbance of the analyte together with the control solution (also called the sample solution):

$$A_{sample} = \lg \frac{I_0}{I_{sample}} .$$

Since absorbances of components are additive (Figure 13.9), the absorbance of the analyte is:

$$A = A_{sample} - A_{control} = \lg \frac{I_0}{I_{sample}} - \lg \frac{I_0}{I_{control}} = \lg \frac{I_{control}}{I_{sample}} .$$

UV/vis spectrophotometers (Figure 13.10) typically require the following components:

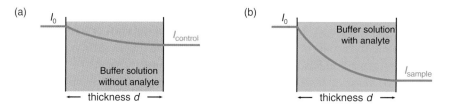

(a) I_0 $I_{control}$ Buffer solution without analyte — thickness d →

(b) I_0 Buffer solution with analyte I_{sample} — thickness d →

Figure 13.9 Absorbances are additive. The absorbance of the analyte is obtained by measuring (a) a control and (b) a sample solution.

- a light source
- a monochromator
- a sample cell/cuvette
- a detection system for transmitted radiation.

The light source is a tungsten filament bulb for the visible part of the spectrum and a deuterium bulb for the UV region. Since the emitted light consists of many different wavelengths, a **monochromator**, consisting of either a prism or a rotating metal grid of high precision called a **grating**, is placed between the light source and the sample. Wavelength selection can also be achieved by using coloured filters as monochromators that absorb all but a certain limited range of wavelengths. This limited range is called the **bandwidth** of the filter. Filter-based wavelength selection is used in **colorimetry**, a method with moderate accuracy, but best suited for specific colorimetric assays (typically carried out in **plate readers**) where only certain wavelengths are of interest. If wavelengths are selected by prisms or gratings, the technique is called **spectrophotometry**.

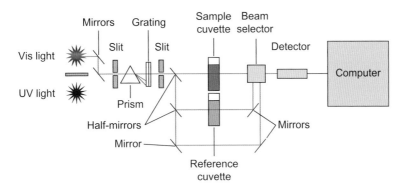

Figure 13.10 Optical arrangements in a dual-beam spectrophotometer. Either a prism or a grating constitutes the monochromator of the instrument. Optical paths are shown as green lines.

A prism splits the incoming light into its components by refraction. Refraction occurs because radiation of different wavelengths travels along different paths in a medium of higher density. In order to maintain the principle of velocity conservation, light of shorter wavelength (higher speed) must travel a longer distance (i.e. blue sky effect). At a grating, the splitting of wavelengths is achieved by diffraction. Diffraction is a reflection phenomenon that occurs at a grid surface, in this case a series of engraved fine lines. The distance between the lines has to be of the same order of magnitude as the wavelength of the diffracted radiation. By varying the distance between the lines, different wavelengths are selected. This is achieved by rotating the grating perpendicular to the optical axis. The resolution achieved by gratings is much higher than that available with prisms. Nowadays instruments almost exclusively contain gratings as monochromators as they can be reproducibly made in high quality by photoreproduction.

The bandwidth of a colorimeter is determined by the filter used as the monochromator. A filter that appears red to the human eye is transmitting red light and absorbs almost any other (visual) wavelength. This filter would be used to examine blue solutions, as these would absorb red light. The filter used for a specific colorimetric assay is thus made of a colour complementary to that of the solution being tested. Theoretically, a single wavelength is selected by the monochromator in spectrophotometers, and the emergent light is a parallel beam. Here, the bandwidth is defined as twice the half-intensity bandwidth. The bandwidth is a function of the optical slit width. The narrower the slit width, the more reproducible are measured absorbance values. In contrast, the sensitivity becomes less as the slit narrows, because less radiation passes through to the detector.

Since borosilicate glass and normal plastics absorb UV light, such cuvettes can only be used for applications in the visible range of the spectrum (up to 350 nm). For UV measurements, quartz cuvettes need to be used. However, disposable plastic cuvettes have been developed that allow for measurements over the entire range of the UV/vis spectrum.

UV/vis spectrophotometers are usually dual-beam spectrometers, where the first channel contains the sample and the second channel holds the control (buffer)

Example 13.2 **ESTIMATION OF MOLAR EXTINCTION COEFFICIENTS**

Question In order to determine the concentration of a solution of the peptide

MAMVSEFLKQ AWFIENEEQE YVQTVKSSKG GPGSAVSPYP TFNPSS

in water, the molar absorption coefficient needs to be estimated.

Answer The molar extinction coefficient ε is a characteristic parameter of a molecule and varies with the wavelength of incident light. Because of useful applications of the Beer –Lambert law, the value of ε needs be known for many molecules being used in biochemical experiments.

Very frequently in biochemical research, the molar extinction coefficient of proteins is estimated using incremental ε_i values for each absorbing protein residue (chromophore). Summation over all residues yields a reasonable estimation for the extinction coefficient. The simplest increment system is based on values from Gill and von Hippel.[1] The determination of protein concentration using this formula only requires an absorption value at $\lambda = 280$ nm. Increments ε_i are used to calculate a molar extinction coefficient at 280 nm for the entire protein or peptide by summation over all relevant residues in the protein:

Residue	ε_i (280 nm) in dm^3 mol^{-1} cm^{-1}
Cys$_2$	120
Trp	5690
Tyr	1280

For the peptide above, one obtains:

$$\varepsilon = 1 \times \varepsilon_{Trp} + 2 \times \varepsilon_{Tyr} = (1 \times 5690 + 2 \times 1280) \text{ dm}^3 \text{ mol}^{-1} \text{ cm}^{-1} = 8250 \text{ dm}^3 \text{ mol}^{-1} \text{ cm}^{-1}$$

1. Gill S. C. and von Hippel P.H. (1989) Calculation of protein extinction coefficients from amino acid sequence data. *Analytical Biochemistry* 182, 319–326. Erratum: *Analytical Biochemistry* (1990) 189, 283.

---

for correction (Figure 13.10). The incoming light beam is split into two parts by a half-mirror. One beam passes through the sample, the other through a control (blank, reference). This approach obviates any problems of variation in light intensity, as both reference and sample would be affected equally. The measured absorbance is the difference between the two recorded transmitted beams of light. Depending on the instrument, a second detector measures the intensity of the incoming beam, although some instruments use an arrangement where one detector measures the incoming and the transmitted intensities alternately. The latter design is better from an analytical point of view as it eliminates potential variations between the two detectors.

Alternatively, one can record the control spectrum first and use this as an internal reference for the sample spectrum. The latter approach has become very popular, as many spectrometers in laboratories are computer-controlled, and baseline correction can be carried out using the software by simply subtracting the control from the sample spectrum.

13.2.4 Applications

The usual procedure for (colorimetric) assays is to prepare a set of standards and produce a plot of concentration versus absorbance called a calibration curve. This should be linear as long as the Beer–Lambert law applies. Absorbances of unknowns are then measured and their concentration interpolated from the linear region of the plot. It is important that one never extrapolates beyond the region for which an instrument has been calibrated, as this potentially introduces enormous errors.

To obtain good spectra, the maximum absorbance should be approximately 0.5, which corresponds to concentrations of about 50 μM (assuming $\varepsilon = 10\,000$ dm^3 mol^{-1} cm^{-1}).

Qualitative and Quantitative Analysis

The absorption spectrum is the plot of absorbance versus the wavelength (or the frequency). The wavelength of maximal absorption λ_{max} is the wavelength for which a maximum is observed in the absorption spectrum and at which the sample possesses a corresponding molar extinction coefficient of

$$\varepsilon_{max} = \frac{A_{max}}{c \times d} \qquad \text{(Eq 13.6)}$$

These wavelengths and molar extinction coefficients can allow identification of a given compound in a given solvent. However, such identification is rather limited and qualitative analysis is only possible for systems where appropriate features and parameters are known, such as identification of certain classes of compounds both in the pure state and in biological mixtures (e.g. protein-bound).

Most commonly, this type of spectroscopy is used for quantification of biological samples either directly or via colorimetric assays. In many cases, proteins can be quantified directly using their intrinsic chromophores, tyrosine and tryptophan. Protein spectra are acquired by scanning from 500 to 210 nm. The characteristic features in a protein spectrum are a band at 278/280 nm and another at 190 nm (Figure 13.7). The region from 500 to 300 nm provides valuable information about the presence of any prosthetic groups or coenzymes. Protein quantification by single wavelength measurements at 260 and 280 nm only should be avoided, as the presence of larger aggregates (contaminants or protein aggregates) gives rise to considerable Rayleigh scatter that needs to be corrected for (Figure 13.7).

Example 13.3   **DETERMINATION OF CONCENTRATIONS**

Question   1. The concentration of an aqueous solution of a protein is to be determined
assuming:
   a. the molar extinction coefficient ε is known
   b. the molar extinction coefficient ε is not known.
   2. What is the concentration of an aqueous solution of a DNA sample?

Answer 1.a The protein concentration of a pure sample can be determined by using the
Beer–Lambert law. The absorbance at 280 nm is determined from a protein
spectrum, and the molar extinction coefficient at this wavelength needs to
be experimentally determined or estimated:

$$\rho^* = \frac{A \times M}{\varepsilon \times d}$$

where ρ^* is the mass concentration in mg cm^{-3} and M the molar mass of
the assayed species in g mol^{-1}.

1.b Alternatively, an empirical formula known as the Warburg–Christian formula can be used without knowledge of the value of the molar extinction
coefficient:

$$\rho^* = \left(1.52 \times A_{280} - 0.75 \times A_{260}\right) \text{ mg cm}^{-3}$$

Other commonly used applications to determine the concentration of protein in a sample make use of colorimetric assays that are based on chemicals (folin, biuret, bicinchoninic acid or Coomassie Brilliant Blue) binding
to protein groups. Concentration determination in these cases requires a
calibration curve measured with a protein standard, usually bovine serum
albumin.

2. As we have seen above, the genetic bases have absorption bands in the
UV/vis region. Thus, the concentration of a DNA sample can be determined
spectroscopically. Assuming that a pair of nucleotides has a molecular mass
of $M = 660$ g mol^{-1}, the absorbance A of a solution with double-stranded
DNA at 260 nm can be converted to mass concentration ρ^* by:

$$\rho^* = 50 \text{ µg cm}^{-3} \times A_{260}.$$

The ratio A_{260}/A_{280} is an indicator of the purity of the DNA solution and
should be in the range 1.8–2.0.

Difference Spectra

The main advantage of difference spectroscopy is its capacity to detect small absorbance changes in systems with high background absorbance. Difference spectra can be obtained in two ways: either by subtraction of one absolute absorption spectrum from another, or by placing one sample in the reference cuvette and another in the test cuvette (see Section 13.2.3). Difference spectra have three distinct features as compared to absolute spectra:

1. Difference spectra may contain negative absorbance values
2. Absorption maxima and minima may be displaced and the extinction coefficients are different from those in peaks of absolute spectra
3. There are points of zero absorbance, usually accompanied by a change of sign of the absorbance values. These points are observed at wavelengths where both species of related molecules exhibit identical absorbances (isosbestic points), and which may be used for checking for the presence of interfering substances.

Common applications for difference UV spectroscopy include the determination of the number of aromatic amino acids exposed to solvent, detection of conformational changes occurring in proteins, detection of aromatic amino acids in the active sites of enzymes, and monitoring of reactions involving 'catalytic' chromophores (prosthetic groups, coenzymes).

Spectrophotometric and Colorimetric Assays

For biochemical assays testing for time- or concentration-dependent responses of systems, an appropriate readout is required that is coupled to the progress of the reaction (reaction coordinate). Therefore, the biophysical parameter being monitored (readout) needs to be coupled to the biochemical parameter under investigation. Frequently, the monitored parameter is the absorbance of a system at a given wavelength, which is monitored throughout the course of the experiment. Preferably, one should try to monitor the changing species directly (e.g. protein absorption, starting product or generated product of a reaction), but in many cases this is not possible and a secondary reaction has to be used to generate an appropriate signal for monitoring. A common application of the latter approach is the determination of protein concentration by Lowry or Bradford assays, where a secondary reaction is used to colour the protein. The more intense the colour, the more protein is present. These assays are called colorimetric assays and a number of those commonly used are listed in Table 13.1.

13.3 CIRCULAR DICHROISM SPECTROSCOPY

13.3.1 Principles

In Section 13.1.3 we have already seen that electromagnetic radiation oscillates in all possible directions and that it is possible to preferentially select waves oscillating in a single plane, as applied for fluorescence polarisation. The phenomenon first known as mutarotation (described by Lowry in 1898) became manifest in due course as a special property of optically active isomers allowing the rotation of plane-polarised light. Optically active isomers are compounds of identical chemical composition and topology, but whose mirror images cannot be superimposed; such compounds are called chiral.

Polarimetry and Optical Rotatory Dispersion

Polarimetry essentially measures the angle through which the plane of polarisation is changed after linearly polarised light is passed through a solution containing a chiral substance. Optical rotatory dispersion (ORD) spectroscopy is a technique that measures this ability of a chiral substance to change the plane-polarisation as a function of the wavelength. The angle α between the plane of the resulting linearly polarised light against that of the incident light is dependent on the refractive index for left (n_{left}) and right (n_{right}) circularly polarised light (Figure 13.6). The refractive index can be calculated as the ratio of the speed of light *in vacuo* and the speed of light in matter. After normalisation against the amount of substance present in the sample (thickness of sample/cuvette length d, and mass concentration ρ'), a substance-specific constant $[\alpha]$ is obtained that can be used to characterise chiral compounds.

Circular Dichroism

In addition to changing the plane of polarisation, an optically active sample also shows unusual absorption behaviour. Left- and right-handed polarised components of the incident light are absorbed differently by the sample, which yields a difference in the absorption coefficients $\Delta\varepsilon = \varepsilon_{left} - \varepsilon_{right}$. This difference is called circular dichroism (CD). The difference in absorption coefficients $\Delta\varepsilon$ (i.e. CD) is measured in units of dm^3 mol^{-1} cm^{-1}, and is the observed quantity in CD experiments. Frequently, results from protein CD experiments are reported as ellipticity θ (Figure 13.6). Normalisation of θ similar to the ORD yields the molar ellipticity:

$$\theta = \frac{M \times \Theta}{10 \times \rho^* \times d} = \frac{\ln 10}{10} \times \frac{180°}{4\pi} \times \Delta\varepsilon \qquad \text{(Eq 13.7)}$$

It is common practice to display graphs of CD spectra with the molar ellipticity θ in units of $1° \ cm^2 \ dmol^{-1} = 10° \ cm^2 \ mol^{-1}$ on the abscissa (Figure 13.11). Three important conclusions can be drawn:

1. ORD and CD are the manifestation of the same underlying phenomenon
2. If an optically active molecule has a positive CD, then its enantiomer will have a negative CD of exactly the same magnitude
3. The phenomenon of CD can only be observed at wavelengths where the optically active molecule has an absorption band.

The Chromophores of Protein Secondary Structure

In Section 13.2.1, we saw that the peptide bond in proteins possesses UV absorption bands in the area of 190–220 nm. The carbon atom vicinal to the peptide bond (the Cα atom) is asymmetric and a stereogenic centre in all amino acids except glycine. This chirality induces asymmetry into the peptide bond chromophore. Because of the serial arrangement of the peptide bonds making up the backbone of a protein, the individual chromophores couple with each other. The (secondary) structure of a polypeptide thus induces an 'overall chirality' which gives rise to the CD phenomenon of a protein in the wavelength interval 190–260 nm.

With protein circular dichroism, the molar ellipticity θ is typically normalised further with respect to the number of residues to yield the mean residue ellipticity θ_{res} (also called MRE), owing to the fact that the chromophores responsible for the chiral absorption phenomenon are the peptide bonds. Therefore, the number of chromophores of a polypeptide in this context is equal to the number of residues less one. Because of the Beer–Lambert law (Equation 13.5), the number of chromophores is proportional to the magnitude of absorption, i.e. in order to normalise the spectrum of an individual polypeptide for reasons of comparison, the CD has to be scaled by the number of peptide bonds.

13.3.2 Instrumentation and Practicalities

The basic layout of a CD spectrometer follows that of a single-beam UV absorption spectrometer. However, owing to the nature of the measured effects, an electro-optic modulator, as well as a more sophisticated detector are needed.

Generally, left and right circularly polarised light passes through the sample in an alternating fashion. This is achieved by an electro-optic modulator, which is a crystal that transmits either the left- or right-handed polarised component of linearly polarised light, depending on the polarity of the electric field that is applied by alternating currents. The photomultiplier detector produces a voltage proportional to the ellipticity of the resultant beam emerging from the sample. The light source of the spectrometer is continuously flushed with nitrogen to avoid the formation of ozone and help to maintain the lamp.

CD spectrometry involves measuring a very small difference between two absorption values that are large signals. The technique is thus very susceptible to noise and measurements must be carried out carefully. Some practical considerations involve having a clean quartz cuvette, and using buffers with low concentrations of additives. While this is sometimes tricky with protein samples, reducing the salt concentrations to values as low as 5 mM helps to obtain good spectra. Also, filtered solutions should be used to avoid any turbidity of the sample that could produce scatter. Saturation of the detector must be avoided, this becoming more critical with lower wavelengths. Therefore, good spectra are obtained in a certain range of protein concentrations only where enough sample is present to produce a good signal and does not saturate the detector. Typical protein concentrations are 0.03–0.3 mg cm^{-3}.

In order to calculate molar ellipticities θ, and thus be able to compare the CD spectra of different samples with each other, the concentration of the sample must be known. Provided the protein possesses sufficient amounts of UV/vis-absorbing chromophores, it is thus advisable to subject the CD sample to a protein concentration determination by UV/vis, as described in Section 13.2.4.

13.3.3 Applications

The main application for protein CD spectroscopy is the verification of the adopted secondary structure. The application of CD to determine the tertiary structure is limited, owing to the inadequate theoretical understanding of the effects of different parts of the molecules at this level of structure. In its simplest implementation, information

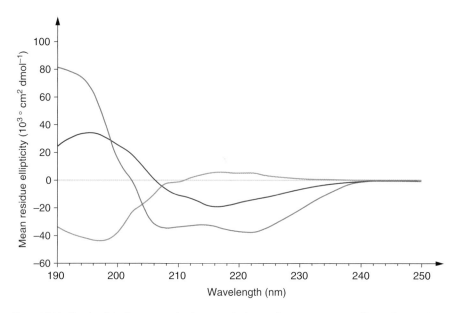

Figure 13.11 Circular dichroism spectra for three standard secondary structures according to Fasman. An α-helical peptide is shown in turquoise, a peptide adopting β-strand (extended) structure in magenta, and a random coil peptide in green.

about the relative weight of secondary-structure elements (Section 2.2.2) present in a folded protein, the CD spectra of the three fundamental secondary structure types α-helix, β-strand and random coil (Figure 13.11) are subjected to a linear combination analysis to best fit the experimentally observed CD spectrum. In other words, the sum of each of the three 'standard spectra', weighted by a factor between 0 and 1, should reconstitute the observed CD spectrum. The weighting factors then represent the percentages of each secondary structure element in the folded protein.

Rather than analysing the secondary structure of a 'static sample', different conditions can be tested. For instance, some peptides adopt different secondary structures when in solution or membrane-bound. The comparison of CD spectra of such peptides in the absence and presence of small unilamellar phospholipid vesicles shows a clear difference in the type of secondary structure. Measurements with lipid vesicles are tricky, because their physical extensions give rise to scatter. Other options in this context include CD experiments at lipid monolayers, which can be realised at synchrotron beam lines, or by the use of optically clear vesicles (reverse micelles).

CD spectroscopy can also be used to monitor changes in secondary structure within a sample over time. Frequently, CD instruments are equipped with temperature-control units and the sample can be heated in a controlled fashion. As the protein undergoes its transition from the folded to the unfolded state, the CD at a certain wavelength (usually 222 nm) is monitored and plotted against the temperature, thus yielding a thermal denaturation curve that can be used for stability analysis.

Further applications include the use of circular dichroism of an observable for kinetic measurements using the stopped-flow technique (see Section 13.8).

Example 13.4   **CIRCULAR DICHROISM SPECTROSCOPY**

Question

(1) In what wavelength range would you expect to find the CD spectrum for compound (a)?

(2) Sketch the CD spectrum for the protein shown in (b).

(a)   (b)

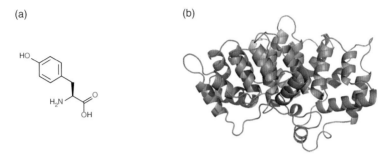

Answer

(1) The CD spectrum for any compound is in the same wavelength range as its UV absorption spectrum. The compound in question is tyrosine, which has an absorption maximum at approximately 280 nm. The CD spectrum for the tyrosine side chain is thus in the near-UV region 260–300 nm.

(2) The spectrum should indicate almost exclusively α-helical secondary structure content. One therefore expects a CD spectrum with two clear minima (208 nm and 222 nm) of similar amplitude. The maximum mean residue ellipticity (MRE) is strong (> 20 000° cm² dmol⁻¹).

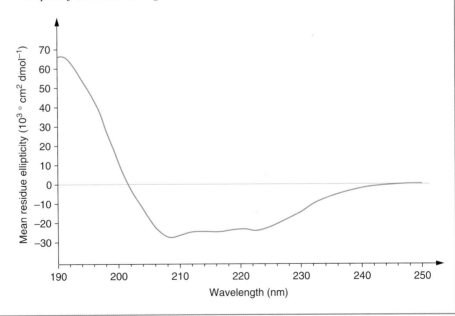

13.4 INFRARED AND RAMAN SPECTROSCOPY

13.4.1 Principles

Within the electromagnetic spectrum (Figure 13.2), the energy range below the UV/vis is the infrared region, encompassing the wavelength range of about 700 nm to 25 μm, and thus reaching from the red end of the visible to the microwave region. The absorption of infrared light by a molecule results in transition to higher levels of vibration (Figure 13.3).

For the purpose of this discussion, the bonds between atoms can be considered as flexible springs, illustrating the constant vibrational motion within a molecule (Figure 13.12). Bond vibrations can thus be either **stretching** or **bending** (deformation) actions. Theory predicts that a molecule with N atoms will have a total of $[3N-6]$ fundamental vibrations ($[3N-5]$, if the molecule is linear): $[2N-5]$ bending, and $[N-1]$ stretching modes (Figure 13.13).

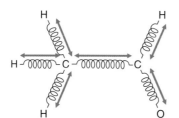

Figure 13.12 Possible stretching vibrations in acetaldehyde.

Infrared and Raman spectroscopy (see below) give similar information about a molecule, but the criteria for the phenomena to occur are different for each type. For asymmetric molecules, incident infrared light will give rise to an absorption band in the infrared spectrum, as well as a peak in the Raman spectrum. However, as shown in Figure 13.13, symmetric molecules, such as CO_2, that possess a centre of symmetry, show a selective behaviour: bands that appear in the infrared spectrum do not appear in the Raman spectrum, and vice versa:

- An infrared spectrum arises from the fact that a molecule absorbs incident light of a certain wavelength which will then be 'missing' from the transmitted light. The recorded spectrum will show an absorption band.
- A Raman spectrum arises from the analysis of scattered light (see also Section 14.3.5). The largest part of an incident light beam passes through the sample (transmission). A small part is scattered isotropically, i.e. uniformly in all directions (Rayleigh scatter), and possesses the same wavelength as the incident beam. The Raman spectrum arises from the fact that a very small proportion of light scattered by the sample will have a different frequency from the incident light. As different vibrational states are excited, energy portions will be missing, thus giving rise to peaks at lower frequencies than the incident light (Stokes lines). Notably, higher frequencies are also observed (anti-Stokes lines); these arise from excited molecules returning to ground state. The emitted energy is dumped onto the incident light, which results in scattered light of higher energy than the incident light.

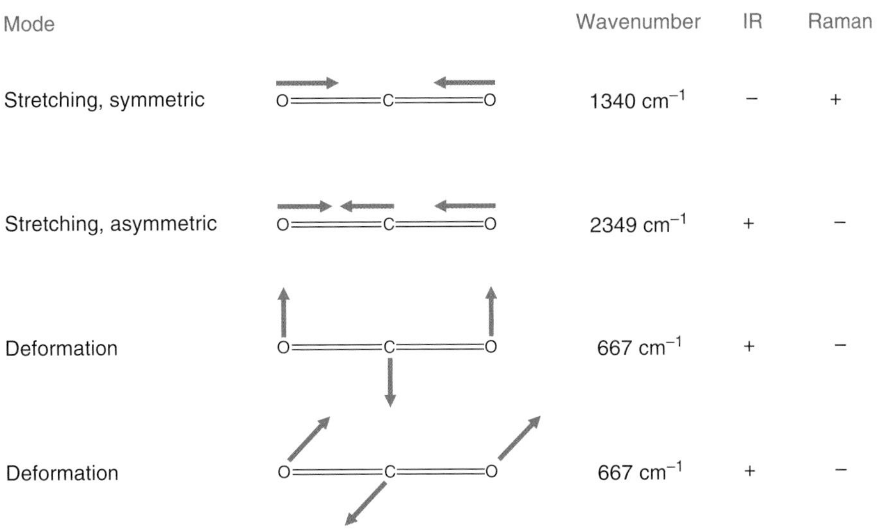

Mode		Wavenumber	IR	Raman
Stretching, symmetric	O=C=O	$1340\ \mathrm{cm}^{-1}$	–	+
Stretching, asymmetric	O=C=O	$2349\ \mathrm{cm}^{-1}$	+	–
Deformation	O=C=O	$667\ \mathrm{cm}^{-1}$	+	–
Deformation	O=C=O	$667\ \mathrm{cm}^{-1}$	+	–

Figure 13.13 Normal vibrational modes for CO_2. For symmetric molecules that possess a centre of symmetry, bands that appear in the IR do not appear in the Raman spectrum. A linear molecule with N atoms possesses $(3N - 5)$ different normal vibrational modes; thus, CO_2 possesses the $3 \times 3 - 5 = 4$ modes shown.

The criterion for a band to appear in the infrared spectrum is that the transition to the excited state is accompanied by a change in dipole momentum, i.e. a change in charge displacement. Conversely, the criterion for a peak to appear in the Raman spectrum is a change in polarisability (the distortion of the electron distribution in the molecule) of the molecule during the transition.

Infrared Spectroscopy

The fundamental frequencies observed are characteristic of the functional groups concerned, hence the term fingerprint. Figure 13.14 shows the major bands of an FT–IR spectrum of the drug phenacetin. As the number of functional groups increases in more complex molecules, the absorption bands become more difficult to assign. However, groups of certain bands regularly appear near the same wavelength and may be assigned to specific functional groups. Such group frequencies are thus extremely helpful in structural diagnosis. A more detailed analysis of the structure of a molecule is possible, because the wavenumber $\tilde{\nu}$ associated with a particular functional group varies slightly, owing to the influence of the molecular environment. For example, it is possible to distinguish between C–H vibrations in methylene (–CH_2–) and methyl groups (–CH_3).

Raman Spectroscopy

The assignment of peaks in Raman spectra usually requires consideration of peak position, intensity and form, as well as depolarisation. This allows identification of the type of symmetry of individual vibrations, but not the determination of structural elements of a molecule. The depolarisation is calculated as the ratio of two intensities with perpendicular and parallel polarisations (see Section 13.1.3) with respect to the incident

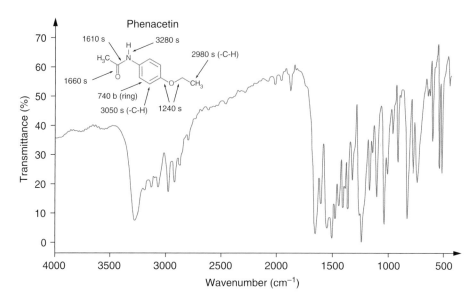

Figure 13.14 FT–IR spectrum of phenacetin, historically the first synthetic fever reducer to go on the market. Bands at the appropriate wavenumbers (in cm⁻¹) are shown, indicating the bonds with which they are associated, and the type (s, stretching; b, bending).

beam. The use of lasers as the light source for Raman spectroscopy easily facilitates the use of linearly polarised light. Practically, the Raman spectrum is measured twice; in the second measurement, the polarisation plane of the incident beam is rotated by 90°.

13.4.2 Instrumentation

The most common source for infrared light is white-glowing zircon oxide or the so-called globar made of silicium carbide with a glowing temperature of 1500 K. The beam of infrared light passes a monochromator and splits into two separate beams: one runs through the sample, the other through a reference made of the substance the sample is prepared in. After passing through a splitter alternating between both beams, they are reflected into the detector. The reference is used to compensate for fluctuations in the source, as well as to cancel possible effects of the solvent. Non-covalent materials must be used for sample containment and in the optics, as these materials are transparent to infrared. All materials need to be free of water, because of the strong absorption of the O–H vibration.

Analysis using a Michelson interferometer enables Fourier transform infrared spectroscopy (FT-IR). The entire light emitted from the source is passed through the sample at once, and then split into two beams that are reflected back onto the point of split (interferometer plate). Using a movable mirror, path length differences are generated between both beams, yielding an interferogram that is recorded by the detector. The interferogram is related to a conventional infrared spectrum by a mathematical operation called Fourier transformation (see also Figure 14.5). Historically, liquid or suspension samples were delivered to the IR instrument held as layers between NaCl planes or solids were pressed into KBr discs or prepared in thick suspensions (mulls) such as nujol. Most contemporary instruments combine the FT-IR technique with a sample probing based on attenuated total reflection (ATR). ATR employs the features of an evanescent field, the same technology used in surface plasmon resonance (Section 14.6.1). The infrared light enters the internal reflection element (a glass prism similar to the one described for SPR) and probes the sample on the surface of the prism. The fact that the sample is probed multiple times has the advantage of yielding stronger absorbances. ATR-FT-IR has become a very popular method since small amounts of solid and liquid samples can be measured conveniently without lengthy preparation.

For Raman spectroscopy, aqueous solutions are frequently used, since water possesses a rather featureless weak Raman spectrum. The Raman effect can principally be observed with bright, monochromatic light of any wavelength; however, light in the visible region of the spectrum is normally used due to few unwanted absorption effects. The ideal light source for Raman spectrometers is therefore a laser. Because the Raman effect is observed in light scattered off the sample, typical spectrometers use a 90° configuration.

13.4.3 Applications

The use of infrared and Raman spectroscopy has typically been in chemical and biochemical research of small-molecule compounds such as drugs, metabolic intermediates and substrates. Examples are the identification of synthesised compounds, or identification of sample constituents (e.g. in food) when coupled to a separating method such as gas chromatography (GC-IR).

However, FT–IR is also used for analysis of peptides and proteins. The peptide bond gives rise to nine characteristic bands, named amide A, B, I, II, III, ..., VII. The amide I $(1600–1700 \text{ cm}^{-1})$ and amide II $(1500–1600 \text{ cm}^{-1})$ bands are the major contributors to the protein infrared spectrum. Both bands are directly related to the backbone conformation and have thus been used for assessment of the secondary structure of peptides and proteins. The interpretation of spectra of molecules with a large number of atoms usually involves deconvolution of individual bands and second-derivative spectra.

Time-resolved FT–IR enables the observation of protein reactions at the submillisecond timescale. One of the first applications of this technique was the investigation of the light-driven proton pump bacteriorhodopsin. For instance, the catalytic steps in the proton-pumping mechanism have been validated with time-resolved FT–IR, and involve transfer of a proton from the Schiff base $(R_1R_2C=N-R_3)$ to a catalytic aspartate residue, followed by reprotonation of a second catalytic aspartate residue.

13.5 FLUORESCENCE SPECTROSCOPY

13.5.1 Principles

Fluorescence is an emission phenomenon where an energy transition from a higher to a lower state is accompanied by radiation. Only molecules in their excited forms are able to emit fluorescence; thus, they have to be brought into a state of higher energy prior to the emission phenomenon.

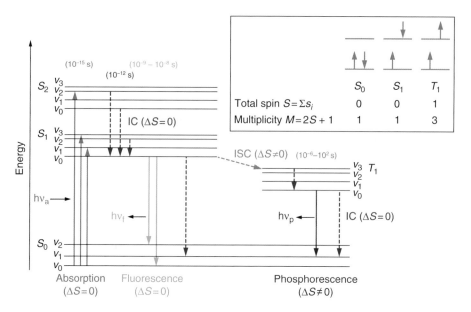

Figure 13.15 Jablonski diagram. Shown are the electronic ground state (S_0), two excited singlet states (S_1, S_2) and a triplet state (T_1). Only select vibrational levels (v) are illustrated. Solid vertical lines indicate radiative transitions, dotted lines show non-radiative transitions. Inset: Explanation of total spin S and multiplicity M. The total spin S is calculated as the sum of the individual electron spins. The multiplicity M is obtained as either 1 (singlet state) or 3 (triplet state).

We have already seen in Section 13.1.2 that molecules possess discrete states of energy. Potential energy levels of molecules have been depicted by different Lennard–Jones potential curves with overlaid vibrational (and rotational) states (Figure 13.3). Such diagrams can be abstracted further to yield Jablonski diagrams (Figure 13.15).

In these diagrams, energy transitions are indicated by vertical lines. Not all transitions are possible; allowed transitions are defined by the selection rules of quantum mechanics. A molecule in its electronic and vibrational ground state (S_0v_0) can absorb photons matching the energy difference of its various discrete states. The required photon energy has to be higher than that required to reach the vibrational ground state of the first electronic excited state (S_1v_0). The excess energy is absorbed as vibrational energy ($v > 0$), and quickly dissipated as heat by collision with solvent molecules. The molecule thus returns to the vibrational ground state (S_1v_0). These relaxation processes are non-radiating transitions from one energetic state to another with lower energy, and are called internal conversion (IC). From the lowest level of the first electronic excited state, the molecule returns to the ground state (S_0) either by emitting light (fluorescence) or by a non-radiative transition. Upon radiative transition, the molecule can end up in any of the vibrational states of the electronic ground state (as per quantum mechanical rules).

If the vibrational levels of the ground state overlap with those of the electronic excited state, the molecule will not emit fluorescence, but rather revert to the ground state by non-radiative internal conversion. This is the most common way for excitation energy to be dissipated and is why fluorescent molecules are rather rare. Most molecules are flexible and thus have very high vibrational levels in the ground state. Indeed, most fluorescent molecules possess fairly rigid aromatic rings or ring systems. The fluorescent group in a molecule is called a fluorophore.

Since radiative energy is lost in fluorescence as compared to the absorption, the fluorescent light is always at a longer wavelength than the exciting light (Stokes shift). The emitted radiation appears as a band spectrum, because there are many closely related wavelength values dependent on the vibrational and rotational energy levels attained. The fluorescence spectrum of a molecule is independent of the wavelength of the exciting radiation and has a mirror image relationship with the absorption spectrum. The probability of the transition from the electronic excited to the ground state is proportional to the intensity of the emitted light.

The fluorescence properties of a molecule are determined by properties of the molecule itself (internal factors), as well as the environment of the protein (external factors). The fluorescence intensity emitted by a molecule is dependent on the lifetime of the excited state. The transition from the excited to the ground state can be treated like a first-order decay process, i.e. the number of molecules in the excited state decreases exponentially with time. In analogy to kinetics, the exponential coefficient k_r is called the rate constant and is the reciprocal of the lifetime: $\tau_r = k_r^{-1}$. The lifetime is the time it takes to reduce the number of fluorescence-emitting molecules to $1/e$ of the original number, and is proportional to λ^3.

The effective lifetime τ of excited molecules, however, differs from the fluorescence lifetime τ_r since other, non-radiative processes also affect the number of molecules in the excited state. τ is dependent on all processes that cause relaxation: fluorescence emission, internal conversion, quenching, fluorescence resonance energy transfer, reactions of the excited state and inter-system crossing.

The ratio of photons emitted and photons absorbed by a fluorophore is called the **quantum yield** Φ (Equation 13.8). It equals the ratio of the rate constant for fluorescence emission k_r and the sum of the rate constants for all six processes mentioned above.

$$\Phi = \frac{N_{em}}{N_{abs}} = \frac{k_r}{k} = \frac{k_r}{k_r + k_{IC} + k_{ISC} + k_{reaction} + k_Q \times [Q] + k_{FRET}} = \frac{\tau}{\tau_r} \qquad \text{(Eq 13.8)}$$

The quantum yield is a dimensionless quantity, and, most importantly, the only absolute measure of fluorescence of a molecule. Measuring the quantum yield requires comparison with a fluorophore of known quantum yield. In biochemical applications, this measurement is rarely done. Most commonly, the fluorescence emissions of two or more related samples are compared and their relative differences analysed.

13.5.2 Instrumentation

Fluorescence spectroscopy works most accurately at very low concentrations of emitting fluorophores. UV/vis spectroscopy, in contrast, is least accurate at such low concentrations. One major factor adding to the high sensitivity of fluorescence

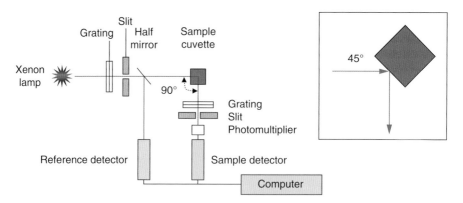

Figure 13.16 Schematics of a spectrofluorimeter with 'T' geometry (90°). Optical paths are shown as blue lines. Inset: Geometry of front-face illumination.

applications is the spectral selectivity. Due to the Stokes shift, the wavelength of the emitted light is different from that of the exciting light. Another feature makes use of the fact that fluorescence is emitted in all directions. By placing the detector perpendicular to the excitation pathway, the background of the incident beam is reduced.

The schematics of a typical spectrofluorimeter are shown in Figure 13.16. Two monochromators are used, one for tuning the wavelength of the exciting beam and a second one for analysis of the fluorescence emission. Due to the emitted light always having a lower energy than the exciting light, the wavelength of the excitation monochromator is set at a lower wavelength than the emission monochromator. The better fluorescence spectrometers in laboratories have a photon-counting detector yielding very high sensitivity. Temperature control is required for accurate work as the emission intensity of a fluorophore is dependent on the temperature of the solution.

Two geometries are possible for the measurement, with the 90° arrangement most commonly used. Pre- and post-filter effects can arise owing to absorption of light prior to reaching the fluorophore and the reduction of emitted radiation. These phenomena are also called **inner filter effects** and are more evident in solutions with high concentrations. As a rough guide, the absorption of a solution to be used for fluorescence experiments should be less than 0.05. The use of microcuvettes containing less material can also be useful. Alternatively, the front-face illumination geometry (Figure 13.16 inset) can be used, which obviates the inner filter effect. Also, while the 90° geometry requires cuvettes with two neighbouring faces being clear (usually, fluorescence cuvettes have four clear faces), the front-face illumination technique requires only one clear face, as excitation and emission occur at the same face. However, front-face illumination is less sensitive than 90° illumination.

13.5.3 Applications

There are many and highly varied applications for fluorescence despite the fact that relatively few compounds exhibit the phenomenon. The effects of pH, solvent composition and the polarisation of fluorescence may all contribute to structural elucidation.

Measurement of fluorescence lifetimes can be used to assess rotation correlation coefficients and thus particle sizes. Non-fluorescent compounds are often labelled with fluorescent probes to enable monitoring of molecular events. This is termed **extrinsic fluorescence**, as distinct from intrinsic fluorescence, where the native compound exhibits the property. Some fluorescent dyes are sensitive to the presence of metal ions and can thus be used to track changes of these ions in in-vitro samples, as well as whole cells.

Since fluorescence spectrometers have two monochromators, one for tuning the excitation wavelength and one for analysing the emission wavelength of the fluorophore, one can measure two types of spectrum: excitation and emission. For fluorescence **excitation spectrum** measurement, one sets the emission monochromator at a fixed wavelength (λ_{em}) and scans a range of excitation wavelengths, which are then recorded as the ordinate (x-coordinate) of the excitation spectrum; the fluorescence emission at λ_{em} is plotted as the abscissa. Measurement of **emission spectra** is achieved by setting a fixed excitation wavelength (λ_{exc}) and scanning a wavelength range with the emission monochromator. To yield a spectrum, the emission wavelength λ_{em} is recorded as the ordinate and the emission intensity at λ_{em} is plotted as the abscissa.

Intrinsic Protein Fluorescence

Proteins possess two intrinsic fluorophores: tryptophan and tyrosine. The latter has lower quantum yield and its fluorescence emission is almost entirely quenched when it becomes ionised, or is located near an amino or carboxyl group or a tryptophan residue. Intrinsic protein fluorescence is thus usually determined by tryptophan fluorescence, which can be selectively excited at 295–305 nm. Illumination at 280 nm excites tyrosine and tryptophan fluorescence and the resulting spectra might therefore contain contributions from both types of residues.

The main application for intrinsic protein fluorescence aims at conformational monitoring. We have already mentioned that the fluorescence properties of a fluorophore depend significantly on environmental factors, including solvent, pH, possible quenchers, neighbouring groups, etc.

A number of empirical rules can be applied to interpret protein fluorescence spectra:

- As a fluorophore moves into an environment with less polarity, its emission spectrum exhibits a hypsochromic shift (λ_{max} moves to shorter wavelengths) and the intensity at λ_{max} increases.
- Fluorophores in a polar environment show a decrease in quantum yield with increasing temperature. In a non-polar environment, there is little change.
- Tryptophan fluorescence is quenched by neighbouring protonated acidic groups.

When interpreting effects observed in fluorescence experiments, one has to consider carefully all possible molecular events. For example, a compound added to a protein solution can cause quenching of tryptophan fluorescence. This could come about by binding of the compound at a site close to the tryptophan (i.e. the residue is surface-exposed to a certain degree), or due to a conformational change induced by the compound.

The comparison of protein fluorescence excitation and emission spectra can yield insights into the location of fluorophores. The close spatial arrangement of fluorophores

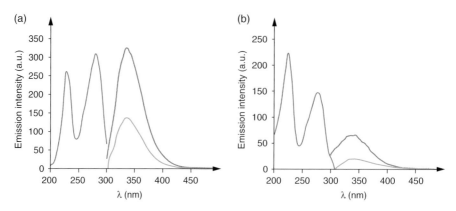

Figure 13.17 Comparison of fluorescence excitation and emission spectra can yield insights into internal quenching. Excitation spectra with emission wavelength 340 nm are shown in dark green. Emission spectra with excitation wavelength 295 nm are shown in light green; emission spectra with excitation wavelength 280 nm are grey. (a) PDase homologue (*Escherichia coli*). The excitation spectrum (λ_{em} = 340 nm) and the emission spectrum (λ_{exc} = 280 nm) show comparable emission intensities. (b) CPDase (*Arabidopsis thaliana*); in this protein, the fluorophores are located in close proximity to each other, which leads to the effect of intrinsic quenching, as obvious from the lower intensity of the emission spectrum as compared to the excitation spectrum.

within a protein can lead to quenching of fluorescence emission; this might be seen by the lower intensity of the emission spectrum when compared to the excitation spectrum (Figure 13.17).

Extrinsic Fluorescence

Frequently, molecules of interest for biochemical studies are non-fluorescent. In many of these cases, an external fluorophore can be introduced into the system by chemical coupling or non-covalent binding. Some examples of commonly used external fluorophores are shown in Figure 13.18. Three criteria must be met by fluorophores in this context. First, it must not affect the mechanistic properties of the system under investigation. Second, its fluorescence emission needs to be sensitive to environmental conditions in order to enable monitoring of the molecular events. And last, the fluorophore must be tightly bound at a unique location.

Common non-conjugating extrinsic chromophores for proteins are 1-anilino-8-naphthalene sulfonate (ANS) or SYPRO® Orange, which emit only weak fluorescence in a polar environment, i.e. in aqueous solution. However, in a non-polar environment, e.g. when bound to hydrophobic patches on proteins, its fluorescence emission is significantly increased and the spectrum shows a hypsochromic shift: λ_{max} shifts from 475 nm to 450 nm. Such fluorescence dyes are called **amphiphilic dyes** and are valuable tools for assessment of the degree of non-polarity. They can also be used in competition assays to monitor binding of ligands and prosthetic groups. SYPRO® Orange is frequently used in a technique called differential scanning fluorimetry (DSF) whereby the binding of small organic molecules to proteins is assessed by monitoring of the thermal stability of a protein in the absence and presence of potential ligands.

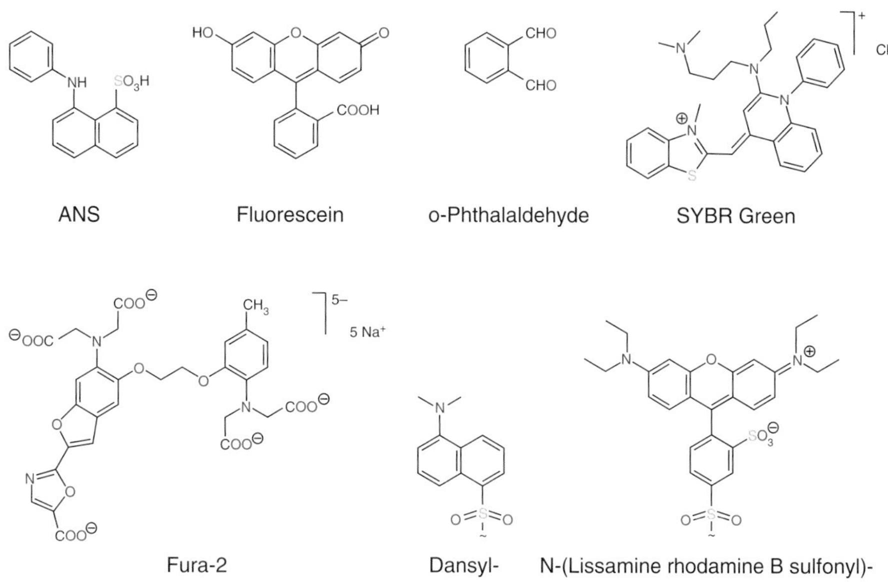

Figure 13.18 Structures of some extrinsic fluorophores. Fura-2 is a fluorescent chelator for divalent and higher-valent metal ions (Ca^{2+}, Ba^{2+}, Sr^{2+}, Pb^{2+}, La^{3+}, Mn^{2+}, Ni^{2+}, Cd^{2+}).

Reagents such as fluorescamine, o-phthalaldehyde or 6-aminoquinolyl-N-hydrox-ysuccinimidyl carbamate have been very popular conjugating agents used to derivatise amino acids for analysis. o-Phthalaldehyde, for example, is a non-fluorescent compound that reacts with primary amines and β-mercaptoethanol to yield a highly sensitive fluorophore.

Metal-chelating compounds with fluorescent properties are useful tools for a variety of assays, including monitoring of metal homeostasis in cells. Widely used probes for calcium are the chelators Fura-2, Indo-1 and Quin-1. Since the chemistry of such compounds is based on metal chelation, cross-reactivity of the probes with other metal ions is possible.

The intrinsic fluorescence of nucleic acids is very weak and the required excitation wavelengths are too far in the UV region to be useful for practical applications. Numerous extrinsic fluorescent probes spontaneously bind to DNA and display enhanced emission. While in earlier days ethidium bromide was one of the most widely used dyes for this application, it has nowadays been replaced by SYBR® Green, as the latter probe poses fewer hazards for health and the environment and has no teratogenic properties, unlike ethidium bromide. These probes bind DNA by intercalation of the planar aromatic ring systems between the base pairs of double-helical DNA. Their fluorescence emission in water is very weak and increases about 30-fold upon binding to DNA.

Labelling with Fluorescent Proteins

The labelling of target proteins with native fluorescent proteins presents a useful approach to study various processes in vivo. This approach requires the generation of fusion proteins at the genetic level, whereby a fluorescent protein such as green fluorescent protein (GFP) is attached N- or C-terminally to the target protein sequence.

The fluorescence label enables visualisation of the target protein in the context of fluorescence microscopy (see Section 11.2.5), but also monitoring of protein expression in preparative biochemistry, in particular when expressing membrane proteins. Further development of fluorescent proteins with different excitation and emission properties, for example red fluorescent protein (RFP) and others, allows for real-time and multi-colour imaging – a methodology that is often referred to as living colour. One example of protein engineering in this context is a genetically altered form of GFP that physically changes structure upon exposure to light of 413 nm. The excitation and emission wavelengths of this photo-activatable GFP are essentially identical to those of native GFP (λ_{exc} = 488 nm, λ_{em} = 509 nm); however, the fluorescence emission intensity is increased 100-fold after an 'activating' light pulse at 413 nm.

Quenching

In Section 13.5.1, we saw that the quantum yield of a fluorophore is dependent on several internal and external factors. One of the external factors with practical implications is the presence of a **quencher**. A quencher molecule decreases the quantum yield of a fluorophore by non-radiating processes. The absorption (excitation) process of the fluorophore is not altered by the presence of a quencher. However, the energy of the excited state is transferred onto the quenching molecules. Two kinds of quenching processes can be distinguished:

1. Dynamic quenching, which occurs by collision between the fluorophore in its excited state and the quencher
2. Static quenching whereby the quencher forms a complex with the fluorophore. The complex has a different electronic structure compared to the fluorophore alone and returns from the excited state to the ground state by non-radiating processes.

It follows intuitively that the efficacy of both processes is dependent on the concentration of quencher molecules. The mathematical treatment for each process is different, because of two different chemical mechanisms. Interestingly, in both cases the degree of quenching, expressed as I_0/I, is directly proportional to the quencher concentration. For **collisional (dynamic) quenching**, the resulting equation has been named the **Stern–Volmer equation**:

$$\frac{I_0}{I} - 1 = k_Q \times [Q] \times \tau_0 \qquad \text{(Eq 13.9)}$$

The Stern–Volmer equation relates the degree of quenching (expressed as the ratio of I_0 to I) to the molar concentration of the quencher [Q], the lifetime of the fluorophore τ_0, and the rate constant of the quenching process k_Q. In the case of **static quenching** (Equation 13.10), (I_0/I) is related to the equilibrium constant K_a that describes the formation of the complex between the excited fluorophore and the quencher, and the concentration of the quencher. Importantly, a plot of (I_0/I) versus [Q] yields for both quenching processes a linear graph with a y-intercept of 1.

$$\frac{I_0}{I} - 1 = K_a \times [Q] \qquad \text{(Eq 13.10)}$$

Thus, fluorescence data obtained by intensity measurements alone cannot distinguish between static or collisional quenching. The measurement of fluorescence lifetimes or the temperature/viscosity dependence of quenching can be used to determine the kind of quenching process. It should be added that both processes can also occur simultaneously in the same system.

The fact that static quenching is due to complex formation between the fluorophore and the quencher makes this phenomenon an attractive assay for binding of a ligand to a protein. In the simplest case, the fluorescence emission being monitored is the intrinsic fluorescence of the protein. While this is a very convenient titration assay when validated for an individual protein–ligand system, one has to be careful when testing unknown pairs, because the same decrease in intensity can occur by collisional quenching.

Highly effective quenchers for fluorescence emission are oxygen, as well as the iodide ion. Use of these quenchers allows surface mapping of biological macromolecules. For instance, iodide can be used to determine whether tryptophan residues are exposed to solvent.

Fluorescence Resonance Energy Transfer (FRET)

Fluorescence resonance energy transfer (FRET) was first described by Förster in 1948. The process can be explained in terms of quantum mechanics by a non-radiative energy transfer from a donor to an acceptor chromophore. The requirements for this process are a reasonable overlap of emission and excitation spectra of donor and acceptor chromophores, close spatial vicinity of both chromophores (10–100 Å), and an almost parallel arrangement of their transition dipoles. Of great practical importance is the correlation:

$$\text{FRET} \sim \frac{1}{R_0^6} \qquad \text{(Eq 13.11)}$$

showing that the FRET effect is inversely proportional to the distance between donor and acceptor chromophores, R_0.

The FRET effect is particularly suitable for biological applications, since distances of 10–100 Å are of the order of the dimensions of biological macromolecules. Furthermore, the relation between FRET and distance allows for measurement of molecular distances and makes this application a kind of 'spectroscopic ruler'. If a process exhibits changes in molecular distances, FRET can also be used to monitor the molecular mechanisms.

The high specificity of the FRET signal allows for monitoring of molecular interactions and conformational changes with high spatial (1–10 nm) and temporal (<1 ns) resolution. The possibility of localising and monitoring cellular structures and proteins in physiological environments makes this method especially attractive. The effects can be observed even at low concentrations (as low as single molecules), in different environments (different solvents, including living cells), and observations may be done in real time.

In most cases, different chromophores are used as donor and acceptor, presenting two possibilities to record FRET: either as donor-stimulated fluorescence emission of

Example 13.5   **FRET APPLICATIONS IN DNA SEQUENCING AND INVESTIGATION OF MOLECULAR MECHANISMS**

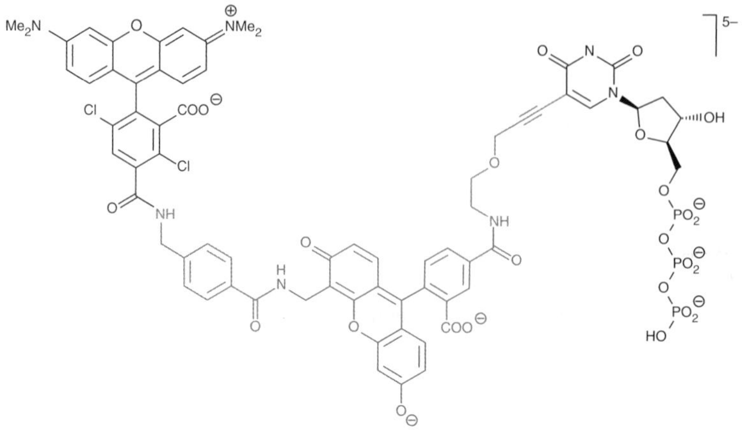

Caption: Structure of one of the four BigDye™ terminators, ddT-EO-6CFB-dTMR. The moieties from left to right are: 5-carboxy-dichloro-rhodamine (FRET acceptor), 4-aminomethyl benzoate linker, 6-carboxy-4′-aminomethylfluorescein (FRET donor), propargyl ethoxyamino linker, dUTP.

BigDyes™ are a widely used application of FRET fluorophores (see figure above). Since 1997, these fluorophores are generally used as chain termination markers in automated DNA sequencing. As such, BigDyes™ are in major part responsible for the great success of genome projects.

In many instances, FRET allows monitoring of conformational changes, protein folding, as well as protein–protein, protein–membrane and protein–DNA interactions. For instance, the three subunits of T4 DNA polymerase holoenzyme arrange around DNA in a torus-like geometry. Using the tryptophan residue in one of the subunits as a FRET donor and a coumarine label conjugated to a cysteine residue in the adjacent subunit (FRET acceptor), the distance change between both subunits was be monitored and seven steps involved in opening and closing of the polymerase were identified. Other examples of this approach include studies of the architecture of *Escherichia coli* RNA polymerase, the calcium-dependent change of troponin and structural studies of neuropeptide Y dimers.

the acceptor or as fluorescence quenching of the donor by the acceptor. However, the same chromophore may be used as donor and acceptor simultaneously; in this case, the depolarisation of fluorescence is the observed parameter. Since non-FRET stimulated fluorescence emission by the acceptor can result in undesirable background fluorescence, a common approach is usage of non-fluorescent acceptor chromophores.

Other applications include tandem dyes used, for example, in flow cytometry (see Section 8.4). FRET-based assays may be used to elucidate the effects of new substrates for different enzymes or putative agonists in a quick and quantitative manner. Furthermore, FRET detection might be used in high-throughput screenings (see Section 24.5.3), which makes it very attractive for drug development.

Bioluminescence Resonance Energy Transfer (BRET)

Bioluminescence resonance energy transfer (BRET) uses the FRET effect with native fluorescent or luminescent proteins as chromophores. The phenomenon is observed naturally, for example with the sea pansy *Renilla reniformis*. It contains the enzyme luciferase, which oxidises luciferin (coelenterazin) by simultaneously emitting light at λ_{exc} = 480 nm. This light directly excites green fluorescent protein (GFP), which, in turn, emits fluorescence at λ_{em} = 509 nm.

While nucleic acids have been the main players in the genomic era, the post-genomic/proteomic era focusses on gene products, the proteins. New proteins are being discovered and characterised, others are already used within biotechnological processes. For the classification and evaluation of enzymes and receptors in particular, reaction systems can be designed such that the reaction of interest is detectable quantitatively using FRET donor and acceptor pairs.

For instance, detection methods for protease activity can be developed based on BRET applications. A protease substrate is fused to a GFP variant on the N-terminal side and dsRED on the C-terminal side. The latter protein is a red fluorescing FRET acceptor and the GFP variant acts as a FRET donor. Once the substrate is cleaved by a protease, the FRET effect is abolished. This is used to directly monitor protease activity. With a combination of FRET analysis and two-photon excitation spectroscopy it is also possible to carry out a kinetic analysis.

A similar idea is used to label human insulin receptor in order to quantitatively assess its activity. The insulin receptor is a glycoprotein with two α and two β subunits, which are linked by dithioether bridges. The binding of insulin induces a conformational change and causes a close spatial arrangement of both β subunits. This, in turn, activates the tyrosine kinase activity of the receptor.

In pathological conditions such as diabetes, the tyrosine kinase activity is different from that in healthy conditions. Evidently, it is of great interest to find compounds that stimulate the same activity as insulin. By fusing the β subunit of human insulin receptor to *Renilla reniformis* luciferase and yellow fluorescent protein (YFP), a FRET donor–acceptor pair is obtained, which reports the ligand-induced conformational change and precedes the signal transduction step. This reporter system is able to detect the effects of insulin and insulin-mimicking ligands in order to assess dose-dependent behaviour.

Fluorescence Recovery After Photobleaching (FRAP)

If a fluorophore is exposed to high-intensity radiation, it may be irreversibly damaged and lose its ability to emit fluorescence. Intentional bleaching of a fraction of fluorescently labelled molecules in a membrane can be used to monitor the motion of labelled molecules in certain (two-dimensional) compartments (Figure 13.19). Moreover, the time-dependent

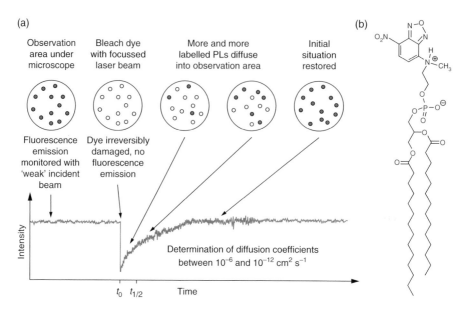

Figure 13.19 (a) Schematic of a FRAP experiment. Time-based monitoring of fluorescence emission intensity enables determination of diffusion coefficients in membranes. (b) A commonly used fluorescence label in membrane FRAP experiments: chemical structure of phosphatidylethanolamine conjugated to the fluorophore NBD.

monitoring allows determination of the diffusion coefficient. A well-established application is the use of phospholipids labelled with NBD (e.g. NBD-phosphatidylethanolamine, Figure 13.19b) which are incorporated into a biological or artificial membrane. The specimen is subjected to a pulse of high-intensity light (photobleaching), which causes a sharp drop of fluorescence in the observation area (Figure 13.19a). Re-emergence of fluorescence emission in this area is monitored as unbleached molecules diffused into the observation area. From the time-dependent increase of fluorescence emission, the rate of diffusion of the phospholipid molecules can be calculated. Similarly, membrane proteins, such as receptors or even proteins in a cell, can be conjugated to fluorescence labels and their diffusion coefficients determined.

Fluorescence Polarisation

Fluorescence polarisation is a property of fluorescent molecules. The polarisation phenomenon has been described in Section 13.1.3. Its measurement gives information on the orientation and mobility of fluorescent molecules. Absorption of polarised light by a chromophore is highest when the plane of polarisation is parallel to the **absorption dipole moment** μ_A (Figure 13.20).

More generally, the probability of absorption of exciting polarised light by a chromophore is proportional to $\cos^2 \theta$, with θ being the angle between the direction of polarisation and the absorption dipole moment. Fluorescence emission, in contrast, does not depend on the absorption dipole moment, but on the **transition dipole moment** μ_E. Usually, μ_A and μ_E are tilted against each other by about 10° to 40°. The probability of emission of polarised light at an angle ϕ with respect to the

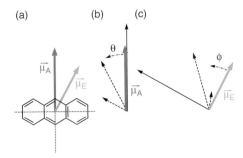

Figure 13.20 (a) The absorption dipole moment μ_A (describing the probability of photon absorption) and transition dipole moment μ_E (describing the probability for photon emission) for any chromophore are usually not parallel. (b) Absorption of linearly polarised light varies with $\cos^2 \theta$ and is at its maximum parallel to μ_A. (c) Emission of linearly polarised light varies with $\sin^2 \theta$ and is highest at a perpendicular orientation to μ_E.

transition dipole moment is proportional to $\sin^2 \phi$, and thus at its maximum in a perpendicular orientation.

As a result if the chromophores are randomly oriented in solution, the polarisation P (Equation 13.3) is less than 0.5. It is thus evident that any process that leads to a deviation from random orientation will give rise to a change of polarisation. This is certainly the case when a chromophore becomes more static. Furthermore, one needs to consider **Brownian motion**. If the chromophore is a small molecule in solution, it will be rotating very rapidly. Any change in this motion due to temperature changes, changes in viscosity of the solvent, or binding to a larger molecule, will therefore result in a change of polarisation.

Experimentally, this can be achieved in a fluorescence spectrometer by placing a polariser in the excitation path in order to excite the sample with polarised light. A second polariser is placed between the sample and the detector with its axis either parallel or perpendicular to the axis of the excitation polariser. The emitted light is either partially polarised or entirely unpolarised. This loss of polarisation is called fluorescence depolarisation.

Fluorescence Cross-Correlation Spectroscopy

With **fluorescence cross-correlation spectroscopy**, the temporal fluorescence fluctuations between two differently labelled molecules can be measured as they diffuse through a small sample volume. Cross-correlation analysis of the fluorescence signals from separate detection channels extracts information of the dynamics of the dual-labelled molecules. Fluorescence cross-correlation spectroscopy has thus become an essential tool for the characterisation of diffusion coefficients, binding constants, kinetic rates of binding and determining molecular interactions in solutions and cells.

Fluorescence Microscopy, High-Throughput Assays

Fluorescence emission as a means of monitoring is a valuable tool for many biological and biochemical applications. We have already seen the use of fluorescence monitoring in DNA sequencing (see Example 13.5); the technique is inseparably tied in with the success of projects such as genome deciphering.

Fluorescence techniques are also indispensable methods for cell biological applications with fluorescence microscopy (see Section 11.2.5). Proteins (or biological macromolecules) of interest can be tagged with a fluorescent label such as GFP from the jellyfish *Aequorea victoria* or RFP from *Discosoma striata*, if spatial and tempo-ral tracking of the tagged protein is desired. Alternatively, the use of GFP spectral variants such as cyan fluorescent protein (CFP) as a fluorescence donor and yellow fluorescent protein (YFP) as an acceptor allows investigation of mechanistic questions by using the FRET phenomenon. Specimens with cells expressing the labelled proteins are illuminated with light of the excitation wavelength, and then observed through a filter that excludes the exciting light and only transmits the fluorescence emission. The recorded fluorescence emission can be overlaid with a visual image computation-ally, and the composite image then allows for localisation of the labelled species. If different fluorescence labels with distinct emission wavelengths are used simultane-ously, even colocalisation studies can be performed (see Section 11.2.5).

Time-Resolved Fluorescence Spectroscopy

The emission of a single photon from a fluorophore follows a probability distribution. With time-correlated single-photon counting, the number of emitted photons can be recorded in a time-dependent manner following a pulsed excitation of the sample. By sampling the photon emission for a large number of excitations, the probability dis-tribution can be constructed. The time-dependent decay of an individual fluorophore species follows an exponential distribution, and the time constant is thus termed the lifetime of this fluorophore. Curve fitting of fluorescence decays enables the iden-tification of the number of species of fluorophores (within certain limits), and the calculation of the lifetimes for these species. In this context, different species can be different fluorophores or distinct conformations of the same fluorophore.

13.5.4 Phosphorescence

An associated phenomenon to fluorescence illustrated in Figure 13.15 is phosphores-cence, an emission that arises from a transition from a triplet state (T_1) to the electronic (singlet) ground state (S_0). The molecule gets into the triplet state from an electronic excited singlet state by a process called intersystem crossing (ISC). The transition from singlet to triplet is quantum-mechanically not allowed and thus only happens with low probability in certain molecules where the electronic structure is favourable. Such mol-ecules usually contain heavy atoms. The rate constants for phosphorescence are much longer and phosphorescence thus happens with a long delay and persists even when the exciting energy is no longer applied. Instrumentation design operates with this property by incorporating a delay between exciting and measuring phosphorescence emission. Phosphorescence emission is observed at low temperature or in highly viscous media.

13.6 LUMINOMETRY

In the preceding section, we mentioned the method of bioluminescence resonance energy transfer (BRET) and its main workhorse, luciferase. Generally, fluorescence phenomena depend on the input of energy in the form of electromagnetic radiation. However, emission of electromagnetic radiation from a system can also be achieved by prior excitation in the course of a chemical or enzymatic reaction. Such processes are summarised as luminescence. Luminometry is not strictly speaking a spectrophotometric technique, but is included here due to its importance in the life sciences.

13.6.1 Principles

Luminometry is the technique used to measure luminescence, which is the emission of electromagnetic radiation in the energy range of visible light as a result of a reaction. Chemiluminescence arises from the relaxation of excited electrons transitioning back to the ground state. The prior excitation occurs through a chemical reaction that yields a fluorescent product. For instance, the reaction of luminol with oxygen produces 3-aminophthalate, which possesses a fluorescence spectrum that is then observed as chemiluminescence. In other words, the chemiluminescence spectrum is the same as the fluorescence spectrum of the product of the chemical reaction.

Bioluminescence describes the same phenomenon, only the reaction leading to a fluorescent product is an enzymatic reaction. The most commonly used enzyme in this context is certainly luciferase (see Section 23.3.2). The light is emitted by an intermediate complex of luciferase with the substrate ('photoprotein'). The colour of the light emitted depends on the source of the enzyme and varies between 560 nm (greenish yellow) and 620 nm (red) wavelengths. Bioluminescence is a highly sensitive method, due to the high quantum yield of the underlying reaction. Some luciferase systems work with almost 100% efficiency. For comparison, the incandescent light bulb loses about 90% of the input energy to heat. Because luminescence does not depend on any optical excitation, problems with autofluorescence in assays are eliminated.

13.6.2 Instrumentation

Since no electromagnetic radiation is required as a source of energy for excitation, no light source and monochromator are required. Luminometry can be performed with a rather simple set-up, where a reaction is started in a cuvette or mixing chamber, and the resulting light is detected by a photometer. In most cases, a photomultiplier tube is needed to amplify the output signal prior to recording. Also, it is fairly important to maintain strict temperature control, as all chemical, and especially enzymatic, reactions are sensitive to temperature.

13.6.3 Applications

Chemiluminescence
Luminol and its derivatives can undergo chemiluminescent reactions with high efficiency. For instance, enzymatically generated H_2O_2 may be detected by the emission of light at 430 nm wavelength in the presence of luminol and microperoxidase (see also Section 6.3.8).

Competitive binding assays may be used to determine low concentrations of hormones, drugs and metabolites in biological fluids. These assays depend on the ability of proteins, such as antibodies and cell receptors, to bind specific ligands with high affinity. Competition between labelled and unlabelled ligand for appropriate sites on the protein occurs. If the concentration of the protein, i.e. the number of available binding sites, is known, and a limited but known concentration of labelled ligand is introduced, the concentration of unlabelled ligand can be determined under saturation conditions when all sites are occupied. Exclusive use of labelled ligand allows the determination of the concentration of the protein and thus the number of available binding sites.

During the process of phagocytosis by leukocytes, molecular oxygen is produced in its singlet state (see Section 13.5.1, Figure 13.15) which exhibits chemiluminescence. The effects of pharmacological and toxicological agents on leukocytes and other phagocytic cells can be studied by monitoring this luminescence.

Bioluminescence

Firefly luciferase is mainly used to measure ATP concentrations. The bioluminescence assay is rapidly carried out with accuracies comparable to spectrophotometric and fluorimetric assays. However, with a detection limit of 10^{-15} M, and a linear range of 10^{-12} to 10^{-6} M ATP, the luciferase assay is vastly superior in terms of sensitivity. Generally, all enzymes and metabolites involved in ATP interconversion reactions may be assayed with this method, including ADP, AMP, cyclic AMP and the enzymes pyruvate kinase, adenylate kinase, phosphodiesterase, creatine kinase, hexokinase and ATP sulfurase (see also Sections 10.1.1, 23.3.2). Other substrates include creatine phosphate, glucose, GTP, phosphoenolpyruvate and 1,3-diphosphoglycerate.

The main application of bacterial luciferase is the determination of electron transfer cofactors, such as nicotine adenine dinucleotides (and phosphates) and flavin mononucleotides in their reduced states, for example NADH, NADPH and $FMNH_2$. Similar to the firefly luciferase assays, this method can be applied to a whole range of coupled redox enzyme reaction systems. The enzymatic assays are again much more sensitive than the corresponding spectrophotometric and fluorimetric assays, and a concentration range of 10^{-9} to 10^{-12} M can be achieved. The NADPH assay is by a factor of 20 less sensitive than the NADH assay.

13.7 ATOMIC SPECTROSCOPY

So far, all methods have dealt with probing molecular properties. In Section 13.1.2, we discussed the general theory of electronic transitions and said that molecules give rise to band spectra, but atoms yield clearly defined line spectra. In atomic emission spectroscopy (AES), these lines can be observed as light of a particular wavelength (colour). Conversely, black lines can be observed against a bright background in atomic absorption spectroscopy (AAS). The wavelengths emitted from excited atoms may be identified using a spectroscope with the human eye as the 'detector' or with a spectrophotometer.

13.7.1 Principles

In a spectrum of an element, the absorption or emission wavelengths are associated with transitions that require a minimum of energy change. In order for energy changes to be minimal, transitions tend to occur between atomic orbitals (which correspond to

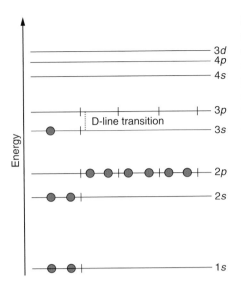

Figure 13.21 Energy levels of atomic orbitals in the sodium atom. Each atomic orbital can be occupied by electrons following the rules of quantum chemistry until the total number of electrons for that element is reached (in the case of sodium: 11 electrons). The energy gap between the 3s and the 3p orbitals in the sodium atom is such that it can be overcome by absorption of orange light.

statistical and spatial localisation of electrons around the atom nucleus; see also Section 17.3.3) close together in energy terms. For example, excitation of a sodium atom and its subsequent relaxation gives rise to emission of orange light ('D-line') due to the transition of an electron from the 3s to the 3p orbital and its return (Figure 13.21).

Electron transitions in an atom are limited by the availability of empty orbitals. Filling orbitals with electrons is subject to two major rules:

- One orbital can be occupied with a maximum of two electrons
- The spins of electrons in one orbital need to be paired in an anti-parallel fashion (Pauli principle).

Together, these limitations mean that emission and absorption lines are characteristic for an individual element.

13.7.2 Instrumentation

In general, atomic spectroscopy is not carried out in solution. In order for atoms to emit or absorb monochromatic radiation, they need to be volatilised by exposing them to high thermal energy. Usually, nebulisers are used to spray the sample solution into a flame or an oven. Alternatively, the gaseous form can be generated by using inductively coupled plasma (ICP). The variations in temperature and composition of a flame make standard conditions difficult to achieve. Most modern instruments thus use an ICP.

Atomic emission spectroscopy (AES) and atomic absorption spectroscopy (AAS) are generally used to identify specific elements present in the sample and to determine their concentrations. The energy absorbed or emitted is proportional to the number of atoms in the optical path. Strictly speaking, in the case of emission, it is the number of excited atoms that is proportional to the emitted energy. Concentration determination with AES or AAS is carried out by comparison with calibration standards.

The presence of sodium results in high backgrounds and is usually measured first. Then, a similar amount of sodium is added to all other standards. Excess hydrochloric acid is commonly added, because chloride compounds are often the most volatile salts. Calcium and magnesium emission can be enhanced by the addition of alkali metals and suppressed by addition of phosphate, silicate and aluminate, as these form non-dissociable salts. The suppression effect can be relieved by the addition of lanthanum and strontium salts. Lithium is frequently used as an internal standard. For storage of samples and standards, polyethylene bottles are used, since glass can absorb and release metal ions, and thus impact the accuracy of this sensitive technique.

Cyclic analysis may be performed that involves the estimation of each interfering substance in a mixture. Subsequently, the standards for each component in the mixture are doped with each interfering substance. This process is repeated two or three times with refined estimates of interfering substance, until self-consistent values are obtained for each component.

Flame instability requires experimental protocols where determination of an unknown sample is bracketed by measurements of the appropriate standard, in order to achieve the highest possible accuracy.

Biological samples are usually converted to ash prior to determination of metals. Wet ashing in solution is often used, employing an oxidative digestion similar to the Kjeldahl method (see Section 5.2.5).

13.7.3 Applications

Atomic Emission and Atomic Absorption Spectrophotometry

The contemporary instrumentation found in analytical laboratories consists of ICP-AES instruments that can detect less than 1 ppm of each of the common elements with the exception of alkali metals. The relative precision is about 1% in a working range of 20–200 times the detection limit of an element. In cases where special attention is applied, precision may be improved to 0.2%. Sodium and potassium are assayed at concentrations of a few ppm using simple filter photometers. The modern emission spectrophotometers allow determination of about 20 elements in biological samples, the most common being calcium, magnesium and manganese.

AES and AAS have been widely used in analytical chemistry, such as environmental and clinical laboratories. Frequently, however, these techniques are replaced by the use of ion-selective electrodes (see Sections 2.4, 10.2.2).

Atomic Fluorescence Spectroscopy

Despite being limited to only a few metals, the main importance of atomic fluorescence spectroscopy (AFS) lies in the extreme sensitivity. For example, zinc and cadmium can be detected at levels as low as 0.1–0.2 ppb. AFS uses the same basic set-up as AES and AAS. The atoms are required to be vapourised by one of three methods (flame, electric, ICP). The atoms are excited using electromagnetic radiation by directing a light beam into the vapourised sample. This beam must be intense, but is not required to be monochromatic, since only the resonant wavelengths will be absorbed, leading to fluorescence (see Section 13.5).

13.8 RAPID MIXING TECHNIQUES FOR KINETICS

Rapid mixing techniques are required for processes with short timescale, such as fast kinetics (Section 23.3.3), ligand binding and protein folding. These techniques involve turbulent mixing of two or more solutions in order to achieve homogeneous solutions in which processes can be monitored on a microsecond timescale.

In the continuous flow method, separate solutions of the enzyme and substrate are introduced from syringes, each of 10 cm³ maximum volume, into a mixing chamber typically of 100 mm³ capacity (Figure 13.22). The mixture is then pumped through a narrow tube that is illuminated by a light source and monitored by a photomultiplier detector. The technique uses relatively small amounts of reactants and is limited only by the time required to mix the two reactants. The **stopped-flow** method is a variant of the continuous flow method. Shortly after the reactants emerge from the mixing chamber, the flow is stopped and the detector triggered to continuously monitor the change in an experimental parameter such as absorbance or fluorescence (Figure 13.22). Special flow cells are used together with a detector that allows readings to be taken 180° to the light source for absorbance, transmittance or circular dichroism measurements, or at 90° to the source for fluorescence, fluorescence anisotropy or light-scattering measurements. Yet another variation is the **quenching method**, where reactants from the mixing chamber are treated with a quenching agent from a third syringe. The quenching agent, such as trichloroacetic acid, stops the reaction, which is then monitored by an appropriate analytical method for the build-up of intermediates. By varying the time between mixing the reactants and adding the quenching reagent, the kinetics of this build-up can be studied. A disadvantage of this approach is that it uses larger amounts of reactants, since the kinetic data are acquired from a series of studies rather than by following one reaction for a period of time.

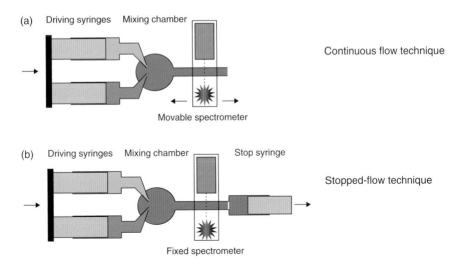

Figure 13.22 Schematic comparison of the experimental set-up for continuous flow (a) and stopped-flow (b).

13.9 SUGGESTIONS FOR FURTHER READING

13.9.1 Experimental Protocols

Beekes M., Lasch P. and Naumann D. (2007) Analytical applications of Fourier transform-infrared (FT–IR) spectroscopy in microbiology and prion research. *Veterinary Microbiology* **123**, 305–319.

Benevides J.M., Overman S.A. and Thomas G.J. Jr. (2004) Raman spectroscopy of proteins. *Current Protocols in Protein Science*, Chapter 17, Unit 17.8, Wiley Interscience, New York, USA.

Brand L. and Johnson M. L. (2008) Fluorescence spectroscopy. *Methods in Enzymology*, Vol. 450, Academic Press, New York, USA.

Ganim Z., Chung H.S., Smith A.W., *et al.* (2008) Amide I two-dimensional infrared spectroscopy of proteins. *Accounts of Chemical Research* **41**, 432–441.

Langowski J. (2008) Protein-protein interactions determined by fluorescence correlation spectroscopy. *Methods in Cell Biology* **85**, 471–484.

Patterson G.H. and Lippincott-Schwartz J. (2002) A photoactivatable GFP for selective photolabeling of proteins and cells. *Science* **297**, 1873–1877.

Sauer K. (2008) Biochemical spectroscopy. *Methods in Enzymology*, Vol. 246, Academic Press, New York, USA.

Scopes R.K. and Smith J.A. (2015) Analysis of Proteins. *Current Protocols in Molecular Biology*, Chapter 10, Wiley Interscience, New York.

Wen Z.Q. (2007) Raman spectroscopy of protein pharmaceuticals. *Journal of Pharmaceutical Sciences* **96**, 2861–2878.

13.9.2 General Texts

General Biophysics

Hofmann A., Simon A., Grkovic T. and Jones M. (2014) *Methods of Molecular Analysis in the Life Sciences*, Cambridge University Press, Cambridge, UK.

Hoppe W., Lohmann W., Markl H. and Ziegler H. (1982) *Biophysics*, 2nd Edn., Springer Verlag, Berlin, Germany.

van Holde K.E., Johnson, C. and Shing Ho P. (2006) *Principles of Physical Biochemistry*, 2nd Edn., Prentice Hall, Upper Saddle River, NJ, USA.

Circular Dichroism Spectroscopy

Fasman G.D. (1996) *Circular Dichroism and the Conformational Analysis of Biomolecules*, Plenum Press, New York, USA.

Fluorescence Spectroscopy

Lakowicz J.R. (1999) *Principles of Fluorescence Spectroscopy*, 2nd Edn., Kluwer/Plenum, New York, USA.

13.9.3 Review Articles

Adams S.T. and Miller S.C. (2014) Beyond D-luciferin: expanding the scope of bioluminescence imaging in vivo. *Current Opinion in Chemical Biology* 21, 112–120.

Czar M.F. and Jockusch R.A. (2015) Sensitive probes of protein structure and dynamics in well-controlled environments: combining mass spectrometry with fluorescence spectroscopy. *Current Opinion in Structural Biology* 34, 123–134.

Domingos S.R., Hartl F., Buma W.J. and Woutersen S. (2015) Elucidating the structure of chiral molecules by using amplified vibrational circular dichroism: from theory to experimental realization. *ChemPhysChem* 16, 3363–3373.

Pescitelli G., Di Bari L. and Berova N. (2014) Application of electronic circular dichroism in the study of supramolecular systems. *Chemical Society Reviews* 43, 5211–5233.

13.9.4 Websites

General biophysics

lectureonline.cl.msu.edu/%7Emmp/applist/Spectrum/s.htm (accessed May 2017)

Absorption spectroscopy

phys.educ.ksu.edu/vqm/html/absorption.html (accessed May 2017)

UV/vis spectroscopy

teaching.shu.ac.uk/hwb/chemistry/tutorials/molspec/uvvisab1.htm (accessed May 2017)

Infrared spectroscopy

orgchem.colorado.edu/Spectroscopy/irtutor/tutorial.html (accessed May 2017)

Raman spectroscopy

www.kosi.com/na_en/products/raman-spectroscopy/raman-technical-resources/raman-tutorial.php (accessed May 2017)

Atomic spectroscopy

zebu.uoregon.edu/nsf/emit.html (accessed May 2017)

14 Basic Techniques Probing Molecular Structure and Interactions

ANNE SIMON AND JOANNE MACDONALD

14.1 INTRODUCTION

Determining the structure of molecules and the interactions between molecules is critical for understanding biochemical processes. All experimental techniques discussed in the following sections revolve around investigations of the three-dimensional structure of molecules as well as the elucidation of interactions between different groups of molecules. As such, some of these techniques are characterised by a somewhat higher level of complexity in undertaking and are often employed at a later stage of biochemical characterisation.

Techniques probing the thermodynamics of a system (such as isothermal titration calorimetry) or atomic/molecular structure and interactions without requirements for washing or flow (NMR spectrosocopy, magnetic resonance imaging, X-ray diffraction, light scattering) can be summarised as direct techniques. Alternatively, molecular interactions can be analysed by applying tracers to the molecules themselves, resulting in molecular switch techniques. Last, there are in vitro methods that combine the use of analytical techniques (such as the spectroscopic techniques outlined in Chapter 13) with physical sampling – these can be classified as indirect techniques.

The analysis of molecular interactions can result in the development of a biosensor: a device that is composed of a biological element and a physico-chemical transduction part that converts signal reception by the biological entity into a physical quantifiable response. Such technology gives rise to the field of biosensing.

14.2 ISOTHERMAL TITRATION CALORIMETRY

14.2.1 Principles

Isothermal titration calorimetry (ITC) enables study of the thermodynamics of molecules binding to each other. This is a general method for studying the thermodynamics of any binding (association) process in solution. It detects and quantifies small heat changes associated with the binding and has the advantages of speed, accuracy and no requirement for either of the reacting species to be chemically modified or immobilised. The apparatus consists of a pair of matched cells (sample and reference) of approximately 2 cm³ volume contained in a microcalorimeter (Figure 14.1a). One of the reactants (say the enzyme preparation) is added to the sample cell and the ligand (substrate, inhibitor or effector) is added via a stepper-motor-driven syringe. The mixture is stirred to ensure homogeneity. The reference cell contains an equal volume of reference liquid. A constant power of less than 1 mW is applied to the reference cell. This directs a feedback circuit activating a heater attached to the sample cell. The addition of the ligand solution causes a heat change due to the binding process and the dilution of both the enzyme and ligand preparations. If the reaction is exothermic, less energy is required to maintain the cell at constant temperature. If the reaction is endothermic, more energy is required. The power required to maintain a constant temperature is recorded as a series of spikes as a function of time (Figure 14.1b). Each spike is integrated to give μcal s⁻¹ and summed to give the total heat exchange per injection. The study is repeated with a series of increasing ligand concentrations and control experiments carried out replacing the ligand with buffer solution to allow the heat exchange (ΔH) associated solely with the addition of ligand to be calculated. A plot is then made of enthalpy change against the molar ratio of the ligand to enzyme. The plot is hyperbolic, from which it is possible to calculate enthalpy, free energy and entropy changes associated with the ligand binding and hence the dissociation constant, K_d, and stoichiometry of binding n. For a discussion of particular strengths and weaknesses of this technique in comparison to other methods please refer to the suggestions in the Further Reading section (14.7).

14.2.2 Applications

Since ITC investigates thermodynamic parameters and kinetic information on molecular interaction processes in solution, a diverse range of interactions in biological systems can be studied, including proteins, peptides, DNA, carbohydrates, lipids, small molecules and cells. The technique allows label-free measurements of thermodynamics of binding reactions and results guide direct applications in many fields (life sciences, drug discovery, etc.). In particular, ITC has been successfully used for determination of rates of enzymatic reactions and their use in investigation of enzyme kinetics (Section 23.3.4). A further important application of ITC has been in the study of the interconversion of protein conformations and the elucidation of the mechanism of allostery (Section 23.2.4). An incremental increase of a component or an inhibitor addition allows the study of reaction mechanisms, but can also be used as an efficient tool to screen compound libraries of molecules. Combined with other techniques, for

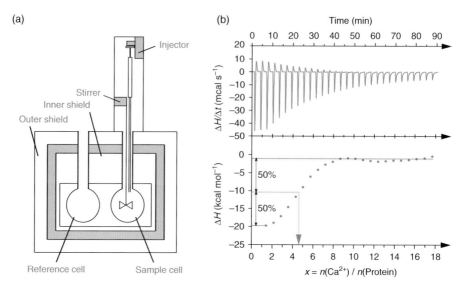

Figure 14.1 (a) Schematics of an isothermal titration calorimeter (ITC). (b) ITC data obtained for titrating calcium ions into a solution of a calcium-binding protein. The top panel shows the raw data. The area underneath each injection peak is equal to the total heat released for that injection. These data can be integrated to yield a plot of the integrated heat against the molar ratio of ligand added to protein (bottom panel). The dotted and dashed lines illustrate determination of stoichiometry at the point of inflection of the binding isotherm. In this example, the determination of $x = 4.6$ indicates a stoichiometry of five calcium ions per protein molecule.

example chromatography, ITC may also be used to identify a target protein for a particular ligand within a biomolecular mixture.

14.3 TECHNIQUES TO INVESTIGATE THE THREE-DIMENSIONAL STRUCTURE

The experimental techniques discussed in the following sections revolve around investigations of the three-dimensional structure of molecules, as well as the elucidation of interactions between different groups of molecules in solution.

14.3.1 Basic Principles

Principles of Nuclear Magnetic Resonance

Conceptually, the nucleus of an atom can be described as a spherical particle that is spinning. This infers a kinetic moment of rotation called nuclear spin (I), which is associated with this movement. The nucleus of an atom is constituted by protons and neutrons, and has a net charge that is normally compensated by the extra-nuclear electrons. The number of all nucleons (A) is the sum of the number of protons (Z) and the number of neutrons (N). Depending on the number of nucleons and protons, the nucleus may possess a half-integer spin (for example $I = \frac{1}{2}$ for hydrogen), an integer spin (for example $I = 1$ for deuterium) or zero (for example $I = 0$ for carbon).

The nuclear charge is due to its protons, and since magnetism arises from the motion of charged particles, a nuclear magnetic moment arises from the nuclear spin. The number of protons (Z) and the number of nucleons (A) determine whether a nucleus will exhibit magnetism. Carbon-12 (^{12}C), for example, consists of six protons ($Z = 6$) and six neutrons ($N = 6$) and thus has $A = 12$. Z and A are even, and therefore the ^{12}C nucleus possesses no nuclear magnetism. Another example of a nucleus with no residual magnetism is oxygen-16 (^{16}O). All other nuclei with Z and A being uneven possess residual nuclear magnetism.

In chemical bonds of a molecule, the negatively charged electrons also possess spins controlled by strict **quantum rules**. A bond is constituted by two electrons occupying the appropriate molecular orbital. According to the **Pauli principle**, the two electrons must have opposite spins, leading to the term **paired electrons**. Each of the spinning electronic charges generates a magnetic effect, but in electron pairs the effect is almost self-cancelling and results in a very small value of the magnetic susceptibility, which is of the order of -10^{-6} g^{-1}. This **diamagnetism** is a property of all substances, because they all contain the minuscule magnets, i.e. electrons. Diamagnetism is temperature independent.

If an electron is unpaired, there is no counterbalancing opposing spin and the observed magnetic susceptibility is much higher and of the order of $+10^{-3}$ to $+10^{-4}$ g^{-1}. This effect by **unpaired electrons** exceeds the 'background' diamagnetism, and gives rise to **paramagnetism**. Free electrons can arise in numerous cases. The most notable example is certainly the paramagnetism of metals such as iron, cobalt and nickel, which are the materials that permanent magnets are made of. The paramagnetism of these metals is called **ferromagnetism**. In biochemical investigations, systems with free electrons (radicals) are frequently used as probes.

The way in which a substance behaves in an externally applied magnetic field allows us to distinguish between dia- and paramagnetism. A paramagnetic material is attracted by an external magnetic field, while a diamagnetic substance is rejected. This principle is employed by the **Guoy balance**, which allows quantification of magnetic effects. A balance pan is suspended between the poles of a suitable electromagnet supplying the external field. The substance under test is weighed in air with the current switched off. The same sample is then weighed again with the current (i.e. external magnetic field) on. A paramagnetic substance appears to weigh more, and a diamagnetic substance appears to weigh less.

For either electronic or nuclear magnets, two possible energy states exist in the presence of an **external magnetic field** (Figure 14.2). In the low-energy state, the field generated by the spinning charged particle is parallel to the external field. Conversely, in the high-energy state, the field generated by the spinning charged particle is anti-parallel to the external field. When enough energy is input into the system to cause a transition from the low- to the high-energy state, the condition of resonance is satisfied. Energy must be absorbed as a discrete dose (quantum) hv, where h is the Planck constant and v is the frequency (see Equation 14.1). The quantum energy required to fulfil the resonance condition and thus enable transition between the low- and high-energy states may be quantified as:

$$h \times v = g \times \beta \times B \qquad\qquad \text{(Eq 14.1)}$$

where g is a constant called the **spectroscopic splitting factor**, β is the magnetic moment of the electron (termed the Bohr magneton) and B is the strength of the applied external magnetic field. The frequency ν of the absorbed radiation is a function of the paramagnetic species β and the applied magnetic field B. Thus, either ν or B may be varied with the same effect.

With appropriate external magnetic fields, the frequency of applied radiation for electron paramagnetic resonance (EPR) is in the microwave region, and for nuclear

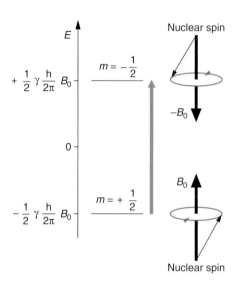

Figure 14.2 Energy levels of a proton in the magnetic field $\bar{B}_0$. The nuclear spin of a nucleus is characterised by its magnetic quantum number m (which can take the values $-I$, $-I\pm1$, ..., I-1, I). For protons ($I = \frac{1}{2}$), m can only adopt $+\frac{1}{2}$ and $-\frac{1}{2}$. The corresponding energies are calculated by $-m\gamma h/(2\pi)B_0$, where γ is a constant characteristic for a particular nucleus, h is the Planck constant, and B_0 is the strength of the magnetic field $\bar{B}_0$.

magnetic resonance (NMR) in the region of radio frequencies. In both techniques, two possibilities exist for determining the absorption of electromagnetic energy (i.e. enabling the resonance phenomenon):

- constant frequency ν is applied and the external magnetic field B is swept
- constant external magnetic field B is applied and the appropriate frequency ν is selected by sweeping through the spectrum.

For technical reasons, the more commonly used option is a sweep of the external magnetic field. After the absorption of energy by the nucleus in the range of the radio frequencies, the system returns the initial equilibrium by the process of **relaxation**. This includes firstly the return of the nuclear spins to their low-energy levels and secondly the loss of magnetisation. The process is characterised by two time constants called spin–spin relaxation and spin–lattice relaxation. Spin–spin relaxation is attributed to a spin exchange between two nuclei in proximity, whereas spin–lattice relaxation is due to magnetic interactions between the nuclear spin and ions around atoms (see Section 14.3.2).

Principles of Light Diffraction

The interaction of electromagnetic radiation with matter causes the electrons in the exposed sample to oscillate. The accelerated electrons, in turn, will emit radiation of the same frequency as the incident radiation, called the secondary waves. The super-position of waves gives rise to the phenomenon of interference. Depending on the displacement (phase difference) between two waves, their amplitudes either reinforce or cancel each other out. The maximum reinforcement is called constructive interference, cancelling is called destructive interference. The interference gives rise to dark and bright rings, lines or spots, depending on the geometry of the object causing the diffraction. Diffraction effects increase as the physical dimension of the diffracting object (aperture) approaches the wavelength of the radiation. When the aperture has a periodic structure, for example in a diffraction grating, repetitive layers or crystal lattices, the features generally become sharper. Bragg's law (Figure 14.3) describes the condition that waves of a certain wavelength will constructively interfere upon partial reflection between surfaces that produce a path difference only when that path difference is equal to an integral number of wavelengths. From the constructive interferences, i.e. diffraction spots or rings, one can determine dimensions in solid materials.

Since the distances between atoms or ions are on the order of 10^{-10} m (1 Å), diffraction methods used to determine structures at the atomic level require radiation in the X-ray region of the electromagnetic spectrum, or beams of electrons or neutrons with a similar wavelength. While electrons and neutrons are particles, they also possess wave properties with the wavelength depending on their energy (de Broglie hypothesis or wave-particle duality). Accordingly, diffraction can also be observed using electron and neutron beams. However, each method also has distinct features, including the penetration depth, which increases in the series: electrons → X-rays →neutrons.

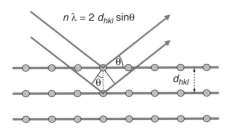

$$n \lambda = 2\, d_{hkl}\, \sin\theta$$

Figure 14.3 Bragg's law. Interference effects are observable only when radiation interacts with physical dimensions that are approximately the same size as the wavelength of the radiation. Only diffracted beams that satisfy the Bragg condition are observable (constructive interference). Diffraction can thus be treated as selective reflection. n is an integer ('order'), λ is the wavelength of the radiation, d_{hkl} is the spacing between a set of lattice planes (identified by the Miller indices h, k and l) and θ is the angle between the incident/reflected beam and the lattice plane.

Principles of Light Scattering

The scattering of light can yield a number of valuable insights into the properties of macromolecules, including the molecular mass, dimensions and diffusion coefficients, as well as association/dissociation properties and internal dynamics. Often, the light sources used for scattering techniques are lasers (see Section 13.1.4) as their special properties of high monochromaticity, narrow focus and strong intensity make them ideally suited for light-scattering applications, such as flow cytometry (see Chapter 8).

The incident light hitting a macromolecule is scattered in all directions, with the intensity of the scatter being only about 10^{-5} of the original intensity. The scattered light is measured at angles higher than 0° and less than 180°. Most of the scattered light possesses the same wavelength as the incident light; this phenomenon is called elastic light scattering.

When the scattered light has a wavelength higher or lower than the incident light, the phenomenon is called inelastic light scattering. Inelastic light scattering is observed in the infrared region of the electromagnetic spectrum (Figure 13.2) and is relevant to Raman spectroscopy (Section 13.4).

14.3.2 Nuclear Magnetic Resonance (NMR) Spectroscopy

Most studies in organic chemistry involve the use of ^{1}H as the nucleus to be probed, but NMR spectroscopy with ^{13}C, ^{15}N and ^{31}P isotopes is also frequently used, particular in biochemical studies. The resonance condition in NMR is satisfied in an external magnetic field of several hundred millitesla (mT), with absorptions occurring in the region of radio waves for resonances of the nuclei (e.g. ^{1}H frequency ~40 MHz). The actual field scanned is small compared with the field strengths applied, and the radio frequencies absorbed are specifically stated on such spectra.

Similar to other spectroscopic techniques discussed earlier, the energy input in the form of electromagnetic radiation promotes the transition of 'entities' from lower- to higher-energy states (Figure 14.2). In the case of NMR, these entities are the nuclear magnetic spins, which populate energy levels according to quantum chemical rules. After a certain time-span, the spins will return from the higher to the lower energy level, a process that is known as relaxation.

The energy released during the transition of a nuclear spin from the higher- to the lower-energy state can be emitted as heat into the environment and is called spin–lattice relaxation. This process happens with a rate of T_1^{-1}, and T_1 is termed the longitudinal relaxation time, because of the change in magnetisation of the nuclei parallel to the field. The transverse magnetisation of the nuclei is also subject to change over time, due to interactions between different nuclei. The latter process is thus called spin–spin relaxation and is characterised by a transverse relaxation time T_2.

The molecular environment of a proton governs the value of the applied external field at which the nucleus resonates. This is recorded as the chemical shift (δ) and is measured relative to an internal standard, which in most cases is tetramethylsilane (TMS; $(H_3C)_4Si$) because it contains 12 identical protons. The chemical shift arises from the applied field inducing secondary fields of about 0.15–0.2 mT at the proton by interacting with the adjacent bonding electrons.

- If the induced field opposes the applied field, the latter will have to be at a slightly higher value for resonance to occur. The nucleus is said to be shielded, the magnitude of the shielding being proportional to the electron-withdrawing power of proximal substituents.
- Alternatively, if the induced and applied fields are aligned, the latter is required to be at a lower value for resonance. The nucleus is then said to be deshielded.

Usually, deuterated solvents such as $CDCl_3$ are used for sample preparation of organic compounds. For peptides and proteins D_2O is the solvent of choice. Because the stability of the magnetic field is critical for NMR spectroscopy, the magnetic flux needs to be tuned, e.g. by locking with deuterium resonance frequencies. The use of deuterated solvents thus eliminates the need for further experiments.

The chemical shift is plotted along the x-axis, and measured in ppm instead of the actual magnetic field strengths. This conversion makes the recorded spectrum independent of the magnetic field used. The signal of the internal standard TMS appears at $\delta = 0$ ppm. The type of proton giving rise to a particular band may thus be identified by the resonance peak position, i.e. its chemical shift, and the area under each peak is proportional to the number of protons of that particular type. Figure 14.4 shows an ^{1}H NMR spectrum of ethyl alcohol, in which there are three methyl protons, two methylene protons and one alcohol-group proton. The peak areas are integrated, and show the proportions $3 : 2 : 1$. Owing to the interaction of bonding electrons with like or different spins, a phenomenon called **spin–spin coupling** (also termed **scalar** or ***J*-coupling**) arises that can extend to nuclei four or five bonds apart. This results in the splitting of the three bands in Figure 14.4 into several finer bands (hyperfine splitting). The hyperfine splitting yields valuable information about the near-neighbour environment of a nucleus.

NMR spectra are of great value in elucidating chemical structures. Both qualitative and quantitative information may be obtained. The advances in computing power have made possible many more advanced NMR techniques. Weak signals can be enhanced by running many scans and accumulating the data. Baseline noise, which is random, tends to cancel out, whereas the signal increases. This approach is known as computer averaging of transients or **CAT scanning**, and significantly improves the signal-to-noise ratio.

Despite the value and continued use of such 'conventional' ^{1}H NMR, much more structural information can be obtained by resorting to pulsed input of radio frequency energy, and subjecting the output to Fourier transform. This approach has given rise to a wide variety of procedures producing multi-dimensional spectra, ^{13}C and other odd-isotope NMR spectra and the determination of multiplicities and scan images.

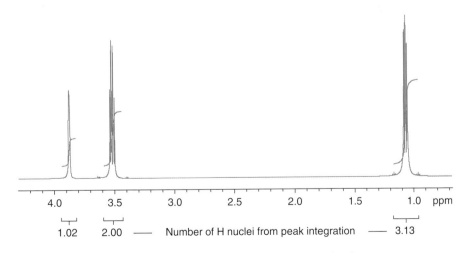

Figure 14.4 ^{1}H NMR spectrum of ethyl alcohol (H$_3$C–CH$_2$–OH) with integrated peaks.

Pulse-Acquired and Fourier Transform Methods

In conventional NMR spectroscopy, the electromagnetic radiation (energy) is supplied from the source as a continuously changing frequency over a pre-selected spectral range (continuous wave method). The change is smooth and regular between fixed limits. Figure 14.5a illustrates this approach. During the scan, radiation of certain energy in the form of a sine wave is recorded. By using the mathematical procedure of Fourier transformation, the time domain can be resolved into a frequency domain. For a single-frequency sine wave, this procedure yields a single peak of fixed amplitude. However, because the measured signal in NMR is the re-emission of energy as the nuclei return from their high-energy into their low-energy states, the recorded radiation will decay with time, as fewer and fewer nuclei will return to the ground state. The signal measured is thus called the free induction decay (FID). Figure 14.5b shows the effect of the FID on the corresponding Fourier transform. The frequency band broadens, but the peak position and the amplitude remain the same. The resolved frequency peak represents the chemical shift of a nucleus resonating at this energy.

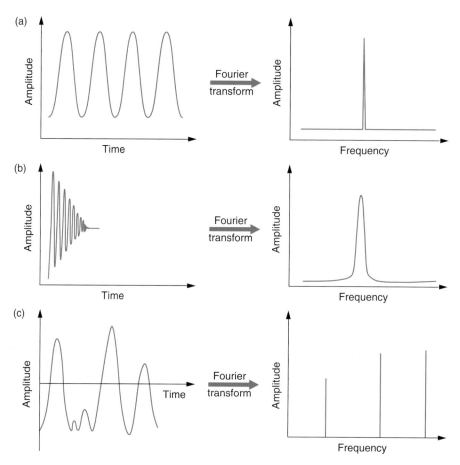

Figure 14.5 Diagrammatic representation of the Fourier transformation of (a) a single-frequency sine wave, (b) a single frequency FID and (c) a three-sine-wave combination.

Alternatively, the total energy comprising all frequencies between the fixed limits can be put in all at the same time (pulse-acquired NMR). This is achieved by irradiating the sample with a broadband pulse of all frequencies in one go. The output will measure all resonance energies simultaneously and will result in a very complicated interference pattern. However, Fourier transformation is able to resolve this pattern into the constituting frequencies (Figure 14.5c).

In the presence of an external magnetic field, nuclear spins precess around the axis of that field with the so-called Larmor frequency. The vector sum of all nuclear magnetic moments yields a magnetisation parallel to the external field, i.e. a longitudinal magnetisation. When a high-frequency pulse is applied, the overall magnetisation is forced further off the precession by a pulse angle. This introduces a new vector component to the overall magnetisation that is perpendicular to the external field; this component is called transverse magnetisation. The FID measured in pulse-acquired spectra is, in fact, the decay of that transverse magnetisation component.

Nuclear Overhauser Effect

It has already been mentioned above that nuclear spins generate magnetic fields that can exert effects through space, for example as observed in spin–spin coupling. This coupling is mediated through chemical bonds connecting the two coupling spins. However, magnetic nuclear spins can also exert effects in their proximal neighbourhood via dipolar interactions. The effects encountered in the dipolar interaction are transmitted through space over a limited distance on the order of 0.5 nm or less. These interactions can lead to nuclear Overhauser effects (NOEs), as observed in a changing signal intensity of a resonance when the state of a near neighbour is perturbed from the equilibrium. Because of the spatial constraint, this information enables conclusions to be drawn about the three-dimensional geometry of the molecule being examined.

^{13}C NMR

Due to the low abundance of the ^{13}C isotope, the chance of finding two such species next to each other in a molecule is very small (see Chapter 9). As a consequence, ^{13}C–^{13}C couplings (homonuclear couplings) do not arise. While ^{1}H–^{13}C interactions (heteronuclear coupling) are possible, one usually records decoupled ^{13}C spectra, where all bands represent carbon only. ^{13}C spectra are thus much simpler and cleaner when compared to ^{1}H spectra. However, the main disadvantage is the fact that multiplicities in these spectra cannot be observed, i.e. it cannot be decided whether a particular ^{13}C is associated with a methyl (-CH_3), a methylene (-CH_2-) or a methyne (=CH-) group. Some of this information can be regained by irradiating with an off-resonance frequency during a decoupling experiment. Another routinely used method is called distortionless enhancement by polarisation transfer (DEPT), where sequences of multiple pulses are used to excite nuclear spins at different angles, usually 45°, 90° or 135°. Although interactions have been decoupled, in this situation the resonances exhibit positive or negative signal intensities dependent on the number of protons bonded to the carbon. In DEPT-135, for example, a methylene group yields a negative intensity, while methyl and methyne groups yield positive signals.

Multi-Dimensional NMR

As we learned above, the observable in pulse-acquired Fourier transform NMR is the decay of the transverse magnetisation, called free induction decay (FID). The detected signal thus is a function of the detection time t_2. Within the pulse sequence, the time t_1 (evolution time) describes the time between the first pulse and signal detection. If t_1 is systematically varied, the detected signal becomes a function of both t_1 and t_2, and its Fourier transform comprises two frequency components. The two components form the basis of a two-dimensional spectrum.

Correlated 2D-NMR spectra show chemical shifts on both axes. Utilising different pulse sequences leads to different methods, such as correlated spectroscopy (COSY), Nuclear Overhauser Effect Spectroscopy (NOESY), etc. Such methods yield the homonuclear ^{1}H couplings. The 1D-NMR spectrum now appears along the diagonal and long-range couplings between particular nuclei appear as off-diagonal signals (Figure 14.6).

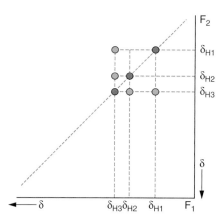

Figure 14.6 Schematics of a correlated 2D-^{1}H NMR spectrum. H3 couples with H2 and H1. H1 and H2 show no coupling.

Summary of NMR Parameters

The parameters obtained from NMR spectra used to derive structural determinants of a small molecule or protein are summarised in Table 14.1.

Table 14.1 NMR-derived structural parameters of molecules

Parameter	Information	Example/Comment
Chemical shift	Chemical group, secondary structure	^{1}H, ^{13}C, ^{15}N, ^{31}P
J-couplings (through bond)	Dihedral angles	${}^3J$(amide-H, Hα), ${}^3J$(Hα, Hβ), ...
NOE (through space)	Interatomic distances	< 0.5 nm
Solvent exchange	Hydrogen bonds	Hydrogen-bonded amide protons are protected from H/D exchange, while the signals of other amides disappear quickly
Relaxation, line widths	Mobility, dynamics, conformational/chemical exchange, torsion angles	The exchange between two conformations, but also chemical exchange, gives rise to two distinct signals for a particular spin
Residual dipolar coupling	Torsion angles	^{1}H–^{15}N, ^{1}H–^{13}C, ^{13}C–^{13}C, ...

Instrumentation

In an analytical NMR instrument, two sets of coils are used to generate and detect radio frequencies (Figure 14.7). Samples in solution are contained in sealed tubes that are rotated rapidly in the cavity to eliminate irregularities and imperfections in sample distribution. In this way, an average and uniform signal is reflected to the receiver to be processed and recorded. In solid samples, the number of spin–spin interactions is greatly enhanced due to intermolecular interactions that are absent in dissolved samples due to translation and rotation movements. As a result, the resonance signals broaden significantly. However, high-resolution spectra can be obtained by spinning the solid sample at an angle of 54.7° (**magic angle spinning**). The sophisticated pulse sequences necessary for multi-dimensional NMR require a certain geometric layout of the radio frequency coils and sophisticated electronics. Advanced computer facilities are needed for operation of NMR instruments, as well as analysis of the acquired spectra.

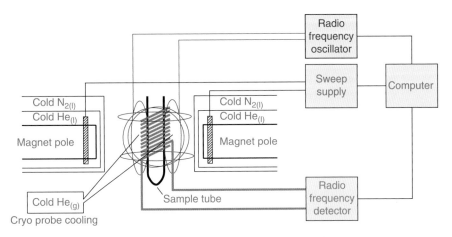

Figure 14.7 Schematic diagram of an NMR spectrometer with cryo probe. The probe coil is held at 25 K by circulating cold helium gas. Cryo probes allow a significant increase in the sensitivity compared to room temperature probes. The magnet needs to be cooled to liquid helium temperature (4 K) to achieve the state of superconductance.

Molecular Structure Determination of Small Molecules

Traditionally, NMR spectroscopy is the main method of structure determination for organic compounds. The chemical shift provides a clue about the environment of a particular proton or carbon, and thus allows conclusions as to the nature of functional groups. Analysis of the chemical shifts therefore informs about the presence of certain functional groups in a molecule. In contrast, spin–spin interactions provide information as to how protons are linked with the carbon skeleton. For structure determination, the spin–spin interactions (fine structure) thus provide crucial information, because they allow reconstruction of the bonding network within the molecule. Furthermore, analysis of the relative signal intensity enables calculation of the number of protons contributing to a particular signal.

Solution Structure of Proteins and Peptides

The structures of proteins up to a mass of about 50 kDa can be determined with biomolecular NMR spectroscopy. The development of magnets with very high field strengths (currently 900 MHz) continues to push the size limit. The preparation of proteins or selected domains for NMR requires recombinant expression and isotopic labelling to enrich the samples with ^{13}C and ^{15}N; ^{2}H labelling might be required as well. Sample amounts in the order of 10 mg used to be required for NMR experiments; however, the introduction of cryo probe technology (see Figure 14.7) has reduced the sample amount significantly (0.1 mg quantities with microprobes). By cooling the radio frequency coil and sections of the pre-amplifiers in liquid helium at near absolute-zero temperatures, cryo probes result in up to 4× higher signal-to-noise ratios compared to a room-temperature probe. This sensitivity gain translates into a substantial reduction (up to 20×) in data collection time.

Heteronuclear multi-dimensional NMR spectra need to be recorded for the assignment of all chemical shifts (^{1}H, ^{13}C, ^{15}N). For inter-proton NOEs, ^{13}C- and ^{15}N-edited 3D-NOESY spectra are required. The data acquisition can take several weeks, after which spectra are processed (Fourier transformation) and improved with respect to digital resolution and signal-to-noise. Assignment of chemical shifts and inter-atomic distances is carried out with the help of software programs. All experimentally derived parameters are then used as restraints in a molecular dynamics or simulated annealing structure calculation. The resulting protein NMR structure is an ensemble of structures, all of which are consistent with the experimentally determined restraints, but converge to indicate the same fold.

14.3.3 Magnetic Resonance Imaging (MRI)

The basic principles of NMR can be applied to imaging of living organisms and has a central role in routine clinical imaging of large-volume soft tissues. Because the proton is one of the more sensitive nuclides and is present in all biological systems abundantly, ^{1}H resonance is used almost exclusively in the clinical environment. The most important compound in biological samples in this context is water. It is distributed differently in different tissues, but constitutes about 55% of body mass in the average human. In soft tissues, the water distribution varies between 60% and 90%. In NMR, the resonance frequency of a particular nuclide is proportional to the strength of the applied external magnetic field. If an external magnetic field gradient is applied, then a range of resonant frequencies is observed, reflecting the spatial distribution of the spinning nuclei.

The number of spins in a particular defined spatial region gives rise to the spin density as an observable parameter. This measure can be combined with analysis of the principal relaxation times (T_1 and T_2). The imaging of flux, as either bulk flow or localised diffusion, adds considerably to the options available. In terms of whole-body scanners, the entire picture is reconstructed from images generated in contiguous slices. MRI can be applied to the whole body or specific organ investigations of the head, thorax, abdomen, liver, pancreas, kidney and musculoskeletal regions (Figure 14.8). The use of contrast agents with paramagnetic properties has enabled

Example 14.1   **ASSESSING PROTEIN CONFORMATIONAL EXCHANGE BY NMR**

Question   Identification of protein–protein interaction sites is crucial for understanding the basis of molecular recognition. How can such sites be identified?

Answer   Apart from providing the absolute three-dimensional structure of molecules, NMR methods can also yield insights into protein interactions by mapping. In a technique called saturation transfer difference NMR (STD-NMR), protein resonances can be selectively saturated. One then calculates the ^{1}H NMR difference spectrum of the ligand by subtracting the saturation experiment from the ligand spectrum without saturation. Intensities of protons in close contact with the ligand appear enhanced in the difference spectrum, allowing the identification of chemical groups within the ligand that interact with the protein. Using titration experiments, this technique also allows determination of binding constants.

Beyond mapping the flexibility of residues in known protein binding sites, NMR techniques can also be used to identify novel binding sites in proteins. Protein motions on the time scale of microseconds to milliseconds are accessible to NMR spectroscopy, and the diffusion constants for rotation around the three principal axes x, y and z (called rotational diffusion tensors) can be determined. The principal axes are fixed in the protein, and the principal components, as well as their orientation, can be derived from analysis of the ratio of the spin–spin and spin–lattice relaxation times T_2/T_1. Analysing these values for the protons of the rigid amide (CO–NH) groups allows characterisation of the conformational exchange of proteins.

Residues constituting the ligand-binding interface often experience a different environment in the bound state compared to the free state. The amide signals of these residues are thus broadened due to exchange between these two environments when the free and bound states are in equilibrium.

This approach has been successfully applied to identify the amino acids at the binding site of a 16 kDa protein that binds to and regulates the 251 kDa hydroxylase of the methane mono-oxygenase protein system. The free and bound forms of the regulatory protein exchange on the time scale of milliseconds.

Other examples include the identification of specific sites involved in the weak self-association of the N-terminal domain of the rat T-cell adhesion protein CD2 (CD2d1) using the concentration dependence of the T_2 values.

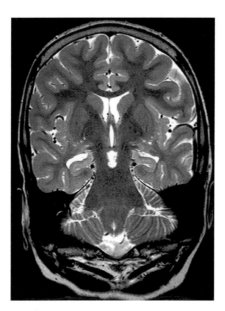

Figure 14.8 Magnetic resonance imaging: 2-mm thick coronal T_2 weighted fast spin echo image at the level of the *foramina monroi*, connecting the anterior horns of the lateral ventricles with the third ventricle. The sequence consisting of 40 images was acquired at a field strength of 3 tesla and generated 0.47 × 0.64 × 2 mm voxels. (Image courtesy of Professor H. Urbach, University of Bonn.)

investigation of organ function, as well as blood flow, tissue perfusion, transport across the blood–brain barrier and vascular anatomy. Resolution and image contrast are major considerations for the technique and subject to continuing development. The resolution depends on the strength of the magnetic field and the availability of labels that yield high signal strengths. MRI instruments used for clinical imaging typically operate with field strengths of up to 3 T, but experimental instruments can operate at more than 20 T, allowing the imaging of whole live organisms with almost enough spatial and temporal resolution to follow regenerative processes continuously at the single-cell level. Equipment cost and data acquisition time remain important issues. However, according to current knowledge, MRI has no adverse effects on human health, and thus provides a valuable diagnostic tool, especially due to the absence of the hazards of ionising radiation.

14.3.4 X-Ray Diffraction

As X-rays are diffracted by electrons, the analysis of X-ray diffraction datasets produces an **electron density map** of a crystal. Since hydrogen atoms have very little electron density, they are not usually determined experimentally by this technique. Unfortunately, the detection of light beams is restricted to recording the intensity of the beam only, leading to the so-called **structure factors**. Other properties, such as polarisation, can only be determined with rather complex measurements. The phase of the light waves is systematically lost in the measurement; this phenomenon has thus been termed the **phase problem**, owing to the essential information contained in the phase in diffraction and microscopy experiments. The X-ray diffraction data can be used to calculate the amplitudes of the three-dimensional Fourier transform of the electron density. Only together with the phases can the electron density be calculated, in a process called Fourier synthesis. Different methods to overcome the phase problem in X-ray crystallography have been developed, including:

- Molecular replacement, where phases from a structurally similar molecule are used
- Experimental methods that require incorporation of heavy element salts (multiple isomorphous replacement)
- Experimental methods where methionine has been replaced by seleno-methionine in proteins (multi-wavelength anomalous diffraction)
- Experimental methods using the anomalous diffraction of the intrinsic sulfur in proteins (single wavelength anomalous diffraction)
- Direct methods, where a statistical approach is used to determine phases. This approach is limited to very-high-resolution datasets and is the main method for small-molecule crystals, as these provide high-quality diffraction with relatively small numbers of reflections.

The electron density map calculated from combining the experimental observations (structure factors) and phases then needs to be interpreted by assigning individual atoms and constructing a model. Depending on the resolution and quality of the data, automated model-building of protein structures by computer software is possible, but in many cases manual adjustments are still required and actuated with the help of molecular graphics. In an iterative refinement process, the adjusted model is subjected to a minimisation procedure that aims to achieve the best possible agreement with the observed structure factors, while also considering a dictionary of allowed geometries of bonding, similar to molecular mechanics force fields (see Section 17.4.1).

Instrumentation

X-rays for chemical analysis are commonly obtained by rotating anode generators (in-house) or synchrotron facilities (Figure 14.9). In rotating anode generators, a rotating metal target is bombarded with high-energy (10–100 keV) electrons that knock out core electrons. An electron in an outer shell fills the hole in the inner shell and emits the energy difference between the two states as an X-ray photon. Common targets are copper, molybdenum and chromium, which have strong distinct X-ray emissions at 1.54 Å, 0.71 Å and 2.29 Å, respectively, that is superimposed on a continuous spectrum known as Bremsstrahlung (see Figure 14.9b). In synchrotrons, electrons are accelerated in a ring, thus producing a continuous spectrum of X-rays. Here, monochromators are required to select a single wavelength.

Single-Crystal Diffraction

A crystal is a solid in which atoms or molecules are packed in a particular arrangement within the unit cell, which is repeated indefinitely along three principal directions in space. Crystals can be formed by a wide variety of materials, such as salts, metals, minerals and semiconductors, as well as various inorganic, organic and biological molecules.

    A crystal grown in the laboratory is mounted on a goniometer and exposed to X-rays produced by rotating anode generators (in-house) or a synchrotron facility. A diffraction pattern of regularly spaced spots known as reflections is recorded on a detector (Figure 14.9d), most frequently image plates (phosphor screens that store exposure to X-rays; the information can be retrieved by a process called

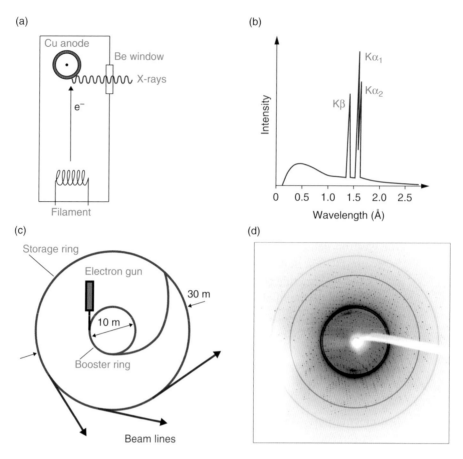

Figure 14.9 Instrumentation for X-ray diffraction. (a) The most common X-ray sources are rotating anode tubes, from which X-rays with material-specific wavelengths can be obtained. The spectrum obtained from a Cu anode is shown in (b). (c) Particle storage rings producing synchrotron radiation are also very common sources of X-rays. With synchrotrons, UV and X-ray light of any wavelength is obtained. (d) From single-crystal X-ray diffraction, discrete diffraction spots are obtained, each of which encodes a particular diffraction angle θ (see Figure 14.3) as well as an intensity.

photostimulable luminescence) or CCD cameras for proteins and movable proportional counters for small molecules.

An incident X-ray beam is diffracted by a crystal such that beams at specific angles are produced, depending on the X-ray wavelength, the crystal orientation and the structure of the crystal (i.e. unit cell).

To record a dataset, the crystal is gradually rotated and a diffraction pattern is acquired for each distinct orientation. These two-dimensional images are then analysed by identifying the appropriate reflection for each lattice plane and measuring its intensity, measuring the cell parameters of the unit cell and determining the appropriate space group. If information about the phases is available, these data can then be used to calculate a three-dimensional model of the electron density within the unit cell, using the mathematical method of Fourier synthesis. The positions of the atomic nuclei are then deduced from the electron density by computational refinement and manual intervention using molecular graphics.

Fibre Diffraction

Historically, fibre diffraction was of central significance in enabling the determination of the three-dimensional structure of DNA by Crick, Franklin, Watson and Wilkins. Certain biological macromolecules, such as DNA and cytoskeletal components, cannot be crystallised, but form fibres. In fibres, the axes of the long polymeric structures are parallel to each other. While this can be an intrinsic property, for example in muscle fibres, in some cases the parallel alignment needs to be induced. As fibres show helical symmetry, by analysing the diffraction from oriented fibres one can deduce the helical symmetry of the molecule, and in favourable cases the molecular structure. Generally, a model of the fibre is constructed and the expected diffraction pattern is compared with the observed diffraction.

Two classes of fibre diffraction patterns can be distinguished. In crystalline fibres (e.g. the A form of DNA), the long fibrous molecules pack to form thin microcrystals randomly arranged around a shared common axis. The resulting diffraction pattern is equivalent to taking a long crystal and spinning it about its axis during the X-ray exposure. All Bragg reflections are recorded at once. In non-crystalline fibres (e.g. B form of DNA), the molecules are arranged parallel to each other, but in a random orientation around the common axis. The reflections in the diffraction pattern are now a result of the periodic repeat of the fibrous molecule. The diffraction intensity can be calculated via Fourier–Bessel transformation, replacing the Fourier transformation used in single-crystal diffraction.

Powder Diffraction

Powder diffraction is a rapid method to analyse multi-component mixtures without the need for extensive sample preparation. Instead of using single crystals, the solid material is analysed in the form of a powder where, ideally, all possible crystalline orientations are equally represented.

From powder diffraction patterns, the inter-planar spacings d of the lattice planes (Figure 14.3) are determined and then compared to a known standard or to a database (Powder Diffraction File by the International Centre for Diffraction Data or the Cambridge Structural Database) for identification of the individual components. Powder diffraction is typically used in the analysis of small organic molecules and inorganic samples.

X-Ray Single Molecule Diffraction

With the advent of so-called X-ray free electron lasers (which are in fact linear electron accelerators), very short pulses of extremely intense X-rays can be generated and their interaction with tiny nanocrystals, or eventually single molecules, produces distinct scatter that can be detected with latest-generation CCD technologies. It has thus become possible to determine atomic and molecular structures of very small crystalline samples or individual objects of nano-scale size (cells, viruses, nanostructures). Furthermore, the pulsing of X-rays at the femtosecond time scale also allows serial structure elucidation and thus the monitoring of conformational processes and reactions in the samples.

14.3.5 Light-Scattering Techniques

Elastic (Static) Light Scattering

Elastic light scattering is also known as Rayleigh scattering and involves measuring the intensity of light scattered by a solution at an angle relative to the incident laser beam. The scattering intensity of macromolecules is proportional to the squared molecular mass, and thus ideal for determination of the molecular mass, since the contribution of small solvent molecules can be neglected. In an ideal solution, the macromolecules are entirely independent of each other, and the light scattering can be described as:

$$\frac{I_\theta}{I_0} \sim R_\theta = P_\theta \times K \times c \times M \qquad \text{(Eq 14.2)}$$

where I_θ is the intensity of the scattered light at angle θ, I_0 is the intensity of the incident light, K is a constant proportional to the squared refractive index increment, c is the concentration and R_θ the Rayleigh ratio. P_θ describes the angular dependence of the scattered light.

For non-ideal solutions, interactions between molecules need to be considered. The scattering intensity of real solutions has been calculated by Debye and takes into account concentration fluctuations. This results in an additional correction term comprising the second virial coefficient B which is a measure for the strength of interactions between molecules:

$$\frac{I_\theta}{I_0} \sim R_\theta = P_\theta \times K \times c \times M \times \left(\frac{1}{1 + 2 \times B \times c \times M} \right) \qquad \text{(Eq 14.3)}$$

Determination of Molecular Mass With Multi-Angle Light Scattering

In solution, there are only three methods for absolute determination of molecular mass: membrane osmometry, sedimentation equilibrium centrifugation (Section 12.5.3) and light scattering. These methods are absolute, because they do not require any reference to molecular mass standards. In order to determine the molecular mass from light scattering, three parameters must be measured: the intensity of scattered light at different angles, the concentration of the macromolecule and the specific refractive index increment of the solvent. As minimum instrumentation, this requires a light source, a multi-angle light scattering (MALS) detector, as well as a refractive index detector. These instruments can be used in batch mode, but can also be connected to an HPLC to enable online determination of the molecular mass of eluting macromolecules. The chromatography of choice is size-exclusion chromatography (SEC; see Section 5.8.3), and the combination of these methods is known as SEC-MALS. Unlike conventional SEC, the molecular mass determination from MALS is independent of the elution volume of the macromolecule. This is a valuable advantage, since the retention time of a macromolecule on the size-exclusion column can depend on its shape and conformation.

Quasi-Elastic (Dynamic) Light Scattering: Photon Correlation Spectroscopy

While intensity and angular distribution of scattered light yields information about molecular mass and dimension of macromolecules, the wavelength analysis of scattered light allows conclusions as to the transport properties of macromolecules. Due to rotation and translation, macromolecules move into and out of a very small region in the solution. This Brownian motion happens at a time scale of microseconds to milliseconds, and the translation component of this motion is a direct result of diffusion, which leads to a broader wavelength distribution of the scattered light compared to the incident light. This analysis is the subject of dynamic light scattering, and yields the distribution of diffusion coefficients of macromolecules in solution.

The diffusion coefficient is related to the particle size by an equation known as the **Stokes–Einstein relation**. The parameter derived is the **hydrodynamic radius** R_{hydro} (also called the **Stokes radius**), which is the size of a spherical particle that would have the same diffusion coefficient in a solution with the same viscosity (see Section 12.2). Most commonly, data from dynamic light scattering are presented as a distribution of hydrodynamic radius rather than wavelength of scattered light.

Notably, the hydrodynamic radius describes an idealised particle and can differ significantly from the true physical size of a macromolecule. This is certainly true for most proteins, which are not strictly spherical and their hydrodynamic radius thus depends on their shape and conformation.

In contrast to SEC, dynamic light scattering measures the hydrodynamic radius directly and accurately, while the former method relies on comparison with standard molecules and several assumptions.

Applications of dynamic light scattering include determination of diffusion coefficients and assessment of protein aggregation, and can aid many areas *in praxi*. For instance, the development of 'stealth' drugs that can hide from the immune system or certain receptors relies on the PEGylation of molecules. Since conjugation with PEG (polyethylene glycol) increases the hydrodynamic size of the drug molecules dramatically, dynamic light scattering can be used for product control and as a measure of the efficiency of the drug.

Inelastic Light Scattering: Raman Spectroscopy

When the incident light beam hits a molecule in its ground state, there is a low probability that the molecule is excited and occupies the next higher vibrational state (Figure 13.3). The energy needed for the excitation is a defined increment, which will be missing from the energy of the scattered light. The wavelength of the scattered light is thus increased by an amount associated with the difference between two vibrational states of the molecule (**Stokes shift**). Similarly, if the molecule is hit by the incident light in its excited state and transitions to the next lower vibrational state, the scattered light has higher energy than the incident light, which results in a shift to lower wavelengths (**anti-Stokes shift**). These lines constitute the Raman spectrum. If the wavelength of the incident light is chosen such that it coincides with an absorption band of an electronic transition in the molecule, there is a significant increase in the intensity of bands in the Raman spectrum. This technique is called resonance Raman spectroscopy (see also Section 13.4).

Small-Angle Scattering

The characteristics of molecules at larger size scales are fundamentally different from those at atomic scales. While atomic-scale structures are characterised by high degrees of order (e.g. crystals), on the micro-scale, matter is rarely well organised and composed of rather complex building blocks (i.e. shapes). Consequently, sharp diffraction peaks are observed in X-ray diffraction from single crystals (probing the structures at atomic scale), but diffuse patterns are obtained from X-ray scattering from biological molecules or nanostructures.

Earlier in this section, we learned that incident light scattered by a particle in the form of Rayleigh scattering has the same frequency as the incident light; it is thus called elastic light scattering. Such techniques use a combination of visible light and molecules, so that the dimension of the particle under study is smaller than the wavelength of the light. When using light of shorter wavelengths such as X-rays, the overall dimension of a molecule is large as compared to the incident light. Electrons in the different parts of the molecule are now excited by the incident beam with different phases. The coherent waves of the scattered light therefore show an interference that is dependent on the geometrical shape of the molecule. As a result:

- in the forward direction (at 0°), there is no phase difference between the waves of the scattered light, and one observes maximum positive interference, i.e. highest scattering intensity
- at small angles, there is a small but significant phase difference between the scattered waves, which results in diminished scattering intensity due to destructive interference.

Small-angle X-ray (SAXS) or neutron scattering (SANS) are experimental techniques used to derive size and shape parameters of large molecules. Both X-ray and neutron scattering are based on the same physical phenomenon, i.e. scattering due to differences in scattering mass density between the solute and the solvent, or indeed between different molecular constituents. An advantage for protein structure determination is the fact that samples in aqueous solution can be assessed.

Experimentally, a monodisperse solution of macromolecules is exposed to either X-rays (wavelength $\lambda \approx 0.15$ nm) or thermal neutrons ($\lambda \approx 0.5$ nm). The intensity of the scattered light is recorded as a function of momentum transfer q:

$$q = 4\pi \times \frac{\sin \theta}{\lambda} \qquad\qquad \text{(Eq 14.4)}$$

Notably, $(2 \times \theta)$ is the angle between the incident and the scattered radiation. Due to the random positions and orientations of particles, an isotropic intensity distribution is observed that is proportional to the scattering from a single particle averaged over all orientations. In neutron scattering, the contrast (squared difference in scattering length density between particle and solvent) can be varied using H_2O/D_2O mixtures or selective deuteration to yield additional information. At small angles, the scattering curve is a rapidly decaying function of q, and essentially determined by the particle shape. Fourier transformation of the scattering function yields the so-called size distribution function, which is a histogram of inter-atomic distances. Comparison of the size distribution function with the particle form factor of regular geometrical bodies allows conclusions as to the shape of the scattering particle. Analysis of the scattering function further allows determination of:

- the radius of gyration R_g, which is the average distance of the atoms from the centre of gravity of the molecule (see Section 2.2.4)
- the mass of the scattering particle from scattering in the forward direction.

Shape Restoration

With the help of computer programs, it is possible to calculate low-resolution three-dimensional structures from the one-dimensional scattering data obtained by small-angle scattering. As a low-resolution technique, the structural information is restricted to the shape of the scattering molecules. Furthermore, the scattering data do not imply a single, unique solution. The reconstruction of three-dimensional structures might thus result in a number of different models. One approach is to align and average a set of independently reconstructed models, thus obtaining a model that retains the most persistent features.

14.4 SWITCH TECHNIQUES

Molecular interactions can be analysed directly, without requiring either washing or flow, using methods that direct switch-like behaviours within the molecules. These methods are often very convenient, as they can be performed in solution with minimal procedural steps.

Switch-like behaviour can be enacted through changes in molecular states associated with the pairing of two molecules. This switch-like behaviour can be adapted into assays that enables analysis of either inter-molecular or intra-molecular interactions.

14.4.1 Principles

The distance between molecules can be assessed by labelling with pairs of molecules that produce a change in molecular state when they are brought into close proximity. This can be used to confirm that two parts within one molecule interact, brought about, for example, by a conformational change (intra-molecular analysis; Figure 14.10a). The effects of potential inhibitors or enhancers (both allosteric and competitive) can also be evaluated. Similarly, this methodology can be used to test whether two different molecules interact, and to measure the affinity of that interaction by measuring the amount of change or signalling that occurs (inter-molecular analysis; Figure 14.10b).

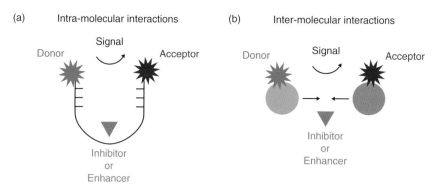

Figure 14.10 Mechanism of using molecule pairs that act as donors and acceptors for a change in molecular state that results in a molecular signal for analysis of both (a) intra-molecular interactions (U-shaped molecule) or (b) inter-molecular interactions. An additional ligand can be screened for inhibition or enhancement of the signalling event.

14.4.2 FRET and Static Quenching

The classic example of switch-like activity is fluorescence resonance energy transfer (FRET), described in Section 13.5.3, where a fluorescent molecule (the donor) has its excitation light absorbed if the acceptor molecule is in close proximity. More recently, mechanisms such as static quenching (Section 13.5.3) have been identified to produce a larger switch-like effect, where the acceptor molecule changes the properties of the donor such that it does not act as a fluorophore, reducing signal-to-noise ratios of the binding events. This mechanism is the basis of several recent molecular analysis tools, since FRET and static quenching measurements effectively serve as a molecular ruler to determine the distance between two biomolecules. For example, the molecular beacon (see below) is used to measure the hybridisation of DNA molecules, and is emerging as a biosensing tool for real-time measurement of analytes. FRET has also been applied to study the structure and conformation of proteins, the spatial distribution and assembly of proteins, receptor/ligand interactions, immune assays, structure and confirmation of nucleic acids, real-time PCR assays and single nucleotide polymorphism (SNP) detection, nucleic acid hybridisation events, distribution and transport of lipids, membrane fusion assays, membrane potential sensing and fluorogenic protease assays.

The switch-like activity of FRET and static quenching has been adopted by molecular engineers to develop miniaturised computing devices, through the generation of molecular binding events that are controlled by logical computing operations such as AND, NOT and OR. Demonstrations of these molecular computers include an automaton made from DNA that can play tic-tac-toe interactively with a human opponent, as well as systems that operate as programmable chemical reaction networks and artificial neural networks.

Molecular Beacons

Molecular beacons are FRET-labelled single-stranded hairpin-shaped oligonucleotide probes (see Figure 14.11), originally developed by Tyagi and Kramer in 1996. The loop region of the beacon is designed to bind specifically to a target nucleic acid. Thus, in the presence of the target nucleic acid, the stem region of the beacon becomes separated, releasing the fluorophore from the quencher such that active fluorescence can be observed. The molecular dynamics of hybridisation enables near-instantaneous real-time reporting on the presence or absence of the desired target. The reactions are also very specific, enabling discrimination between target sequences that differ by a single nucleotide substitution (e.g. for SNP detection). Applications thus also include real-time nucleic acid detection, real-time PCR quantification and allelic discrimination, and clinical/diagnostic applications have been reported. In addition, the use of multiple beacons with multiple fluorophore/quencher combinations enables multiplexing of detection, where multiple targets can be analysed in a single system.

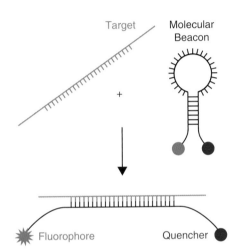

Figure 14.11 Molecular beacons consist of a stem (5–7 bp), and a loop (18–30 base pairs) and are labelled at the 5′ and 3′ ends by fluorophores and quenchers, respectively. Fluorophore and quencher combinations can include fluorescein and Black Hole quencher 1; Dabcyl, Cyanine 3 or Texas Red and Black Hole Quencher 2 or Dabcyl; Cyanine 5 and Black Hole Quencher 3 or Dabcyl.

14.4.3 ALPHA-Screen®

ALPHA screening is a trade-marked switch-like assay technology sold by PerkinElmer that uses a donor bead containing phthalocyanine, which produces a singlet oxygen (see Section 13.5.1) when exposed to light at 680 nm. This singlet oxygen interacts with an acceptor bead containing thiophene derivatives, producing emission at 600 nm. When the two beads are distant (>200 nm) the reactive oxygen decays and no signal is generated. ALPHA stands for <u>a</u>mplified <u>l</u>uminescent <u>p</u>roximity <u>h</u>omogeneous <u>a</u>ssay, and is characterised by high sensitivities, low signal/background ratios, and the ability to study both low- and high-affinity interactions. ALPHA screening has been demonstrated to be useful in the analysis of affinity interactions of small binding proteins, studies on phosphorylation and oligomerisation, and other enzymatic and binding reactions. Numerous kits for various assays are available via the manufacturers' websites.

14.4.4 Differential Scanning Fluorimetry

Another method that uses fluorophores in switch-like behaviour is differential scanning fluorimetry (DSF; see also Section 13.5.3). The method is used to monitor thermal transitions of biomolecules, and exploits the properties of amphiphilic dyes, such as SYPRO® Orange or the thiol-reactive dye

7-diethylamino-3-(4′-maleimidylphenyl)-4-methylcoumarin (CPM), to be highly fluorescent in certain chemical microenvironments (such as hydrophobic binding pockets), but not in general aqueous solutions. The method can be used to study biomolecule conformational stability, including folding and unfolding at different temperatures. The fluorescence intensity of the dye is plotted as a function of temperature, enabling the apparent melting temperature (T_m) of the biomolecule to be determined. This is particularly useful for determining folding and unfolding in complex multi-domain proteins. In addition, it can also be used in screening assays to identify low-molecular-weight ligands that bind and stabilise purified proteins.

14.5 SOLID-PHASE BINDING TECHNIQUES WITH WASHING STEPS

14.5.1 Principles

Solid-phase binding assays are a generic class of biosensing technique that can be used for the study of molecular binding events. They can also be used to quantify the concentration of molecules required for detection of the binding event. Molecules of interest (ligands) are attached to a surface, and then a series of reagents are added to determine if binding occurs between different ligands. Washing is performed between reagent addition steps to remove unbound molecules from the reaction.

Biomolecules (ligands) that can be studied using solid-phase binding assays are often antibodies with epitopes specific to antigens of interest (see Section 7.3), however the technique is applicable to the analysis of interactions between any class of biomolecule, including proteins, peptides, small molecules, lipids, aptamers (single-stranded DNA or RNA that can bind to particular target molecules) and nucleic acids in general. These techniques are useful for in vitro study of binding such as protein–protein interactions, protein–peptide interactions, or even protein–nucleic acid interactions. However, each ligand must have one additional binding site (apart from the binding site of interest) to enable attachment to either a solid support (ligand 1) or a labelling tracer molecule (ligand 2).

14.5.2 Analytical Techniques

The sensitivity to which binding can be determined in solid-phase binding assays is highly dependent on the physical format in which the assay is performed, the buffering conditions surrounding the molecules, and particularly the combination of tracer and analytical technique used to observe the binding events. Indeed, the tracer molecule largely directs the analytical technique required to observe the binding. This may be a radioactive label (e.g. radioimmunoassays; Section 7.3.5), which is highly sensitive, but technically difficult to perform due to maintaining safety of the operator. Other systems, such as the enzyme-linked immunosorbent assay (ELISA; Sections 7.3.1 and 7.5) use an enzyme that produces a colorimetric or luminescence product: colorimetric assays reagents are relatively inexpensive and the output can be viewed by eye, ablating the need for expensive readers; however, luminescence is much more sensitive, resulting in expansion of the linear range of the assay by several orders of magnitude. Other systems use fluorophores, which are measured spectroscopically (outlined in Chapter 13). Similarly, molecules that operate at different parts of the electromagnetic spectrum, such as infrared, can be used. Alternatively, molecules that afford the release of electrons can also be used, such as the production of hydrogen peroxide by peroxidase; this enables direct coupling with electrical sensing devices, for the production of electrochemical biosensors (see potentiometry in Section 10.2.2).

14.5.3 Assay Formats

Assays to determine binding between two ligands can be performed using several different formats. In the direct assay (Figure 14.12a), one ligand (circle) is adhered to the surface, and the second ligand (star) is linked directly to the tracer. In the indirect assay (Figure 14.12b), the tracer is placed on a third binding molecule known to bind to the second ligand. In the sandwich assay (Figure 14.12c), the binding of the first ligand to the surface is directed through an additional binding partner. Alternatively, a competitive assay can be run (Figure 14.12d), where a known amount of the second ligand bound to the tracer competes for binding to the first ligand with an unknown amount of the second ligand that may be present in a sample. Extending these principles further can lead to three-component (or higher) assays (Figure 14.13), where a third ligand (triangle), often a small molecule, is used to inhibit binding between two other ligands. This inhibition can occur competitively (Figure 14.13a), or non-competitively, through binding to an allosteric site on either the first or the second ligand (Figure 14.13b). Alternatively, the third ligand may enhance binding between the other two ligands, which is called cooperativity, and this enhancement can be due to forming a direct bridge that links the two molecules (Figure 14.13c) or through binding on an allosteric site on one of the ligands that changes the shape of the molecule to more favourably participate in the binding event (Figure 14.13d). See also Section 23.2.3 for concepts on allostery and cooperativity.

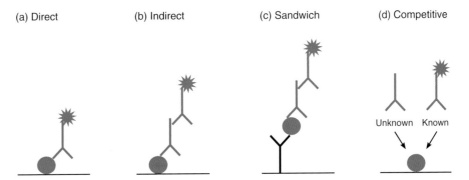

(a) Direct   (b) Indirect   (c) Sandwich   (d) Competitive

Unknown   Known

Figure 14.12 Different configurations for solid-phase binding assays with two ligands (shown as circle and star, respectively).

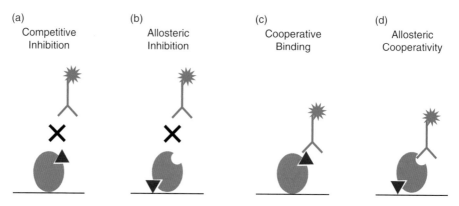

Figure 14.13 Schematic illustration of solid-phase binding assays with three components to probe an effector.

14.6 SOLID-PHASE BINDING TECHNIQUES COMBINED WITH FLOW

14.6.1 Surface Plasmon Resonance

Surface plasmon resonance (SPR) is a surface-sensitive method for monitoring small changes in the refractive index or the thickness of thin films. It is mainly used for monitoring the interaction of two components (e.g. ligand and receptor), one of which is immobilised on a sensor chip surface (Figure 14.14a), such as a hydrogel layer on a glass slide, via either biotin–avidin interactions or covalent coupling using amine or thiol reagents similar to those used for cross-linking to affinity chromatography resins (see Section 5.8.1). Typical surface concentrations of the bound protein component are in the range of 1–5 ng mm^{-2}. The sensor chip forms one wall of a microflow cell so that an aqueous solution of the ligand can be pumped at a continuous, pulse-free rate across the surface. This ensures that the concentration of ligand at the surface is maintained at a constant value. Environmental parameters such as temperature, pH and ionic strength are carefully controlled, as is the duration of exposure of the immobilised component to the ligand. Replacing the ligand solution by a buffer solution enables investigation of the dissociation of bound ligand.

Binding of ligand to the immobilised component causes an increase in mass at the surface of the chip; vice versa, dissociation of ligand causes a reduction in mass. These mass changes, in turn, affect the refractive index of the medium at the surface of the chip, the value of which determines the propagation velocity of electromagnetic radiation in that medium.

Plasmon is a term for a collection of conduction electrons in a metal or semiconductor. Excitation of a plasmon wave requires an optical prism with a metal film of about 50 nm thickness. Total internal reflection (TIR) occurs when a light beam travelling through a medium of higher refractive index (e.g. glass prism with gold-coated surface) meets an interface with a medium of lower refractive index (e.g. aqueous sample) at an angle larger than the critical angle. TIR of an incident light beam at the prism–metal interface elicits a propagating plasmon wave by leaking an electrical

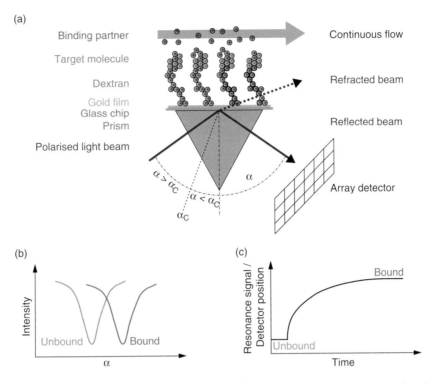

Figure 14.14 Principle of a surface plasmon resonance (SPR) measurement. (a, b) The angle α of total internal reflection shifts if material is bound on the sample side of the SPR chip. (c) A time-dependent trace or sensorgramme is recorded to monitor binding of material to the chip in the course of an experiment. Note that kinetic parameters can be determined from the sensorgramme.

field intensity, called an **evanescent field wave**, into the medium of lower refractive index where it decays at an exponential rate and effectively only travels one wavelength.

Since the interface between the prism and the medium is coated with a thin layer of gold, incident photons excite a vibrational state of the electrons in the conducting band of the metal. In thin metal films this propagates as a longitudinal vibration. The electrons vibrate with a resonance frequency (hence the term 'resonance') that is dependent on metal and prism properties, as well as the wavelength and the angle of the incident beam. Excitation of the plasmon wave leads to a decreased intensity of the reflected light. Thus, SPR produces a dip in the reflected light intensity at a specific angle of reflection (Figure 14.14b). The propagating surface plasmon wave enhances the amplitude of the evanescent field wave, which extends into the sample region.

When binding to the chip occurs, the refractive index on the sample side of the interface increases. This alters the angle of incidence required to produce the SPR effect and hence also alters the angle of reflected light. The change in angle brings about a change in detector position, which can be plotted against time to give a sensorgramme reading (Figure 14.14c). The angle is expressed in resonance units (RU), such that 1000 RU corresponds to a change in mass at the surface of the chip of about 1 ng mm^{-2}.

Since in SPR instruments the angle and the wavelength of the incident beam are constant, a shift in the plasmon resonance leads to a change in the intensity of the reflected beam. The shift is restricted locally and happens only in areas where the optical properties have changed. The use of an array detector as compared to a single detector cell therefore allows for the creation of an SPR image. SPR can detect changes in the refractive index of less than 10^{-4} or changes in layer heights of about 1 nm. This enables not only the detection of binding events between biomolecules, but also binding at protein domains or changes in molecular monolayers with a lateral resolution of a few μm. For SPR, light of wavelengths between infrared (IR) and near-infrared (NIR) may be used. In general, the higher the wavelength of the light used, the better the sensitivity, but at lower lateral resolution. Vice versa, if high lateral resolution is required, red light is used because the propagation length of the plasmon wave is approximately proportional to the wavelength of the exciting light.

Applications

The SPR technique enjoys frequent use in modern life science laboratories, due to its general applicability and the fact that there are no special requirements for the molecules to be studied (label-free), such as fluorescent properties, spectral labels or radio labels. It can even be used with coloured or opaque solutions.

Generally, all two-component binding reactions can be investigated, which opens a variety of applications in the areas of drug design (protein–ligand interactions), as well as mechanisms of membrane-associated proteins (protein–membrane binding) and DNA-binding proteins. SPR has thus successfully been used to study the kinetics of receptor–ligand, antibody–antigen and protein–protein interactions. The method is extensively used in proteomic research and drug development.

SPR Imaging

The focus of SPR imaging experiments has shifted in recent years from characterisation of ultrathin films to analysis of biosensor chips, especially affinity sensor arrays. SPR imaging can detect DNA–DNA, DNA–protein and protein–protein interactions in a two-dimensional manner. The detection limit for such biosensor chips is on the order of 10^{-9} M (nM) to 10^{-15} M (fM). Apart from the detection of binding events as such, the quality of binding (low affinity, high affinity) can also be assessed by SPR imaging. Promising future applications for SPR imaging include peptide arrays that can be prepared on modified gold surfaces, which could prove useful for assessing peptide–antibody interactions. The current time resolution of less than 1 s for an entire image also allows for high-throughput screenings and *in situ* measurements.

Mass-Spectrometry-Coupled Biosensors

Mass spectrometry (MS; see Chapters 15 and 21) can be used for identification of molecules that are interacting with each other. The technique of biomolecular interaction analysis by mass spectrometry (BIA-MS, also called SPR-MS) combines an SPR bio-sensor with mass spectrometry. The SPR biosensor allows the selection, concentration and quantification in real time of biomolecules interactions at the molecular level. SPR imaging allows multi-analyte analysis through SPR protein microarrays and a simul-taneous examination of several protein biomarkers present in a single experiment with minimisation of reagents and analytes used, and miniaturised design. The hyphenated mass spectrometry analysis allows identification of interacting molecules. The usual ways of sample ionisation in MS are the matrix-assisted laser desorption ionisation (MALDI) or the electrospray ionisation (ESI) methods. SPR-MALDI-MS allows the direct analysis of the sample surface and ESI-MS approaches uses elution of the sample from the surface. SPR-MS is particularly useful for protein–protein interaction studies and ligand fishing; compound screening can also be undertaken. Further improvements on the contemporary SPR technology to improve detection and quantification of bio-molecular interactions at even smaller scales include silicon evanescent waveguides. For a similar concept, see also microchip-electrophoresis-coupled mass spectrometry (MCE-MS) in Section 6.6.

14.6.2 Lateral Flow and Microfluidics Systems

A microfluidics system offers a confined space for handling and controlling volumes of liquid in a continuous liquid stream, droplets or bubbles into microchannels in the nanolitre to millilitre range.

Principles

The heart of a microfluidic device is the microchannel, which is where reaction, sep-aration, or detection takes place. The flow of fluid through a cylindrical microfluidic channel is characterised by the Reynolds number:

$$Re = \frac{l \times v \times \rho}{\eta} \qquad \text{(Eq 14.5)}$$

where l is the characteristic length of the channel, v is the velocity of flow, ρ is the fluid density and η is the viscosity of the fluid. The geometry of the channel may be circular, square, rectangular or many other shapes, and there are various adjustments that can be applied to Equation 14.5 to fit a given shape.

The flow of molecules through a microchannel can be actuated by several mechanisms:

- Diffusion, where molecules move along a concentration gradient
- Convection, where fluids are moved by heat transfer
- Electrokinetics, where the walls of a microchannel have an electric charge that moves ions towards an electrode of opposite polarity
- Pressure, where pumps displace the liquid, either by peristaltic means via squeezing fluids in the desired direction, or centrifugal forces via rotation to accelerate fluids outward from the centre of the device.

Fabrication

Microfluidic channels can be fabricated using various substrates, including glass (silica) or polymers such as polydimethylsiloxane (PDMS). Channels can be formed by photolithography, where a patterned photomask exposes certain areas of a device to curing agents such as UV light to produce hardened areas. Alternatively, channels can be formed by etching that carves channels from a solid surface (e.g. by using the tip of an atomic force microscope, AFM), or by layering materials such as wax via inkjet printing.

Complete flow devices are composed of elements of sample preparation, fluid control and measurements. The performance of the resulting design relies on a large number of factors. Channel dimensions, flow velocities, transport and reaction rate are critical to the microfluidics operation. Additionally, the biosensing part requires consideration of surface functionalisation and measurement methods. Detection methods may include fluorescence, surface plasmon resonance and mass spectrometry. The final sensitivity and detection limit are defined according to the targeted application and the detection method used. Other issues such as portability, shelf-life, lifespan, cost and reuse of design are all important factors in commercial development of microfluidic devices.

Applications

The majority of applications are in the biomedical field for clinical or home uses, but also in food monitoring and biotechnological processes with detection of analytes such as glucose, lactate and fructose (see also Section 10.2.2). Nucleic acids can be detected by incorporating PCR within the microfluidics system. Biochips have been designed to perform two-dimensional electrophoresis, transcriptome analysis, liquid chromatography for proteins and DNA, and protein analysis, particularly in screening conditions for protein crystallisation. Microfluidics systems have potential impact commercially for a broad range of applications in multiplexing, automation and lab-on-chip technologies. Other examples include microchip electrophoresis (MCE), discussed in Section 6.6.

One of the simplest forms of microfluidics is the lateral flow assay (Figure 14.15), a paper-based device where molecules are embedded into several reaction zones within a solid phase, and liquids are wicked through the device by capillary force. This device is often used for analysis of antibodies binding to an antigen (analyte), but could be used for determining any binding interaction. The sensitivity of the device is limited, as many molecules are absorbed by the paper material as it progresses through the device. However, the simplicity, portability, low cost and speed of the assay is revolutionising the at-home diagnostics industry. Initially developed to detect human chorionic gonadotropin (hCG) by Unipath in 1988 (home pregnancy testing), the device has subsequently been used to detect parasites, bacteria, viruses, cells, toxins, hormones and biological markers.

14.6.3 Fluorescence-Activated Cell Sorting

Fluorescence-activated cell sorting (FACS) is another powerful technique that uses flow, and can be used to characterise molecules and molecular interactions on the surface of cells. The principles of this technique are outlined in Sections 7.8 and 8.3. In this case, a fluorescently labelled probe (e.g. an antibody or an oligonucleotide)

(a)   Loading of analyte

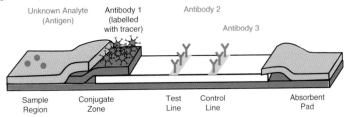

(b)   Binding of unknown analyte to labelled antibody

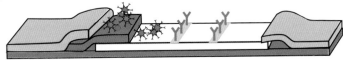

(c)   A specific analyte is captured by antibody 2,
        antibody 3 directly binds to tracer-labelled antibody

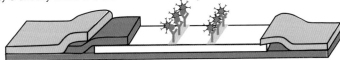

(d)   Accumulation of tracer leads to coloured precipitate

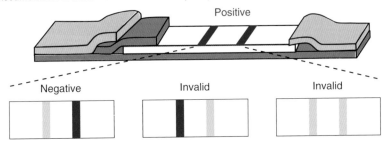

Figure 14.15 Working schematics of a typical lateral flow device. Sample containing analyte is added to a sample pad (cellulose acetate), where it is wicked by capillary action through a series of embedded reaction zones. An antibody-labelled tracer molecule binds to the analyte in the conjugate zone, and the complex moves across a nitrocellulose membrane and is captured by antibodies pre-deposited in a test line. Accumulation of the tracer molecule (e.g. gold nanoparticles) results in the appearance of a coloured precipitate at the test line. Unbound particles are removed by continued capillary flow to the absorbent pad. A control line binds directly to the tracer molecule, indicating that the reagents have wicked correctly through the device, regardless of the presence of analyte. See also Section 7.5.

indicates the presence or absence of cells with a specific biomarker of interest. The number of cells labelled by the probe are counted and cells that contain the biomarker can be sorted from those that do not. This technique is highly useful in immunology, where subpopulations of cells containing specific biomarkers can be captured and sorted, such as B Cells (CD19+) from T cells (CD3+), or cell-mediated T helper cells (CD4+) from cytotoxic T cells (CD8+).

14.7 SUGGESTIONS FOR FURTHER READING

14.7.1 Experimental Protocols

Majka J. and Speck C. (2007) Analysis of protein–DNA interactions using surface plasmon resonance. *Advances in Biochemical Engineering and Biotechnology* **104**, 13–36.

Mueller M., Jenni S. and Ban N. (2007) Strategies for crystallization and structure determination of very large macromolecular assemblies. *Current Opinion in Structural Biology* **17**, 572–579.

Skinner A.L. and Laurence J.S. (2008) High-field solution NMR spectroscopy as a tool for assessing protein interactions with small molecule ligands. *Journal of Pharmaceutical Science* **97**, 4670–4695.

Wlodawer A., Minor W., Dauter Z. and Jaskolski M. (2008) Protein crystallography for non-crystallographers, or how to get the best (but not more) from published macromolecular structures. *FEBS Journal* **275**, 1–21.

Zhou X., Kini R.M. and Sivaraman J. (2011) Application of isothermal titration calorimetry and column chromatography for identification of biomolecular targets. *Nature Protocols* **6**, 158–165.

14.7.2 General Texts

Biophysical Methods

Hofmann A., Simon A., Grkovic T. and Jones M. (2014) *Methods of Molecular Analysis in the Life Sciences*, Cambridge University Press, Cambridge, UK.

Biosensors

Cooper M.A. (2010) *Label-free Biosensors: Techniques and Applications*, Cambridge University Press, Cambridge, UK.

Calorimetry

Feig A.L. (2016) Calorimetry. *Methods in Enzymology*, Vol. 567, Academic Press, Burlington, MA, USA.

Nuclear Magnetic Resonance

Friebolin H. (2010) *Basic One- and Two-Dimensional NMR Spectroscopy*, Wiley-VCH, Weinheim, Germany.

Protein Crystallography

Rhodes G. (2006) *Crystallography Made Crystal clear*, 3rd Edn., Academic Press, Burlington, MA USA.

14.7.3   Review Articles

Anker J.N., Hall W.P., Lyandres O., *et al.* (2008) Biosensing with plasmonic nanosensors. *Nature Materials* **7**, 442–453.

Blamire A.M. (2008) The technology of MRI: the next 10 years? *British Journal of Radiology* **81**, 601–617.

Campbell C.T. and Kim G. (2007) SPR microscopy and its applications to high-throughput analyses of biomolecular binding events and their kinetics. *Biomaterials* **28**, 2380–2392.

Hickman A.B. and Davies D.R. (2001) Principles of macromolecular X-ray crystallography. *Current Protocols in Protein Science*, Chapter 17, Unit 17.3, Wiley Interscience, New York, USA.

Ishima R. and Torchia D.A. (2000) Protein dynamics from NMR. *Nature Structural Biology* **7**, 740–743.

Leatherbarrow R.J. and Edwards P.R. (1999) Analysis of molecular recognition using optical biosensors. *Current Opinion in Chemical Biology* **3**, 544–547.

Lipfert J. and Doniach S. (2007) Small-angle X-ray scattering from RNA, proteins, and protein complexes. *Annual Reviews of Biophysical and Biomolecular Structure* **36**, 307–327.

McDermott A. and Polenova T. (2007) Solid state NMR: new tools for insight into enzyme function. *Current Opinion in Structural Biology* **17**, 617–622.

Miao J., Ishikawa T., Shen Q. and Earnest T. (2008) Extending X-ray crystallography to allow the imaging of noncrystalline materials, cells, and single protein complexes. *Annual Reviews in Physical Chemistry* **59**, 387–410.

Mross S., Pierrat S., Zimmermann T. and Kraft M. (2015) Microfluidic enzymatic biosensing systems: a review. *Biosensors and Bioelectronics* **70**, 376–391.

Neumann T., Junker H.D., Schmidt K. and Sekul R. (2007) SPR-based fragment screening: advantages and applications. *Current Topics in Medicinal Chemistry* **7**, 1630–1642.

Neylon C. (2008) Small angle neutron and X-ray scattering in structural biology: recent examples from the literature. *European Biophysics Journal* **37**, 531–541.

Phillips K.S. and Cheng Q. (2007) Recent advances in surface plasmon resonance based techniques for bioanalysis. *Analytical and Bioanalytical Chemistry* **387**, 1831–1840.

Putnam C.D., Hammel M., Hura G.L. and Tainer J.A. (2007) X-ray solution scattering (SAXS) combined with crystallography and computation: defining accurate macromolecular structures, conformations and assemblies in solution. *Quarterly Reviews in Biophysics* **40**, 191–285.

Renaud J.-P., Chung C.-W., Danielson U.H., *et al.* (2016) Biophysics in drug discovery: impact, challenges and opportunities. *Nature Reviews Drug Discovery* **15**, 679–698.

Spiess H.W. (2008) NMR spectroscopy: pushing the limits of sensitivity. *Angewandte Chemie International Edition* **47**, 639–642.

Stiger E.C.A., de Jong G.J. and van Bennekom W.P. (2013) Coupling surface-plasmon resonance and mass spectrometry to quantify and to identify ligands. *Trends in Analytical Chemistry* **45**, 107–120.

Vuignier K., Schappler J., Veuthey J.-L. Carrupt, P.-A. and Martel, S. (2010) Drug-protein binding: a critical review of analytical tools. *Analytical and Bioanalytical Chemistry* **398**, 53–66.

14.7.4 Websites

NMR spectroscopy

www.cis.rit.edu/htbooks/nmr/ (accessed May 2017)

teaching.shu.ac.uk/hwb/chemistry/tutorials/molspec/nmr1.htm (accessed May 2017)

X-ray diffraction

ww1.iucr.org/cww-top/edu.index.html (accessed May 2017)

www.ruppweb.org/Xray/101index.html (accessed May 2017)

Small-angle scattering

www.ncnr.nist.gov/programs/sans/tutorials/index.html (accessed May 2017)

www.embl-hamburg.de/workshops/2001/EMBO (accessed May 2017)

Surface plasmon resonance

www.sprpages.nl/ (accessed May 2017)

www.biacore.com/ (accessed May 2017)

TaqMan Molecular Beacons

www.genelink.com/newsite/products/mbintro.asp (accessed May 2017)

15 Mass Spectrometric Techniques

SONJA HESS AND JAMES I. MACRAE

15.1 INTRODUCTION

Mass spectrometry (MS) is an extremely valuable analytical technique in which the molecules in a test sample are converted to gaseous ions that are subsequently separated in a mass analyser according to their **mass-to-charge (m/z) ratio** and then detected. The **mass spectrum** is a plot of the relative abundances of the ions at each m/z ratio. Note that it is the mass-to-charge ratio (m/z), and *not* the actual mass, that is measured. For example, if a biomolecule is ionised in positive ion mode, the instrument measures the m/z after the addition of one proton (i.e. 1.0072772984 Da for exact mass or 1.0078 Da for average mass). Similarly, for a biomolecule ionised in negative ion mode, an m/z after the loss of one proton is measured.

The essential features of all mass spectrometers are:

- Generation of ions in the gas phase
- Separation of ions in a mass analyser
- Detection of each species of a particular m/z ratio.

Several techniques exist to generate ions and are discussed below. Of these, the development of **electrospray** ionisation (ESI; Section 15.2.4) and **matrix-assisted laser desorption ionisation** (MALDI; Section 15.2.5), has effectively expanded the detectable mass range, enabling the measurement of almost any biomolecule. Mass analysers separate ions by use of either a magnetic or an electric field (Section 15.3); detectors produce a measurable signal in the form of either a voltage or a current that can be transformed by a computer into data to be analysed (Section 15.4). The symbol M_r is used to designate relative molecular mass. As a relative measure, M_r has no units.

The treatment of mass spectrometry in this chapter will be rather non-mathematical and non-technical. Mass spectrometry has a wide array of applications, including drug discovery and the sciences of proteomics (Chapter 21) and metabolomics (Chapter 22).

This chapter will focus on the fundamental principles of mass spectrometry. The intention is to give an overview of the different types of instrumentation that are available and discussion of their applications, complementary techniques and the advantages/disadvantages of each system. Sample preparation and data analysis will also be covered. A further reading list is provided, covering more technical and mathematical aspects of mass spectrometry.

15.1.1 Components of a Mass Spectrometer

In essence, all mass spectrometers consist of the following three components (Figure 15.1):

- An ion source to convert neutral molecules into gas-phase ions. Common ion sources include electrospray ionisation (ESI), matrix-assisted laser desorption ionisation (MALDI), electron impact ionisation (EI) and direct chemical ionisation (CI), among others.
- A mass filter/analyser to separate the gas-phase ions by their m/z ratios. Examples of these include quadrupoles, time-of-flight (TOF) detectors, ion traps, Orbitrap™ and ion cyclotron resonance Fourier transform (FT) detectors (these latter two are termed image current detectors).
- A detector. Common detectors include electron multipliers, conversion dynodes, and image current detectors. Ions detected are recorded, processed and visualised using a computer.

To ensure a free flight path for the gas-phase ions and to minimise collisions between ions and air molecules, mass spectrometers are operated under very high vacuum (10^{-9} torr or 10^{-7} Pa). This vacuum is generated using a combination of turbomolecular pumps, diffusion pumps and rotary vane pumps.

To allow introduction of a sample, mass spectrometers must be equipped with a sample inlet. For MALDI, the sample is introduced to the mass spectrometer on a sample plate. For ESI, EI and CI, sample introduction is through an insertion probe. Introduction of a sample straight into the mass spectrometer in this manner is known as direct infusion. However, mass spectrometry is often combined with chromatography to enable separation of biomolecules before ionisation. These techniques are

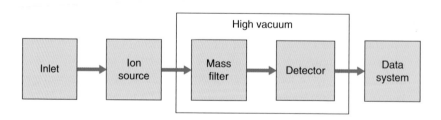

Figure 15.1 Basic components of a mass spectrometer.

collectively termed hyphenated mass spectrometry and include high-performance liquid chromatography and gas chromatography coupled to mass spectrometry (HP LC-MS and GC-MS, respectively).

15.2 IONISATION

Ions may be produced from a neutral molecule by removing an electron or adding a proton to produce a positively charged cation, or by adding an electron or removing a proton to form a negatively charged anion. These ionic forms of neutral molecules are termed molecular ions, or precursor ions, respectively. If the precursor ions are disintegrated into a number of smaller ions, fragment ions (or product ions) are generated. This process is termed fragmentation and is an important principle in identification of the precursor molecules.

While both positive- and negative-ion mass spectrometry are often used, for simplicity the remainder of this section will mainly focus on positive-ion MS, since this is more common and has wider applications. The principles of separation and detection of positively and negatively charged ions are essentially the same.

15.2.1 Electron Impact Ionisation (EI)

Electron impact ionisation (also termed electron ionisation, EI) is widely used for the analysis of metabolites, pollutants and pharmaceutical compounds, and is the main ionisation technique coupled to gas chromatography in GC-MS. The electrons for EI are produced by heating a metal filament with an electric current running through it. These electrons are accelerated to 70 eV (the electron volt, eV, is a measure of energy) and focussed to allow passage into the ionisation chamber, into which the neutral molecules of the sample to be analysed (the analyte) are also introduced. Sample ionisation occurs in this high vacuum chamber when the electrons collide with neutral molecules of the analyte (Figure 15.2). Interaction with the analyte results in either loss of an electron from the substance (to produce a cation) or electron capture (to produce an anion). Importantly, the analyte must be in the vapour state in the EI source, which generally limits the applicability to molecules below around 450 Da. EI is a hard ionisation technique and often results in complete, or near complete, fragmentation of the analyte. This may or may not be desirable, depending on the information required.

Chemical bonds in organic molecules are formed by the pairing of electrons. Ionisation resulting in a positively charged cation requires loss of an electron from one of these bonds (i.e. the electron bombardment of EI effectively knocks out an electron from the bond) and leaves the bond with a single unpaired electron. This constitutes a radical as well as a cation, and hence representation as $M^{\bullet+}$, where the $+$ sign indicates the ionic state and $\bullet$ the radical. Conversely, electron capture results in the addition of an unpaired electron and therefore a negatively charged radical anion, hence the symbol $M^{\bullet-}$. Radical ions generated under the conditions of electron bombardment are relatively unstable, limiting EI to smaller molecules. Their energy in excess of that required for ionisation has to be dissipated by forming fragment ions. A typical EI spectrum thus contains mostly product ions, with minimal or no precursor ions.

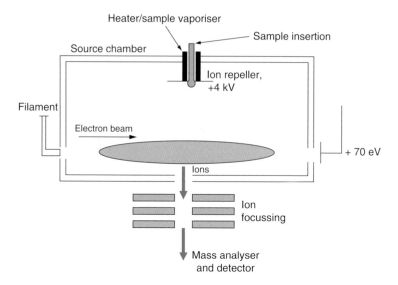

Figure 15.2 Electron impact source. Electrons are produced by thermionic emission from a filament of tungsten or rhenium. The filament current is typically 0.1 mA. Electrons are accelerated toward the ion source chamber (held at a positive potential equal to the accelerating voltage) and acquire energy equal to the voltage between the filament and the source chamber – typically 70 eV. The electron trap is held at a fixed positive potential with respect to the source chamber. Gaseous analyte molecules are introduced into the path of the electron beam where they are ionised. Due to the positive ion repeller voltage and the negative excitation voltage that produce an electric field in the source chamber, the ions leave the source through the ion exit slit and are analysed.

For the production of a radical cation, since it is not known where either the positive charge or the unpaired electron precisely reside within the molecule, it is the convention to place $^{\bullet}{}^{+}$, and $^{-}$ signs outside the abbreviated bracket sign (1) when depicting an ion (see Figure 15.3).

When the precursor ion fragments, one of the products carries the charge and the other the unpaired electron, i.e. it splits into a radical and an ion. The product ions are therefore true ions and not radical ions. The radicals produced in the fragmentation process are neutral species and as such cannot be analysed by the mass spectrometer and are instead pumped away by the vacuum system. Only charged species are accelerated out of the source and into the mass analyser. It is also important to recognise that almost all possible bond breakages can occur and any given fragment will arise both as an ion and a radical. The distribution of the charge and unpaired electron, however, is by no means equal. It depends entirely on the thermodynamic stability of the products of fragmentation. Furthermore, any fragment ion may break down further (ultimately until single atoms are obtained) and hence few ions of a particular type may survive. As such, smaller m/z ions are generally more abundant in EI spectra than those of a higher m/z.

A simple example is given by n-butane ($CH_3CH_2CH_2CH_3$) and some of the major fragmentations are shown in Figure 15.3a. The resultant EI spectrum is shown in Figure 15.3b.

15.2.2 Chemical Ionisation (CI)

Chemical ionisation (CI) is used for a range of samples similar to those for EI and, like EI, is often used as the ion source in GC-MS. It is particularly useful for the determination of molecular masses, as high-intensity molecular ions are produced due to less fragmentation. CI is therefore regarded as a **soft ionisation** technique and can give rise to much cleaner spectra. The source is very similar to the EI source, but contains a suitable reagent gas such as methane (CH_4) or ammonia (NH_3) that is initially ionised by EI. The high gas pressure in the source results in ion–molecule reactions between reagent gas ions (such as CH_4^+ and NH_3^+), some of which react with

(a)

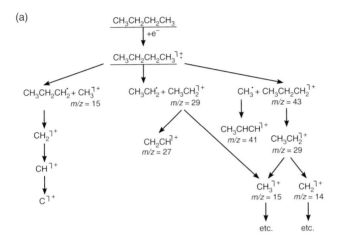

(b)

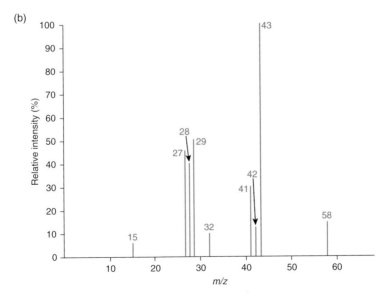

Figure 15.3 Fragmentation pathways (a) and the EI spectrum (b) of *n*-butane. In the spectrum, the relative abundance is plotted from 0 to 100% where the largest peak is set at 100% (base peak). Spectra represented in this way are said to be **normalised**.

the analyte to produce analyte ions. The mass differences from the neutral parent compounds therefore correspond to reagent gas adducts of the molecule. Since CI is essentially a chemical reaction between two substrates (the reagent gas ion and the analyte), the flow of reagent gas determines the degree of fragmentation of the analyte. Varying this flow can therefore reveal more (or less) information about the precursor molecule.

15.2.3 Fast Atom Bombardment (FAB)

At the time of its development in the early 1980s, fast atom bombardment (FAB) was the first method that enabled the analysis of biomolecules in solution without prior derivatisation. This important advancement was possible due to the fact that FAB, like CI above, is a soft ionisation technique, leading to the formation of ions with low internal energies and little consequent fragmentation. The sample is mixed with a relatively involatile, viscous matrix such as glycerol, thioglycerol or *m*-nitrobenzyl alcohol. This mixture, placed on a probe, is introduced into the source housing and bombarded with an ionising beam of neutral atoms (such as argon, helium, xenon) at high velocity. Molecular ion species arise as either protonated or deprotonated entities ($[M + H]^+$ or $[M - H]^-$, respectively), allowing positive and negative ion mass spectra to be determined. Other charged adducts can also be formed, such as sodium $[M + Na]^+$ and potassium $[M + K]^+$ adducts.

15.2.4 Electrospray Ionisation (ESI)

Electrospray ionisation (ESI) is a soft ionisation technique that enables the analysis of large intact (underivatised) biomolecules, such as proteins, DNA, glycerolipids and glycans. This advancement in ionisation technique revolutionised biology and, consequently, resulted in the award of the Nobel Prize for Chemistry to John Fenn in 2002. As the name implies, electrospray ionisation involves the production of ions by spraying a solution of the analyte through a needle into an electrical field. This creates an aerosol with very small charged droplets of solvent containing analyte. Under conditions of electrospray, the solvent evaporates from the charged droplets. When these charged droplets shrink, Coulombic repulsion of the charges increase until they reach the 'Rayleigh limit', the point at which the repulsion is stronger than the surface tension, leading to droplet 'explosion'. This creates a charged analyte ion that is free of solvent and introduced into the mass spectrometer. It is therefore important for solvent evaporation to occur quickly and efficiently and so introduction of a curtain or sheath gas (usually nitrogen) around the spray needle may be used to promote this. A diagrammatic representation of the ESI source is shown in Figure 15.4.

While most ESI sources are fed by an HPLC system, direct infusion experiments can also be carried out. For direct infusion, a sample is dissolved in a typical solvent (e.g. acetonitrile/water (1:1) or methanol/water (1:1), usually containing 1% acetic acid or 0.1% formic acid) and loaded in a syringe that is connected to the source. The organic acid present in the solvent provides the protons for the ionisation of the analyte and assists in droplet formation/evaporation. At low pH, basic groups become ionised. For

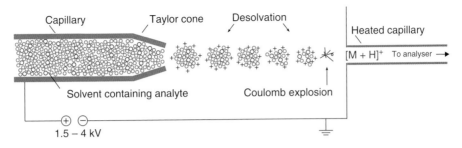

Figure 15.4 Electrospray ionisation source. The Taylor cone generates ESI creates very small droplets of solvent containing analyte by **nebulisation** as the sample is introduced into the source through a fine capillary. The solvent evaporates in the high-vacuum region as the spray of droplets enters the source. As a result of the strong electric field acting on the surface of the sample droplets and electrostatic repulsion, droplet size decreases and eventually single species of charged analyte (free of solvent) remain. These may have multiple charges depending on the availability of ionisable groups.

example, in peptides the basic side groups of Arg, Lys, His and the N-terminal amino group become ionised.

In ESI, smaller molecules, such as small peptides, usually produce a mixture of singly, doubly and triply charged ions. However, in the case of larger biomolecules, multiply charged ions are frequently formed, resulting in m/z ratios that are sufficiently small to be observed in the quadrupole analyser (see Section 15.3.1). Thus masses of large intact proteins, DNA and organic polymers can also be measured in electrospray MS, even if the mass is larger than the usual 2000 or 3000 m/z limit, since the charge state z is the denominator in the unit m/z. For example, proteins are normally analysed in positive ion mode, where charges are introduced by addition of protons. The number of basic groups in a protein (mainly the side chains of Arg, Lys, His and the N-terminus) determines the maximum number of charges carried by the molecule. The distribution of basic residues in many proteins is such that multiple peaks – one for each $[M + n\,H]^{n+}$ ion – are centred on $m/z \approx 1000$. In Figure 15.5 depicts an ESI spectrum of a large protein with a mass of 102 658 ± 7 Da, forming a typical envelope of multiply charged ions (labelled from +106 to +74). For the species with 100 protons, i.e. with 100 positive (H^+) charges, the charge number z is 100, and hence the mass-to-charge ratio $m/z = 1027.6$. This species is therefore denoted as $[M + 100\,H]^{100+}$. The resulting mass of the analyte molecule M based on this peak alone is therefore $M = (1027.6\ Da \times 100) - (100 \times 1.0078\ Da) = 102\ 659.2$ Da. When the data are processed for all multiply charged peaks in this manner (deconvolution), the average for each set of peaks gives a good accuracy mass determination of the neutral molecule, resulting in a single peak (in this example, $M = 102\ 658 \pm 7$ Da).

Nanospray (NanoESI)

A significant development in ESI MS has been the introduction of nanoflow electrospray ionisation (nanoESI), which significantly increases the sensitivity of the analysis due to very low flow rates. MS (and in particular LC-MS) had previously

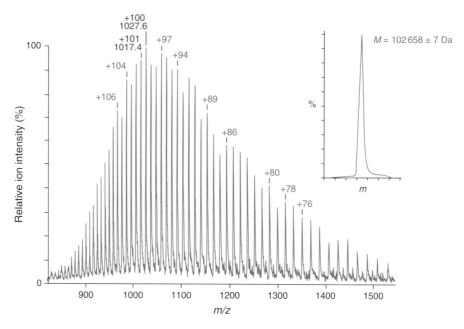

Figure 15.5 An ESI spectrum depicting how the mass of a large intact protein can be accurately measured by electrospray MS. The ion species are annotated by charge state, e.g. with +99, +100, +101 charges. Purple ion species are shown with their associated m/z values and are referred to in the main text. The inset image shows the deconvoluted spectrum.

been limited to microflow rates (i.e. microlitres per minute), but the development of nanoflow (i.e. nanolitres per minute) has led to lower sample consumption and higher signal-to-noise ratios. Depending on the ESI or nanoESI source (and the mass analysers these sources are connected to), the required sample concentration can be as low as a few attomoles per microlitre (amol μl^{-1}).

Most modern nanospray applications are performed in dynamic experiments, where capillary columns with small diameters (i.e. internal diameter less than 100 μm) are used to achieve analyte separation at low flow rates. In high-pressure and ultra-high-pressure liquid chromatography, such low flow rates can be achieved directly. Alternatively, a stream with an initially high flow rate can be split to achieve the desired reduced flow rate. For example, a stream splitter can be used to divert a large (99–99.9%) proportion of the solvent flow from the pump to waste, allowing the remainder to pass through the column to be analysed in a more sensitive manner.

15.2.5 Matrix-Assisted Laser Desorption Ionisation (MALDI)

Matrix-assisted laser desorption ionisation (MALDI) produces gas-phase protonated ions by excitation of the sample molecules using the energy of a laser transferred via a UV light-absorbing matrix. There is a variety of conjugated organic compounds that can be used to form the matrix. Choice of MALDI matrix is determined by class

Example 15.1   **PROTEIN MASS DETERMINATION BY ESI**

Question   A protein was isolated from human tissue and subjected to a variety of investigations. Relative molecular mass determinations gave values of approximately
12 000 Da by size-exclusion chromatography and 13 000 Da by gel electrophoresis. After purification, a sample was subjected to electrospray ionisation
mass spectrometry and the following data were obtained.

m/z	773.9	825.5	884.3	952.3	1031.3
Abundance (%)	59	88	100	66	37

Given that:

$$n_2 = \frac{m_1 - 1}{m_2 - m_1}$$

and $M = n_2 \times (m_2 - 1)$ and assuming that the only ions in the mixture arise by
protonation, deduce an average molecular mass for the protein by this method.

Answer M by exclusion chromatography = 12 000 Da

M by gel electrophoresis = 13 000 Da

Taking ESI peaks in pairs yields:

$m_1 - 1$	$m_2 - m_1$	n_2	$m_2 - 1$	M (Da)	z
951.3	79.0	12.041	1030.3	12 406.6	12
883.3	68.0	12.989	951.3	12 357.1	13
824.5	58.8	14.022	883.3	12 385.7	14
772.9	51.6	14.978	824.5	12 349.9	15

The average mass can now be obtained by the mean of the four observations:

$$M_{aver} = \frac{\sum M}{4} = \frac{49\,499.3 \text{ Da}}{4} = 12\,375 \text{ Da}$$

Note: Relative abundance values are not required for the determination of the
mass.

of molecules to be analysed in the sample. Examples of matrix compounds and their
application for particular biomolecules are shown in Table 15.1. These are designed
to maximally absorb light at the wavelength of the laser, typically a nitrogen laser
of 337 nm or a neodymium/yttrium-aluminium-garnet (Nd-YAG) laser at 355 nm.

Table 15.1 **Examples of MALDI matrix acids**

Matrix acid	Structure	Application
α-Cyano-4-hydroxycinnamic acid (CHCA)		Peptides < 10 kDa (glycopeptides)
Sinapinic acid (3,5-dimethoxy-4-hydoxycinnamic acid) (SA)		Proteins > 10 kDa
2,5-Dihydroxybenzoic acid (DHB) (gentisic acid)		Neutral carbohydrates, synthetic polymers (oligos)
'Super DHB', mixture of 10% 5-methoxysalycilic acid (2-hydroxy-5-methoxybenzoic acid) with DHB		Proteins, glycosylated proteins
3-Hydroxypicolinic acid		Oligonucleotides
2-(4′-Hydroxybenzeneazo)-benzoic acid (HABA)		Oligosaccharides

The sample (1–10 pmol mm^{-3}) is mixed with an excess of the matrix and dried onto a target plate where they cocrystallise upon drying. Pulses of laser light of a few nanoseconds duration cause rapid excitation and vaporisation of the crystalline matrix, resulting in ejection of matrix and analyte ions into the gas phase (Figure 15.6). This generates a plume of matrix and analyte ions that are subsequently analysed in a mass analyser, often a TOF analyser (see Section 15.3.3).

Typically a few hundred pulses of laser light are used, each with a duration of a few nanoseconds, and multiple ion plume mass analyses are accumulated to build up a satisfactory spectrum. In order to target optimal areas of the sample matrix for application of the laser, a camera positioned over the sample plate allows tracking and moving of the laser beam to find so-called sweet spots, where the composition of cocrystallised matrix and sample is optimal for good sensitivity.

In the MALDI ionisation process, the neutral matrix MH serves as an absorbing medium for the ultraviolet light, converting the incident laser energy into molecular electronic energy, for both desorption and ionisation. After ionisation, the matrix ions MH$^+$ can transfer H$^+$ ions to the neutral sample analyte A, which subsequently becomes charged ([AH]$^+$):

$$MH + A \longrightarrow MH^+ + A \longrightarrow M + [AH]^+ \qquad \text{(Eq 15.1)}$$

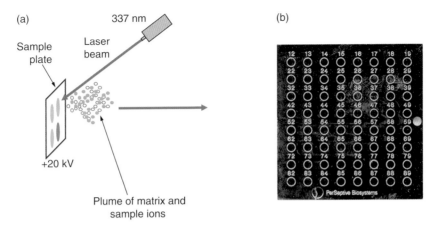

Figure 15.6 MALDI ionisation mechanism and MALDI-TOF sample plate. (a) The sample is mixed, in solution, with a 'matrix' – the organic acid in excess of the analyte (usually in a ratio between 1000:1 to 10 000:1) – and transferred to the MALDI plate. An ultraviolet (UV) laser is directed to the sample with a beam of a few micrometres diameter for desorption. The laser pulse is absorbed by the matrix molecules, causing rapid heating of the region around the area of impact and electronic excitation of the matrix. The immediate region of the sample explodes into the high vacuum of the mass spectrometer, creating gas-phase protonated molecules of both the acid and the analyte. The ionised matrix molecules then serve as proton donors for the analyte. (b) A MALDI sample plate. Numbered circles represent spots onto which the matrix/sample mixture is added and dried.

Sample Concentration for MALDI Experiments

Maximum sensitivity is achieved in MALDI–TOF if samples are diluted to a particular concentration range. If the sample concentration is unknown, a dilution series may be needed to produce a satisfactory sample/matrix spot of suitable concentration on the MALDI plate. Peptides and proteins often give optimal spectra from 0.1 to 10 pmol μl^{-1}. Some proteins, particularly glycoproteins, may yield better results at concentrations up to 10 pmol μl^{-1}, while for oligonucleotides, better spectra are obtained between 10 and 100 pmol μl^{-1}. Polymers usually require a concentration in the region of 100 pmol μl^{-1}.

One clear advantage of MALDI-TOF analyses is the ability to produce large mass ions, with high sensitivity and considerable tolerance to salt contaminants. MALDI is a soft ionisation method that does not produce abundant amounts of fragmentation. Since the molecular ions are produced with little fragmentation, it is a valuable technique for examining larger molecules, such as large glycans and proteins (see Figures 15.7, 15.8).

MALDI Sample Plates

MALDI sample plates are often made of stainless steel and include 96 circles or 'wells' for matrix/sample application. These plates are therefore ideal for multiple sample analysis where close external calibration is used (i.e. a compound(s) of known molecular mass is placed on an adjacent spot to calibrate the instrument). A further advantage of these plates is the ease of visual inspection of matrix crystallisation on the plate surface.

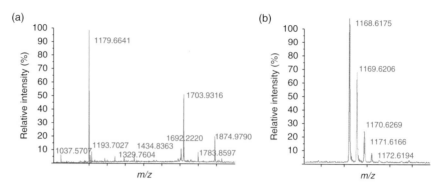

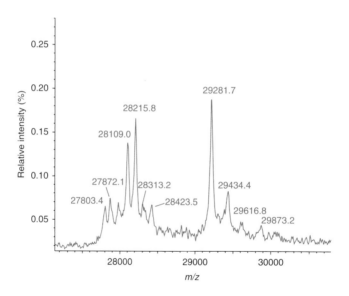

Figure 15.7 Two examples of MALDI–TOF peptide spectra. (a) Spectrum from a protein digest mixture. (b) Expansion of a small part of a spectrum showing ^{13}C-containing forms (see Section 21.4). Due to the high concentrations of the matrix molecules, MALDI-TOF spectra usually have large amounts of 'matrix ions' in the lower m/z ranges (not shown).

Figure 15.8 MALDI-TOF spectrum depicting various protein isoforms. The spectrum contains almost exclusively singly charged ions, representing the molecular ion species of the constituent proteins.

Alternatively, Teflon-coated 384-well plates allow increased sensitivity in instances where sample concentration is crucial. Due to the very small diameter of the spots, it is difficult to accurately spot samples manually and so these plates are typically used in conjunction with automated sample spotting. Only in the centre of each spot is the surface of the plate exposed. In this way, the sample does not spread over the whole surface but rather concentrates itself into the centre of each spot as it dries. Gold-coated plates with 2 mm wells (see Figure 15.6b) have similar favourable properties in that they limit the spread of sample when used with organic solvents, e.g. tetrahydrofuran

(THF) preparations for polymers. These plates also allow on-plate reactions within each well, for example with thiol-containing reagents that bind to the gold surface.

MALDI Imaging

One developing field in which MALDI is a central technique is that of mass spectrometry imaging (MSI). MSI allows 'imaging' of a sample (e.g. a biological tissue) by building a profile of the molecules detected in different regions of the tissue. In simple terms, this involves a frozen slice of tissue being added to a MALDI plate and coated with a sprayed matrix. The MALDI laser can be routinely adjusted to 20 µm so that the mass spectrometric analysis of these plates yields relatively high-resolution 'maps' of the tissues (Figure 15.9). The major advantage for using MALDI imaging over other imaging techniques (e.g., Chapter 11) is that one can analyse thousands of analytes (including metabolites, lipids, drugs, proteins and peptides) at the same time without the need to introduce a label that might either affect the biophysical properties of the analytes or itself possess non-specific properties. However, MALDI imaging requires a substantial amount of time to acquire an image. It should be kept in mind, then, that sample disintegration/degradation is a key concern when planning such an experiment.

15.2.6 Other Ionisation Techniques

The ionisation techniques above are the most commonly used in laboratories. Many other ionisation methods also exist, and new ones are constantly being developed. They can be grouped into plasma and chemical ionisation, thermal ionisation, neutral desorption-extractive electrospray ionisation, reactive electrospray-assisted laser

(a)

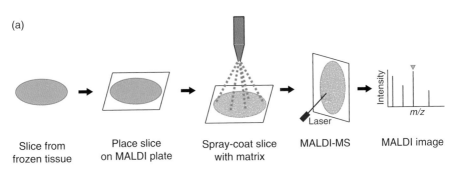

| Slice from frozen tissue | Place slice on MALDI plate | Spray-coat slice with matrix | MALDI-MS | MALDI image |

Figure 15.9 MALDI imaging. (a) The entire slice from the frozen tissue is placed on a MALDI plate and coated with matrix spray. The plate is then raster sampled and data are assembled to generate a high-resolution picture of the detected ions. (b) MALDI imaging of a cryostat section of a mouse kidney. Shown are the inner region (medulla), the inner and outer cortex, with the yellow spots being the glomeruli. P-PE is plasmenyl phosphatidylethanolamine and PA is phosphatidic acid. The analysis was carried out on a Bruker 15T solariX FT-ICR MS with a spatial resolution of 15 µm in negative ion mode; mass resolving power: 100 000 at m/z 700, total number of pixels: 126 509. Selected ion images of m/z 749.5128 and m/z 774.5444 (representing phospholipids with a C_{40} chain and five or six double bonds, respectively) annotated by mass accuracy (comparison with LipidMaps) are shown. (Image provided courtesy of Jeff Spraggins, The National Resource for Imaging Mass Spectrometry, Vanderbilt University.)

Figure 15.9 (Continued)

desorption/ionisation, charged droplet, and acoustic techniques. What they all have in common is that particular molecular interactions lead to a charge transfer at the sample surface. The interested reader can learn more about these techniques by following references that can be found at the end of this chapter.

15.3 MASS ANALYSERS

Once ions are created and leave the ion source, they pass into a mass analyser, the function of which is to separate the ions and to measure their masses. It is important to remember that the actual discrimination in the analyser is against the mass-to-charge ratio (m/z), rather than the mass, for each ion. There are many different types of mass analyser, each exploiting different combinations of physical and chemical properties of the analyte ions in order to separate them for detection. There is no 'perfect' mass analyser, each having their own advantages and limitations, and so choosing one for a project should be based on the particular application, and the required information and level of performance.

15.3.1 Quadrupole Mass Spectrometer

The central component of quadrupole analysers is a set of four parallel cylindrical rods (Figure 15.10). To each pair of rods, a combination of a direct current (DC) voltage and a superimposed radio frequency (RF) voltage are applied, creating a continuously varying electric field along the length of the analyser. At a given voltage, ions of a particular mass-to-charge ratio resonate and are accelerated down the analyser towards the detector. All ions of other mass-to-charge ratios are 'non-resonant', so impact on the rods, do not pass through the quadrupole and hence are not detected. In other words, at any given moment, ions of a particular mass-to-charge are allowed

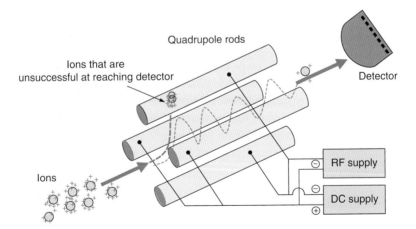

Figure 15.10 Quadrupole analyser. The fixed (DC, direct current) and oscillating (RF, radio frequency) fields cause the ions to undergo wave-like trajectories through the quadrupole filter. For a given set of fields, only certain trajectories are stable, which only allows ions of specific m/z to travel through to the mass analyser.

to pass through the quadrupole analyser and, as the voltages applied to the quadrupoles vary, ions of different mass-to-charge ratios can pass through. In this way, by ramping up and cycling the voltages applied to the quadrupole rods, the analyser 'scans' through a large range of mass-to-charge ratios. Quadrupoles are coupled to a detector that counts the ions that pass through the analyser. Knowledge of the voltages applied when an ion is detected allows us to determine the mass-to-charge ratio of that ion. These are the basic principles of scanning mass spectrometry. Quadrupoles can routinely analyse m/z values of up to 3000, which is extremely useful for biological MS since, as we have seen, proteins and other biomolecules normally give a charge distribution of m/z that is centred below this value (see Figure 15.5).

Quadrupoles are often coupled together to allow exquisite analysis of precursor ions. A common such set-up is a triple quadrupole analyser (QQQ or QqQ). This analyser consists of two scanning quadrupoles and, between these, a collision quadrupole – hence 'triple quadrupole'. The central quadrupole (which is actually a misnomer, as it is usually a hexapole or even octapole) is known as the collision quadrupole (or cell) and, through application of additional voltages and inert gas, ions can be fragmented. This process is termed collision-induced dissociation (CID) and the resultant product ions are analysed in the third quadrupole. The implications and uses of this powerful technique, termed tandem MS or MS/MS, are discussed in Section 21.3.

Note that in hybrid instruments, quadrupoles can be combined with other mass analysers, such as a TOF (Section 15.3.3) or an Orbitrap™ (Section 15.3.5), depending upon the application required.

15.3.2 Ion-Trap Mass Spectrometer

An ion trap (Figure 15.11) is essentially a device into which ions are transferred and subsequently measured. The process of trapping and analysis occurs almost simultaneously, within milliseconds of each other. After each measurement, the trap is refilled with new ions arriving from the source. The time it takes to refill the trap is called the cycle time. Ion-trap mass spectrometers have wide applications in analysis of peptides and small molecules, and can be useful when key analyte ions are of low abundance. Ion-trap MS can also provide structural information and, in the case of peptides, amino-acid sequence information. In addition to MS/MS experiments, resultant fragment ions can be fragmented further to reveal extra structural information. In theory, these ions can be fragmented again and again to reduce precursor ions down to their constituent parts – such experiments are termed MS^n, where n is the number of rounds of fragmentation (see Figure 15.12).

15.3.3 Time-of-Flight (TOF) Mass Spectrometer

A time-of-flight (TOF) mass spectrometer is a mass analyser that theoretically has an unlimited mass range. Ions are accelerated by an electric field to the same kinetic energy and are separated according to mass as they fly through a 'flight tube'. The flight tube can be a simple single tube with a detector at one end, where ions are detected at different times, depending on their mass-to-charge ratio.

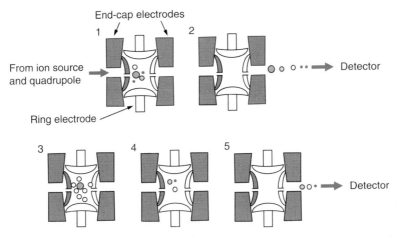

Figure 15.11 Diagram of an ion trap. The ion trap contains three hyperbolic electrodes, which form a cavity in a cylindrical device of around 5 cm diameter in which the ions are trapped (stored) and subsequently analysed. Each end-cap electrode has a small hole in the centre. Ions produced from the source enter the trap through a quadrupole and the entrance end-cap electrode. Potentials are applied to the electrodes to trap the ions (diagrams 1 and 2). The ring electrode has an alternating potential of constant RF (radio frequency), but variable amplitude. This results in a three-dimensional electrical field within the trap. The ions are trapped in stable oscillating trajectories that depend on the potentials and the m/z of the ions. To detect these ions, the potentials are varied, resulting in the ion trajectories becoming unstable, and the ions are ejected in the axial direction out of the trap in order of increasing m/z into the detector. A very low pressure of helium is maintained in the trap, which 'cools' the ions into the centre of the trap by low-speed collisions that normally do not result in fragmentation. These collisions merely slow the ions down so that during scanning, the ions leave quickly in a compact packet, producing narrower peaks with better resolution. For MS/MS, all the ions are ejected except those of a particular m/z ratio that has been selected for fragmentation (see diagrams 3, 4 and 5). The steps are: (3) selection of precursor ion, (4) collision-induced dissociation of this ion (see Section 21.3.1), and (5) ejection and detection of the fragment ions.

More commonly, the flight tube has a reflector (or reflectron) within it. The reflector is a type of ion mirror that results in higher resolution in TOF instruments. The reflector increases the overall path length for an ion and it corrects for minor variations in the energy spread of ions of the same mass. Both effects improve resolution. The device has a gradient electric field and the depth to which ions will penetrate this field, before reversal of direction of travel, depends upon their energy. Higher-energy ions will travel further and lower-energy ions a shorter distance. The flight times thus become focussed, while neutral fragments are unaffected by the deflection. Figure 15.13 shows a diagrammatic representation of a TOF reflector.

TOF mass spectrometers often exist as hybrid mass spectrometers, most commonly as Q-TOF instruments, where an isolating quadrupole and a collision cell are placed in front of the TOF tube, as shown in Figure 15.14. This set-up allows for a variety of experiments. For example, ions can be filtered through the quadrupole, passed through the collision cell and precursor mass-to-charge ratios determined with high accuracy in the TOF. Subsequently, the voltage in the collision cell can be increased

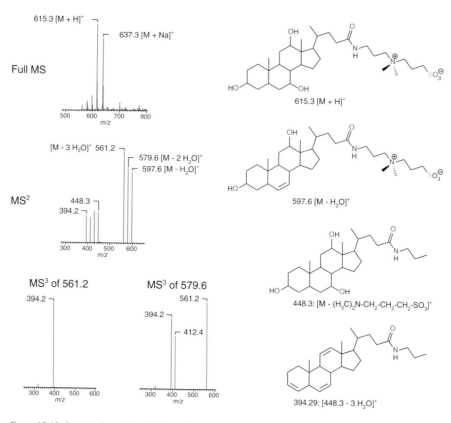

Figure 15.12 Structural analysis, MS^n in an ion trap. In this example of a steroid-related compound, the structure can be analysed when the [M + H]$^+$ ions at m/z = 615.3 are retained in the ion trap. These ions are subjected to collision-induced dissociation (CID; see Section 21.3.1), which results in the loss of the quaternary ammonium tail and partial loss of some hydroxyl groups in the tandem MS (MS^2) experiment. The major fragment ions (561.2 and 579.6) are selected for further CID (MS^3) which leads to subsequent losses of more hydroxyl groups from the steroid ring.

to induce dissociation (i.e. fragmentation) of the analytes before TOF detection. Furthermore, instead of selecting a specific ion at a time, large mass ranges can be passed through the detector in a first scan and detected as non-fragmented species. In a subsequent scan, all fragmented ions can be detected. While this technology requires sophisticated algorithms to interpret the collected data, it is a powerful tool that has been used to great success in a number of fields, such as proteomics, and drug and doping analysis.

15.3.4 Fourier Transform Ion Cyclotron Resonance Mass Spectrometer

Fourier transform ion cyclotron resonance (FT-ICR) mass spectrometry is a powerful technique in the analysis of a wide range of biomolecules. It is currently the most accurate mass spectrometric technique and can achieve very high mass resolution. A resolving power of 10^6 or higher is attainable with most instruments. The instrument

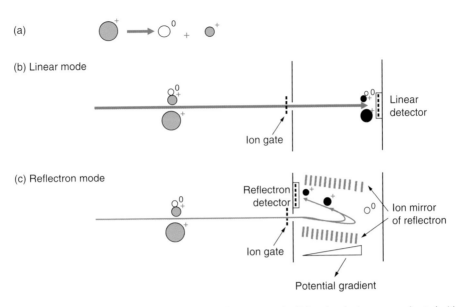

Figure 15.13 Principles of time-of-flight (TOF) MS. After entering the flight tube, the ions are accelerated with the same potential at a fixed point and a fixed initial time and will separate according to their mass-to-charge ratios. Lighter ions travel faster to the detector than the heavier ions. This time of flight can be converted to mass. A reflector extends the flight path and corrects for minor variations in the energy spread of the ions of the same mass.

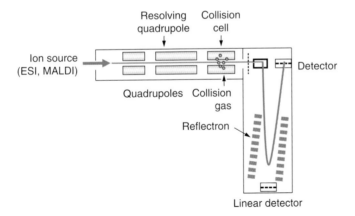

Figure 15.14 Schematic diagram of a typical hybrid quadrupole TOF-MS. A variety of sources can be coupled to a TOF analyser, such as ESI, EI and MALDI sources.

also allows tandem MS analyses to be performed. FT-ICR MS is based on excitation of ions using radio frequency signals, while orbiting in a superconducting magnetic field. As a result, the ions circulate at their cyclotron frequencies, which is inversely proportional to their m/z ratio and the strength of the applied magnetic field. The

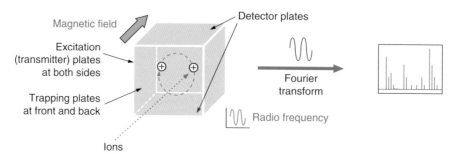

Figure 15.15 Schematic diagram of the Fourier transform ion cyclotron resonance (FT-ICR) instrument. The technique involves trapping, excitation and detection of ions to produce a mass spectrum. The trapping plates maintain the ions in orbit and are shown at the front and back in the schematic. The excitation (or transmitter) plates where the radio frequency (RF) pulse is given to the ions are shown at each side and the detector plates that detect the image current (to be Fourier transformed) are shown at the top and bottom. The ions are focussed and transferred into the analyser cell under high vacuum. The analyser cell is a type of ion trap within a spatially uniform strong magnetic field, which constrains the ions in a circular orbit. The orbit frequency is determined by the mass, charge and velocity of the ion. While the ions are in these stable orbits between the detector electrodes, they will not elicit a measurable signal. By applying a particular RF signal of duration of a few milliseconds, ions of a given m/z are excited to a wider orbit, thus giving rise to a detectable image current. Different frequencies are used to select different m/z ratios. After excitation, the ions relax back to their previous orbits and high sensitivity can be achieved by repeating this process many times. The time-dependent image current is subjected to Fourier transform in order to obtain the component angular frequencies, which correspond to the m/z of the different ions. The mass spectrum is therefore determined to a very high mass resolution, since frequency can be measured more accurately than any other physical property.

circulating ions produce a detectable image current in the cell in which they are trapped. This image current is then mathematically processed (Fourier transformed) to obtain the component frequencies of the different ions, which correspond to their m/z (Figure 15.15). Since the resolving power of FT-ICR MS is proportional to the strength of the applied magnetic field, superconducting magnets (7–14.5 tesla) are required. Although these instruments are relatively expensive, the high-resolution spectra obtained using FT-ICR allow accurate measurement of mass-to-charge ratios – crucial in correct identification of biomolecules and their fragments.

15.3.5 Orbitrap Mass Spectrometer

The Orbitrap™ mass analyser is a lower-cost alternative to the high-magnetic-field FT-ICR mass spectrometer. Ions are trapped in the Orbitrap™, where they undergo harmonic ion oscillations, along the axis of an electric field (see Figure 15.16). The frequency of the axial oscillations, which can be detected as an image current, is inversely proportional to the m/z value of the analyte. Similar to FT-ICR measurements, Fourier transforms are performed to obtain the mass spectrum. The instrument has high mass resolution (up to 500 000), high mass accuracy (<1 ppm), an m/z range of up to 6000 and a dynamic range greater than 10^4 with subfemtomol sensitivity.

(a)

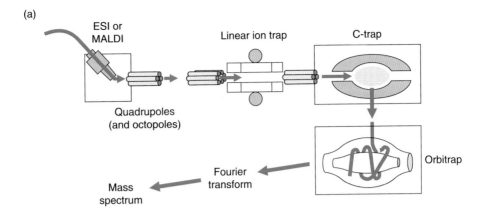

(b)

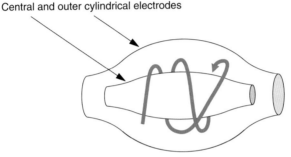

Figure 15.16 Simplified schematic of a classic hybrid Orbitrap™ mass spectrometer with a linear ion trap. Ions, generated in the ion source (ESI or MALDI), are initially transferred to the linear ion trap using radio frequency guiding by quadrupoles (and octopoles). The ions are either stored or passed through the linear ion trap to the C-trap where they are squeezed into a small cloud. These bundled ion packets are then injected into the Orbitrap™ analyser.

Several hybrid and even tribrid versions of the Orbitrap™ mass spectrometer exist, where they are combined with quadrupole or ion-trap mass analysers, or both, enabling a number of unique, high-resolution mass spectrometric experiments to be performed.

15.3.6 Other Analysers

The mass spectrometers above cover most of the instruments commonly found in laboratories, although others also exist. While not discussed in detail here, it is worth mentioning these since they are used for certain applications. Sector mass spectrometry (most commonly magnetic sector MS) involves application of a magnetic field in a direction perpendicular to the direction of ion motion. This causes ions to move in an arc and, in combination with an electric sector and often additional magnetic sectors, allows separation of ions based on m/z ratio. Accelerator mass spectrometry (AMS) makes use of ions accelerated to incredibly high energies before analysis. In this process, electrons are stripped away from negatively charged ions, which thus become positively charged. These high-energy ions move through the rest of the accelerator at high speeds and are then detected by single-ion counting. This technique efficiently suppresses molecular (and some atomic) isobars, allowing separation of low-level isotopes from a neighbouring abundant mass (e.g. separation of ^{14}C from ^{13}C and ^{12}C). Therefore, this technique has particular use in radiocarbon dating.

15.4 DETECTORS

The ions separated by the mass analyser arrive on the surface of a detector, generating an electric current. This current is amplified and processed by a computer. The resulting total ion current (TIC) is the sum of the current carried by all the ions being detected at any given moment. This is a very useful parameter to measure during on-line, chromatography-coupled MS. Counts of individual ions that make up the TIC is an important measurement in determination of the identities (and relative quantities) of biomolecules. In order for these ions to be effectively measured, the signal generated by an ion is strengthened by passage through an electron multiplier, amplifying the signal.

15.4.1 Electron Multiplier and Conversion Dynode

Electron multipliers are the most commonly used detectors for mass spectrometers. They are frequently combined with a conversion dynode – a device that converts the ion beam from the mass analyser into electrons in direct proportion to the number of bombarding ions. A positively or negatively charged ion hits the conversion dynode held at high potential, and causes the emission of secondary particles containing secondary ions, electrons and neutral particles (see Figure 15.17). These secondary

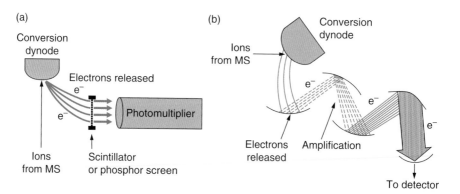

Figure 15.17 Conversion dynode and electron multiplier. (a) Ions hit the conversion dynode and emit secondary particles, including electrons. (b) Each ion strikes the conversion dynode, converting them to electrons. These electrons then travel to the next, higher-voltage dynode. The secondary electrons from the conversion dynode are accelerated and focussed onto another dynode, which itself emits secondary electrons. In short, each electron produces several secondary electrons and signal amplification continues through the 'cascading effect' of secondary electrons from dynode to dynode, finally resulting in a measurable current at the end of the electron multiplier. The cascade of electrons continues until a sufficiently large current for detection is obtained. A series of up to 10–20 dynodes (set at different potentials) provides a signal amplification gain of 10^6 or 10^7.

electrons are accelerated into the dynodes of the electron multiplier. They strike the dynodes with sufficient energy to dislodge electrons, which pass further into the electron multiplier, colliding with the dynodes held at higher potential and thus producing more and more electrons. This cascade of electrons amplifies the signal and continues until a sufficiently large current for detection is obtained.

15.5 OTHER COMPONENTS

15.5.1 Ion-Mobility-Mass Spectrometry

Ion-mobility-mass spectrometry (IMS) is becoming an important method for the separation of molecules based on their 'shape'. In ion mobility experiments, ionised molecules enter a drift tube. The drift tube normally has a uniform electric field held at atmospheric or reduced pressure and this field propels entering ions, where they interact with neutral drift molecules contained within the tube. Ions of the same mass-to-charge are separated by shape, or collisional cross-section, with physically smaller molecules travelling through the system faster than larger molecules, due to reduced interaction with the drift molecules. While having a number of applications, common uses are found in structural studies (for example in observation of protein/peptide confirmations, or in the separation of isobaric lipid compounds) or in security screening at airports and other facilities to test for the presence of oxidising agents and/or explosives.

15.6 SUGGESTIONS FOR FURTHER READING

15.6.1 Experimental Protocols

Hess S. (2013) Sample preparation guide for mass spectrometry-based proteomics. *Current Trends in Mass Spectrometry* **7**, 12–17.

15.6.2 General Texts

Siuzdak G. (2006) *The Expanding Role of Mass Spectrometry for Biotechnology*, 2nd Edn., MCC Press, San Diego, CA, USA.

15.6.3 Review Articles

Breitling R., Pitt A.R. and Barrett M.P. (2006) Precision mapping of the metabolome. *Trends in Biotechnology* **24**, 543–548.

Chen H., Gamez G. and Zenobi R. (2009) What can we learn from ambient ionization techniques? *Journal of the American Society for Mass Spectrometry* **20**, 1947–1963.

Eliuk S. and Makarov A. (2015) Evolution of Orbitrap mass spectrometry instrumentation. *Annual Review of Analytical Chemistry (Palo Alto Calif)* **8**, 61–80.

Glish G.L. and Burinsky D.J. (2008) Hybrid mass spectrometers for tandem mass spectrometry. *Journal of the American Society for Mass Spectrometry* **19**, 161–172.

Hah S.S. Henderson P.T., and Turteltaub K.W. (2009) Recent advances in biomedical applications of accelerator mass spectrometry. *Journal of Biomedical Science* **16**, 54.

Marshall A.G. and Hendrickson C.L. (2008) High-resolution mass spectrometers. *Annual Review of Analytical Chemistry (Palo Alto Calif)* **1**, 579–599.

Miller P.E. and Denton M.B. (1986) The quadrupole mass filter: Basic operating concepts. *Journal of Chemical Education* **63**, 617–622.

Norris J.L. and Caprioli R.M. (2013) Analysis of tissue specimens by matrix-assisted laser desorption/ionization imaging mass spectrometry in biological and clinical research. *Chemical Reviews* **113**, 2309–2342.

15.6.4 Websites

MassBank (Japan)
www.massbank.jp/?lang=en (accessed May 2017)

Mass Spectrometry Data Center (NIST)
chemdata.nist.gov/ (accessed May 2017)

m/zCloud (HighChem LLC, Slovakia)
www.mzcloud.org/ (accessed May 2017)

16 Fundamentals of Bioinformatics

CINZIA CANTACESSI AND ANNA V. PROTASIO

16.1 INTRODUCTION

Over the last decade, we have witnessed a massive expansion in demand for and access to low-cost high-throughput sequencing of nucleic acids, which can be predominantly attributed to the advent and establishment of so-called next–generation sequencing technologies (NGS; see Section 20.2.2). The availability and progressively decreasing costs of such technologies has been accompanied by an ever-increasing number of nucleic acid and protein sequences being deposited in public repositories and, in turn, by the need to draw biologically meaningful information or interpretations from these data. As a consequence, the discipline of bioinformatics has become instrumental in many areas of biology and, in particular, molecular biology.

Bioinformatics can be defined as a 'fusion' of biology and informatics, which includes applied mathematics, computer sciences, information technology and statistics. This multi-disciplinary field of research includes two major components, one aimed at developing computational tools and algorithms to facilitate storage, analysis and manipulation of sequence data, and one aimed at applying such tools to the discovery of new biological insights on the organism(s) under consideration. Researchers involved in the field of bioinformatics comprise both algorithm- or software-developers and end–users. While the main interest of the first group lies in writing sequence analysis programs and tools (programming, often called 'coding'), the second wishes to apply these tools to answer questions of biological relevance. Within this latter group, experienced end-users often download and maintain programs on their private personal computers or servers, analyse a number of sequences (thousands to millions) simultaneously, have a working knowledge of programming languages and are therefore skilled in the use of command-line-based software. On the other hand, occasional users mainly deal with a limited number of sequences and thus prefer the use of 'user-friendly', web-server-based tools, which often offer a reduced set of options and a limited capacity when compared with the corresponding downloadable software packages. This chapter intends to provide an overview of the basic methods and bioinformatics resources available for the analysis of nucleic acids and protein sequences, and is primarily addressed to occasional users. For details on bioinformatic analyses of large-scale sequence datasets, such as those generated by NGS technologies, the reader is referred to Chapter 20, while programmatic or script access to programs will require more advanced programming skills, for example using the Python language (see Chapter 18).

The field of bioinformatics relies heavily on information databases; hence the most relevant for the field of life sciences are presented in this chapter. The public sequence databases discussed below have been selected based on: (i) their long-standing reputation as key resources for bioinformatics analyses of sequence data over the last decades and (ii) the fact that they are consistently maintained and regularly updated. To the best of our knowledge, the tools and databases presented in this chapter are the most up-to-date versions at the time of writing. It should, however, be noted that due to the ever-evolving nature of the bioinformatics discipline and its heavy reliance on the World Wide Web, many resources are subject to future changes.

This chapter, as others in this book, contains worked examples aimed at helping the reader to put theory into practice. In order to fully benefit from these examples, access to a computer with internet connection and software installation rights is recommended.

16.2 BIOLOGICAL DATABASES

Databases represent well-structured and accessible portals to the vast amount of biological information produced by researchers worldwide. While early public databases were constructed mainly for storage of DNA and protein sequences, modern databases are highly integrated and interactive, and cover various aspects of the biological sciences. The most prominent public databases for the biological sciences are listed in Table 16.1. Inevitably, this is a current snapshot, as new databases are continuously created to accommodate the needs of users and requirements of funding agencies to make data publicly available and accessible to both researchers and the general public. Users should always read the 'Help' and/or 'About' sections of databases, which contain information on stored data and the purpose of the database itself. This helps to evaluate whether a given database is the right resource for the user query to be undertaken. It is also important to check the latest release or update of the database; there are many new online databases launched every year, but only a few are regularly reviewed.

Table 16.1 **A selection of the most significant and widely used databases in the fieldof biological sciences**

Name	Websiteª	Description
GenBank(NCBI)	www.ncbi.nlm.nih.gov/	International nucleotide sequence databases and repositories
ENA(EMBL-EBI)	genbank/www.ebi.ac.uk/ena	
DDBJ	www.ddbj.nig.ac.jp/	
UniProt	www.uniprot.org/	Protein database, sequence and functional annotation
Ensembl	www.ensembl.org/	Vertebrate and eukaryotic genomes
Ensembl genomes	www.ensemblgenomes.org/	Genome-scale data for bacteria, protists, fungi, plants and invertebrate metazoa
InterPro	www.ebi.ac.uk/interpro/	Functional analysis of protein sequences
Pfam	pfam.xfam.org/	Manually curated collection of protein domain families

Note: ªAll websites accessed May 2017.

A fairly comprehensive description of publicly available databases for the biological sciences has been compiled by the National Center for Biotechnology Information (NCBI).

16.2.1 Primary and Secondary Databases

There are primary and secondary public databases. Primary databases focus on providing a repository of experimental data and are populated with entries submitted by experimental scientists. Data submission can be as little as an oligonucleotide (~20 nucleotides) or as large as the whole human genome. GenBank from the NCBI (National Center for Biotechnology Information in Maryland, USA), ENA (European Nucleotide Archive) from EMBL (European Molecular Biology Laboratory, with headquarters in Germany) and DDBJ (DNA Data Bank of Japan) from NIG (National Institute of Genetics, Japan) are included in the International Nucleotide Sequence Database Collaboration (INSDC) consortium, and they represent the three main primary nucleotide databases. Conversely, UniProt (the Universal Protein resource) from a consortium of international laboratories, is a primary protein database. These databases also work as public repositories and it is common practice for experimental researchers to submit raw and/or compiled data in such repositories pior to publication in scientific journals. Primary databases are often grouped into consortia, which allow sharing of information. In this way, one entry submitted to NCBI can be seen in DDBJ and ENA; this process is often referred to as mirroring. Primary databases offer the user search tools to query the contents; this will be discussed in Section 16.4.

Secondary databases focus on compiling certain types of information from the primary databases and offering this back to the user, often providing additional analysis or resources. For example, Pfam and InterPro, both from EMBL-EBI, provide tools for the classification of protein sequences into families according to the presence of conserved domains. Like most primary databases, secondary databases also provide the user with a range of tools for accessing the stored contents.

Nowadays, primary and secondary databases are well and widely integrated with each other, for instance via the use of sequence accession numbers. These are unique identifiers assigned by the database to each submitted sequence or entry, thus allowing users to unequivocally identify each individual entry and corresponding information. In cases where multiple submissions exist for a given entry (for instance the same amino-acid sequence submitted by multiple researchers), databases may link these sequences via their accession numbers.

16.2.2 Reference Sequence Database RefSeq

Given the large number of redundant sequences submitted to all the repository databases, a non-redundant database was necessary that could summarise and centralise all records found for a given gene. The RefSeq collection hosted by NCBI provides such a resource, covering all natural biomolecules from plasmids to higher eukaryotes that are submitted to any of the members of INSDC. Each RefSeq entry summarises sequence information gathered across several data sources and provides a unique platform where genetic and functional annotation are linked. Much of the RefSeq database content is based on manual curation (see below), which adds extraordinary value to this repository. RefSeq also hosts collections of genes of medical interest, together with their annotations and is therefore a starting point in the study of many genetic disorders.

16.2.3 Database Curation

While most public databases and repositories are automatically updated, the accuracy of information linked to each individual entry is the responsibility of the submitter. Additionally, some analyses on large amounts of data are solely carried out by dedicated software (with little or no human intervention) and then presented to the user as database entries. It is thus important to carefully evaluate the information stored in databases and always treat automated annotations cautiously and critically. In some cases, human intervention is important, whereby an expert analyses and presents data to the public – in a similar way as a museum curator does with items in a collection; this is called manual curation. By way of example, the differences between automatically and manually curated databases are illustrated by the two genome data projects Ensembl and HAVANA. Ensembl is the major public repository for whole-genome sequences and was created primarily to allow fast access to the human genome. Large stretches of DNA sequence were submitted to Ensembl by the Human Genome Project Consortium and genes were automatically annotated using computer software. These are now accessible through the Ensembl website. However, given the importance of the human and mouse genomes, there is great interest among scientists, medical doctors and the community in having a reliable and well-curated database of the collection of human genes. HAVANA is an initiative by the Wellcome Trust Sanger Institute that aims to provide manual annotation to relevant genes in the human and other vertebrate genomes. These manually curated data are made available via the VEGA website. Finally, and illustrative of the highly integrated nature of databases, Ensembl shows findings produced by VEGA (also called HAVANA genes) and vice versa.

16.2.4 Literature Databases

Besides nucleotide and protein sequence data, databases can also store scientific literature; such databases represent the modern version of books and journal libraries, and have revolutionised the way research articles and bibliographies in general are accessed. PubMed is the literature database from the NCBI, and it is probably the most widely used database of this kind in the life sciences. As with nucleotide and protein sequences, each publication is given an accession number, called the PubMed ID (or PMID). Most bibliography managing software can handle these accession numbers to retrieve data from PubMed and automatically populate relevant bibliographic information, saving the user valuable time and minimising the chance of typing errors.

16.3 BIOLOGICAL DATA FORMATS

In bioinformatics, nucleotide and protein sequences are written as a string of letters, each representing one nucleotide (A = adenine, T = thymine, G = guanine, C = cytosine; U = uracil; see Table 0.11) in the case of DNA/RNA, or one amino acid (M = methionine, P = proline, etc; see Table 0.9) in the case of proteins. It is often necessary to attach additional information to a given sequence, such as its given name, the organism from which it originates, sequence length, etc. These strings of characters can be written in a variety of formats. While some formats can hold only one sequence (single-sequence formats), others can hold two or more sequences in the same file (multiple-sequence formats), either one sequence after the other or as an alignment (see Section 16.4).

The importance of keeping sequence information in a particular format or layout arises from the use of algorithms. While humans can ign^ore C H A N G E S *in format* and still understand the meaning of a message, computer algorithms expect data in a given format, and incorrect formatting can result in the program not performing its functions as expected. In the following sections, we will discuss some of the key formats in which nucleotide and protein sequences can be presented and the additional data that can be included in these formats. Such data files are generally ASCII files, i.e. they contain human-readable text and can be opened and edited using any text editor. Most of the file names are written with a particular extension, which is appended with a dot, for example `myfile.fasta` or `myfile.fa` to designate a particular format (here, the FASTA format). This denomination is purely organisational and many software will analyse the data format irrespective of the extension (i.e. based on the actual formatting of the text within the file).

16.3.1 FASTA

The FASTA format (typical file extensions: '.fasta' or '.fa') is one of the most widely used formats in bioinformatics. It is usually a single-sequence format beginning with a 'greater than' ('>') symbol, followed by a single line descriptor of variable length, the header. This usually contains all relevant information linked to the sequence, such as organism of origin, sequence accession number (a unique identifier linked to the sequence when deposited in a public database, see Section 16.2), sequence description, features and comments, which can be separated by spaces or symbols, such as dashes ('-'), underscores ('_') and pipe symbols ('|'). The sequence appears on the following line, directly following the header (Figure 16.1). Blank lines in a FASTA file are ignored by (most) bioinformatics applications.

The FASTA format can also be used as a multiple-sequence format (multi-FASTA), simply by listing one sequence after the other in a single file. This format is widely accepted by many applications, for instance those performing both single- and multiple-sequence alignments, such as BLAST (see Section 16.4.1).

```
>gi|170285762|emb|AM922248.1| Eimeria tenella ITS2, isolate T8, clone Et3
TTAACAACTCCTACTAGTAGGCCATGCTGCTGTCAGTCTCTGTTCCTTGTGGTCCTGTGAGGGTTCGGCGATGCTGCCGACAGAAGTGAGTGG
TTTGCTCGTTTCTGTTTTGTGTCGCGGAATTTTTTCGGTTCACCAAAGGGGAGGTAGAAGCATGTTTGGTTTCATTTGAGTGTCGTTGCATTG
GTATTGAAGGAGATGCGGCGTCTCTCGAAATTGTTGTCGGCAGCGGTGCTGTGTGTCTGCACAGTGTGCCGTTTTACATGCCTGTGCTTTCTA
TAGTGCCGTCGTATGCTCCTTTCATTCGGAAAGAGAGAGATACGGTGGTTGTATTTTATGCAACATTGTTTGTCTCATTCTGGACGAATGTTT
TGAGCAGGGCTAGGGCGAGGTATAATAGTGCATGGGTATGCGACAACGTGAAACGACATATAGTACACGGCACCATGCACGTGTTGCATGCGT
CGTTTTTTTTCGGTATTACACATGTATGTATAGACCTGAAATCAGT
```

Figure 16.1 Example of FASTA formatted sequence information.

16.3.2 GenBank File

The GenBank format is the standard sequence format used by the GenBank database, an open-access and large collection of annotated nucleotide and protein sequences submitted by external users and maintained, updated and curated by the National Center for Biotechnology Information (NCBI) of the United States (see Section 16.2). This format includes pre-defined fields that each contain specific information linked to the sequence (the annotation section), starting with the word 'LOCUS', and the sequence itself (the sequence section), beginning with 'ORIGIN' and ending with a line including the symbol '//' (Figure 16.2). For more details on information linked to each entry field in the GenBank format, the reader may consult www.ncbi.nlm.nih.gov/Sitemap/samplerecord.html (accessed May 2017).

16.3.3 EMBL File

Similar to the GenBank format, the EMBL flat file is the standard format used by the European Molecular Biology Laboratory (EMBL). This format begins with an identifier, marked with 'ID', followed by several annotation lines; the start of the sequence is marked by 'SQ' and the end by '//' (Figure 16.3).

In keeping with the FASTA format, GenBank and EMBL formats can hold multiple sequences (of the same format) in one single file. Note that the GenBank and EMBL formats allow much more information to be stored at the cost of space. The choice of one or other format typically depends on the particular needs of the user and their project.

```
LOCUS         AM922248                  511 bp DNA      linear     INV 27-NOV-2008
DEFINITION    Eimeria tenella ITS2, isolate T8, clone Et3.
ACCESSION     AM922248
VERSION       AM922248.1  GI:170285762
KEYWORDS      Parasite; Coccidia; Poultry
SOURCE        Eimeria tenella
  ORGANISM    Eimeria tenella
              Eukaryota; Alveolata; Apicomplexa; Conoidasida; Coccidia;
              Eucoccidiorida; Eimeriorina; Eimeriidae; Eimeria.
REFERENCE     1
  AUTHORS     Cantacessi,C., Riddell,S., Morris,G.M., Doran,T., Woods,W.G.,
              Otranto,D. and Gasser,R.B.
  TITLE       Genetic characterization of three unique operational taxonomic
              units of Eimeria from chickens in Australia based on nuclear spacer
              ribosomal DNA
  JOURNAL     Vet. Parasitol. 152 (3-4), 226-234 (2008)
  PUBMED      18243560
REFERENCE     2   (bases 1 to 511)
  AUTHORS     Cantacessi,C.
  TITLE       Direct Submission
  JOURNAL     Submitted (05-DEC-2007) Cantacessi C., Department of Veterinary
              Sciences, University of Melbourne, 250 Princes Highway Werribee,
              3030, AUSTRALIA
FEATURES               Location/Qualifiers
     source            1..511
                       /organism="Eimeria tenella"
                       /mol_type="genomic DNA"
                       /isolate="T8"
                       /isolation_source="chicken faeces"
                       /host="Gallus gallus"
                       /db_xref="taxon:5802"
                       /clone="Et3"
                       /country="Australia:Queensland"
                       /PCR_primers="fwd_name: WW2, fwd_seq: acgtctgtttcagtgtct,
                       rev_name: WW4r, rev_seq: aaattcagcgggtaacctcg"
     misc_feature      1..511
                       /note="internal transcribed spacer 2, ITS2"
ORIGIN
        1 ttaacaactc ctactagtag gccatgctgc tgtcagtctc tgttccttgt ggtcctgtga
       61 gggttcggcg atgctgccga cagaagtgag tggtttgctc gtttctgttt tgtgtcgcgg
      121 aatttttcg gttcaccaaa ggggaggtag aagcatgttt ggtttcattt gagtgtcgtt
      181 gcattggtat tgaaggagat gcggcgtctc tcgaaattgt tgtcggcagc ggtgctgtgt
      241 gtctgcacag tgtgccgttt tacatgcctg tgctttctat agtgccgtcg tatgctcctt
      301 tcattcggaa agagagagat acggtggttg tattttatgc aacattgttt gtctcattct
      361 ggacgaatgt tttgagcagg gctagggcga ggtataatag tgcatgggta tgcgacaacg
      421 tgaaacgaca tatagtacac ggcaccatgc acgtgttgca tgcgtcgttt tttttcggta
      481 ttacacatgt atgtatagac ctgaaatcag t
//
```

Figure 16.2 Example of GenBank file format.

16.3.4 CLUSTAL

This format (typical file extension: '.aln') includes multiple sequences in one file, either interleaved as an alignment or one after the other. This format is automatically generated by the program CLUSTAL W (or its web-based version CLUSTAL X, see Section 16.4), and it can be used as an input format for phylogenetic algorithms (see Section 16.6.5). The format starts with the word 'CLUSTAL', followed by an empty line and one or more sequence blocks. Each sequence is identified by a name, followed by a space and a string of characters. When the CLUSTAL format includes a sequence alignment, dots can be placed directly under the first sequence in each block or, alternatively, a line of stars ('*') under each sequence block, to indicate nucleotide or amino-acid conservation amongst the sequences in the file; dashes ('-') indicate deletions (Figure 16.4; see Section 16.4.3).

```
ID   AM922248; SV 1; linear; genomic DNA; STD; INV; 511 BP.
XX
AC   AM922248;
XX
DT   18-MAR-2008 (Rel. 95, Created)
DT   27-NOV-2008 (Rel. 97, Last updated, Version 2)
XX
DE   Eimeria tenella ITS2, isolate T8, clone Et3
XX
KW   Coccidia; Poultry; Parasite.
XX
OS   Eimeria tenella
OC   Eukaryota; Alveolata; Apicomplexa; Conoidasida; Coccidia; Eucoccidiorida;
OC   Eimeriorina; Eimeriidae; Eimeria.
XX
RN   [1]
RP   1-511
RA   Cantacessi C.;
RT   ;
RL   Submitted (05-DEC-2007) to the INSDC.
RL   Cantacessi C., Department of Veterinary Sciences, University of Melbourne,
RL   250 Princes Highway Werribee, 3030, AUSTRALIA.
XX
RN   [2]
RX   DOI; 10.1016/j.vetpar.2007.12.028.
RX   PUBMED; 18243560.
RA   Cantacessi C., Riddell S., Morris G.M., Doran T., Woods W.G., Otranto D.,
RA   Gasser R.B.;
RT   "Genetic characterization of three unique operational taxonomic units of
RT   Eimeria from chickens in Australia based on nuclear spacer ribosomal DNA";
RL   Vet. Parasitol. 152(3-4):226-234(2008).
XX
DR   MD5; 50e29af8c61cb173f7bcaef4c20fe7aa.
XX
FH   Key             Location/Qualifiers
FH
FT   source          1..511
FT                   /organism="Eimeria tenella"
FT                   /host="Gallus gallus"
FT                   /isolate="T8"
FT                   /mol_type="genomic DNA"
FT                   /country="Australia:Queensland"
FT                   /isolation_source="chicken faeces"
FT                   /clone="Et3"
FT                   /PCR_primers="fwd_name: WW2, fwd_seq: acgtctgtttcagtgtct,
FT                   rev_name: WW4r, rev_seq: aaattcagcgggtaacctcg"
FT                   /db_xref="taxon:5802"
FT   misc_feature    1..511
FT                   /note="internal transcribed spacer 2, ITS2"
XX
SQ   Sequence 511 BP; 102 A; 94 C; 145 G; 170 T; 0 other;
     ttaacaactc ctactagtag gccatgctgc tgtcagtctc tgttccttgt ggtcctgtga        60
     gggttcggcg atgctgccga cagaagtgag tggtttgctc gtttctgttt tgtgtcgcgg       120
     aattttttcg gttcaccaaa ggggaggtag aagcatgttt ggtttcattt gagtgtcgtt       180
     gcattggtat tgaaggagat gcggcgtctc tcgaaattgt tgtcggcagc ggtgctgtgt       240
     gtctgcacag tgtgccgttt tacatgcctg tgctttctat agtgccgtcg tatgctcctt       300
     tcattcggaa agagagagat acggtggttg tattttatgc aacattgttt gtctcattct       360
     ggacgaatgt tttgagcagg gctagggcga ggtataatag tgcatggta  tgcgacaacg       420
     tgaaacgaca tatagtacac ggcaccatgc acgtgttgca tgcgtcgttt tttttcggta       480
     ttacacatgt atgtatagac ctgaaatcag t                                      511
//
```

Figure 16.3 Example of EMBL file format.

16.3.5 NEXUS

This format (typical file extensions: '.nex' or '.nxs') is one of the most frequently used input formats for phylogenetic algorithms and, in particular, for phylogenetic reconstructions using Bayesian inference methods (see Section 16.6). It begins with the wording '#nexus', followed by blocks containing commands. Each block begins with 'BEGIN block_name;' and ends with 'end;'. A simple NEXUS file is composed of the taxa block, which includes information linked to the sequence data in the file (such as number of sequences and length in base pairs (bp) of the alignment) and the data block, which includes the alignment (Figure 16.5). Details on the commands used to build phylogenetic trees using the NEXUS format are given in the section on algorithms for phylogenetics (Section 16.6.5).

```
CLUSTAL W (1.74) multiple sequence alignment

EbST1   CTTGATCATC TCAAGAAGTA CA-------A CCGTCATAAG CATTTTT--T ACAATGCTGT
EbST2   ......TT.T .......A.. ..CTACCAC. G.A.G...T. TGCC.GGAG. .GGT...CA.
EbST3   .......T.T .......A.. ..CTACCAC. G.A.G...T. TGCC.GGAG. .GGT...CA.
EbST4   .......T.T ...-...A.. ..CTACCAC. G.A.G...T. TGCC.GGAG. .GGT...CA.

EbST1   TAATGATGTT ATAATCTATA TGAGGATGCG TCGTGTGAAG TTTTTTTTTT GGGGTTTTTA
EbST2   .....G...G G.GG.T.C.T .TG---CT.. ...C.A.GGA .A....CCGA T.C..CG.-G
EbST3   ..G..G...G G.GG.T.C.T .TG---CT.. ...C.ATGGA .A...CCCGA T.C..CG.-G
EbST4   ..G..G...G G.GG.T.C.T .TG---CT.. ...C.ATGGA .A...CCCGA T.C..CG.-G
```

Figure 16.4 Example of CLUSTAL alignment format.

```
>#nexus
begin data;
dimensions ntax=6 nchar=60;
format datatype=dna interleave=yes gap=- missing=?;
matrix
Cow      ATGGCATATCCCATACAACTAGGATTCCAAGATGCAACATCACCAATCATAGAAGAACTA
Carp     ATGGCACACCCAACGCAACTAGGTTTCAAGGACGCGGCCATACCCGTTATAGAGGAACTT
Chicken ATGGCCAACCACTCCCAACTAGGCTTTCAAGACGCCTCATCCCCCATCATAGAAGAGCTC
Human    ATGGCACATGCAGCGCAAGTAGGTCTACAAGACGCTACTTCCCCTATCATAGAAGAGCTT
Loach    ATGGCACATCCCACACAATTAGGATTCCAAGACGCGGCCTCACCCGTAATAGAAGAACTT
Mouse    ATGGCCTACCCATTCCAACTTGGTCTACAAGACGCCACATCCCCTATTATAGAAGAGCTA
;
end;
```

Figure 16.5 Example of NEXUS alignment format.

16.3.6 PHYLIP

Together with the previous format, the PHYLIP format (typical file extensions: '.phy' or '.ph') is also widely used as an input multiple-sequence format in software aimed to determine phylogenetic relationships amongst sequences (e.g. the PHYLIP software, see Section 16.6.5). A PHYLIP file begins with a line indicating the number of sequences in the file, followed by the length (in bp) of the alignment. This is followed on the next line by the alignment block (Figure 16.6).

```
6 59
Cow      ATGGCATATCCCATACAACTAGGATTCCAAGATGCAACATCACCAATCATAGAAGAACTA
Carp     ATGGCACACCCAACGCAACTAGGTTTCAAGGACGCGGCCATACCCGTTATAGAGGAACTT
Chicken ATGGCCAACCACTCCCAACTAGGCTTTCAAGACGCCTCATCCCCCATCATAGAAGAGCTC
Human    ATGGCACATGCAGCGCAAGTAGGTCTACAAGACGCTACTTCCCCTATCATAGAAGAGCTT
Loach    ATGGCACATCCCACACAATTAGGATTCCAAGACGCGGCCTCACCCGTAATAGAAGAACTT
Mouse    ATGGCCTACCCATTCCAACTTGGTCTACAAGACGCCACATCCCCTATTATAGAAGAGCTA
```

Figure 16.6 Example of PHYLIP alignment format.

16.3.7 Sequence Format Converters

Given the incessant sharing of sequence data amongst databases and the requirements of a range of bioinformatics software for specific input-file formats, users frequently face the need to convert one format into another. Free online tools have been developed for this purpose. An example of these is the EMBOSS Seqret tool, which includes, amongst several features for rapid sequence manipulation, that of automatically converting a wide range of sequence formats into one or more selected by the user via a drop-down menu. In addition, many user-friendly, open access programs for sequence analysis and manipulation (for example, Molecular Evolutionary Genetics Analysis, MEGA) offer the option of automatically saving single sequences and/or sequence alignments in formats required as inputs by other applications (Figure 16.7).

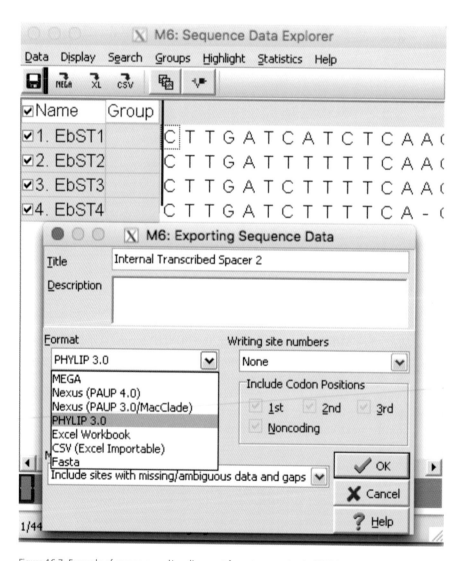

Figure 16.7 Example of sequence and/or alignment format conversion in MEGA.

16.4 SEQUENCE ALIGNMENT AND TOOLS

Every time a new sequence is discovered, scientists face the challenge of assigning an identity, i.e. identifying specific features in the sequence that can inform them of its origin, structure and/or function. This process is also known as **sequence annotation**, and will be discussed in detail in Section 16.5. The first step in sequence annotation always involves comparisons with other nucleic acids or protein sequences available in public databases (see Section 16.2) via sequence alignments, the process by which each character (nucleotide or amino-acid residue) in a given sequence is compared with those in one (pairwise alignment) or more (multiple-sequence alignment) other sequences to establish residue–residue correspondence. Sequence alignments form the basis of database searching (see Section 16.2), as well as of algorithms designed to infer evolutionary relationships between two or more sequences, a topic which will be discussed in Section 16.6.

When two or more sequences contain a certain number of common residues (which share biochemical properties such as charge and hydrophobicity), they are said to be similar. In cases where such similar sequences share an evolutionary origin via a common ancestor, they are also **homologous**. It is important to stress that the definition of homology, unlike that of similarity, is based on the determination of common evolutionary relationships between two or more aligned sequences, and is therefore not solely based on correspondence of residues.

Sequence alignments can generally be categorised into local and global alignments. While local alignments aim to find one or more regions in two or more sequences that can be aligned with one another, global alignments aim to align the sequences along their entire length. A variation of global alignments are genomic alignments that aim to align large genomic regions up to a few megabases long (see Chapter 20). Commonly, local alignments are used in database searching, i.e. when a query nucleotide or amino-acid sequence is compared against sequence data in public databases; conversely, global alignments are used when two or more sequences, of approximately the same length, are closely related. Examples of widely used bioinformatic tools for performing local alignments are FASTA, and its direct evolution, the Basic Local Alignment Search Tool (BLAST) family of programs, whereas CLUSTAL (in its varieties CLUSTAL W and CLUSTAL X) is possibly one of the most user-friendly, best-known programs for performing global alignments. Given the frequency with which these two tools are used in a range of bioinformatic applications, we will discuss them in more detail in the following paragraphs.

16.4.1 BLAST

BLAST is by far the most widely used local alignment tool for sequence similarity search. Most public sequence databases offer users the possibility of using web-based versions of BLAST to compare their own input sequence with available sequence data. BLAST uses a practical, heuristic approach for sequence comparison, i.e. it aims to find a satisfactory, but time-efficient solution to the problem, rather than an optimal but time-consuming result. This aspect is particularly crucial in bioinformatics, since users are often faced with the challenge of having to compare hundreds to millions of sequences simultaneously. In BLAST, an input sequence is called a query, whereas matching sequences are called subjects, hits or targets. First, short matching sequence fragments between query and hit are identified (seeding); when these matches are identified, a score is assigned to the alignment. This is based on a pre-defined scoring matrix and depends on the nature (nucleotide or protein) of the sequences being investigated. Then, the alignment is progressively extended in both directions in order to increase the score (in case of matching adjacent residues) or decrease it (in case of non-matching residues). When the score drops below a given threshold, the alignment process is halted and the results are reported. Scoring matrices for both DNA and protein sequences, as well as a set of default parameters (e.g. score threshold after which an alignment attempt is halted), are embedded in BLAST, and therefore they do not require setting by individual users. In most online sequence databases, BLAST results are presented as a series of alignments; those with the highest scores (also called **maximum scoring pairs**, MSPs) are shown at the top, directly below the query sequence. In addition to the score, other parameters indicative of the robustness of each individual alignment are shown. These include the e-value (expectation value) and the **percentage identity** of target and template sequence. The first indicates the number of alignments with scores equal to or above the threshold that are expected to occur by chance; reliable alignments are characterised by low e-values (for instance, in cases of two nearly identical sequences, the e-value indicated will be very low and often rounded to 0.0). The percentage identity refers to the occurrence of the same residues at the same position throughout the entire length of the two sequences, expressed as a percentage (Figure 16.8).

Different tools within the BLAST package (see Table 16.2) offer users the possibility of using a nucleotide or amino-acid sequence as a query against either DNA- or protein-sequence databases. In particular, in blastn and blastx, a nucleotide sequence is used as a query against a DNA- and protein-sequence database, respectively. Conversely, in blastp, an amino-acid sequence is used as query against a protein database. In tblastn, an amino-acid sequence is used as a query against a databases of proteins conceptually translated from nucleotide sequences in all six frames (see Section 16.5.1); in tblastx, an amino-acid sequence conceptually translated from a nucleotide sequence in all six frames is used as a query against a protein database.

Table 16.2 **The BLAST family of programs and corresponding query, subject and database**

Query	Subject	Program	Database type
Nucleotide	Nucleotide	blastn	DNA
Nucleotide	Amino acid	blastx	Protein
Amino acid	Amino acid	blastp	Protein
Amino acid	Translated nucleotide	tblastn	DNA
Translated nucleotide	Amino acid	tblastx	Protein

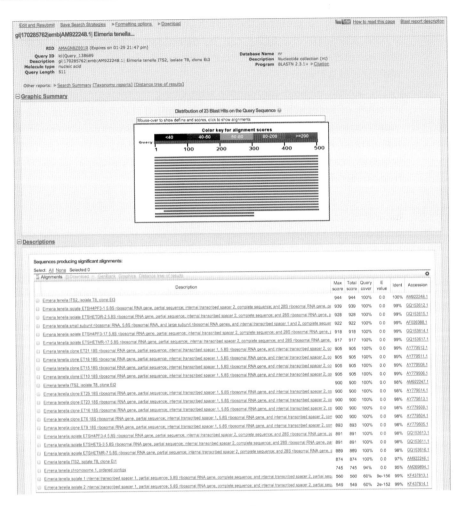

Figure 16.8 Example of BLAST output from the NCBI database.

16.4.2 BLAT and other tools

BLAT, an adaptation of BLAST, is a sequence-similarity search algorithm that was developed in the 2000s to reduce the processing time and to accommodate querying large amounts of sequence data. It provides researchers with a rapid tool to align large stretches of sequences to whole genomes, performing 500 times faster than BLAST. Its improved speed is based on the generation of an indexing catalogue of the subject sequences. The indexing step, albeit slow, is generated once for each of the sequences (often whole genomes) and is saved. The subsequent alignment steps use this index to find the best matching sequences. In the most widely used web-based BLAT server at the University of California at Santa Cruz (UCSC), the user can select the subject genome from a drop-down menu, as well as the version of the genome assembly. Because BLAT relies on an indexing step to perform the search, not all species are represented on this web server. Alternatively, BLAT can be downloaded locally as stand-alone software and instructed to perform the indexing step on a selected sequence prior to running the BLAT search.

The advent of high-throughput sequencing technologies and the subsequent vast amounts of sequencing data generated required the development of even faster and more accurate similarity searching tools. These algorithms are commonly known as mappers or mapping algorithms. Their main purpose is to assign millions of short sequence reads (often 100–200 bp) to a location in a given genome. This is discussed in more detail in Chapter 20.

16.4.3 CLUSTAL

CLUSTAL is a user-friendly tool for performing global alignments of two or more nucleotide or amino-acid sequences. The program uses a heuristic, progressive alignment method, i.e. conducts a series of pairwise alignments between each possible pair of sequences and assigns a score to each of these alignments. Starting from the best scoring alignment, more sequences are progressively added (using the neighbour–joining method, see Section 16.6.4) to produce a global alignment, including all sequences. Identical residues in two or more aligned sequences are called matches; however, residues that appear to have been inserted between matches to generate a longer sequence are called insertions; the 'missing' residues in the corresponding aligned sequence(s) are called gaps or deletions, and they are generally indicated with a dash symbol '-' (for an example of a CLUSTAL alignment, see Figure 16.4). Generally, the alignment between two or among many closely related sequences should be characterised by the smallest number possible, or ideally the absence, of insertions/deletions.

16.4.4 Hidden Markov Models (HMMs)

When dealing with two or more conserved sequences, these can be aligned to a sequence profile or model, a process that largely relies on the use of profile-hidden Markov models or profile-HMMs. Hidden Markov models (HMMs) are statistical, probabilistic methods aimed to generate optimal sequence alignments, where matches are maximised, while gaps are minimised. They are widely used in a range of different disciplines, from speech recognition to alignments of multiple protein sequences. In biology, HMMs provide flexible tools for sequence alignments, where mismatches are permitted without penalisation, depending on the differences/similarities between the seed sequences from which the model originates.

Figure 16.9 shows a schematic representation of the steps involved in the generation of a profile-HMM (pHMM). First, a sequence alignment is presented (a) and an ungapped model (b) is computed based on the frequency at which each nucleotide occurs at every position in the sequence. Match states M1–M5 each represent a frequency or probability. For example, in match state M2 of the ungapped HMM, the probability of finding a 'G' is 1 in 5 or 0.2, while in M4 the probability of finding the

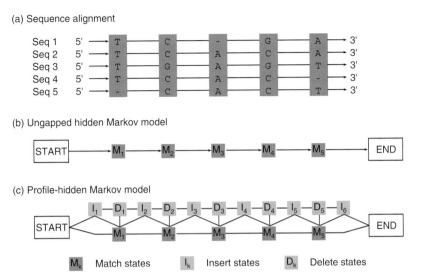

(a) Sequence alignment

Seq 1	5' →	T	C	-	G	A	→ 3'
Seq 2	5' →	T	C	A	C	A	→ 3'
Seq 3	5' →	T	G	A	G	T	→ 3'
Seq 4	5' →	T	C	A	C	-	→ 3'
Seq 5	5' →	-	C	A	C	T	→ 3'

(b) Ungapped hidden Markov model

START → M₁ → M₂ → M₃ → M₄ → M₅ → END

(c) Profile-hidden Markov model

I₁ – D₁ – I₂ – D₂ – I₃ – D₃ – I₄ – D₄ – I₅ – D₅ – I₆

START   M₁ M₂ M₃ M₄ M₅   END

Mₖ Match states   Iₖ Insert states   Dₖ Delete states

Figure 16.9 Schematic representation of HMM generation (modified from Yoon, 2009). For a detailed explanation see main text.

same nucleotide is 2 in 5 or 0.4. More complex situations arise for the calculation of these probabilities in cases where insertions and deletions occur at a particular position. In a pHMM (Figure 16.9c), an enhanced probabilistic description of the 'insert' and 'delete' state probabilities is provided.

pHMMs are currently one of the most widely used tools for classification of proteins. Starting from a selected number of representative protein sequences, their non-conserved regions are trimmed or ignored, while their highly conserved domains are used as seeds for the pHMM. When querying protein classification databases, this approach is the most frequently used in order to search for conserved amino-acid sequence regions. Examples of such classification databases include InterProScan (see Section 16.5.4) and Pfam. The latter provides manually curated seed alignments for the generation of pHMMs.

A commonly used tool for representing the probability of a given residue occurring in a specific position in a seed alignment used to generate a pHMM are sequence logos. In this graphic representation, the height of nucleotide (or amino acid) is directly proportional to the frequency of its conservation in a given position, whereas stacks of letters represent the relative frequency of the corresponding nucleotides or amino acids in that position. In the example shown in Figure 16.10, 'T' and 'A' at positions 1 and 3 are highly conserved, while 'C' in position 2 has a higher frequency than 'C' in position 4. One limitation of this type of representation is that insertions and deletions are not clearly shown; for this purpose, alternative logo generators have been developed, in which the width of each character represents the frequency of insertions/deletions (HMM logos). A convenient tool to generate sequence logos is the Weblogo service.

Figure 16.10 Graphic representation of a profile-HMM generated using Weblogo.

16.4.5 Protein Sequence Alignments

Since protein sequences consist of a more complex alphabet of characters compared to nucleotide sequences, and the relationships between residues are determined by a range of biochemical and biophysical properties, more sophisticated algorithms and tools are often used for amino-acid sequence alignments. These include the HMMs discussed above, which are particularly useful when comparing pairs of relatively dissimilar sequences that may be difficult to align using a traditional method such as BLAST.

Indeed, while two or more proteins may differ more or less substantially in sequence, they may share similar secondary structures, i.e. helices, strands or coils (see Section 2.2.2). The presence of specific strings of amino acids determines the occurrence of such structural features; therefore, algorithms designed to predict protein secondary structures often form the basis of structure-based protein alignment tools. Such programs build on the prediction of secondary-structure elements in amino-acid strings that are dissimilar in sequence to guide and/or improve protein sequence alignments (e.g. PSIPRED, SBAL). In structure-guided amino-acid sequence alignments, the ordered secondary-structure elements (helices, strands) are mapped onto the amino-acid string. A popular mapping method indicates helices and strands in different background colours, thus providing an accessible visualisation of the structure mapping (see Figure 16.11). A more detailed account of available software for the prediction of protein sequence features of biological relevance is given in Section 16.5.5.

```
>Ctg1    MNNYMS--LSKYSQLAWQSSNRIGCVVVFCWNSWTVVVCEYNPGGDLPGEVIYDEGDPCTEDSHCQCPGCVCSRDEALCI    218
>Ctg2    MQVFNRG-VGHYTQMVWQSSKKLGCGVKVCNN-FVLAGCQYQRRGNLLGANIYDKGNTCSM---CNCPQCRCDPRVGLCD    217
>Ctg3    MQVFNRG-VGHYTQMVWQSSKKLGCGVKVCNN-FVLAGCQYQRRGNLLGANIYDKGNTCSM---CNCPQCRCDPRVGLCD    216
>Ctg4    MQVFNRG-VGHYTQMVWQSSRKIGCGVKVCEK-FVLAGCQYQKRGNLIGADIYEKGNICSM---CNCQQCRCNKSEGLCD    213
>Ctg5    EQVFNRG-VGHYTQMVWQSSRRIGCGVKVCAK-FVLAGCQYQKRGNLIGADIYEKGDVCSM---CNCQQCRCNKKEGLCD    202
>Ctg6    MEVFNRG-VGHYTQVVWQLSNKIGCAVEWCSD-MTLAACEYDSAGNYLGQLIYEIGDPCKKNEDCKCTQCVCSSE-----    223
>Ctg7    QEVYDRG-VGHYTQLAWQTSDRIGCAIQWCPS-MTLVGCEYNPTGNNHNQLIYDIGDPCTTDEDCQCTGCT---------    189
>Ctg8    TPYNEGKPIQMFTRMAWATSDLMGCGVASCGD-YNSVVCRYKPGGNNLYEQLYMKGTPCSA---CPGDMFCTTSM-----    211
```

Figure 16.11 Example of structure-based amino-acid sequence alignment. In this alignment produced by SBAL, helices are mapped in green and strands in red, while cysteine residues are shown with yellow highlights.

Example 16.1   **APPRAISAL OF A GENE PRODUCT**

1. Go to the NCBI database search page at www.ncbi.nlm.nih.gov and search for the term 'daf-38'.
   a. Notice the number of entries in different databases; which database has the largest number of entries?

2. Click on the 'Nucleotide' database.
   a. What nucleotide entries are listed?
   b. Which one is the most likely to give information about a gene/mRNA/protein?

3. Click on 'daf-38, mRNA'.
   a. Which format is used to present the information?
   b. What can you infer about this mRNA/protein function (hint: look for the 'product' description)?
   c. Who made the submission of this sequence?
   d. Where could you find publications related to this entry?
   e. What other databases are cross-referenced in this entry (hint: look for the tag 'db_xref')?

4. Click the 'FASTA' link at the top of the page.
   a. What type of sequence is displayed?

5. Click 'Run BLAST' on the right-hand-side menu. This will automatically place the entry in the BLAST search window (in the form of an accession number rather than the FASTA sequence). By default, this feature performs a search using the blastn algorithm. Scroll to the bottom of the page and select the 'Highly similar sequences (megablast)' from the 'Program Options' section. Next to the BLAST button, click on 'Show results in a new window'. Now click 'BLAST'.
   a. Be patient, this search can take up to 30 seconds; the page will update automatically.
   b. Once results are shown, study the graphical representation of BLAST hits.
   c. Was it possible to identify similar mRNAs/proteins in other species? What about another genus (i.e. not *Caenorhabditis*)?
   d. Based on an *e*-value cut-off of 10^{-5}, how many entries can be considered significant BLAST hits?

6. Return to the BLAST submission page. This time, select 'Somehow similar sequences (blastn)' from the 'Program Options' section. Run BLAST.
   a. How does the results page change in comparison with the previous search?
   b. Are these last settings more or less stringent than those previously used?

7. Return to the BLAST submission page and choose a BLAST program suitable to use a nucleotide query against protein database as subject. If in doubt, consult Table 16.2 in this chapter. Run BLAST leaving all options as default.
   a. Once results are shown, study the graphical representation of the BLAST hits. Are there more or less hits than in the first BLAST search?
   b. Scroll down to the 'Descriptions' section and select the entry with the lowest e-value that is not of the same genus of the query sequence. Click on the description – this will automatically take you to the sequence alignment.
   c. What are the main difference between the query and the subject?

8. Return to the 'Descriptions' section. Notice the 'Download' tab in this section. Select at least 6 but no more than 10 hits/sequences using the selection boxes on the left of each entry. Use the download option, choosing the option 'FASTA – complete sequences' from the drop-down menu. Choose a destination on your computer and save the file.

9. Open another tab in your internet browser and navigate to the ClustalO server at www.ebi.ac.uk/Tools/msa/clustalo.
   a. Click on 'Choose a file' and navigate to the folder where you saved the FASTA file from the previous step.
   b. Run ClustalO.
   c. Study the alignments and identify the regions with high similarity.
   d. Which region is more highly conserved, the N-terminal or the C-terminal region?

16.5 ANNOTATION OF PREDICTED PEPTIDES

16.5.1 Conceptual Translation

The process of converting the DNA into protein is known as translation and it is carried out by the ribosome and messenger RNA in the cell cytoplasm. The message is encoded in three-letter words called codons and the set of rules that dictates which codon encodes for which amino acid is famously known as the genetic code (see Table 0.11). Using the genetic code, it is possible to predict the amino-acid sequence that is encoded in a given stretch of DNA; this is called conceptual translation. For each strand of DNA there are three possible reading frames with the protein translation starting at the first, second or third letter, and reading three-letter words thereafter. Figure 16.12 shows the three possible reading frames for a stretch of DNA.

In the example shown in Figure 16.12, only the first and third frames will produce an open reading frame (ORF), that is an uninterrupted stretch of amino-acid sequence. The second frame contains a stop in translation ('-'), encoded by one of the three possible stop or termination codons (i.e. TAG, TAA, TGA). In addition, because

```
ATGGCATATCCCATACAACTAGGATTCCAAGAT
M  A  Y  P  I  Q  L  G  F  Q  D    1st reading frame
W  H  I  P  Y  N  -  D  S  K       2nd reading frame
   G  I  S  H  T  T  R  I  P  R    3rd reading frame
```

Figure 16.12 Representation of conceptual translation of a short nucleotide sequence. The first reading frame represents the amino-acid sequence resulting from reading the DNA sequence starting from the first letter (ATG, GCA, and so on). Second and third reading frames are produced by starting on the second and third letters respectively (TGG, CAT and so on for the second reading frame and GGC, ATA for the third reading frame).

DNA is double-stranded, any given sequence can be read either in the forward or the reverse direction. Hence, if the direction of the sequence to be conceptually translated is unknown, there are six possible ways or frames in which the message can be read. These correspond to three frames for the forward strand (5'→3' direction) and three for the reverse strand (3'→5'). A number of web-based tools are available for the conceptual translation of DNA, such as Translate by the Swiss Institute for Bioinformatics (SIB) through its portal ExPASy or the transeq tool in the EMBOSS suite.

16.5.2 Homology-Based Annotation

Often, the inference of function based on sequence similarity is driven by the presence of conserved protein domains (three-dimensional entities of a protein; see Section 2.2.2). For example, in the case of enzymes, their catalytic site commonly locates to a specific three-dimensional (3D) motif of the protein structure. If this 3D motif or structure is perturbed, for example by changing key amino acids in its sequence, these changes can have a dramatic impact on the function of the enzyme. Because the sequence is essential for the maintenance of the 3D structure, the same (or highly similar) sequence is present in other proteins that carry the same function. A good example of conserved protein domains is found in homologous proteins (see Sections 16.4, 17.4.5).

16.5.3 Gene Ontology

With the increasing number of DNA and protein sequences submitted to databases, the problem of defining and/or identifying the function of gene products (in most cases proteins) arises. For example, consider homologous genes (see Section 16.4 for the definition of homology), that are likely to encode for proteins with identical or similar functions. Such homologous genes are annotated with the same function in public databases. Gene ontology (GO) is the use of a controlled vocabulary of terms, also called GO terms, to describe such gene products, allowing the creation of relationships between gene products and a common language to describe them. Created by an international consortium that comprises a large number of researchers across many disciplines, the first GO terms were assigned to sequences based on experimental data. Nowadays, the assignment of GO terms to a given sequence relies heavily on phylogenetic and evolutionary relationships with highly similar genes. The resulting gene descriptions are meaningful to users and can be understood by software algorithms. Gene ontology assigns terms in three categories:

- Cellular component (CC)
- Molecular function (MF)
- Biological process (BP).

Every GO term is assigned an identification number (similar to the accession numbers described earlier in Section 16.2.1) starting with the prefix 'GO:' followed by a number and a controlled vocabulary description of the function. For example, GO:0008134 is the identification number for the term 'transcription factor binding'. In addition, each GO term has at least one, but very frequently many, child and ancestor relationships with other terms.

In the example shown in Figure 16.13, the GO term 'transcription factor binding' (GO:0008134) is a child of 'protein binding', which itself is a child of 'binding'. At the same time, 'transcription factor binding' has many child terms, such as 'activating transcription factor binding' and 'RNA polymerase I transcription factor binding'. Quite obviously, the higher up in the family tree of terms the more general the descriptions are and the lower in the tree the more specific the function description becomes.

There are many web-based tools for searching and assigning GO terms to genes and gene products. AmiGO is the main access point provided by the GO consortium. At the same time, GO terms are also represented in many of the primary and secondary databases described earlier (see Section 16.2.1). For instance, protein entries in Uniprot include GO terms as part of their annotation.

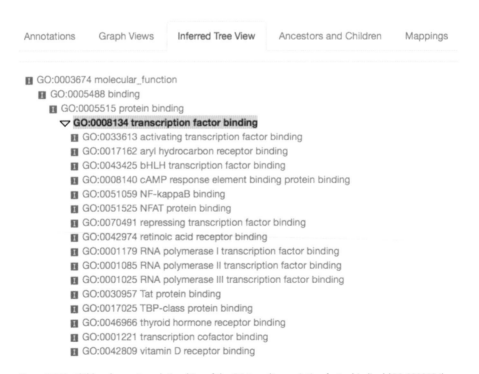

Figure 16.13 Child and ancestor relationships of the GO term 'transcription factor binding' (GO:0008134).

16.5.4 InterProScan

InterProScan is a software package that provides a one-step solution for comprehensive annotation of protein sequences based on information available in databases (Figure 16.15). It has been developed and is currently maintained by members of the InterPro consortium. A single InterProScan search is equivalent to 11 different searches in 11 distinct databases. InterProScan can be used to search for protein domains and/or signatures in whole genomes, as well as in individual sequences. However, the InterProScan web service only supports the submission of one sequence at a time. Its main strength consists in providing the user with non-redundant results; redundancy in results is a regular problem when running multiple (different) prediction algorithms on the same sequence. Once a query has been submitted, the results page reports a schematic representation of the localisation of the different conserved domains and signatures across the query sequence, as well as details of protein-family membership and associated GO terms. The results page is highly interactive, and the user can directly access information linked to different domains and signatures identified in the query sequence via the numerous links provided. Because InterPro compiles searches from various databases, results are presented without a significance value (e-value, see Section 16.4). In cases where the latter is required, it is advisable to submit the query sequence to the database of choice (Pfam, SMART, etc).

16.5.5 Prediction of Protein Features

The Center for Biological Sequence Analysis based in Denmark provides a very comprehensive toolkit of algorithms used for the prediction of a variety of features of nucleic and amino-acid sequences, including those associated with protein localisation, sorting and post-translational modifications; many of these predictions are based on pHMMs (see Section 16.4.4). Here, we will focus on a brief description of a few servers for the prediction of selected protein features.

Signal Peptides

Following protein synthesis in the cell cytoplasm, newly formed polypeptides must be moved to their final destination, either within or outside the cell. This sorting mechanism is possible due to the presence of short amino-acid motifs (15–30 amino acids) called signal peptides (SPs), which are found at the N-terminus of the protein sequence. Thus, signal peptides act as tags determining the localisation of the newly synthesised protein. Following protein transport to its final destination, signal peptides can be either retained or cleaved by peptidases. In bacteria, signal peptides direct the protein across the plasma membrane, while in eukaryotes, proteins with signal peptides travel across the endoplasmic reticulum (ER) and through the Golgi apparatus. However, exceptions exist: not all proteins with a signal peptide will be transported extra-cellularly and not all extra-cellular proteins require a signal peptide in order to reach their final localisation outside the cell. Details on the latter mechanism, known as alternative secretion, are beyond the scope of this chapter.

Similar to signal peptides, some eukaryotic proteins include amino-acid motifs that act as retention signals (= target peptides); the presence of such signals leads to the

Figure 16.14 Example of an InterProScan web service result with Uniprot accession number G4VHT1. The results page is interactive and provides convenient access to detailed information on the conserved domains identified by the algorithms embedded in InterProScan. The links are listed on the right-hand side, next to the graphical representation of the domain architecture. The InterPro accession numbers (listed on the left and starting with the prefix 'IPR') link to domain descriptions compiled by the InterPro database. At the bottom of the page, and where applicable, a list of predicted GO terms is presented.

transport of the corresponding protein to specific subcellular organelles, such as chloroplasts in plants (= cTP signals) or mitochondria (= mTP signals) in animals. As for signal peptides, retention signals can also be cleaved by dedicated peptidases.

The conserved nature of signal and target peptides can be used to predict the sorting behaviour of proteins. For example, signal peptides are characterised by the presence of three discrete regions within its ~30 amino-acid sequence: a positively charged N-terminus, a hydrophobic middle region and a C-terminus characterised by the presence of uncharged polar residues. This region is followed by a peptidase cleavage site, which is also conserved across amino-acid sequences containing signal peptides. Despite minor differences across organisms, the overall topology of signal peptides (such as hydrophobic regions or predominance of positively charged residues) is maintained. These known features can be used by a prediction algorithm as a model; the software is then able to detect the presence of model features in a given query sequence. The more model features are provided, the more accurate the software is in detecting such signatures. HMMs are frequently used for the prediction of SPs, cTPs and mTPs (see Section 16.4.4), either alone or in combination with other prediction algorithms, such as neural networks and support vector machines.

TargetP is a widely used algorithm for the prediction of such protein features, available as a web-based tool from the CBS website. It combines a set of prediction algorithms for mitochondrial, chloroplastic, secretory pathways and peptidase cleavage sites. An example output is presented in Figure 16.15.

TargetP 1.1 Server - prediction results

Technical University of Denmark

```
### targetp v1.1 prediction results ################################
Number of query sequences:  1
Cleavage site predictions included.
Using NON-PLANT networks.
```

Name	Len	mTP	SP	other	Loc	RC	TPlen
Sequence	480	0.890	0.039	0.135	M	2	6
cutoff		0.000	0.000	0.000			

Explain the output. Go back.

Figure 16.15 Example of a result page from TargetP. The numbers in columns headed by 'mTP', 'SP' and 'other' represent the calculated probability of the occurrence of a mitochondrial target peptide (mTP), a signal peptide (SP) or other motif in the sequence under consideration. 'Loc' indicates the predicted localisation (in this particular case, 'M' indicates mitochondrial localisation). 'RC' stands for reliability class and 'TPlen' indicates the predicted cleavage site. See www.cbs.dtu.dk/services/TargetP-1.1/output.php (accessed May 2017) for further details.

Transmembrane Domains

In addition to being transported to specific cellular and extra-cellular sites by means of signal peptides, some proteins can be embedded in or cross a cellular membrane. These proteins are known as transmembrane (TM) proteins and they are characterised by the presence of specific sets of amino acids forming the TM domain – the part of the protein that is anchored to or embedded in the membrane. Given the highly hydrophobic nature of lipid membranes, the amino acids found in the TM domain are also primarily hydrophobic. Therefore, by analysing the hydrophobicity of a given protein sequence, it is possible to predict which protein regions are extra-cellular, cytoplasmic or embedded in the cellular membrane.

   A widely used tool for the prediction of TM domains in a protein sequence is TMHMM. It uses HMMs (see Section 16.4.5) to calculate the probability of a given stretch of amino acids embedded in a membrane. The TMHMM output (see Figure 16.16) includes a table showing a breakdown of the regions of the polypeptide given in amino-acid residue

```
# WEBSEQUENCE Number of predicted TMHs:    7
# WEBSEQUENCE Exp number of AAs in TMHs:   151.67665
# WEBSEQUENCE Exp number, first 60 AAs:    1.90502
# WEBSEQUENCE Total prob of N-in:          0.03897
WEBSEQUENCE        TMHMM2.0        outside       1     62
WEBSEQUENCE        TMHMM2.0        TMHelix      63     85
WEBSEQUENCE        TMHMM2.0        inside       86     96
WEBSEQUENCE        TMHMM2.0        TMHelix      97    119
WEBSEQUENCE        TMHMM2.0        outside     120    133
WEBSEQUENCE        TMHMM2.0        TMHelix     134    153
WEBSEQUENCE        TMHMM2.0        inside      154    173
WEBSEQUENCE        TMHMM2.0        TMHelix     174    196
WEBSEQUENCE        TMHMM2.0        outside     197    218
WEBSEQUENCE        TMHMM2.0        TMHelix     219    241
WEBSEQUENCE        TMHMM2.0        inside      242    261
WEBSEQUENCE        TMHMM2.0        TMHelix     262    284
WEBSEQUENCE        TMHMM2.0        outside     285    303
WEBSEQUENCE        TMHMM2.0        TMHelix     304    326
WEBSEQUENCE        TMHMM2.0        inside      327    375
```

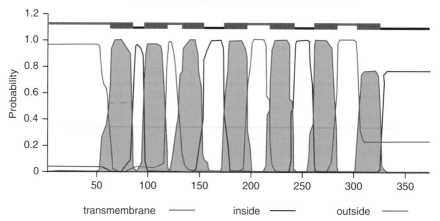

Figure 16.16 TMHMM output of GPER1_HUMAN. In this example, the polypeptide submitted to TMHMM Server 2.0 has seven TM domains (red regions in graph). The graph plots the probability of each amino acid being part of a TM domain based on the overall sequence (y-axis) versus the amino-acid residue position/number in the sequence (x-axis).

numbers (start and end) and which particular region has the highest probability of placement: non-membrane regions 'inside' or 'outside' the cell, as well as the membrane region 'TMhelix'.

Post-Translational Modification: Phosphorylation and Glycosylation

Following translation, some proteins undergo substantial remodelling before they can perform their biological functions. These processes are known as post-translational modifications, i.e. changes that occur after protein translation. In contrast, modifications occurring during translation are known as cotranslational modifications. Often, these modifications are necessary for protein functionality, but they can also be used by the cell as signalling mechanisms and thus turn on or turn off certain cellular processes. One example of a post-translational modification is the cleavage of signal or target peptides (see above). However, most commonly, post-translational modifications are conjugation reactions catalysed by enzymes, and may include addition of carbohydrate groups (glycosylation) and phosphate groups (phosphorylation), carried out by glycosylases and phosphatases, respectively. The CBS website offers a number of programs for predicting the presence/absence of signals indicative of post-translational modification events. Available training sets are often species-specific and are limited to bacterial, eukaryotic and/or human proteins. For example, NetNGlyc provides an algorithm for the identification of N-linked glycosylation sites in human proteins, while NetPhosYeast is used for the prediction of serine and threonine phosphorylation sites in yeast proteins.

16.5.6 Kyoto Encyclopedia of Genes and Genomes (KEGG)

The Kyoto Encyclopedia of Genes and Genomes (KEGG) is a highly integrated network of different databases that provides users with high-level information on metabolic and biochemical pathways in a range of organisms. This network aims to provide a functional annotation tool based on orthology. Two or more genes (in different species) are defined as orthologues if they have evolved from a common ancestral gene as a consequence of a speciation event; orthologue genes share common functions. The KEGG database therefore allows inference of functions of a given query gene by comparative analyses with known orthologues. One of the strengths of this database is the utilisation of intuitive graphical diagrams to recreate and illustrate different metabolic and biochemical pathways (Figure 16.17). In its most recent release, the KEGG can be used to find metabolic and biochemical pathways in whole genomes. This process consists of two steps: first, the algorithm assigns a molecular function to each gene based on KEGG orthology and then pathways are drawn based on the presence/absence of orthologous genes in the KEGG database.

16.5.7 Clusters of Orthologous Groups (COG)

The Cluster of Orthologous Groups of proteins (COG) facilitates the functional annotation of microbial genomes. The main strength of this database lies in the fact that it encompasses complete genomes, thus allowing a thorough examination of the complete sets of genes available for a given organism. In many

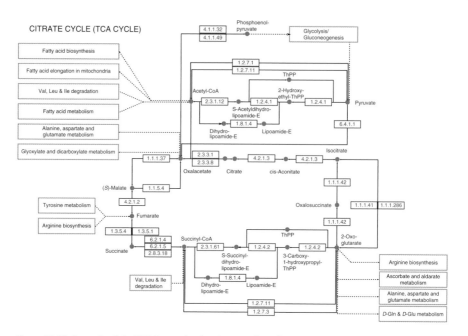

Figure 16.17 Example of the KEGG entry for the citrate cycle pathway map. Gene products are represented by red squares and molecular interactions are denoted as black arrows. Interactions with other pathways (yellow boxes) are also included in this diagram and indicated by dashed arrows. Products of metabolic reactions are represented by filled circles.

aspects, this database provides similar tools to those offered by other protein domain databases (such as InterPro or Pfam). However, data stored in the COG databases are primarily full-length bacterial proteins, whereas other databases mainly store sequence data from humans and model organisms, such as mouse or fruit fly.

16.6 PRINCIPLES OF PHYLOGENETICS

16.6.1 Introduction

Phylogenetics is a branch of bioinformatics that aims to study the evolutionary relationships between biological entities (often called taxa) using simple diagrams known as trees. While these relationships may be investigated using a variety of biological data (e.g. morphological characters), the most widely used type of data in contemporary phylogenetics are nucleotide and/or amino-acid sequences (molecular phylogenetics). Indeed, over the course of evolution, genes and their products accumulate mutations (nucleotide or amino-acid substitutions) that can inform about the history of the organisms under consideration and their relationships with those more or less distant. It is, however, worthwhile to note that phylogenetic relationships established via the analysis of individual

Example 16.2   **PREDICTION OF PROTEIN FEATURES**

1. Return to the GenBank entry page for daf-38, mRNA (see Example 16.1) or navigate to the entry under accession number NM_070486 in the nucleotide database of NCBI.

2. Click the Uniprot accession number. This will take you to the Uniprot database entry for the protein encoded in this gene.

3. Study the Gene Ontology annotation. Based on the descriptions, can you infer any additional functions for this protein? What is the most likely cellular localisation of this protein?

4. Scroll to the 'Family and Domains' section. Find the link to the Pfam entry and click on the accession number. Study the Pfam entry for this conserved domain.
   a. Which names are reported in Pfam for this conserved domain?

5. Go back to the Uniprot entry and scroll to the section 'Sequence'.
   a. How many amino acids constitute this polypeptide?
   b. Click on the FASTA link, the FASTA sequence should display.

6. Using another tab or window of your web browser, navigate to InterProScan (www.ebi.ac.uk/interpro/) and paste the FASTA sequence from the previous step into the search box and run the analysis.
   a. According to the InterPro results, which protein family does this sequence belong to?
   b. Does this correlate with the Pfam classification?

7. Using another tab or window of your web browser, navigate to TargetP (www.cbs.dtu.dk/services/TargetP/) and paste the FASTA sequence from the previous step and run the analysis.
   a. Is there a predicted cellular localisation for this protein?

8. Using another tab or window of your web browser, navigate to TMHMM (www.cbs.dtu.dk/services/TMHMM/) and paste the same FASTA sequence; run the analysis.
   a. How many transmembrane helices are found?
   b. According to the graphical representation, do all transmembrane helices have the same likelihood to be embedded in the cellular membrane (hint: probabilities are plotted on the y-axis)?
   c. How do these results compare to those suggested by Pfam and/or InterProScan?

genes and/or gene products only provide significant information on the evolutionary history of said genes or proteins, and not necessarily on the species from which the sequences are derived. In order to address species evolution by molecular phylogenetics, data integration from several trees constructed from a number of genes and/or proteins is necessary. Within the scope of this text, we will only address aspects linked to single gene or protein phylogeny. Additional available resources on the principles of species phylogeny are suggested below (Further Reading, Section 16.7).

The application of molecular phylogenetics to the study of gene/protein evolutionary relationships is based on the assumption that the molecular markers of choice, i.e. the genes or proteins used to construct the phylogenetic tree, are present in all species under consideration as homologous sequences (see Section 16.4), i.e. they share a common ancestor and have diverged over the course of evolution. While, ultimately, a particular molecular marker is chosen by the user, some key criteria are usually considered when selecting a nucleotide or protein sequence for such studies. These criteria are mainly linked to the:

- purpose of the study
- characteristics of the candidate molecular markers.

As a general principle, since gene evolution occurs more rapidly than protein evolution, the former are widely used for phylogenetic studies involving closely related organisms (e.g. mammals or plants). However, despite using the same genetic code (see Section 16.5.1, Table 0.11), codon usage (i.e. the differences in frequency of codons in a coding stretch of DNA) may vary even between relatively closely related organisms, which may lead to biases not necessarily linked to evolutionary events. For this reason, and especially in the case of more divergent species, amino-acid sequences can be chosen instead for the construction of phylogenetic trees. Key examples of genes frequently used in molecular phylogenetic studies are:

- Nuclear ribosomal genes: present in all living organisms, they are characterised by the presence of both highly conserved and highly variable regions. Examples include the 18S rRNA of the small ribosomal subunit (SSU) and the 5S, 5.8S and 25/28S rRNAs of the large subunit (LSU) in eukaryotes, and the 16S rRNA of the SSU and the 5S and 23S rRNA of the LSU in prokaryotes.
- Mitochondrial genes: characterised by frequent rearrangement events and presence of large stretches of non-coding DNA in between coding regions, they are frequently used in species-level phylogenetics. Key examples are the cytochrome oxidase I and II (COI I/II), which are characterised by slow evolutionary rates when compared to other mitochondrial genes.
- Elongation factor 1 alpha protein-coding genes: genes encoding ubiquitous proteins extensively used for inferring phylogenetic relationships amongst distant taxa.

The integration of information obtained from two or more phylogenetic reconstructions based on different gene and/or protein markers often proves useful for resolution of complex phylogenetic relationships.

16.6.2 Glossary

Before delving into the description of the methods and algorithms used in phylogenetics, it is useful to look at a number of key terms that are used to describe a typical tree (Figure 16.18). The lines in the tree are called branches, while the leaves at the end of each branch are the taxa or, alternatively, operational taxonomic units. If the lengths of the individual branches are scaled to represent the evolutionary divergence of each branch relative to the others, the tree is called a phylogram. In contrast, if this information is not included in the tree and the taxa are aligned in a column or row, the tree is referred to as a cladogram. The pattern by which different branches are connected to each other is referred to as tree topology. The point at which two branches connect is called a node, and represents the common ancestor from which the connected taxa descend. A group of taxa that includes all descendants of a common ancestor is called a clade. In particular, a clade is known as monophyletic when it includes the most recent common ancestor and its descendants; it is called paraphyletic when it includes the most recent ancestor and some, but not all, of its descendants. A clade is further defined as polyphyletic when it includes a group of taxa, but not their most recent common ancestor (Figure 16.18b).

A node on a tree that has two immediate descendants is called a dichotomy; conversely, when the descendants are more than two, the node is defined as a polytomy. In addition, a tree can be unrooted if a common ancestor is not defined, or rooted when all taxa represented in the tree can be related to a common ancestor (Figure 16.19).

Since, in phylogenetics, the common ancestor of a group of taxa is often unknown, the root of a tree can be defined using a taxon known as an outgroup, i.e. a sequence homologous to those representing the other taxa in the tree (the ingroup), but sufficiently separate from an evolutionary perspective. For instance, if the ingroup is made up of sequences from birds, a homologous sequence from a reptile could be used as an outgroup, provided that there is sufficient evidence that this has separated from the bird homologues prior to them separating from each other. In cases where a robust outgroup sequence is unavailable, the tree can be rooted at midpoint, meaning that the root is placed at the midpoint of the two most divergent groups (based on overall branch length). However, in this case, it is assumed that both branches evolved equally from the root. This assumption is based on the molecular clock hypothesis, which states that sequence evolution occurs at constant rates and the divergence time can therefore be calculated based on the relative length of the branches of the tree.

16.6.3 Substitution Models

Following the choice of an appropriate molecular marker, the next step towards building a phylogenetic tree is to generate an alignment of sequences representing the ingroup (and, if available, the outgroup) taxa (see previous section). Accurate

(a)

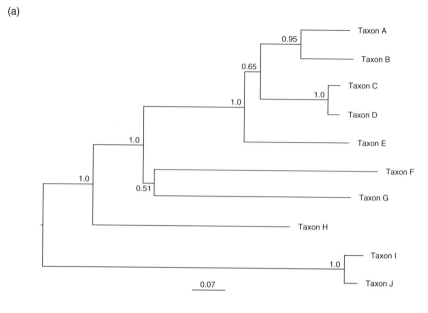

(b)

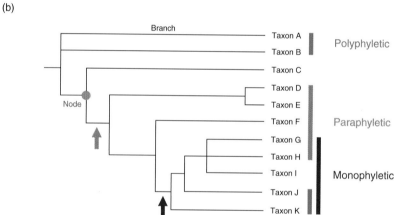

Figure 16.18 (a) A simple cladogram. In this example, the scale at the bottom represents a quantification of genetic change. The lengths of individual branches indicates this value for individual taxons; it is usually calculated based on the number of nucleotide or amino-acid substitutions divided by the length of the sequence. (b) A simple phylogram.

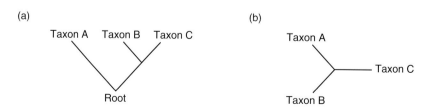

Figure 16.19 Examples of (a) rooted and (b) unrooted phylogenetic trees.

alignments are a prerequisite for the generation of accurate phylogenetic trees and, while the use of automatic software is useful (see Sections 16.4 and 16.4.5), manual adjustments are recommended in order to optimise and refine the alignment. It also needs to be decided whether whole alignments are to be used for phylogenetic reconstructions, or whether unambiguous fragments need to be extracted in order to reduce the possibility of errors or biases. Following the alignment, the next step is to choose an appropriate substitution model. In phylogenetics, this is defined as the process by which the characters in a sequence change into another set of characters. For instance, in a given sequence, an 'A' is replaced by a 'C'; however, this substitution may have occurred in one single step (A→C) or in multiple steps (A→T→G→C, or as C→G→C, the latter is known as a reverse substitution). In addition, two or more sequences may be characterised by the presence of the same nucleotide at a given position, not because of common ancestry, but rather as a result of convergent evolution, i.e. the process by which two or more sequences with separate ancestries evolve towards the same final state. This latter process is known as homoplasy and, together with reverse substitutions, it complicates the overall process of trying to calculate evolutionary distances between two or more sequences. Substitution models are statistical tools used to correct for these biases; each of these models treat substitutions at each position differently, and therefore the choice of the model to use is based on a series of assumptions. The key substitution models used in DNA phylogenetics are:

- Jukes–Cantor model: assumes that all substitutions occur at the same rate, and that all nucleotides are substituted with the same frequency
- Kimura two-parameter model: assumes that transitions (i.e. the substitution from purine to purine or pyrimidine to pyrimidine; A↔G or C↔T) occur more frequently than transversions (i.e. purine to pyrimidine or vice versa).

While equivalents of these models can be used to correct for multiple substitutions in amino-acid sequence alignments, the point accepted mutation (PAM) and Jones–Taylor–Thornton (JTT) matrices are normally used for protein phylogenies. These describe the relative probability of one amino acid being replaced by another. In brief, a PAM matrix implies a certain degree of mutation, which is measured in PAM units. For instance, a 2-PAM matrix implies the occurrence of evolutionary events which will involve 2% of amino acids in the sequences.

16.6.4 Building a Phylogenetic Tree

Methods to build phylogenetic trees can be classified into two major categories, i.e. the phenetic distance-based methods and the cladistic character-based methods. The first methods rely on measurements of distances among sequences, i.e. the sequence data are transformed into pairwise distances based on their similarity, and the resulting matrix is used to construct the phylogenetic tree. Conversely, the character-based methods rely on the direct use of the aligned characters (i.e. nucleotides or amino acids) to infer phylogenetic relationships among sequences. The key distance- and character-based methods used in phylogenetics are reviewed in the following paragraphs. For detailed explanations of the algorithms employed by each of these methods, we would like to refer the reader to Section 16.7.

Distance-Based Methods

These include the Unweighted Pair Group Method Using Arithmetic Average (UPGMA) and the neighbour-joining (NJ) methods. With the UPGMA method, the pair of sequences characterised by the smallest distance between them (based on a pre-loaded distance matrix) are first clustered together and a midpoint root is placed between them. Then, a new distance matrix is calculated which now considers the first cluster as a single taxon. A new taxon characterised by the smallest distance from the original cluster is added and the process is repeated until all ingroup taxa are placed in the tree. The last remaining taxon, characterised by the largest distance from all other taxa, is considered as the outgroup and is therefore used to root the tree. The main advantage of using the UPGMA method is certainly the speed of calculation. However, the principles behind this method imply that all ingroup taxa evolve at identical rates (unweighted distances), thus potentially producing inaccurate tree topologies.

While the NJ method is similar to UPGMA in that it also relies on the use of progressively reduced distance matrices for phylogenetic reconstructions, the evolutionary rates of individual taxa are considered unequal and thus require the use of correction values (r-value or transformed r-value) which account for this discrepancy. With NJ, a tree is initially built that includes all ingroup taxa joined onto a single node. Then, pairs of taxa are selected based on corrected distances, and the tree is progressively resolved. The NJ method is unable to autonomously define an outgroup taxon and its selection is left to the user. One disadvantage of the NJ method is the fact that, unlike with character-based methods (see below), only one tree topology is tested, thus disregarding the possibility that a given ingroup sequence may be equally distant to two or more other sequences being tested. This can lead to biases in the reconstruction of the initial tree. Therefore, to overcome this limitation, a generalised NJ method can be used, in which multiple trees are generated from a variety of initial clusters, and the best tree is finally selected based on the topology that best fits the evolutionary distances of the sequences under consideration.

Character-Based Methods

These include the widely used maximum parsimony (MP), maximum likelihood (ML) and Bayesian inference (BI) methods. The basic principle of the maximum parsimony methods is that simpler hypotheses are preferable to complicated ones. Therefore, in MP analyses, trees with the smallest number of substitutions are chosen as the best option to describe the phylogenetic relationships amongst taxa. A tree reconstruction using MP thus begins with the search of all possible trees, and the selection of the tree characterised by the minimum number of mutations required to reconstruct the phylogeny. The first step towards building a tree with the MP method is the selection of the so-called informative sites. These are positions in the sequences where at least two different characters occur, which each of them occurring at least twice. Conversely, the non-informative sites which, like the name suggests, are sites in which mutations do not occur or occur only once, have limited value in the calculations on the phylogenetic relationships amongst taxa, and are therefore discarded from further analyses. Once informative sites are selected, the number of substitutions occurring for each of these is calculated and summed up. Finally, the tree with the smallest number of changes is selected as the best tree. A variation of the MP method is the weighted parsimony that takes into account the fact that some substitutions occur more frequently than others (e.g. transitions versus transversions, see above). While the principles behind the parsimony methods are simple and logical, the use of this strategy may yield inaccurate tree topologies when, for instance, episodes of convergent evolution occur frequently in the dataset under consideration, or when dealing with very divergent sequences.

Similar to MP is the maximum likelihood method, which selects the most likely tree to best represent the phylogenetic relationships amongst the sequences. Unlike MP, however, it does not only select informative sites, but instead calculates the likelihood of a given ancestor evolving into the sequences under consideration. This likelihood is calculated based on a substitution model, e.g. the Jukes–Cantor model for nucleotide sequences (see Section 16.6.3). A tree reconstruction using the ML method begins with the calculations of the likelihood of all possible tree topologies, and of all possible ancestors. The assumption here is that all characters in the alignment have evolved independently; therefore calculations are made for each site in the sequences. Once the likelihood of each branch pattern is calculated, these are summed up, and the tree with the highest likelihood score is finally selected. The fact that the ML method takes into account the whole sequence, rather than a subset of characters, explains why this method is generally considered more accurate than MP. However, the reliability of a phylogenetic tree constructed using the ML method is strictly dependent upon the choice of a robust substitution model. In addition, it is computationally intensive, and its use is limited to small datasets or the availability of high-performance computing facilities.

In order to overcome these issues, other methods have been developed. Amongst these, one of the most popular is the Bayesian inference method. This method essentially consists of finding the set of trees that are most likely to best represent the data by adjusting previous expectations based on new observations. This methodology is also known as posterior probability, which is defined as the probability of a given event occurring, calculated in light of the occurrence of previous events. The BI method therefore aims to select the trees with the highest posterior probability. A BI phylogenetic reconstruction begins with the search of all possible trees, all equally likely to best represent the data (prior probability). Then, a conditional likelihood is computed, which is the frequency of all mutations observed in the alignment. These two parameters are then used to calculate the final posterior probability and select the set of trees predicted to best represent the data. The search for the best set of trees is performed by random sampling using an algorithm known as **Markov chain Monte Carlo** (MCMC; see also Section 17.4.5). Since the BI method looks for a group of trees, rather than the best tree (see MP and ML methods), it is computationally less intensive than the methods described above and, importantly, provides the user with an improved possibility of accurately identifying true phylogenetic relationships amongst the sequences under consideration.

Assessing the Robustness of a Phylogenetic Tree

After constructing a phylogenetic tree, the user inevitably faces the challenge of trying to establish the reliability of the relationships that have been inferred, both at the single branch/clade and at the overall tree level. Methods have been developed to address this issue, and these include bootstrapping and jack-knifing (at the branch level), as well as conventional statistical analyses at the tree level. Statistical support is represented as numbers placed at each node, from 0 to 100% for bootstrap and jack-knife, and from 0 to 1 for posterior probabilities.

The **bootstrapping** assessment method introduces slight modifications into the original dataset (for example by replacing some characters), and then repeats sampling after each modification in order to obtain an estimate of the robustness of the original inference. This process is usually repeated hundreds or thousands of times and new phylogenetic trees are constructed in each round. The number of repetitions is at the discretion of the user. The larger the number of repetitions, the more accurate the branch support; however, performing thousands of repetitions can be time-consuming and computationally intensive.

Jack-knifing is another method to assess the reliability of individual branches/ clades. Here, half of the sequences of the original dataset are randomly chosen and removed, and the phylogenetic analyses are repeated using the remaining set of sequences (which is different each time). While this method is substantially less computationally intensive than bootstrapping, it is often considered less accurate, since it only considers partial datasets. In addition, these datasets differ from one another, and the repeated analyses can therefore not be considered true replicates.

Since the Bayesian inference method randomly samples trees over thousands or millions of repeats via the MCMC method, this procedure is considered an evaluation of the robustness of each assigned node. The posterior probability values thus directly serve as an assessment criterion. However, since random sampling occurs around a subset of near-best trees, posterior probability values are normally higher than bootstrap values, even when calculated on identical sequence datasets.

16.6.5 Phylogenetic Analysis Software

A wide range of software is available for phylogenetic analyses of sequence data. These include both commercial and open-source programs. In addition, a number of sequence analysis and editing tools (e.g. MEGA) often incorporate algorithms for phylogenetics. Amongst the most widely used commercial tools for phylogenetics is PAUP (Phylogenetic Analysis Using Parsimony). This program incorporates a vast range of available models and options, all of which can be controlled by the user in each single step of the phylogenetic calculations, which allows optimisation of the calculation parameters based on the exact nature of the sequence data under consideration. However, for this same reason, PAUP requires an in-depth knowledge of phylogenetic algorithms, and may not appear friendly to some users. PHYLIP (Phylogenetic Inference Package) is a command-line-based, open-source software incorporating algorithms for inferring both distance- and character-based phylogenies; there is also a web-based version. The open-source software MrBayes is one of the most widely used software for Bayesian inference phylogenetic analyses and requires nucleotide or amino-acid sequence alignments in NEXUS file format (see Section 16.3.5, Figure 16.5). Amongst the most popular software for tree visualisation and editing are TreeView and FigTree.

The following code is an example input script for MrBayes. The input consists of a nucleotide sequence alignment in NEXUS format. Besides the taxa and the data block (see Section 16.3.5), the file contains a tree block with detailed user-provided instructions on the parameters to be used for the construction of a phylogenetic tree (see mrbayes.sourceforge.net/Help/commref_mb3.1.txt; accessed May 2017). The consensus tree in this example is based on the 50-majority rule, i.e. a tree containing all clusters that occur in at least 50% of all trees generated.

```
# Script 16.1
# nexus
begin data;
 dimensions ntax=10 nchar=80;
 format datatype=dna interleave=yes gap=- missing=?;
 matrix
Seq1    CTGTCCCACTCTAAGTCCAG CAATGAGACTGGTATCCTGG ACATGGCCCAAAGAGGGTGA
Seq2    CTGCCCCACCCTAAGTCCAA CACTGAGTACGGTTGTTTGG AAATGGCCCAAGGAGGGTGA
Seq3    CTGCTCCGCCCTAAGTCCAA CACTGAGTACGGTTGTTTGG AAATGGCCCATTGAGGGTGA
Seq4    CTGCTCCGCCCTAAGTCCAA CACTGAGTACGGTTGTTTGG AAATGGCCCATTGAGGGTGA
Seq5    CTGCTCCACCCTAAGTCCAA CACTGAGTACGGTTGTTTGG AAATGGCCCAGGGAGGGTGA
Seq6    CTGCTCCACCCTAAGTCCAA CACTGAGTACGGTTGTTTGG AAATGGCCCAGGGAGGGTGA
Seq7    CTGCTCCACCCTAAGTCCAA CAATGAGTACGGTAGTACGG ACATGGCCCACAGAGGGTGA
Seq8    CTGCTCCACCCTAAGTCCAA CACTGAGTACGGTTGTTTGG AAATGGCCCAAGGAGGGTGA
Seq9    CTACTCCACCCTAAGTCCAA CACTGAGTGCGGTTGTATGG ACATGGCCCAGGGAGGGTGA
Seq10   CTGCCCCATCCTAAGTCCAA CACTGAGTACGGTTGTTTGG ACATGGCCCAAGGAGGGTGA
 ;

begin mrbayes;
# Seq 1 is used as outgroup
outgroup 1;
log start filename=users_filename.log replace;
```

```
# general model of DNA substitution
# with gamma-distributed rate variation across sites
lset nst=6 rates=gamma;
# quiet execution
set autoclose = yes;
# Markov chain Monte Carlo (mcmcp)
# 10000000 cycles (ngen), sample every 100 cycles (samplefreq),
# 4 chains (nchains),
# print information every 1000 cycles (printfreq)
# keep information on branch length (savebrlens)
mcmcp ngen=10000000 samplefreq=100 nchains=4 printfreq=1000
savebrlens=yes filename=users_filename;
mcmc;
# produce a summary of the trees produced (sumt)
# discard 1000 samples before calculating statistics (burnin)
# use 50-majority (halfcompat) to produce consensus tree (contype)
sumt filename=users_filename burnin=1000 contype=halfcompat;
log stop;
end;
```

Example 16.3   **CALCULATION OF PHYLOGENETIC TREES**

Question   Construct a phylogenetic tree of 12 related nucleotide sequences.

1. Using your Internet browser of choice, open the NCBI nucleotide database at www.ncbi.nlm.nih.gov/nucleotide (accessed May 2017).

2. Retrieve and download the nucleotide sequences with the following accession numbers in FASTA format: AM922237.1, AM922238.1, AM922239.1, AM922240.1, AM922241.1, AM922242.1, AM922243.1, AM922244.1, AM922245.1, AM922246.1, AM922247.1, AM922248.1 (on each sequence page, click on 'FASTA' to access the FASTA format directly).

3. Review the nature of these sequences by browsing scientific literature linked to this sequence data. Make notes of the features of these sequences.

4. Create a FASTA multiple sequence file using a text editor such as NotePad (Windows), gEdit (Linux) or TextEdit (iOS) and save the file as exercise.fasta.

5. Copy and paste the multiple sequence file in Clustal Omega at www.ebi .ac.uk/Tools/msa/clustalo/ (accessed May 2017) and perform a nucleotide sequence alignment by selecting 'Clustal w/o numbers' as output format.

6. Once the alignment is completed, it will appear on the computer screen. Carefully review the result.

7. Click 'Send to ClustalW2_Phylogeny'.

8. Produce a phylogenetic tree using the different options provided.

9. Describe the output tree.

10. Download and install MEGA (www.megasoftware.net/; accessed May 2017).

11. Once installed, open MEGA and load the multiple sequence FASTA file by clicking on 'File > Open a File/Session > exercise.fasta'.

12. Select the 'Align' option.

13. Click on 'Alignment > Align by ClustalW'; leave alignment parameters as default and click 'OK'.

14. Once completed, export the alignment in MEGA format and save as 'exercise.meg'; input title of the data is at user's discretion. Click 'No' at the question 'Protein-coding nucleotide sequence data?'.

15. Click 'File > Open a File/Session > exercise.meg'; select 'Nucleotide sequences'.

16. Visualise and analyse the alignment by clicking on the 'TA' window.

17. Click on 'Phylogeny > Construct/Test Maximum Likelihood tree'.

18. Carefully review parameters and select appropriate tests based on infor-mation collected in point 3. Which phylogeny test? Which substitution model?

19. Once the parameters are set, click 'Compute'.

20. Carefully review and describe the phylogenetic tree generated.

21. Compute additional trees (e.g. character- and distance-based trees) by modifying the parameters provided. What differences can be noted and how would you explain them?

Answer   The choice of phylogenetic methods (i.e. character- or distance-based) and of the substitution model to be used during the phylogenetic reconstructions will greatly influence the topology of the trees obtained. In this case, sequences represent a highly variable, non-coding region of the ribosomal DNA, i.e. the internal transcribed spacer (ITS). These regions are known to accumulate mutations, therefore inter-specific variability is high. For this reason, a phylo-genetic reconstruction using the ML method is recommended.

16.7 SUGGESTIONS FOR FURTHER READING

16.7.1 Experimental Protocols

Hall B.G. (2011) *Phylogenetic Trees Made Easy: A How-To Manual,* 4th Edn., Sinauer Associates, Sunderland, MA, USA.

Madden T. (2013) The BLAST Sequence Analysis Tool. In: *The NCBI Handbook,* 2nd Edn., www.ncbi.nlm.nih.gov/books/NBK153387/ (accessed May 2017).

Mark T. (2009) Manual annotation of the mouse genome: HAVANA and EUCOMM. *Nature Precedings.* doi: 10.1038/npre.2009.3181.1.

McWilliam H., Li W., Uludag M., Squizzato S., Park Y.M., Buso N., Cowley A.P. and Lopez R. (2013) Analysis tool web services from the EMBL-EBI. *Nucleic Acids Research* 41, W597-W600.

Schneider T.D. and Stephens R.M. (1990) Sequence logos: a new way to display consensus sequences. *Nucleic Acids Research* 18, 6097–6100.

Schuster-Böckler B., Schultz J. and Rahmann S. (2004) HMM Logos for visualization of protein families. *BMC Bioinformatics* 5, 7.

16.7.2 General Texts

Attwood T. and Parry-Smith D. (2001) *Introduction to Bioinformatics,* 1st Edn., Benjamin Cummings, San Francisco, CA, USA.

Dear P. (2007) *Bioinformatics,* 3rd Edn., Scion Publishing Ltd, Banbury, UK.

Xiong J. (2006) *Essential Bioinformatics,* 1st Edn., Cambridge University Press, Cambridge, UK.

16.7.3 Review Articles

García-Campos M.A., Espinal-Enríquez J. and Hernández-Lemus E. (2015) Pathway Analysis: State of the Art. *Frontiers in Physiology* 6, 383.

Yoon B.J. (2009) Hidden Markov models and their applications in biological sequence analysis. *Current Genomics* 10, 402–415.

16.7.4 Websites

AmiGO (tools for gene ontology)
amigo.geneontology.org/amigo (accessed May 2017)

BLAST (Basic Local Alignment Search Tool)
blast.ncbi.nlm.nih.gov/Blast.cgi (accessed May 2017)

BLAT (sequence similarity search)
genome.ucsc.edu/cgi-bin/hgBlat (accessed May 2017)

Center for Biological Sequence Analysis
www.cbs.dtu.dk/services/ (accessed May 2017)

CLUSTAL (sequence alignments)
www.clustal.org/ (accessed May 2017)
www.ebi.ac.uk/Tools/msa/clustalo/ (accessed May 2017)

Cluster of Orthologous Groups of Proteins (COG)
www.ncbi.nlm.nih.gov/COG/ (accessed May 2017)

EMBOSS Seqret (sequence conversion)
www.ebi.ac.uk/Tools/sfc/emboss_seqret/ (accessed May 2017)

EMBOSS transeq (conceptual translation)
www.ebi.ac.uk/Tools/webservices/services/st/emboss_transeq_rest (accessed May 2017)

Ensembl (software system for production and maintaining automatic annotation of genomes)
www.ensembl.org/ (accessed May 2017)

European Nucleotide Archive
www.ebi.ac.uk/ena (accessed May 2017)

ExPASy Translate (conceptual translation)
web.expasy.org/translate/ (accessed May 2017)

FASTA alignment tool
www.ebi.ac.uk/Tools/sss/fasta/ (accessed May 2017)

FigTree (phylogenetic tree visualisation and editing)
tree.bio.ed.ac.uk/software/figtree/ (accessed May 2017)

InterProScan
www.ebi.ac.uk/interpro/search/sequence-search (accessed May 2017)

Kyoto Encyclopedia of Genes and Genomes (KEGG)
www.genome.jp/kegg/ (accessed May 2017)

MEGA: Molecular Evolutionary Genetics Analysis (sequence conversion)
www.megasoftware.net/ (accessed May 2017)

MrBayes (Bayesin inference phylogenetic analysis)
morphbank.ebc.uu.se/mrbayes/ (accessed May 2017)

MUSCLE (multiple alignment software)
www.drive5.com/muscle/ (accessed May 2017)

NCBI database overview
www.ncbi.nlm.nih.gov/guide/all/#databases (accessed May 2017)

NetNGlyc (N-linked glycosylation sites in human proteins)
www.cbs.dtu.dk/services/NetNGlyc/ (accessed May 2017)

NetPhosYeast (phosphorylation sites in yeast proteins)
www.cbs.dtu.dk/services/NetPhosYeast/ (accessed May 2017)

PAUP (Phylogenetic Analysis Using Parsimony)
paup.csit.fsu.edu/ (accessed May 2017)

PHYLIP (Phylogenetic inference package)
evolution.genetics.washington.edu/phylip.html (accessed May 2017)
evolution.genetics.washington.edu/phylip/phylipweb.html (accessed May 2017)

PSIPRED (secondary structure and fold prediction based on amino-acid sequence)
bioinf.cs.ucl.ac.uk/psipred/ (accessed May 2017)

SBAL (structure-based amino-acid sequence alignment tool)
www.structuralchemistry.org/pcsb/ (accessed May 2017)

STRING: functional protein association networks
string-db.org/ (accessed May 2017)

Structure Function Linkage Database
sfld.rbvi.ucsf.edu/django/ (accessed May 2017)

TargetP (prediction of signal and target peptides)
www.cbs.dtu.dk/services/TargetP/ (accessed May 2017)

TMHMM (prediction of transmembrane domains)
www.cbs.dtu.dk/services/TMHMM/ (accessed May 2017)

TreeView (phylogenetic tree visualisation and editing)
darwin.zoology.gla.ac.uk/~rpage/treeviewx/ (accessed May 2017)

UGENE (software package with many useful tools for work involving sequences)
ugene.net/ (accessed May 2017)

VEGA (Vertebrate Genome Annotation) browser
vega.sanger.ac.uk/ (accessed May 2017)

Weblogo
weblogo.berkeley.edu/ (accessed May 2017)

17 | Fundamentals of Chemoinformatics

PAUL TAYLOR

17.1 INTRODUCTION

Chemoinformatics (or as it is also known, cheminformatics) basically relates to the storage and manipulation of chemical data. Many of the techniques gathered together under this title pre-date its introduction, and perhaps the term was coined to explicitly signal that a similar discipline to bioinformatics exists within the realm of theoretical chemistry. The term is also heavily related to drug discovery, given that those with the largest commercial interest in systematic computational analysis of molecules are drug companies. This area is explicitly addressed elsewhere in this book (see Chapter 24) and so here we restrict ourselves to the computational representation of molecules, the derivation of useful molecular properties, application of the techniques to biological systems, and the storage and comparison of those data.

Chemoinformatics can be used on chemicals of any elemental composition, however it is most well developed for small organic molecules, i.e. those compounds that are mostly made from combinations of carbon, hydrogen, nitrogen, oxygen and sulfur, with the addition of small amounts of other elements, most notably the halogens (fluorine, chlorine, bromium, iodine) and, particularly for biological chemistry, phosphorous. While small positively charged metal ions are commonly encountered in compounds (often as counter ions to organic acids) larger metal ions are comparably rare.

Chemoinformatics calculations can address features of single molecules; however, they become invaluable when dealing with a set of molecules among which certain features need to be compared. Like most computational processes, it is of course possible to perform many of the calculations described here by hand, albeit many of these are tedious and repetitive in nature. Once automated, chemoinformatics allows the comparison of molecules to/within large datasets, and many calculations specifically involve comparison of one molecule to another. The concept of similarity is central to chemoinformatics studies, as often molecules classed as similar can be expected to exhibit similar physiological behaviours, and particularly bind to macromolecules in a similar way. This can often be done by direct comparison of the molecular structures (if they differ only by a single atom or bond); however, the ideal is to find molecules that possess similar properties, but come from distinct chemical families.

Finally, it should always be borne in mind that chemoinformatics is a branch of theoretical chemistry. A great many of the techniques used in this context rely on approximations to the exact physics occurring in the molecule. One should therefore be aware of the particular approximations for any of these techniques, in particular to appreciate when its predictions become uncertain. Ultimately, some degree of experimental verification must be available for calculations made.

17.1.1 Two-Dimensional Versus Three-Dimensional Data

Many of the original techniques used in chemoinformatics were originally developed to use the information contained in the classical chemical structural drawing of a molecule. A chemical compound is indeed a three-dimensional (3D) object, that often exists in several slightly different forms, but with a particular spatial arrangement of its chemical functionalities. As well as the obvious whole-body motions, groups connected by rotatable bonds are moving relative to one another. These rotations are moderated by steric barriers, which are caused by internal close atomic contacts, intermolecular van der Waals forces and electrostatic interactions. The recurring question for a 3D representation is thus: does this represent a likely conformation of the molecule? And even if it does, are there other more likely conformations? Consequently, representations based on two dimensions are inherently physically less relevant than 3D ones. A key point in favour of two-dimensional (2D) representations is that they involve no choice of conformer, and as such, for a given set of compounds, a suitable choice of 2D representation may lead to the compounds being more directly comparable to each other.

17.1.2 Uses in Experimental Design and Results Analysis

When looking at a pathway or a mechanism, one often faces questions such as, "How can we probe this further?" Chemoinformatics provides an economical way to identify chemical probes for the reaction in question. For example, it is possible to identify a range of compounds similar to known substrates and cofactors that can then be used to investigate the reaction. If we know that the reaction changes with a solution property (e.g. pH), then we can appraise from a theoretical perspective what impact the changes in that property has on the state of the molecules of interest. For any experiment where we have a measure of binding affinity, or generally activity, for a set of molecules, we can calculate a range of properties for each molecule and attempt to correlate them with these activities (see Figure 17.1). Clearly, the further away from the direct measurement of the relative strengths of binding between molecules (or generally activities) we go, the more unlikely we are to get a clear correlation. But even in situations where a very indirect effect of the binding event is measured, it is

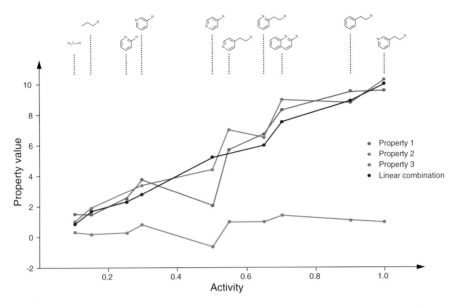

Figure 17.1   Correlation of properties and activities of chemical compounds in a set. Three properties have been calculated for each individual compound in the set and plotted against the activity of that compound (all values are in arbitrary units). A linear combination of the weighted properties shows a good correlation to the activity. The relationship derived for this particular set of compounds (training set) can be used to predict activity for chemically related compounds based on the three calclulated properties.

often possible to generate a correlation and use this as a tool to predict activity for new compounds.

17.1.3   Uses in Hypothesis Generation

It is in this area that chemoinformatics can be genuinely innovative, allowing the researcher to explore multiple different mechanisms for a reaction and formulate possible explanations for the primary observation in order to motivate and guide further research. In the past, this might have required recourse to specialist groups with high-performance computing facilities. However, the modern PC is literally the equivalent of the supercomputer of a generation ago, and although chemoinformatics software has not evolved at anything like the same rate as hardware, it is often possible to learn how to use it with sufficient application. As in any computing discipline (see also Chapter 18), the software will certainly produce incorrect output if it is presented with faulty inputs. It is thus of utmost importance to carefully check inputs and results for physical reasonableness; for example, in a similarity calculation, an important first question to ask is if the search structure was actually a physically meaningful molecule. An issue with interpretation of chemoinformatics results is often that while these approaches can produce very convincing tables and figures, one has to keep in mind that these calculations are based on approximations and assumptions. Only if one can be certain that the inputs conform to these assumptions, is it appropriate to have some confidence in the outputs and use them to plan actual experiments that can validate your hypothesis.

17.2 COMPUTER REPRESENTATIONS OF CHEMICAL STRUCTURE

The classical way to represent a chemical structure is the **chemical formula** (also called empirical formula); phenylalanine, for example, possesses the formula $C_9H_{11}N_1O_2$. However, since this formula can represent multiple different molecules, chemists normally draw the molecule. These drawings are two-dimensional representations of structure (see Section 17.2.2) and show how individual atoms are connected. Typically, when a chemist draws an organic molecule (almost all molecules involved in biological systems belong to this class), unnamed atoms are carbon atoms. Also, the hydrogen atoms are often excluded and multiple bonds are indicated by multiple lines (Figure 17.2).

There are a number of things to note about this. First, the aromatic benzene ring is shown as alternating single and double bonds (the so-called Kekulé structure), but it could alternatively be shown as a hexagon with a ring within (the so-called aromatic structure), indicating the delocalised π bond. Second, the carboxylic acid is shown with its formal bonding, and third, the bond to the nitrogen is shown as a wedge, indicating the stereochemistry at the Cα atom. The stereochemistry shown in Figure 17.2 is the naturally occurring form L-phenylalanine (if using the D/L nomenclature

(a)

(b)

Figure 17.2 Two-dimensional drawing of L-phenylalanine. (a) All hydrogen atoms omitted. (b) With polar hydrogen atoms.

Example 17.1   **GENERATION OF A TWO-DIMENSIONAL CHEMICAL REPRESENTATION**

1. Using a web browser, navigate to the JSME home page at peter-ertl.com/jsme/.

2. Draw the molecule 2-fluoro-butane.

   a. This is most simply done by using the chain tool shown as the ⌇.

   b. Hold the mouse button down and drag until you have a chain of four atoms. Note that the number in the status bar of the application shows the number of bonds (3).

   c. Then use the singe bond addition (the ——) to add a single atom to the second carbon of the chain.

   d. Now use the F element symbol to modify the newly added atom from the default carbon to fluorine.

by Fischer for chirality), or (*S*)-phenylalanine (when using the *R*/*S* nomenclature by Cahn–Ingold–Prelog). There are many computational tools to perform this type of sketch, including ChemDraw and BIOVIA Draw, and within web applications that require chemical input there will usually be a window to perform sketches, based around the JME Molecular Editor or MarvinSketch. Given the increasing issues surrounding Java security, Javascript-based solutions like Ertl's JSME are also appearing.

Formally, a diagram such as in Figure 17.2 is described as a **molecular graph**, with the atoms being the nodes of the graph, bonds being edges. There is a large body of research dealing with graphs, their similarity and the discovery of the **largest common subgraph** between graphs (in our case the largest single piece of structure in common), and also with graphs that have annotated nodes and edges – this work can be applied to the task of representing molecular structure.

While these applications are useful in allowing human input of chemical information, the actual formats used to exchange data between applications are based around defined file formats for each molecule. The structure of the file format varies with the level of detail required.

17.2.1 Formula Representations

The most compact representations of a chemical molecule are the line notations, where only information about the elemental compositions and their connectivity is encoded. Possibly the most well known of these formats is the simplified molecular input line entry system (SMILES) format originally designed by the US EPA and extended and popularised by Daylight Chemical Information Systems. In this system, each element symbol is considered as being bonded to the previous, with branched chains indicated by '()' and ring systems indicated by numbers marking the opening and closing member of the ring. For a straightforward system like *n*-butane the SMILES string is trivial, `CCCC`. Isobutane, or methylpropane is denoted as `CC(C)C`, so the propane chain is shown as `CCC` with the methyl branch being the `(C)`. Double bonds are shown with '='; for example, methylpropene is coded as `CC(C)=C`. The SMILES notation for cyclohexane is `C1CCCCC1`, where the numbers indicate chains joining. Note that the identically connected benzene is represented by `c1ccccc1`, the lower case indicating aromaticity. Our original example of *L*-phenylalanine is `N[C@@H](Cc1ccccc1)C(=O)O`. The stereochemical information is encoded within `[C@@H]` which indicates that the following groups, when viewed from the nitrogen towards the stereogenic carbon, are the substituents (H, R-group and carboxylic acid) arranged in clockwise order. Conversely, `[C@H]` would indicate that they are arranged anticlockwise. SMILES strings are a useful and compact way to represent structures.

Alternative similar representations include the InChI (International Chemical Identifier) format by the International Union of Pure and Applied Chemistry (IUPAC). An advantage of the InChI system is that the format is not proprietary and software with an open-source type licence is available to process it, and unique identifiers can be generated from InChI (so-called InChI Keys) strings via a hashing algorithm, in principle allowing for fast similarity searching.

One drawback is that identical structures can be represented in multiple ways; this is normally addressed by producing a **canonical representation**, where the atoms

Example 17.2 **GENERATION OF SMILES REPRESENTATIONS**

1. Continuing with 2-fluoro-butane from Example 23.1, we can generate a SMILES string for the current drawing on the JSME home page (peter-ertl. com/jsme/) simply by pressing the 'Get smiles' button in the 'Export' section of that web page.

2. Now modify the terminal carbon next to the fluorine of 2-fluoro-butane to a bromine atom. Then indicate the stereochemistry of the carbon atom with the two halogen substituents by using the wedge bond icon, producing a diagram like

3. Generate the SMILES string for the new molecule by pressing the 'Get smiles' button. Note how the stereochemical information has been introduced in the SMILES string.

are sorted by rules to produce an input independent of atom ordering. This has been a consistent issue with chemical representations, even though the problem was recognised early in the development of computational representation by Morgan. Canonicalisation algorithms are beyond the scope of this chapter, however one should not attempt to compare molecules with these representations, unless one has full confidence that the linear method encoding software produces consistent results for any version of a molecule that is subjected to it.

17.2.2 Connection Tables

A second form of representation is the connection table. There are many variants of this, with one of the oldest and frequently used being the MDL (now BIOVIA) MOL format. This format constitutes the key section of a large number of formats. At its core, it is essentially a codified chemical drawing.

Using again phenylalanine as an example, the MOL format is illustrated in Figure 17.3. Box A contains three lines and is largely commentary; line 1 is normally a title, line 2 describes the software that generated the file, and line 3 can be a general comment. Box B consists of only one line and informs about the counts, where the first two numbers indicate the number of atoms and the number of bonds in the entry, the fifth gives the number of M property records. V2000 identifies the version the file is compliant with. V2000, is the most common format, but there is a new v3000 format, which is useful if the atom or bond count exceeds 999. In box C, there are x-, y- and z-coordinates for each atom. As z is consistently 0.0000 this file codes for a two-dimensional structure, and the x- and y-coordinates represent a plot of the molecule, where the coordinates have no absolute dimension. In box D, the bonds within

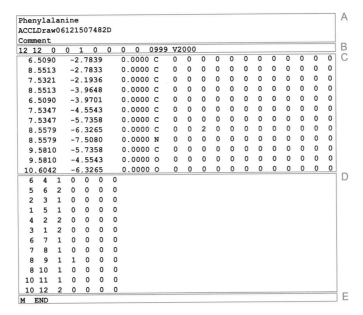

Figure 17.3 A MOL format representation of phenylalanine. Note that the z-coordinate shows 0.0000 throughout, hence this represents a two-dimensional structure. Boxes are shown for illustration purposes only.

the molecule are defined. Each bond is represented by the atom number of the two atoms being bonded, followed by the formal bond order and the stereochemistry. For example, for the bond line 8 9 1 1 0 0 0 0, this means:

- 8: atom 8 (carbon with $x = 8.5579$, $y = -6.3265$, $z = 0.0000$) is bonded to
- 9: atom 9 (nitrogen with $x = 8.5579$, $y = -7.5080$, $z = 0.0000$);
- 1: and the bond is a single bond,
- 1: where the bond is coming outwards from the page, exactly as seen in the original sketch (see Figure 17.2).

The last box (E) indicates the end of the molecular property records, by use of the sentinel line M END.

It is also possible to represent three-dimensional data using this type of format. In doing so, we are moving from a simple connection table of atoms to a description of atomic coordinates within a Cartesian coordinate system. Almost universally, the coordinates are represented in ångström units, although occasionally some formats (e.g. the GROMACS format) may use nanometres (1 nm = 10 Å). In some formats, derived from structures determined by X-ray crystallography (see Section 14.3.4), the coordinates may be defined in terms of fractions of the unit cell edge. Since not all crystallographic unit cells possess orthogonal axes (like the Cartesian coordinate system), this requires that the cell geometry is given, and the calculation of the Cartesian coordinates (in Ångstom) becomes more complex when the angles in the unit cell do not equal 90°.

In introducing SD-formatted files (SDF files), which are derived from the MOL file format, we once again use the example of L-phenylalanine, this time represented as a three-dimensional structure and added hydrogen atoms (Figure 17.4). SDF files can hold multiple MOL entries, separated by $$$$. In addition, it is possible to add so-called properties that are added by using particular property tags > <... . >. In the example shown in Figure 17.4, a SMILES string for the encoded molecule is included, preceded by the tag > <UNIQUE_SMILES> Property tags can be used to introduce arbitrary data into the file and often store molecular properties like molecular mass or logP (see Section 24.2.2). The property entry is terminated by a blank line.

The MOL/SDF file format is very rich in its ability to encode chemical information for use in database systems, which reflects its origins in large-scale chemical

```
C9H11NO2
APtclcactv06121508333D 0    0.00000     0.00000      1

 23 23  0  0  1  0  0  0  0  0 0999 V2000
    3.0135   -1.2385   -0.3791 C   0  0  0  0  0  0  0  0  0  0  0  0
    3.0417    1.0912    0.1725 C   0  0  0  0  0  0  0  0  0  0  0  0
    3.6954   -0.0402   -0.2785 C   0  0  0  0  0  0  0  0  0  0  0  0
    1.7064    1.0242    0.5237 C   0  0  0  0  0  0  0  0  0  0  0  0
    1.6780   -1.3053   -0.0288 C   0  0  0  0  0  0  0  0  0  0  0  0
    1.0243   -0.1739    0.4222 C   0  0  0  0  0  0  0  0  0  0  0  0
   -0.4313   -0.2471    0.8055 C   0  0  0  0  0  0  0  0  0  0  0  0
   -1.2970    0.0553   -0.4192 C   0  0  2  0  0  0  0  0  0  0  0  0
   -1.0764    1.4442   -0.8440 N   0  0  0  0  0  0  0  0  0  0  0  0
   -2.7497   -0.1377   -0.0675 C   0  0  0  0  0  0  0  0  0  0  0  0
   -3.2758   -1.3723   -0.0422 O   0  0  0  0  0  0  0  0  0  0  0  0
   -3.4408    0.8190    0.1916 O   0  0  0  0  0  0  0  0  0  0  0  0
    3.5231   -2.1214   -0.7357 H   0  0  0  0  0  0  0  0  0  0  0  0
    3.5746    2.0272    0.2518 H   0  0  0  0  0  0  0  0  0  0  0  0
    4.7389    0.0119   -0.5518 H   0  0  0  0  0  0  0  0  0  0  0  0
    1.1957    1.9081    0.8764 H   0  0  0  0  0  0  0  0  0  0  0  0
    1.1452   -2.2414   -0.1077 H   0  0  0  0  0  0  0  0  0  0  0  0
   -0.6601   -1.2470    1.1743 H   0  0  0  0  0  0  0  0  0  0  0  0
   -0.6386    0.4845    1.5866 H   0  0  0  0  0  0  0  0  0  0  0  0
   -1.0269   -0.6202   -1.2309 H   0  0  0  0  0  0  0  0  0  0  0  0
   -1.3383    2.0305   -0.0657 H   0  0  0  0  0  0  0  0  0  0  0  0
   -0.0812    1.5547   -0.9683 H   0  0  0  0  0  0  0  0  0  0  0  0
   -4.2079   -1.4961    0.1834 H   0  0  0  0  0  0  0  0  0  0  0  0
  4  6  1  0  0  0
  5  6  2  0  0  0
  2  3  1  0  0  0
  1  5  1  0  0  0
  2  4  2  0  0  0
  1  3  2  0  0  0
  6  7  1  0  0  0
  7  8  1  0  0  0
  8  9  1  1  0  0
  8 10  1  0  0  0
 10 11  1  0  0  0
 10 12  2  0  0  0
  1 13  1  0  0  0
  2 14  1  0  0  0
  3 15  1  0  0  0
  4 16  1  0  0  0
  5 17  1  0  0  0
  7 18  1  0  0  0
  7 19  1  0  0  0
  8 20  1  0  0  0
  9 21  1  0  0  0
  9 22  1  0  0  0
 11 23  1  0  0  0
M  END
> <UNIQUE_SMILES>NC(CC1=CC=CC=C1)C(O)=O

$$$$
```

Figure 17.4 An SD-formatted (SDF) file representation of L-phenylalanine. In contrast to Figure 17.3, the z-coordinates are now populated, thus making this a representation of a three-dimensional structure.

information systems. For many applications, however, it is somewhat ambiguous, especially, if one intends to use the structure in a molecular forcefield application, the description of the structure is insufficiently detailed, since it is not possible to represent hybrid bonding and non-integral charges. Many molecular modelling packages have more complex formats, and one of the more popular was originally designed by the company Tripos and confusingly also known as MOL format, although now it is more common to use the later MOL2 format. This format is particularly useful for larger molecules or molecular assemblies, so we introduce it by illustrating the representation of the tripeptide VFG, where phenylalanine is now part of a polypeptide (Figure 17.5).

```
@<TRIPOS>MOLECULE
peptide_vfg.pdb
 22 22 0 0 0
SMALL
GASTEIGER
Energy = 0

@<TRIPOS>ATOM
      1   N      34.7390   18.9610  -11.0420 N.3      2   VAL2      -0.1231
      2   CA     33.9030   17.9980  -10.3330 C.3      2   VAL2       0.1692
      3   C      34.8000   17.3120   -9.2940 C.2      2   VAL2       0.2592
      4   O      35.7590   16.6050   -9.6650 O.2      2   VAL2      -0.2718
      5   CB     33.1400   17.0340  -11.2320 C.3      2   VAL2       0.0266
      6   CG1    32.2510   16.0840  -10.4340 C.3      2   VAL2       0.0020
      7   CG2    32.2940   17.7140  -12.2900 C.3      2   VAL2       0.0020
      8   N      34.4910   17.5460   -8.0380 N.am     3   PHE3      -0.1962
      9   CA     35.1850   16.9030   -6.9180 C.3      3   PHE3       0.1771
     10   C      34.7420   15.4410   -6.7710 C.2      3   PHE3       0.2620
     11   O      33.5250   15.1620   -6.8620 O.2      3   PHE3      -0.2715
     12   CB     34.9670   17.6320   -5.5940 C.3      3   PHE3       0.0574
     13   CG     35.9440   18.7370   -5.3750 C.ar     3   PHE3      -0.0200
     14   CD1    35.6660   20.0500   -5.7980 C.ar     3   PHE3      -0.0042
     15   CD2    37.0000   18.5570   -4.4730 C.ar     3   PHE3      -0.0042
     16   CE1    36.5770   21.0760   -5.5680 C.ar     3   PHE3      -0.0003
     17   CE2    37.8690   19.5890   -4.1570 C.ar     3   PHE3      -0.0003
     18   CZ     37.6360   20.8730   -4.6660 C.ar     3   PHE3      -0.0000
     19   N      35.7240   14.6390   -6.3310 N.am     4   GLY4      -0.1954
     20   CA     35.3660   13.2800   -5.8700 C.3      4   GLY4       0.1869
     21   C      34.9240   13.4200   -4.4150 C.2      4   GLY4       0.2309
     22   O      35.3030   14.4030   -3.7810 O.2      4   GLY4      -0.2863
@<TRIPOS>BOND
     1      7      5      1
     2      5      6      1
     3      5      2      1
     4      1      2      1
     5      2      3      1
     6      4      3      2
     7      3      8     am
     8      8      9      1
     9      9     10      1
    10      9     12      1
    11     11     10      2
    12     10     19     am
    13     19     20      1
    14     20     21      1
    15     14     16     ar
    16     14     13     ar
    17     12     13      1
    18     16     18     ar
    19     13     15     ar
    20     18     17     ar
    21     15     17     ar
    22     21     22      2
```

Figure 17.5 A Tripos MOL2 file representation of the tripeptide VFG.

Comparison of the MOL2 format as illustrated in Figure 17.5 with the MOL (Figure 17.3) and SDF (Figure 17.4) formats shows that the three formats are quite similar. There is still a counts line that holds information about the number of atoms and bonds in each section, but the information about the molecule is now clearly divided up, which is achieved by keywords of the format @<TRIPOS>XXX so that the MOL2 file can be easily parsed. The molecule block, initiated by the keyword @<TRIPOS>MOLECULE, holds general information about the molecule, including a name. In this case `peptide_vfg.pdb`. Note the enhancements over the MOL format in the atom block (@<TRIPOS>ATOM):

- each atom has a distinct name rather than just an element symbol
- each atom is given an enhanced type descriptor, in this case the atom type within the Tripos force field
- there is a substructure ID number and name, typically used to describe polymers like proteins or DNA
- there is also a partial charge for each atom.

We can see from the molecule header block that these charges have been calculated using the basic Gasteiger bond increment method. Similarly, the bonds block (@<TRIPOS>Bond) has an enhanced bond type.

Finally, we should consider a very common filetype used in macromolecular data, which is strictly a connection table, but one where the connections are largely inferred. Although more modern formats based on the crystallographic information file (CIF) are becoming more widely used, the basic format for interchanging macromolecular information is still the Protein Data Bank or PDB format. As with the CIF or macromolecular CIF (mmCIF) formats, a lot of information about the experiment used to generate the structural data is included. For chemoinformatics applications, one is usually just interested in the element and position data. For illustration, the VFG tripeptide is shown in PDB format in Figure 17.6.

```
ATOM     10  N   VAL A   2      34.739  18.961 -11.042  1.00 19.96           N
ATOM     11  CA  VAL A   2      33.903  17.998 -10.333  1.00 18.10           C
ATOM     12  C   VAL A   2      34.800  17.312  -9.294  1.00 19.39           C
ATOM     13  O   VAL A   2      35.759  16.605  -9.665  1.00 22.14           O
ATOM     14  CB  VAL A   2      33.140  17.034 -11.232  1.00 16.81           C
ATOM     15  CG1 VAL A   2      32.251  16.084 -10.434  1.00 21.92           C
ATOM     16  CG2 VAL A   2      32.294  17.714 -12.290  1.00 19.46           C
ATOM     17  N   PHE A   3      34.491  17.546  -8.038  1.00 19.89           N
ATOM     18  CA  PHE A   3      35.185  16.903  -6.918  1.00 17.43           C
ATOM     19  C   PHE A   3      34.742  15.441  -6.771  1.00 15.70           C
ATOM     20  O   PHE A   3      33.525  15.162  -6.862  1.00 18.52           O
ATOM     21  CB  PHE A   3      34.967  17.632  -5.594  1.00 17.94           C
ATOM     22  CG  PHE A   3      35.944  18.737  -5.375  1.00 16.78           C
ATOM     23  CD1 PHE A   3      35.666  20.050  -5.798  1.00 15.97           C
ATOM     24  CD2 PHE A   3      37.000  18.557  -4.473  1.00 19.95           C
ATOM     25  CE1 PHE A   3      36.577  21.076  -5.568  1.00 17.32           C
ATOM     26  CE2 PHE A   3      37.869  19.589  -4.157  1.00 17.65           C
ATOM     27  CZ  PHE A   3      37.636  20.873  -4.666  1.00 17.91           C
ATOM     28  N   GLY A   4      35.724  14.639  -6.331  1.00 16.79           N
ATOM     29  CA  GLY A   4      35.366  13.280  -5.870  1.00 16.34           C
ATOM     30  C   GLY A   4      34.924  13.420  -4.415  1.00 11.91           C
ATOM     31  O   GLY A   4      35.303  14.403  -3.781  1.00 16.23           O
END
```

Figure 17.6 The tripeptide VFG in PDB format.

As illustrated in Figure 17.6, the section that defines the structure of the tripeptide VFG comprises of lines that start with the **ATOM** keyword. For each atom (shown as a consecutive number and an atom type), there is information as to which amino-acid residue it belongs to, the chain identifier and the residue number. This is followed by the x-,y- and z-coordinates, and then the site occupation factor (SOF; also called occupancy) and the atomic (or thermal) displacement factor (ADF; also called B-factor). Both of these latter parameters are derived from the crystallographic experiment, the SOF indicating the fraction of total unit cells in which that atom is present and the B-factor indicating the extent of thermal movement associated with that atom. Note that the example shown in Figure 17.6 is derived from an experimental structure and thus has experimental values, while the example shown in Figure 17.7 is converted from a calculated structure and thus has arbitrary values (typically set to **0.00**). Often in situations where PDB data are derived from some type of calculation, the B-factor field can be used to present a quality or score associated with that atom.

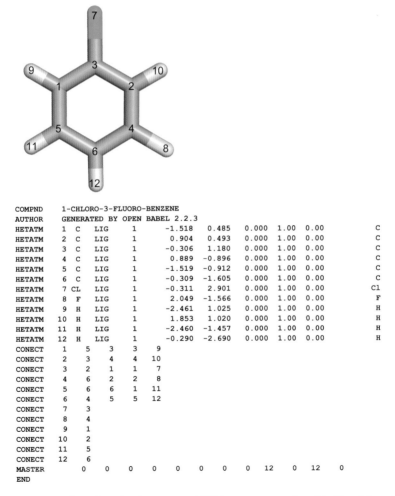

```
COMPND    1-CHLORO-3-FLUORO-BENZENE
AUTHOR    GENERATED BY OPEN BABEL 2.2.3
HETATM    1  C    LIG    1       -1.518    0.485   0.000  1.00  0.00           C
HETATM    2  C    LIG    1        0.904    0.493   0.000  1.00  0.00           C
HETATM    3  C    LIG    1       -0.306    1.180   0.000  1.00  0.00           C
HETATM    4  C    LIG    1        0.889   -0.896   0.000  1.00  0.00           C
HETATM    5  C    LIG    1       -1.519   -0.912   0.000  1.00  0.00           C
HETATM    6  C    LIG    1       -0.309   -1.605   0.000  1.00  0.00           C
HETATM    7  CL   LIG    1       -0.311    2.901   0.000  1.00  0.00           Cl
HETATM    8  F    LIG    1        2.049   -1.566   0.000  1.00  0.00           F
HETATM    9  H    LIG    1       -2.461    1.025   0.000  1.00  0.00           H
HETATM   10  H    LIG    1        1.853    1.020   0.000  1.00  0.00           H
HETATM   11  H    LIG    1       -2.460   -1.457   0.000  1.00  0.00           H
HETATM   12  H    LIG    1       -0.290   -2.690   0.000  1.00  0.00           H
CONECT    1    5    3    3    9
CONECT    2    3    4    4   10
CONECT    3    2    1    1    7
CONECT    4    6    2    2    8
CONECT    5    6    6    1   11
CONECT    6    4    5    5   12
CONECT    7    3
CONECT    8    4
CONECT    9    1
CONECT   10    2
CONECT   11    5
CONECT   12    6
MASTER    0    0    0    0    0    0    0    0   12    0   12    0
END
```

Figure 17.7 Three-dimensional structure and PDB format representation of 1-chloro-2-fluoro-benzene.

In the PDB format, the connectivity within each molecule can be formally defined by a dictionary where connectivity is assumed from the atom names used. However, for common macromolecules such as peptides and proteins this can be assumed, as these are made up of individual amino acids as building blocks. Therefore, most software that parse and interpret PDB files derive the connectivity from inter-atomic distances 'on-the-fly'. The PDB format can of course also accommodate arbitrary molecules where the connectivity is not necessarily known *a priori* and additional connectivity information can be supplied (identified by a **CONECT** command at the start of each line). This is illustrated for 1-chloro-2-fluoro-benzene in Figure 17.7. The first **CONECT** line, for example, codes for

CONECT   1   5   3   3   9

- make a bond from atom 1 to atom 5
- make a bond from atom 1 to atom 3
- make another bond from atom 1 to atom 3 (i.e. there is now a double bond between atoms 1 and 3)
- make a bond from atom 1 to atom 9.

17.2.3   Matrix Representation of Molecular Graphs

The simplest matrix representation of atomic connectivity is the **adjacency matrix**. This is a square diagonally symmetric matrix where each element *i,j* represents the connectivity between atoms *i* and *j*. This can be just a simple binary yes/no version (represented by 1 and 3) or encode the actual bond order, in which cases numbers larger than 1 are also possible.

Illustrated for the example of phenylalanine in Figure 17.8, a simple matrix like the adjacency matrix represents all the σ bonds in the structure. Trivially, a sum of the matrix yields twice the total bond count and the sum of a row gives the number of bonds for that atom. Therefore, simple arithmetic can yield information about the level of cyclisation in the molecule.

(a)   (b)

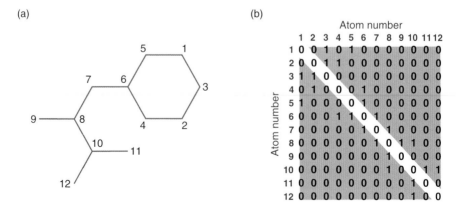

Figure 17.8 (a) Arbitrary numbering of atoms in phenylalanine. (b) Adjacency matrix derived for phenylalanine using the numbering in (a).

A more complex matrix tabulates the total smallest number of bonds between any two atoms in the molecule. This is called a **connectivity matrix** and is illustrated for phenylalanine in Figure 17.9. The sum of the elements in either of the symmetric halves is the **Wiener index**, a molecular descriptor for compactness; it thus informs about the level of branching in the molecule. Note that the Wiener index, like many other descriptors, is calculated based only on the non-hydrogen atoms in a molecule. Other similar descriptors weighted by atomic properties can be calculated from this matrix.

(a)   (b)

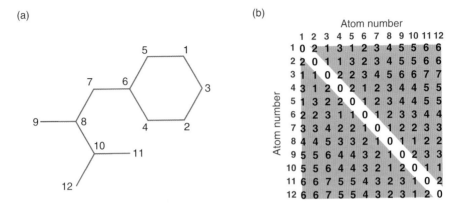

Atom number

	1	2	3	4	5	6	7	8	9	10	11	12
1	0	2	1	3	1	2	3	4	5	5	6	6
2	2	0	1	1	3	2	3	4	5	5	6	6
3	1	1	0	2	2	3	4	5	6	6	7	7
4	3	1	2	0	2	1	2	3	4	4	5	5
5	1	3	2	2	0	1	2	3	4	4	5	5
6	2	2	3	1	1	0	1	2	3	3	4	4
7	3	3	4	2	2	1	0	1	2	2	3	3
8	4	4	5	3	3	2	1	0	1	1	2	2
9	5	5	6	4	4	3	2	1	0	2	3	3
10	5	5	6	4	4	3	2	1	2	0	1	1
11	6	6	7	5	5	4	3	2	3	1	0	2
12	6	6	7	5	5	4	3	2	3	1	2	0

(Atom number — vertical axis label)

Figure 17.9 (a) Arbitrary numbering of atoms in phenylalanine. (b) Connectivity matrix derived for phenylalanine using the numbering in (a).

Example 17.3   **CALCULATION OF THE WIENER INDEX**

Question   Calculate the Wiener index for the following molecule:

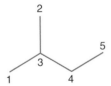

Answer   We can simply sum the product of the number of atoms on each side of each bond:

Bond 1–3	atom 1	atoms 2, 3, 4, 5	1×4	=	4
Bond 3–2	atoms 1, 3, 4, 5	atom 2	4×1	=	4
Bond 3–4	atoms 1, 3, 2	atoms 4, 5	3×2	=	6
Bond 4–5	atoms 1, 3, 2, 4	atom 5	4×1	=	4
				Total	18

Alternatively, we can generate the connectivity matrix and sum the elements of either of the symmetric halves:

$$
\begin{array}{ccccc}
0 & 2 & 1 & 2 & 3 \\
2 & 0 & 1 & 2 & 3 \\
1 & 1 & 0 & 1 & 2 \\
2 & 2 & 1 & 0 & 1 \\
3 & 3 & 2 & 1 & 0 \\
\end{array}
\qquad
\begin{aligned}
0 + 2 + 1 + 2 + 3 &= 8 \\
0 + 1 + 2 + 3 &= 6 \\
0 + 1 + 2 &= 3 \\
0 + 1 &= 1 \\
0 &= \underline{0} \\
&\ \ 18
\end{aligned}
$$

The Wiener index for 2-methyl-butane is thus 18.

17.2.4 Meta Data Derived From Structures

The Wiener index introduced in the above section is an example of a molecular descriptor, a type of meta data wholly dependent on connectivity data. Different descriptors depend on different types of data, but all of them are dependent on the topology and elemental composition of the molecule. Whereas the Wiener index is a topological descriptor, other descriptors seek to describe the flexibility of the molecule (based on number of rotatable bonds in the molecule) and the level of branching. At a first glance, the number of rotatable bonds and the branching may seem similar, but the descriptors in fact describe different aspects of a molecule. For example, both pentane and 2,2-dimethylpropane have the same number of rotatable bonds, but the different level of branching leads to substantial differences in the overall selections of molecular shape (Figure 17.10).

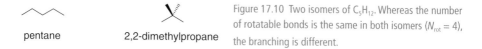

pentane   2,2-dimethylpropane

Figure 17.10 Two isomers of C_5H_{12}. Whereas the number of rotatable bonds is the same in both isomers ($N_{rot} = 4$), the branching is different.

A major reason to use descriptors is that they allow comparison of molecules more easily than the full connectivity matrices. As can be seen in the simple case of the Wiener index, descriptors do not vary according to the atom ordering inside the molecule. Of course, a single descriptor is unlikely to offer a meaningful similarity discrimination; however, linear combinations of descriptors may correlate well enough with a particular activity to give some level of predictive power (see Figure 17.1). Furthermore, certain descriptors might have special relevance to a particular physiological process (e.g. oral bioavailability, blood–brain barrier permeability) and these can be used as filters to exclude non-viable compounds from a particular study.

A second form of meta data is the molecular fingerprint. Molecular fingerprints can be based on almost any recognisable feature in a molecule, and consist of a sequence of bits. The bit is 1 when the feature in question is within the molecule, and 0 when it is not (see example in Figure 17.11).

Fingerprints can be of arbitrary length, and two molecules are compared by a logical AND of the fingerprints. Various schemes can be used to provide such similarity

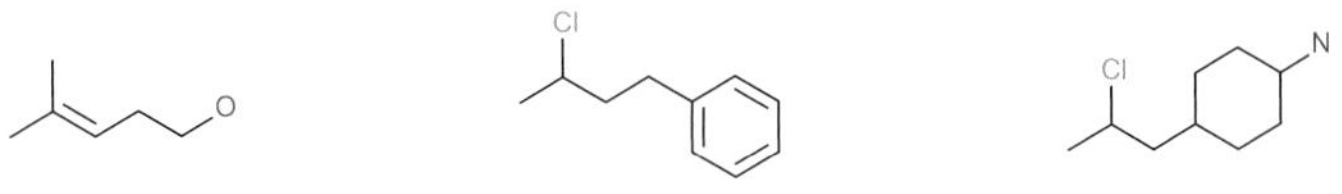

	Has C=C bond	Has carbonyl function	Has alcohol function	Has amine function	Has nitro function	Contains halogen	Contains non-aromatic ring	Has 6-membered aromatic ring
A	1	0	1	0	0	0	0	0
B	0	0	0	0	0	1	0	1
C	0	0	0	1	0	1	1	0

Figure 17.11 Illustration of a molecular fingerprint consisting of eight criteria for three molecules A, B and C.

scores, with the best known being the Tanimoto coefficient. To calculate this coefficient, one counts the number of bits set in one fingerprint (a), the number of bits set in the other (b) and the number of bits set in the logical of both (c); the Tanimoto coefficient is then calculated as per:

$$\text{Tanimoto cofficient} = \frac{c}{a + b - c} \qquad \text{(Eq 17.1)}$$

Simple methods of fingerprinting code for the presence of particular elements (e.g. halogens) or fragments like the amide function. More complex systems might code for connected paths of atoms.

The molecule shown in Figure 17.12 could code for

- the 2-atom paths Cl-C, C-O
- the 3-atom paths Cl-C-O, C-O-C
- the 4-atom path Cl-C-O-C.

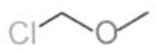

Figure 17.12 Chloromethyl-methyl-ether.

Not all of these might be coded for in a particular fingerprint, since the bit string to represent all keys is potentially huge. However, a hashing function that compresses the string at the expense of accuracy could be used. Searching for similar fingerprints can be a very fast activity, a large number can be held in a modern computer memory and the bit counting and logical AND functions are computationally easy to perform.

17.2.5 Format Conversion

As can be seen from the examples given above, format conversion can be a non-trivial exercise. It is usually quite easy to go from a format with greater detail to one with less information, but the opposite may prove more challenging. There are several open-source conversion tools that will produce correct format versions of different molecules and even provide 2D–3D conversion. However, care should be exercised when using these tools, in particular when any form of extended atom type description is being generated (e.g. writing a format that includes force-field types) and the input information is purely elements and their formal bond orders. Similarly, any conversion from a two- to a three-dimensional format involves assumptions about the molecular geometry, and although a low-energy conformation will likely be generated during conversion, it cannot be regarded as the inevitable outcome (i.e. the true energy minimum). While numerous software packages exist that will do this, for those interested in examining different file formats and interconversions, two open-source projects should be considered: first OpenBabel, which offers a format converter, a 2D–3D converter and a C++ programming interface, as well as other tools; second, for those interested in developing their own software, the Chemistry Development Kit is a very useful Java library that can read and write many formats, also has 2D–3D conversion abilities and can generate molecular descriptors, amongst many other capabilities. Bindings to the package for Python (see Chapter 18) are also available. You can experiment with format conversion with the Online SMILES Translator and Structure File Generator offered by the National Cancer Institute (NCI, NIH).

17.3 CALCULATION OF COMPOUND PROPERTIES

Frequently, one needs to calculate properties of compounds, because methods to measure these properties are time consuming (e.g. partition coefficients) or not immediately accessible (e.g. partial charges). In case of partial charges, the resulting dipole moment of a molecule needs to be determined; despite such physical measurements being possible, they require a lot of time and dedicated equipment. The advantage of chemoinformatics is that large sets of such properties can be calculated and used to select subsets of compounds with a favourable combination of properties.

17.3.1 Simple Formula-Derived Properties

These are the trivially calculable properties like molecular mass, formal charge, counts of atom types, and hydrogen bond donors and acceptors. Despite being simple, descriptors like these are often very powerful, with many of them occurring in the empirically derived rules that are used in compound selection for drug or agrochemical discovery applications (e.g. Lipinski rules; see Section 24.2.2).

17.3.2 Solubility and Partition Coefficients

One of the most common issues that needs to be addressed is the question of whether a compound is soluble in water, particularly if it is to be used in a biological context. There is little point in investing much computational and experimental effort, if finally, when synthesised, it becomes clear that the maximum aqueous concentration is so small as to make the compound completely unusable for its intended application. Computational approaches to this problem tend to be an automated and systematised version of chemists' intuition. Most chemists can look at a formula and have some inherent feeling for its possible solubility. This estimate is based on the ratio of hydrophobic parts to polar groups (hydroxyl groups, acids and amines etc.). Various computational approaches reduce the molecule to a series of fragments or analyse the topology and use a training set to produce a correlation between the fragments and experimentally observed values.

This leads on to the partition coefficient, or more usually in the biological context the octanol/water partition coefficient. The partition coefficient is the logarithm of the ratio of the concentration of a compound in one solvent to its concentration in a second immiscible solvent; in our case, the immiscible solvents are octanol and water.

$$\text{Partition coefficient} = \log P = \frac{c(\text{compound in octanol})}{c(\text{compound in water})} \qquad \text{(Eq 17.2)}$$

The more positive this number is, the more the compound favours octanol to water. Compounds with a negative logP are likely to be soluble. Further, in a biological context, we can infer something about the organ or cellular location of a compound, and, more importantly, the propensity of the compound to bind in a hydrophobic pocket on a protein. The archetypal logP calculation method is cLogP, based on the so-called Pomona method of logP estimation (see also Section 24.2.2). This is essentially a summation of fragment partition coefficients with various complex correction values added to produce closer agreement to a test set. Other similar methods have been developed, some of which attempt to deal gracefully with the problem of what to do if the molecule subjected to calculation contains a fragment that does not correspond to any in the cLogP library. As in many chemoinformatics algorithms, it is important to know the extent of molecular space that is covered by the technique and be aware of any inaccuracies that will occur in the calculated properties should computation be conducted on a molecule for which the method has not been adequately parametrised.

17.3.3 Partial Charges and Molecular Orbital Calculations

An important aspect of a molecule is the distribution of charge across the atoms in a molecule. In simple terms, we know that in solution a carboxylic acid dissociates (see Section 2.3.2) into a positively charged proton and a negatively charged counter ion; in the counter ion, the negative charge is not located on a single oxygen atom, but is spread over the entire carboxylic acid. In fact, the charge is distributed, to some extent, over the entire molecule, and even in neutral species there is charge separation between atoms of differing electronegativities. Obviously, when molecules interact with each other, this charge distribution is important; an arrangement of molecules that has charges of the same sign in close proximity is clearly much less stable than one where complementarity of charge occurs.

The fractional charges within a molecule are called partial charges, to differentiate them from the integral formal charges assigned to entire molecules. Most frequently, calculated partial charges assigned to atoms are those from the method of Gasteiger and Marsilli. This method starts with a formal charge assignment and then repeatedly applies incremental movements of charge across bonds until a consistent charge distribution is achieved. This method was original proposed for a very limited set of compounds, but is applied routinely to general organic compounds; however, the resulting partial charges are often inaccurate.

More accurate partial charges can be derived from molecular orbital (MO) methods. These methods build molecular orbitals from linear combinations of atomic orbitals.

Normally, the atomic orbitals are represented by approximations of orbitals from the hydrogen atom. There are usually several layers of approximation, using hydrogen-type orbitals in multi-electron systems. In other words, only the hydrogen atom can be treated rigorously, but the atomic orbitals of heavier atoms need to be approximated. The functions used to approximate atomic orbitals are called basis sets, and the number of actual functions in any basis set can increase, depending on the level of sophistication. For example, a common simple basis set is STO-3G, which stands for Slater-type orbitals, approximated by three Gaussian functions, and these are generally the form used in semi-empirical MO calculations. Many more complex basis sets are possible and these can be used in more rigorous *ab initio* calculations. The greater the number of contributing functions to a basis set, the better the approximation of the atomic orbitals becomes, but at the cost of requiring many more calculations and thus computing time. This places a limit on the overall size of molecule that can be dealt with; however, more recent developments such as fragment molecular theory can tackle larger systems.

Molecular orbital theory is a complex topic, with many factors to consider when deciding exactly what basis set and calculation technique to use for a particular simulation. In the context of partial charges, those derived from almost any MO calculation are more accurate than the various bond increment techniques, and thus produce a corresponding increase in the quality of any subsequent calculation. Due to substantial computing time, MO theory can be useful to obtain parameters for small organic molecules, but the size limits imposed by the complexity of the calculation severely limit its usefulness in biological systems (see Section 17.4.4).

17.4 MOLECULAR MECHANICS

Molecular mechanics is a molecular simulation method where the basic simulation unit is the atom. Atoms are modelled as points with a collection of potential energy terms between them. The simplest term is the bond potential (E_{bond}). This models a bond as a classical spring that behaves as a harmonic oscillator, with an equilibrium distance R_{eq} and a force constant k. This is, of course, a simplification. In a real bond, the function is anharmonic, and if stretched sufficiently, the bond will break; but within the range of bond stretches observed in intact molecules, the harmonic potential is almost identical to the genuine bond-stretching function. Since the harmonic function is considerably less challenging to compute, molecular mechanics models typically apply this approximation, hence limiting these simulations to the behaviour of intact molecules. It is thus also important to keep in mind that molecular mechanics energies are only reasonable in near equilibrium atomic arrangements. Comparison of the harmonic function with the Morse potential, which better models the actual bond function, shows that they are almost identical, close to the equilibrium distance (Figure 17.13).

For any arrangement for which we know the connectivity between the atoms, we can compute the overall change from the equilibrium bond energy just by summing all the energies from each bond and combining them into E_{bond}. For example, in the molecule 4-fluoro-pentan-2-one (Figure 17.14) there are 15 bonds to consider.

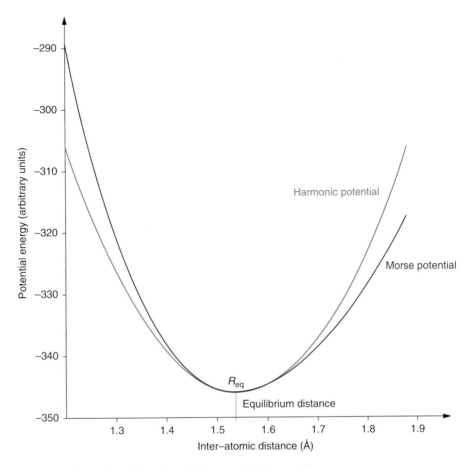

Figure 17.13 Comparison of harmonic and Morse bond stretch potentials.

Figure 17.14 4-Fluoro-pentan-2-one.

Each bond connecting different elements has a particular equilibrium distance and force constant. Therefore, one needs to find values for C–H (single), C–C (single), C–F (single) and C=O (double) bonds. Similarly, **bond angles** can be treated as harmonic functions, with five different potentials (combined into E_{angle}) required for the molecule shown in Figure 17.14 (H–C–H, H–C–C, H–C–F, C–C–F, C–C=O). Further, there is a term required to deal with **torsion angles** (E_{torsion}), that will be periodic in nature. In addition, there are terms for **non-bonded interactions**, comprising of van der Waals forces (E_{vdW}) and electrostatic interactions ($E_{\text{electrostatic}}$). The **van der Waals interactions** can be quite accurately modelled by the Lennard–Jones potential (Figure 17.15). It is also sometimes called the 6–12 potential since it arises from the combination of a 6th power attractive term and a 12th power repulsive term.

The van der Waals term needs to be calculated for every pairwise non-bonded interaction in the molecule, so the number of calculations increases approximately with the square of the number of atoms in the system.

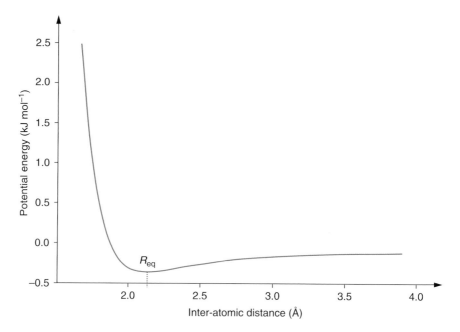

Figure 17.15 The Lennard–Jones potential is used to model van der Waals interactions between two non-bonded atoms.

Similarly, the electrostatic term ($E_{electrostatic}$) is taken across all pairwise interactions, and for a given pair of atoms i and j is of the form:

$$E_{ij} = \frac{Q_i \times Q_j}{4\pi \times \varepsilon_0 \times \varepsilon_r \times r_{ij}} \qquad \text{(Eq 17.3)}$$

where Q_i and Q_j are the charges of atoms i and j, respectively, and r_{ij} is the distance between both atoms; ε_r represents the dielectric constant for a material and ε_0 is the vacuum permittivity. In the biological context, ε_r is a value appropriate for material in solution and the entire term ($4\pi \times \varepsilon_0 \times \varepsilon_r$) can be collapsed into a single constant.

Importantly, whereas the van der Waals effect decreases by the distance to the sixth power and so is quite short range, electrostatic forces scale linearly with distance and so are much more long range.

Finally, the potential energy function for any arrangement of atoms is obtained as a combination of the individual terms introduced above:

$$E_{tot} = E_{bond} + E_{angle} + E_{torsion} + E_{vdW} + E_{electrostatic} \qquad \text{(Eq 17.4)}$$

17.4.1 Molecular Mechanics Force Fields

All of the terms introduced above require particular parameters describing the features of atoms involved in forming a molecule, i.e. the equilibrium bond distance and force constant for any pair of two atoms, etc. When discussing file formats (see Section 17.2.5), we noted that file formats for molecular modelling systems feature extended atom and bond types. These are required to allow identification of the correct molecular mechanics parameters. Clearly, a C–C single bond between sp^3-hybridised carbon atoms is different from a C–C single bond between sp^2-hybridised carbon atoms. This large collection of parameters and the accompanying equations are normally referred to as the molecular force field, as it defines the forces between atoms. Some force fields are only parameterised for certain classes of molecules, and, in general, few force fields have parameters for the heavier elements. Some force fields may include so-called special terms that provide an out-of-plane correcting force; this is often required as the routine parameters frequently do not produce flat delocalised electron systems like aromatic rings. Other special terms correct the chirality volume, which is used to ensure the correct enantiomer in chiral systems. Neither of these terms has a physical equivalent in real-world systems and ideally should be vanishingly small in systems that are at equilibrium. Many force fields used for biological systems remove the explicit hydrogen atoms, and instead include them in the heavier atoms they are attached to; these are referred to as unified-atom force fields. Importantly, when special terms are being used, the primary function of molecular mechanics force fields is to ensure a meaningful geometry of bonded atoms rather than to produce accurate energies.

17.4.2 Energy Evaluation

A commonly used application of molecular mechanics is to evaluate the relative stability of two atomic assemblies. One can evaluate the energy of each assembly (where each is a conformation of the same system, or each is a possible complex with different components) to discover which is the most likely of the two. This is based on the fundamentals of a chemical equilibrium such as:

$$A \rightleftharpoons B \qquad \text{(Eq 17.5)}$$

that has an excess of the lower-energy species, and the exact ratio of the amounts of species A and B is governed by the difference in the Gibbs free energy change (ΔG), which is defined as:

$$\Delta G = \Delta H - T \times \Delta S \qquad \text{(Eq 17.6)}$$

Using molecular mechanics (and the associated force field) one can estimate the relative enthalpies of species A and B and thus estimate the enthalpy difference ΔH. If one (incorrectly) assumes that there was no overall change in entropy in the system shown in Equation 17.6, then one can suggest that the lower enthalpy state is more stable than the one with higher enthalpy. A common situation where this approach is applied is in examining two protein–ligand complexes and evaluating which is the more stable of the two. The result indicates which of the two evaluated protein–ligand complexes is more likely to form in a solution containing the protein and the two ligands. From the calculated energy difference one can then go on to estimate the relative abundance of each species. One should always bear in mind, however, that the assumption of similar entropies for the two complexes is not always appropriate, and might thus lead to incorrect estimates.

17.4.3 Minimisation

A second use of force fields is to take any assembly of atoms, and produce an arrangement of the same atoms in the same topology, but with a lower energy, and ultimately an assembly that possesses the lowest energy possible. Since the topology of the assembly is to be conserved, covalent bonds can neither be made nor broken, but non-covalent features such as electrostatic interactions, hydrogen bonds and van der Waals interactions may be varied. This problem constitutes a minimisation problem, with the target function shown in Equation 17.4. Minimisation of a function is a well-understood area, with several different types of algorithms available. While a full description of minimisation methods is beyond the scope of this text, an understanding of the types of method is useful for critical analysis and technique selection.

The simplest methods involve repeated evaluations of the function, with the most well-known of these being the simplex method. The advantage is that nothing more than a method to evaluate the energy at any given atomic configuration is required. However, the rate at which the resulting energies converge towards the minimum energy is very poor and the number of energy calculations is large.

The next, and most commonly used, are so-called first-order methods. These make use not just of the energy function itself, but also its derivative with respect to each atomic coordinate. The derivative can be calculated numerically, so each coordinate is moved by a small amount $\vec{d}$ and the partial derivative for that coordinate is $\Delta E / \vec{d}$ (note that $\vec{d}$ is a three-dimensional vector). For N atoms this requires $(3 \times N)$ energy calculations per derivative calculation. A better, but more complex, technique is to derive an analytical (as opposed to numerical) derivative for each term in the potential energy function. Then, for each coordinate only those terms need be calculated that involve just this coordinate. Judicious use of non-bonded-term cut-off distances can also reduce the extent of the calculation. Once the derivative vector is determined, the minimum energy must be in the direction 'downhill' of the current position; so if one searches along this vector the lower-energy conformation should be found. The derivative calculation is then done at the new position, and the process repeated until the resulting energies have converged. This technique is called steepest-descent minimisation, and is much faster than the simplex method. Despite this improvement, under certain circumstances this method can still take many iterations to reach the minimum; a typical analogy is to think of a skier moving zig-zag down a hill, rather than skiing straight down to the centre of the valley. An improvement of the steepest-decent method is the conjugate-gradient approach, where each new downhill vector is combined with the previous vector to produce faster convergence.

There also exist methods that use not only the derivative of the function to be minimised, but also the second-order derivative. In theory, these can detect the minimum in a single operation, but only work well in smoothly varying functions. Since molecular potential energy functions contain a large number of non-smoothly varying features (e.g. they become infinitely large when atoms get too close or bonds stretch too far), these second-order methods become unstable and fail. Therefore, such methods are only applied to systems that have previously undergone minimisation by one of the other methods above.

No matter which method is applied, one thing should always be clear: the final result of a minimisation depends on the starting point. All of the methods described above will only find an atomic assembly with an energy that is lower than the energy at the initial starting point.

By way of example, let the potential energy of an atomic assembly be represented as a function of the conformation, as illustrated in Figure 17.16. If the starting conformation is any particular colour, then minimisation with the methodologies introduced above can only get to the minimum of that colour. For instance, if the starting conformation is in the orange section of the graph (D), a minimiser has no way to get to the global minimum E. In molecular terms, this means that if one is modelling a cyclohexane ring, and start with it perfectly flat, a minimiser will keep it flat, as it must go through a higher-energy state to get to the more realistic boat or chair conformer (see Figure 2.3 in Chapter 2). Similarly, a flat *trans*-polypeptide chain will not be minimised into a correctly folded protein, as it must transit through a higher-energy state to get there.

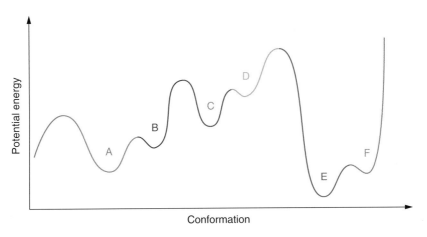

Figure 17.16 Schematic illustration of the potential energy of an atomic assembly as a function of the conformation.

17.4.4 Molecular Dynamics

In molecular dynamics, one uses the potential energy function to simulate the movement of atoms in a molecule over time. In the previous section, we saw how the derivative of the energy function can be used to produce a minimised energy conformation from any starting conformation. Since force is the first derivative of the potential energy with respect to the space coordinate, one can use these derivatives as forces acting on the atoms. With these forces, it is possible to solve Newton's equations of motion for the system, and over a short time step (typically of the order of a femtosecond) obtain velocities and new positions. If this is done repeatedly, one can eventually produce a simulation of atomic movement. Much of the software in this field is designed towards macromolecular simulation, although they will work with small-molecule systems, too.

Example 17.4 **MOLECULAR FORCE FIELD AT WORK**

Using the Molecule Calculator web application, you can see a molecular force field at work.

1. Using a web browser, navigate to the MolCalc website (molcalc.org/).

2. The initial molecule shown is methane, but other starting molecules are available from the molecule list.

3. Change one of the hydrogen atoms to another atom type, e.g. nitrogen. Click on the N button, and then on one of the hydrogen atoms. This should produce the molecule methyl-amine (H_2N-CH_3).

4. Then click on 'Drag' to activate the drag function; using the left mouse button, drag the C–N bond away from its equilibrium distance.

5. The force field will then be used to calculate a low-energy conformation starting from the distorted molecule. If this doesn't happen automatically, click the 'Optimise' button.

However, usually a lot of preparation is required to produce suitable definitions of the molecular topology within the context of the force field to be used.

Perhaps the most commonly used software systems in this context are Amber, Gromacs and NAMD. The molecule to be simulated must first be minimised, as any large mechanics terms will produce correspondingly high forces and velocities in the system, with distortion and poor geometry propagating out from the problem area throughout the molecule. Although simulations were often done just with the molecule of interest (= *in vacuo*), the simulations can be carried out with water present, so that the results have some relevance to conformations in biological systems, with appropriate counter ions introduced to produce a net charge of zero. Typically, a very large number of water molecules are introduced to a regularly shaped box around the molecule. A technique known as **periodic boundary conditions** is then used so that any solvent molecule (and indeed any atom) leaving the box is reintroduced to the simulation at the opposing face. If this technique was not used, then water would diffuse away from the protein during the simulation. In order to screen the protein from interactions with itself, the water barrier has to be quite large, and this introduces a large number of extra atoms to be simulated. Given the small time step that must be used, the scale of the calculation becomes very large indeed. Techniques are used to reduce the requirement to explicitly calculate the energy and derivative between each pair of atoms, originally using a cut-off function, but more commonly now using the **particle mesh Ewald approximation**, where short-range interactions are summed directly, and the longer-range interactions replaced by a Fourier space summation. Even with such techniques, a molecular dynamics run can be quite intensive, with even a small protein requiring substantial processor time. Most programmes are now adapted to use the multiple processor cores of modern computer systems, and many are able to use the massively parallel computing architecture found in modern graphics display cards, so that a relatively reasonably priced desktop computer can perform useful simulations in days rather than months.

The primary output from a dynamics calculation is the trajectory, which gives the atomic positions of all the atoms in the system. There is usually little interest in the many solvent molecules simulated beyond the first or second hydration shell of the molecule. After a trajectory has been obtained, it is thus required to devise some useful parameters to be analysed for the given system. Many analyses can be performed on the trajectory to see if it contains useful information; the most basic and frequently applied is to assess the usefulness of the calculation by evaluating the root mean square deviation from the initial atomic configuration. This is a measure of how well the atoms resemble the starting conformation, and the expected behaviour is that the simulation will diverge rapidly from the starting conformation, but then stabilise over time. If the system has stabilised, then other measurements, for example formation of intermolecular hydrogen bonds or correlations between distant parts of the molecule, can then be used to form testable hypotheses about the system. Again, it needs to be stressed that the trajectory represents a model of the molecular movements over time, and if one of the initial assumptions is incorrect (e.g. poor choice of force field or incorrect initial molecular model) then that model may be rather meaningless. The utility of the model lies in its ability to either suggest a hypothesis that can be tested by laboratory experiments, or to explain a set of observations that have already been made.

17.4.5 Related Methods

Simulated Annealing

As discussed in the previous section, minimisation of an energy function using a force field can only produce a minimum-energy conformer downhill from the starting conformer; it is not clear whether the attained local minimum represents a globally low-energy conformation. One way to search for low-energy conformers is with a molecular dynamics calculation. If the energy of the system is increased by adding a small random increase to the velocity of each atom over time, this simulates heating of the system under study. With sufficient energy added, one can possibly overcome any conformational energy barriers in the system. By sampling the trajectory at this increased temperature, and then cooling it by slowly removing energy, an annealing process is simulated that mimics processes used to remove stresses in materials. This approach is known as simulated annealing and can be used to sample all the useful conformers of a molecule, eventually producing the conformer at the global minimum. Although this method samples conformational space and is more likely to find a global minimum, one can never be sure that it has indeed been found.

Monte Carlo Methods

Monte Carlo methods also aim at finding the global minimum, albeit with a more stochastic approach. Like in the other methods introduced above, an initial model is processed through many states, but there is no clear ordered relationship between one state and the next. In molecular Monte Carlo simulations, the most common procedure is the Metropolis method, and the output is not a trajectory across time, but rather an ensemble of states that collectively can provide a statistical model of the protein. The approach still involves a molecular model with its potential energy calculated from a force field. Also, when simulating a model in a biological context, there is a solvent box and periodic boundary condition, as described in Section 17.4.4. A minimised molecular model is used as starting state and small movements are then applied randomly to atoms, yielding a new conformer. If the calculated potential energy of this new conformation is lower than that of the starting state, it is kept and becomes the new current state. In order to increase the efficiency of this approach in the quest to find the global minimum, certain conformers with higher energy than the current state are accepted. For this purpose, if the energy of a new state is higher than that of the current state, the Boltzmann factor:

$$F = e^{-\frac{|E_1 - E_2|}{k \times T}}$$ (Eq 17.7)

is calculated for the two states. Then, a random number R between 0 and 1 is chosen and if $F > R$, the new state is accepted as the current one. In the schematic energy diagram shown in Figure 17.16, this allows sampling in an uphill direction. If this was done for a very large number of trials, then a population of different states is obtained where the number of conformers in a particular state reflects the probability of finding the system in that state.

Protein Model Generation

A common problem encountered in many studies is the absence of an experimental protein structure. The severity of this problem varies. At its simplest, the structure of a very close relative of the protein has been determined, and a few simple point mutations can be introduced with a molecular graphics tool and the new model minimised. A more challenging case is constituted by a situation where the structure of the general fold is available, but the target protein possesses multiple mutations, insertions and deletions. In this case, a comparative protein modelling approach can be used in order to obtain a structural model of the target protein (often called a homology model). Finally, there is the worst case where there is very little homologous structural information available in the databases of experimental structures, and one needs to fall back to *ab initio* approaches. The central paradigm of protein structure prediction is that for any given naturally occurring sequence there exists a stable fold, and the expressed linear amino-acid chain will spontaneously fold correctly. However, research in the last 20 years has shown that intrinsically disordered proteins are far more common than initially thought. But if a target sequence falls outside that group, one expects that it would be relatively simple to predict the structure; at least the gross fold of an unknown sequence or even exact side-chain conformations.

The most successful class of programs in this area are the comparative modelling programs. In this method, 3D structural information from the Protein Data Bank (PDB) is used to provide the bulk of the information, by locating structurally conserved regions (SCRs) of sequence from donor structures, filling in the short loop sections between these SCRs and then finally engaging in some form of side-chain conformer optimisation. There are several fully automated web-based services that use this approach, as well as other systems, where more input is required from the user. It has been observed that the final error of the model is largely a combination of the errors present in the donor structures, combined with the errors in initial sequence alignment. Despite being a matter of debate, more accurate models are arguably produced by performing the best initial sequence selection and alignment possible, and subsequently use of one of the less-automated software products.

Various techniques exist for using the information from the SCRs, ranging from the bulk copy of coordinates to the more sophisticated generation of spatial restraints from the input model. The spatial restraint technique readily allows the incorporation of more data, from either the donor models or from molecular-mechanics-based data. A typical model building sequence would then involve a sequence search for relevant sections of good-quality donor structure within the Protein Data Bank, with multiple donor structures for each section of sequence. Whereas it is possible to proceed with this initial alignment, better results tend to be achieved by using multiple sequence alignments that consider the target and donor sequences, since alignments based on multiple sequences are likely to increase the overall quality of the alignment. The sequence alignment and the relevant parts of the donor structures can then be passed to the modelling software and multiple conformations of the target structure are computed. A final important part of the process is model validation. The modelling software typically return some indication of how well the produced structures fit its own internal metrics, but in addition, some of the more general protein structure evaluation tools should be used. Protein structure validation software evaluate geometric properties of the protein model and show how the model fits within the statistical ranges of known experimental structures. Typical checks include:

- Assessments of the model's bond lengths and angles which are required to be in acceptable ranges
- Compliance of the backbone torsion angles (ϕ and φ; see Section 2.2.1) with allowed regions of a Ramachandran plot
- Evaluation of whether the ratio of polar to apolar contacts for a residue lies within the range observed in experimental structures.

A second technique for protein structure modelling is *ab initio* model generation. In the most successful of these techniques (the approach used by the Rosetta software) the term '*ab initio*' is a slight misnomer, as the initial starting points for the models are derived from observed distributions of conformers in short runs of sequence. In this approach, the protein conformation is optimised from an initial extended conformation, with an initial fragment of the extended chain chosen and substituted by one of the conformers available in the starting library. The move from extended to folded conformation is performed, not by direct changes in the Cartesian coordinates, but by altering the torsion angles in the chain to match those of the insert. Monte Carlo methods are used to optimise the insertions, in an attempt to avoid getting trapped in a local minimum associated with that part of the folding process. This small-scale interaction selection from a database source is combined with a larger molecular-mechanics-based force field with specialised terms that can be used or ignored as the model begins to approach a folded state. This reflects the requirement to move the structure initially to a roughly folded state, which is then optimised, using more routine potential-energy terms to achieve a reasonable model of a folded protein. Again, the final model should be checked for validity using external tools, and if multiple models are produced, then the one with the best validity scores can be selected. *Ab initio* folding programs have been effective with small proteins, and these techniques are finding further uses in the solution of experimental structures, where the observed data is insufficient to solve the structure.

17.5 DATABASES

Databases are an area where classic computational techniques have made an enormous impact on biochemical research. Very early in the history of their development, both the structure and sequence elucidation areas saw the advantages of sharing data as widely as possible. Initially, this was with the development of standardised data formats that could be collated and periodically made available to others in the field, and later with the development of bespoke database and searching tools. Databases also became very important to the pharmaceutical industry as a way of documenting their extensive compound collections. With the development of the internet and the concomitant improvement of web browsers and servers, it has become clear that databases coupled to web sites are one of the more effective ways of sharing data and techniques with other research groups.

17.5.1 Database Types

Any attempt to classify the databases currently in use would be fairly arbitrary and would fail to include at least one important database. However, one division that is obvious is between databases that derive their holdings from other data sources, and those that act as primary repositories for experimental data. Four major databases of interest to the biological modelling community, and which perform this repository function, are the European Nucleotide Archive (ENA) for gene sequences, UniProt for protein sequence data (see Sections 16.2, 18.3.3), the Protein Data Bank for macromolecular data and the Cambridge Crystallographic Data Centre (CCDC) for small-molecule crystal structures. Beyond these data repositories, there are large numbers of meta databases that provide useful data and tools such as Pfam, CATH and SCOP (see Sections 2.2.3, 16.2). Finally, for chemoinformatics applications where one needs access to small-molecule data and properties beyond the experimentally derived, there are PubChem and ZINC (see Section 24.3.2).

The template used to develop many early commercial websites – the Linux, Apache, MySQL and php (or perl) stack of programs (known as the LAMP stack) – maps quite well to the format of a chemoinformatics project, with the database being the location for results storage and web pages allowing less technically adept group members to query and manipulate the data. Website development is a very rapidly changing area, with a constant flow of innovative techniques and add-ins, but the data–view–action model of website design is still quite useful within a chemoinformatics context.

17.5.2 Curation and the Update Problem

As mentioned earlier, often it becomes clear during the lifetime of a chemoinformatics or computational biology project that it could benefit from the creation of a database and website that allows others to use the amassed information (or developed software), and, importantly, act as a focal point for a publication dealing with that project. The creation of such a website, while not trivial, is relatively simple. However, in many projects this ignores two factors, the initial curation and the updating of the database. If the data source is a limited set of experimental results, then these problems can be quite small. However, if the information is derived from a larger database, or even involves data mining from publications, then the curation and update problems immediately become a major issue. The **curation problem** is due to the screening process for depositing data and published papers not being perfect. There are many examples of poor-quality data that have made their way into the public domain. This is not even a result of professional misconduct, but rather genuine errors in the data or results that are not reproducible. It may also be that the source database being used is correct, but its content is skewed in a particular direction. For example, any attempt to naively derive statistical data from the Protein Data Bank, will lead to a result skewed towards lysozyme over any random structure, since lysozyme occurs very frequently in the database. This particular version of the problem is well understood and there are well-developed procedures to take an unbiased sample from the database; however, these results are still biased towards soluble globular proteins and away from membrane bounds just by the nature of the experimental techniques employed. In order for a database to be as correct and unbiased as possible, one must curate the input data to remove possibly incorrect data and ensure that the entire set fairly represents the range of data possible. For a database with tens of input sources this can be done manually, but if one is dealing with thousands then the data acquisition and processing methods must deal with this. This is linked with the second problem: the **database update**. What happens when the person looking after the database has to move on? If only that person really understands the process of adding new data, the database itself has become effectively fixed and incapable of update. Thus, when planning a database, ongoing addition and curation of data should be built into the very design so that the resource is robust enough to be used by a non-specialist.

17.6 SUGGESTIONS FOR FURTHER READING

17.6.1 Experimental Protocols

Basak S.C., Niemi G.J. and Veith G.D. (1990) A graph-theoretic approach to predicting molecular properties. *Mathematical and Computer Modelling* 31, 511–516.

Dalby A., Nourse J.G., Hounshell W.D., *et al.* (1992) Description of several chemical structure file formats used by computer programs developed at Molecular Design Limited. *Journal of Chemical Information and Computer Sciences* 32, 244–255.

Eswar N., Eramian D., Webb B., Shen M.Y. and Sali A. (2008) *Protein structure modeling with MODELLER*. In *Structural Proteomics: High-Throughput Methods*, Kobe B., Guss M., Huber T., Eds., Humana Press, New York, USA, pp. 145–159.

Gasteiger J. and Marsili M. (1980) Iterative partial equalization of orbital electro-negativity: a rapid access to atomic charges. *Tetrahedron* **36**, 3219–3228.

Heller S.R., McNaught A., Stein S., Tchekhovskoi D. and Pletnev I.V. (2013) InChI – the worldwide chemical structure identifier standard. *Journal of Cheminformatics* **24**, 7.

Leo A. and Weininger D. (1984) *CLOGP Version 3.2 User Reference Manual*, Medicinal Chemistry Project, Pomona College, Claremont, CA, USA.

Mulliken R.S. (1955) Electronic population analysis on LCAO-MO molecular wave functions. I. *The Journal of Chemical Physics* **23**, 1833–1840.

Weininger D. (1988) SMILES, a chemical language and information system. 1. Introduction to methodology and encoding rules. *Journal of Chemical Information and Computer Sciences* **28**, 31–36.

17.6.2 General Texts

Morgan H.L. (1965) The generation of a unique machine description for chemical structures: a technique developed at chemical abstracts service. *Journal of Chemical Documentation* **5**, 107–113.

Todeschini R. and Consonni V. (2009) *Molecular descriptors for chemoinformatics*. In *Methods and Principles in Medicinal Chemistry*, Book 41, Wiley-VCH, Weinheim, Germany.

Willett P. (2011) *Similarity searching using 2D structural fingerprints*. In *Chemoinformatics and Computational Chemical Biology*, Bajorath J., Ed., Humana Press, New York, USA, pp. 133–158.

17.6.3 Review Articles

Berman H.M. (2007) The protein data bank: a historical perspective. *Acta Crystallographica Section A: Foundations of Crystallography* **64**, 88–95.

Bonneau R. and Baker D. (2001) Ab initio protein structure prediction: progress and prospects. *Annual Review of Biophysics and Biomolecular Structure* **30**, 173–189.

Christen M., Van Gunsteren W.F. (2008) On searching in, sampling of, and dynamically moving through conformational space of biomolecular systems: a review. *Journal of Computational Chemistry* **29**, 157–166.

Gasteiger J. (2014) Solved and unsolved problems of chemoinformatics. *Molecular Informatics* **33**, 454–457.

Geppert H., Vogt M. and Bajorath J. (2010) Current trends in ligand-based virtual screening: molecular representations, data mining methods, new application areas, and performance evaluation. *Journal of Chemical Information and Modeling* **50**, 205–216.

Karplus M., McCammon J.A. (2002) Molecular dynamics simulations of biomolecules. *Nature Structural and Molecular Biology* **9**, 646–952.

Raha K., Peters M.B., Wang B., *et al.* (2007) The role of quantum mechanics in structure-based drug design. *Drug Discovery Today* **12**, 725–731.

Van Gunsteren W.F., Luque F.J., Timms D. and Torda A.E. (1994) Molecular mechanics in biology: from structure to function, taking account of solvation. *Annual Review of Biophysics and Biomolecular Structure* 23, 847–863.

17.6.4 Websites

Amber Molecular Dynamics Package
ambermd.org/ (accessed May 2017)

Avogadro (an advanced molecule editing and visualisation software)
avogadro.cc/ (accessed May 2017)

BIOVIA Draw (molecule editor)
accelrys.com/products/collaborative-science/biovia-draw/ (accessed May 2017)

BKchem (a Python molecule editor)
bkchem.zirael.org/ (accessed May 2017)

Cambridge Crystallographic Data Centre (CCDC)
www.ccdc.cam.ac.uk/ (accessed May 2017)

cDraw (a Java chemical drawing software)
www.structuralchemistry.org/pcsb/cdraw.php (accessed December 2017)

ChemDraw (molecule editor)
www.cambridgesoft.com/software/overview.aspx (accessed May 2017)

Chemistry Development Kit (Java libraries for chemoinformatics)
sourceforge.net/projects/cdk/ (accessed May 2017)

Gromacs (molecular dynamics package)
www.gromacs.org/ (accessed May 2017)

International Chemical Identifier (InChI)
www.iupac.org/inchi/ (accessed May 2017)

JChemPaint (the molecule editor of the Java Chemistry Development Kit CDK)
jchempaint.github.io/ (accessed May 2017)

JSME Molecule Editor (a Javascript-based molecule editor)
www.peter-ertl.com/jsme/ (accessed May 2017)

MarvinSketch (a Java-based molecule editor)
www.chemaxon.com/products/marvin/marvinsketch/ (accessed May 2017)

Modeller (comparative modelling software)
salilab.org/modeller/ (accessed May 2017)

Molecule Calculator (MolCalc)
molcalc.org/ (accessed May 2017)

NAMD Scalable Molecular Dynamics

www.ks.uiuc.edu/Research/namd/ (accessed May 2017)

Online SMILES Translator and Structure File Generator
cactus.nci.nih.gov/translate/ (accessed May 2017)

OpenBabel (format conversion)
www.openbabel.org (accessed May 2017)

Rosetta (algorithms for computational modelling and analysis of protein structures)
www.rosettacommons.org/software/ (accessed May 2017)

Simplified Molecular Input Line Entry System (SMILES)
www.daylight.com/smiles/ (accessed May 2017)

18 | The Python Programming Language

TIM J. STEVENS

18.1 INTRODUCTION

A biologist will often turn to computer programming in situations where the amount or the complexity of data is too much to be sensibly handled by spreadsheets, and where no other, more specialised, software exists. Often only a relatively simple program needs to be written to get something useful from biological data, which would otherwise not be available. For biologists, the task of writing a computer program can sometimes seem like a significant barrier, but once the basic programming skills are learned then many possibilities are enabled. This chapter offers an introduction to the Python language and gives some concrete examples of programs that may be useful in molecular biology. However, there is not space to cover all aspects of the language and many of its finer details. For this we recommend further reading, but nonetheless hope this chapter serves to illustrate the basics and to show what is possible.

Python is one of the most popular programming languages and is becoming an increasingly attractive option for the biologist. It is a high-level, general-purpose language that is well supported and relatively easy to learn; indeed it is now taught in mainstream UK schools. Also, it has a large number of external modules, including many relating to mathematics, science and biology. Python is easy to install, if it isn't already installed as standard, and runs on almost all kinds of computer system. In this chapter we will show some of the features and capabilities of Python 3 and then apply this to several example programs to illustrate the sort of things that can be achieved for molecular biology. Python version 2, should you need to work with that instead, is very, very similar and most programs are easily transferred (e.g. using the conversion program 2to3 supplied with Python), although the two versions are not 100% compatible.

Even if you don't intend to use Python in the long-run or for all programming work, it nonetheless serves as a good starting point to learn some of the major principles of many modern computing languages. Python is a high-level language like Perl, Matlab and R, which is directly interpreted when a program is run; there is no distinct compilation step. Though Python programs are not as quick to run as equivalents written in strongly typed, compiled languages (C, C++, FORTRAN, Java, etc.) there are fast modules to perform common numerical operations, such as NumPy, which allows array-based operations similar to Mathematica and Matlab. Also, it is possible to connect a Python program to a module written in a fast compiled language. An easy way to do this is to use Numba, a Python module that allows you to mix Python with compiled code. Scripts discussed in this chapter are also available online at www .cambridge.org/hofmann.

18.2 GETTING STARTED

The main Python language releases are available at www.python.org though it is typically already installed on Macintosh OSX and Linux computers. For Windows users, installation requires a file download and a simple double click. Another, and perhaps better, way to obtain the Python language is to install a package like Anaconda (www .anaconda.com), which will not only provide Python, but also a number of useful external mathematical and scientific modules at the same time, and which work together properly. Otherwise such modules have to be downloaded and installed separately, though when using the Anaconda package, missing modules may be easily installed using the 'conda' command. For this chapter we will assume that the following external modules are installed: SciPy (which also provides NumPy and Matplotlib), Pandas, PySam and Ruffus.

Python code is usually saved in a text file that is later run by the Python interpreter. Lines of Python code also can be entered manually at the >>> command prompt that appears when you type the `python` command, and this can be handy for quick and small jobs. However, for the examples in this chapter we will assume a file-based approach is taken. Although it is up to the user to decide what the names of Python code files should be, they typically end with `.py`. Often the program will be run by issuing a command of the form `python my_code.py`, though if you are using a graphical environment look for `Run` or a similar option in the menu.

18.2.1 Basic Code Structure

A very simple Python program is illustrated below. Note that all the commands start at the beginning of their line, i.e., they have no spaces or indentations before them. Indentation is a very important part of the Python language syntax and it is assumed to be consistent in each block of code; that's how blocks are defined, as illustrated later. Spaces that are internal to each line, such as those next to the equals sign here are much less critical; indeed you could have more or no spaces here. The program consists of the six lines:

```
# Script 18.1
mass = 3.4
volume = 1.8
density = mass/volume        # Division
```

```
density = round(density, 3)   # Round to three decimal places
print(density)
```

What this tells the Python interpreter to do is to assign a value of 3.4 to the label called `mass` and assign 1.8 to `volume`. The `density` is a new value that is calculated from the other values: one divided by the other, as indicated by the '/' symbol. Note that there are comments, from the # symbol onwards, which is only for a human to read and will not be interpreted as Python. Next the density is re-assigned to a new value which is its old value rounded to three decimal places: the old value and the number of decimal places are stated in the parenthesis of `round()`. Finally we use the `print()` function, which has the effect of displaying the underlying numeric value of the density to screen. Each labelled item, which is more correctly called a **variable**, is a way of attaching a name to a piece of data, and as the term suggests, the actual value that is being referred to can change while the remainder of the program remains the same, i.e. we could calculate the density in the same way for different masses and volumes. Note that we are free to choose any name for our variables within certain rules: names can contain numbers, letters and underscore ('_') only but may not start with a number. The use of `round()` and `print()` are examples of **functions**: named tasks performing a specific set of operations on input data, specified inside the parenthesis and which often generate some output. Many functions are inbuilt into Python (see python.org for full documentation), but as we illustrate below it is possible to write new functions inside a Python program.

The '/' symbol used for division is an example of one of the inbuilt mathematical symbols (operators) in Python. The full complement of these is given in Table 18.1.

Table 18.1 **Mathematical operators in programming languages**

Operation	Description	Example
x + y	Addition	revenue = profit + expenses
x - y	Subtraction	income = profit - taxes
x * y	Multiplication	area = volume * height
x / y	Division	mean = (x + y + z) / 3.0
x // y	Floor division: divide and round down.	a = 13.0 // 5.0 # a is 2.0
x % y	Modulus: remainder after division.	a = 13 % 5 # a is 3
-x	Negate the value of x.	a = -5 * 3 # a is -15
x ** y	Raise to the power, i.e. x^y.	a = 2 ** 3 # a is 8

Note that there are also related operators that modify the value of a variable in-place. For example x `*=` 3 means x is assigned to a new value which is triple its old value: equivalent to x = x * 3. Similarly x += 3 means add 3 to x and x -= 3 means subtract 3 from x.

18.2.2 Simple Data Types

There are several basic data types in Python (see Table 18.2). The simplest of these include numbers, as already illustrated, values for true or false and a special **None** value.

Table 18.2 **Basic data types in Python**

Data type	Description	Example	Converter
Integer	Whole numbers	`x = 128`	`int()`
Floating point	Numbers with decimal points and/ or power of ten exponent (scientific notation)	`x = 5e-3` `y = 12.00`	`float()`
Complex	Numbers with real and imaginary parts	`x = 1.0-2j;` `y = 1j`	`complex()`
Boolean	Truth values `True` and `False`	`x = True;` `y = False`	`bool()`
None	A special value `None` for nothingness/ undefined	`x = None`	

Integer numbers are specified without any decimal point or exponent and represent an exact whole number. Floating point numbers represent a number of significant digits and an exponent, though this is often implicit. The scientific notation, with a power of ten specified after an `e` in the number is optional, and a given number can often be written in many different ways, e.g. `0.001`, `1e-3` and `0.1e-2` are the same value. Floating point numbers have limited precision, i.e. they cannot represent all fractions precisely, as one might expect with ⅓ etc. However, some floating-point calculations create an error in the least significant digits (caused by the underlying system being binary), which can occasionally cause problems for the programmer if not expected:

```
print(3.0 * 7.1)   # Gives 21.299999999999997 not 21.3 !
```

Calculations with integers often give integer results, though division of two integers gives a floating point result (this was different in Python 2). Calculations involving any floating point number tend to give a floating point result.

```
# Script 18.2
x = 2                      # Integer
y = 5                      # Integer
z = 3.0                    # Floating point
print(x * y)               # 10 - Integer
print(x / y)               # 0.4 - Floating point due to division
print(x + z)               # 5.0 - Floating point
t = type(x * y - z)        # Get a value's data type
print(t)                   # float
```

Each entity in Python is an object with a given type; it is a rich description of the data rather than just the plain underlying value. As shown above, the data type can be inspected directly using the `type()` function. The rich, object-oriented nature of Python is revealed by the dot syntax, which accesses values and functions that belong to an item. In the following example, the documentation text is accessed for an integer object using .__doc__, where the double underscores are a hint this is a special inbuilt value:

```
# Script 18.3
x = 1                   # An arbitrary integer
print(x.__doc__)  # Print Python documentation for this object
```

Objects can be converted between one data type and another with the relevant conversion function, though naturally this can cause rounding when creating an integer from a float, e.g. int(7.3) gives integer 7. The Boolean data type can be created directly by using the special keywords True and False or by using the bool() function, which treats zeros, empty containers (see below) and the None object as false, and everything else as true.

```
# Script 18.4
x = True
print(x)
print(bool(0.0))   # False
print(bool(7))     # True
```

Many True and False values in Python arise as a result of a comparison operation. For example, when testing if one number is smaller than another:

```
# Script 18.5
x = 2
print(x > 5)     # False
y = x < 3
print(y)         # True
```

The Boolean operations and, or and not can be used to combine multiple comparisons. Note, as shown below, that testing equality uses double equals signs ==, which is easily confused with the single sign used for assignment, and that != is the test for not equals.

```
# Script 18.6
x = 1
y = -1
x > y and y > 1  # False - second comparison is False
x != 2 or y == 1 # True - first comparison is True as x is 1
not (y > x)      # True - comparison is False; not False is True
```

The complement of general comparison operators in Python is specified in the Table 18.3. In Python 3, inequality comparisons can only be made between values of a comparable type (Python 2 is different in this regard; you can compare anything, whether it is meaningful or not).

Table 18.3 **Comparison operations in Python**

Operator	Description	Example	
==	Tests whether two values are equal	`x = 3`	
		`x == 3.0`	`# True`
!=	Tests whether two values are not equal	`x = 3`	
		`x != 3`	`# False`
>	Tests whether the first value is greater than the second	`x = 2**10`	`# 2 to the power 10`
		`x > 1024`	`# False equal not more`
<	Tests whether the first value is smaller than the second	`x = 2**10`	
		`x < 1025`	`# True`
>=	Tests whether the first value is greater than or equal to the second	`x = 2**10`	
		`x >= 1024`	`# True`
<=	Tests whether the first value is less than or equal to the second	`x = 2**10`	
		`x <= 512`	`# False`
is	Tests whether two values represent the same Python object	`3 == 3.0`	`# True`
		`3 is 3.0`	`# False`
is not	Tests whether two values represent different Python objects	`3 is not 3.0`	`# True`

The keyword operators `is` and `is not` are of note because they compare whether two items are the same Python object, not whether their values are the same. As we illustrate in Table 18.3, an integer and floating point number can be equal in value but they are two different objects, with different data types. The `is` comparison is often used with `None`, to detect if something is defined, which is distinct from whether it has a zero value.

18.2.3 Text Strings

Text in Python is represented by the String data type and can be specified in code using single or double quotes, which distinguishes the characters from unquoted parts of Python.

```
# Script 18.7
x = 'Hello'
y = "world"
print(x,y)   # Print both values to screen
```

There is also a triple-quote syntax which allows text strings to flow over multiple lines:

```
# Script 18.8
x = "'This text flows from one line
to the next line inside triple quotes'"
print(x)
```

Text strings can be created from any value with `str()`, like when using `print()`, and where appropriate they may be converted to other data types:

```
x = str(1.23)                 # x is text '1.23'
y = int('     007     ')   # y is the integer 7
```

Some symbolic operators work with strings, though naturally in a textual way, even if the characters happen to represent digits:

```
x = '1' + '99'     # '199' - concatenate characters
y = 'abc' * 3      # 'abcabcabc'
```

Many operations are performed on strings using inbuilt functions that belong to the Python object, as accessed using the dot notation, noting that the original data remains unaffected:

```
x = 'abc'          # x is a string object
y = x.upper()      # y is 'ABC' x remains 'abc'
```

Strings can be considered to be arrays of characters, and accordingly have a length. Character or substring elements can be accessed using square brackets, by specifying an appropriate index (starting at zero) or a range:

```
# Script 18.9
x = 'abcde'
print(len(x))     # 5
print(x[0])       # First character 'a' at index 0
print(x[-1])      # Last character 'e'
print(x[1:3])     # 'bcd' range; from index 1 to 3
```

Some elements of a textual data are not printable as normal character glyphs, such as new lines or tab stops. These are represented in Python strings using special escape codes that start with '\'. For example '\n' means new line:

```
print('Hello\nworld') # Starts a new line before 'world'
```

The full list of character escape codes is listed in Table 18.4. It is notable that Python has a concept of raw strings where these codes are not used: raw strings have an 'r' before their quotes, e.g. r'Hello\nworld' has actual '\' and 'n' characters in the middle, not a new line.

There are many inbuilt functions (methods) that are associated with strings, as listed in the full Python documentation, and many of these will be used in later examples.

```
x = ' Hello\n'                # Remove edge whitespace characters
y = x.strip()                 # 'Hello'
                              # List of separate words
z = x.split('Val Gly Lys') # ['Val','Gly','Lys']
```

One of the most important of these functions is format(), which allows variable values to be inserted inside a text string. The values are inserted into the string at the {} positions, replacing the bracket section. The brackets can optionally have a number, specifying which item to insert and also a format specification after ':'

Table 18.4 **Character escape codes in Python**

Code	Description	Example
\\	A backslash character, which needs to be forced when the following character would otherwise form an escape code.	`text = '\\title'` Text is '\title' and does not have a tab code (\t) inside.
\'	A single quote, which may need to be escaped in situations where it should not be considered as the start or end of a string.	`text = 'Don\'t do that!'` Note: Not required when a string is defined with double quotes.
\"	A double quote, which may need to be escaped in situations where it should not be considered as the start or end of a string.	`text = "Shout \"Help!\"loudly."` Note: Not required when a string is defined with single quotes.
\n	A newline (linefeed) control character. Used to separate lines of text in UNIX and Linux based computers.	`text = 'Line A\nLine B\n'` Text value is split into two lines on Linux and UNIX machines.
\r	A carriage return control character. Used in combination with \n on Windows-based systems to separate lines of text.	`text = "Line A\r\nLine B\r\n"` Text is split into two lines on Windows machines.
\t	A tab character, providing indentation with white space to pre-set stop points.	`text = 'Col 1\tCol 2\tCol 3\n'` Tabs indent to form three columns.
\u····	Specifies a Unicode character using a 16-bit hexadecimal value.	`text = '\u03b1-helix'` Creates 'α-helix', e.g. for graphical displays.
\x··	Specifies a character using a hexadecimal value.	`text = '\x48\x65\x6C\x6C\x6f'` Text is hexadecimal code for 'Hello'.

which specifies how to represent the item, e.g. to specify the number of decimal places for a floating point or scientific notation.

```
# Script 18.10
name = 'Mary'
x = 0.12
y = 34121.0
a = 'Hello {}'.format(name)          # name replaces brackets
print(a)                             # 'Hello Mary'
b = 'X {:.5 f} Y {:.2e}'.format(x, y) # 5 dp float 2 dp sci
print(b)                             # 'X 0.12000 Y 3.41e4'
c = '{:3d},{:3d}'.format(12,7)       # pad digits to 3 chars
print(c)                             # '  12,  7'
```

18.2.4 Collections

Collections are Python objects that can contain other Python objects; they are containers for organising data. For example, a list of various objects, specified with square parenthesis:

```
x = True
y = ['GCAT', 4.9, x]        # a list of 3 items, incl. the value of x
z = []                      # z is an empty list
```

Containers can contain the simple data types already discussed and also other containers (within certain rules) and they can also be empty. The basic inbuilt types of container in Python are summarised in Table 18.5.

Lists are specified with square brackets and can be accessed by numeric indices and ranges, like strings, with the first item being at index 0. The value of a specific index or range can also be set.

```
# Script 18.11
x = ['a', 'b', 'c', 'd']
print(x[2])             # 'c' - the character at index 2
print(x[1:])            # ['b', 'c', 'd'] - from index 1 to the end
x[0] = 'z'              # Character at index 0 is set to 'z'
del x[2]                # Deletes item at index 2
print(x)                # ['z', 'b', 'd']
y = [[4,7], [9,6]]      # A list containing two other lists
print(y[1])             # [9,6] - the sub-list at y index 1
print(y[1][0])          # 9 - item 0 from sub-list 1
```

Collections can be made from other collections using the appropriate creation function, e.g. `list()` to make list and `set()` to make a set. More generally however,

Table 18.5 **Container types in Python**

Type	Description	Example	Converter
List	A modifiable, ordered list of items	`x = ['cat', 'dog', 'pig']` # a list of three strings	`list()`
Tuple	An unmodifiable, ordered list of items	`x = (0.742, 0.159)` # a tuple of two floats	`tuple()`
Set	A modifiable, unordered collection of unique items	`x = {9, 1, 7, 2, 5}` # a set of five numbers	`set()`
Frozenset	An unmodifiable, unordered collection of items	`x = frozenset([1,2,3])`	`frozenset()`
Dictionary	A modifiable, unordered collection of items accessed by unique keys	`x = {'A':1, 'C':3, 'E':5}` # a dictionary with three # key:value pairs	`dict()`

they can be created from Python objects that are iterable: things that can generate a sequence of items. For example, as shown below, `range()` generates a sequence of integer numbers up to a specified limit (a start number and step could also be specified), which is then used to make a list.

```
# Script 18.12
x = list('PQRST')      # Create list from text string
print(x)               # ['P','Q','R','S','T'] - list of characters
y = list(range(10))    # A list of the range from 0 up to <10
print(y)               # [0,1,2,3,4,5,6,7,8,9]
```

Collections can be unpacked into individual variables if the number of items assigned matches the collection size:

```
# Script 18.13
x = [4, 1, 0]                  # Three items
a, b, c = x                    # a is 4, b is 1, c is 0
x = ['a', 'b', 'c', 'd']
print(len(x))                  # 4 - number of items in list
print('c' in x)                # True - string 'c' is in the list
print(2 in x)                  # False - number 2 is not in list
```

As with all the main Python collections, lists have a size, accessible with `len()`, and membership tests are performed using the `in` operator.

Tuples, specified with round brackets, are similar to lists and have indexed items etc. However, they cannot be modified: their contents are defined when they are created. This is useful because tuples can be used as keys in a dictionary (see below) while lists cannot.

```
# Script 18.14
x = (9, 3, 1, 0)       # Create a tuple
y = (2,)               # Tuple with one item (needs trailing comma)
z = ()                 # Empty tuple
print(x[-1])           # Print the last item - 0
x[0] = 6               # Does not work! Tuples cannot be changed!
w = list(x)            # Create equivalent list from a tuple
w[0] = 6               # List copy can be changed
```

Sets are specified with curly brackets. They do not contain repeated values and the items do not have an order, and so cannot be accessed by index. They have helpful associated set operations (intersection, union, difference, disjoint test, subset test etc.), and allow for membership tests.

```
# Script 18.15
x = {1,2,3,4,3,2,1}    # x is {1,2,3,4} - duplicates ignored
y = set([3,4,5,6])     # y created from a list
print(len(y))          # 4
```

```
print(1 in y)          # False; 1 is not in y
print(x & y)           # {3,4} - intersection, items in both
print(x | y)           # {1,2,3,4,5,6} - union, items in either
```

Dictionaries are lookup tables containing `key:value` pairs, they are also specified with curly brackets, but can be distinguished from sets by ':' joining the pairs. The values in a dictionary are accessed by key, not by index, and each key is used only once. Dictionary keys are typically strings or numbers, but more generally they are restricted to items that are not modifiable (so you cannot use lists or other dictionaries as keys), but the values being referred to can be any data type. The following example is a dictionary where the keys are strings (DNA base codes) and the values are numbers (nucleotide masses).

```
# Script 18.16
d = {"G":329.21, "C":289.18, "A":313.21, "T":314.19}
print(d['A'])         # 313.21 - value associated with 'A'
print(len(d))         # 4 - number of key:value pairs
print(d.keys())       # Just keys 'G', 'A', 'T', 'C'
print(d.values())     # Just values 329.21, 313.21, 314.19, 289.18
```

If a key is already present in the dictionary then a simple assignment of the form `dict[key] = value` is used to change the value associated with that key. However, if a key was not already present this kind of assignment will add a new `key:value` pair. Existing keys cannot be changed directly, but it is possible to remove a `key:value` pair using `del` and add the same value back again with a different key.

```
# Script 18.17
d['T'] = 304.19    # Change the value of an existing item
d['U'] = 291.08    # Add a new key:value pair
print(len(d))      # 5 - dictionary has keys G,C,A,T,U
del d['U']         # Delete key U and its value from dictionary
```

The collection types have a number of inbuilt functions (methods) that are accessed with the dot syntax, some of which are listed below. Naturally the functions available to a given collection are appropriate to its type, e.g. sets do not have functions that refer to positional indices.

```
# Script 18.18
x = ['Mon', 'Tue', 'Wed']      # A list of strings
y = ['Fri', 'Sat', 'Sun']      # And another
x.append('Thu')                # Add a single new item to end
x.extend(y)                    # Add items from another collection
x.sort()                       # Sort contents alphabetically
x.remove('Sun')                # Remove an item
x.index('Sat')                 # Positional index of an item
```

```
# Script 18.19
s = {'G', 'C', 'A', 'T'}  # A set with 4 strings
t = {'N', 'R', 'Y'}
s.add('U')                     # Add a single item (if not present)
s.update(t)              # Add new items from another collection
```

18.2.5 Control Code

Lines of Python code are generally executed in sequential order, one after the other. However, there are situations where we wish to deviate from this paradigm, for example to repeat a section of code several times in a loop, or to only execute a block of code under certain conditions. Accordingly, there are a set of keyword commands including if, for, while, try and def that are used to control the execution of a subsequent code block, which is indented relative to the keyword. For example, the if statement is used to only perform operations if a particular condition is met (if the value of an expression is logically true).

```
# Script 18.20
x = 3
if x < -1 or x > 1:      # Run the next indented lines only when true
    x *= 2
    print('Value was doubled')
print('Value is:', x)    # Always executed, not in indented block
```

It should be noted that any number of spaces can be used for indentation of the controlled block, but it must be consistent. Four spaces are generally recommended, though we sometimes use two here for space reasons. The if statement can also have a number of elif clauses and a final else clause, each with their own block: elif does further conditional tests if all preceding ones failed, and else marks the block that is run if no condition is met.

```
# Script 18.21
if x > 0:
    print('Positive')
elif x < 0:              # Checked if first condition was false
    print('Negative')
else:                    # If all fails
    print('Zero')
```

When testing whether the expression after an if statement is true it is obvious what happens in situations like x < 10 or y == 5, given that these comparisons generate True or False (Boolean) objects. However, any object in Python can be tested for truth. A few specific objects are deemed to be logically false: False, None, 0, 0.0, empty string '' and empty collections, while almost everything else is deemed to be true.

```
# Script 18.22
x = 'abc'
if x:                  # Test innate truth of x
    print('true')  # This prints; a non-empty string is true
else:
    print('false')
```

Repetitive loops can be created with a **for** statement or a **while** statement. The **for** loop extracts items from a collection (or other iterable object, like a string of characters) and assigns a loop variable (x in the below example) to each value in turn, repeating execution of the indented block of code each time. Here a list of numbers is defined and x is repeatedly assigned to the value of each item in the list, which is printed out and then added to the total in each cycle:

```
# Script 18.23
total = 0
data = [1,4,9,25,36]
for x in data:        # x is first 1, then 4, then 9 etc.
    print(x)          # Current value of x in this cycle
    total += x        # Add current value of x to total
```

It is often convenient to use the enumerate() function with a **for** loop. This allows the loop to iterate over both numbers for the items (usually the positional indices) and their actual values. Below the number and value (i and x respectively) are stated as separate variables in the **for** line, using the collection unpacking syntax illustrated previously:

```
# Script 18.24
text = 'AGCAGTAGACGAACAT'      # String of characters
for i, x in enumerate(text):  # Extract index and character value
    print(i, x)                # Print index and value for each cycle
```

A **while** loop repeats a block of code while a certain condition evaluates to be true, and so it is important to make sure that the condition is eventually false (on the command-line, Ctrl+C keys can be used to stop an 'infinite' loop).

```
# Script 18.25
x = 1
while x < 1000:  # Repeat the indented block while this is true
    print(x)
    x *= 2       # Double the value
print(x)         # 1024 >= 1000 - final value stopped the loop
```

Loops of both kinds can be skipped, for the remainder of their block, using **continue** and stopped entirely with **break**. Also, control statements (with indented blocks) can appear inside one another:

```
# Script 18.26
t = 0
data = [3, -1, 2, -5, None, 9, -2]
for x in data:
    if x < 0:
        continue           # Skip the remainder of 'for' loop
    elif x is None:
        break              # Quit entirely
    t += x * x             # Otherwise do a calculation
```

In keeping with the topic of loops, there is another kind in Python that does not have an indented syntax. It is not a general-purpose loop like the ones described above, rather it is a means of constructing a collection (list, tuple, set, dictionary) where a kind of **for** loop is specified inside the collection's brackets. There is also an option to add a filter to the construction of the collection using an internal **if** statement. Considering the following example for constructing a list:

```
# Script 18.27
squares = []
for x in range(1, 10):
    squares.append(x*x) # 1, 4, 9, 16, 25, 36 …
```

This could be equivalently written as a list comprehension, effectively building a list from the inside, where the item that enters the list appears before the **for**:

```
squares = [x*x for x in range(1,10)]
```

Adding a filter is achieved using an **if** after the **for** section(s). Also, changing the bracket type changes the type of collection constructed. Curly brackets {} could be used to construct a set or a dictionary, depending on whether a **key:value** dictionary specification is made.

```
# Generate a tuple with squares of odd x
odd_sq = (x*x for x in range(1,10) if x % 2 == 1)
# Dictionary of index:character pairs
{i:x for i,x in enumerate('ABCDEF')}
```

A **try: except:** block is used to catch and deal with illegal circumstances. The code in a **try** block is run and if a problem occurs an **except** block of code may be run if a particular kind of error (a type of **Exception** object) is detected. In this way we can prevent the program from failing and sensibly handle an error. If we choose, the original error can be retriggered using **raise()**.

```
# Script 18.28
x = 1
y = 0
try:              # Run the following block and check for failure
    w = x / y
    # y was zero
```

```
    except ZeroDivisionError as err:
        # Warn, but otherwise ignore
        print('divided by zero, continuing')
    # Other, unspecified problem
    except Exception as err:
        # Trigger the error, do not continue
        raise(err)
```

The keyword **def** is used to define functions, i.e. user-created subroutines. In essence these are a specification for a named bit of code. Defining a function is distinct from running (or calling) a function, but once defined, a function can be called into action any number of times; achieved by using its name with parentheses (which convey any input data). Naturally the general idea is that functions represent reusable code, performing the same operation in many different places, albeit for different input data. The following function called my_calc, takes two input values (two arguments) that are labelled as x and y inside the function and which do not have any specific values inside the definition. The definition involves specifying a calculation, the result of which, z is passed back from the function at the **return** statement.

```
def my_calc(x, y):
    z = x * x - y * y
    return z
```

Once the definition exists, the name my_calc can be used on two input values, which fill the x and y slots in the function, generating a value for z, which is then what is output from the function and, in this case, printed:

```
print(my_calc(4,5)) # -9
print(my_calc(3,2)) # 5
```

There is significant flexibility with the input arguments of Python functions. They can have defaults, for when they are not explicitly specified, and there is freedom to use named or unnamed arguments. Named input arguments can appear in any order, though they must be stated after any unnamed ones, which fill slots in order, as shown above. The next example is a modification of the previous function that uses a default value of 1 for y on the **def** line. Calling the function is then illustrated with and without specifying an explicit value for y, using the named and unnamed conventions:

```
# Script 18.29
# x is mandatory, y defaults to 1 if not given
def my_calc(x, y=1):
    z = x * x - y * y
    return z
a = my_calc(7)              # x is 7, y defaults to 1
b = my_calc(x=2, y=2)       # Name both arguments
c = my_calc(y=9, x=-1)      # Name arguments in a different order
d = my_calc(3, y=-2)        # Unnamed arguments (x is 3) come first
print(a, b, c, d)
```

Decorators are a relatively recent addition to Python and allow a modifying statement to be added to the start of a function with an '@' syntax. In essence this wraps one function with another, which can modify and inspect both its input and output.

```
@decorator_func
def my_calc(x, y=1):
    z = x * x - y * y
    return z
```

For the sake of brevity, creating decorator functions will not be discussed; however, their use will be shown in later examples.

18.2.6 Object Classes

All the Python objects (items of data) used thus far have been of the standard types. However, it is possible to create custom Python objects using the **class** keyword. This creates a named prototype that connects data values and bound functions (methods) together in an organised way. A class can often be thought of as equivalent to a table in a database. For reasons of space this aspect of Python will not be discussed in great detail in this chapter. However, a simple example is provided below that illustrates a rudimentary **Person** object. The **class** block contains the definitions of two functions. It defines the __init__() method which, because it has a special name, is run any time an object of that type is created: here its task is to associate the input **name** and **age** with **self**; a special variable that represents the run-time object. A second method **get_first_name**() is defined that extracts and passes back the first part of the full name, which was stored as **self.name**.

```
# Script 18.30
class Person():
  # This indented block is in the class definition
  # name, age: values specified when object is made
  def __init__(self, name, age):
    self.name = name        # Link input value name to the object
    self.age = age          # Link input value age to the object

  # This indented block is a second, custom function
  def get_first_name(self):
    names = self.name.split()  # self refers to run-time object
    return names[0]            # Give back first word
p1 = Person('Lisa Simpson', 8)      # Make object of Person class
p2 = Person('Bart Simpson', 10)     # Make another
print(p1.age, p2.age)       # Values linked to objects - 8, 10
print(p1.get_first_name())  # Run a linked function - gives 'Lisa'
```

As illustrated above, two objects were made using the **Person** prototype, thus creating two different instances of that class: stored as variables **p1** and **p2**. The methods and simple attributes that were associated with the `self` are available to each instance using the dot syntax, e.g. `p1.name` and `p1.get_first_name()`. Here, the `self` value stated in **class** definition represents the object stated before the dot, i.e. `p1`, and is not passed via the parentheses.

18.2.7 Modules

Some functions, like `len()` or `int()`, are available at any time in Python. However, functions must often be imported into a program from a separate module before they can be used. There are three basic sources of modules: those that automatically come as part of every Python installation (the Standard Library), those that require a separate installation (such as NumPy or BioPython) and those that are specific to the user. If a module is accessible to Python it can be used via the **import** keyword and the various components of the module are referred to with dot syntax:

```
# Script 18.31
import math              # Import the inbuilt mathematics module
print(math.e)           # An attribute representing e
print(math.exp(2.0))    # Use the exponent (eˣ) function
```

It is also possible to locally use a different name for the module using the **import..**
as.. syntax:

```
# Script 18.32
import math as m    # Import as a different name
print(m.log(2.0))   # Use the logarithm function
```

Alternatively, specific components of a module may be imported using a **from..**
import.. syntax:

```
# Script 18.33
from math import sqrt, cos    # Import named module components
x = sqrt(3/2)                 # Use the square root function
print(cos(x))                 # Use the cosine function
```

The `math` module used above contains several commonly used mathematical constants and functions. There are various other commonly used libraries in the standard set, some of which we illustrate below. However, for a full listing see the documentation at python.org. The `sys` module relates to the run-time Python environment:

```
# Script 18.34
import sys
print(sys.argv) # List of words typed at the command line
print(sys.path) # Directory search path for Python modules
sys.exit()      # Quit the Python program
```

The random module is use to generate pseudo-random numbers:

```
# Script 18.35
import random
random.seed(3)                          # Set the random number seed
# A random integer between 1 to 10
print(random.randint(1,10))
# A random float between 0.0 and 2.0
print(random.uniform(0.0, 2.0))
data = [1,2,3,4,5]
random.shuffle(data)                    # Shuffle list order
```

The os module is for things that depend on which particular operating system (e.g. Windows, OSX, Linux) is running. Much of this relates to use of file systems, i.e. dealing with file paths, directories, permissions etc.

```
os.chdir(dir_name)   # Change the current working directory
os.mkdir(dir_name)                      # Make a new directory
os.listdir(dir_name)                    # Get a list of directory contents
os.remove(file_path)                    # Delete a file
os.rename(old_path, new_path) # Move a file to a new location
```

It is notable that os doesn't handle copying files; this is done with the shutil module. Within the os module of particular importance is the os.path submodule, which handles the text strings that represent locations within the file system.

```
file_path = '/home/user/test.py'
os.path.exists(file_path)       # True if the file path exists
# Split into [leading, end] parts
os.path.split(file_path)        # ['/home/user','test.py']
os.path.isdir(file_path)        # True if a directory, else False
# Join 'usr' and 'local' with directory separator: 'usr/local'
os.path.join('usr','local')
# Split into path and file extension
os.path.splitext('folder/file.txt') # ['folder/file', '.txt']
```

The re module is used for regular expressions; pattern matching in text strings. Often the functions generate a special match object that can be interrogated, for example to find where the pattern was found and what the actual substring was.

```
# Script 18.36
import re
# Make a regular expression object with one or more digits
pattern = re.compile('\d+')
text = 'A 123 B 456'                    # A string to look in
match_obj = pattern.search(text) # Match inside string
print(match_obj.group())                # '123' - matching substring
print(match_obj.start)                  # 2 - position of match
# Substitute all matches with '**'
```

```
text_2 = pattern.sub(text, '**')
print(text_2)                      # 'A ** B **'
hits = pattern.findall(text)       # List of matching sub-strings
print(hits)                        # ['123', '456']
```

A selection of commonly used modules from the Python Standard Library is given below (Table 18.6). Full documentation of these and the other modules mentioned above can be found at python.org.

Table 18.6 **Frequently used standard modules in Python**

Module	Description
argparse	A module that helps interpret command-line options/arguments, i.e. information typed after the name of a program, as available in sys.argv.
copy	Create a new Python object by copying an existing object. Can create shallow or deep copies; whether any object contained by an object is itself also copied.
datetime	Provides date, time, timedelta and datetime objects to represent temporal information. Deals with daylight savings, date formatting, time string interpretation etc.
glob	Provides file name fetching using UNIX-like wild cards, i.e. patterns that include '*' and '?' rather than regular expressions.
ftplib	Used to send and receive files using the File Transfer Protocol.
gzip, bz2, zipfile, tarfile	Modules that deal with creating and extracting compressed and/or archived files.
http	Used to send and receive information across the Internet using the Hypertext Transport Protocol. A lower-level library than urllib.
multiprocessing	Run Python code as separate, parallel, processes/jobs on multiple core/processor systems.
pickle	Converts Python object data to and from a text string (serialisation) which may be saved to or loaded from a file system.
platform	Used to get information about the current computer and its architecture.
shutil	Performs higher-level file operations, such as copying files, copying trees of files and finding executable files.
sqlite3	Interaction with SQLite : a lightweight file-based SQL database.
string	Provides some functions not directly available to string objects. Useful for accessing particular sets of characters such as white space, punctuation, digits etc.
subprocess	Run an external program as a separate job/process and connect any input/output data streams.
time	Basic time-related functions using numbers and strings. Can be used to time program execution and to pause execution (time.sleep()).

Table 18.6 **(cont.)**

Module	Description
`threading`	Run Python code in separate threads, which will not run concurrently on multiprocessor systems (use `multiprocessing` for that). Can be useful to process intermittent data streams.
`urllib`	Used to send and receive information across the Internet. Higher-level and often more convenient than `httplib`. Handles web proxies, redirection, passwords, cookies etc. Often used to interact with web services and databases.
`zlib`	Used to compress data into more compact representation, using the zlib algorithm. Can be useful for caches and undo functions.

Writing custom modules is easily achieved in Python: in general, a normal file containing Python code can be imported as a module. There are a few caveats to this, but most importantly the Python system needs to know where to look in the file system to find a module. Modules are found by looking in a series of directories, the search path, for file names that match the attempted import. Some of these will be in standard locations for the Python installation, but search directories can be added at any time via `sys.path`. Considering a file called `my_module.py` that resides in the directory '/home/user/my_mods/' and which has the following contents:

```
# Script 18.37
CONSTANT = 1.0545718e-34    # Example constant
def my_calc(x, y):          # Example function
    return x*x - 2*x*y - y*y
```

The directory containing the file can be appended to `sys.path` so that we can import `my_module` in a different Python program and thus access its variables and functions. It is notable that the file extension ('.py' here) is omitted from the module name when it is imported.

```
import sys
print(sys.path)    # Current module search path
# Directory with my_module.py
sys.path.append('/home/user/my_mods')
# Import custom module (no '.py' extension)
import my_module

# Use variable from my_module (no.py) Python file
print(my_module.CONSTANT)
# Use function from my_module (no.py) Python file
print(my_module.my_calc(9,-1))
```

It is often convenient to more permanently add the locations of custom Python modules to the search path by adding them to the PYTHONPATH environment variable (generally set in the operating system), so that `sys.path` automatically contains the required locations whenever any Python program is run. If instead the module was located inside a subdirectory of the one in `sys.path`, such as '/home/user/ my_mods/my_examples/', then the module could be imported as follows, using dot syntax:

```
import my_examples.my_module    # Import from sub-directory
```

Before Python version 3.3, to make this subdirectory import work, a file called '__init__.py', which is usually blank or contains only '**pass**', must be present in the 'my_examples' directory.

When a module is imported, its contents are run as Python. While this is usually no problem for defining constants and functions etc., sometimes the file may contain code that should only be run when the file is used directly and not when it is imported. To overcome this, the special internal Python __name__ variable can be inspected. This will be the string '__main__' if the code is run as the main program, but will otherwise be the name of the module. Hence this can simply be checked, to make sure the code is not imported as a module.

```
# Script 18.38
def my_calc(x, y):              # Example function
    return x*x-2*x*y-y*y
if __name__ == '__main__':      # Is the code run as a main program?
    print(my_calc(2,3))         # Run test code only when it is main,
    print(my_calc(-1,0))        # no module imports in this block
```

Though the standard libraries are extensive and custom modules allow great flexibility, there are a large number of external modules that are really useful for molecular biology, helping one to build programs using pre-written and well-tested code. A small selection of popular general-purpose external modules useful for molecular biology is given Table 18.7.

18.2.8 Storing Data

A key operation in most programs is reading and saving data to and from disk, or another storage file system. In Python, we can choose to handle all of the writing explicitly in a program. However, if there is a module that handles a particular data format then that is often a good choice instead. For example, we can use BioPython modules to read most common bioinformatics formats and use Pandas to read and write comma-separated value (CSV) files and Excel spreadsheets. Note that the term 'CSV format' is often also used for other delimiter-separated formats that employ different field delimiters, including tabs (\t) or other character delimiters such as semicolon or |.

File operations revolve around a file object. This is typically generated using the inbuilt **open()** function, given the location of the data on the file system. Once the file object is created, various functions can be used to read and write data; the object is the Python interface to the stored data. In the next example a file object is created, and all of its data read. The use of 'r' is to specify that the file is opened for reading, though this is the default when not specified. The data may be read in its entirety using **read()**:

Table 18.7 **Select external modules often used in Python programming**

Module	Description
NumPy	Numeric Python with a highly functional multi-dimensional array object and associated functionality for linear algebra, pseudo-random numbers, Fourier transforms, etc.
BioPython	A collection of tools for computational biology. Provides modules to work with and manipulate many common bioinformatics format files relating to sequences, alignments, phylogenetics, molecular structures, etc.
PySam	The Python interface to SAMtools, which allows reading, writing and manipulation of BAM and SAM format files used in high-throughput DNA sequence mapping. Also deals with variant call VCF/BCF files.
HTSeq	A module for the analysis of high-throughput DNA sequencing data. Deals with the common informatics formats, including FASTQ, SAM, BAM, BED and GFF. Has specialised data structures to handle large genomic array and genomic interval data.
PIL	The Python Imaging Library, which handles loading and saving image data in a large number of different file formats. Provides many functions to manipulate image data: enhance, filer and mask, etc.
Pandas	Manipulation and analysis of data tables with an organisation similar to spreadsheets and relational databases. Provides many advanced functions for data manipulation and file access (including MS Excel, CSV and plain text).
Pymol, UCSF Chimera	Popular molecular structure viewers that may be imported as Python modules to produce graphics for protein structure, etc.
SciPy	Extensive scientific and engineering library, building upon NumPy with modules for statistics, integration, optimisation, signal processing, etc.
matplotlib	A plotting library to create many different types of charts and graphs from numeric data, with customisable graphical styles.
scikit-learn	A library for machine learning, clustering, regression, dimensionality reduction etc. that is built upon SciPy and NumPy.

```
# Script 18.39
# Location of data in file system
path = 'examples/data_file.txt'
# Create file object for reading
file_obj = open(path, 'r')
# Read all the data as a string
data = file_obj.read()
print(data)
```

If the data are reread then nothing results because the file object remembers the last read point and this refers to the end of the data after the first complete read. If we want to change this, we can use **seek()** to go to a particular position (in bytes, with 0 being the start). When the file is no longer needed, it is closed for further use.

```
# Script 18.40
print(file_obj.read())    # Empty, nothing to read at end of file
file_obj.seek(0)          # Point at start of data
```

```
print(file_obj.read())      # Read everything
file_obj.close()            # Close file, no further operations
```

Recently, it has become common practice to use the `with.. as..` syntax to deal with file objects. The `with` block creates a managed context, which in this case means that the file object is only open within the block and closed automatically after

```
with open(path) as file_obj:
    print(file_obj.read())
```

There are other functions to read the file in terms of lines, separated with \n (or \r if universal read mode `rU` mode is used). All lines can be read at once using `readlines()`, or `readline()` can fetch a single line.

```
with open(path, 'rU') as file_obj:
    print(file_obj.readline())     # First line
    print(file_obj.readline())     # Second line
    print(file_obj.readlines())    # All remaining lines as a list
```

Subsequent `readline()` calls give each line in turn; the pointer to the file data picks up at the end of the previous line. However, it is often more convenient to treat the file object as an iterator and loop through it, i.e. as if it were a list of lines:

```
# Script 18.41
file_obj = open(path, 'r')
for line in file_obj:        # Loop through all lines in file
    print(line)
```

Many files that are commonly read by Python are essentially stored as text. Consequently, any numbers must be properly interpreted as such, appropriately creating proper `int` or `float` objects, if we want to use the data for calculations etc. It is fairly common to have lines where the data are separated by spaces or tabs, such as:

```
chr    mb_size    proteins
1      249.250    2012
2      243.199    1203
3      198.022    1040
```

The following code will read this, given an appropriate file name, and calculate the total for the `mb_size` column:

```
# Script 18.42
total = 0
with open(f_name) as file_obj:
    file_obj.readline()          # Read first header line; not used
    for line in file_obj:        # Go through each remaining line
        data = line.split()      # Split line at space into a list
        size = float(data[1])    # Make a number from column [1]
        total += size            # Increase count
```

For writing to a file, mode w or mode a must be used: opening with w mode initially writes a blank file (deleting any previous data) while a appends to the end of a file. To do the actual writing, data are simply passed to the write() function, in one or more parts. If we want to save the data as lines, the appropriate newline characters (\n etc.) need to be added.

```python
file_obj = open(f_name, 'w') # Open file object in writing mode
x = 1
while x < 1000:               # A loop to generate many lines
    line = '{}\n'.format(x)   # Create the string for each line
    file_obj.write(line)      # Write each line
    x *= 2
file_obj.close()
```

When dealing with files it is common practice to accept the name of the file to use at the time when the program is run. On the command line, this is easily achieved by specifying the name of the file after the Python script name. What was entered at the command line is then accessed using the sys.argv list. For example, if the following is typed at the operating system command prompt (>):

```
> python programFile.py data/inputFile.txt
```

then the name of the files can be captured in the following way, noting that the first item in the sys.argv list is the Python script itself, so it is the second item (index 1) that we usually want.

```python
# Script 18.43
import sys                        # Access the sys module
py_script = sys.argv[0]           # 'programFile.py'
data_file = sys.argv[1]           # 'data/inputFile.txt'
# Make file object and read its data
data = open(data_file).read()
```

When dealing with common data file formats there may be a module available that deals with file access to handle all of the reading and writing for you and interpret the data properly. For example the BioPython module can read and write many bioinformatics formats. In the next example, we import the SeqIO module. This can read from an open file object using its parse() function to generate sequence record objects (named protein below) from which data is accessed via the dot syntax. In this case the data format is specified as FASTA, common for protein or nucleic acid sequences:

```python
# Script 18.44
# Load BioPython module; must be installed
from Bio import SeqIO
file_name = "examples/demoSequences.fasta"  # Location of data
file_obj = open(file_name, "rU")   # Universal read mode
```

```
# Go through each entry
for protein in SeqIO.parse(file_obj, 'fasta'):
    print(protein.id)                # The ID of seq record
    print(protein.seq)               # The actual sequence
file_obj.close()
```

Next, the ability of the pandas module to read comma-separated value (CSV) files is demonstrated using the read_csv() function. Similar read_excel() and read_sql_table() functions, for Excel spreadsheets and SQL databases, respectively, also exist.

```
# Script 18.45
import pandas
data_set = pandas.read_csv('in.csv', sep='\t',
                           header=None, names=['A','B','C'])
for col in data_set:               # Loop through column names
    for value in data_set[col]:    # Column names act as keys
        print(col, value)
```

Though modules like pandas can read and write data to and from SQL databases there are also modules to directly handle SQL databases. There are different modules for the different SQL implementations, e.g. sqlite3 to handle the file-based SQLite and MySQLdb for MySQL. Such modules are used to open a database connection and submit SQL statements (a completely different language to Python). Here a connection is made to a MySQL database:

```
import MySQLdb                       # Module must be installed
connection = MySQLdb.connect(db='my_db_name', host='127.0.0.1',
                user='my_username', passwd='my_password')
```

A connection to an SQLite database just needs the file location:

```
import sqlite3 # Installed as standard
connection = sqlite3.connect('genome_example.sqlite')
```

Given a connection, a cursor object can then be created, that acts as a point of entry into the database, with the ability to submit SQL statements.

```
# Script 18.46
cursor = connection.cursor()   # Make a cursor

make_table_smt = """           # This multi-line string is SQL,
CREATE TABLE Genome (          # it specifies a Genome table
   build_id VARCHAR(12),       # with various columns
   genus TEXT NOT NULL,
   species TEXT NOT NULL,
```

```
   strain TEXT,
   num_chromsomes INT,
   mb_length FLOAT,
   pubmed_id VARCHAR(12),
   PRIMARY KEY (build_id));
"""
cursor.execute(make_table_smt) # Execute SQL string
cursor.close()                 # The cursor is finished
connection.commit() # Commit the changes to the database
```

Naturally, data can be put into a database table with the appropriate (INSERT and UPDATE) statements and specifying the data that comes from Python; as the second input argument in execute(). When doing a large number of identical operations (albeit for different data) it is possible to use the executemany() function on a whole list of data, i.e. for multiple rows. In the below example the ? character is used as a placeholder for the inserted data, which is the convention for sqlite3. Other implementations can use different characters, for example MySQLdb uses %s instead.

```
# Script 18.47
cursor = connection.cursor() # A new cursor into the database
# Create the INSERT statement using "\" to split over two lines
add_genome_smt = 'INSERT INTO Genome (build_id, genus, ' \
   'species,strain, num_chromsomes) VALUES (?, ?, ?, ?, ?)'
# Data to add
mouse_data = ['GRCm38', 'Mus', 'musculus', 'C57BL/6', 22]
# Add one row
cursor.execute(add_genome_smt, mouse_data)
# Data to add
multi_data = [['GRCh38', 'Homo', 'sapiens', None, 25]
              ['GRCh37', 'Homo', 'sapiens', None, 25],
              ['BDGP6', 'Drosophila', 'melanogaster', '2057', 7]]
# Add several rows
cursor.executemany(add_genome_smt, multi_data)
# Instructions for an update
set_size_smt = 'UPDATE Genome SET mb_length=? WHERE build_id=?'
# Alter two rows
cursor.execute(set_size_smt, [142.573, 'BDGP6'])
cursor.execute(set_size_smt, [3482.010, 'GRCm38'])
```

Reading single rows of data can be done by using fetchone() on the result of a SELECT query, and all rows can be fetched with fetchall().

```
get_organism_smt = 'SELECT genus, species FROM Genome ' \
   'WHERE build_id=?'
```

```
genus, species = \
  cursor.execute(get_organism_smt, ['GRCh37']).fetchone()
```

Given that reading all rows takes memory, it is common practice to loop through the result of an `execute()`, handling rows one-by-one:

```
smt = 'SELECT build_id, genus, species, mb_length FROM Genome'
for build, g, s, sz in cursor.execute(smt):
    print(build, g, s, sz)
cursor.close()
```

An alternative means of working with SQL databases, without having to create SQL statements as strings, is to use a module like `sqlalchemy`. This is an object-relational mapper: it creates Python objects to represent the various components of the database so that data access and manipulation are done with a normal object-oriented Python syntax.

18.2.9 Numeric Python

Though mathematical operations can be performed in regular Python, the NumPy module is often faster and more convenient. Most of this revolves around `ndarray` objects, which store arrays of data of a specified type (usually, but not limited to, numeric types). Such arrays can be used in a similar way to lists: they contain an ordered sequence of values and can be used in loops etc. However, an important idea with NumPy arrays is that operations can be performed on the array as a whole, rather than looping through all the component elements. This means that a fast internal implementation can be used. Also, it results in syntax similar to matrix algebra, where each variable is an entire array.

An `ndarray` can have a number of different dimensions/axes, so that it can represent vectors, matrices, tensors etc. Arrays can be created from other collections, though the result will be of a uniform data type, which may be specified, but is otherwise determined from the input.

```
# Script 18.48
# Import the function to make ndarrays
from numpy import array
x = array([[1,2,3],[4,5,6]])    # Make array from list of lists
print(x.shape)                  # (2,3) - rows x columns
print(x.size)                   # 6 - elements in total
print(x.ndim)                   # 2 - two axes
print(x.dtype)                  # int - whole numbers
y = array(x)                    # Copy an array
# Convert to 3 x 2
y.reshape((3,2))                # [[1,2], [3,4], [5,6]]
# Make array forced as floating point
z = array((3,5,1,2), float)
```

There is an index and range (slicing) syntax that is similar to that used with lists and strings etc., but which separates the different axis specifications with a comma, such as `data[i,j]` for a 2-dimensional array. This differs from regular Python, where separate brackets are needed for each sublist, e.g. using `data[i][j]`. Accessing with multiple brackets will work with NumPy arrays, but will be slower, as it makes an intermediate array.

```
# Script 18.49
# Flat array of five float numbers
x = array([5.1, 6.2, 7.3, 8.4, 9.5])
x[2]              # 7.3 - number at index 2
# Slice range makes new array
x[1:4]           # array([6.2, 7.3, 8.4])
y = array([[1,2,3], [4,5,6]]) # Two-dimensional array
y[1,2]           # 6 - row one, column two
y[0]             # array([1, 2, 3]) - entire row zero
y[0,:]           # array([1, 2, 3]) - row zero, all columns (same)
y[:,2]           # array([3, 6]) - all rows, column two
y[-1,:]          # array([4, 5, 6]) - last row, all columns
y[:,1:]          # array([[2, 3],[5, 6]]) - column one onwards
y[1,0:2]         # array([4, 5]) - row one, first two columns
y[::-1,:]        # array([[4, 5, 6],[1, 2, 3]]) - reversed rows
y[:,(2,1,0)] # array([[3, 1, 2],[6, 4, 5]]) - new column order
```

There are various functions to create particular kinds of filled arrays, e.g. full of zeros or with linear increments.

```
# Script 18.50
import numpy as np
a = np.zeros((2,3))          # 2 x 3 array full of 0.0
b = np.ones((3,2), int)      # 3 x 2 array full of 1
c = np.full((2,2), 7.0)      # 2 x 2 array full of 7.0
d = np.identity(3)           # 3 x 3 identity matrix
# From 1.0 up to <3.0 in steps of 0.5
e = np.arange(1.0, 3.0, 0.5) # [1.0, 1.5, 2.0, 2.5]
```

Many common operations are applied in an element-wise manner, i.e. to each value individually, and operations can work between two arrays if they have compatible sizes.

```
x = np.array([1.0, 2.0, 3.0])
y = np.array([3.0, 4.0, 5.0])
x + 2    # array([3.0, 4.0, 5.0]) - Add 2 to all values
y / 2    # array([1.5, 2.0, 2.5]) - Divide all values by 2
x + y    # array([4.0, 6.0, 8.0]) - i.e. 1+3, 2+4, 3+5
```

```
x * y     # array([3.0, 8.0, 15.0])
x - y     # array([-2.0, -2.0, -2.0])
x / y     # array([0.33333333, 0.5, 0.6])
```

NumPy provides equivalent functions to the `math` module, which do the same thing, except to operate on all the elements of an array (and they also work on single numbers). As well as NumPy arrays, the functions will accept regular Python lists or tuples as input, but an array is created.

```
angles = np.array([30.0, 60.0, 90.0, 135.0])
radians = np.radians(angles)
cosines = np.cos(radians)     # array([0.866, 0.50, 0.0, -0.707])
np.log([10.0, 2.71828, 1.0]) # array([2.302585, 1.0, 0.0])
np.exp([2.302585, 1.0, 0.0]) # array([10.0, 2.71828, 1.0])
```

There are various methods inbuilt into an array object, for example to calculate sums, extrema, mean and standard deviation. These may be applied to the whole array or only along a specific axis, via the `axis` argument.

```
x = np.array([[3,6],
              [2,1],
              [5,4]])
x.min()            # 1                      - minimum value
x.max()            # 6                      - maximum value
x.max(axis=0)      # array([5,6])           - maximum value row
x.sum()            # 21                     - summation of all
x.sum(axis=1)      # array([9, 3, 9])       - add columns together
x.mean()           # 3.5                    - the mean value of all
x.mean(axis=1)     # array([4.5, 1.5, 4.5]) - mean of each row
```

Comparisons between arrays generate Boolean arrays giving **True** or **False** for each elemental comparison. When doing this it is handy to know which values are true: `nonzero()` will get a tuple of non-zero index arrays than can be used to modify the elements that were true in the comparison. In a similar manner `argsort()` can be used to get the indices of an array in value order, which can then be used for sorting.

```
x = np.array([1, 8, 9])
y = np.array([5, 0, 3])
z = x > y               # array([False, True, True])
idx = z.nonzero()       # Indices where z is true (x value > y value)
x[idx]                  # Values from x that were greater

# Array of indices in order of increasing values
idx = y.argsort()  # ([1, 2, 0])
y[idx]                  # [[0, 3, 5]] - values in increasing order
```

There are many functions that relate to matrix operations and linear algebra, for example to calculate various products (inner, outer, cross), to transpose and find inverse matrices:

```
x = np.array(((1,1),(1,0)))
y = np.array(((0,1),(1,1)))
# Dot product (matrix multiplication)
z = np.dot(x, y)          # array([[1, 2], [0, 1]])

x = np.array([[1, 2, 3], [4, 5, 6]])
# Transpose (swap rows with cols)
y = x.transpose()         # array([[1, 4],[2, 5],[3, 6]])

x = numpy.array(((1,1),(1,0)))
# Inverse
y = numpy.linalg.inv(x)   # array([[0., 1.], [1., -1.]])
```

NumPy has a number of handy submodules, such as `fft` for Fourier transforms and `random` for pseudo-random number generation

```
# Script 18.51
from numpy import arange, exp, fft, random, pi
l = 0.04
w = 0.1
t = arange(0.0, 100.0, 1.0) # A range of time values
# Wave equation using complex numbers
x = exp(2j*pi*w*t) * exp(-l*t)
# Fast Fourier transform array of wave
y = fft.fft(x)
# Uniform sample 100 values in [0, 5)
a = random.uniform(0.0, 5.0, 100)
# Sample normal distribution: mean 0.0, std. dev. 2.5
b = random.normal(0.0, 2.5, 100)
```

NumPy installations include the `matplotlib` module, which provides the ability to make graphs:

```
# Script 18.52
from matplotlib import pyplot
from numpy import arange, random
x_vals = arange(1000)
y_vals = random.poisson(5, 1000)
# Make scatter plot
pyplot.scatter(x_vals, y_vals, s=4, marker='*')
# Save graph as an image
pyplot.savefig("TestGraph.png", dpi=72)
pyplot.show()   # Show on screen
```

SciPy is an extensive library that builds upon the NumPy arrays and their functions. It provides specialised scientific functionality for areas, including further linear algebra, optimisation, integration, interpolation, signal processing, image processing and differential equations. Below, a simple example is given, showing the `ndimage` submodule, which is useful for reading and writing image pixel data.

```
# Script 18.53
from scipy import ndimage
img_file = 'examples/my_image.png'
pixmap = ndimage.imread(img_file)   # Read image data as array
height, width, channels = pixmap.shape
red_channel = pixmap[:,:,0]          # Color channels are last axis
green_channel = pixmap[:,:,1]
```

18.3 EXAMPLES

Given the introduction to some of the features of the Python language, and a few of its modules, the remainder of this chapter will work through a few short example programs for molecular biology. To keep the examples short and simple, they are not as extensive as they could be. For example, there is no sanity check for DNA sequences, to make sure they only contain the expected letters. However, all of the examples are working Python that provide real functionality.

18.3.1 Reverse Complement Sequence

This example finds the reverse complement of a DNA sequence, pairing G with C, A with T and their corresponding reciprocals. The program is written as a function that takes an input sequence `seq` and gives back (returns) a reverse complement sequence `rev_comp`. The sequence is assumed to be an iterable collection of characters like a list or a string. After the definition of the function (the indented block of code) we call the function with some test input data to check it works properly.

```
# Script 18.54
# Define function name and input argument
def rev_complement(seq):
    # DNA letter mapping dictionary
    rev_map = {'A':'T','T':'A','G':'C','C':'G'}
    # Reverse complement string is initially empty
    rev_comp = ''
    # Go through each seq letter in turn
    for x in seq:
        # Look up complementary letter in dictionary
        y = rev_map[x]
        # Grow the rev. comp. String by adding new letter to the start
        rev_comp = y + rev_comp
    return rev_comp # Rev. comp string sent as output
```

```
# A test DNA one-letter sequence
f_seq = 'AGCATAAGAATAGCAGCAGCGCGA'
# Run function on test sequence
r_seq = rev_complement(f_seq)
# Print original and reverse complement
print(f_seq, r_seq)
```

18.3.2 Calculate Sequence Identity

The next example calculates the percentage of sequence identity for two input sequences seq_a and seq_b. The sequences should be aligned and carry '-' characters to represent alignment gaps. It should be noted that the sequences are not assumed to be the same length and because of this we calculate n as the minimum sequence length, which gives a range of sequence positions that are valid for both inputs.

```
# Script 18.55
# Define function taking two input arguments
def calc_seq_ident(seq_a, seq_b):
  # Find minimum input sequence length
  n = min(len(seq_a), len(seq_b))
  # Starting identity count is zero
  count = 0.0
  # Loop through valid position indices
  for i in range(n):
    a = seq_a[i] # Letter at index i for first seq
    b = seq_b[i] # Letter at index i for second seq
    # Test if letters are same and not a gap
    if a == b and a != '-':
      # .. if they are increase count by one
      count += 1.0
  # Calculate and send back identity as % total
  return 100.0 * count/n

seq1 = 'ALIGDPVENTS'
seq2 = 'ALIGN-MENTS'
x = calc_seq_ident(seq1, seq2) # Run on some test sequences
print('Seq identity:', x, '%') # x is 72.7
```

18.3.3 Download an Informatics File from a URL

The following function illustrates how data can be fetched over the Internet using the urlib module which does the hard work of finding the requested file and downloading its data. The url variable contains most of the Internet address

location, but needs a specific database entry identifier (db_id) to be inserted into
the address (using format()) before the connection can be opened to the correct
file. There is a slight complication in this function as the data are downloaded in
their raw binary form and we have to specifically convert it to a Python string
object via decode(), which interprets the data using a standard (UTF-8) character
encoding scheme.

```
# Script 18.56
import os
# Define function with 3 arguments, file_name is optional
def download_db_id(db_id, url, file_name=None):
  # Import urlopen() function
  from urllib.request import urlopen
  # Makes object to handle link
  response = urlopen(url.format(db_id))
  data = response.read()       # Read data from URL
  data = data.decode('utf-8') # Interpret data as plain text
  # If save file was specified
  if file_name:
    # ... create file object for writing
    file_obj = open(file_name, 'w')
    file_obj.write(data) # Save all data to file
    file_obj.close()
  return data # Hand back data from function
```

The function is tested by downloading a protein structure from the PDB and a
FASTA format sequence from the UniProt database. The URLs for these databases are
given below and include '{}', indicating where the identifier code for the specific
database entry will be added. It is notable that these variables use the uppercase convention as they are acting as constants and lie outside the functions.

```
# Get data from the PDB and save as a file
PDB_URL = 'http://www.rcsb.org/pdb/cgi/export.cgi/' \
          '{}.pdb?format=PDB&compression=None'
download_db_id('1AFO', PDB_URL, '1AFO.pdb')
# Get FASTA data from Uniprot, no save
UP_FASTA_URL = 'http://www.uniprot.org/uniprot/{}.fasta'
data = download_db_id('P02788', UP_FASTA_URL)
print(data)   # Show downloaded data
```

The next example combines the file download in the above example with use of
BioPython. The SeqIO module from BioPython will be used to handle the FASTA
format data obtained from the UniProt database, which saves us from having to interpret the data format ourselves. The function takes a database entry identifier db_id
and downloads the FASTA formatted data, as illustrated above, before automatically

parsing the data and sending back the (first or only) sequence as a sequence record object created by BioPython. This object has handy attributes that can be accessed via the dot syntax to get at the actual sequence etc.

```
# Script 18.57
def get_uniprot_seq_record(db_id):
  from Bio import SeqIO
  # Add extension to ID to make file name
  file_name = db_id + '.fasta'
  if not os.path.exists(file_name):
    download_db_id(db_id, UNIPROT_FASTA_URL, file_name)
  file_obj = open(file_name) # Open file for reading
  for seq_record in SeqIO.parse(file_obj, 'fasta'):
    return seq_record # Give back first record encountered
sr = get_uniprot_seq_record('P18754')
print(sr.id, sr.seq)
```

The next example shows how we can use Python to run an external non-Python program. We will use Python to create the input data, run the program with various options and collect the output data without needing to write any files. This means that the external program can be wrapped and controlled as if it were a Python function. Here the external program is BLAST for sequence-based homology detection (finding protein and DNA sequences that are similar because they are related by evolution). The example assumes that the BLAST program is installed on the computer and that a BLAST sequence database is available (this may be created from sequence files using the `makeblastdb` program). The mandatory input to the function is a biological sequence (a Python string), the location of the BLAST database (without any file extension) and the type of BLAST program to run: `blastp, blastn` etc. Two optional parameters control the e-value (the significance threshold for sequence hits) and the number of processor cores to use (i.e. to run the job in parallel). Running the BLAST program is handled by `Popen` from the `subprocess` module. This creates an object with all of the required command-line parameters and connects the input and output to Python (using `PIPE`). The input data is prepared as a FASTA format version of the sequence and sent directly to the `Popen` object, which, by using `communicate()`, sends back output and error data. The output data is then interpreted as lines of text, converted into proper number objects, where appropriate, and placed in a list.

```
# Script 18.58
# Define function; the first 3 arguments are mandatory
def blast_search(seq, database, program,
                 e_cutoff=1e-10, cpu_cores=None):

  # Imports for running external jobs
  from subprocess import Popen, PIPE
  if not cpu_cores: # CPU count parameter
    import multiprocessing
```

```python
  # Use all avail. CPU cores
  cpu_cores = multiprocessing.cpu_count()

# Make a list of all options and params to run BLAST
# Output format 6 means table output
# Numeric parameters are converted to strings with str()
cmd_args = [program, '-outfmt', '6', '-db', database,
              '-num_threads', str(cpu_cores),
              '-evalue', str(e_cutoff)]
# Try to create a process to run BLAST, but catch any error
try:
  proc = Popen(cmd_args, stdin=PIPE, stdout=PIPE)
except Exception as err: # Something went wrong
  print('BLAST command failed')
  print('Command used: "%s"'% ' '.join(cmd_args))
  print(err)
  return [] # Send back an empty list if BLAST failed

# FASTA format string template
template = '>UserQuery\n{}\n'
# Insert the sequence in template
in_data = template.format(seq.upper())
# Remove any special unicode characters
in_data.encode('utf-8')

# Send input get output
out_data, err_data = proc.communicate(in_data)
# Interpret output as text
results_str = out_data.decode('utf-8')

results = [] # Start with empty results
# Split output into lines
for line in results_str.split('\n'):
  data = line.split() # Split line into sub-list

  # If line was not blank
  if data:
    # Some columns are integers
    for col in range(3,10):
      data[col] = int(data[col])

    # Other columns are float
    for col in (2,10,11):
      data[col] = float(data[col])

    results.append(data) # Add row of data to results

return results # Send back results
```

To test the function, we specify the location of the BLAST database files (without any file extension) and use the sequence obtained above to search the BLAST database using the `blastp` program, which compares protein sequences.

```
db = '/data/tmd/uniprot-Chordata'  # BLAST database: must exist
blast_hits = blast_search(sr.seq, db, 'blastp')# Run on seq used
                                               before
for hit in blast_hits:                 # Go through each row in
                                       output list
   print(hit[1], hit[10], hit[11])   # Print some data columns
```

Another way to handle the BLAST results is to use the `pandas` module to create a highly functional `DataFrame` object to represent the table of hits. This will label the table columns with informative names and allow the data to be accessed in a similar manner to a dictionary, yielding a `Series` object for each column that has functionality, very similar to a NumPy array.

```
# Script 18.59
from pandas import DataFrame, Series, ExcelWriter
from matplotlib import pyplot
import os

# Column/key names
cols = ('query_id', 'subject_id', '%_identity',
        'alignment_length','mismatches', 'gap_opens', 'q_start',
        'q_end', 's_start', 's_end', 'evalue', 'bit_score')
# Get results list
hits = blast_search(seq_record.seq, db, 'blastp')
# Convert to Pandas table
data_set = DataFrame(hits, columns=cols)
# Show data types of the columns
print(data_set.dtypes)

# Fetch a column
column = data_set['%_identity']
# numpy-like functions
print(column.min(), column.mean(), column.max())

# Conditionally select indices
rows = data_set['%_identity'] > 75
# Show selected indices
print(data_set[rows])

# Create a new column from existing column series
data_set['offset'] = Series(data_set['q_start']- \
   data_set['s_start'])
# Save as tab-separated file
data_set.to_csv('BLAST_results.csv', '\t')
```

```
# Write to Excel
excel_writer = ExcelWriter('BLAST_results.xlsx')
data_set.to_excel(excel_writer, sheet_name='Sheet1')
excel_writer.save()

# Plot a graph directly from a column Series (uses matplotlib)
ax = column.plot('hist', bins=50,
   title='Seq identity distribution')
ax.set_xlabel('% Identity') # Set x-axis label
ax.set_ylabel('Count')      # Set y-axis label
pyplot.show()               # Display graphics on screen
```

18.3.4 Principal Component Analysis with NumPy

The below function illustrates the use of the NumPy module and performs a principal component analysis (PCA): treating input data as vectors, it finds the orthogonal directions in the data of maximal variance (the Eigenvectors of the covariance matrix). This is often used on high-dimensionality data to create simpler representations that still preserve the most important features. The function takes two arguments, the input data, to be provided as a list of vectors, and the number of principal components to extract. There is a small complication in this function as `linalg.eig()` outputs a matrix (p_comp_mat below) where each Eigenvector is a column, rather than a row. This orientation is useful for applying the matrix as a transformation, as we demonstrate below. However, in the code it means the matrix is sorted and selected on its last axis (using `[:,:n]` etc.). Therefore, the transpose, `pcomps.T` is used when extracting the two vectors.

```
# Script 18.60
from numpy import random, dot, cov, linalg, concatenate, array
from matplotlib import pyplot
def get_principal_components(data, n=2):
   data = array(data)        # Convert input to array
   mean = data.mean(axis=0)  # Mean vector
   # Centre all vectors and transpose
   centred_data = (data - mean).T
   covar = cov(centred_data)  # Covariance matrix
   # Get Eigenvalues and Eigenvectors
   evals, evecs = linalg.eig(covar)
   # Eigenvalue indices by decreasing size
   indices = evals.argsort()[::-1]
   # Sort Eigenvectors according to Eigenvalues
   evecs = evecs[:,indices]
   # Select required principal components
   p_comp_mat = evecs[:,:n]
   return p_comp_mat
```

The function is tested using some random 3D data. Here, the `random` module from NumPy is used to create three clusters of vector points. Initially each has the same mean (0.0) and standard deviation (0.5), but the last two clusters are transposed by adding an offset vector. The clusters are concatenated together (along the long axis) to create the final test dataset with three "blobs".

```
size = (100, 3)   # 100 points times 3 dimensions
d1 = random.normal(0.0, 0.5, size)
d2 = random.normal(0.0, 0.5, size) + array([4.0, 1.0, 2.0])
d3 = random.normal(0.0, 0.5, size) + array([2.0, 0.0, -1.0])
test_data = concatenate([d1, d2, d3], axis=0)
pcomps = get_principal_components (test_data, n=2) # Run PCA
```

The principal components are given back as a matrix, albeit in transposed form:

```
pc1, pc2 = pcomps.T # Extract the two PC vectors from columns
for x, y, z in (pc1, pc2):
  # Scale values so they can be seen better on graph
  x *= 10
  y *= 10
  # Plot PC x, y as line from origin (0,0)
  pyplot.plot((0, x), (0, y))
# Extract x and y values from transpose
x, y, x = test_data.T
pyplot.scatter(x, y, s=20, c='#0080FF', marker='o', alpha=0.5)
pyplot.show()
```

The principal component matrix can be used to transform the test data. Here the first two principal components are used as new *x*- and *y*-axis directions, illustrating that the transformation gives a better separated 2D view of the data.

```
transformed = dot(test_data, pcomps)
x, y = transformed.T
pyplot.scatter(x, y, s=20, c='#FF0000', marker='^')
pyplot.show()
```

18.3.5 Statistics Using SciPy

SciPy provides objects representing random variables for a variety of different statistical distributions. These are created by specifying the particular parameters for the distribution (e.g. mean and standard deviation for normal/Gaussian). Here, a Poisson distribution is illustrated. A random variable object is created from which we access the probability mass function pmf(). This generates the probabilities of the input values, according to the probability distribution.

```
# Script 18.61
from scipy.stats import norm, poisson
from numpy import arange
from matplotlib import pyplot
```

```
poisson_rand_var = poisson(2.0)   # Random variable object
# Value to plot probabilities for
x_xals = arange(0, 10, 1)
# Get probabilities from the distribution
y_vals = poisson_rand_var.pmf(x_xals)
pyplot.plot(x_xals, y_vals, color='black')   # Make line plot
pyplot.show()   # Show on screen
```

The next SciPy statistics example uses the random variable objects in a different way, to perform a tailed test, as would be done to estimate a *p*-value: the probability of obtaining a value from a given random distribution with stated parameters. The function performs the test on an input array of numbers for a normal distribution with given mean value, `mv` and standard deviation, `std`. There is an option to state if we want to do a one- or two-tailed test, i.e. consider only values on the same side of the mean or both sides.

```
# Script 18.62
from numpy import array
from scipy.stats import norm
def normal_tail_test(values, mv, std, one_sided=True):
  # Normal distribution random variable object
  norm_rv = norm(mv, std)
  # Calculate differences from distribution mean
  diffs = abs(values-mv)
  # Use cumulative density function
  result = norm_rv.cdf(mv-diffs)
  if not one_sided:   # Two-tailed test
    # Distribution is symmetric: double the area
    result *= 2
  return result
```

The function can be tested given some parameters and test values. In this case, the values could represent the heights of male humans.

```
mean = 1.76
std_dev = 0.075
values = array([1.8, 1.9, 2.0])
result = normal_tail_test(values, mean, std_dev, one_sided=True)
print('Normal one-tail', result)      # [0.297, 0.03097, 0.000687]
```

18.3.6 Plotting DNA Read Quality Scores from FASTQ Files

The following examples move on to illustrate handling high-throughput sequencing data. First we will work with FASTQ sequence read format and will handle this directly in standard Python, though in the second example, an external module, `pysam` is used to deal with the file reading etc.

The objective for the following function is to read a stated number of quality scores from each sequence read of a FASTQ file and plot these as a graph. Naturally, the function takes the name of the file to read as an input argument, as well as a number of nucleotide positions to look at: this could be the whole read length or shorter. The operation of the function is fairly simple as it reads the lines of the file and collects quality scores. However, because the FASTQ format consists of four lines for each entry (annotation, sequence, annotation and qualities) the lines are read in groups of four using `readline()` and a `while` loop. There is a slight complication because the data can be shorter than our requested read length (some sequences terminate early). In such cases we extend the quality data with a padding string containing spaces that allows all quality data to be analysed in the same way (the padding spaces give zero scores). It is notable that the quality scores in FASTQ files are stored as characters; one for each position in the DNA read sequence. These characters are converted to numeric values by the NumPy `fromstring()` function; each character is converted into its ASCII code number and the smallest value (`32` for a space) is subtracted to get scores ranging from zero.

```
# Script 18.63
from numpy import array, zeros, fromstring, int8, arange
from matplotlib import pyplot
def plot_fastq_qualities(fastq_file, read_len=100):
    file_obj = open(fastq_file)        # Open FASTQ file for reading
    qual_scores = []                   # Initial scores are empty list
    line = file_obj.readline()         # Read first line (a header)
    pad = ' ' * read_len               # Spaces (to pad short data)
    while line:  # Continue looping while there are more lines
        seq = file_obj.readline()      # Sequence line; not used
        head = file_obj.readline()     # Second header line; not used
        codes = file_obj.readline()    # Quality score code line; used
        codes = codes.strip()          # Remove trailing newline char
        if len(codes) < read_len:      # If quality code string too short
            codes += pad               # Extend with spaces
        codes = codes[:read_len]       # Chop string to desired length
        # Convert string to numbers
        qvals = fromstring(codes, dtype=int8)
        qvals -= 32  # Set lowest possible value to zero
        # Add quality scores for this seq to list
        qual_scores.append(qvals)
        line = file_obj.readline()     # Read next header line
    qual_scores = array(qual_scores)   # Convert list to array
    positions = arange(1, read_len+1)  # Array of seq. positions
    # Find mean for each position
    ave_scores = qual_scores.mean(axis=0)
```

```
  # Standard deviation for each position
  st_devs = qual_scores.std(axis=0)
  # Plot error bars of standard dev. height at score positions
  pyplot.errorbar(positions, ave_scores, yerr=st_devs,
    color='red')
  # Plot a line graph of scores on top
  pyplot.plot(positions, ave_scores, color='black', linewidth=2)
  pyplot.show()
# Test FASTQ file location
fastq_file = '/data/My_DNA_sample.fastq'
plot_fastq_qualities(fastq_file)
```

The next high-throughput sequencing example uses SamTools (importing the pysam module) to get paired sequence read data, i.e. two ends of the same DNA fragment, from a BAM format file. This format contains information about where DNA sequences have been mapped (by alignment) to the larger sequences of chromosomes within a reference genome. The task for this function is to find what are known as discordant pairs of DNA reads. These are where the mapped chromosome positions are further apart than expected given the size range of DNA fragments that were sent for sequencing. This function finds paired reads that map either far apart on the same chromosome or to completely different chromosomes. Such occurrences can indicate places where there have been rearrangements in the sequenced genome compared to the reference.

```
# Script 18.64
def find_discordant_read_pairs(sam_file, max_sep=2e3,
  verbose=True):
  # Import inside the function, for a change
  from time import time
  from pysam import Samfile, AlignmentFile
  # Record start time (seconds since 1/1/1970)
  start_time = time()
  # Split input file ext.
  root_name, ext = os.path.splitext(sam_file)
  # Out file has new ending
  out_file = root_name + "_discordant.bam"
  # Open BAM file for reading
  in_sam = Samfile(sam_file, 'rb')
  # Make new (empty) BAM file, base headers on input file
  out_sam = AlignmentFile(out_file, "wb", template=in_sam)
  # Fetch chromosome names
  chromo_names = in_sam.references
  # Some message templates
  msg_cis = 'Long range: {}:{} - {}:{}'
  msg_trans = 'Interchromosome: {}:{} - {}:{}'
```

```
n = 0                         # Number of discordant read pairs found
for align in in_sam:          # Go through each input alignment record
   chr_a = align.tid          # Primary chromosome of alignment
   pos_a = align.pos          # Primary base pair position
   chr_b = align.rnext        # Secondary chromosome
   pos_b = align.pnext        # Secondary position
   if chr_a == chr_b:                 # If both chromosomes the same
      delta = abs(pos_b-pos_a)        # Get separation of read ends
      if delta > max_sep:             # If too big
         # Write alignment record to output BAM
         out_sam_file.write(align)
         n += 1                       # Increase count
         if verbose:   # Check option to print a message
            chr_name = chromo_names[chr_a]
            print(msg_cis.format(chr_name, pos_a, chr_name, pos_b))
   else:                              # Chromosomes were different
      # Write alignment record to output BAM
      out_sam_file.write(align)
      n += 1                          # Increase count
      if verbose:                     # Check option to print a message
         chr_name_a = chromo_names[chr_a]
         chr_name_b = chromo_names[chr_b]
         print(msg_trans.format(chr_name_a, pos_a, chr_name_b,
            pos_b))
in_sam.close()
out_sam.close()
# Time taken: current time minus start
delta_time = time()-start_time
# Write out some useful information
print('Written {:d} records to BAM file: {}'.format(n,
   out_file_name))
print('Time taken: {:.1f} s'.format(delta_time))
# Test process on a _paired_ BAM file
bam_file = '/data/my_sample.bam'
find_discordant_read_pairs(bam_file, verbose=False)
```

18.3.7 Creating a File Processing Pipeline with Ruffus

This example demonstrates the construction of a data processing pipeline using the Ruffus module (the name Ruffus comes from a species of snake). The idea behind this module is that if a process can be broken down into component stages, each of which involve reading one kind of file and writing another, then the file transitions can be formally described and tracked. If the task at hand can be split into multiple independent files for a stage then that part of the pipeline can be run in parallel on a multi-processor computer. The example demonstrated here for a pipeline involves stages that have already been discussed above: FASTA format sequence files will be downloaded, BLAST will be run on each sequence and the table of BLAST output will be converted into an Excel spreadsheet. The components of this pipeline are functions that are mostly based on code previously described, although they are functions that always work from an input file to generate an output file.

The component functions form the stages of a pipeline by being wrapped by `ruffus` functions: this involves the `@ruffus.func_name` decorator syntax, which is easy to add just before the definition of an existing function. What these wrapper functions do is to track which files are being input and output from the function. The Ruffus system then knows which stages of the pipeline to run according to which files have not yet been made and which need to be remade if they have new inputs (newer than any previous output). This check-pointing approach to the files is made at each stage and means that the pipeline can be stopped and will restart where it last left off. There are a variety of different Ruffus decorator functions that represent different network topologies; whether there is a one-to-one, one-to-many, many-to-one, etc. relationship between the input and output files of each stage. Here the example is kept simple using `@ruffus.originate` to start the pipeline on a given number of FASTA files, and then following with `@ruffus.transform`, which tracks the one-to-one transformation of files (UniProt FASTA to BLAST to Excel).

Initially some imports are made and some variables are defined. The `call()` function is imported to run BLAST and is a somewhat simpler way of running the program than the `Popen` shown before. This simpler approach can be used because we are letting BLAST read and write normal files and don't have to take special measures to send the data directly to and from Python. The inputs to the pipeline are a list of FASTA file names that are derived from the UniProt identifiers, and these filenames match exactly what can be downloaded.

```
# Script 18.65
# Function to run an external program
from subprocess import call
from urllib.request import urlopen
# Location of BLAST database
BLAST_DATABASE = '/data/my_blast_db'
UNIPROT_URL = 'http://www.uniprot.org/uniprot/{}'
# Create a list of FASTA file names from UniProt IDs
uniprot_id_list = ['P68431', 'P62805', 'P0C0S8', 'Q96A08']
fasta_files = [x + '.fasta' for x in uniprot_id_list]
```

The first stage of the pipeline does not accept input file names from a previous stage so we use the `@originate` decorator function to specify that files are only generated by this stage.

```
@ruffus.originate(fasta_files)    # First stage, makes FASTA files
def get_uniprot_seq(fasta_file): # Downloads a named FASTA file
# Insert file name into URL
url = UNIPROT_URL.format(fasta_file)
response = urlopen(url)   # Access remote location
data = response.read().decode('utf-8')   # Read data as text
file_obj = open(fasta_file, 'w')   # Open file for writing
file_obj.write(data)   # Save downloaded data
file_obj.close()
```

The next stage takes the FASTA files and runs BLAST on each, creating a text table file which we end with '.blast'. Because this stage does the conversion of one file to another we use the @ruffus.transform decorator to track the Ruffus pipeline. The decorator function has three inputs: the function that ran previously, so that it can catch the output files from the previous stage; an indication of what kind of files are going to be accepted as input, here stating they will have '.fasta' as a suffix, and that this will be replaced; and lastly a substitution string to make an output file name from an input file name. The input to the run_blast() function is fairly simple: just the input fasta_file, the output table_file and various BLAST parameters. What all this means is that although the run_blast function makes one input from one output, the Ruffus pipeline will use this function on every FASTA file downloaded by the previous stage.

```
@ruffus.transform(get_uniprot_seq, ruffus.suffix(".fasta"),
   ".blast")
def run_blast(fasta_file, table_file, e_cutoff=1e-10,
   cpu_cores=1):
   cmd_args = ['blastp', 2018-query', fasta_file, '-out',
               table_file, '-outfmt', '6', '-num_threads',
               str(cpu_cores), '-db', BLAST_DATABASE, '-evalue',
               str(e_cutoff)]
   call(cmd_args) # Run external program with command parameters
```

The final stage in the pipeline takes the BLAST output (which we named '.blast') and creates an Excel spreadsheet for each set of results. The use of @ruffus.trans-form is similar to before: we are just making a transition between different file extensions and a different function is creating the output files.

```
@ruffus.transform(run_blast, ruffus.suffix(".blast"), ".xlsx")
def make_spreadsheet(table_file, xl_file):
   # A tuple containing spreadsheet column headings
   columns = ('query_id', 'subject_id', '%_identity',
              'alignment_length', 'mismatches', 'gap_opens',
              'q_start', 'q_end', 's_start', 's_end', 'evalue',
              'bit_score')
   # Data type for each column in sheet
```

```
data_types = [str, str, float, int, int, int,
              int, int, int, int, float, float]
# Pair columns with data types
data_types = zip(columns, data_types)
# Make dictionary of col_name:dtype
data_types = dict(data_types)
# Make a data table object with pandas; the dtype dictionary
# allows the correct conversion of the BLAST table to numbers
data_frame = pandas.read_table(table_file, dtype=data_types,
                               names=columns, header=None)
# Make Excel file
excel_writer = pandas.ExcelWriter(xl_file)
# Write data to sheet
data_frame.to_excel(excel_writer, sheet_name='Sheet1')
```

With the pipeline of functions complete we can inspect its status and then run it, in this case on four processors, i.e. so four protein sequences are run through the pipeline at the same time.

```
ruffus.pipeline_printout()
ruffus.pipeline_run(multiprocess=4,)
ruffus.pipeline_printout()
```

The pipeline will not run anything if we try to start it again: it detects all the required output files exist and so considers the job complete. However, if an input file is removed, the pipeline will run again, but only to regenerate the missing part of the output.

```
ruffus.pipeline_run(multiprocess=4,)
# Complete: will not re-run
import os
# Delete one of the first stage files
os.unlink("P68431.fasta")
# The pipeline runs for the missing part
ruffus.pipeline_run()
```

Lastly, instead of downloading sequences, the input to the pipeline could be specified as a single large FASTA file containing multiple protein sequences. In this case the pipeline would be started by splitting the big FASTA file into small files with one protein sequence in each. The rest of the pipeline can then operate as before. This kind of one-to-many file transformation can be achieved with the split_fasta function defined below. It is wrapped with the Ruffus @ruffus.split decorator, which tracks that the big file is used to make multiple files ending with '.fasta' (the '*' matches the leading part of the file name):

```
@ruffus.split('big_input.faa', '*.fasta')
def split_fasta(big_fasta_file)
```

```
for line in open(big_fasta_file):     # Each line in big file
  # If a header line starting with '>'
  if line[0] == '>':
    # Split proper UniProt annotations
    annos = line.split('|')
    # ID is the first annotation item
    uid = annos[1]
    fasta_file = uid + '.fasta'        # Make file name from ID
    file_obj = open(fasta_file, 'w')   # Open file for this ID
  file_obj.write(line) # Line written to open file for current ID
  # Will be used for sequence lines until new header line is read
```

18.4 SUGGESTIONS FOR FURTHER READING

18.4.1 Experimental Protocols

Kremer L.P., Leufken J., Oyunchimeg P., Schulze S. and Fufezan C. (2016) Ursgal, universal Python module combining common bottom-up proteomics tools for large-scale analysis. *Journal of Proteome Research* 15, 788–794.

18.4.2 General Texts

Stevens T.J. and Boucher, W. (2015) *Python Programming for Biology: Bioinformatics and Beyond*, 1st Edn., Cambridge University Press, Cambridge, UK.

18.4.3 Review Articles

Bassi S. (2007) A primer on python for life science researchers. *PLoS Computational Biology* 3, e199.

18.4.4 Websites

Scripts discussed in this chapter
www.cambridge.org/hofmann

Anaconda (Python Data Science Platform)
www.anaconda.com/ (accessed May 2017)

Python Software Foundation
www.python.org/ (accessed May 2017)

19 Data Analysis

JEAN-BAPTISTE CAZIER

19.1 INTRODUCTION

19.1.1 Data, Data, Data

Data are big, data are hard, data are experimental; data are everywhere.

Converting data into (useful) information is a challenge, but a necessary step to not only justify data collection but also appraise processes and arrive at conclusions. This chapter strives to reach such an ambitious goal. Beyond the provision of analytical methods, it aims to highlight potential pitfalls and tips for success.

We shall start by contemplating the very essence of data analysis, discussing the theoretical concepts and relevant approaches required for a particular analysis. Given the diversity in complexity and nature, we will consider whether it is possible, or desirable, to separate a part from the whole: does it make sense to look only at a subset of data to draw a conclusion? Such a question is just as valid when separating one particular dataset from another, for example when pulling apart a graphical representation from statistical analysis. A figure can only illustrate a result, not quantify its significance. Very importantly, when designing an experiment and collecting data one should always remember the words of R.A. Fisher, the eminent statistician and geneticist: 'To call in the statistician after the experiment is done may be no more than asking him to perform a post-mortem examination: he may be able to say what the experiment died of.'

Then we will get into the practicalities of data handling, particularly the actual manipulation of the data into a format required for further analysis. It may not be the most glamorous subject, but it is essential to enable the appropriate and efficient use of the data collected. Just like a well-designed experiment, well-organised data can make its collection, visualisation and analysis a smooth and enjoyable experience. Data reformatting presents frequent mishaps, including an inadvertent loss, an erroneous characterisation, a mix-up and even an outright corruption. Unfortunately, such misfortune happens all too often; so being prepared will not only minimise the impact, but help capture any eventual problem earlier, easing its fix. We will also emphasise that data handling should be directly linked to its graphical representation and expected analysis on the platform of choice. We will therefore discuss the optimal use of spreadsheets, R and Python languages (see also Chapter 18).

Once data are properly gathered and organised, only then is it time to examine it. Graphical representation of datasets should occur prior to the analysis itself, to aid not only understanding the data structure, but essentially performing a first-pass sanity check for gaps, outliers or batch effects. Since the characteristics of the collected data should be known and the data well organised, the 'natural' way to examine it should come naturally; a digital version of Boileau's 'Ce qui se conçoit bien s'énonce claire-ment – Et les mots pour le dire arrivent aisément.' (What is well understood is enunci-ated clearly – and its expression is easily done.) We will illustrate several ways to use the R and Python platforms to plot data in a relevant fashion. Early exploration of the data should first confirm that the experiment succeeded, and the right data have been collected. It also presents an early chance to spot potential outliers requiring further attention before they affect the whole analysis. Finally, it should 'give a feel' for the trend in the experimental data through the recognition of patterns.

Finally, we will tackle the all-important data analysis itself using 'real-life' datasets. A variety of essential statistical tests such as t-test, ANOVA, Fisher's exact tests or chi-square tests will be covered with examples in both Python and R. We will intro-duce the linear and logistic models where graphical representation meets analysis. But, most importantly, the concepts of Type I and Type II errors, as well as power and distributions, will be presented in the context of exemplary data. Scripts discussed in this chapter are also available online at www.cambridge.org/hofmann.

19.1.2   A Holistic Approach

In a world of Big Data, it is tempting to gather as much data as possible, frequently including 'any' data and 'just-in-case' data. More information is always great, but more data does not necessarily translate into more information. Furthermore, large amounts of data come at a price, even if data are available at ever lower costs, whether by generation or collection from outside sources. All these data need to be carefully organised, otherwise there await not only spurious results, but information can easily become invisible in a plethora of uninformative data. 'More data' translates as much into 'more information' as it increases the chances of drawing incorrect conclusions. Problems can arise in a number of different guises:

- Error: Data are never perfect and some error in generation, collection or transcription can easily occur, especially if handled manually.
- Bias: Collection of data is affected by other factors in a systematic fashion, e.g. a specific machine or technician.
- Corruption: Some of the data are cut or only partially represented.
- Mix-up: Values are exchanged between samples or data types; the former, being harder to spot, can have more dramatic effects.

- Incompleteness: Not all the information can be collected for all samples and therefore the information can be sparse, leading to irrelevance or eventual bias.

19.1.3 Necessary Rigour

In order to avoid, or at least minimise, potential issues with data, some simple rules should be followed:

- Define what the data are to be used for; for example, is it worth collecting the outside temperature on the day of experiment?
- Define what the collected data correspond to and how they are captured.
- Make sure that sufficient data are collected to reach a conclusion.
- Define what the data should be like: its type, range, etc.; for example, nucleobase representation of DNA should be expected to be A for adenine, C for cytosine, G for guanine or T for thymine only, not a numerical value, nor a colour.
- Make use of electronic, duplicated storage rather than manual collection.
- Keep the data together with clear documentation of the above points.
- Introduce meta data into datasets; for example, consider self-describing data in formats such as JSON, XML or HDF5.
- Minimise the manual manipulations in favour of automatic protocols.

Once the data are well organised, and their structure clearly understood, it is much easier to explore and look for relevant conclusions to be drawn from statistically sound analyses.

19.2 DATA REPRESENTATIONS

19.2.1 Data Formatting

Data obtained from biochemical experiments can be extremely diverse and its representation and storage should be appropriate. One would not expect gender to be stored in a similar fashion to height, or an MRI scan. Still, once data are generated they need recording to enable any future use. Formatting of data according to a defined set of rules is an efficient way to avoid some of the above-mentioned issues.

Definition

Formatting is the first essential step, which often occurs without being noticed. By default, most information appears to be collected in **alpha-numeric** style, i.e. both letters, from A to Z, and numbers, from 0 to 9, are stored as text, as well as some additional appropriate characters. While this is very convenient for a human to read and understand, it can be hard to standardise. The need for organised representation of data often leads to **data wrangling**, where hours, if not days, can be spent putting the data in the 'right format', such as preparing for input to any given tool. Furthermore, this approach is far less efficient to perform than the actual analysis. The recording of human-readable data is based on a character encoding scheme known as the American Standard Code for Information Interchange (ASCII). In order to store an ASCII character in binary code used by computers, a pattern of eight binary digits (bits), consisting of '0' and '1', are required. This conversion is costly in terms of computational cycles required.

Quantitative versus Qualitative Data

Data can be categorical (qualitative) or numerical (quantitative). For example, gender can be collected as 'Male'/'Female', 'Man'/'Woman', 'M'/'F' or 'M'/'W', depending on the time or person collecting the data. While such data collection is relatively easy to transcribe a posteriori, it is an unnecessary complication. In fact, having free-text recording for such qualitative information is not ideal and better done by selecting from a pre-defined list: either 'Male' or 'Female', not forgetting 'Unknown' or 'Not Available'. In some instances, categorical data might be hard to define a priori (e.g. hair colour), but frequently it would be most beneficial for such instances to make use of pre-defined lists.

Quantitative data should only be stored as numerical values. However, it is important to clarify how these values are presented and whether there should be boundaries. For example, without further explanation, a value of 0.8 for percentage can potentially represent 0.8% or 80%. In particular instances, one might want to avoid confusion by restricting data entry to a certain range. A recent report in a UK tabloid tells the story of a man who had to get a specific medical examination in order to prove his employability as he (erroneously) recorded his height to be 17 cm instead of 176 cm.

Precision

Both quantitative and qualitative measures benefit from an appropriate level of precision. Snow can be described by 1, 10 or even 100 words depending on the language used and the context. However, it is not always relevant to describe a type of snow that occurs only at the South Pole when working with general weather in Europe.

Precision is even more essential when dealing with quantitative values as it is easy to miss useful information, and even easier to gather useless data. When comparing small effects, it can be essential to gather further decimals that are unlikely to be relevant for a person's height. Similarly, reporting a p-value should not necessarily focus on the first three digits to highlight a value below 0.05, since in some cases it is important to compare results with p-values of 0.001 to those with p-values of 0.000001; in this example, reporting the first three digits would only provide '0.00' and '0.00'. One option for storage of such information is to perform a logarithmic transformation, in order to focus on the information about how small a value is. In the above case, this transformation yields $\lg 0.001 = -3$ and $\lg 0.000001 = -6$, making it much easier to store and compare the transformed data, '-3' and '-6'.

19.2.2 The Illusions of the Spreadsheets

The spreadsheet solutions seem to be able to do it all, from data handling, formatting and processing to visualisation. It is commonly, and rightfully, used to capture summary information such as budget from small amounts of well-organised data. Interactive tools to capture 'means' and 'sums' provide a convenient way to explore the data. Since the results of experiments are intrinsically provided in a neat table format, the ubiquitous use of these convenient tools is quite understandable.

At the same time, the convenience of spreadsheets all too often gives rise to indiscriminate use and blind acceptance of results, creating a dangerous illusion of correct data. The scientific world, as well as industry relying heavily on numerical processing, such as insurance and statistics businesses, are susceptible to this illusion, and faulty formula or cut-and-paste errors frequently result in mistakes. Indeed, a range of tools is available within spreadsheet software, but many are irrelevant, or inappropriate for scientific analysis. Most importantly, the concept itself can be dangerous for all types of applications, considering the following aspects:

- Size: Working with small amounts of data is fine, but Big Data, such as results from genomics studies will generate more data points than the generous 1048576 rows by 16384 columns that the spreadsheet format can handle.
- Formatting: The automatic formatting ability of certain software should be welcome to enforce the previously mentioned organisation of data. However, it often comes with unpredictable effects, such as a 1/1 genotype being transformed automatically into the 1st of January, leading to the whole column of genotypes to be treated as dates.
- Analytical tools: The functions available are somewhat limited; they include sum, mean and sometimes standard deviation. Proper, yet simple, statistical tests (e.g. Fisher's exact test) are not available.
- Scripting: Often, a scripting language such as Visual Basic is present in spreadsheet programs, and is welcome because there is no other way to perform complex operations going beyond the limited list of tools provided interactively.
- Plotting: Spreadsheets allow one to quickly visualise data, but have a limited range of plots; for example, a box plot chart, or density plots, do not exist as standard.
- Recycling: The tendency to manipulate the data interactively does not help the need for traceability of an analysis, and even less the ability to rerun an analysis when one value is changed or a row added.
- Accident: The worst danger of all is an accidental change of values while moving around the data. Putting the input and output data together, in the same worksheet, makes it impossible to spot such accidental changes and correct them.

Notably, there have been recent efforts to integrate R as an alternative processing language to compensate for the problems of scripting, tools and plotting. Such efforts further highlight the limitations of spreadsheets and indicate the optimal solution, which is to use dedicated and well-established software, such as R, for analysis of large datasets.

19.2.3 Big Data

What is true for any data is obviously true for Big Data. No matter how big the dataset is, some specific points are worth highlighting, in particular, because the dimensions of the 'big Vs' are (still) growing:

- Volume: The large size of Big Data is obvious, but deserves a mention. The genomic revolution in the past decade is an obvious candidate where we moved from a few markers representing the genome to every single base pair.

- Variety: Big Data does not necessarily correspond to a lot of the same data. Increasingly, Big Data is made up of accumulations of various types, such as in the growing field of 'omics, which includes data from genomes, transcriptomes, proteomes, metabolomes, microbiomes, etc.
- Variability: The data collected can vary dramatically in terms of quality and reproducibility. Not all data types can be treated equally.
- Velocity: Big Data grow fast and more data are easily generated. This applies not only to volume and variety, but variability as well.
- Value: Not all data are useful, or worth collecting. In this context, 'value' can just as well mean a price as an effort (time spent).
- Vagaries: There are no limits in the ability to accumulate data. Health monitor devices are just one example adding to the behavome (the totality of an individual's behaviours) and exposome (the totality of environmental exposures of an individual) to the large 'omics family.

19.2.4 Infrastructure

Big Data is creating new challenges for the information technology (IT) infrastructure as the increased number of transfers from storage devices to processing units are putting resources under stress.

Storage

More data obviously mean more space required for storage. The byte is the unit of digital information and consists of eight bits (see Section 19.2.1). While we are often manipulating files of the size of kB (kilobytes of data, 10^3 bytes) to MB (1000 kB, 10^6 bytes) for images and presentations, Big Data generally refers to working with data on the TB (terabyte, 10^{12} bytes) scale and larger. Many data scientists have already entered the world where petabyte (10^{15} bytes) is a common measure, while some plan for the exabyte (10^{18} bytes) world. Not only does it require expensive large-scale facilities to store the data, it also becomes cumbersome to transport, and therefore duplicate. New challenges arise in the form of decisions on whether/what to back up, and how to timetable computing, since data cannot easily be transferred from one computing centre to another.

Computing

Computing of large datasets can be testing for the existing IT infrastructure. In the best scenario, Big Data corresponds to more of the same, in which case it just requires more central processing unit (CPU) resources, and either takes proportionally longer to process, or allows for a parallel split of the computational work across machines. For example, data across the genome can sometimes be split into individual chromosomes, each separately processed and then recombined. This is, however, not always possible, since most computing clusters have a maximum amount of time allowed to process a job (so-called walltime). Such limitations are usually put in place to keep rogue processes under control, but the Big Data revolution can make such safeguarding tools obsolete. If computational processes cannot be subdivided into smaller-size tasks, they will either require more random access memory (RAM) or use swap memory; since swap space is located on hard drives as opposed the the physical memory provided by RAM, swap memory has a slower access time than physical memory. Utilising large amounts of RAM, if available, can save a lot of computing time.

Algorithms

One of the smartest ways to handle the deluge of data and minimise its impact on storage and compute requirements is to improve the analytical methods used. Coming back to the representation of data (see Section 19.2.1), it is possible to increase the efficiency of data handling transitioning from text (alpha-numeric) to binary storage. For example, rather than storing the text patterns 'Male' and 'Female' which take 4+6 letters (32 or 48 bits, respectively), it is possible to reduce it to a binary 0/1 format ('0': male, '1': female) which is an order of magnitude smaller to store (1 bit if stored as Boolean, 8 bits if stored as Integer; see Section 18.2.2 for data types). Such efficient conversion disconnects the human reader from being able to directly interpret the data, but has the positive side effect of enforcing its clear definition.

A very positive algorithmic development related to Big Data includes the recognition of patterns otherwise hidden from view. This is typically what machine learning achieves, a common form of so-called artificial intelligence. Still, such methods are not without issue, as model-free approaches can lead to over-fitting, which typically occurs when conclusions are too closely linked to a specific dataset and not reproducible when another identical experiment is performed. A famous such case is Google Flu Trends which achieved world-wide fame in 2009 for its ability to predict the flu epidemic using purely data from its search engine. Unfortunately, this success could not be repeated in the same survey the very next year.

19.2.5 Analytical Languages

Since each tool has its strengths and weaknesses, one can enter an interminable debate on the right tool to use. The choice of analytical tool is very much dependent on the field, type of application and user preference. The following briefly introduces some commonly used and freely available tools.

AWK (awk, mawk, nawk, gawk)

AWK uses a text-driven approach. It is a very light processing tool and easy to use in the command line. It has been in existence for decades and is typically used for processing very long files which can be treated line by line.

```
# Script 19.1
# Processing of the ASCII-file file.tab with tab-separated values
awk '
    BEGIN {FS="\t"}
    (NR>1){
        if ($1>2) {print $1,$2-$4} else {print $1,$3}
        n=n+1
    }
    END {print "This is the End after processing "n" lines."}
' file.dat
```

Python

Python is a relatively new language that has become increasingly popular in the field of Big Data and bioinformatics (see also Chapter 18). Originally conceived as object-oriented, it is extremely clear to write, read and modify. Python is modular in nature and a large range of modules are available; of note are the very useful Pandas module to manipulate data and SciPy to perform analysis and generate graphics.

```
# Script 19.2
# Calculate the sum of squares of data in an array
s = 0
data = [3, -1, 2, -5, None, 9, -2]
for d in data:
    if not d is None:
        s += d * d
print('The sum of squares is ',s)
```

R

R has been designed for statistical analysis as a freely available open-source incarnation of Splus. It can handle complex structures, perform the most complex of statistical tests and generate publication-quality graphics. Being an advanced software, it is more challenging to master than a spreadsheet application, but easy to modify. It is ideal for all data science applications, but is particularly useful for the biological sciences, as numerous applications have been developed in R and are easily imported. These libraries are organised and are available in the BioConductor package manager (see Section 19.2.6).

```
# Script 19.3
# Plotting a normal distribution
> x<-seq(-4,4,length=100)
> dx<-dnorm(x,mean=0,sd=1)
> plot(x,dx,type="l",main="Normal Distribution")
> p1<-seq(-4,-1.5,length=100)
> dp1<-dnorm(p1)
> p2<-seq(1.5,4,length=100)
> dp2<-dnorm(p2)
> polygon(c(-4,p1,-1.5),c(0,dp1,0),col="gray")
```

In the following sections, we will demonstrate some applications of R; AWK has no graphical abilities, and Python is covered in Chapter 18.

19.2.6 Getting Started With R

Installation

R is a freely available, open-source effort led by the Comprehensive R Archive Network (CRAN) which has made the software environment available across different platforms (Windows, Mac, Linux). You can download ready-to-install binaries from their website, which also hosts a wealth of documentation and additional packages. Of particular interest is R-Studio, which is a user-friendly interface or integrated desktop environment (IDE), allowing for simultaneous coding in R, examination of data and results.

Essentials

R can be run from the command line where text is typed after a prompt (>); while a script might be presented as

```
> print("Hello World!")
```

it means you simply need to type `print("Hello World!")` after the prompt and hit the Enter key. Comments are written after the # character and are not taken into account; for example:

```
> print("Hello World!") # Print Hello World to the screen
```

Here are a few essential commands for use within R:

```
> print("this text")  # Print "this text" to the screen
> q()                  # Quit R without saving anything
> help(print)          # Instant help on the function print
# Execute the file script.R as if it was typed as command line
> source("script.R")
# Give instructions on how to use the function
> ?function_name
```

Packages

R itself is an open processing environment that supports the development of new packages. These consist of libraries of data and/or scripts developed for a specific use. While the most basic functions such as `mean` or `t.test` are built into R, there is no limit to what one might want to use. If you are not a cutting-edge statistician and you do not find the function you are looking for in R, chances are someone else has already developed and shared their work in an R repository. Datasets and functions defined in a package are accessible only once it is loaded with the `library()` function:

```
# Script 19.4
# List packages available, not necessarily active
> library()
> data(agriculture) # Attempt to load the Agriculture dataset
Warning message:
In data(agriculture) : dataset 'agriculture' not found
# Run bannerplot function
> bannerplot(agnes(agriculture), main = "Bannerplot")
Error: could not find function "bannerplot"
> library(cluster) # Load the cluster library and try again
> data(agriculture)
> bannerplot(agnes(agriculture), main = "Bannerplot")
```

Sometimes the chosen package is not installed on your system, but it is available from the CRAN repository and installed using the R package manager. If you are connected to the World Wide Web, it is therefore straightforward to retrieve the package content to get it installed, allowing for it to be loaded:

```
# Script 19.5
> library(pwr) # Attempt to load the power tool library, pwr
Error in library(pwr) : there is no package called 'pwr'
> install.packages("pwr") # Install the "pwr" package
Installing package into '/Users/cazierj/Library/R/3.1/library'
(as 'lib' is unspecified)
--- Please select a CRAN mirror for use in this session ---
trying URL
'http://www.stats.bris.ac.uk/R/bin/macosx/mavericks/contrib/3.1/
pwr_1.1-3.tgz'
Content type 'application/x-gzip' length 56256 bytes (54 Kb)
opened URL
==========================================================
downloaded 54 Kb

The downloaded binary packages are in
     /var/folders/fy/downloaded_packages
> library(pwr)
> ?pwr
pwr-package {pwr} R Documentation
Basic Functions for Power Analysis pwr

Description

Power calculations along the lines of Cohen (1988) using in
particular the same notations for effect sizes. Examples from
the book are given.
[...]
```

For biochemical applications, there is a dedicated branch of package development, BioConductor, which has its own repository and simplified interface. It can be loaded simply by sourcing (i.e. executing) an R script available on the internet:

```
# Script 19.6
# Execute the online script biocLite.R
> source("http://bioconductor.org/biocLite.R")
Installing package into '/Users/cazierj/Library/R/3.1/library'
(as 'lib' is unspecified)
trying URL
'http://www.bioconductor.org/packages/3.0/bioc/bin/macosx/
mavericks/contrib/3.1/BiocInstaller_1.16.5.tgz'> biocLite("mQTL.
NMR") # Install the specific package mQTl.NMR, and all its
dependencies automatically
BioC_mirror: http://bioconductor.org
Using Bioconductor version 3.0 (BiocInstaller 1.16.5), R version
3.1.1.
```

```
Installing package(s) 'mQTL.NMR'
also installing the dependencies 'GenABEL.data', 'qtl', 'GenABEL',
'outliers'

trying URL
'http://www.stats.bris.ac.uk/R/bin/macosx/mavericks/contrib/3.1/
GenABEL.data_1.0.0.tgz'
Content type 'application/x-gzip' length 2417502 bytes (2.3 Mb)
opened URL
==================================================
downloaded 2.3 Mb

trying URL
'http://www.stats.bris.ac.uk/R/bin/macosx/mavericks/contrib/3.1/
qtl_1.39-5.tgz'
Content type 'application/x-gzip' length 5514460 bytes (5.3 Mb)
opened URL
==================================================
downloaded 5.3 Mb

trying URL
'http://www.stats.bris.ac.uk/R/bin/macosx/mavericks/contrib/3.1/
GenABEL_1.8-0.tgz'
Content type 'application/x-gzip' length 3769294 bytes (3.6 Mb)
opened URL
==================================================
downloaded 3.6 Mb

trying URL
'http://www.stats.bris.ac.uk/R/bin/macosx/mavericks/contrib/3.1/
outliers_0.14.tgz'
Content type 'application/x-gzip' length 51326 bytes (50 Kb)
opened URL
==================================================
downloaded 50 Kb

trying URL
'http://bioconductor.org/packages/3.0/bioc/bin/macosx/mavericks/
contrib/3.1/mQTL.NMR_1.0.0.tgz'
Content type 'application/x-gzip' length 1774669 bytes (1.7 Mb)
opened URL
==================================================
downloaded 1.7 Mb

The downloaded binary packages are in
    /var/folders/fy/downloaded_packages

> ??mQTL.NMR # Look at the built-in documentation of the package
```

```
mQTL.NMR-package {mQTL.NMR}     R Documentation
Metabolomic Quantitative Trait Locus mapping for 1H NMR data
```

```
Description
```

mQTL.NMR provides a complete mQTL analysis pipeline for 1H NMR data. Distinctive features include normalisation using most-used approaches, peak alignment using RSPA approach, dimensionality reduction using SRV and binning approaches, and mQTL analysis for animal and human cohorts.
```
[...]
```

19.2.7 Manipulating Data With R

Simple Data Types

Data can be typed from the command line or loaded directly from files. As organised and structured data are essential in R, many default, yet conservative, rules are applied on the fly.

```
# Script 19.7
> x<-3          # Set x to hold the value 1
> print(x)      # Display its value with the print() command
[1] 3
> x             # Or even simply by stating its name
[1] 3
> x+x           # Simple operations are straight forward
[1] 6
> x*x
[1] 9
> sqrt(x)       # even the square root, using a built-in function
[1] 1.732051
> y<- -3.4
> y
[1] -3.4
> x-y           # Calculate the difference between numbers x and y
[1] 6.4
> t<-"Text"     # Set t to hold the String "Text"
> t
[1] "Text"
> x-t           # Calc. of the difference between x and t fails
Error in x - t : non-numeric argument to binary operator
> typeof(x)     # x is a number with double precision
[1] "double"
> typeof(t)     # t is a string with character type
[1] "character"
> z<- TRUE      # Set a BOOLEAN value TRUE to z
> z
[1] TRUE
> x-z           # The difference between x and z provides an answer
[1] 2
```

```
> typeof(z)
[1] "logical"
> typeof(x-z)  # The BOOLEAN 'TRUE' is transformed into value 1
[1] "double"   # double 3 - 1 = 2
```

More complex constructs are available, for example vectors; these are single entities of an ordered collection of data of the same data type. To set up a vector named Courier New consisting of a list of numbers, the concatenate function c() is used as per

```
# Script 19.8
# Create a vector of length 5 with set values
> v <- c(10, 1.6, 1, 2.6, 1.7)
> v
[1] 10.0 1.6 1.0 2.6 1.7
> w<-1:10    # Create a vector of length 10
> w
[1] 1 2 3 4 5 6 7 8 9 10
> v+1         # One can operate on all numbers
[1] 11.0 2.6 2.0 3.6 2.7
> v*w         # Even between vectors
[1] 10.0 3.2 3.0 10.4 8.5 60.0 11.2 8.0 23.4 17.0
> sqrt(v)    # Or more complex operation like square root
[1] 3.162278 1.264911 1.000000 1.612452 1.303840
> mean(v)    # Average across the values of the vector
[1] 3.38
> sd(v)       # Standard Deviation
[1] 3.744596
> sum(v)      # Sum
[1] 16.9
> length(v) # Length of the vector
[1] 5
```

Advanced Data Types

The advanced data types include matrices and a more complex data frame that allows different types to be stored together.

Arrays are multi-dimensional generalisations of vectors, i.e. a matrix with an index. An array is a vector of vectors and can therefore be indexed by two or more indices and can be printed in special ways:

```
# Script 19.9
# Create a 2x3 array containing values from 2 to 7
> m<-array(2:7,dim=c(2,3))
> m
     [,1] [,2] [,3]
[1,] 2 4 6
[2,] 3 5 7
> m[1,3]         # Access a single cell by index
[1] 6
```

```
> m[1,]          # Access a single row
[1] 2 4 6
> m[,c(1,3)]    # Access a pair of columns
     [,1] [,2]
[1,]  2    6
[2,]  3    7
```

Factors are the categorical equivalent of the vector and are ideal for storing qualitative rather than quantitative data.

```
# Script 19.10
> hair<-c("black","fair","black","black","brown","fair","white",
"brown","white","bald")
> f_hair<-factor(hair)   # Make a factor from this vector
> f_hair
[1] black fair black black brown fair white brown white bald
Levels: bald black brown fair white
> levels(f_hair)          # The categories are the levels
[1] "bald" "black" "brown" "fair" "white"
# There are no levels in the original vector
> levels(hair)
NULL
```

Lists are generalised vectors and allow elements that are not of the same type with a named value. Lists are a convenient way to return the results of a statistical computation.

```
# Script 19.11
> res <- list(test_name=c("t.test","Fisher_exact"), pval=0.001,
df=3, err=c(0.4,0.7,0.9)) # List of 4 components
> res
$test_name
[1] "t.test" "Fisher_exact"
$pval
[1] 0.001
$df
[1] 3
$err
[1] 0.4 0.7 0.9
> res$pval                   # Access the full element pval
[1] 0.001
> res$err                    # Access the full element err
[1] 0.4 0.7 0.9
> res$test_name[1]           # Access an element in an element
[1] "t.test"
```

Data frames are the most complex, most useful and therefore most frequently used data type. Data frames are like a matrix, but with columns of different types. They can be seen as data matrices with one row per observational unit, but with (possibly) both numerical and categorical variables. Experiments are often described with data frames: treatments are categorical, while response is numerical.

File handling

Large amounts of data should not be entered manually, but loaded directly from files. Just like Python and the Pandas module, R provides various tools to get input from data stored in files.

The function `read.table()` is the obvious tool to read structured data into a data frame. If we have the following data contained in a file named `table.dat` it can be loaded as per:

SampleID	Gender	Weight	Age	Affected
S1	Male	66.3	20	TRUE
S2	Male	80.2	21	FALSE
S3	Female	53.1	29	TRUE
S4	Female	21.3	18	TRUE
S5	Female	54.6	12	FALSE
S6	Female	78.1	23	TRUE

```
# Script 19.12
# Read the table considering the first row as data
> read.table("table.dat")
          V1     V2     V3  V4        V5
1 SampleID Gender Weight Age  Affected
2         S1   Male   66.3  20      TRUE
3         S2   Male   80.2  21     FALSE
4         S3 Female   53.1  29      TRUE
5         S4 Female   21.3  18      TRUE
6         S5 Female   54.6  12     FALSE
7         S6 Female   78.1  23      TRUE
```

or

```
# Script 19.13
# Read the table and use the first row to name the columns
> read.table("table.dat",header=T)
  SampleID Gender Weight Age Affected
1       S1   Male   66.3  20     TRUE
2       S2   Male   80.2  21    FALSE
3       S3 Female   53.1  29     TRUE
```

```
4        S4 Female    21.3  18      TRUE
5        S5 Female    54.6  12      FALSE
6        S6 Female    78.1  23      TRUE
```

If some data are missing, an error is returned. As the fact that data are missing is recognised when the number of field separators (e.g. tabulation signs, \t) is less than expected, this can be alleviated if a field separator is given where an element in the set is missing. In this case the missing field will be set with NA status.

SampleID	Gender	Weight	Age	Affected
S1	Male	66.3	20	TRUE
S2	Male	80.2	21	FALSE
S3	Female	53.1	29	TRUE
S4	Female	21.3	18	TRUE
S5	Female	54.6		FALSE
S6	Female	78.1	23	TRUE

```
# Script 19.14
# Read the table with missing entries
> read.table("table2.dat")
Error in scan(file, what, nmax, sep, dec, quote, skip, nlines,
na.strings,   :
  line 6 did not have 5 elements

> t<-read.table("table2.dat",header=T,sep="\t")
> print(t)
  SampleID Gender Weight Age Affected
1       S1   Male   66.3  20     TRUE
2       S2   Male   80.2  21    FALSE
3       S3 Female   53.1  29     TRUE
4       S4 Female   21.3  18     TRUE
5       S5 Female   54.6  NA    FALSE
6       S6 Female   78.1  23     TRUE

> typeof(t)     # The data are interpreted as a list of various
[1] "list"       # subtypes
> typeof(t$Age)
[1] "integer"
> typeof(t$Affected)
[1] "logical"
> typeof(t$Gender)
[1] "integer"
> summary(t)    # Obtain a summary of the values per type
```

```
SampleID      Gender        Weight                Age              Affected
S1:1       Female:4    Min.    :21.30    Min.    :18.0    Mode :logical
S2:1       Male  :2    1st Qu.:53.48    1st Qu.:20.0    FALSE:2
S3:1                   Median :60.45    Median :21.0    TRUE :4
S4:1                   Mean   :58.93    Mean   :22.2    NA's :0
S5:1                   3rd Qu.:75.15    3rd Qu.:23.0
S6:1                   Max.   :80.20    Max.   :29.0
                                        NA's    :1
```

The function `scan()` is a more primitive tool, but allows for more flexibility in the data stored in a file.

Built-in data can be found in packages alongside the functions, as seen previously and is useful for testing the use of a function.

```
# Script 19.15
> data(euro)   # Load the built-in Euro conversion values
> euro
        ATS            BEF            DEM            ESP            FIM
FRF            IEP
   13.760300     40.339900      1.955830   166.386000      5.945730
 6.559570      0.787564
        ITL            LUF            NLG            PTE
1936.270000     40.339900      2.203710   200.482000
> data(package="cluster")   # Datasets accessible from packages
Data sets in package 'cluster':

agriculture               European Union Agricultural Workforces
animals                   Attributes of Animals
chorSub                   Subset of C-horizon of Kola Data
flower                    Flower Characteristics
plantTraits               Plant Species Traits Data
pluton                    Isotopic Composition Plutonium Batches
ruspini                   Ruspini Data
votes.repub               Votes for Republican Candidate in
                          Presidential Elections
xclara                    Bivariate Data Set with 3 Clusters
# Built-in data can be loaded from a package
> data(flower, package="cluster")
> data(cars)
> dim(cars)
[1] 50  2
> cars[1:10,]    # A subset of the dataset can be used
    speed dist
1      4    2
2      4   10
3      7    4
4      7   22
```

```
5      8    16
6      9    10
7     10    18
8     10    26
9     10    34
10    11    17
> summary(cars)    # and relevant content displayed
      speed                dist
  Min.    : 4.0    Min.    :  2.00
  1st Qu.:12.0    1st Qu.: 26.00
  Median :15.0    Median : 36.00
  Mean    :15.4    Mean    : 42.98
  3rd Qu.:19.0    3rd Qu.: 56.00
  Max.    :25.0    Max.     :120.00
```

19.2.8 Plotting Data

It is essential to use the appropriate graphical representation of data. Often, there are different ways in which data can be represented in order to convey a message. However, the goal is always to highlight the relevant information about the data in an instant. Although the following is not an exhaustive list, it summarises some specific types of plots that are common in the scientific field:

- Scatter plots are the most common representation of simple data when an output is a function of input, whether it is categorical or not; they allow one to spot a trend or an outlier quickly.
- Histograms are most useful when dealing with large datasets, combining the frequency of occurrence of individual phenomena into an overall distribution.
- Box plots are extremely common in biochemical sciences where the distribution of a measure can be compared between categories, or treatments. Such plots present the distributions of the values on the y-axis with the standard error, and the categories on the x-axis.
- Heatmaps have proven very popular to cluster data between two main factors, highlighting patterns. Combined with a dendrogram, this type of representation is very good at highlighting categories that cluster together. The use of colours to represent a continuous measurement is an alternative to 3D representation, but there lies a danger in the choice of grading that may mislead conclusions.
- Pie charts might be very popular, but they have been shown time and again to be misleading. For example, the visual perception of a percentage is biased, and small groups (with a low percentage) are almost invisible.

The choice of the appropriate representation depends both on the type of data (categorical, quantitative or mixed) and on the message it is intended to convey.

Plotting Data With R

Not only is the use of the appropriate graphical representation important to present the desired results, it should also not mislead the reader. Graphical representations are often closely linked to some analysis. The following examples highlight the use of the different type of plots introduced above (with the exception of pie charts) using R.

(a) Cars dataset   (b) Ruspini dataset   (c) Titanic dataset

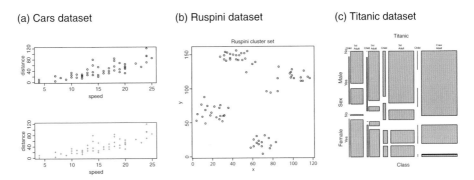

Figure 19.1 Illustration of scatter plots using the above examples of (a) Cars, (b) Ruspini and (c) the Titanic dataset. The latter dataset provides information on the fate of passengers on the fatal maiden voyage of the ocean liner Titanic, summarised according to economic status (class), sex, age and survival.

Scatter Plots

In a scatter plot, data are ordered automatically according to quantities plotted on the two axes; the provided list of data thus does not necessarily have to be ordered (Figure 19.1). The easiest way to plot values of two variables against each other with default parameters is to use the `plot()` function.

```
# Script 19.16
# Cars dataset visualised as scatter plot
> data(cars)
> dim(cars)
[1] 50 2
> names(cars)
[1] "speed" "dist"
> par(mfrow=c(2,1)) # Split the graphic in 2 rows and 1 column
> plot(cars)
> plot(cars,pch="+",xlab="Speed",ylab="Distance",col="red")
# Or use several parameters
```

```
# Script 19.17
# Ruspini dataset visualised as scatter plot
> library(cluster)
> head(ruspini) # Another set of x vs y data
    x   y
1   4   53
2   5   63
3  10   59
4   9   77
5  13   49
6  13   69
> plot(ruspini, main="Ruspini cluster set")
# Highlights a clustering pattern
```

```
# Script 19.18
# Titanic dataset visualised as scatter plot
> plot(Titanic,main="Titanic")
```

Histograms

Histograms provide a visual interpretation of numerical data by indicating the number of data points that lie within a range of values. These range of values are called classes or bins. The frequency of the data that falls in each class is depicted by the use of a bar. The higher that the bar is, the greater the frequency of data values in that bin. Notice that, unlike a bar chart (which is typically used for categorical data), there are no gaps between the histogram bars, because a histogram represents a continuous dataset. Nevertheless, some bars might have zero height due to the absence of a particular event.

```
# Script 19.19
# The Cars dataset visualised as a histogram
> par(mfrow=c(2,1))
> hist(cars$speed)
> hist(cars$speed,breaks=15,col="gray")
> abline(v=21,col="red")
```

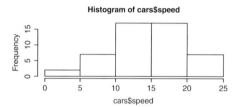

Histogram of cars$speed

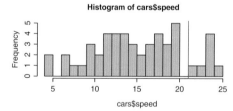

Histogram of cars$speed

Figure 19.2 Two different histogram plots of the Cars dataset.

When constructing histograms, one needs to decide whether scores are rounded up or rounded down at the boundaries of bins. Also, a suitable bin width needs to be chosen such that the bins are not too thin or too wide. A small bin width results in resolution of too much individual data and makes it difficult to see the underlying pattern. If the bin width is chosen too large, too much individual information is merged, again making it difficult to recognise underlying trends in the data (Figure 19.2).

When a frequency distribution is normally distributed, we can find out the probability of a score occurring by standardising the scores, known as standard scores (or Z-scores). Z-scores are expressed in terms of standard deviations from their means; therefore, they have a distribution with a mean of 0 and a standard deviation of 1.

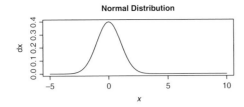

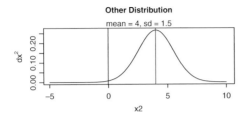

Figure 19.3 A normal distribution (top) has a mean of 0 and a standard deviation of 1. In contrast, distributions with non-zero mean and a standard deviation different from 1 are not called 'normal' (bottom).

Applying a Z-transformation to a dataset thus results in a shifting of the mean value to 0 and a scaling such that a standard deviation of 1 is achieved (Figure 19.4).

```
# Script 19.20
# A normal distribution with a mean of 0 and std. dev. of 1
# Split the display in 2 rows and 1 column
> par(mfrow=c(2,1))
> x<-seq(-5,10,length=100)    # Interval from 5 to 10
> dx<-dnorm(x)                # Default normal distribution
# Plot distribution
> plot(x,dx,type="l",main="Normal Distribution")
```

```
# Script 19.21
# Another distribution with mean of 4 and std. dev. of 1.5
> x2<-x
> dx<-dnorm(x2,mean=4,sd=1.5)
# Plot distribution
> plot(x2,dx2,type="l",main="Other Distribution")
> abline(v=4,col="red")         # Draw vertical line in red at mean
> abline(v=0,col="blue")        # Draw vertical line in blue at 0
> mtext(text="mean=4, sd=1.5")  # Include extra text on top
```

We can apply such a Z-transformation to the speed data of the Cars dataset:

```
# Script 19.22
# Z-transformation of the Cars dataset
> par(mfrow=c(2,1))
> s<-cars$speed
> zs<-(s-mean(s))/sd(s)
> hist(cars$speed,breaks=10,col="gray",xlim=c(-5,25),main="Speed
Distribution",xlab="Speed")
> hist(zs,breaks=10,xlim=c(-5,25),main="Z Speed Distribution",
xlab="Z transform of Speed")
```

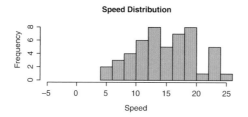

Figure 19.4 Histogram representation of the Cars dataset before (top) and after a Z-transformation (bottom).

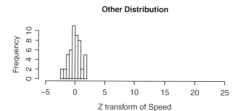

Box Plots

In this type of representation, the bottom and the top of the box indicate the first and third quartiles; the band inside the box shows the position of the second quartile (the median). The lines extending vertically from the boxes are called whiskers and indicate the variability outside the upper and lower quartiles (Figure 19.5).

```
# Script 19.23
# Box plot of a dataset of smoking-related deaths
# breslow dataset available from
# https://vincentarelbundock.github.io/Rdatasets/datasets.html
> data(breslow)
# Smoking = 1, No Smoking = 0; y is number of related deaths
> breslow[1:5,]
age smoke n y ns
1 40 0 18790 2 0
2 50 0 10673 12 0
3 60 0 5710 28 0
4 70 0 2585 28 0
5 80 0 1462 31 0
> boxplot(y~smoke,data=breslow,main="Smoking-related deaths",
xlab="Smoking",ylab="Number of related deaths")
```

Heatmaps

A heatmap visualises data where the individual values contained in a matrix are represented as colours (Figure 19.6). The implementation in R closely follows the protocols described by Eisen and colleagues, and also draws dendrograms of the cases and variables using correlation similarity metric and average linkage clustering.

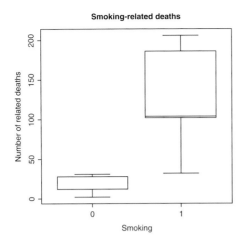

Smoking-related deaths

Figure 19.5 Illustration of a box plot with whiskers using the Breslow dataset of smoking-related deaths.

```
# Script 19.24
# Heatmap representation of the data with automatic construction
# of dendrogram to reflect the organisation of the data
> par(mfrow=c(2,1))
> x<-as.matrix(swiss) # Investigate the fertility in Swiss cantons
> head(x)
              Fertility Agriculture Examination Education Catholic
Infant.Mortality
Courtelary        80.2        17.0          15        12     9.96
22.2
Delemont          83.1        45.1           6         9    84.84
22.2
Franches-Mnt      92.5        39.7           5         5    93.40
20.2
Moutier           85.8        36.5          12         7    33.77
20.3
Neuveville        76.9        43.5          17        15     5.16
20.6
Porrentruy        76.1        35.3           9         7    90.57
26.6
> heatmap(as.matrix(swiss))

# Script 19.25
# Same plot with different colour scheme and added colour on the
# sides from a Rainbow palette
> rc<-rainbow(nrow(x),start=0, end =.3)
> cc<-rainbow(ncol(x),start=0, end =.3)
> heatmap(x,col=cm.colors(256),scale="column",RowSideColors=rc,
ColSideColors=cc,xlab="Specification variables",
ylab="Cantons",main="Heatmap")
```

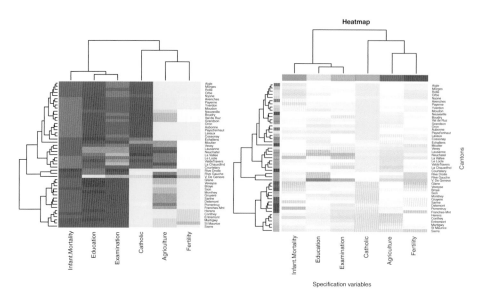

Figure 19.6 Visualisation of Swiss fertility and socioeconomic indicators as a heatmap, using the above examples.

More Graphics in R: ggplot2

The package ggplot2 is a complete plotting system for R. It provides the ability to produce complex multi-layered graphics while taking care of fiddly details such as drawing legends (Figure 19.7). Mastering the ggplot2 language is somewhat more advanced than basic R usage, but excellent reference and tutorial resources are available (see Further Reading section 19.5.4).

```
# Script 19.26
# A few official ggplot2 examples
> install.packages("ggplot2")
Installing package into '/Users/cazierj/Library/R/3.1/library'
(as 'lib' is unspecified)
> library(ggplot2)


# Script 19.27
# Create factors with value labels
mtcars$gear <-factor(mtcars$gear,levels=c(3,4,5),
  labels=c("3gears","4gears","5gears"))
mtcars$am <- factor(mtcars$am,levels=c(0,1),
  labels=c("Automatic","Manual"))
mtcars$cyl <-factor(mtcars$cyl,levels=c(4,6,8),
  labels=c("4cyl","6cyl","8cyl"))
```

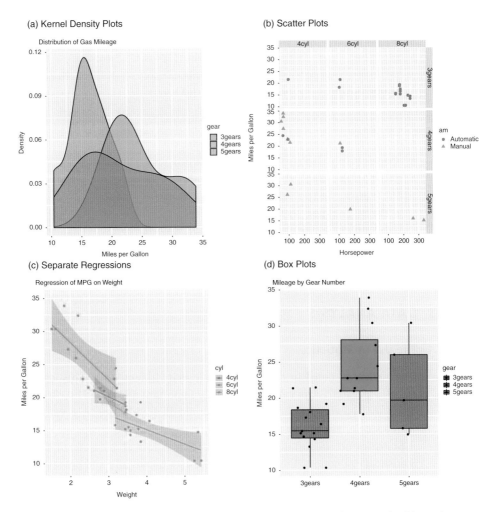

Figure 19.7 Illustration of data representations generated with ggplot2 using above examples. (a) Kernel density plots, (b) scatter plots, (c) separate regressions and (d) box plots.

```
# Script 19.28
# Kernel density plots for mpg
# grouped by number of gears (indicated by color)
> qplot(mpg, data=mtcars, geom="density", fill=gear,alpha=I(.5),
    main="Distribution of Gas Milage", xlab="Miles Per Gallon",
    ylab="Density")


# Script 19.29
# Scatter plots of mpg vs. hp
# for each combination of gears and cylinders
# in each facet, transmission type is represented by shape and colour
> qplot(hp, mpg, data=mtcars, shape=am, color=am,facets=gear~cyl,
    size=I(3), xlab="Horsepower", ylab="Miles per Gallon")
```

```
# Script 19.30
# Separate regressions of mpg on weight for each number of cylinders
> qplot(wt, mpg, data=mtcars, geom=c("point","smooth"), method="lm",
    formula=y~x, color=cyl, main="Regression of MPG on Weight",
    xlab="Weight", ylab="Miles per Gallon")


# Script 19.31
# Box plots of mpg by number of gears
# observations (points) are overlayed and jittered
> qplot(gear, mpg, data=mtcars, geom=c("boxplot","jitter"),
    fill=gear, main="Mileage by Gear Number", xlab="",
    ylab="Miles per Gallon")
```

These are just a few of the most common ways to represent data. Many more types of applications are possible and further examples are available for browsing in the R Graph Gallery.

19.3 DATA ANALYSIS

19.3.1 The Use of Statistics

Leek and Peng noted in their 2015 article that "[t]here is no statistic more maligned than the *p*-value". A *p*-value is genuinely important and full of essential information. Being that powerful a tool, it should be handled with care. A *p*-value is associated with a single test, not an experiment, even if the test is a composite of many tests.

Statistics are important as they represent a formal measure of the experiment. Statistical evaluation allows us to analyse data impartially and prevent us from being misled by spurious results, personal bias or, even worse, common sense. Albeit intuitive decisions and conclusions guided by common sense can help us discard ludicrous ideas, such pre-factual approaches are rather restrictive and the chances are that we would still believe that the Earth was flat, and the Sun rotated around us.

The relevance of statistics in the context of data analysis, and the *p*-value in particular, lies in its close relationship to key concepts such as those of false-positives and false-negatives (a.k.a. alpha and beta values, or Type I and Type II errors; see Section 19.3.9). We shall look into details of these concepts alongside power, which is most relevant to the design of an experiment, for example to inform about how many measurements need to be taken in order to have an acceptable false-positive and false-negative rates.

19.3.2 Analytical Considerations and Experimental Error

Many biochemical investigations involve the quantitative determination of the concentration and/or amount of a particular component (the analyte) present in a test sample. For example, in studies of the mode of action of enzymes, trans-membrane transport and cell signalling, the measurement of a particular reactant or product is investigated as a function of a range of experimental conditions and the data used to calculate kinetic or thermodynamic constants. These, in turn, are used to deduce details of the mechanism of the biological process taking place. Irrespective of the experimental rationale for undertaking such quantitative studies, all quantitative experimental data must first be questioned and validated in order to give credibility to the derived data and the conclusions that can be drawn from them. This is particularly important in the field of clinical biochemistry, in which quantitative measurements on a patient's blood and urine samples are used to aid a clinical diagnosis and monitor the patient's recovery from a particular disease.

Thorough appraisal requires that the experimental data be assessed and confirmed as an acceptable estimate of the true values by the application of one or more standard statistical tests. Evidence of the validation of quantitative data by the application of such tests is required by the editors of refereed journals for the acceptance for publication of draft research papers. The following sections will address the theoretical and practical considerations behind these statistical tests.

Selecting an Analytical Method

The nature of the quantitative analysis to be carried out will require a decision to be taken based on the analytical technique to be employed. A variety of analytical methods may be capable of achieving the desired analysis and the decision to select one may depend on a variety of issues. These include:

- the availability of specific pieces of apparatus
- the precision, accuracy and detection limits of the competing methods
- the precision, accuracy and detection limit acceptable for the particular analysis
- the number of other compounds present in the sample that may interfere with the analysis
- the potential cost of the method (particularly important for repetitive analysis)
- the possible hazards inherent in the method and the appropriate precautions needed to minimise risk
- the published literature method of choice
- personal preference.

The most common biochemical quantitative analytical methods are visible, ultraviolet and fluorimetric spectrophotometry, chromatographic techniques such as HPLC and GC coupled to spectrophotometry or mass spectrometry, ion-selective electrodes and immunological methods such as ELISA. Once a method has been selected, it must be developed and/or validated using the approaches discussed in the following sections. If it is to be used over a prolonged period of time, measures will need to be put in place to ensure that there is no drift in response. This normally entails an internal quality-control approach using reference test samples covering the analytical range that are measured each time the method is applied to test samples. Any deviation from the known values for these reference samples will require the whole batch of test samples to be reassayed.

The Nature of Experimental Errors

Every quantitative measurement has some uncertainty associated with it. This uncertainty is referred to as the experimental error, which is a measure of the difference between the true value and the experimental value. The true value normally remains unknown except in cases where a standard sample (i.e. one of known composition) is being analysed. In other cases, it has to be estimated from the analytical data by the methods that will be discussed later. The consequence of the existence of experimental errors is that the measurements recorded can be accepted with a high, medium or low degree of confidence, depending upon the sophistication of the technique employed, but seldom, if ever, with absolute certainty. Experimental error may be of two kinds: systematic error and random error.

Systematic Error

Systematic errors (also called determinate errors) are consistent errors that can be identified and either eliminated or reduced. They are most commonly caused by a fault or inherent limitation in the apparatus being used, but may also be influenced by poor experimental design. Common causes include the misuse of manual or automatic pipettes, the incorrect preparation of stock solutions, and the incorrect calibration and use of pH meters. Such errors may be constant (i.e. have a fixed value irrespective of the amount of test analyte present in the test sample under investigation) or proportional (i.e. the size of the error is dependent upon the amount of test analyte present). Thus, the overall effect of the two types in a given experimental result will differ. Both of these types of systematic error have three common causes:

- Analyst error: This is best minimised by thorough training and/or by the automation of the method.
- Instrument error: This may not be eliminable and hence alternative methods should be considered. Instrument error may be electronic in origin or may be linked to the composition of the sample.
- Method error: This can be identified by comparison of the experimental data with that obtained by the use of alternative methods.

Identification of Systematic Errors

Systematic errors are always reproducible and may be positive or negative, i.e. they increase or decrease the experimental value relative to the true value. The crucial characteristic, however, is that their cause can be identified and corrected. There are four common means of identifying this type of error:

- Use of a control sample (often called a 'blank'): This is a sample that you know contains none of the analytes under test so that if the method gives a non-zero answer then it must be responding in some unintended way. The use of blank samples is difficult in cases where the composition of the test sample is complex, for example, serum.
- Use of a standard reference sample: This is a sample of the test analyte of known composition so the method under evaluation must reproduce the known answer.
- Use of an alternative method: If the test and alternative methods give different results for a given test sample then at least one of the methods must have an in-built flaw.
- Use of an external quality assessment sample: This is a standard reference sample that is analysed by other investigators based in different laboratories employing the same or different methods. Their results are compared and any differences in excess of random errors (see below) identify the systematic error for each analyst. The use

of external quality assessment schemes is standard practice in clinical biochemistry laboratories (see Section 10.2.3).

Random Errors

Random errors (also called indeterminate errors) are caused by unpredictable and often uncontrollable inaccuracies in the various manipulations involved in the method. Such errors may be variably positive or negative, and are caused by factors such as difficulty in the process of sampling, random electrical noise in an instrument or by the analyst being inconsistent in the operation of the instrument or in recording readings from it.

Standard Operating Procedures

The minimisation of both systematic and random errors is essential in cases where the analytical data are used as the basis for a crucial diagnostic or prognostic decision, as is common, for example, in routine clinical biochemical investigations and in the development of new drugs. In such cases, it is normal for the analyses to be conducted in accordance with standard operating procedures (SOPs) that define in full detail the quality of the reagents, the preparation of standard solutions, the calibration of instruments and the methodology of the actual analytical procedure that must be followed. In addition, a well-written SOP will allow any investigator to produce identical results, eliminating operator-induced error.

Assessment of the Performance of an Analytical Methodology

As introduced in Section 2.7, analytical methods can be characterised by a number of performance indicators that define how the selected method performs under specified conditions. The assessment of precision and accuracy is discussed in the following sections.

19.3.3 Assessment of Precision

After a quantitative study has been completed and an experimental value for the amount and/or concentration of the test analyte in the test sample obtained, the experimenter must ask the question, 'How confident can I be that my result is an acceptable estimate of the true value?' (i.e. is it accurate?). An additional question may be, 'Is the quality of my analytical data comparable with that in the published scientific literature for the particular analytical method?' (i.e. is it precise?). Once the answers to such questions are known, a result that has a high probability of being correct can be accepted and used as a basis for the design of further studies, whilst a result that is subject to unacceptable error can be rejected. Unfortunately, it is not possible to assess the precision of a single quantitative determination. Rather, it is necessary to carry out analyses in replicate (i.e. the experiment is repeated several times on the same sample of test analyte) and to subject the resulting dataset to some basic statistical tests.

If a particular experimental determination is repeated numerous times and a graph constructed of the number of times a particular result occurs against its value, it is normally bell-shaped with the results clustering symmetrically about a mean value (see Figure 19.3). This type of distribution is called a Gaussian or normal distribution. In such cases, the precision of the dataset is a reflection of random error. However, if the plot is skewed to one side of the mean value, then systematic errors have not been eliminated. Assuming that the dataset is of the normal distribution type, there are three statistical parameters that can be used to quantify precision: standard deviation, coefficient of variation and variance.

These three statistical terms are alternative ways of expressing the scatter of the values within a dataset about the mean (μ), calculated by summing their total value and dividing by the number of individual values (N):

$$\mu = \frac{1}{N} \times \sum_{i=1}^{N} x_i \qquad \text{(Eq 19.1)}$$

Each term has its individual merit, but all of them measure the width of the normal distribution curve such that the narrower the curve, the smaller the value of the term and the higher the precision of the analytical dataset.

Standard Deviation

The standard deviation (σ) of a dataset is a measure of the variability of the population from which the dataset was drawn. It is calculated by use of Equation 19.2 or 19.3:

$$\sigma = \sqrt{\frac{1}{N-1} \times \sum_{i=1}^{N} (x_i - \mu)^2} \qquad \text{(Eq 19.2)}$$

$$\sigma = \sqrt{\frac{1}{N-1} \times \left[\sum_{i=1}^{N} x_i^2 - \frac{\left(\sum_{i=1}^{N} x_i\right)^2}{N} \right]} \qquad \text{(Eq 19.3)}$$

$(x_i - \mu)$ is the difference between an individual experimental value (x_i) and the calculated mean (μ) of the individual values. Since these differences may be positive or negative, and since the distribution of experimental values about the mean is symmetrical, they would cancel each other if they were simply added together. The differences are therefore squared, then summed to give consistent positive values (known as the sum of squares). To compensate for the square operation, the square root of the resulting calculation has to be taken to obtain the standard deviation.

Standard deviation has the same units as the actual measurements and this is one of its attractions. The mathematical nature of a normal distribution curve (see Figure 19.3) is such that 68.2% of the area under the curve (and hence 68.2% of the individual values within the dataset) are within one standard deviation on either side of the mean. Similarly, 95.5% of the area under the curve is within two standard deviations and 99.7% within three standard deviations. Exactly 95% of the area under the curve falls between the mean and 1.96 standard deviations. The **precision** (or imprecision) of a dataset is commonly expressed as ±1 standard deviation of the mean.

The term $(N - 1)$ is called the **degrees of freedom** of the dataset and is an important variable. The initial number of degrees of freedom possessed by a dataset is equal to the number of results (N) in the set. However, when another quantity characterising the dataset, such as the mean or standard deviation, is calculated, the number of degrees of freedom of the set is reduced by 1; and by 1 again for each new derivation made. Many modern calculators and computers include programs for the calculation of standard deviation. However, some use variants of Equation 19.2 in that they use N as the denominator rather than $(N - 1)$ as the basis for the calculation. If N is large (for example, greater than 30), then the difference between the two calculations is small, but if N is small (and certainly if it is less than 10), the use of N rather than $(N - 1)$ will significantly underestimate the standard deviation. This may lead to false conclusions being drawn about the precision of the dataset. Thus, for most analytical biochemical studies it is imperative that the calculation of standard deviation is based on the use of $(N - 1)$.

Coefficient of Variation

The coefficient of variation (CV) (also known as relative standard deviation) of a dataset is the standard deviation expressed as a percentage of the mean, as shown in Equation 19.4.

$$CV = \frac{\sigma}{\mu} \times 100\% \qquad\qquad \text{(Eq 19.4)}$$

Since the mean (μ) and standard deviation (σ) have the same units, the coefficient of variation is simply a percentage. This independence of the unit of measurement allows methods based on different units to be compared. The majority of well-developed analytical methods have a coefficient of variation within the analytical range of less than 5% and many, especially automated methods, of less than 2%.

Variance

The variance of a dataset is the mean of the squares of the differences between each value (x_i) and the mean (μ) of the values. It is also the square of the standard deviation, hence the symbol σ^2.

$$\sigma^2 = \frac{1}{N} \times \sum_{i=1}^{N} (x_i - \mu)^2 \qquad\qquad \text{(Eq 19.5)}$$

It has units that are the square of the original units and this makes it appear rather cumbersome, which explains why standard deviation and coefficient of variation are the preferred ways of expressing the variability of datasets. The importance of variance will be evident in later discussions of ways to make a statistical comparison of two datasets.

Example 19.1   **ASSESSMENT OF THE PRECISION OF GLUCOSE DETERMINATION IN SERUM**

Question   Suppose that two serum samples, A and B, were each analysed 20 times for serum glucose by the glucose oxidase method (see Section 10.2.3) such that sample A gave a mean value of 2.00 mM with a standard deviation of ±0.10 mM and sample B a mean of 8.00 mM and a standard deviation of ±0.41 mM. What is the precision of the results for both samples based on the (i) the standard deviation, and (ii) the coefficient of variation?

Answer   The standard deviation and coefficient of variation as measures of precision each have their merits.

(i) On the basis of the standard deviation values it might be concluded that the method had given a better precision for sample A (0.10 mM) than for B (0.41 mM).

(ii) However, this ignores the absolute values of the two samples. If this is taken into account by calculating the coefficient of variation, the two values are 5.0% and 5.1%, respectively. Based on the CV, the method had shown the same precision for both samples.

This illustrates the fact that standard deviation is an acceptable assessment of precision for a given dataset, but if it is necessary to compare the precision of two or more datasets, particularly ones with different mean values, then the coefficient of variation should be used.

Example 19.2   **ASSESSMENT OF THE PRECISION OF AN ANALYTICAL DATA SET**

Question   Five measurements of the fasting serum glucose concentration were made on the same sample taken from a diabetic patient. The values obtained were 2.3, 2.5, 2.2, 2.6 and 2.5 mM. Calculate the precision of the dataset.

Answer   Precision is normally expressed either as one standard deviation of the mean or as the coefficient of variation of the mean. These statistical parameters therefore need to be calculated.

Mean (Equation 19.1):

$$\mu = \frac{2.2 \text{ mM} + 2.3 \text{ mM} + 2.5 \text{ mM} + 2.5 \text{ mM} + 2.6 \text{ mM}}{5} = 2.42 \text{ mM}$$

Calculation of the relevant terms:

x_i	$(x_i - \mu)$	$(x_i - \mu)^2$	x_i^2
(mM)	(mM)	(mM^2)	(mM^2)
2.2	−0.22	0.0484	4.84
2.3	−0.12	0.0144	5.29
2.5	+0.08	0.0064	6.25
2.5	+0.08	0.0064	6.25
2.6	+0.18	0.0324	6.75
Σ 12.1	0.00	0.108	29.39

Standard deviation:

Using Equation 19.2, this yields:

$$\sigma = \sqrt{\frac{1}{5-1} \times 0.108 \ \text{mM}^2} = 0.164 \ \text{mM}$$

and with Equation 19.3:

$$\sigma = \sqrt{\frac{1}{5-1} \times \left[29.39 \ \text{mM}^2 - \frac{(12.1 \ \text{mM})^2}{5}\right]} = \sqrt{\frac{1}{4} \times 0.108 \ \text{mM}^2} = 0.164 \ \text{mM}.$$

Coefficient of variation:

This is calculated as per Equation 19.4:

$$CV = \frac{0.164 \ \text{mM}}{2.42 \ \text{mM}} \times 100\% = 6.78\%$$

Discussion: In this case, it is easier to appreciate the precision of the dataset by considering the coefficient of variation. The value of 6.78% is moderately high for this type of analysis. Automation of the method would certainly reduce it by at least half. Note that it is legitimate to quote the answers to these calculations to one more digit than was present in the original dataset. In practice, it is advisable to carry out the statistical analysis on a far larger dataset than that presented in this example.

19.3.4 Assessment of Accuracy

Population Statistics

Whilst standard deviation and coefficient of variation give a measure of the variability of a dataset, they do not quantify how well the mean of the dataset approaches the true value. To address this issue, it is necessary to introduce two concepts in relation to population statistics: the confidence limit and confidence interval.

In the previous section, we defined the mean μ and the standard deviation σ, which can be determined for a dataset where every member of a population is sampled. If a dataset is made up of a very large number of individual values so that N is a large number, the entire dataset may be examined by randomly chosen samples. In these cases, the mean and standard deviation become estimated parameters, which is indicate by $\bar{x}$ (sample mean) instead of μ (population mean), and s (sample standard deviation) instead of σ (population standard deviation).

The two population parameters μ and σ are the best estimates of the true values since they are based on the largest number of individual measurements, so that the influence of random errors is minimised. In practice, the population parameters are seldom measured for obvious practical reasons and the sample parameters have a larger uncertainty associated with them. The uncertainty of the sample mean ($\bar{x}$) deviating from the population mean (μ) decreases with the reciprocal of the square root of the number of values in the dataset, i.e. $1/\sqrt{N}$ (see Equation 19.6).

Thus, to decrease the uncertainty by a factor of two, the number of experimental values would have to be increased fourfold. Similarly, to decrease the uncertainty by a factor of 10, the number of measurements would need to be increased 100-fold. The nature of this relationship emphasises the importance of evaluating the acceptable degree of uncertainty of the experimental result before the design of the experiment is completed and the practical analysis begun. Modern automated analytical instruments recognise the importance of multiple results by facilitating repeat analyses at maximum speed. It is good practice to report the number of measurements on which the mean and standard deviation are based, as this gives a clear indication of the quality of the calculated data.

Confidence Intervals, Confidence Limits and the Student's t Factor

Accepting that the population mean (μ) is the best estimate of the true value, the question arises 'How can I relate my experimental sample mean ($\bar{x}$) to the population mean?' The answer is by using the concept of confidence. Confidence level expresses the level of confidence, expressed as a percentage, that can be attached to the data. Its value has to be set by the experimenter to achieve the objectives of the study. The confidence interval is a mathematical statement relating the sample mean to the population mean. A confidence interval gives a range of values about the sample mean such as to include the population mean with a given probability (determined by the confidence level). The relationship between the two means is expressed in terms of the standard deviation of the sample set (s), the square root of the number of values (N) in the dataset and a factor known as Student's t:

$$CI = \bar{x} \pm \frac{t \times s}{\sqrt{N}} \qquad\qquad (\text{Eq } 19.6)$$

where CI is the confidence interval of the population mean (μ), x (with a line on the top as earlier) is the measured mean, s is the measured standard deviation, N is the number of measurements and t is Student's t factor. The term $s/\sqrt{N}$ is known as the standard error of the mean and is a measure of the precision of the sample mean. Unlike standard deviation, the standard error depends on the sample size and will fall as the sample size increases. The two measurements are sometimes confused, but in essence, standard deviation should be used if we want to know how widely scattered the measurements are and standard error should be used if we want to indicate the uncertainty around a mean measurement.

The confidence level can be set at any value up to 100%. For example, it may be that a confidence level of only 50% would be acceptable for a particular experiment. However, a 50% level means that there is a one in two chance that the sample mean is not an acceptable estimate of the population mean. In contrast, the choice of a 95% or 99% confidence level would mean there was only a one in 20 or a one in 100 chance, respectively, that the best estimate had not been achieved. In practice, most analytical biochemists choose a confidence level in the range 90–99% and most commonly 95%. Student's t is a way of linking probability with the size of the dataset and is used in a number of statistical tests. Student's t values for varying sample sizes of a dataset (and hence with the varying degrees of freedom) at selected confidence levels are available in statistical tables. Some values are shown in Table 19.1. The numerical value of t is equal to the number of standard errors of the mean that must be added and subtracted from the mean to give the confidence interval at a given confidence level. Note that as the sample size (and hence the degrees of freedom) increases, the confidence levels converge. When N is large and we wish to calculate the 95% confidence interval, the value of t approximates to 1.96 and some texts quote Equation 19.6 in this form. The term Student's t factor may give the impression that it was devised specifically with students' needs in mind. In fact, 'Student' was the pseudonym of a statistician, by the name of William Sealy Gossett, who in 1908 first devised the term and who was not permitted by his employer to publish his work under his own name.

Table 19.1 **Select values of Student's *t***

Degrees of freedom	Confidence level (%)					
	50	90	95	98	99	99.9
2	0.816	2.920	4.303	6.965	9.925	31.598
3	0.765	2.353	3.182	4.541	5.841	12.924
4	0.741	2.132	2.776	3.747	4.604	8.610
5	0.727	2.015	2.571	3.365	4.032	6.869
6	0.718	1.943	2.447	3.143	3.707	5.959
7	0.711	1.895	2.365	2.998	3.500	5.408
8	0.706	1.860	2.306	2.896	3.355	5.041
9	0.703	1.833	2.262	2.821	3.250	4.798
10	0.700	1.812	2.228	2.764	3.169	4.587
15	0.691	1.753	2.131	2.602	2.947	4.073
20	0.687	1.725	2.086	2.528	2.845	3.850
30	0.683	1.697	2.042	2.457	2.750	3.646

Example 19.3   ASSESSMENT OF THE ACCURACY OF AN ANALYTICAL DATASET

Question   Calculate the confidence intervals at the 50%, 95% and 99% confidence levels of the fasting serum glucose concentrations given in Example 19.2.

Answer   Accuracy in this type of situation is expressed in terms of confidence intervals that express a range of values over which there is a given probability that the true value lies.

As previously calculated, the mean is 2.42 mM and the standard deviation is 0.161 mM. Inspection of Table 19.1 reveals that for four degrees of freedom (the number of experimental values minus one) and a confidence level of 50%, $t = 0.741$ so that the confidence interval for the population mean is given by:

$$CI = 2.42 \text{ mM} \pm \frac{0.741 \times 0.161 \text{ mM}}{\sqrt{5}} = (2.42 \pm 0.05) \text{ mM}$$

For the 95% confidence level and the same number of degrees of freedom, $t = 2.776$, hence the confidence interval for the population mean is given by:

$$CI = 2.42 \text{ mM} \pm \frac{2.776 \times 0.161 \text{ mM}}{\sqrt{5}} = (2.42 \pm 0.20) \text{ mM}$$

For the 99% confidence level and the same number of degrees of freedom, $t = 4.604$, hence the confidence interval for the population mean is given by:

$$CI = 2.42 \text{ mM} \pm \frac{4.604 \times 0.161 \text{ mM}}{\sqrt{5}} = (2.42 \pm 0.33) \text{ mM}$$

Discussion: These calculations show that there is a 50% chance that the population mean lies in the range 2.37 – 2.47 mM, a 95% chance that the population mean lies within the range 2.22 – 2.62 mM and a 99% chance that it lies in the range 2.09 – 2.75 mM. Note that as the confidence level increases, the range of potential values for the population mean also increases. You can calculate for yourself that if the mean and standard deviation had been based on 20 measurements (i.e. a fourfold increase in the number of measurements) then the 50% and 95% confidence intervals would have been reduced to (2.42 ± 0.02) mM and (2.42 ± 0.07) mM, respectively. This re-emphasises the beneficial impact of multiple experimental determinations, but at the same time highlights the need to balance the value of multiple determinations against the accuracy with which the experimental mean is required within the objectives of the individual study.

Criteria for the Rejection of Outlier Experimental Data

A very common problem in quantitative biochemical analysis is the need to decide whether or not a particular result is an outlier and should therefore be rejected before the remainder of the dataset is subjected to statistical analysis. It is important to identify such data as they have a disproportionate effect on the calculation of the mean and standard deviation of the dataset. When faced with this problem, the first action should be to check that the suspected outlier is not due to a simple experimental or mathematical error. Once the suspect figure has been confirmed, its validity is checked by application of Dixon's Q-test. Like other tests to be described later, the test is based on a null hypothesis, namely that there is no difference in the values being compared. If the hypothesis is proved to be correct, then the suspect value cannot be rejected. The suspect value is used to calculate an experimental rejection quotient, Q_{exp}. This quotient is then compared with tabulated critical rejection quotients, Q_{table}, for a given confidence level and the number of experimental results (Table 19.2). If Q_{exp} is less than Q_{table}, the null hypothesis is confirmed and the suspect value should not be rejected, but if it is greater than Q_{table}, then the value can be rejected. The basis of the test is the fact that in a normal distribution, 95.5% of the values are within the range of two standard deviations of the mean. In setting limits for the acceptability or rejection of data, a compromise has to be made on the confidence level chosen. If a high confidence level is chosen, the limits of acceptability are set wide and therefore there is a risk of accepting values that are subject to error. If the confidence level is set too low, the acceptability limits will be too narrow and therefore there will be a risk of rejecting legitimate data. In practice, a confidence level of 90% or 95% is most commonly applied. The Q_{table} values in Table 19.2 are based on a 95% confidence level.

The calculation of Q_{exp} is based upon Equation 19.7 that requires the calculation of the separation of the questionable value from the nearest acceptable value (gap) coupled with knowledge of the range (difference between maximum and minimum) covered by the dataset:

$$Q = \frac{gap}{range} \qquad\qquad\qquad \text{(Eq 19.7)}$$

Table 19.2 **Values of Q for the rejection of outliers**

Number of observations	Q (95% confidence)
4	0.83
5	0.72
6	0.62
7	0.57
8	0.52

Example 19.4   **IDENTIFICATION OF AN OUTLIER EXPERIMENTAL RESULT**

Question   If the dataset in Example 19.2 contained an additional value of 3.0 mM, could this value be regarded as an outlier point at the 95% confidence level?

Answer   Using the Q-test, we can determine whether the value of 3.0 mM classifies as an outlier. The closest value to the value in question (3.0 mM) is 2.6 mM. The *gap* is thus:

$$gap = 3.0 \text{ mM} - 2.6 \text{ mM} = 0.4 \text{ mM}.$$

The *range* is calculated to be

$$range = 3.0 \text{ mM} - 2.2 \text{ mM} = 0.8 \text{ mM}.$$

With Equation 19.7 we obtain for the experimental Q:

$$Q_{exp} = \frac{0.4 \text{ mM}}{0.8 \text{ mM}} = 0.5$$

Using Table 19.2 for six data points shows that Q_{table} is equal to 0.62. Since Q_{exp} is smaller than Q_{table}, the point should not be rejected as there is a more than 95% chance that it is part of the same dataset as the other five values.

It is easy to show that an additional data point of 3.3 mM rather than 3.0 mM would give a Q_{exp} of 0.64 and could be rejected.

19.3.5 Validation of an Analytical Method

A t-test in general is used to address the question as to whether or not two datasets have the same mean. Both datasets need to have a normal distribution and equal variances. There are three types:

- Unpaired t-test: Used to test whether two datasets have the same mean.
- Paired t-test: Used to test whether two datasets have the same mean, where each value in one set is paired with a value in the other set.
- One-sample t-test: Used to test whether the mean of a dataset is equal to a particular value.

Each test is based on a **null hypothesis**, which holds that there is no difference between the means of the two datasets. The tests measure how likely the hypothesis is true. The attraction of such tests is that they are easy to carry out and interpret.

Analysis of a Standard Solution: One-Sample *t*-Test

Once the choice of the analytical method to be used for a particular biochemical assay has been made, the normal first step is to carry out an evaluation of the method in the laboratory. This evaluation entails the replicated analysis of a known standard solution of the test analyte and the calculation of the mean and standard deviation of the resulting dataset. Then, the question, 'Does the mean of the analytical results agree with the known value of the standard solution within experimental error?' is asked. To answer this question a one-sample *t*-test is applied.

In the case of the analysis of a standard solution, the calculated mean and standard deviation of the analytical results are used to calculate a value of Student's t (t_{calc}) using Equation 19.8:

$$t_{calc} = \frac{|\text{known value} - \bar{x}|}{s} \times \sqrt{N} \qquad \text{(Eq 19.8)}$$

This value is then compared with tabled values of t (t_{table}) for the particular degrees of freedom of the dataset and at the required confidence level (Table 19.1).

These tabled values of t represent critical values that separate the border between different probability levels. If t_{calc} is greater than t_{table}, the analytical results are deemed not to be from the same dataset as the known standard solution at the selected confidence level. In such cases, the conclusion is therefore drawn that the analytical results do not agree with the standard solution and hence that there are unidentified errors in them. There would be no point in applying the analytical method to unknown test analyte samples until the problem has been resolved.

Example 19.5 **VALIDATING AN ANALYTICAL METHOD**

Question A standard solution of glucose is known to be 5.05 mM. Samples of it were analysed by the glucose oxidase method (see Section 10.2.3 for details) that was being used in the laboratory for the first time. A calibration curve obtained using least mean square linear regression was used to calculate the concentration of glucose in the test sample. The following experimental values were obtained: 5.12, 4.96, 5.21, 5.18 and 5.26 mM. Does the experimental dataset for the glucose solution agree with the known value within experimental error?

Answer It is first necessary to calculate the mean and standard deviation for the set and then to use it to calculate a value for Student's *t*.

Applying Equation 19.1 yields the mean:

$$\mu = \frac{5.12 \text{ mM} + 4.96 \text{ mM} + 5.21 \text{ mM} + 5.18 \text{ mM} + 5.26 \text{ mM}}{5} = 5.15 \text{ mM}$$

This allows for calculation of the relevant terms to calculate the standard deviation

x_i (mM)	$(x_i - \mu)$ (mM)	$(x_i - \mu)^2$ (mM^2)	x_i^2 (mM^2)
5.12	−0.03	0.0009	26.21
4.96	−0.19	0.0361	24.60
5.21	0.06	0.0036	27.14
5.18	0.03	0.0009	26.83
5.26	0.11	0.0121	27.67

Σ	25.73	−0.02	0.0536	132.46

according to Equation 19.2:

$$\sigma = \sqrt{\frac{1}{5-1} \times 0.0536 \text{ mM}^2} = 0.12 \text{ mM}$$

and according to Equation 19.3:

$$\sigma = \sqrt{\frac{1}{5-1} \times \left[132.46 \text{ mM}^2 - \frac{(25.73 \text{ mM})^2}{5}\right]} = \sqrt{\frac{1}{4} \times 0.05342 \text{ mM}^2} = 0.12 \text{ mM}$$

Now applying Equation 19.8 to give t_{calc}:

$$t_{calc} = \frac{\left|5.05 \text{ mM} - 5.15 \text{ mM}\right|}{0.12 \text{ mM}} \times \sqrt{5} = 1.863$$

From Table 19.1 at the 95% confidence level with four degrees of freedom, t_{table} = 2.776. The calculated t value is therefore less than t_{table}, and the conclusion can be drawn that the measured mean value does agree with the known value.

Using Equation 19.4, the coefficient of variation for the measured values can be calculated to be 2.33%.

Comparison of Two Competitive Analytical Methods: Unpaired t-Test

In quantitative biochemical analysis, it is frequently helpful to compare the performance of two alternative methods of analysis in order to establish whether or not they give the same quantitative result within experimental error. To address this need, each method is used to analyse the same test sample using replicated analysis. The mean and standard deviation for each set of analytical data is then calculated and a Student's t-test applied. In this case, the t-test measures the overlap between the datasets such that the smaller the value of t_{calc} the greater the overlap between the two datasets. This is an example of an unpaired t-test.

In using the tables of critical t values, the relevant degrees of freedom is the sum of the number of values in the two datasets minus 2 (i.e. $N_1 + N_2 - 2$). The larger the number of degrees of freedom, the smaller the value of t_{calc} needs to be to exceed the critical value at a given confidence level. The formulae for calculating t_{calc} depend on whether or not the standard deviations of the two datasets are the same. This is often obvious by inspection, revealing whether or not the two standard deviations are similar.

However, if in doubt, an F-test, named in honour of Sir Ronald Aylmer Fisher, can be applied. An F-test is based on the null hypothesis that there is no difference between the two variances. The test calculates a value for F (F_{calc}), which is the ratio of the larger of the two variances to the smaller variance. It is then compared with critical F values (F_{table}) available in statistical tables or computer packages (Table 19.3). If the calculated value of F is less than the table value, the null hypothesis is proved and the two standard deviations are considered to be similar. If the two variances are of the same order, then Equations 19.9 and 19.10 are used to calculate t_{calc} for the two datasets. If not, Equations 19.11 and 19.12 are used.

$$t_{calc} = \frac{|\overline{x_1} - \overline{x_2}|}{s_{pooled}} \times \sqrt{\frac{N_1 \times N_2}{N_1 + N_2}} \qquad \text{(Eq 19.9)}$$

$$s_{pooled} = \sqrt{\frac{s_1^2 \times (N_1 - 1) + s_2^2 \times (N_2 - 1)}{N_1 + N_2 - 2}} \qquad \text{(Eq 19.10)}$$

$$t_{calc} = \frac{\overline{x_1} - \overline{x_2}}{\sqrt{\frac{s_1^2}{N_1} + \frac{s_2^2}{N_2}}} \qquad \text{(Eq 19.11)}$$

$$\text{Degrees of freedom} = \frac{\left(\frac{s_1^2}{N_1} + \frac{s_2^2}{N_2}\right)^2}{\frac{\left(\frac{s_1^2}{N_1}\right)^2}{N_1 + 1} + \frac{\left(\frac{s_2^2}{N_2}\right)^2}{N_2 + 1}} - 2 \qquad \text{(Eq 19.12)}$$

where x_1 and x_2 are the calculated means of the two methods, s_1^2 and s_2^2 are the calculated standard deviations of the two methods, and N_1 and N_2 are the number of measurements in the two methods.

Table 19.3 **Critical values of *F* at the 95% confidence level**

Degrees of freedom for Set 2	Degrees of freedom for Set 1							
	2	3	4	6	10	15	30	∞
2	19.0	19.2	19.2	19.3	19.4	19.4	19.5	19.5
3	9.55	9.28	9.12	8.94	8.79	8.70	8.62	8.53
4	6.94	6.59	6.39	6.16	5.96	5.86	5.75	5.63
5	5.79	5.41	5.19	4.95	4.74	4.62	4.50	4.36
6	5.14	4.76	4.53	4.28	4.06	3.94	3.81	3.67
7	4.74	4.35	4.12	3.87	3.64	3.51	3.38	3.23
8	4.46	4.07	3.84	3.58	3.35	3.22	3.08	2.93
9	4.26	3.86	3.63	3.37	3.14	3.01	2.86	2.71
10	4.10	3.71	3.48	3.22	2.98	2.84	2.70	2.54
15	3.68	3.29	3.06	2.79	2.54	2.40	2.25	2.07
20	3.49	2.10	2.87	2.60	2.35	2.20	2.04	1.84
30	3.32	2.92	2.69	2.42	2.16	2.01	1.84	1.62
∞	3.00	2.60	2.37	2.10	1.83	1.67	1.46	1.00

At first sight these four equations may appear daunting, but closer inspection reveals that they are simply based on variance (s^2), mean ($\bar{x}$) and number of analytical measurements (N) and that the mathematical manipulation of the data is relatively easy.

Comparison of Two Competitive Analytical Methods: Paired *t*-Test

A variant of the previous type of comparison of two analytical methods based upon the analysis of a common standard sample is the case in which a series of test samples is analysed once by the two different analytical methods. Here, there is no replication of analysis of any test sample by either method. The *t*-test can then be applied to the differences between the results of each method for each test sample. The formula for calculating t_{calc} for such a paired *t*-test is given by Equation 19.13:

$$t_{calc} = \frac{|\bar{d}|}{s_d} \times \sqrt{N}, \text{ with } s_d = \sqrt{\frac{\sum_{i=1}^{N}(d_i - \bar{d})^2}{N-1}} \qquad \text{(Eq 19.13)}$$

where d_i is the difference between the paired results, d (with a line on the top as in the equation) is the mean difference between the paired results, N is the number of paired results and s_d is the standard deviation of the differences between the pairs.

Example 19.6 **COMPARISON OF TWO ANALYTICAL METHODS USING REPLICATED ANALYSIS OF A SINGLE TEST SAMPLE**

Question A sample of fasting serum was used to evaluate the performance of the glucose oxidase and hexokinase methods for the quantification of serum glucose concentrations (for details see Section 10.2.3). The following replicated values were obtained: for the glucose oxidase method 2.3, 2.5, 2.2, 2.6 and 2.5 mM and for the hexokinase method 2.1, 2.7, 2.4, 2.4 and 2.2 mM.

Establish whether or not the two methods gave the same results at a 95% confidence level.

Answer Glucose oxidase method:

Mean and standard deviation are known from Example 2: μ_1 = 2.42 mM, σ_1 = 0.164 mM, σ_1^2 = 0.026 mM^2.

Hexokinase method:

Using the standard formulae, we can calculate the mean, standard deviation and variance for this dataset; this yields: μ_2 = 2.36 mM, σ_2 = 0.230 mM, σ_2^2 = 0.053 mM^2.

Comparison:

We can then apply the F-test to the two variances to establish whether or not they are the same:

$$F_{calc} = \frac{0.053 \text{ mM}^2}{0.026 \text{ mM}^2} = 2.04$$

F_{table} for the two sets of data, each with four degrees of freedom and for the 95% confidence level is 6.39 (Table 19.3). Since F_{calc} is less than F_{table} we can conclude that the two variances are not significantly different. Therefore, using Equations 19.9 and 19.10 we can calculate that:

$$S_{pooled} = \sqrt{\frac{0.026 \text{ mM}^2 \times (5-1) + 0.053 \text{ mM}^2 \times (5-1)}{5+5-2}} = \sqrt{\frac{0.316 \text{ mM}^2}{8}}$$

$$= 0.199 \text{ mM}$$

$$t_{calc} = \frac{|2.42 \text{ mM} - 2.36 \text{ mM}|}{0.199 \text{ mM}} \times \sqrt{\frac{5 \times 5}{5+5}} = 0.477$$

Using Table 19.1 at the 95% confidence level and for eight degrees of freedom, t_{table} is 2.306. Thus, t_{calc} is far less than t_{table} and so the two sets of data are not significantly different, i.e. the two methods have given the same result at the 95% confidence level.

Example 19.7   **COMPARISON OF TWO ANALYTICAL METHODS USING DIFFERENT TEST SAMPLES**

Question   Ten fasting serum samples were each analysed by the glucose oxidase and the hexokinase methods. The following results were obtained:

Glucose oxidase method	(mM)	1.1	2.0	3.2	3.7	5.1	8.6	10.4	15.2	18.7	25.3
Hexokinase method	(mM)	0.9	2.1	2.9	3.5	4.8	8.7	10.6	14.9	18.7	25.0

Do the two methods give the same results at the 95% confidence level?

Answer   Before addressing the main question, note that the ten samples analysed by the two methods were chosen to cover the whole analytical range for the methods. To assess whether or not the two methods have given the same result at the chosen confidence level, it is necessary to calculate a value for Student's t and to compare it to the tabulated value for the $(N - 1) = 9$ degrees of freedom in the study. In order to determine t_{calc} using Equation 19.13, we first need to calculate the value of s_d.

The step-by-step calculation of relevant terms is shown in the following table

Glucose oxidase method (mM)	Hexokinase method (mM)	d_i (mM)	$d_i - \bar{d}$ (mM)	$(d_i - \bar{d})^2$ (mM)
1.1	0.9	0.2	0.08	0.0064
2.0	2.1	-0.1	-0.22	0.0484
3.2	2.9	0.3	0.18	0.0324
3.7	3.5	0.2	0.08	0.0064
5.1	4.8	0.3	0.18	0.0324
8.6	8.7	-0.1	-0.22	0.0484
10.4	10.6	-0.2	-0.32	0.1024
15.2	14.9	0.3	0.18	0.0324
18.7	18.7	0.0	-0.12	0.0144
25.3	25.0	0.3	0.18	0.0324
		$\bar{d} = 0.12$		$\Sigma\left(d_i - \bar{d}\right)^2 = 0.3560$

but can also be done conveniently in R:

```
# Script 19.32
> g<-c(1.1,2.0,3.2,3.7,5.1,8.6,10.4,15.2,18.7,25.3)
> h<-c(0.9,2.1,2.9,3.5,4.8,8.7,10.6,14.9,18.7,25.0)
> d<-g-h
> d
[1]  0.2 -0.1  0.3  0.2  0.3 -0.1 -0.2  0.3  0.0  0.3
> mean(d)
[1] 0.12
> sum((d-mean(d))**2)
[1] 0.356
```

The standard deviation s_d can then be calculated as per:

$$s_d = \sqrt{\frac{0.356 \text{ mM}^2}{10-1}} = 0.199 \text{ mM}$$

or using R:

```
> sd(d)
[1] 0.1988858
```

This delivers the calculated t value as per the following equation:

$$t_{calc} = \frac{0.12 \text{ mM}}{0.199 \text{ mM}} \times \sqrt{10} = 1.907$$

and the corresponding calculation in R looks like:

```
> t.test(g,h,paired=TRUE)

        Paired t-test

data: g and h
t = 1.908, df = 9, p-value = 0.08875
alternative hypothesis: true difference in means is not
equal to 0
95 percent confidence interval:

 -0.02227432 0.26227432
sample estimates:
mean of the differences
0.12
```

Using Table 19.1, t_{table} at the 95% confidence level and for nine degrees of freedom is 2.262. Since t_{calc} is smaller than t_{table}, the two methods do give the same results at the 95% confidence level. Inspection of the two datasets shows that the glucose oxidase method gave a slightly higher value for seven of the ten samples analysed.

An alternative approach to the comparison of the two methods is to plot the two datasets as an x-y- plot and to carry out a regression analysis of the data. If this is done using the hexokinase data as x and the glucose oxidase data as the y variable, the following results are obtained:

slope: $m = 1.0016$; intercept: $t = 0.1057$; correlation coefficient: $R = 0.9997$.

The slope of very nearly one confirms the similarity of the two datasets, whilst the small positive intercept on the y-axis confirms that the glucose oxidase method gives a slightly higher, but insignificantly different, value to that of the hexokinase method.

19.3.6 Calibration Methods

Quantitative biochemical analyses often involve the use of a calibration curve produced by the use of known amounts of the analyte using the selected analytical procedure. A **calibration curve** is a record of the measurement (absorbance, peak area, etc.) produced by the analytical procedure in response to a range of known quantities of the standard analyte. It involves the preparation of a standard solution of the analyte and the use of a range of aliquots in the test analytical procedure. It is good practice to replicate each calibration point and to use the mean ±1 standard deviation for the construction of the calibration plot. Inspection of the compiled data usually reveals a scatter of the points about a linear relationship. The method of **least mean squares linear regression** is the most common way of fitting a straight line to data, typically either with dedicated statistical software or in basic applications with spreadsheet programs. Importantly, one needs to appreciate that the accuracy of the values for slope and intercept that it gives are determined by experimental errors of the x and y values of the individual data points.
   The mathematical expression for a straight line is

$$y = m \times x + t \tag{Eq 19.14}$$

and the regression analysis results in numerical values for the slope m and the y-intercept t, both of which characterise the best-fit straight line for the experimental dataset. This is then used to construct the calibration curve and subsequently to analyse the test analyte(s). The results of any regression analysis are reported together with a quantity describing the **goodness-of-fit**, most commonly the **correlation coefficient** R. The stronger the correlation between the two variables, the closer the value of R to either +1 or −1.

Most commonly, the correlation coefficient is reported as its squared value, and quoted to four decimal places. Values for good correlation commonly exceed an R^2 of 0.99.

In the routine construction of a calibration curve, a number of points have to be considered:

- Selection of standard values: A range of standard analyte amounts/concentrations should be selected to cover the expected values for the test analyte(s) in such a way that the values are equally distributed along the calibration curve. Test samples should not be estimated outside this selected range, as there is no evidence that the regression analysis relationship holds outside the range. It is good practice to establish the analytical range and the limit of detection for the method (see also Section 2.7). It is also advisable to determine the precision (standard deviation) of the method at different points across the analytical range and to present the values on the calibration curve. Such a plot is referred to as a precision profile. It is common for the precision to decrease (standard deviation to increase) at the two ends of the curve and this may have implications for the routine use of the curve. For example, the determination of testosterone in male and female serum requires the use of different methods, since the two values (reference range 10–30 nM for males, < 3 nM for females) cannot be accommodated with acceptable precision on one calibration curve.
- Use of a **control sample** (blank): This is a sample in which no standard analyte is present (see also Section 19.3.2). A control sample should always be included in the experimental design when possible (it will not be possible, for example, with analyses based on serum or plasma). Any experimental value, e.g. absorbance, obtained for the control must be deducted from all other measurements. This may be achieved automatically in spectrophotometric measurements by the use of a double-beam spectrophotometer in which the blank sample is placed in the reference cell or automatic baseline correction of the previously measured control sample in single-beam instruments.
- Type of correlation: It should not be assumed that all calibration curves are linear. They may be curved and be best represented by other, non-linear functions, such as a quadratic equation or logarithmic function.
- **Recalibration**: A new calibration curve should be constructed on a regular basis. It is not acceptable to rely on a calibration curve produced on a much earlier occasion.

19.3.7 Internal Standards

An additional approach to the control of time-related minor changes in a calibration curve and the quantification of an analyte in a test sample is the use of an internal standard. An ideal internal standard is a compound that has a molecular structure and physical properties as similar as possible to the test analyte and which gives a similar response to the analytical method as the test analyte. This response, expressed on a unit quantity basis, may be different from that for the test analyte, but provided that the relative response of the two compounds is constant, the advantages of the use of the internal standard are not compromised. Quite commonly, the internal standard is a structural or geometrical isomer of the test analyte.

Practically, a known fixed quantity of the standard is added to each test sample and analysed alongside the test analyte by the standard analytical procedure. The resulting response for the standard and the range of amounts or concentrations of the test analyte is used to calculate a relative response for the test analyte and used in the construction of the calibration curve. The curve therefore consists of a plot of the relative response to the test analyte against the range of quantities of the analyte.

Internal standards are commonly used in liquid and gas chromatography since they help to compensate for small temporal variations in the flow of liquid or gas through the chromatographic column. In such applications, it is essential that the internal standard responses (e.g. migration time on chromatography material) are near to, but distinct from, the test analyte. If the analytical procedure involves preliminary sampling procedures, such as solid-phase extraction, it is important that a known amount of the internal standard is introduced into the test sample at as early a stage as possible and is therefore taken through the preliminary procedures. This ensures that any loss of the test analyte during these preliminary stages will be compensated by identical losses to the internal standard so that the final relative response of the method to the two compounds is a true reflection of the quantity of the test analyte.

19.3.8 Getting the Right Conclusions

Statistics versus Data Manipulations

Sometimes, statistics can be misused and the results presented would therefore not reflect the reality. In the famous case of Potti versus Baggerly in 2011, lauded results were pulled apart by careful and accurate use of statistics. Not only were there errors in the data handling that were not spotted, but the unwise use of statistics hid the truth. This episode was one of the worst scandals of research fabrication: a demonstration of what not to do.

At the beginning of the millennium, Potti and Nevis published a seminal article on personalised medicine using genetic microarrays. Unfortunately, these results did not seem to hold under scrutiny from Baggerly and Coombes, who spent years analysing the data and trying to reproduce the published results. Despite them finding many possible data-handling errors, ranging from data corruption to sample mix-up, they could only partially reproduce the published graphs and results. Only by employing forensic statistical analysis did they reach the conclusion that the results were not just a mere mistake, but an outright fraud. This bitter case ended in numerous retractions, bad publicity and the end of careers. Such a downfall was not the result of bad luck; as it turned out, other warning signs of dubious claims, including observations by a whistleblower, have since been uncovered (and ignored by institutional officials).

Clearly, to help reduce the chances of similar misconduct happening, the recent initiatives of more open data sharing need to be continued and intensified. Open data, as defined by the Open Definition are data that:

- are available and can be accessed in a convenient and modifiable form, at virtually no cost
- are provided with the permission to reuse and redistribute
- allow universal participation without restrictions.

Importantly, such efforts also address the important question of inter-operability, meaning the ability of diverse systems and organisations to work together (inter-operate) and inter-mix different datasets. The importance of this latter aspect is illustrated by a recent mistake that led to the claim that humans across the whole of Africa carried DNA inherited from Eurasian immigrants. The error occurred when genetic variants in the ancient Ethiopian sample were compared with those in the reference human genome. Due to incompatibility between two software packages used, some variants were removed from the analysis – this could have been prevented by a computer script that was not deployed. However, thanks to the openness of the authors, this problem found a swift resolution and positive outcome.

Simpson's Paradox

It should be emphasised that drawing wrong conclusions from statistics does not necessarily reflect malign intentions. Simpson's paradox is a good example of such a case where 'over-analysing' data can lead to erroneous conclusions.

In 1973, the University of California, Berkeley, was accused of sex discrimination as the graduate school was enrolling a disproportionate amount of male students compared to their female counterpart (44% versus 35%). Considering the sample size involved, there is little doubt that such difference would reach statistical significance. Still, the case for discrimination was lost, owing to statistical analysis. A thorough analysis of the data revealed an altogether different picture, whereby the difference between the number of admitted male and female students in each of the 85 departments was not statistically significant, apart from 10 instances (6 against male students, 4 against female students). Why is there such a difference between these two statistical approaches? The answer lies in the sizes of the different departments, which are not directly comparable. The larger departments are over-represented in comparison to smaller ones (Table 19.4). But it is not a case of a biased selection, rather a biased application: it turned out that female students applied more to the competitive departments than their male counterparts. While there is no gender bias in these competitive departments, it disproportionally affects the overall female student success rate.

Therefore, the apparent statistical test to apply in this case, a t-test on pools of male and female students, would actually not reflect the data correctly and therefore lead to wrong conclusions.

Table 19.4 **Gender-specific admission at the university level, in its six largest departments, and their combination**

| Student | Overall | Department | | | | | | Combined |
		A	B	C	D	E	F	
Male Applicants	8442	825	560	325	417	191	373	2691
Admitted	44%	62%	63%	36%	33%	26%	7%	30%
Female Applicants	4321	108	25	593	375	393	341	1835
Admitted	35%	82%	67%	34%	34%	23%	6%	46%

19.3.9 Further Methods

There are many more statistical tests and concepts available than those discussed in the examples above, and it is a statistician's job to make sure there are new and evolving statistical approaches to handle the ever-increasing amounts of data generated by improved technology. There remain a few more specific tests and concepts that a life scientist is likely to come across; these are discussed below.

ANOVA

The ANOVA family of statistical models is a natural extension of the t-test to multiple conditions developed by R.A. Fisher. Most of what one needs to know about ANOVA appears in its name: ANalysis Of VAriance. When several groups need to be compared, one can be tempted to perform multiple individual t-tests between pairs of groups. However, such an approach does not make full use of the data structure and can even be seriously misleading (due to an increased overall error rate when conducting many individual t-tests). In contrast, ANOVA tests are the appropriate way to tackle such datasets and come essentially in two flavours:

- One-way ANOVA is used when considering a single measure (e.g. expression value of a given gene) across balanced groups (e.g. differing diets as treatment) to compare whether there is a significant difference between them. One-way ANOVA is an omnibus test statistic (testing whether the explained variance in a dataset is overall significantly greater than the unexplained variance) and cannot tell which specific groups were statistically significantly different from each other, only that at least two groups were. To determine which specific groups differed from each other, one needs to use a post hoc test.
- Two-way ANOVA is used if another factor is evaluated simultaneously, i.e. the above treatment has two components such as diet and exercise level. Note that the numbers of factors can be increased and is referred as three-, four-way ANOVA etc.

There are numerous extensions of ANOVA to address specific problems. The two most common are:

- ANCOVA, which focusses on COVAriance rather than variance to integrate for a more general model than the linear, linking ANOVA to regression.
- MANOVA is a multi-variate ANOVA, which includes the covariance of the outcome of the univariate ANOVA.

Counting

Most of the tests mentioned so far have been dealing with continuous values as they naturally arise from experimental measurements. However, sometimes variables are discrete and even binary in nature, in which case it is more relevant to count and compare the number of occurrences in various groups. Typically, a contingency table can be represented as 2×2 or 3×2 table, where all experiments are put in a specific cell.

Fisher's exact test compares all possible combinations of the total counts across the various cells. With very large counts, such tests can become extremely costly and time-consuming. One therefore tends to use **Pearson's chi-square distribution** as an approximation, which leads to an alternative p-value. Note that while chi-square testing is very efficient with large sample sizes, it is very inaccurate if a single cell value is below 5 and should therefore not be used.

Truth and Lies: Hypothesis Testing

Statistics provides a robust framework to handle the truth, but, more importantly, the lack of it. Biochemical experiments are rarely perfect and it is therefore essential to be able to evaluate **false positive** as much as **false negative** results; in statistics, these are usually referred to as Type I (alpha) and Type II error (beta), respectively. These definitions are a core part of the framework of **hypothesis testing** and, as such, are an essential component of statistics. False positives and false negatives occur when we do not perform a correct inference of true positive and true negative values (see Table 19.5).

Statistical tests are defined by opposition: the (null) hypothesis (H_0) that everything occurs by chance is tested experimentally against an alternative hypothesis (H_1). Importantly, one can never fully reject a null hypothesis, and therefore p-values (see below) aim to be as small as possible, but very rarely zero.

p-value

The ubiquitous p-value is defined as the probability of obtaining a result equal to or 'more extreme' than what was actually observed, when the null hypothesis is true; in other words, how likely we are to capture a false positive. It therefore corresponds to the acceptable level of false positive results and is based on a threshold for the Type I error, the alpha value. In the life sciences, the p-value should typically be lower than 0.05; in physics, it should typically be lower than 0.01, due to the nature of the experiments performed. The p-value is computed as the area under a normal probability density distribution (see Figure 19.8); such a distribution can be calculated with R as in the following example:

Table 19.5 **Table of errors**

| | | Null hypothesis H_0 | |
		True	False
Test of H_0	Reject	False positive Type I error alpha	True positive Correct inference
	Fail to Reject	True negative Correct inference	False negative Type II error beta

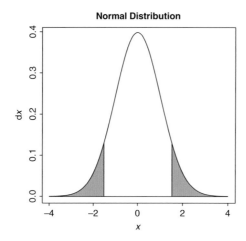

Figure 19.8 Plotting the probability density (y-coordinate), computed under the null hypothesis, of each outcome (x-coordinate) yields a probability density function. In a normally distributed probability density function, the total area under the curve equals 1. The p-value is the area under the normal probability density distribution past the observed data point.

```
# Script 19.33
# p-value in the normal probability density distribution
> x<-seq(-4,4,length=100)
> dx<-dnorm(x,mean=0,sd=1)
> plot(x,dx,type="l",main="Normal Distribution")
> p1<-seq(-4,-1.5,length=100)
> dp1<-dnorm(p1)
> p2<-seq(1.5,4,length=100)
> dp2<-dnorm(p2)
> polygon(c(-4,p1,-1.5),c(0,dp1,0),col="gray")
> polygon(c(1.5,p2,4),c(0,dp2,0),col="gray")
> abline(h=0)
```

Example 19.8   **COUNTING STATISTICS TO ANALYSE RESULTS OF A TREATMENT EXPERIMENT**

Question   A total of 100 samples have been subjected to two different treatments (A and B) and analysed for the presence or absence of an effect. In treatment regime A, 24 affected and 26 unaffected samples were observed. In treatment regime B, there were 35 affected and 15 unaffected samples. Is there a significant difference in the number of affected versus unaffected samples?

Answer   The contingency table for this counting statistics can be constructed as follows:

	Treatment A	Treatment B	Total per group
Affected	24	35	24 + 35 = 59
Unaffected	26	15	26 + 15 = 41
Total per treatment	24 + 26 = 50	35 + 15 = 50	24 + 26 + 35 + 15 = 100

Using R and Fisher's exact test for this experiment, the double-sided p-value can be determined as per:

```
# Script 19.34
> F<-matrix(c(24,26,35,15),nrow=2);
> F
     [,1] [,2]
[1,]   24   35
[2,]   26   15
> fisher.test(F)
    Fisher's Exact Test for Count Data
data: F
p-value = 0.04143
alternative hypothesis: true odds ratio is not equal to 1
95 percent confidence interval:
 0.1599061 0.9685788
sample estimates:
odds ratio
 0.3994095
```

The corresponding chi-square test can be performed as follows:

```
> chisq.test(F)
    Pearson's Chi-squared test with Yates' continuity cor-
rection
data: F
X-squared = 4.1339, df = 1, p-value = 0.04203
```

Fisher's exact test yields a p-value of 0.04143; the chi-square test yields the slightly different p-value of 0.04203, since it is an approximation. Nevertheless, both values indicate statistical significance at 95%.

Power

Whilst the p-value addresses the avoidance of false positives (Type I errors), the power is concerned with avoiding false negative results (Type II errors, beta value). In other words, it is used to denote how well one does in missing desired results. The power is defined as (1 − beta) and will increase with the ability to accurately reject the null hypothesis. This can be achieved with an increased (expected) effect size (defined as the difference between the mean of the treatment group and the mean of the control group, divided by the standard deviation of the control group). However, biological effects are rarely under our control, so when designing an experiment, the main handle on increased power is an increased sample size. Since one can calculate ahead of time how many samples are needed, this is an efficient way to gain meaningful

Table 19.6 **Power applications in the R pwr package**

R function	Test statistic	Example
pwr.t.test	*t*-test (one-, two- or paired-samples)	pwr.t.test(n=25,d=0.75,sig. level=.01,alternative="greater")
pwr.t2n.test	*t*-test (two samples with unequal *N*)	pwr.t2n.test(n1=10,n2=15,d=0.6,sig. level=.05)
pwr.anova.test	Balanced one-way ANOVA	pwr.anova.test(k=5,f=.25,sig. level=.05,power=.8)
pwr.chisq.test	Chi-square test	pwr.chisq.test(w=2,N=200,df=15,sig. level=2)
pwr.r.test	Correlation	pwr.r.test(r=0.8,sig.level=2, power=.8)

results. While in clinical settings, the power is often aimed to be above 0.8 (or 80%, i.e. 20% of missed hits is acceptable), this is different in research settings, where a more stringent threshold of typically 0.99 (or 99%, i.e. 1% of missed hits is acceptable) is applied. Some useful power applications in the R pwr package are summarised in Table 19.6.

19.4 CONCLUSION

Statistics should be our best, yet least-compromising, friend. It should be there to protect ourselves from our own prejudice. We should not be afraid of the challenges it poses, since we need to be ready to shake our data, and question our models to ensure the chosen approaches and derived conclusions are robust. If we do not do it ourselves, someone else will, publicly, and it will be most embarrassing, even if no deception was intended. Being subject to peer review goes beyond papers, it includes experimental design, data collection and method choice – it is the very nature of science. We need to welcome statistics and embrace it; how else could someone dare to claim correlation between educational attainment and height?

19.5 SUGGESTIONS FOR FURTHER READING

19.5.1 Experimental Protocols

Bickel P.J., Hammel E.A. and O'Connell J.W. (1975) Sex bias in graduate admissions: data from Berkeley. *Science* 187, 398–404.
Dougherty D. (1997) *Sed & Awk*, 2nd Edn., O'Reilly Media, Sebastopol, CA, USA.
Eisen M.B., Spellman P.T., Brown P.O. and Botstein D. (1998) Cluster analysis and display of genome-wide expression patterns. *Proceedings of the National Academy of Sciences USA* 95, 14863–14868.

Okbay A., Beauchamp, J.P., Fontana, M.A., *et al.* (2016) Genome-wide association study identifies 74 loci associated with educational attainment. *Nature* 533, 539–542.

Wickham H. (2009) *ggplot2: Elegant Graphics for Data Analysis*, Springer, New York, USA.

19.5.2 General Texts

McCandless D. (2000) *Information is Beautiful*, Collins, New York, USA.

19.5.3 Review Articles

Lazer D., Kennedy R., King G. and Vespignani A. (2014) The parable of Google Flu: traps in Big Data analysis. *Science* 343, 1203–1205.

Leek J.T. and Peng R.D. (2015) Statistics: p values are just the tip of the iceberg. *Nature* 520, 612.

Leek J.T. and Peng R.D. (2015) Opinion: Reproducible research can still be wrong: adopting a prevention approach. *Proceedings of the National Academy of Sciences USA* 112, 1645–1646.

19.5.4 Websites

Scripts discussed in this chapter
www.cambridge.org/hofmann

Keith Baggerly: The importance of reproducible research in high-throughput biology
www.youtube.com/watch?v=7gYIs7uYbMo (accessed May 2017)

Cookbook for R (Winston Chang)
www.cookbook-r.com/ (accessed May 2017)

ggplot2
ggplot2.org/ (accessed May 2017)

The Comprehensive R Archive Network
cran.r-project.org/ (accessed May 2017)

The Open Data Handbook
opendatahandbook.org/ (accessed May 2017)

The Open Definition
opendefinition.org/ (accessed May 2017)

The R Graph Gallery
www.r-graph-gallery.com/ (accessed May 2017)

The R Project for Statistical Computing
www.r-project.org/ (accessed May 2017)

20 Fundamentals of Genome Sequencing and Annotation

PASI K.KORHONEN AND ROBIN B.GASSER

20.1 INTRODUCTION

Advanced nucleic acid sequencing and bioinformatic technologies allow the investigation of genomes and transcriptomes, and thus provide useful tools to investigate the molecular biology, biochemistry and physiology of organisms.

This chapter describes genome-sequencing methodologies, approaches and algorithms used for genome assembly and annotation. Compared with the long-established biochemical methodologies, the sequencing and drafting of genomes is constantly evolving as new bioinformatics tools become available. Despite technological advances, there have been considerable challenges in the sequencing, annotation and analyses of previously uncharacterised eukaryotic genomes. In this chapter, we will therefore discuss some bioinformatic workflows that have proven to be efficient for the assembly and annotation of complex eukaryotic genomes, and give a perspective on future research toward improving genomic and transcriptomic methodologies.

A genome project starts with the assembly and annotation of a draft genome from experimentally determined DNA sequences referred to as reads (nucleotide sequence regions). The success of an assembly depends on the sequencing technology and the assembly algorithms used, as well as the quality of the data. Typically, short-read, shotgun assemblies do not lead to complete chromosomal assemblies of eukaryote genomes, mainly because of challenges in resolving regions with repeats, a lack of uniform read coverage linked to a suboptimal quality of genomic DNA or poor library construction. Clearly, the quality of a genomic assembly is critical for subsequent gene predictions, and the quality of gene prediction is crucial to achieve an acceptable annotation.

The annotation of a genome is typically divided into structural and functional phases. For structural annotation, genomic features, such as genes, RNAs and repeats, are predicted, and their composition and location in the genome inferred. For gene prediction, results of both *ab initio* and evidence-based predictions (from mRNA, cDNA and/or proteomic data) are often combined. Functional annotation (also called functional prediction) assigns a potential function to a gene or genome element. In general, functions are predicted using similarity searches, structural comparisons, phylogenetic approaches, genetic interaction networks and machine-learning approaches. The following sections cover commonly used algorithms and recent methods for genome assembly, the prediction of protein-encoding genes and the functional annotation of such genes and their products.

20.2 GENOMIC SEQUENCING

The two main approaches used are whole-genome and hierarchical shotgun sequencing. For whole-genome shotgun sequencing, total genomic DNA is fragmented and sequenced, and the sequence reads are then computationally assembled into a draft genome. In contrast, in hierarchical shotgun sequencing, the genome is first divided into segments that are used to prepare a library by cloning (for example a bacterial artificial chromosome, BAC; see Section 4.12). BAC clones from the library are then individually sequenced and computationally assembled. Hierarchical shotgun (short-read) sequencing is more suited to genomes with high repeat contents than whole-genome shotgun sequencing. In contrast, whole-genome shotgun sequencing (long-read) methods are suited to resolve long repeat regions.

20.2.1 Sanger Sequencing

Dideoxy sequencing (Sanger sequencing) was one of the first methods used to sequence DNA. This 'first-generation sequencing' method was used for the BAC-coupled hierarchical shotgun sequencing of the human genome, which took more than 10 years to complete. Parallel to this publicly funded genome project, J. Craig Venter and coworkers undertook a competing effort employing paradigm shifting, whole-genome shotgun technology. In spite of the major progress achieved in these two human genome projects and the ability to produce high-quality read data ($\leq$ 900 bases), Sanger sequencing has been laborious, costly and slow for genomic sequencing. In the original Sanger sequencing method, the DNA sample is divided into four separate polymerase chain reactions (PCR; see Section 4.10), all of which contain the four standard deoxynucleotides (dATP, dGTP, dCTP, dTTP – collectively called dNTPs) and the DNA polymerase. To each reaction only one of the four dideoxynucleotides (ddATP, ddGTP, ddCTP, ddTTP – collectively called ddNTPs) are added. The ddNTPs terminate elongation of the DNA strand; they are added at an approximately 100-fold lower concentration than the corresponding dNTPs, allowing for enough fragments to be produced, while still transcribing the complete sequence. The DNA fragments resulting from PCR are thermally denatured and separated by size using gel electrophoresis (Section 6.2.2) in four different lanes (one for each ddNTP). The relative positions of the different bands among the four lanes, consecutively from bottom to top, are then used to read the DNA sequence. The introduction of ddNTPs with fluorescence labels enabled automated high-throughput sequencing, such that gel electrophoresis needed to be replaced by capillary electrophoresis. A further improvement of this technology, referred to as **dye-terminator sequencing**, used different fluorescence labels for each of the four ddNTPs (Figure 20.1). This development allowed for PCR-based sequencing to be carried out in a single reaction mix rather than four separate reactions. Dye-terminator sequencing is still used as a routine method for laboratory DNA sequence determination.

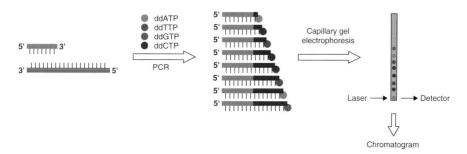

Figure 20.1 Schematic of contemporary dye-terminator sequencing. The use of fluorescently labelled dideoxynucleotides (see also Example 13.5) allows optical readout following the separation of DNA fragments by capillary gel electrophoresis.

20.2.2 Next-Generation Sequencing

With the Sanger sequencing method being fairly slow and laborious, next-generation sequencing (NGS), second-generation or massively parallel sequencing technologies, were developed. In contrast to sequencing a single DNA fragment, the DNA to be sequenced is typically prepared in the form of random fragments, with defined oligonucleotide sequences at either end that are captured on a flow cell by surface-bound complementary oligonucleotides (Figure 20.2). In the first step, each bound fragment is amplified to produce a so-called clonal cluster that contains ~1000 copies – a sufficient quantity for detection by the CCD camera. For the next step, sequencing reagents, including fluorescently labelled dNTPs are added. The flow cell is continuously illuminated with a laser for excitation and scanned by the CCD camera to record fluorescence emission. As individual bases are incorporated, the fluorescence emission from each cluster is recorded. As each individual type of dNTP is labelled with a distinct fluorescence label, the emission wavelength serves to identify the type of base being incorporated (so-called **sequencing by synthesis**). As this process is carried out across millions of fragments, it is regarded as massively parallel sequencing.

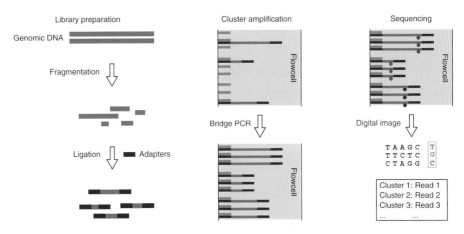

Figure 20.2 Schematic workflow of sequencing by synthesis. A library of fragments is prepared from genomic DNA and ligated to specialised adapters used to hybridise to complementary adapters on the flow cell surface. Each bound fragment is amplified into a clonal cluster. For sequencing, reversibly binding nucleotides with fluorescence labels are used and imaged for each cluster in real time in order to produce reads.

Successfully employed second-generation technologies include Genome Analyzer by Illumina, Supported Oligonucleotide Ligation and Detection (SOLiD) by Applied Biosystems and Genome Sequencer by Roche. Since their launch, these NGS platforms have evolved substantially, in terms of chemistry, read length, throughput and cost. The common aspect of these technologies is the iterative sequencing of PCR-amplified templates in wash-and-scan flow cells, producing read lengths of 35–700 bases. Differences exist in the methodologies of cluster generation, as well as the sequencing step. Both SOLiD and Genome Sequencer use emulsion PCR technology for DNA template preparation, whereas Genome Analyzer uses bridge PCR for the same purpose. In emulsion PCR, DNA sequences are replicated on a bead surface within tiny aqueous bubbles floating in an oil solution. In contrast, bridge PCR utilises single-stranded, adapter-ligated fragments bound to the surface of the flow cell; the free end of a ligated fragment 'bridges' to a complementary oligonucleotide on the surface.

The sequencing by synthesis method has also been implemented using different approaches. While Illumina's MiSeq and HiSeq instruments use a four-colour sequencing-by-synthesis method with reversible terminators, Applied Biosystems' SOLiD employs pyrosequencing. This approach monitors the release of pyrophosphate (diphosphate, $P_2O_7^{4-}$) upon incorporation of a new nucleotide into the strand being sequenced rather than the nucleotide itself. The released pyrophosphate is converted by the enzyme ATP sulfurylase to ATP, which in turn acts as a substrate for the luciferase-mediated conversion of luciferin to oxyluciferin (bioluminescence; see Section 13.5.3) and thus can be analysed spectroscopically. Finally, SOLiD uses oligonucleotide ligation, with two-base encoding and four-colour imaging. In this method, oligonucleotides of a fixed length with labels according to the sequenced position are annealed and ligated; the preferential ligation for matching sequences results in a signal informative of the nucleotide at that position.

Advantages and Challenges

Advantages of NGS include speed and relatively low cost of sequencing. A large genome (≥2 Gb) can now be sequenced for less than US$1000. However, limitations are the high cost of the sequencers themselves and the short length of sequence reads. These short-read datasets typically result in fragmented genome assemblies, which is particularly evident in eukaryotic genomes replete with long, highly repetitive repeat regions. Using short-read data only, the genome assembly algorithms cannot resolve long repeat regions or extend contigs (a set of overlapping DNA segments that form a consensus region of the nucleotide sequence) to chromosomal contiguity due to the multitude of similar anchor sequences in repeat regions. In order to address this problem, data from external libraries (for example, paired-end read libraries, mate pair libraries or Chicago libraries) bridge genomic regions using a pair of reads relating to a known insert size between the pairs. Typically, insert sizes range from 20 to 800 bases for paired end reads and from 2 to 40 kb for mate-pair libraries.

Third-Generation Sequencing

Third-generation sequencing supersedes and complements NGS by producing long reads (≤ 100 kb) from single DNA molecules from tiny amounts of genomic DNA template at reduced cost and throughput time. At present, some second-generation sequencers, such as HiSeq, MiSeq, the Moleculo Technology service from Illumina and GS Junior from Roche compete with third-generation methods. Currently available third-generation sequencing include Ion Torrent (Life Technologies), PacBio real-time sequencer (Pacific BioSciences) and Nanopore (e.g. MinION by Oxford Nanopore Technologies) sequencing technologies.

Ion Torrent can produce reads covering more than 400 bases by using a complementary metal-oxide semiconductor (CMOS) integrated circuit, named ion chip, to recognise a pH change caused by the release of hydronium ion (H_3O^+) side-products from a polymerase attaching nucleotides to the template sequence. Ion Torrent might be applicable to single molecule sequencing, but, currently, the DNA template is amplified by emulsion PCR. Because of this limitation and the short read length, Ion Torrent might be categorised as a second-generation sequencing technology.

The PacBio real-time sequencer platform is based on the use of a single-molecule real-time (SMRT) sequencing cell and produces read lengths larger than 10000 bases. Although PacBio sequencing is attractive, the cost of sequencing and the reported error rate of sequences (13–15%) are fairly high. However, using high sequence coverage (more than 100 times), a resultant sequence can be relatively error-free, and the technology can be used to assemble genomic regions that cannot be resolved employing NGS methods. Importantly, used in combination with short-read data, this technology can achieve substantially improved genomic assemblies.

Nanopore-based methods can sequence a single DNA molecule without the need for an intervening PCR amplification or chemical labelling step, or the need for optical instrumentation to identify the chemical label, and have potential for high-throughput, low-cost sequencing. The principle of such methods is that a molecule is passed through a pore, driven either by a salt gradient, an electric field or by an enzyme, thereby modulating measurable ionic flow. This flow has particular characteristics for each base, and allows the sequence in the DNA strand to be recorded. The technical challenges of this approach include blockages in the pores, enzyme dissociation from the nucleic acid strand and inaccuracy of base recognition. Currently, only Oxford Nanopore has been able to produce long reads, with error rates of 5–40%.

20.3 ASSEMBLY OF GENOMIC INFORMATION

The DNA sequence reads produced by sequencers are usually accompanied by a quality score for each nucleotide. The challenge is to correctly assemble these reads computationally to produce multiple, long contiguous sequences (contigs), with the goal of resolving the sequences to chromosomal contiguity. Computer algorithms used for assembly include the overlap, layout, consensus (OLC) (first-generation), de Bruijn graph-based (second-generation), greedy algorithms and overlap, correct, assemble (OCA; a recent hierarchical genome assembly method) (Table 20.1).

Table 20.1 **Comparison of genome assembly algorithms**

Name	Algorithm[a]	Distinguishing feature	Correction[b]	Data support	Read length	Year
SEQAID	OLC	First graph based	–	Sanger	Long	1984
TIGR	OLC	k-mer alignment	–	Sanger	Long	1995
CAP3	OLC	Paired-end read support	–	Sanger	Long	1999
Celera	OLC	Unitig concept	–	Sanger	Long	2000
EULER	DB	DBG repeat handling	–	Solexa, Illumina, Roche 454, Sanger	Short	2001
Newbler	OLC	454 Life Sciences assembler	–	Roche 454	Long	2005
SSAKE	Greedy	First short read assembler	–	Solexa, Illumina	Short	2007
SHARCGS	Greedy	Read filtering	–	Solexa, Illumina	Short	2007
CABOG	OLC	Sanger and SOLiD support	–	Roche 454, Sanger	Short+ long	2008
Velvet	DB	Read threading	–	Solexa, Illumina	Short	2008
ALLPATHS-LG	DB	Large genome support	–	Solexa, Illumina	Short	2008
ABySS	DB	Full parallel processing	–	Solexa, Illumina	Short	2009
SOAP denovo	DB	Compact graph structure	–	Solexa, Illumina	Short	2010
SPAdes	DB	Paired assembly graphs	Yes	Solexa, Illumina	Short	2012

Table 20.1 **(cont.)**

Name	Algorithm[a]	Distinguishing feature	Correction[b]	Data support	Read length	Year
SGA	OLC	FM-indexed[c] string graph	–	Solexa, Illumina	Short	2012
MaSuRCA	OLC/DB	Parallel k-mer counting (Jellyfish 2011)	–	Illumina	Short	2013
PBcR	OCA	First to support long-read data using hierarchical genome assembly method overlap, correct, assemble (OCA). The workflow was used in SMRT portal	No	Sanger, Solexa, Illumina, PacBio RS, Roche 454	Short+ long or long	2012
HGAP	OCA	Provides preassembled low error-rate long-reads for a long-read capable assembler. Workflow adopted for SMRT portal with Celera 8.1 assembler	No	PacBio RS	Long	2013
Falcon	OLC	Diploid genome assembly using string graphs		PacBio RS	Long	2014
Canu	OCA	Single cell long-read. Uses MHAP in error correction	No	PacBio RS, Oxford Nanopore	Long	2015
MaSuRCA	OLC/DB	Hybrid short and long-read method to create megareads used to assemble 22 Gb genome of *Pinus taeda*		Illumina, PacBio RS, Oxford Nanopore	Short+ long	2016
ABruijn	DB	De Bruijn graph-based long-read assembler No error correction required.	No	PacBio RS, Oxford Nanopore	Long	2016

Notes: [a]DB: de Bruijn graph; OCA: overlap, correct, assemble; OLC: overlap, layout, consensus. [b]Error correction for long reads required. [c]FM-index is a compressed full-text substring index based on the Burrows–Wheeler algorithm that is also used in digital archive compression such as bzip2.

The general workflow of a genome assembly pipeline is schematically outlined in Figure 20.3.

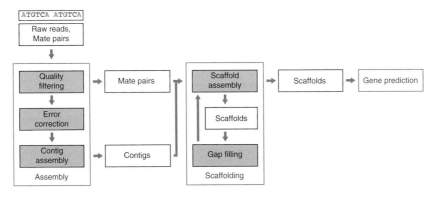

Figure 20.3 A simplified flowchart illustrating the genome assembly pipeline for short-read data.

20.3.1 Graph Theory

The major challenge in assembling genomic information in the correct order revolves around the problem of having millions of individual pieces to assemble, like in a jig-saw puzzle, but only that some pieces may be missing, others may be malformed, and there may also be pieces mixed in from another puzzle (e.g. contaminating sequences). Such problems are computationally often addressed using graph theory. A graph comprises a collection of vertices (called **nodes**) and edges that connect individual nodes (Figure 20.4). If there is no distinction between the nodes associated with each edge, the graph is called undirected. If the edges are directed from one node to another, the graph is called a directed graph.

One can take different approaches to walk through a graph. A **Eulerian path** visits every edge exactly once, whereas a **Hamiltonian path** visits every node exactly once. In practice, it is much more difficult to construct a Hamiltonian path. The relationships between DNA sequence reads can be represented as a graph, where the nodes are individual reads and an edge connects two nodes if their reads have an overlap.

Undirected Graph   Directed Graph

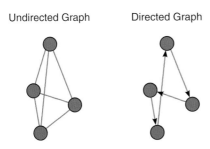

Figure 20.4 Graphs comprise a collection of vertices (nodes) and edges. In undirected graphs, there is no particular order in the association of nodes. In directed graphs, the edges possess a particular direction.

20.3.2 OLC Algorithms

Assembly algorithms used for data produced using first-generation sequencers are composed of overlap, layout and consensus phases (see Table 20.1). The overlap, layout, consensus strategy first approximates the pairwise overlaps between reads and then computes a graph structure (see Figure 20.5). To implement this overlap stage, all reads need to be aligned in a pairwise manner. However, this is not computationally feasible using traditional Needleman–Wunsch or Smith–Waterman dynamic programming alignment algorithms. To circumvent this issue of all versus all pairwise alignments, reads are usually first transformed into so-called k-mers (= all ordered subsequences of length k among all reads). Then, to identify the high-confidence pairwise matches, the alignment can be applied to a subset of reads with identical k-mers. This approach substantially reduces the number of pairwise alignments needed to create an overlap graph. Ideally, all reads in the resultant graph would be used once, and a Hamiltonian path would reveal the order of the reads, thus reconstructing the genomic sequence. However, this is not computationally feasible and, instead, the reads are multi-aligned to the optimised layout (with the relative position of reads along the genome determined) and consensus sequences are then established from these alignments by majority-voting.

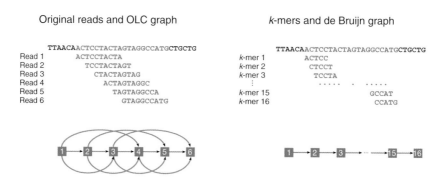

Figure 20.5 Comparison of reads and k-mers, and different graph approaches. Left: Six reads were generated for a genomic region and are laid out in order along the original sequence. The evaluation of overlaps of a minimum of five bases results in an overlap graph shown at the bottom. Most nodes have more than one ingoing or outgoing edge, making the reconstruction of the original sequence very computing-intense. Right: The original reads have been chopped into k-mers ($k = 5$). The evaluation of overlaps with a minimum of ($k - 1$) bases results in a de Bruijn graph with most nodes having only one incoming and one outgoing edge. Such a graph can be used to construct a Eulerian path.

20.3.3 Greedy and de Bruijn Algorithms

Second-generation sequencers produce enormous numbers of reads, which required the development of new assembly algorithms (Table 20.1). Such large datasets create challenges for OLC-based assemblers, but programs such as Newbler and CABOG have resolved this issue for pyrosequence data. The first programs to support NGS data were the greedy assembler implementations SSAKE and SHARCGS, in which the contigs are iteratively extended by best-matching reads. These assembly algorithms use a simple but effective strategy in that they greedily join together the reads that are most similar to each other. A problem with such greedy approaches is that the best match of a read in one contig might lead to a better assembly if this read were assigned to another contig.

To avoid computationally demanding pairwise and multiple alignment phases of OLC-based implementations, the de Bruijn graph fragment assembly method was soon applied to second-generation assemblers. In short, instead of representing full reads in an overlap graph, a de Bruijn graph represents all k-mers from these reads. First, a suitable k-mer length is chosen and the reads are split into patterns of length k.

A directed graph is then constructed, such that an edge points from an individual k-mer to the k-mer whose last $(k - 1)$ nucleotides overlap with the first $(k - 1)$ bases of the current k-mer (see Figure 20.5). This graph is then processed and 'pruned' to a Eulerian graph to identify a Eulerian walk (crossing each edge once) in order to reconstruct the genome. The use of the de Bruijn graph reduces the algorithmic phases to graph construction, graph simplification and error removal.

20.3.4 The Challenge of Assembling Across Repeat Regions

The assembly of repeat regions using first- and second-generation assembly algorithms is often a struggle. The accurate assembly of a repeat region becomes challenging when the length of the repeat is longer or equal to the length of the reads. To partially overcome this challenge, mate pairs or jumping pairs (paired reads with long insert sizes) that span repeat regions can be employed. However, the use of third-generation sequencing and assemblers should overcome this obstacle in many cases.

20.3.5 Assemblers for Long-Read Data

For third-generation sequencers, which are capable of producing significantly longer reads than first- and second-generation sequencers, the computational effort in assembler implementations concentrates on enhanced read pre-processing to eliminate numerous sequencing errors. This pre-processing results in relatively small numbers of long sequences of high quality, thus stimulating the reuse of OLC-based algorithms to avoid the artificial shortening of read lengths (resulting in k-mers) encountered using de Bruijn graph-based assemblers. Therefore, this approach has the capacity to assemble and resolve long repeat regions. The first implementation to support third-generation data was PacBio corrected Reads (PBcR) embedded in the Celera Assembler, in which high-quality short reads produced by second-generation sequencers (see Section 20.2.2) are used to resolve error-prone long-read data produced using third-generation sequencers. The assembler used was based on the OLC-based CABOG Celera implementation (Table 20.1). Many second-generation assembler implementations (e.g. SPAdes and MaSuRCA) now support hybrid assemblies from both short- and long-read datasets, and some implementations (e.g., HGAP and MHAP) allow self-correction of long-read data without the need for second-generation reads. To date, at least one de Bruijn graph-based assembler (ABruijn) is capable of producing high-quality genome assemblies using uncorrected long-read data. Instead of haploid genome assemblies, long-read data also paves the way to achieving diploid assemblies by means of string graphs (Falcon).

20.4 PREDICTION OF GENES

The assembled genomic sequence data are of little value without further bioinformatic processing. Gene prediction, also called structural annotation or gene finding, aims to identify structural elements in a genomic region that represent a gene. In eukaryotes, these elements consist of 5'- and 3'-untranslated regions (UTRs), introns, exons and respective protein-coding sequences (CDSs), and splice sites (Figure 20.6).

Methods for gene prediction can be extrinsic (based on a similarity search) or intrinsic (*ab initio*). While extrinsic techniques align transcriptomic, protein-sequence and/or other evidence datasets for gene prediction, intrinsic methods use statistical patterns to identify gene regions in a genomic sequence. Predicted gene-element data are typically represented by a unified General Feature Format (GFF) specification. A general layout of gene prediction and annotation is illustrated in Figure 20.7.

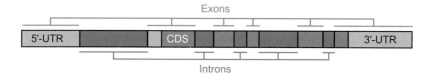

Figure 20.6 Structural elements of a gene. Exons encompass both transcribed untranslated regions (UTRs: 5'-UTRs, 3'-UTRs) and coding sequences (CDSs), whereas introns between exons are regions that are spliced from the gene product.

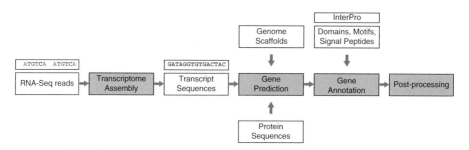

Figure 20.7 A general pipeline for gene prediction and functional annotation.

20.4.1 Extrinsic Methods

For these methods, evidence data (i.e. transcriptomic and/or protein sequences) are aligned to the genomic sequence, and genes are predicted based on the alignment success. In order to provide a means of accurately predicting a gene structure, full-length cDNA sequences are preferred, but sometimes their availability is limited. Therefore, fragmented transcriptomic sequences, assembled from RNA-Seq data (see Section

4.17.1), and protein sequences of taxonomically closely related species are also used. The mRNA sequences are typically derived from the species under investigation and thus match the genomic sequence (so-called native alignment), whereas the protein sequences of closely related species are not expected to match exactly conceptually translated genomic sequences (so-called trans alignment).

Alignment inaccuracies, the fragmented nature of evidence (mRNA or protein sequence) data and splice variants from genes are challenges for extrinsic methods. Some alignment algorithms (for example EST_GENOME, AAT, Exonerate) use BLAST (see Section 16.4.1) to produce seed alignments. These seeds are then extended using different dynamic programming variants (Needleman–Wunsch or Smith–Waterman algorithms). Pair HMM aligners, such as GeneWise and Pairagon, align evidence data accurately across exons and introns, but at the cost of large computation times. Of these aligners, Exonerate processes data relatively rapidly, is capable of aligning both nucleotide and protein sequences, and is, thus, relatively widely used for alignment tasks.

20.4.2 Intrinsic Methods

These methods are divided into consensus (signal sensor) and non-consensus (content sensor) categories. Consensus-based methods predict known nucleotide patterns (signals) in gene elements, such as splice sites, start and stop codons, and the Kozak consensus sequence (relates to the initiation of translation; see Section 4.5.7). Methods utilising the weighed matrix method (WMM), such as position weight matrix (PWM), weigh-ed array model (WAM), maximal dependence decomposition (MDD) and windowed weight array model (WWAM), are used to recognise these signals. In brief, WMM calculates the signal probability and assumes that individual nucleotides are independent. In contrast, WAM assumes dependencies between adjacent nucleotides; MDD implements a decision tree of WMMs, extending the dependency considerations across non-adjacent nucleotides; and WWAM assumes dependencies across three consecutive nucleotides and averages related conditional probabilities among five consecutive nucleotides.

Non-consensus methods use nucleotide composition (content) to recognise gene elements and sequence areas, such as coding and non-coding regions. Hidden Markov models (HMMs; see Section 16.4.4) using hexamer sequence composition has proven to be the most successful discriminator between these two regions, when predicting nucleotide-by-nucleotide. However, to extend the prediction capability of this single nucleotide approach to versatile gene elements and even complete gene structures, the prediction algorithms are enhanced with so-called three-period, fifth-order generalised HMMs (GHMMs), in which hexamer sequences are used, together with the built-in knowledge of codon structure, to ensure the preservation of a reading frame. For instance, the programs GENSCAN and GeneMark-ES use a GHMM-based three-period, fifth-order Markov chain model. To further improve predictions, interpolated Markov models (IMMs), in which Markov models of different orders are interpolated, are implemented in gene finders such as AUGUSTUS and GlimmerHMM. *Ab initio* prediction algorithms (see Table 20.2) have been enhanced using information from syntenic

Table 20.2 **Key *ab initio* gene prediction algorithms**

Name	Sensor Content[a]	Signal[b]	Features	Year
GENSCAN	GHMM	WMMs, MDD	First to predict multiple genes	1997
GeneID	GHMM	PWM	DP gene structure prediction	2000
TWINSCAN	GHMM	–	Conservation probability	2001
AUGUSTUS	IMM	WWAM, HMM	Improved intron model	2003
GlimmerHMM	IMM	WMMs, MDD	Uses GeneSplicer and GlimmerM	2004
SNAP	GHMM	WMMs, WAM	User-friendly training procedure	2004
GeneMark-ES	HSMM	–	Unsupervised training	2005
N-SCAN	GHMM	–	Phylogenetic Bayes networks	2006
CONTRAST	CRF	SVM	Combines signals to gene structure	2007
mGene	SVM	SVM	Structure prediction by Hidden semi-Markov support-vector machines (HSMSVMs)	2009

Notes: [a]GHMM: generalised hidden Markov model; IMM: interpolated Markov model; HSMM: hidden semi-Markov model; CRF: conditional random field; SVM: support-vector machine. [b]HMM: hidden Markov model; MDD: maximal dependence decomposition; PWM: position weight matrix; SVM: support-vector machine; WAM: weighed array model; WMM: weighed matrix method; WWAM: windowed weight array model.

(= colocalised) regions among multiple genomes; however, it is advisable to employ genomes from taxonomically closely related species.

To create functional prediction models, *ab initio* predictors have to be trained using reliable training datasets, which are specific to each genome. If training data are not available, parameter values for prediction models can be estimated by predicting genes first using suboptimal parameter values (for example, copied from prediction models for closely related species, inferred from the structures of core eukaryotic genes (CEGs) or obtained from unsupervised gene prediction programs, such as GeneMark), and then by recalculating new values using these predicted genes.

20.4.3 Combining Gene Predictions From Multiple Sources

Consensus gene predictions are achieved by combining evidence-based and *ab initio* predictions from multiple sources (see Table 20.3), and can provide confident predictions. The first attempt to combine the prediction data from multiple sources was made using the program COMBINER. Here, consensus prediction algorithms are based on both linear and statistical combinations of the prediction data from multiple sources. JIGSAW, the successor of COMBINER, has internal support for *ab initio* predictions with GHMMs and expresses external evidence of structural elements of a gene using feature vectors. These vectors give a weighting coefficient to each prediction source, and dynamic programming (combined with decision trees) is used to establish optimal gene structures.

Table 20.3 **Key combined gene prediction programs**

Name	Features	Year
GenomeScan	GHMM with evidence by blastx	2001
TWINSCAN-EST	Nucleotide evidence data	2006
N-SCAN-EST	Nucleotide evidence data	2006
AUGUSTUS+	Nucleotide evidence data	2006
COMBINER	Linear and statistical combination	2004
Ensembl	Prefers evidence data	2004
JIGSAW	GHMM	2005
GLEAN	LCA	2007
EVIGAN	Bayes network with ML parameter estimation	2008
MAKER2	AED	2008
EVM	Manual weight adjustment of evidence data and gene prediction data sources	2008

Other frameworks include Ensembl, EVIGAN, GLEAN, MAKER2 and EVM. Ensembl prefers evidence-based over *ab initio* predictions, thus achieving high-quality annotations at the cost of sensitivity. EVIGAN predicts gene structures using dynamic Bayes networks, for which the parameters are estimated with the maximum likelihood (ML) method, and GLEAN uses the latent class analysis (LCA) algorithm to give consensus predictions. In GLEAN-based LCA implementation, gene structures are predicted from gene structural elements. MAKER2 uses annotation edit distance (AED) to estimate the share of evidence data for the consensus prediction, therefore offering the advantage of estimating the reliability of any prediction. The successor versions of *ab initio* programs, such as GENSCAN (GenomeScan), TWINSCAN (TWINSCAN-EST), N-SCAN (N-SCAN-EST) and AUGUSTUS (AUGUSTUS+), can also utilise evidence data for gene predictions. EVM accommodates the use of variable amounts of gene-prediction and evidence data, and allows for manual weight adjustment of each data source.

20.5 FUNCTIONAL ANNOTATION

The functional annotation of genes (**functional genomics**) is a large field that also utilises extensive experimentation to describe the function and interactions of genes and gene products. However, in this section, we will only consider computational predictions of the functions of protein-encoding sequences and their inferred products. A simplified work flow is illustrated in Figure 20.7. Computational prediction can rely, for instance, on sequence similarity, clustering and phylogenomics, structure and structural similarity, protein–protein interaction networks and/or machine learning. Commonly used tools in the process of annotation, such as BLAST (Section 16.4.1) and the InterPro software framework (Section 16.5.4), are based on sequence similarity.

The description of a uniform and coherent annotation of protein function(s) is achieved using a common set of rules, defined in unified classification schemes such as Gene Ontology (GO; Section 16.5.3), Enzyme Commission (EC; Section 23.1.1) and Kyoto Encyclopedia of Genes and Genomes (KEGG) BRITE functional hierarchies (see Section 16.5.6). The structure of a protein often relates to its function, but can differ in hierarchy from functional classification schemes. Three classification schemes have been devised for protein structures, namely Structural Classification of Proteins (SCOP), Class, Architecture, Topology, Homologous (CATH) superfamily and Families of Structurally Similar Proteins (FSSP), as discussed in Section 2.2.3.

The success or reliability of functional prediction is influenced by the accuracy of an alignment of homologous characters in two or more sequences. Typically, two randomly generated protein sequences of more than 100 amino acids in length have 10–20% sequence identity. This identity range imposes practical limits on the success of identifying homologous protein sequences, and has raised a concept called the twilight zone (identity range: 15–25%), in which the reliability of the prediction of homologous protein sequences is only 10%. In contrast, at higher sequence identity (> 30%), the reliability of a prediction is 90%.

However, for enzyme pairs that share more than 50% amino-acid sequence identity, less than 30% have identical EC numbers (see Section 23.1.1). To confidently transfer the three digits of an EC annotation scheme, 40% sequence identity is required, and even 60% identity is needed to transfer all four digits at 90% accuracy. On the other hand, only 86% of BLAST matches with e-values (expectation values) of less than 10^{-50} have an identical EC annotation. By contrast, the general structure of a protein is given by its fold (Section 2.2.2) and imposed by the amino acids covering only 3–4% of the sequence. Proteins with a similar structure (structural homologues; see Section 2.2.2) are likely at amino-acid sequence identities of more than 33%, but can potentially be identified, even at less than 15% sequence identity (the so-called midnight zone).

Errors in functional annotation of genes are widespread in current databases. Estimates indicate that 5–63% of gene annotations in public databases are incorrect or misleading, and are propagated via analyses of new genomes. These errors originate from various sources, including genome assembly and gene prediction. Erroneous or incomplete genome assemblies might relate to truncated or chimeric sequences, as well as single nucleotide errors, which can cause inaccurate prediction of genes and gene functions. At present, RefSeq (Section 16.2.2) and UniProt/SwissProt (Section 16.2.1) are the best-curated databases, due to the fact they require multiple lines of experimentally derived evidence.

20.5.1 Annotation Based on Sequence Similarity

Functional annotation methods rely on sequence similarity comparisons with proteins of known function(s). These methods establish similarities either by global alignment of full-length sequences or by local alignments of domains, motifs (partial sequences) or fingerprints (motif sets). The concepts of domain, motif and fingerprint in sequences contained in these databases and functional prediction methods are illustrated in Figure 20.8.

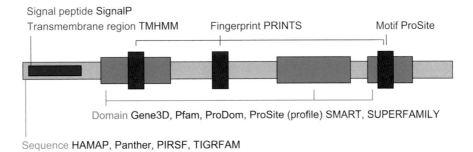

Figure 20.8 Databases for sequence, domains, motifs and fingerprints, as well as methods to predict signal peptides and transmembrane regions are integrated in the InterPro framework.

A widely used approach is to transfer the annotation of a homologous protein identified by gapped BLAST. The sensitivity of BLAST is comparable to that of the Smith–Waterman algorithm, but evolutionarily distant homologues are not always recognised. To address this latter issue, an iterative algorithm using a position-specific score matrix was devised and implemented in the software PSI-BLAST. A score matrix is reconstructed for individual iterations using sequences from previous iterations. BLAST-based methods do not annotate proteins based on features, but rather they transfer an annotation from another protein predicted to be a homologue. In some cases, homologues may only align over a small portion of their overall lengths, which could lead to an erroneous transfer of function. In addition, it is possible that the homologue from which function is being transferred might originally have been wrongly annotated. Because BLAST is also commonly used to predict orthologous proteins (that is, arising from a speciation event) from KEGG databases (see Section 16.5.6), any mis-annotation can result in erroneous predictions of metabolic pathways and protein families.

There are many publicly available databases and methods for the sequence-based prediction of protein functions. These databases and methods are collectively integrated in the widely used InterPro framework (see Section 16.5.4) that supports a range of databases of superfamilies, families, domains, and signal peptide and transmembrane domain predictions.

Databases with curated or predicted domains and motifs deserve detailed consideration, because they are concerned with the functional components of proteins. For example, Pfam has a curated (PfamA) and a computationally generated (PfamB) domain-family database. This database generates clusters of domain families defined using the program ADDA, in which clusters are formed from pairwise comparisons of profiles of domains inferred by penalising splits and partial overlaps in a pairwise, BLAST-aligned protein similarity matrix. In contrast, the Simple Modular Architecture Research Tool (SMART) requires manual intervention during annotation and is linked to a database called Search Tool for the Retrieval of Interacting Genes (STRING). ProDom compares the results from PSI-BLAST against the UniProtKB database and infers domain information from the resultant data. ProDom complements domain databases, such as Pfam, PROSITE and SMART. The SUPERFAMILY resource uses the

SCOP classification scheme (see Section 2.2.3) for inferred protein-domain superfamilies, and assigns gene ontology (GO) terms (see Section 16.5.3) to these families using gene ontology annotation. Gene3D combines both structural and functional information to annotate domains found in sequences in the databases UniProtKB, RefSeq and Ensembl; it uses the structural CATH classification scheme (see Section 2.2.3), and clusters annotated superfamilies into functional subfamilies using GeMMA. PROSITE recognises protein motifs using regular expressions and weight matrix profiles, augmented by the annotation rule database ProRule. The implementation of ProRule increases the reliability by imposing rules, such as essential amino acids in the active sites of an enzyme.

The annotation of proteins encoded in a genome can be inferred by combining results from searches of all of the above, curated public databases containing information/data on protein sequences, domains and motifs. However, currently, these resources are far from complete, and the domains of some genes/proteins annotated in the course of a new genome project are likely to differ significantly from those found in existing databases, or are not yet defined, thus compromising their recognition.

20.5.2 Annotation Based on Clustering and Phylogenetics

Methods for predicting **orthologous** (arisen from speciation) and **paralogous** (arisen from gene duplication) protein groups among multiple species can be divided into those that rely on clustering based on similarity and those that use a phylogenetic approach. The resultant groups are expected to represent proteins with the same or similar functions, and can thus be used for protein annotation (see Table 20.4 for an overview of clustering and phylogenomic annotation algorithms).

Clustering methods, such as OrthoMCL, InParanoid and MultiParanoid, and the databases OrthoDB and Clusters of Orthologous Groups of Proteins use all-versus-all similarity matrices, created based on pairwise alignments of protein sequences using algorithms, such as BLAST, FASTA and Smith–Waterman. Note that the 'FASTA format' is also a commonly used file format used for storing biological sequences (see also Section 16.3.1). The largest publicly available all-versus-all protein sequence similarity score matrix is called Similarity Matrix of Proteins (SIMAP). SIMAP 1 (now discontinued) imported proteins from multiple databases, such as Ensembl, NCBI GeneBank, NCBI RefSeq, UniProt/TrEMBL, UniProt/SwissProt and PDB, and was updated monthly; in the release of January 2015, it encompassed more than 48 million sequences, whereas SIMAP 2 is limited to proteins encoded in complete genome sequences and is not publicly available. SIMAP 2 is employed, for instance, in the database called 'evolutionary genealogy of genes: Non-supervised Orthologous Groups' (eggNOG), in which the proteins are assembled into **in–paralogous** (as opposed to **out–paralogous** and orthologous) groups by comparing sequence similarities within and among clades. In-paralogous genes arise where two lineages share a gene duplication (Figure 20.9). Orthologous proteins amongst these in-paralogous groups are then identified by creating and merging reciprocal best hits among three species.

These clustering methods can be improved by replacing **amino–acid substitution models** (e.g. BLOSUM) with models that better estimate phylogenetic distances, such

Table 20.4 **Clustering and phylogenomic annotation algorithms**

Name	Orthologues and paralogues inferred by	Pairwise similarity matrix	Year
InParanoid	Pairwise reciprocal	BLAST based	2001
OrthoMCL	Markov Clustering algorithm	BLAST based	2003
MultiParanoid	Pairwise reciprocal	Uses InParanoid	2006
OrthoDB	Three-wise reciprocal	Smith–Waterman (3–reciprocal hits)	2013
KOG	Three-wise reciprocal	BLAST based (3–reciprocal hits)	2003
SIMAP 1	–	FASTA and Smith–Waterman; very large	2010
eggNOG	Taxonomy and phylogenetic distance	Uses SIMAP	2008
SYNERGY	Taxonomy and phylogenetic distance	FASTA weighed by phylogenetic distance	2007
PhIG	Taxonomy and phylogenetic distance between families	BLAST weighed by phylogenetic distance	2006
TreeFam	Taxonomy and phylogenetic distance between families	Uses PhIG	2006
PANTHER	Taxonomy and phylogenetic distance between families	Profile HMM	2013
SIMAP 2	–	Proteins from sequenced genomes	2015

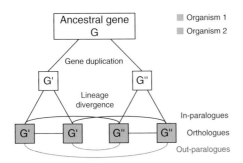

Figure 20.9 Relationships of orthologues, in-paralogues and out-paralogues.

as JTT and WAG, and by reconciliation of a deduced phylogenetic tree of individual genes to the phylogenetic tree of species, in order to predict orthologues and paralogues. This approach has implications in terms of the prediction of protein function, because, in contrast to orthologues, paralogues tend to have divergent functions. The reconciliation is accounted for in methods or databases such as SYNERGY, Phylogenetically Inferred Groups (PhIG), TreeFam and PANTHER.

Databases offered by these methods provide a compelling option to rapidly predict protein function, but often they are limited in their coverage of species and proteins.

Furthermore, using sequence similarity searches to position the query sequence in phylogenetic trees in these databases, in which substitution models and taxonomic information are utilised for construction, seems questionable. Instead, the query sequence should be positioned in a phylogenetic tree using methodologies commensurate with database construction. However, even perfect positioning does not guarantee an accurate prediction of function for the query protein sequence, because homologous proteins do not always have the same function. In summary, although some methods attempt to address this issue by reconciliation of orthologous protein trees with a phylogenetic tree of species, in order to assign proteins as orthologues or paralogues, a more accurate means of identifying conserved functions is needed.

20.5.3 Annotation of Proteins Based on Structure

The sequence-based annotation methods described in Sections 20.5.1 and 20.5.2 predict the function of a protein, provided a well-annotated homologous protein exists. When this is not the case, function can still be predicted by comparing the predicted folds of gene products against structurally similar proteins (structural homologues) in databases such as the Protein Data Bank (PDB). Although this approach can successfully assign a function in many cases, there is a caveat to be considered.

Only ~ 60% of structurally similar proteins without significant sequence similarity share a binding site location; thus, the function inferred from this comparison may not always be correct. Furthermore, increasing numbers of three-dimensional structures in the PDB have been derived from high-throughput structure determination efforts (structural genomics initiatives). As these efforts have only been directed at determining the three-dimensional structures, often functional knowledge about these proteins is lacking.

A second problem arises in cases of convergent evolution. In divergent evolution, the structure and function of even distantly related orthologous proteins are relatively well conserved, therefore supporting structure-based function prediction. However, in convergent evolution, the same function is observed with different folds, thus preventing the use of structural homologues to infer a function.

The widely used sequence-to-structure-to-function annotation paradigm consists of four steps:

1. The search and selection of a three-dimensional template of a protein
2. Model-building from well-aligned template fragments
3. Subsequent quality evaluation of resultant model(s)
4. The prediction of function.

For the first step, *ab initio* structure predictions can be used in the absence of matching structural templates or to infer the structure of non-aligned regions, such as loops. Typically, rapid protein secondary-structure-based profile–profile threading algorithms and meta-predictors (combining multiple threading algorithms) are used to search for and select templates. A model is then built from such templates or template fragments using rigid-body segment matching or spatial constraint methods, and loops are modelled separately either from loop libraries or *ab initio* predictions (see currently used software in Table 20.5). The evaluation of the quality

Table 20.5 **Software for the functional annotation of proteins based on structure**

Name	Pairwise similarity matrix	Year
Phyre2	Fast structure and function prediction	2003, 2015
JESS	Prediction of conserved residues in 3D space	2003
I-TASSER	Winner of CASP[a] competition over many years	from 2006
ConFunc	Predicts conserved residues among GO enrichments	2008
COFACTOR	Function prediction module in I-TASSER	2012

Notes: [a]Critical Assessment of Protein Structure Prediction

of a model is based on various alignment metrics (such as TM-score and root mean square deviation), but these scores are always dependent on similarity comparisons between or among multiple structures, which sometimes compromises their accuracy. Subsequently, the prediction of function is achieved by aligning the predicted final model to known protein structures using structural alignment techniques, such as combinatorial extension, DALI, Vorolign, TM-align and the phenotypic plasticity method (PPM).

To increase the accuracy of structure-based function prediction, conserved amino acids in active and binding sites need to be evaluated. For enzymes, catalytic residues and their location within the protein and orientation in active sites are usually conserved, and do not always associate with structural variation, thereby allowing the functional annotation of distantly related homologues. The identification of conserved residues in protein families is based on multiple sequence alignment. Methods such as evolutionary trace paths extract conserved residues from three-dimensional models and use entropy considerations in the scoring. Other methods, such as ConFunc, predict conserved residues among gene ontology (GO)-specific groups following multiple sequence alignment, and directly assign query sequences to GO domains (see also Section 16.5.3). The program JESS estimates residue conservation of catalytic sites of three-dimensional models, and provides structural templates for use by the Catalytic Site Atlas (CSA; see also Section 23.4.1). Implementations, such as COFACTOR in the I-TASSER framework, first predict function based on the global structure of a protein, estimate conserved residues in a protein and then compare these residues to the catalytic structural templates in CSA, and to the known functional residues in protein templates. In the Phyre2 implementation, confident protein structure and functional residue predictions are submitted to the program 3DLigandSite for the inference of ligand binding sites.

20.5.4 Additional Annotation Methods

Annotation methods other than those described above include experimentally evaluated and computationally predicted protein–protein interaction networks and protein–protein complexes, for which results are found in databases such as DIP and STRING. Promising annotation appears to be achieved when using machine learning and supervised classification methods, and unsupervised clustering methods. These methods can be applied to predict individual features of proteins (for example, domain boundaries, subcellular location and conserved residues), to collectively predict a function based on data integrated from different sources (sequence, structure, transcription, taxonomy, and metabolic and protein–protein interaction networks), or to enhance an existing homology-based annotation. The functional classification of proteins can be evaluated in the annual Critical Assessment of Function Annotation (CAFA) challenge.

20.6 POST-GENOMIC ANALYSES

Genome assembly and annotation provide a foundation for many different molecular and biochemical analyses. In the following section, three examples of post-genomic analyses are given.

20.6.1 Synteny Analyses

Synteny is a concept in which conserved segments (orthologous chromosomal regions) in different genomes preserve their order. Syntenic blocks are defined as nucleotide sequence segments that can be converted to conserved segments by small rearrangements, called microarrangements. Such blocks include orthologous genes or non-coding conserved nucleotide sequence tracts known as anchor-nucleotide regions. The identification of such regions and the subsequent prediction of syntenic blocks are achieved through a genome alignment, which is usually a challenging computational task, and can be complicated by the presence of repeats in genomes. The last step in a synteny analysis is to rearrange syntenic blocks among the genomes and to visually display the result. The rearrangement algorithms are typically heuristic, parsimony-based implementations. For fragmented *de-novo*-assembled genomes, the arrangement of the contigs on chromosomes is not known and therefore adds a level of complexity to the prediction of synteny and the display of results; currently available programs do not cope well with these issues. However, the use of long-read data should decrease fragmentation in an assembly, consequently reducing complexity linked to the computational aspects of synteny. An accurate synteny analysis can assist gene prediction and functional annotation, as well as the prediction of chromosomal duplications, deletions and rearrangements, serving as a basis, for instance, for predicting ancestral genomes. The programs used to resolve and visually display synteny are listed in Table 20.6. These programs either align the genomic sequences using alignment implementations (e.g. LASTZ, blastn and blastp), or require files with genomic positions of anchors.

20.6.2 Transcriptomic Analyses

Presently, RNA sequencing (RNA-Seq) is the preferred technology to investigate transcriptomes linked to *de-novo*-assembled genomes, because it offers genome-wide transcription profiles for any species, and can also identify genes not detected in draft genome assemblies. In addition, unlike hybridisation- and PCR-based microarrays,

Table 20.6 **Key software commonly used to resolve and display synteny**

Software	Comments
GRIMM	Command line; requires anchor file as input
Cinteny	Web service; limited to three user defined species
AutoGraph	Web service; limited to three user defined species
FISH	Limited to synteny for two genomes only
Mummer	Genome-wide alignment; limited to two genomes only
VISTA	Web service; limited support for scaffold assemblies
MCScanx	Command line; requires BLAST results and gene coordinates
OrthoCluster	Command line and web service; requires orthologous relationships for genes
Circos	Visualisation program; supports only circular chromosomes
SyMap	Automated system for identifying and displaying genome synteny alignments

RNA-Seq does not require the use or design of oligonucleotide probes, and platforms used for genomic sequencing can also sequence transcribed RNA molecules from a sample. The resultant RNA-Seq data can then be used to explore transcription qualitatively and quantitatively, in order to detect differentially transcribed genes and gene transcription profiles, and to detect splice junctions. Importantly, RNA-Seq data can be assembled to serve as training or evidence data for the prediction of genes.

RNA-Seq data can have multiple sources of bias, which need to be considered before analyses are conducted. Technical sources of bias can include library size, GC content, template shearing and the transfer of technical errors introduced during library construction. Methods that account for technical biases using data normalisation include SVA and PEER, as well as trimmed mean of M-values (TMM) and DESeq. However, presently, there are no normalisation methods for comparative analyses of transcription among species, and possible biases in datasets are therefore unknown. Transcription profiling and the analysis of differential transcription using RNA-Seq data can be undertaken employing either parametric (e.g. EdgeR and DESeq) or the non-parametric (NOIseq) software. The sensitivity and accuracy of RNA-Seq depends on sequencing depth and the read-mapping method selected for the analyses, but also on the accuracy and completeness of the gene models used for gene prediction.

20.6.3 Metabolic Pathway Analyses

Metabolic pathway and enrichment analyses can be conducted using gene and transcription data encoded in *de-novo*-assembled and annotated genomes. Generally, pathway predictions are highly dependent on accurate enzyme annotations, which emphasises the importance of the original quality of gene predictions and functional annotations.

Table 20.7 **Databases and programs for metabolic pathway analyses**

Pathway identification	Novel pathway prediction	Synthetic pathway prediction
KEGG	IPA™	RouteSearch
MetaCyc	MetaCore™	FMM
BioCyc	Pathway Tools	RetroPath
ConsensusPathDB	PathMiner	
Reactome		

The key metabolic pathway databases and computational tools for this purpose are summarised in Table 20.7. Pathway analyses can be carried out using information available in public databases, such as the Kyoto Encyclopedia of Genes and Genomes (KEGG; see also Section 16.5.6), MetaCyc, BioCyc, ConsensusPathDB and Reactome, and the commercial Ingenuity Pathway Analysis™ (IPA) or MetaCore™ databases employing available analysis software embedded within these databases. Typically, these databases assume that the organism of interest is readily integrated into the database, but are usually not able to accommodate data from an annotated genome from a new species. Only the KEGG database and the MetaCore™ suite allow the tailoring of analyses to an unknown organism. In the KEGG database, genes are first mapped to KEGG orthologues, which can then be used to extract enriched pathways, whereas MetaCore™ supports the construction of a customised database. However, such analyses are limited to identifying known pathways, rather than inferring pathways *ab initio*. The database EuPathDB aims to predict or identify pathogen-specific pathways, but does not support the integration of *de-novo*-assembled genomes. Nonetheless, novel metabolic pathways could be inferred or predicted using the methods implemented in programs such as Pathway Tools and PathMiner. These programs are typically based on subgraph predictions from all-encompassing metabolic networks (reactions and compounds). The construction of complete metabolic networks, which involves numerous manual steps and might take years to achieve, would substantially improve the quality of pathway predictions.

20.7 FACTORS AFFECTING THE SEQUENCING, ANNOTATION AND ASSEMBLY OF EUKARYOTIC GENOMES, AND SUBSEQUENT ANALYSES

A major factor contributing to the quality of the genome assembly is the integrity and amount of genomic DNA isolated for subsequent library construction and sequencing. For some organisms, such as parasites, there are significant limitations in extracting high-quality, high-molecular-weight genomic DNA. If small quantities of such DNA are available, techniques, such as multiple displacement amplification (MDA), can be used to produce up to a billion-fold genomic DNA, even from a single cell. However, this amplification process can lead to a bias toward CG pairing, and some DNA elements may be 'lost', particularly when single cells are used as templates; a few thousand copies of genomic DNA are required to minimise possible amplification bias in MDA. Following amplification, AT-rich areas, such as those found sometimes in repeat regions, might lack the necessary sequencing depth for successful genome assembly. For short-read data produced by second-generation sequencers (see Section 20.2.2), long repeat regions cannot be assembled reliably, and this bias might go unnoticed. However, such amplification bias should be detected using third-generation sequencing approaches, where read length relates directly to the quality of genomic DNA extracted and/or library construction. Other factors that can affect the quality of a genome assembly include:

- high levels of sequence heterogeneity in genes among individuals within some species
- chromatin diminution (a process occurring in early developmental stages of some multi-cellular eukaryotes, in which a considerable fraction of chromatin is eliminated from the genome)
- habitat-specific contamination of isolated genomic DNA.

20.8 CONCLUDING REMARKS

To improve the quality of genome assemblies and annotation, third-generation sequencing technologies can now be employed to produce long-read sequence data to complement the assembly of short-read datasets in enhanced bioinformatic pipelines. Additional components now need to be integrated into annotation pipelines, including extensive post-processing of predicted genes, accurate gene prediction that targets small sets of genes of interest, and the use of phylogenetic, structural and machine-learning prediction methods for functional annotation to account for orphan genes (i.e. previously uncharacterised genes), which are either entirely new or distant homologues to known genes available in current public databases. To further study such orphan genes, advanced bioinformatics methods and/or carefully designed biochemical assays are essential to unravel their functional roles in biological pathways.

20.9 SUGGESTIONS FOR FURTHER READING

20.9.1 Experimental Protocols

Alexeyenko A., Tamas I., Liu G. and Sonnhammer E.L. (2006) Automatic clustering of orthologs and inparalogs shared by multiple proteomes. *Bioinformatics* 22, e9–15.

Altschul S.F., Madden T.L., Schaffer A.A., *et al.* (1997) Gapped BLAST and PSI-BLAST: a new generation of protein database search programs. *Nucleic Acids Research* 25, 3389–3402.

Altun Y., Tsochantaridis I. and Hofmann T. (2003) Hidden Markov support vector machines. In *Twentieth International Conference on Machine Learning (ICML 2003)*, AAAI Press, Menlo Park, CA, USA.

Beier S., Himmelbach A. and Schmutzer T., *et al.* (2016) Multiplex sequencing of bacterial artificial chromosomes for assembling complex plant genomes. *Plant Biotechnology Journal* 14, 1511–1522.

Dale J.M., Popescu L., Karp P.D. (2010) Machine learning methods for metabolic pathway prediction. *BMC Bioinformatics* 11, 15.

Dean F.B., Hosono S., Fang L.H., *et al.* (2002) Comprehensive human genome amplification using multiple displacement amplification. *Proceedings of the National Acadademy of Sciences USA* 99, 5261–5266.

Needleman S.B. and Wunsch C.D. (1970) A general method applicable to the search for similarities in the amino acid sequence of two proteins. *Journal of Molecular Biology* 48, 443–453.

Rothberg J.M., Hinz W., Rearick T.M., *et al.* (2011) An integrated semiconductor device enabling non-optical genome sequencing. *Nature* 475, 348–352.

Smith T.F. and Waterman M.S. (1981) Identification of common molecular subsequences. *Journal of Molecular Biology* 147, 195–197.

20.9.2 General Texts

Mäkinen V., Belazzougui D., Cunial F. and Tomescu A.I. (2015) *Genome-Scale Algorithm Design: Biological Sequence Analysis in the Era of High-Throughput Sequencing*, 1st Edn., Cambridge University Press, Cambridge, UK.

Wang K. (2016) *Next-Generation Sequencing Data Analysis*, CRC Press, Boca Raton, FL, USA.

20.9.3 Review Articles

Alkan C., Sajjadian S. and Eichler E.E. (2011) Limitations of next-generation genome sequence assembly. *Nature Methods* 8, 61–65.

Brent M.R. (2005) Genome annotation past, present, and future: how to define an ORF at each locus. *Genome Research* 15, 1777–1786.

Ekblom R. and Wolf J.B. (2014) A field guide to whole-genome sequencing, assembly and annotation. *Evolutionary Applications* 7, 1026–1042.

Erdin S., Lisewski A.M. and Lichtarge O. (2011) Protein function prediction: towards integration of similarity metrics. *Current Opinion in Structural Biology* 21, 180–188.

Goodwin S., McPherson J.D. and McCombie W.R. (2016) Coming of age: ten years of next-generation sequencing technologies. *Nature Reviews Genetics* 17, 333–351.

Holt R.A. and Jones S.J. (2008) The new paradigm of flow cell sequencing. *Genome Research* 18, 839–846.

Niedringhaus T.P., Milanova D., Kerby M.B. Snyder, M.P. and Barron A.E. (2011) Landscape of next-generation sequencing technologies. *Analytical Chemistry* 83, 4327–4341.

Rost B. (2002) Enzyme function less conserved than anticipated. *Journal of Molecular Biology* 318, 595–608.

Sleator R.D. (2010) An overview of the current status of eukaryote gene prediction strategies. *Gene* 461, 1–4.

Stein L. (2001) Genome annotation: from sequence to biology. *Nature Reviews Genetics* 2, 493–503.

20.9.4 Websites

Critical Assessment of Function Annotation (CAFA)
biofunctionprediction.org (accessed May 2017)

Critical Assessment of Protein Structure Prediction (CASP)
predictioncenter.org/ (accessed May 2017)

General Feature Format (GFF) specification
www.sanger.ac.uk/resources/software/gff/spec.html (accessed May 2017)

Jellyfish mer counter
www.genome.umd.edu/jellyfish.html (accessed May 2017)

Next Generation Sequencing Practical Course (EBI)
www.ebi.ac.uk/training/online/course/ebi-next-generation-sequencing-practical-course/how-take-course (accessed May 2017)

Sequencing by Synthesis (Illumina)
www.illumina.com/technology/next-generation-sequencing/sequencing-technology.html (accessed May 2017)

21 | Fundamentals of Proteomics

SONJA HESS AND MICHAEL WEISS

21.1 INTRODUCTION: FROM EDMAN SEQUENCING TO MASS SPECTROMETRY

Prior to mass spectrometry, **Edman degradation** was the only technique to obtain the sequence information of proteins. Edman sequencing was based on the chemical reaction of the N-terminal amine with phenyl isothiocyanate, leading to a phenylth-iocarbamoyl derivative, which was cleaved upon acidification and determined based on chromatography or electrophoresis. This was a slow process identifying one amino acid per reaction cycle. In addition, it required that the N-terminus of the proteins of interest was not blocked. However, most intact proteins, if they are not processed from a secretory or pro-peptide form, are blocked at the N-terminus, most commonly with an acetyl group. Other amino-terminal blocking includes fatty acylation, such as myristoylation or palmitoylation. Cyclisation of glutamine to a pyroglutamyl res-idue and other post-translational modification to N-termini also occur. In short, all these modifications leave the N-terminal residue without a free proton on the alpha nitrogen, thus Edman chemistry cannot proceed. Nowadays, Edman sequencing plays a minor role in protein analyses and has been surpassed by biological mass spectrom-etry techniques.

This revolution of biological mass spectrometry was largely enabled by the soft ioni-sation techniques ESI and MALDI (see Sections 15.2.4 and 15.2.5) that allowed the large-scale analysis of biomolecules (proteins, peptides, oligonucleotides, oligosaccharides and lipids) and thereby revolutionised the areas of proteomics and metabolomics

(Chapter 22). In contrast to electron impact (EI), these soft ionisation techniques produce molecular ions and only insignificant amounts of fragment ions. Therefore, in order to obtain structural sequence information on biomolecules, tandem MS (or MS/MS) has been developed. Furthermore, the faster speed and sensitivity of tandem MS soon dwarfed the sequencing turnaround available by Edman degradation.

21.2 DIGESTION

The identification of proteins by mass spectrometry usually involves protease cleavage, mostly by trypsin. Owing to the specificity of this protease, tryptic peptides usually have basic groups at the N- and C-termini. Trypsin cleaves after lysine and arginine residues, both of which have basic side chains (an amino and a guanidino group, respectively). This results in a large proportion of high-energy doubly charged positive ions that are easily fragmented. The digestion of the protein into peptides is followed by identification of the peptides by tandem mass spectrometry (Section 21.3). This is commonly referred to as bottom-up or shotgun proteomics.

21.3 TANDEM MASS SPECTROMETRY

Tandem MS refers to the analysis of precursor ions, followed by the dissociation of these precursor ions and analysis of the resulting fragment ions. Thus, for every identifiable species, the molecular weight can be determined from the precursor ion and the sequence can be deduced from its fragment ions. Although often referred to as 'sequencing', technically this is rather a 'matching' to theoretical peptide spectra.

Tandem MS can either happen 'in time' or 'in space'. Tandem 'in time' refers to the analysis of precursor ions, fragmentation and analysis of the fragment ions in the same analyser. The analysers used for this type of methodology include ion traps and ion cyclotron resonance (see Sections 15.3.2 and 15.3.4). In contrast, tandem 'in space' requires a physical separation of the precursor ion analysis (in the first quadrupole), fragmentation (usually in a second quadrupole or multipole) and fragment ion analysis in the third analyser. This is typically achieved in either a triple quadrupole (QQQ), Q-Trap (illustrated in Figure 21.1), Q-TOF, or Q-Exactive, which is a quadrupole coupled to an orbitrap (see Section 15.3.5). The fragmentation requires an inert collision gas (usually helium or argon) and through conversion of the kinetic energy of the ions to vibrational energy, fragmentation of the peptide bonds can occur. This is known as collision–induced dissociation (CID) or collision-activated dissociation (CAD).

21.3.1 Collision-Induced Dissociation

CID is the most commonly used fragmentation technique and cleaves peptide bonds, thereby generating an ammonium ion from the C-terminus (termed y-ion) or an acylium ion (termed b-ion) from the N-terminus (Figure 21.1). Although each collision only produces one ion, the large number of ions that are generated in one analysis yield a distribution of cleaved fragment ions that can be used to identify the sequence of the underlying peptide (Figure 21.2).

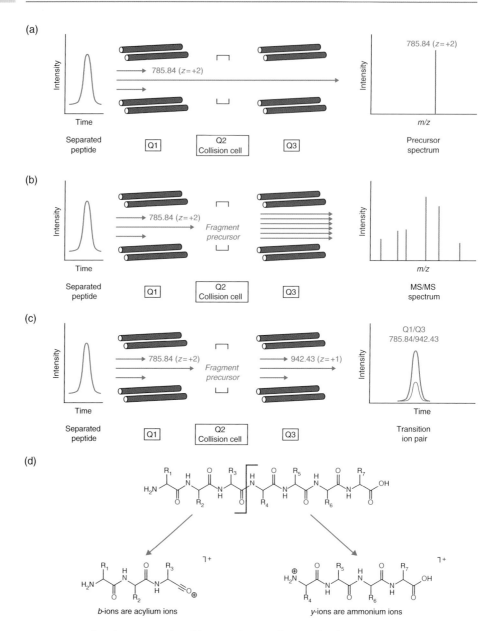

Figure 21.1 Quadrupole MS/MS analysis. (a) An ion of a particular *m/z* value (here 785.84) is selected in the first quadrupole, Q1, and passes through the second and third quadrupoles, Q2 and Q3, where its precursor mass is detected. (b) In the next scan, the same ion is selected in Q1 and passed through Q2, where it is subjected to collision with the collision gas and an additional voltage is applied. All the fragmented ions that are generated in this collision are passed through Q3 to generate an MS/MS spectrum in the Q-Trap mode. (c) If instead of passing all ions through Q3, select ions (here 942.43) are scanned through Q3, transition ion pairs (i.e. the precursor and fragment ions) are recorded in the QQQ mode. (d) The fragmentation depicted here is at the peptide bond and one of the fragments will retain the charge, resulting in either a *y*- or a *b*-ion. Whereas the *y*-ions derive from the C-terminus and thus are ammonium ions, the *b*-ions are derived from the N-terminus and are acylium ions. Each collision produces only one *b*- or *y*-ion. Because many precursor ions are present, fragmentation occurs throughout the peptide bonds, creating *b*- and *y*-ion series (see Figure 21.2).

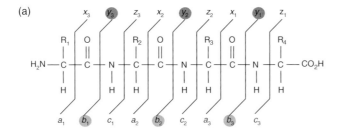

(a)

(b)

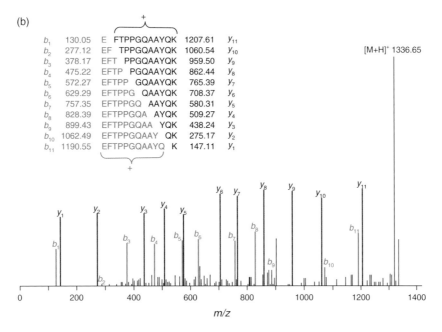

b_1	130.05	E FTPPGQAAYQK	1207.61 y_{11}
b_2	277.12	EF TPPGQAAYQK	1060.54 y_{10}
b_3	378.17	EFT PPGQAAYQK	959.50 y_9
b_4	475.22	EFTP PGQAAYQK	862.44 y_8
b_5	572.27	EFTPP GQAAYQK	765.39 y_7
b_6	629.29	EFTPPG QAAYQK	708.37 y_6
b_7	757.35	EFTPPGQ AAYQK	580.31 y_5
b_8	828.39	EFTPPGQA AYQK	509.27 y_4
b_9	899.43	EFTPPGQAA YQK	438.24 y_3
b_{10}	1062.49	EFTPPGQAAY QK	275.17 y_2
b_{11}	1190.55	EFTPPGQAAYQ K	147.11 y_1

[M+H]⁺ 1336.65

Figure 21.2 Peptide fragment ion nomenclature and tandem MS spectrum of a peptide. (a) Charge may be retained by either the N- or C-terminal fragment, resulting in the a-, b- and c- ion series or x-, y- and z-ion series, respectively. Ions in the b- and y- series frequently predominate in collision-induced dissociation (CID). Corresponding neutral fragments are not detected. (b) The sequence of the peptide from a mutant haemoglobin is: EFTPPGQAAYQK. The figure shows the tandem mass spectrum from CID of the doubly charged [M + 2H]²⁺ precursor, m/z = 668.3. Cleavage at each peptide bond results in the b- or y- ions when the positive charge is retained by the fragment containing the N- or C-terminus of the peptide, respectively (see inset).

Although a wide variety of fragmentations may occur, there is a predominance of peptide bond cleavage, which gives rise to peaks in the spectrum that differ sequentially by the residue mass. The mass differences are thus used to reconstruct the amino-acid sequence (**primary structure**) of the peptide (Table 0.9).

Different series of ions (a, b, c and x, y, z) may be registered, depending on which fragment carries the charge. Ions x, y and z arise by retention of charge on the C-terminal fragment of the peptide. The a-, b-, and c- ion series arise from the N-terminal end of the peptide, when the fragmentation results in retention of charge on these fragments.

Figure 21.2a shows an idealised peptide subjected to fragmentation. Particular series will generally predominate so that the peptide sequence may be deduced from both ends by obtaining complementary data (Figure 21.2b). In addition, ions can arise from side-chain fragmentation.

The protein is identified by searching databases of expected masses from all known peptides from every protein (or translations from DNA) and matching them to theoretical masses from fragmented peptides. The sensitivity of sources such as nanoESI using tandem MS is routinely at the attomolar level.

Example 21.1   **PEPTIDE MASS DETERMINATION (I)**

Question   Consider the following mass spectrometric data obtained for a peptide.

(i) The MALDI spectrum showed two signals at m/z = 1609 and 805.

(ii) There were two significant signals in a positive electrospray mass spectrum at m/z = 805 and 827 in the presence of sodium chloride, the latter signal being enhanced on addition of sodium chloride.

(iii) Signals at m/z = 161.8, 202.0, 269.0 and 403.0 were observed when the sample was introduced into the mass spectrometer via an electrospray ionisation source (without addition of sodium salts).

Use these data to give an account of the ionisation methods used. Discuss the significance of the data and deduce a relative molecular mass for the peptide.

Answer

(i) Signals in the MALDI spectrum were observed at m/z = 1609 and 805. These data could represent the following possibilities:

(a) $m/z = 1609 \equiv [M + H]^+$ when $m/z = 805 \equiv [M + 2H]^{2+}$
and $m/z = 403 \equiv [M + 4H]^{4+}$, giving $M = 1608$ Da

(b) $m/z = 1609 \equiv [2M + H]^+$ when $m/z = 805 \equiv [M + H]^+$
and $m/z = 403 \equiv [M + 2H]^{2+}$, giving $M = 804$ Da

(ii) The distinction between the above options can be made by considering the ESI data in the presence of Na$^+$ ions. This resulted in peaks at m/z = 805 and 827, the latter being enhanced on the addition of sodium chloride. This evidence suggests:

$m/z = 805 \equiv [M + H]^+$

$m/z = 827 \equiv [M + Na]^+$

giving $M = 804$ Da and supports option (b) from the MALDI data.

(iii) The multiply charged ions observed in the electrospray ionisation method allow an average M to be calculated. Using the standard formula (see Example 15.1):

m_1	$m_1 - 1$	m_2	$m_2 - m_1$	$n_2 = \dfrac{m_1 - 1}{m_2 - m_1}$	$m_2 - 1$	$M = n_2 \times (m_2 - 1)$ (Da)	z
269.0	268.0	403.0	134	2.0	402.0	804	2
202.0	201.0	269.0	67	3.0	268.0	804	3
161.8	160.8	202.0	40.2	4.0	201.0	804	4

The molecular mass M is 804 Da, confirming the conclusions from (ii).

Example 21.2 **PEPTIDE MASS DETERMINATION (II)**

Question   An unknown peptide and an enzymatic digest of it were analysed by mass spectrometric and chromatographic methods as follows:

(i) MALDI–TOF mass spectrometry of the peptide gave two signals at $m/z = 3569$ and 1785.

(ii) The data obtained from analysis of the peptide using coupled HPLC–MS operating through an electrospray ionisation source were $m/z = 510.7$, 595.7, 714.6, 893.0 and 1190.3.

Explain what information is available from these observations and determine a molecular mass, using the amino-acid residue mass values from (ii), for the unknown peptide.

Answer

(i) Signals from MALDI–TOF were observed at $m/z = 3569$ and 1785. These data could represent either of the following possibilities:

(a) $m/z = 3569 \equiv [M + H]^+$ when $m/z = 1785 \equiv [M + 2H]^{2+}$, giving $M = 3568$ Da

(b) $m/z = 3569 \equiv [2M + H]^+$ when $m/z = 1785 \equiv [M + H]^+$, giving $M = 1784$ Da

(ii) Electrospray ionisation data represent multiply charged ions. Using the standard formula (see Example 15.1) the mean M may be obtained.

m_1	$m_1 - 1$	m_2	$m_2 - m_1$	$n_2 = \dfrac{m_1 - 1}{m_2 - m_1}$	$m_2 - 1$	$M = n_2 \times (m_2 - 1)$ (Da)	z
893.0	892.0	1190.3	297.3	3.0003	1189.3	3568.3	3
714.6	713.6	893.0	178.4	4.0000	892.0	3568.0	4
595.7	594.7	714.6	118.9	5.0016	713.6	3569.2	5
510.7	509.7	595.7	85.0	5.9964	594.7	3566.1	6

With ΣM = 14271.6 Da, the mean of the four values of the molecular mass can be calculated and yields M = 3567.9 Da. This more precise value confirms the considerations made above.

Example 21.3   **PEPTIDE SEQUENCE DETERMINATION (I)**

Question An oligopeptide obtained by tryptic digestion was investigated both by ESI–MS and MS/MS in positive mode, and gave the following m/z data:

ESI	223.2	297.3								
MS/MS	147	204	261	358	445	592	649	706	803	890

(i) Predict the sequence of the oligopeptide. Use the amino-acid residual mass values in Table 0.9.

(ii) Determine the average molecular mass.

(iii) Identify the peaks in the ESI spectrum.

Note: Trypsin cleaves on the C-terminal side of arginine and lysine.

Answer

(i) The highest mass peak in the MS/MS spectrum is m/z = 890, which represents $[M + H]^+$.

Hence M = 889 Da.

The following mass differences (Δ) between the individual peaks are obtained:

m/z	147	204	261	358	445	592	649	706	803	890
Δ		57	57	97	87	147	57	57	97	87
aa		Gly	Gly	Pro	Ser	Phe	Gly	Gly	Pro	Ser

The mass differences between sequence ions represent the amino-acid (aa) residue masses. The lowest mass sequence ion, $m/z = 147$, is too low for arginine and must therefore represent Lys + OH, which therefore constitutes the C-terminal end of the peptide. The ion series above is thus the y-series, and the sequence in conventional order from the N-terminal end is then:

Ser-Pro-Gly-Gly-Phe-Ser-Pro-Gly-Gly-Lys

(ii) The summation of the mass for all residues yields $\Sigma\, M = 889$ Da, which is a validation of the value of M derived above.

(iii) The m/z values in the ESI spectrum represent multiply charged species and may be identified as follows:

$m/z = 223.2 \equiv [M + 4H]^{4+}$ from $889/223.2 = 3.9$, therefore $z = 4$

$m/z = 297.3 \equiv [M + 3H]^{3+}$ from $889/297.3 = 2.9$, therefore $z = 3$

Remember that z must be an integer and hence values need to be rounded to the nearest whole number.

21.3.2 Electron Transfer Dissociation/Electron Capture Dissociation

In addition to CID, electron-based dissociation techniques such as electron capture dissociation (ECD; used in FT-ICR instruments) and electron transfer dissociation (ETD; used in ion-trap instruments) have been developed. In either case, an electron is transferred to the analyte, ultimately breaking the N-Cα bonds of the generated radical cation species, and producing c- and z-ions (see Figure 21.2). Electron-based fragmentation techniques are the preferred choice for the analysis of post-translational modifications, since they generally preserve those modifications better than CID.

21.4 THE IMPORTANCE OF ISOTOPES FOR FINDING THE CHARGE STATE OF A PEPTIDE

Since the mass detector operates on the basis of mass-to-charge ratio (m/z), the charge of the ions needs to be determined to make a proper mass assignment. With sufficient resolution, this can be accomplished by inspecting the isotopic distribution of the peptide ions. While the natural abundance of ^{13}C is only around 1.1%, with increasing size, there is an increased probability that peptides contain at least one (or few) ^{13}C atoms. The same is true for hydrogen (^{2}H), nitrogen (^{15}N), oxygen (^{17}O), and sulfur (^{33}S). A peptide of 20 residues has approximately equal peak heights of the 'all mono-isotopic peaks' (m/z_{mono}) and those with one additional neutron (m/z_{mono+1}) resulting from ^{13}C, ^{15}N, ^{2}H, ^{17}O or ^{33}S contributions to the isotopic pattern.

A singly charged peptide will therefore show adjacent peaks differing in one mass unit; a doubly charged peptide will show adjacent peaks differing in half a mass unit, and so on (Figure 21.3 and Table 21.1). In the example shown in Figure 21.3, the peptide has a mass calculated from its sequence as 1295.69 Da. The experimentally derived values are, for the singly charged ion, $[(M + H) /1] = 1296.65$ and for the doubly charged ion, $[(M + 2H) /2] = 648.82$.

If the resolution of a given mass spectrum is not sufficient to resolve individual isotope peaks, the average mass is reported. This is still the case with larger polypeptides and proteins. With modern instruments, however, all isotopes can be resolved.

Table 21.1 **Mass differences due to isotopes in multiply charged peptides**

Charge on peptide	Apparent mass	Mass difference between isotope peaks
Single charge	$[(M + H) /1]$	1 Da
Double charge	$[(M + 2H) /2]$	0.5 Da
Triple charge	$[(M + 3H) /3]$	0.33 Da
n charges	$[(M + nH) /n]$	$1/n$ Da

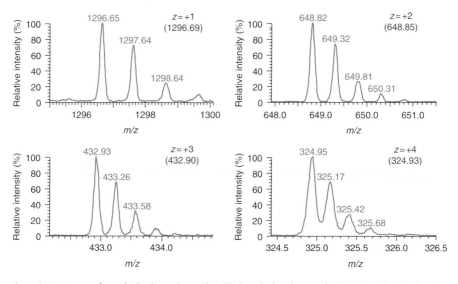

Figure 21.3 Spectra of a multiply charged peptide with the calculated mass of 1295.69 Da. Finding the charge state of a peptide involves zooming in on a particular part of the mass spectrum to obtain a detailed image of the mass differences between different peaks that arise from the same biomolecule, due to isotopic abundance.

Example 21.4 **PEPTIDE SEQUENCE DETERMINATION (II)**

Question Determine the primary structure of the oligopeptide that gave the following, positive mode, MS/MS data:

m/z	150	306	443	530	617

Answer

A: The highest mass peak in the MS/MS spectrum is $m/z = 617$, which represents $[M + H]^+$.

m/z	150		306		443		530		617
Δ		156		137		87		87	
aa		Arg		His		Ser		Ser	

The mass differences (Δ) between sequence ions represent the amino-acid (aa) residue masses. In general, y-ions are more stable and thus, often predominate over more labile b-ions. Since the y-ions are derived from the C-terminus, the sequence needs to be reversed, to be in its conventional order from the N- to the C-terminus. The conventional order for the sequence is then:

Ser-Ser-His-Arg-?

It is important to note that no assignment has been given for the remaining $m/z = 150$. It may not in fact be a sequence ion and more information would be required, such as an accurate molecular mass of the oligopeptide, in order to proceed further. It is, however, possible to speculate as to the nature of this ion. If the $m/z = 150$ ion is the C-terminal amino acid then it would end in -OH and be 17 mass units greater than the corresponding residue mass. The difference between 150 and 17 is 133, which is equal to methionine, so this amino acid is the likely end to the chain:

Ser-Ser-His-Arg-Met

21.5 SAMPLE PREPARATION AND HANDLING

One of the most important and often underestimated aspects of a successful proteomics experiment is proper sample handling. Mass spectrometry is the only biophysical method that consumes the sample. Therefore, contaminations that are introduced to the mass spectrometer influence not only the outcome of the proteomics experiment, but also the mass spectrometer itself, and thus subsequent analyses. In a worst case scenario, a single contamination may require extensive cleaning, including venting the instruments if the contamination travelled down the ion optics of the instrument. This usually leads to serious down time and should therefore be avoided.

Contaminations can come from a variety of sources. Keratin contamination from flakes of skin and hair can be a major problem, particularly when handling gels and slices; therefore gloves and laboratory coats must be worn. Work on a clean surface in a laminar air flow hood, and use of a dedicated box of clean polypropylene micro-centrifuge tubes tested not to leach out polymers, mould release agents, plasticisers, etc. are required.

Electrospray ionisation (ESI) mass analysis is affected, seriously in some cases, by the presence of particular salts, buffers and detergents. MALDI is, in general, much more salt-tolerant than ESI. If samples are analysed by MALDI, on-plate washing can remove buffers and salts and yield a sufficient clean-up. With electrospray, proper sample clean-up must be given particular attention. If it cannot be avoided in the first place, removal of buffer salts, EDTA, DMSO, non-ionic and ionic detergents (e.g. SDS, NP-40, Triton X) etc. can be achieved by using ion-exchange resins, precipitation, and/or dialysis. It should be noted that even as little as 0.1% of SDS, Triton X or NP-40 can lead to an electrospray signal reduction of more than 90%. CHAPS, dodecylmaltoside and similar non-ionic detergents generally show better performance in proteomics experiments and should therefore be used whenever possible. It is not uncommon that proteomics experiments require several iterative optimisation rounds.

Sample clean-up can also be achieved by **pipette tip chromatography**. This consists of a ZipTip®, a miniature C_{18} reverse phase chromatography column, packed in a 10 nm³ pipette tip. The sample, in low or zero organic solvent-containing buffer, is loaded into the tip with a few up-and-down movements of the pipette piston to ensure complete binding of the sample. Since most salts and buffers described above will not bind, the sample is trapped on the reverse phase material, washed with water and eluted with a high concentration of organic solvent (typically 50–75% acetonitrile). This is particularly applicable for clean-up of samples after in-gel digestion of protein bands separated on SDS-PAGE (see Section 6.3.1). Coomassie Brilliant Blue dye is also removed by this procedure. Additionally, the technique can be used to concentrate samples and fractionate a mixture.

Similar miniaturised pipette tip columns have been made commercially, as well as in laboratories around the world to achieve certain sample properties, e.g. using an immobilised metal ion affinity column to enrich for phosphopeptides (see Section 21.6.1).

21.5.1 Digestion Methods

Trypsin is the most commonly used digestion enzyme for proteomics applications. Since tryptic peptides generally have at least two positive charges (the N-terminal amino group, and the side-chain amino groups of lysine or arginine at the C-terminal end), the resulting m/z values are in the proper mass range for MS/MS identification. The amount of trypsin added to a complex mixture needs to be carefully considered. A protease-to-protein ratio of 1:100 to 1:20 is usually recommended. Too much trypsin could lead to an overload of autolytic tryptic peptides, whereas too little trypsin may generate peptides with too many missed cleavages. Both conditions would hinder successful identification of the peptides of interest. For extensive proteolysis of a protein, double digestion with Lys-C and trypsin are usually performed for best results. Commercially available trypsin is either mutated or treated with tosyl-phenylalanyl-chloromethyl-ketone (TPCK) to reduce the abundance of auto-digestion products. However, the existence of autolytic tryptic peptides can also be advantageous when used as an internal calibrant.

Other proteases such as the endoproteinase Glu-C, which cleaves C-terminally to glutamic acids, or Asp-N, which cleaves N-terminally to aspartic acids, have also been used successfully for proteomics analysis. The combination of results from several digestions usually increases sequence coverage, enabling the detection of those peptides that would be missed in a single tryptic digestion, either because they would be too small, too big, or are post-translationally modified (e.g. cross-linked, glycosylated).

21.6 POST-TRANSLATIONAL MODIFICATION OF PROTEINS

Many chemically distinct types of post-translational modifications of proteins are known to occur and regulate the structure and function of proteins. An up-to-date database of the broad chemical diversity of known modifications and the side chains of the amino acids to which they are attached can be found in the UNIMOD database.

The most common modifications studied by mass spectrometry are acetylation (N-terminal and at lysine side chains), methylation, phosphorylation, ubiquitylation, glycosylation and disulfide bonds.

21.6.1 Protein Phosphorylation and Identification of Phosphopeptides

Phosphoryl groups are covalently but reversibly attached to eukaryotic proteins in order to regulate their activity (Section 23.5.1). The main residues arising from phosphorylation are O-phosphoserine (see Figure 21.4), O-phosphothreonine and O-phosphotyrosine, but other amino acids in proteins can also be phosphorylated, thus giving rise to: O-phospho-aspartate, S-phosphocysteine, N-phospho-arginine, N-phosphohistidine, and N-phospholysine.

Global phosphopeptide analyses are routinely carried out for phosphorylation studies, either after enrichment of the protein or, more commonly and more effectively, on the peptide level. Protein-level enrichment is based on molecular mass, liquid chromatographic separation or immuno-affinity. Enriched proteins are digested and analysed by mass spectrometry. Alternatively, proteins can be digested first and then enriched at the peptide level using titanium dioxide (TiO_2), immobilised metal affinity chromatography (IMAC), or immuno-affinity using antibodies against phosphorylated amino acids or peptides. Analysis of modified peptides by mass spectrometry is essential to confirm the exact location and number of phosphorylated residues.

For the identification of phosphorylation sites, electron transfer dissociation (ETD), electron capture dissociation (ECD), collision-induced dissociation (CID) and higher-energy collisional dissociation (HCD, a collision-based dissociation developed for orbitrap mass analysers) have been used. Note that radiolabelled samples cannot

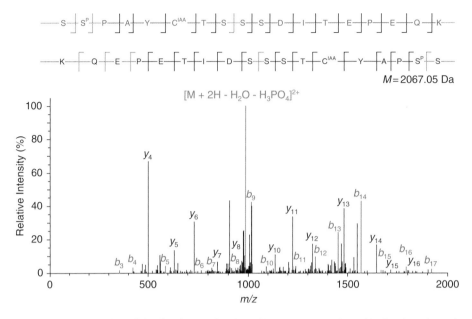

Figure 21.4 MS/MS analysis of the phosphopeptide SS^{P}PAYC^{IAA}TSSSDITEPEQK, where S^P is phosphoserine and C^{IAA} is carbamidomethylated cysteine. The MS^2 spectrum is shown. The parent ion shows a loss of water and of H$_3$PO$_4$, 98 Da. MS^2 data reveal the sequence of the phosphorylated peptide.

be analysed by mass spectrometry, as they would contaminate the entire mass spectrometer and, eventually, laboratory and personnel.

21.6.2 Mass Spectrometry of Glycosylation Sites and Structures of Sugars

The attachment points of *N*-linked (through asparagine) and *O*-linked (through serine and threonine) glycosylation sites and the structures of the complex carbohydrates can be determined by MS. The loss of each monosaccharide unit of distinct mass can be interpreted to reconstruct the glycosylation pattern (see example in Figure 21.5).

The GlycoMod website, part of the ExPASy suite, provides valuable assistance in interpretation of the spectra. GlycoMod is a tool that can predict the possible oligosaccharide structures that occur on proteins from their experimentally determined masses. The program can be used free of charge for derivatised oligosaccharides, as well as for glycopeptides. Another algorithm, GlycanMass, also part of the ExPASy suite, can be used to calculate the mass of an oligosaccharide structure from its composition.

21.6.3 Identification of Disulfide Linkages by Mass Spectrometry

Mass spectrometry is also used to identify the location of disulfide bonds in a protein. Identification of the position of disulfide linkages involves the analysis of the protein by LC-MS (see Section 15.1.1), with disulfide bonds intact, as well as reduced and alkylated (Figure 21.6). These different samples are then digested to yield peptides, using acidic conditions to minimise disulfide exchange. Proteases with active site thiols (e.g. papain, bromelain) should be avoided. Instead, pepsin and cyanogen bromide

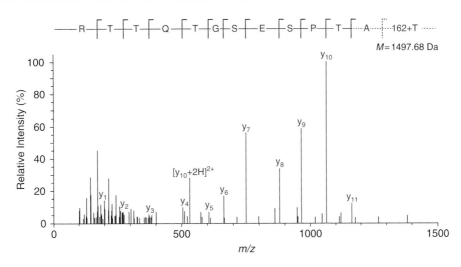

Figure 21.5 MS/MS analysis of the glycopeptide T¹⁶²ATPSESGTQTTR, where T¹⁶² is hexosylated threonine. The MS² spectrum shows the y-ion series. MS² data reveal the sequence of the hexosylated peptide.

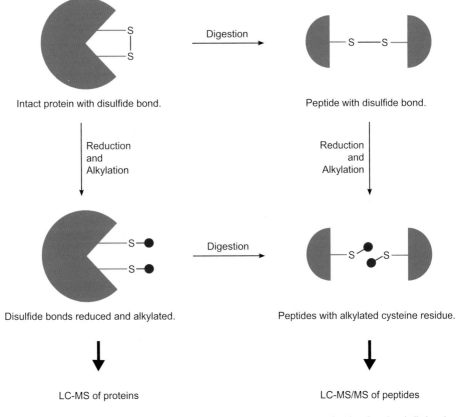

Figure 21.6 Strategy to determine disulfide bonds in proteins. Proteins intact and with reduced and alkylated disulfide bonds are analysed by LC-MS. The comparison enables the determination of the number of disulfide bonds per protein. Non-reduced and reduced/alkylated proteins are digested and analysed by LC-MS/MS to determine their disulfide bond location.

are particularly useful. The disulfide-linked peptide fragments are separated and iden-tified by LC-MS/MS. The separation is repeated after reduction with reagents such as mercaptoethanol and dithiothreitol (DTT) to cleave the –S–S– bonds; to prevent reoxidation, the disulfide bonds are then alkylated. Peptides that initially possessed a disulfide linkage disappear from the spectrum and reappear at the appropriate posi-tions for the individual components.

21.7 ANALYSING PROTEIN COMPLEXES

Mass spectrometry is frequently used to identify partner proteins that interact with a particular protein of interest. Most commonly, interacting proteins are isolated by immunoprecipitation of tagged proteins from cell transfection. An emerging technology for identification of interacting molecules is the so-called **biomolecu-lar interaction analysis by mass spectrometry** (BIA-MS, also called SPR-MS; see Section 14.6.1).

21.7.1 Analysis of Complex Protein Mixtures by Mass Spectrometry

In principle, proteome analysis can be carried out either using an in-gel digestion or an in-solution digestion procedure.

In-gel digestion procedures usually include the following steps:

- Separate proteins by gel electrophoresis, one-dimensional (1D) or two-dimensional (2D)
- Stain the gel to identify bands or spots of interest
- Excise entire gel, or specific gel bands or spots
- Digest proteins in the gel and extract the peptides
- De-salt the peptides
- Conduct mass analysis of the peptides
- Search database.

In addition to in-gel digestions, **in-solution digestion** procedures have successfully been developed. Usually, they require a double digestion of the proteins (using endo-proteinase Lys-C and trypsin) in an appropriate buffer, followed by de-salting, mass analysis and database search.

Both of these procedures have some limitations. For instance, it can be difficult to solubilise all hydrophobic and transmembrane proteins for in-solution digestion. At the same time, it can be difficult to exhaustively extract all peptides from a gel. These two approaches will therefore generate subsets of mutually exclusive peptides. For 2D gels in particular, there is also a limitation on the amount of material that can be loaded on the first dimension of the gel to fully take advantage of the second-dimension separation, although thousands of proteins can be reproducibly separated on one 2D gel from approximately 1 mg of tissue/biopsy specimen or biological fluid. Some proteins are of such low abundance that they can currently not be identified next to the high-abundance proteins. The dynamic range of protein expression can be as high as nine orders of magnitude, which exceeds the dynamic range of even the most advanced instruments currently available. Therefore, it can be very challenging to identify all proteins expressed in a cell, regardless of their copy numbers. However, several thousand proteins can be routinely identified.

21.7.2 Quantitative Analysis of Complex Protein Mixtures by Mass Spectrometry

Most proteomics experiments require the determination of the quantitative changes upon disturbance of the steady-state equilibrium, e.g. due to mutation or drug treatment. Several methods have been developed for this purpose.

Stable Isotope Labelling with Amino Acids in Cell Culture

Alternatives to label-free quantitation (where ion intensities of the peptides are used for quantitation) include stable isotope labelling with amino acids in cell culture (SILAC; Figure 21.7), which is currently the most accurate method available. As the name implies, this technique involves metabolic labelling of protein samples in cell cultures that are grown with different stable-isotopically labelled amino acids such as ^{13}C and/or ^{15}N lysine and arginine. Since trypsin cleaves C-terminally of lysine and arginine residues, this results in one label per tryptic peptide. For instance, a tryptic peptide that is labelled with a heavy arginine (where six ^{12}C are replaced with ^{13}C) results in a mass shift of 6 when compared to a peptide that is unlabelled and only contains the normal isotopes.

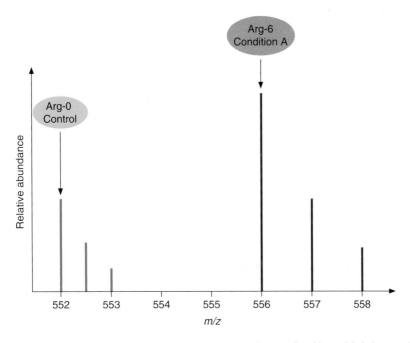

Figure 21.7 Samples labelled with light (Arg-0, with normal isotopes) and heavy labels (Arg-6, where six ^{12}C are replaced with ^{13}C, resulting in a mass shift of 6 when compared to the normal, light isotope) produce a spectrum with two sets of peaks. The heavy-to-light ratio of the monoisotopic peaks (indicated by the black arrows) can be used to quantify both species.

When a treated (cell culture 1, isotope-labelled) and an untreated sample (cell culture 2, no labelling) are mixed, one expects to see two peaks for each tryptic peptide, one light (no label) and one heavy (isotope-labelled) peak. The ratio of these peaks can be used to determine the protein ratio. Because of its accuracy, SILAC has been widely used for the investigation of protein interactions and signalling pathways. A disadvantage of this methodology is the fact that two peaks appear in the mass spectrum per tryptic peptide, therefore increasing the complexity of an already complex mixture, and thus triggering two MS/MS events per peptide for identification.

iTRAQ or TMT

An alternative method for quantitative analysis of protein mixtures involves the use of iTRAQ (isobaric tags for relative and absolute quantification) or TMT (tandem mass tag) reagents. These are sets of isobaric (same mass) reagents that are amine reactive and yield labelled peptides identical in mass and hence also identical in single MS mode. However, they produce strong, diagnostic, low-mass tandem MS signature ions, allowing simultaneous quantitation of multiple samples (Figure 21.8). Currently, duplex, 6plex, 8plex and 10plex kits are commercially available. Information on post-translational modifications is preserved, and since all peptides are tagged isobarically, the signals are not split into heavy and light signals, resulting in higher proteome coverage overall. Depending on the instrument used, both iTRAQ and TMT can suffer from severe interferences that are only overcome by the latest-generation instruments. An advantage of iTRAQ and TMT is that multiple peptides per protein are analysed, improving the statistical significance of the quantitative results.

The procedure involves reduction, alkylation and digestion with trypsin of the protein samples in parallel, in an amine-free buffer system. The resulting peptides are labelled with the iTRAQ or TMT reagents, and then combined. Depending on sample complexity, they may be directly analysed by LC-MS/MS after bulk purification using a cation exchange resin to remove reagent by-products. Alternatively, to reduce overall peptide complexity of the sample mixture, fractionation can be carried out when eluting from a cation exchange column.

Selected-Ion Monitoring (SIM) and Multiple-Reaction Monitoring (MRM)

Selected-ion monitoring (SIM) was initially developed for the sensitive analysis of drugs, as well as their contaminants and metabolites. It is typically set up by selecting ions that are characteristic of a target compound or family of compounds. Since only a small range of m/z values is detected and analysed in this technique, a substantial increase in sensitivity can be achieved. This concept has also been transferred to the area of proteomics, where **multiple-reaction monitoring** (MRM) is commonly performed. During an on-line chromatography–mass-spectrometry analysis, the instruments can be set up to monitor multiple ion masses as the peptides elute successively from the reverse-phase nanoHPLC column. This methodology produces highly reproducible quantitative data from select peptides/proteins to be studied in a specific pathway under multiple conditions. Consequently, the technique has gained popularity in recent years. A related technique, called **single-reaction monitoring** (SRM), is used in clinical settings, such as the analysis of platelets from patients from whom

(a)   Total mass of isobaric tag = 145.1 Da

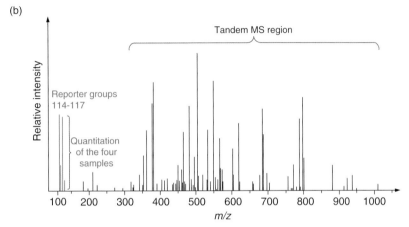

Figure 21.8 (a) iTRAQ reagent structure. The iTRAQ reagents consists of a reporter group (charged after fragmentation), a peptide reactive group (an NHS, N-hydroxysuccinimide), which reacts with amino groups, and a neutral balance group. The isotope labels in the balance group are chosen such that the overall mass of the reagent is 145 Da. iTRAQ reagents with different isotope labels are thus 'isobaric'. The peptide reactive group covalently links an iTRAQ reagent isobaric tag with each lysine side chain and N-terminus of a polypeptide, labelling all peptides in a given sample digest. By combining multiple iTRAQ-labelled digests into one sample mixture, the MS resembles that of an individual sample (assuming the same peptides are present). The balance group ensures that an iTRAQ-labelled peptide displays the same mass, independent of the reporter group mass (114, 115, 116, or 117 Da). The reporter group retains the charge after fragmentation, while the balance group is eliminated by neutral loss. In the case of '8plex' iTRAQ reagents, additional isotopes are employed to yield reporter group masses of 113 to 121 Da (with corresponding masses of balance groups of 192 to 184 Da). Reporter groups with masses of 120 Da and balance groups with masses of 185 Da are not used, as these may be confused with the ammonium ion of phenylalanine. (b) Identification and quantification of proteins using iTRAQ reagents. During MS/MS, along with the usual peptide fragmentation, the isobaric tag cleaves at the sites indicated. As a result of fragmentation, there is neutral loss of the balance group, and the reporter groups are generated, displaying diagnostic ions in the low-mass region between m/z values 114 to 117. Since this region is free of other common ions, quantification of the peak areas of these resultant ions represents the relative amount of a given peptide in the respective sample.

only small amounts of blood can be obtained (often the case with infants or patients with coagulation disorders). Although sophisticated algorithms are required, they are usually supported by the instrument manufacturers. Excellent open-source software (Skyline) is also available.

In addition, one can take advantage of so-called neutral loss scanning, in order to increase the sensitivity for phospho- or glycopeptides. For example, phosphoserine- and phosphothreonine-containing peptides can be identified by a loss of 98 Da due to the β-elimination of H_3PO_4 (see Figure 21.4).

21.8 COMPUTING AND DATABASE ANALYSIS

21.8.1 Identification of Proteins

Large-scale identification of proteins is accomplished by matching peptides that are generated through *in silico* digestion of an appropriate protein database to the measured spectra. As mentioned in Section 21.3, it is important to note that while this matching is sometimes referred to as 'sequencing', it does not involve actual sequencing at all. In fact, this method requires prior knowledge of the sequence information.

Several database search engines have been developed, including, but not limited to, SEQUEST Mascot, Protein Prospector and MaxQuant. The underlying search algorithms vary slightly from one program to another, thus resulting in small differences in their outputs, but at the same false discovery rates. Different protein databases can be searched:

- UniProt is a non-redundant database that contains manually annotated and reviewed records (Swiss-Prot) as well as computer-annotated records (TrEMBL). It is the most commonly used database for protein searches.
- NCBInr is a non-redundant database maintained by NCBI for use with their search tools BLAST and Entrez. The entries have been compiled from several databases such as GenBank CDS translations, PIR, SWISS-PROT, PRF and PDB, making this the largest protein database.
- dbEST is the division of GenBank that contains 'single-pass' cDNA sequences, or expressed sequence tags, from a number of organisms. dbEST is very large and is divided into three sections: EST_human, EST_mouse and EST_others. Because of their size, searches of these databases take far longer than a search of one of the non-redundant protein databases. Therefore, an EST database should only be searched if a protein database search has failed to find a match.
- Decoy databases identical in size to the chosen database, but with random sequences, are commonly used to estimate false discovery rates.

In order to ensure reproducibility, search parameters have to be defined. Under normal circumstances, the number of modifications should be limited. With every additional possible modification, the number of possible matches grows exponentially, as does the search time. Typical standard modifications include carboxy-amidomethylation of cysteine residues as a fixed modification for all iodo-acetamide treated samples, as well as oxidation of methionine and N-terminal acetylation as variable modifications. Should additional post-translational modifications (phosphorylation, ubiquitylation, glycosylation) be of interest, they may be added to the search parameters as well. If a number of large-size peptides are detected in the initial analysis, the missed cleavages parameter, which in a comprehensively digested sample is set to 1 or 2, should be increased. In the interest of computation time, the search should be limited to a particular species or taxon, such as human or mammalia.

Unmatched masses can be searched again, since sometimes unknown or unsearched modifications can prevent a peptide from being identified. Note the **delta p.p.m.** (the difference between the theoretical and the experimental mass of a particular peptide) should be low and consistent. Obviously, this is highly dependent on the mass accuracy and correct calibration of a particular instrument. If high-resolution mass spectrometers (Q-TOF, orbitrap or FT-ICR) are used and an internal calibration has been performed, the **mass accuracy** parameter can be set to 20 ppm or lower, down to the ppb range.

21.9 SUGGESTIONS FOR FURTHER READING

21.9.1 Experimental Protocols

Graham R.L.J., Kalli A., Smith G.T., Sweredoski M.J. and Hess, S. (2012) Tips from the bench: avoiding pitfalls in proteomics sample preparation. *Biomacromolecular Mass Spectrometry* 4, 261–272.

Graham R.L.J., Sweredoski M.J. and Hess S. (2011) Stable Isotope Labeling Of Amino Acids In Cell Culture: An Introduction For Biologists. *Current Proteomics* 4, 1–15.

21.9.2 General Texts

Reinders J. (2016) *Proteomics in Systems Biology: Methods and Protocols (Methods in Molecular Biology)*, 1st Edn., Springer, New York, USA.

Siuzdak G. (2006) *The Expanding Role of Mass Spectrometry for Biotechnology*, 2nd Edn., MCC Press, San Diego, CA, USA.

21.9.3 Review Articles

Aebersold R. and Mann M. (2003) Mass spectrometry-based proteomics. *Nature* 422, 198–207.

Doll S. and Burlingame A. (2015) MS-Based detection and assignment of protein PTMs. *ACS Chemical Biology* 10, 63–71.

Steen H. and Mann M. (2004) The ABC's (and XYZ's) of peptide sequencing. *Nature Reviews Molecular Cell Biology* 5, 699–711.

21.9.4 Websites

dbEST (Protein database)
www.ncbi.nlm.nih.gov/dbEST/ (accessed May 2017)

ExPASy (Expert Protein Analysis System)
www.expasy.org/tools/ (accessed May 2017)

FindMod (Prediction of potential protein post-translational modifications)
web.expasy.org/findmod/ (accessed May 2017)

GlycanMass (Calculation of the mass of an oligosaccharide structure)
web.expasy.org/glycanmass/ (accessed May 2017)

Glyocomod (Prediction of possible oligosaccharide structures)
web.expasy.org/glycomod/ (accessed May 2017)

Mascot (Protein identification from MS spectra)
www.matrixscience.com/search_form_select.html (accessed May 2017)

MaxQuant (Quantitative proteomics with high-resolution MS data)
www.maxquant.org/ (accessed May 2017)

Merck Millipore (Sample clean-up and ZipTips®)
www.millipore.com/catalogue/module/c5737 (accessed May 2017)

NCBInr Database Search with BLAST
http://www.ncbi.nlm.nih.gov/blast.ncbi.nlm.nih.gov/Blast.cgi?PROGRAM=blastp&
PAGE_TYPE=BlastSearch&LINK_LOC=blasthomehttp://www.ncbi.nlm.nih.gov/
(accessed May 2017)

Nest Group (Sample clean-up)
www.nestgrp.com/protocols/protocol.shtml#massspec (accessed May 2017)

Pierce/ThermoFisher (Protein mass spectrometry analysis)
www.thermofisher.com/au/en/home/life-science/protein-biology/protein-mass-
spectrometry-analysis.html (accessed May 2017)

Protein Prospector (Sequence database mining with MS data)
prospector.ucsf.edu/ (accessed May 2017)

Skyline (Targeted mass spectrometry analysis)
skyline.gs.washington.edu/ (accessed May 2017)

UNIMOD (Protein modifications for mass spectrometry)
www.unimod.org/ (accessed May 2017)

UniProt (Protein database)
www.uniprot.org/(accessed May 2017)

22 | Fundamentals of Metabolomics

JAMES I. MACRAE

22.1 INTRODUCTION: WHAT IS METABOLOMICS?

Metabolomics is essentially the measurement of the thousands of low-molecular-weight metabolites present in a biological system and is an important field of post-genomics biology. Metabolites, and metabolite levels, can be regarded as the physiological end-point of cellular responses to genetic factors, protein expression/activity and environmental changes (for example, nutrient availability). As such, measurement of the metabolite repertoire (or metabolome) of a cell, tissue or biofluid can give a precise snapshot of the physiological state of the system at any given moment (Figure 22.1). Along with its sister '-omics' fields of genomics, transcriptomics and proteomics, metabolomics has important roles in understanding cell biology, but it also has wider-reaching applications, such as in environmental research and toxicology, as well as disease biomarker discovery, drug discovery, drug abuse and forensic science, to name but a few.

As is the case with proteomics (see Chapter 21), metabolomics is not simply one process. In fact, metabolomics encompasses a number of specialised fields, each with their own advantages and challenges, and can range from vast clinical cohort (or population) studies requiring large numbers of samples and deep statistical analysis, to intricate study of flux through specific metabolic pathways in an organelle. Such a wide range of applications requires careful consideration of the approaches required – from experimental design and metabolite extraction techniques to data acquisition and data processing/statistics. However, no matter what approach is taken, sample preparation and choice of data acquisition instrumentation is vital for a successful experiment.

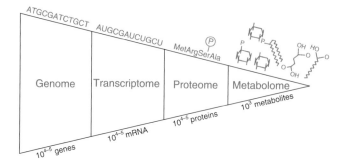

Figure 22.1 Metabolomics triangle: the link between genotype and phenotype. The cartoon depicts the relationship between genes, RNA, proteins and metabolites in a biosystem. While precise numbers vary considerably, a typical genome may contain in the region of $n \times 10^{4-5}$ genes, which are variably transcribed and translated into $n \times 10^{4-5}$ proteins, which in turn impact on only thousands of metabolites. The metabolome therefore provides exquisite detail of the physiological state of a biosystem at any given time.

22.2 SAMPLE PREPARATION

When performing any metabolomics experiment, it is crucial that the samples are prepared in an appropriate fashion. This includes considerations about metabolic quenching, metabolite extraction, sample storage, and preparation of the sample for metabolite analysis (data acquisition).

22.2.1 Metabolic Quenching

One of the key aspects of most metabolomics experiments is the quenching of the metabolism of that cell/tissue/biofluid. Quenching metabolism is the process of halting the metabolic processes of the biosystem as swiftly and reproducibly as possible. This is important for two reasons. Firstly, when we are observing the metabolome of a sample, we wish to view a snap-shot of the metabolism of the system at that precise moment. Without quenching before metabolite extraction, the biosystem will be susceptible to metabolic changes caused by the stresses of being withdrawn from their 'normal' conditions, as well as possible interference by the metabolite extraction procedure itself. For example, removing a culture flask from an incubator set at 37 °C and 5% CO_2 will change both the temperature and the gaseous environment of the culture, causing stress and potential metabolic changes/adaptations in the organism in question. Quenching the metabolism quickly slows down and/or halts metabolic responses in the biosystem, hence minimising metabolic flux and the chance of results being compromised by 'false-positive' stress-response metabolites. Secondly, we must also consider the stability of certain metabolites. Since, in general, higher temperatures result in faster kinetic rates (see Section 23.2.1), unstable metabolites may be rapidly degraded or turned over. Quenching the metabolism helps to minimise this process.

Metabolic quenching can be performed in a number of ways and is largely dependent on the biosystem being analysed, but almost always involves rapidly cooling the sample. One way of quenching the metabolism of small volumes of cultured cells (< 50 ml) is to submerge the sample tube in a dry ice/ethanol slurry until the temperature of the sample reaches as close to zero as possible without freezing (which could lyse any cells, causing loss of metabolites). This method may not always be suitable, however, especially when tracking metabolism over very short time courses. When this is the case, it is more favourable to transfer aliquots of the culture medium to a volume of very cold methanol (often at −20 °C or lower). Both of these methods require that the sample is withdrawn from a culture suspension.

Example 22.1   **EXTRACTION OF METABOLITES FROM *PLASMODIUM FALCIPARUM***

Question   The blood stages of the malaria parasite *Plasmodium falciparum* can be grown within red blood cells in 30 ml cultures at 37 °C and 5% CO_2. Commonly, infected red blood cells account for 10–20% of the total red blood cell population in the culture flask. In order to obtain pure samples of infected red blood cells for metabolite extraction, the culture can be passed through a ferrous wool column in the presence of a magnet. Uninfected red blood cells will pass through the column freely, while cells infected with mature parasites will stick to the column, due to accumulation of a ferrous haemoglobin breakdown product, haemazoin. These can be eluted by removal of the magnet and washing with medium or a suitable saline solution. The process of loading cells onto the column and eluting infected cells takes approximately 30 minutes. What considerations should be made when purifying infected red blood cells for metabolite extraction?

Answer

1. The purification procedure is likely to be stressful to the cells, both physically and nutritionally. It is therefore important to quench the metabolism before loading onto the column.

2. Red blood cells are non-adherent and can easily be suspended in medium. The 30 ml culture can therefore be transferred in the flask to a dry ice/ethanol slurry for rapid quenching.

3. Once quenched, the cells must be kept as cold as possible without freezing. Therefore, the ferrous wool column/magnet apparatus should also be kept close to zero, for example in a 4 °C cold room. The elution/wash solutions should also be ice cold.

4. The uninfected red blood cells may be useful as a control sample and should be reserved and kept under quenched conditions.

When investigating adherent cell lines (for example, fibroblasts), it is often better to quench the cells in the plate/flask. This can be achieved by rapid removal of the culture medium followed by addition of cold methanol (analogous to the above) and/ or transfer of the plate/flask to either an ice–water or liquid nitrogen bath. For the above cases, washing the sample with a cold wash solution (e.g. phosphate-buffered saline, PBS) should be considered for removal of 'contaminating' residual metabolites from the culture medium. For tissue samples, rapid quenching can be particularly challenging – especially when dealing with substantial tissue sizes. In some cases, it may be possible to submerge the tissue directly into liquid nitrogen; for larger tissues, using a freeze–clamp (a clamp pre-chilled in liquid nitrogen) may be a more suitable option to ensure efficient quenching. When dealing with tissue samples, it is often best to lyophilise the sample after quenching to help prevent sample deterioration and aid the extraction process.

22.2.2 Metabolite Extraction

As with quenching, the methods of metabolite extraction are sample-dependent. However, consideration of the metabolites of interest must also be taken into account. When extracting metabolites from a biosystem for the first time, one must discern the most suitable extraction method – i.e. one that extracts the most or most-relevant metabolites in a reproducible manner. For most applications, metabolites are extracted using an organic solvent or combination of solvents.

After quenching (and washing, if applicable) of culture suspensions, centrifugation will result in a pellet of quenched cells from which the metabolites can be extracted. Where the cells are straightforward to lyse, metabolites are commonly extracted by addition of a combination of chloroform, methanol and water. Chloroform is often added first, as this – in combination with vortex agitation/sonication – will help to lyse the cell membranes and denature the enzymes within, acting as a fail-safe for the quenching process. Chloroform will dissolve many apolar metabolites and so methanol and/or water is added to dissolve the polar metabolites present.

While chloroform, methanol and water are probably the most commonly used solvents in general metabolomics, other solvents and additives are also used where specific metabolites are of interest or the cell/tissue sample is particularly difficult to lyse. For example, anti-oxidants may be added to help prevent lipid oxidation. In addition to chemical methods, there are a number of physical methods that can be employed when cells/tissues are resistant to lysis. Common methods include freeze-thaw cycling, agitation in the presence of inert beads (mechanical shearing) and ultrasonication (i.e. using waves with a frequency > 20 kHz). It is also important to consider the extraction vessel. For example, chloroform should be avoided where metabolite extraction is performed directly in culture dishes (e.g. with adherent cells) since the solvent could react with the plastic and leach contaminants from the dish itself. Here, it is preferred to extract metabolites in polar solvents, transfer the extract to a more suitable vessel and add chloroform, if required, at this point.

When extracting from a biofluid (e.g. a culture medium or body fluid), there are no concerns about lysing cells or disrupting tissues and so the aim of metabolite extraction is primarily to dissolve the metabolites while removing any contaminants, such as residual particulate material or proteins, DNA, etc. Particulate material is easily removed through centrifugation and most other potential contaminants can be precipitated and removed by addition of organic solvents.

In determining the most suitable extraction method, reproducibility is a key concern. In this regard, assessment of the coefficient of variation (CV) for each metabolite in a method is calculated (see also Section 19.3.3 and Example 19.2). The CV is in essence the variability (often the standard deviation of the mean) of measured abundances of all metabolites in a sample (Figure 22.2). Where the CV is high, the metabolite abundance is not reproducible and so the extraction method should be improved. However, different extraction methods may affect the CV of metabolites in different ways and so the CVs of the primary metabolite(s) of interest may need to be considered over other metabolites.

22.2.3 Sample Storage

Metabolomics experiments can often be lengthy and take place over a number of days, months, or even years. For example, a cohort study of biomarkers for a particular disease may require samples from the same patient to be obtained over a period of time, or it may be the case that sampling a sufficient number of patients takes place over a long period. Another example might be a comparison of the metabolome of

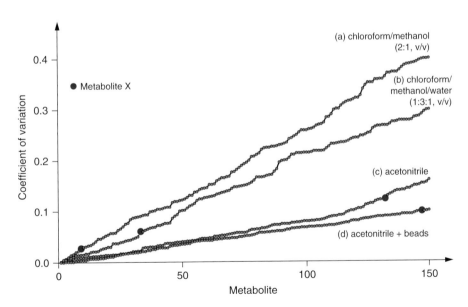

Figure 22.2 A coefficient of variation (CV) plot depicts the relative variation in the abundances of metabolites recovered from identical replicates. This is a measure of the reproducibility of a given method, with shallower gradients depicting less variability (i.e. the method is more reproducible). Metabolites are arrayed on the x-axis in order of CV value (y-axis) for four different extraction methods. The CV is calculated as the quotient of standard deviation and mean of the metabolite abundance and is often depicted as a percentage. The highlighted Metabolite X is referred to in Example 22.2 below.
Note: Ratios given in brackets indicate solvent volume ratios.

Example 22.2   **EXTRACTION OF METABOLITES FROM AN ENCAPSULATED YEAST CELL**

Question   Many single-celled organisms, such as encapsulated yeast cells, protect themselves with a thick cell wall. The method of metabolite extraction used for
studies on other single-cell organisms – organic solvent extraction chloro-
form/methanol/water (1:3:1, v/v) – was applied to encapsulated yeast cells,
but yielded no metabolites when analysed by gas-chromatography-coupled
mass spectrometry (GC-MS). Closer inspection using fluorescence microscopy
revealed that the cells remained fully intact after this treatment. After trialling
a variety of alternative extraction methods, microscopy showed that four of
these appeared to have lysed the cells successfully. Analysis of these extracts
by GC-MS yielded reasonable chromatograms and the coefficient of variation
(CV) was measured for all 150 detectable metabolites (Figure 22.2). From what
you know about the extraction procedures and the CV plot, determine the best
method to use in future experiments. Should the same method be used if the
metabolite of interest is Metabolite X (marked in purple in Figure 22.2)?

Answer

Of the four methods, the method with the best (i.e. lowest) overall CV was
Method D (acetonitrile with beads). However, this method is only slightly better
than Method C (acetonitrile alone). The use of zirconium beads adds complexity
to the method, being both cumbersome to handle, adding to the length of the
extraction procedure, and adding an extra financial cost. With this in mind,
Method C is on balance the best method to choose. However, when using Method
C (or D), metabolite X has a relatively high CV. While Method A (chloroform/
methanol, 2:1 v/v) has the least favourable overall CV plot, the CV for Metabolite
X is better than in the other three methods. Therefore, Method A should be used
when Metabolite X is the only metabolite of interest. However, if other metabolites are important to the study, Method C may still be more suitable.

the same crop species grown in different geographical regions and so samples must
be prepared in different places. Or perhaps the instrument of choice for data acquisition is not available on the day of metabolite extraction and so the samples must be
stored. In all cases, it is important to consider at what stage of the extraction process
the samples should be stored, and how.

   In most cases, once the metabolites have been extracted and separated from cellular/
tissue/particulate debris, the extracts can be stored. Polar metabolites are most commonly
stored in plastic tubes (e.g. Eppendorf tubes) at −80 °C. Lipids, however, can adhere to
such tubes and are preferably stored in glass. Where oxidation is a primary concern,
storage dry and/or under nitrogen may be required. Other classes of compounds, such as
hydrocarbons, may not respond well to freeze–thaw procedures and should instead be

stored at 4 °C. Sample longevity is also important and all metabolites have a shelf-life no matter what storage method is used. This deterioration of metabolites should especially be considered when samples are collected over a long period of time.

22.2.4 Sample Preparation for Data Acquisition

How metabolite extracts are prepared for analysis is dependent on the method of detection. Commonly, when using organic solvent extraction methods as described above, polar and apolar metabolites can be partitioned into separate fractions and analysed individually. Use of internal (added during extraction) and/or external (added post-extraction) standards is important for quality control, quantification of metabolites and normalisation of data. Which standards to use varies depending on the metabolite(s) being measured and the detection method used. Furthermore, extracts often need to be reconstituted in solvents specific to a particular detection method, while chemical derivatisation of metabolites is sometimes required. Some common methods of data acquisition are discussed in the following section.

22.3 DATA ACQUISITION

Being an -omics methodology, metabolomics seeks to identify as many metabolites as possible in a given sample. The most common techniques for metabolomics data acquisition are mass spectrometry and nuclear magnetic resonance, although other methods are also used. Since each method has its own specific application and range of metabolite detection, multiple analytical platforms are often required in order to maximise metabolome coverage.

22.3.1 Mass Spectrometry

Mass spectrometry (MS, discussed in detail in Chapter 15) is a very sensitive and powerful tool for detection, identification and quantification of metabolic features in a given biological sample. Liquid-chromatography-coupled mass spectrometry (LC-MS) and GC-MS are two of the most widely used methods used for metabolomics data acquisition, and can often be considered high-throughput in comparison to alternative techniques.

LC-MS is the most widely used MS technique in metabolite profiling for a number of reasons. First and foremost, in comparison to GC-MS and capillary-electrophoresis MS (CE-MS), LC-MS has the widest range of observable and identifiable metabolites (Figure 22.3). This is largely due to the availability of columns with various interaction chemistries. However, and especially in comparison to GC-MS, LC-MS also has a wider detectable mass range (commonly up to around $m/z = 2000$ for LC-MS, compared to around $m/z = 800$ for GC-MS), a wider dynamic range (i.e. the range of quantifiable abundances between the limit of detection and saturation) and often requires little sample processing before injection into the instrument. In addition, LC-MS can often give a greater amount of information about a particular metabolite than GC-MS. While GC-MS most often involves **electron impact (EI) ionisation** (a hard ionisation technique; Section 15.2.1), LC-MS usually involves the softer process of electrospray ionisation (Section 15.2.4). This allows detection of the molecular ion of the metabolite (i.e. an ionised form of the complete molecule), which can be fragmented to produce diagnostic fragments for identification.

Although LC-MS is very well suited to metabolite profiling, GC-MS is also widely used in metabolomics. GC-MS usually involves sample derivatisation in order to detect metabolites. This process adds a level of specificity to GC-MS experiments and so, in combination with the relative ease of separation of many isomeric molecules and the characteristic metabolite fragmentation fingerprints that are observed when using EI ionisation, identifications of metabolites can be made with comparatively high confidence. Moreover, GC-MS is a relatively cheap, highly robust and reproducible technique, largely due to the low number of operational variables (the ionisation energy and mobile phase is almost always the same from experiment to experiment). The highly reproducible fragmentation afforded by EI is most notably exploited in targeted metabolomics, particularly in metabolic labelling experiments (See Section 22.4.2), and makes GC-MS a more quantitative method than LC-MS. However, important structural/positional information can be lost when using the hard EI, and so the softer technique of chemical ionisation (CI) can be used in order to produce larger fragment ions to address this.

22.3.2 Nuclear Magnetic Resonance (NMR)

NMR (see Section 14.3.1) is a very commonly used method of metabolomics data acquisition. In particular, NMR is most often associated with analysis of biofluids, such as urine, blood plasma and culture medium. NMR profiling is cheaper and can often be higher throughput than equivalent MS profiling and little is needed in the way of sample preparation other than resuspension in an appropriate solvent buffer. This can allow measurement of some metabolites that are otherwise undetectable by

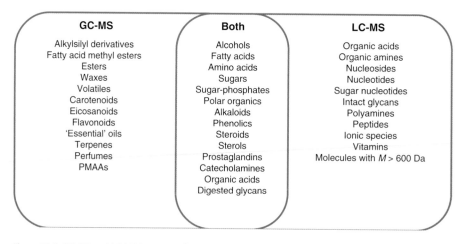

GC-MS	Both	LC-MS
Alkylsilyl derivatives	Alcohols	Organic acids
Fatty acid methyl esters	Fatty acids	Organic amines
Esters	Amino acids	Nucleosides
Waxes	Sugars	Nucleotides
Volatiles	Sugar-phosphates	Sugar nucleotides
Carotenoids	Polar organics	Intact glycans
Eicosanoids	Alkaloids	Polyamines
Flavonoids	Phenolics	Peptides
'Essential' oils	Steroids	Ionic species
Terpenes	Sterols	Vitamins
Perfumes	Prostaglandins	Molecules with $M > 600$ Da
PMAAs	Catecholamines	
	Organic acids	
	Digested glycans	

Figure 22.3 GC-MS and LC-MS have complementary and overlapping metabolite detection. The figure depicts those metabolites observable by each platform. This list is by no means exhaustive and exceptions to class detectability occur.

Note: 'Essential' oils are called essential since they represent the essence of a plant's fragrance; here, 'essential' does not indicate indispensability.

MS, particularly volatile metabolites lost in MS sample preparation procedures such as bicarbonate (HCO_3^-, an important metabolite to measure when following carbon metabolism) and cellular ethanol (a key metabolic end-product of some yeasts and bacteria). Moreover, NMR is potentially more reproducible than even GC-MS and, hence, more quantitative. Like GC-MS, NMR is often used in metabolic flux analysis as it potentially allows assessment (directly or indirectly) of the nature of every atom in a given metabolite. This makes NMR – in particular ^{13}C- and ^{15}N-NMR – a very powerful technique when trying to elucidate the precise nature of metabolic pathways. However, a key drawback of NMR lies in its relatively poor sensitivity when compared to MS. Potentially thousands of metabolites are observable when using LC-MS and hundreds using GC-MS, while NMR can observe only ~100 in a given sample.

22.3.3 Other Detection Methods

While MS and NMR are certainly the predominant methodologies of choice for metabolomics, others are also used. These are often most appropriate in bespoke, specific applications and hence are not always true metabolomics or metabolite-profiling techniques. For example, respirometry involves the direct measurement of cellular or mitochondrial respiration, often in response to addition of drugs or changes in culture conditions (e.g. hypoxia). Respiration is assayed by measurement of oxygen consumption rates in a cell culture, while mitochondrial metabolism is measured by ion production (for example H^+ ions, as extra-cellular pH). This technique can give rapid measurements of metabolic responses and has the potential, when used in conjunction with other metabolomics techniques, to provide an additional layer of understanding to cell metabolism.

22.4 UNTARGETED AND TARGETED METABOLOMICS

Metabolomics studies can be broadly described as being either untargeted or targeted. The choice of strategy is dependent upon the hypothesis being addressed. Different experimental design considerations are required, depending on which type of study is undertaken. The fundamental question to be addressed in experimental design is whether the study requires profiling of many metabolites (untargeted) or the analysis of a specific metabolite or set of metabolites (targeted).

22.4.1 Untargeted Metabolomics

The aim of untargeted metabolomics, or metabolite profiling, is to detect and identify as many metabolites from a sample as possible. Metabolite detection is highly dependent upon the method of data acquisition and most commonly involves MS and NMR techniques (see Section 22.3). As described above, due to the difference in sensitivity and metabolite repertoire of these platforms, metabolite profiling studies often use a combination of instruments/techniques in order to maximise metabolome coverage. However, when performing large and/or complex studies with the sample number and controls required for statistical rigour, it may not always be possible or practical to generate material for more than one analytical method. There is no fault in this, provided the scientist is aware of the limitations of a single-platform approach. In this regard, untargeted experiments result in an instrument-specific metabolite profile rather than the metabolome per se.

In essence, untargeted metabolomics is a simple concept: analysis of as many metabolites in a sample as possible. However, this is by no means straightforward. Many considerations and challenges present themselves to the scientist when designing and performing such a study (Figure 22.4) and can be grouped into three main areas: experimental design, statistical rigour and metabolite identification.

When designing any scientific experiment, it is important to ensure that appropriate controls are included. The biological question should be carefully considered so that the most suitable samples are taken. As discussed in Section 22.2, it is important to ensure that metabolic quenching, extraction, preparation and storage occurs in a reproducible manner. It is important to record as much additional information, termed **meta data**, about the samples as possible. For example, in a large cohort study looking for potential biomarkers of early-onset diabetes in human urine, it would be important to collect any meta data that could potentially influence the results. This covers genotypic, phenotypic and environmental data and includes – but is not limited to – age, gender, ethnicity, smoking habits, medication, diet, weight, lifestyle, height, general health, evidence of genetic predisposition, etc.

Furthermore, the number of samples required to give statistically meaningful results must be considered. The precise number of samples required (and the calculation thereof) depends largely on the statistical tests that will be performed to challenge the principal hypothesis. Of course, in some cases, biological limitations mean that the number of samples for true statistical rigour might not be reached. For example, this may be the case when profiling specific populations of T cells only present in very low numbers, or when looking for biomarkers or drug responses in patients with rare disorders. In these instances, the results may not be significant, but they might still be useful as part of a larger dataset/study and meaningful information (e.g. trends) may be obtained. As long as the scientist is aware of, and correctly reports, the statistical limitations of the experiment, the study can still be of value in directing future research avenues.

Untargeted	Targeted
Looks for all 'features'	Looks at few defined metabolites
Poor for quantification	Good for quantification
Time-consuming peak identification	Stable isotope labelling
Little pathway information	Flux analysis
May highlight novel/unanticipated metabolites/pathways	Detailed pathway information
Large sample numbers required	Misses potentially valuable information
Multiple platforms often required	Generally fewer replicates required
Complex data analysis (gap-filling, statistics, etc.)	Often completed on one platform
	Complex data analysis (labelling, etc.)

Figure 22.4 Targeted versus untargeted metabolomics. The figure depicts some of the key considerations, advantages and disadvantages of targeted and untargeted approaches to metabolomics.

The most time-consuming aspect of untargeted metabolomics is data analysis. Whether acquired on MS or NMR platforms, the basic work flow for data analysis is comparable: (i) data acquisition; (ii) feature finding; (iii) statistical analysis; (iv) feature identification; (v) quantification; (vi) validation. While steps (i) to (iii) can be relatively quick (a few days), steps (iv) to (vi) can take months, years or may not be possible at all.

The below work flow is focussed on MS studies, but many of the same principles can be applied to NMR:

1. Data acquisition: Samples should always be run in a random order, and include suitable quality controls to allow for operation-induced changes (such as metabolite degradation or instrument performance) in the analytical runs themselves. Chromatography should be as robust and reproducible as possible.

2. Feature finding: Many algorithms exist for processing MS data. Feature finding is different to peak finding, since a peak may contain numerous features (or metabolites). Peak thresholds, signal-to-noise ratios, and accounting for ion adducts should be considered. Chromatograms should be aligned and features quantified. Feature abundances can then be calculated and, if necessary, batch-corrected (to minimise false positives due to operational changes) and/or normalisation to internal standards or loading (e.g. cell number, dry weight, volume, etc.).

3. Statistical analysis: This includes methods to determine reproducibility (i.e. quality) of the data, differences and similarities within and between sample groups, significance testing for these differences, and determination of those features responsible (or likely responsible) for the differences observed. This is covered in more detail in Section 22.5.

4. Feature identification: Feature identification can be exceptionally challenging in MS even when running appropriate verified chemical standards and using high-resolution MS, since coeluting metabolites sharing the same m/z and fragmentation patterns may in fact be different molecules. For example, stereo- and geometric isomers are often indistinguishable by the most common LC metabolomics techniques. Leucine and isoleucine may coelute, as can glucose, mannose, galactose and any other hexopyranose. This issue can be addressed by using another platform (in this case, these compounds are easily separated by GC-MS) and so true identification can often only occur when the correct platforms are used. The degree of confidence in a feature assignment can be separated into four levels:

a. Unknown: MS or spectral data do not provide any information to allow feature annotation or chemical classification.

b. Putatively characterised metabolite class: MS or spectral data allow assignment of the feature to a particular class of metabolite, but no more. For example, an ion with $m/z = 357$ might correspond to a sugar-phosphate derivative in GC-MS, but without further details, a precise annotation cannot be made.

c. Putatively annotated: Annotation is based upon MS fragmentation or spectral patterns and/or comparison to available MS or spectral libraries, but with no chemical standard, identification cannot be verified.

d. Identification: The metabolite matches at least two independent orthogonal pieces of data when compared to an authentic standard analysed under the same conditions (e.g. retention time, fragmentation patterns, accurate mass, NMR spectra, etc.).

Of course, this makes identification of a true unknown even more challenging, since one would have to purchase or synthesise the most likely metabolite. There are many software packages available that attempt to putatively identify MS features by comparison with commercial and public database libraries. However, these often rely on the molecule being analysed in precisely the same way as in the database and, nevertheless, any annotations should always be verified with an authentic standard.

5. Quantification: Whether identified or not, a feature can still be quantified. In most untargeted metabolomics projects, relative quantification is more important than absolute quantification. In this regard, after any required batch corrections or normalisations, the abundances of metabolites between samples can be directly compared and expressed as values relative to each other, rather than an absolute quantification (e.g. mM, nmol, mg, etc.). Because of the importance of relative differences, rather than absolute abundances, in untargeted metabolomics, the dynamic range of the data acquisition instrument is of paramount importance. This is one reason why LC-MS and NMR are more favourable than GC-MS in these studies.

6. Validation: This can relate to validation of putative annotations in order to make them true identifications, or it may refer to validation of the result or the hypothesis drawn from the result. Validating a result may simply mean repeating an experiment. Validating a hypothesis would likely require devising further experiments in order to challenge it.

22.4.2 Targeted Metabolomics

The rationale behind targeted metabolomics is to use general metabolomics approaches on focussed areas of the metabolome and metabolism. This can mean a particular metabolite, an individual metabolic pathway or even a collection of pathways. An example of targeted metabolomics is the study of central carbon metabolism, where glycolysis, the tricarboxylic acid (TCA) cycle, pentose phosphate pathway, and other closely related pathways are studied. Despite a number of metabolites and pathways involved, this is defined as a targeted approach.

The most important criteria for targeted metabolomics are quite different from those of untargeted metabolomics. In targeted metabolomics, absolute recovery and quantification accuracy are of great importance, as opposed to dynamic range of detection. Feature identification is essential and is generally more straightforward than in untargeted metabolomics. Extraction and acquisition methods may be altered in order to preferentially detect certain metabolites to the detriment of those less important to the study. Thus, targeted metabolomics may fail to detect potentially interesting information about the metabolism of that biosystem.

However, since targeted metabolomics is focussed, generally fewer replicate samples are required than for untargeted studies, and projects can often be completed using just one analytical platform. Perhaps the most powerful branch of targeted metabolomics is in the ability to generate exquisite detail about the nature of metabolic fluxes through the use of stable-isotope label-incorporation techniques.

22.4.3 Stable-Isotope Labelling

The incorporation of stable-isotope labels is a powerful technique whereby a bio-system is supplied with a nutrient containing stable isotopic forms of an atom (or atoms) and the metabolic passage of that atom is followed using MS or NMR analysis. Nutrients containing ^{13}C atoms are often used, although ^{15}N and ^{2}H are also common. For example, a cell culture may be fed with U-$^{13}C_6$-glucose (see *Note*) or a subject may be given a constant infusion of the same carbon source. In each case, incorporation of ^{13}C atoms is measured in metabolites of downstream path-ways, such as those of glycolysis, the pentose phosphate pathway, the TCA cycle and inositol metabolism, to name but a few. In its simplest guise, measurement of the number of molecules of a particular metabolite containing ^{13}C is an indicator of the metabolic activity of the enzyme(s) leading to its production, i.e. the carbon flux along that specific pathway. By assembling this labelling information into metabolic pathway databases, general overviews of metabolic flux throughout a whole biosystem can potentially be observed. This information can highlight novel or unanticipated metabolic pathways in a biosystem and can provide valuable insight into the nature of the physiology of that system. For true measurements of metabolic flux, time-course measurements of label incorporation and abundance are required.

Note: The nomenclature for labelled molecules is straightforward. In the exam-ple of U-$^{13}C_6$-glucose, '^{13}C' refers to the atom isotope (i.e. the labelled atom), the subscript '6' refers to the number of atoms labelled in the molecule, and the 'U' stands for 'universal' (i.e. in this case, all carbons in the molecule are ^{13}C). Similarly, 1,2-$^{13}C_2$-glucose refers to a glucose molecule containing two ^{13}C atoms and that these atoms are in the 1 and 2 positions of the molecule. This distinction can be important in interpretation of labelling patterns.

Although measurement of stable-isotope label incorporation can be achieved with LC-MS, this approach is most often associated with NMR and GC-MS, due to the readily available structural data of intact molecules (NMR) or fragment ions (GC-MS). However, modern high-resolution LC-MS can now be used to study isotope patterns. In GC-MS analyses, quantifiable measurement of label incorporation into multiple fragment ions is possible, which can allow assessment of both the amount of label incorporation and the precise location(s) of incorporated labelled atoms into the molecule. When quantifying label incorporation, a number of individual steps are involved: (1) identification of the labelled molecule; (2) quantification of the abundance of each **isotopologue** (any ion derived from a particular metabolite con-taining one or more labelled atoms) of the diagnostic ion; (3) correction for natural isotope abundance and (4) calculation of label incorporation into and abundance of the ion/molecule.

1. The mass spectrum of a labelled ion will differ from the unlabelled version and can sometimes be very complex, especially where there are multiple combinations of labelled atoms possible. Although this makes automated identification exceptionally challenging, in most cases, labelled molecules coelute with their unlabelled counterparts and hence their identity can be confirmed by retention time. For example, one diagnostic ion of phosphoenolpyruvate (PEP) after trimethysilyl derivatisation (required for GC-MS analysis) is $m/z = 369$ and contains three carbon atoms. Therefore, when labelling with U-^{13}C$_6$-glucose, this ion becomes $m/z = 372$ due to incorporation of three ^{13}C atoms (each being one mass unit higher than ^{12}C). Library searches therefore do not identify this ion as PEP, but since both species, $m/z = 369$ and $m/z = 372$, coelute, the latter can be identified as ^{13}C$_3$-PEP. Bioinformatics tools for automated identification of label incorporation have been developed – particularly for high resolution LC-MS – and are ever improving, but manual inspection of relevant peaks is still necessary to ensure correct peak assignment.

2. In the above case, measurement of the proportion of PEP ions that are $m/z = 372$ compared to $m/z = 369$ gives the fraction of molecules into which ^{13}C has been incorporated, i.e. the 'mol % label incorporation' (see (4.) below). Therefore, in order to calculate label incorporation, measurement is required of the abundances of the unlabelled ion and each isotopologue ion.

3. However, before calculation of label incorporation can be made, the spectrum must be corrected for the natural abundance of isotopic forms of every atom in the ion. In order to do this, the empirical formula of the ion must be known. This is important since, although the natural abundance of ^{13}C is only 1.08% of all carbon, for silicon – an atom present in a number of GC-MS derivatisation agents – two natural isotopes occur, ^{29}Si and ^{30}Si, accounting for 7.78% of all silicon. This presents a considerable factor to be removed in label incorporation calculations. The process of removal of natural isotope abundance is termed natural isotope correction and can easily be performed by a mathematical algorithm, resulting in the so-called mass isotopomer distribution (MID). An MID shows the corrected abundance values for each isotopologue of the ion and can reveal useful information about the biosynthesis of a metabolite.

4. When calculating abundance of a labelled metabolite, the abundances of all unlabelled and isotopologue ions should be taken into consideration. In this case, the abundance of PEP would be the sum of the ion abundances from $m/z = 369$ to $m/z = 372$. Calculation of label incorporation can be expressed in a number of ways. Firstly, and perhaps most obviously, label incorporation can be expressed as the number of molecules (containing any number of labelled atoms) labelled in a metabolite pool – the 'mol % label'. Alternatively, labelling can be expressed as the number of carbons labelled in a metabolite pool. The differences between the modes of presenting label incorporation are explored in Example 22.3.

Combining stable-isotope labelling data with metabolome data over a time course can be used to elucidate the presence and activity of metabolic pathways in a given biosystem. By observing label incorporation data in different ways, more information about a pathway can be gleaned than by 'total label incorporation' alone. This is explored further in Example 22.4.

Example 22.3   **CALCULATION OF LABEL INCORPORATION INTO CITRIC ACID**

Question   Label incorporation can be expressed in a number of ways. Often, analysis of isotopologues of a metabolite can give deeper insight into its metabolic role. The below spectra show MIDs of an ion corresponding to a six-carbon fragment of citrate obtained after labelling cells with U-$^{13}C_6$-glucose. The spectra show the relative isotopomer abundance after correction for natural isotope abundance. Use these spectra to calculate the label incorporation into citrate. The x-axis indicates the number of ^{13}C atoms in the citrate ion, which contains six carbon atoms; 'M0' indicates the monoisotopic mass containing no ^{13}C atoms, and '$+n$' indicate isotopomers containing n ^{13}C atoms. *Hint*: there are two ways of expressing label incorporation and the results for each may be different.

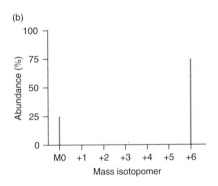

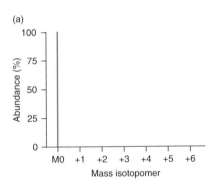

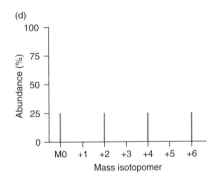

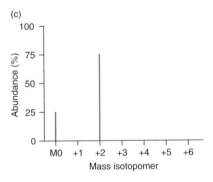

Answer

A helpful way to interpret these spectra is to draw the ions as ball-and-stick figures to highlight labelled and unlabelled carbons. In the below cartoons, the six-carbon citrate ion is depicted using coloured (^{13}C) and non-coloured (^{12}C) circles. Note that the MIDs do not infer the precise positions of the ^{13}C atoms within the molecule; their positions in this figure are illustrative only.

(a)   (b)

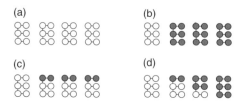

(c)   (d)

Panel (a): there are no isotopologues of the 'M0' ion and hence citrate is unlabelled (0% labelling).

Panel (b): 25% of the citrate ions are 'M0', while 75% are 'M+6'. Therefore, 75%
of the citrate *molecules* are labelled. For the 'M+6' ions, all six carbon atoms
are labelled. Since the ions are either fully labelled or unlabelled, 75% of the
total carbon atoms are labelled.

Panel (c): 25% of the citrate ions are 'M0', while 75% are 'M+2'. Therefore, 75%
of the citrate *molecules* are labelled. For the 'M+2' ions, two of the six carbons
are labelled. Therefore, for every four molecules, six carbon atoms will be
labelled and 18 unlabelled, hence 25% of the *total carbon atoms* are labelled.

Panel (d): 25% of the citrate ions are 'M0', 25% are 'M+2', 25% are 'M+4', and
25% are 'M+6'. Therefore, 75% of the citrate *molecules* are labelled. For every
four molecules, 12 carbon atoms will be labelled and 12 unlabelled, hence 50%
of the total carbon atoms are labelled.

Example 22.4   **METABOLIC LABELLING OF THE TRICARBOXYLIC ACID (TCA) CYCLE**

Question   The below figure shows the corrected MIDs for pyruvate and key tricarboxylic
acid (TCA) metabolites from analysis of a cell culture labelled for 1, 2 and 3
minutes in the presence of U-^{13}C$_6$-glucose. The ions observed contain all carbon atoms for that metabolite (i.e. three for pyruvate, six for citrate, and four
for succinate, fumarate and malate). The x-axis indicates the number of ^{13}C
atoms in the metabolite ion; 'M0' indicates the monoisotopic mass containing
no ^{13}C atoms, with '+n' indicating isotopomers containing n ^{13}C atoms. Using
this information and knowledge of TCA architecture, determine (a) whether the
TCA cycle is operating in this organism, (b) if it is present, in which direction
it is operating and (c) what can be ascertained about anaplerotic contributions
to the pathway (i.e. contributions not directly from acetyl-CoA). *Hint*: The
TCA cycle is entirely located in the mitochondrion. This organism is known to
carboxylate pyruvate to oxaloacetate and convert this to malate in the cytosol.
Abd: abundance.

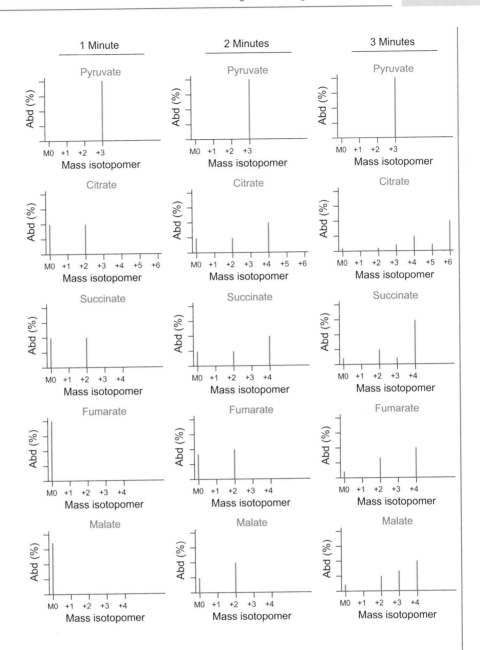

Answer

By representing the isotopologue data as simple ball-and-stick figures (see below), it is easier to interpret the data. The cartoon ball-and-stick depictions of ^{13}C incorporation into TCA metabolites are inferred from the equivalent MID panels above. Unlabelled (^{12}C) carbon atoms (non-coloured) and ^{13}C atoms (coloured) are shown for the predominant isotopomers of each metabolite. Note, in this case, the MIDs do not infer the precise positions of the ^{13}C atoms within the molecule; the positions of the ^{13}C atoms in this figure are illustrative only.

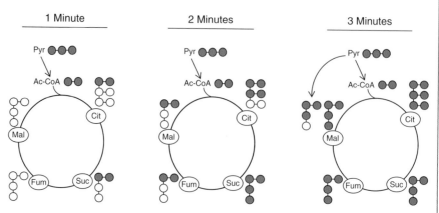

The 1 min time point shows that two ^{13}C atoms have entered citrate and succinate, but no further. This suggests that U-^{13}C$_6$-glucose has been metabolised to ^{13}C$_2$-acetyl-CoA, which has been turned over to citrate, and that this has been metabolised through oxidative TCA reactions to succinate. The 2 min time point shows that the label has now been incorporated into all measurable TCA metabolites and that, in the case of citrate and succinate, four ^{13}C atoms have been incorporated (i.e. a second turn of the TCA cycle has begun). This supports the hypothesis that the TCA cycle is complete and operates in a canonical (oxidative) direction. By 3 minutes, all TCA metabolites show evidence of complete labelling and hence multiple turns of the TCA cycle (i.e. four ^{13}C atoms for succinate, fumarate and malate, and six ^{13}C atoms for citrate). The 3 min time point also shows evidence of metabolites containing three ^{13}C atoms, and that this is greatest in malate. Since ^{13}C$_3$-pyruvate can be converted to ^{13}C$_3$-malate in the cytosol, the presence of three ^{13}C atoms in all TCA metabolites suggests that this malate can enter the mitochondrion and contribute to TCA metabolism in an anaplerotic manner. There is also evidence of citrate containing five ^{13}C atoms, indicative of citrate obtaining three ^{13}C atoms from ^{13}C$_3$-malate anaplerosis and two from ^{13}C$_2$-acetyl-CoA.

22.4.4 Metabolic Flux

By combining label incorporation data with absolute metabolite abundances, the metabolic flux through a pathway can be deduced. Flux measurements are dependent on the ability to accurately measure metabolite abundance. As mentioned before, this, in turn, is dependent on the ability to recover (extract) the metabolite from the biosample, the stability of the metabolite, the abundance and detectability of the metabolite and the dynamic range of the acquisition instrument. These constraints therefore limit flux studies to a relatively small proportion of metabolites in any given experiment. However, missing metabolites in a pathway might not always be a problem in determining the metabolic flux. In Example 22.4, only four of the nine TCA metabolites were observed, and yet conclusions about TCA metabolism could be made.

In order to capture the complex metabolic flux data and integrate it into what is already known about metabolism from genomic, proteomic or biochemical data, specialised computational frameworks have been developed to produce metabolic models. This methodology allows visualisation of the flux data on metabolic maps and can provide information about how metabolic activity is organised in the organism/tissue, the importance of individual metabolic processes under different conditions, and can highlight potential chokepoints in a system. A chokepoint is a reaction that either uniquely consumes or produces a specific metabolite in a pathway. In general, inhibition of chokepoint enzymes is expected to be detrimental to cell physiology and is therefore of interest biologically for drug/genetic therapy and product yield in biofuel production, for example.

Obtaining enough information from stable-isotope labelling studies for generation of a useful metabolic flux map can be an arduous process. Time-course analyses or comparisons of labelling under various growth conditions are often required, but, as mentioned before, data acquisition-based quantification limitations can leave crucial data gaps in our pathways of interest. An alternative approach is to generate genome-scale metabolic models *in silico* using constraint-based modelling (CBM) techniques, such as flux balance analysis (FBA). FBA modelling is based upon pre-existing knowledge of the stoichiometry of metabolic reactions (from genomic, proteomic and biochemical data) and nutritional requirements of that cell/tissue. Constraints are added to the model, such as to limit nutritional input, waste export, simplify cell metabolism to a few metabolic pathways and make assumptions about compartmentalisation and inter-organelle nutrient transport. FBA models can be set to maximise an objective function (e.g. biomass production) and calculate possible flux patterns that would allow the processes required for the objective function to proceed under those constraints. This modelling can also be used for *in silico* prediction of essential genes, i.e. chokepoints. FBA models can be tested and honed using stable-isotope labelling experiments. Once optimised, *in-silico*-predicted essential genes can be targeted for further study. This has been used to great effect, especially in elucidation of essential genes in unicellular organisms (see Further Reading, Section 22.7).

22.5 CHEMOMETRICS AND DATA ANALYSIS

Broadly speaking, data analysis encompasses all processes after data acquisition; the extraction of information from a dataset is termed chemometrics. This begins with initial data extraction (peak picking, peak integration, feature identification) and processing (batch correction, normalisation, quantification, label incorporation, natural abundance correction, isotopologue analyses), through to pattern recognition (principal component analyses, loadings plots, hierarchical cluster analyses) and statistical/visualisation approaches (t-tests, ANOVA analyses, volcano plots, Z-transformations, heatmaps, etc). While data extraction and processing is generally the same for all metabolomics studies, the way in which statistical analyses are employed and visualised may be different for individual studies.

When performing any chemometric analysis, it is prudent to discuss the work with a skilled statistician so that the results obtained will be meaningful. Ideally, this should take place before the study begins, to ensure that enough sample replicates and statistically relevant controls are collected. This is especially important in untargeted metabolomics. The sections below describe some important chemometrics tools used in metabolomics. This list is by no means exhaustive and serves only as a brief introduction to these concepts.

22.5.1 Pattern Recognition Approaches

Pattern recognition is a global term referring to chemometric techniques that seek to cluster or differentiate samples based on their metabolite content. Depending on the method used, these methods can be unbiased (i.e. blind) or use pre-existing knowledge (i.e. meta data) of the samples, and they can seek to cluster samples based on similarities or differences. These techniques are most commonly used in untargeted metabolomics to compare two or more sample groups. However, they can also be applied to targeted studies, generally to assess dataset quality and reproducibility.

Principal component analysis (PCA) is one such method and aims to find the linear combination of features that best separates two or more classes of objects. The statistical process to produce a PCA plot involves orthogonal linear transformation of abundances of all features in a chromatogram (for MS) or spectrum (for NMR) into a smaller set of linearly uncorrelated variables (principal components) that account for the greatest possible variance in the dataset. This means that the first principal component is a vector accounting for the greatest proportion of variability in the data; similarly, the second principal component accounts for the next greatest proportion of variability, and so forth. PCA plots are shown in two or three dimensions with each principal component allocated to one axis (Figure 22.5). The position of each sample in the plot relates to its overall variance factor in each principal component vector. Each PCA analysis has a corresponding loadings plot, which reveals those metabolites contributing most to each principal component and hence the variance in the dataset.

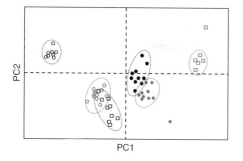

Figure 22.5 Principal component analysis (PCA) plot. The PCA of the data matrix generated by progressive clustering and deconvolution of MS chromatograms. This aims to find the linear combination of features that best separates two or more classes of objects. The x- and y-axes represent the principal components accounting for the greatest (PC1) and second greatest (PC2) variability, respectively. Circles and squares represent individual replicates of six different sample types. Dotted lines show general groupings of each sample type.

Linear discriminant analysis (LDA) is closely related to PCA in that they both transform multi-factorial datasets into linear vectors. However, while PCA seeks to model the variability within an entire dataset, LDA specifically seeks to model differences between the individual groups of the dataset. To put this simply, a PCA addresses the question 'What is the variability in this dataset?', whereas an LDA addresses the question 'These sample groups are different – how are they different and by how much?'.

Hierarchical cluster analysis (HCA) is a statistical method that attempts to show correlations between individual samples by building a hierarchy of clusters. HCAs are most often shown as a dendrogram, with the most similar samples clustering together, and the most similar clusters grouping together (Figure 22.6).

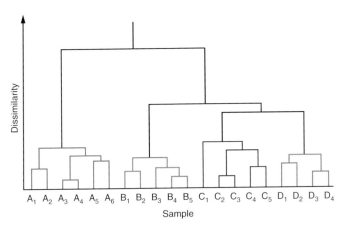

Figure 22.6 Hierarchical cluster analysis (HCA) dendrogram. An HCA is a technique to find clusters of observations within a dataset. In appearance, the dendrogram shows how objects (samples) are similar (or dissimilar) to each other, with those most similar clustering initially in pairs (e.g. samples B_4 and B_5 in the figure), trios (e.g. samples B_3-B_5), groups (e.g. samples A_3-A_6), and ultimately clusters (e.g. samples A_1-A_6). The definition of what constitutes a particular cluster is largely arbitrary and user-defined.

22.5.2 Statistical Approaches

There are multiple ways in which the significance of a particular result can be calculated. Calculation of statistical significance is termed hypothesis testing (see also Section 19.3.9) and the most common methods to do this include t-tests (unpaired and paired; Section 19.3.5) and ANOVA (analysis of variance; Section 19.3.9) analyses. In all cases, these analyses seek to determine whether the difference between two or more mean values is significant or otherwise – in other words, they test the null hypothesis that the mean of two results is the same. Unpaired t-tests are used to determine significance in two unrelated groups (for example, cell type A and cell type B), while paired t-tests are used to determine significance in two related groups (for example, cell type A at time 0 and time 10 minutes). ANOVA is similar to the t-test, but is used when more than two groups are present. Since metabolomics datasets are often large and involve statistical testing of many values, correction for multiple comparisons is often necessary to ensure that null hypotheses are not incorrectly rejected. Whatever hypothesis test method and correction factors are used, the result is always a so-called *p*-value (see Section 19.3.9). This function can assume values between 0 and 1, with small *p*-values (typically less than 0.05, i.e. 5%) suggesting that a result is significant.

Z-transformation plots compare datasets from two samples by comparing the values (e.g. metabolite abundance) of one group (the sample) to the mean of the corresponding values of the other group (control). Rather than plotting fold change, each value is plotted according to its number of standard deviations away from the mean of the control group (Figure 22.7). In this way, a Z-transformation (see also Section 19.2.8) depicts the significance of differences, as opposed to the magnitude of differences. In the example shown in Figure 22.7, those metabolites near the top and bottom of the panel possess more significant differences in abundance than those in the middle of the panel.

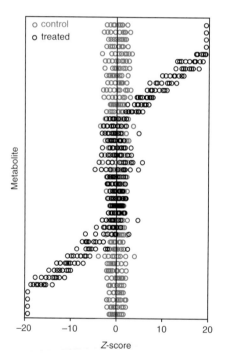

Figure 22.7 A Z-transformation depicts how individual replicates of a group of samples compare to the average of another group (control). This is usually performed in respect to metabolite abundance. In this figure, the mean of the control for each metabolite is plotted along the x-axis and given the value '0' since it is zero standard deviations away from the mean. The positions of the blue circles indicate, for each replicate of the control, the number of standard deviations away from the control mean value. The positions of the black circles indicate, for each replicate of the treated dataset, the standard deviations away from the control mean value. Close grouping suggests that there was low variability (high reproducibility) in that dataset. Z-transformations can often be depicted with additional lines, usually at Z-scores (i.e. standard deviations) of −2 and +2 or −5 and +5, to help depict which data points can be regarded as significant, or otherwise.

Volcano plots are used to compare two metabolomics datasets by plotting the fold change of the mean value of a metabolite (abundance or labelling) against the corresponding *p*-value. Most commonly, the log of the fold change and the negative $\log_{10}$ of the *p*-value are used. In essence, this allows one to visualise those metabolites that have the greatest change in abundance between two samples, and those that are changing most significantly (Figure 22.8). As mentioned in the above discussion of Z-transformations, a significant change does not necessarily equate to a high fold change, and vice versa.

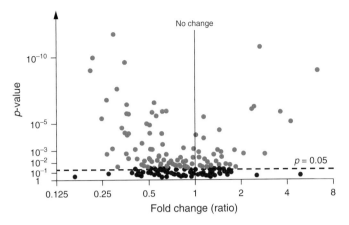

Figure 22.8 A volcano plot shows the relationship between those values that change most and those that are most significant. On the x-axis is shown the fold change of a value (e.g. metabolite abundance) from that of the mean of the control group (or the difference of group means), while the y-axis depicts the p-value of this value after correction for multiple testing (often shown on a logarithmic scale). The significance threshold (most commonly $p = 0.05$) is shown as a dotted line. In such a graph, metabolites (or values) that change most greatly and significantly from the control mean will be plotted in the upper left and right quadrants. Purple data points indicate statistically insignificant values.

22.5.3 Visual Approaches

Visual approaches are simply methods in which matrices of data (such as those from a data spreadsheet table) are depicted in a more-easily viewable and comprehensible manner. In its simplest form, this would be a bar chart or a graph to show, for example, changes in metabolite levels over time. As an extension of this, box plots are used to visualise more information about the reproducibility (quality) of a value by summarising the distribution in a clear manner. A typical box plot indicates the median, the interquartile range (the box) and commonly the most extreme data points within 1.5 interquartile ranges of the interquartile range itself, along with any outliers (Figure 22.9; see also Section 19.2.8).

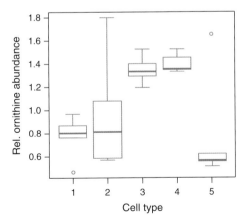

Figure 22.9 Box plots (often called box–whisker plots) depict values (e.g. metabolite abundance) and basic descriptive statistics of those values. In this figure, values are shown for the metabolite ornithine in five different cell types. The thick line represents the median value – this is also known as the second quartile. The lower and upper edges of the box represent the first and third quartiles, respectively, and hence the box depicts the interquartile range of the data values. The lines (or whiskers) can represent several different values, but are most commonly representative of either (a) the minimum and maximum values of the dataset; (b) the lowest and highest points within 1.5 quartile ranges of the lower and upper quartiles, respectively; or (c) one standard deviation away from the mean. Other alternatives also exist. The circles are outlier values that lie outside of the whisker values.

Heatmaps (see also Section 19.2.8) are a clear and simple way of visualising metabolomics data – especially relative abundance or label incorporation data for a number of metabolites and samples. Heatmaps rarely contain any statistical information (e.g. standard deviations, standard error of the mean, p-values, etc.) and so are often accompanied by a more detailed table of results. However, heatmaps can sometimes be used to illustrate global changes in a metabolome in a manner analogous to genome or proteome arrays. In these cases, some degree of statistical analysis is employed. An example of a heatmap is shown in Section 22.6.2.

22.6 FURTHER METABOLOMICS TECHNIQUES AND TERMINOLOGY

22.6.1 Lipidomics

Lipidomics is metabolomics focussed entirely on lipids and lipid-containing molecules. Due to the complex nature of lipid biochemistry, and the almost endless possibilities of lipid structural composition, the study of lipidomics is sometimes regarded as an entirely separate field to metabolomics. Subtle differences in fatty acid chain length and degrees of saturation (i.e. the number of double bonds present in the moiety) can make chromatography very challenging and the resulting chromatograms very complex. Therefore, truly global lipidomics may be unattainable in some samples and so targeted lipidomics, focussing on particular lipid classes, can often be a more favourable approach.

Lipid-containing molecules are often very large and even the simplest structures are generally too large to be analysed in their intact state by GC-MS. To overcome this issue, molecules are often digested into their constituent parts (e.g. fatty acid, glycerol, sugar and/or organic acid moieties). Such analysis can give useful information about the repertoire of fatty-acid moieties (chain lengths, saturation) in the sample, but without prior fractionation fatty acids cannot be assigned to any specific lipid class(es).

LC-MS and NMR are often preferred to GC-MS in lipidomics due to the ability to analyse many lipids in their intact (assembled) form. For example, mature phospholipids (which are important molecules found throughout nature and involved in membrane structure, signal transduction, trafficking and storage, amongst other roles) generally consist of two fatty-acid moieties attached through a glycerol backbone to a polar headgroup. The headgroup and length/saturation of the fatty-acid moieties contribute to their localisation and function and are therefore of interest to determine analytically. Liquid chromatography can partially separate different phospholipid classes (Figure 22.10a) and subsequent MS allows determination of masses of the intact molecules (via m/z values, Figure 22.10b). Fragmentation of each feature produces product ions that can be used to elucidate both the headgroup and the fatty-acid composition of each molecule (Figure 22.10 c,d). Product ions that are unique to a given lipid class (e.g. the ion with $m/z = 241$ of phosphatidylinositol, PI) are amenable to multiple reaction monitoring (MRM) analyses (see Section 21.7.2) and specific detection of that lipid. MRM is therefore a popular method in lipidomics, often complemented by high-resolution mass spectrometry for determination of the accurate mass of each identified molecule. However, without sophisticated and well-planned experimentation, ambiguity in lipid structure can remain, such as the precise location of double bonds in an acyl chain and which lipid moiety is attached to which position in larger molecules. For example, cardiolipin has four acyl chains linked through one glycerol and two glycerophosphate molecules – i.e. nine carbon atoms are theoretically available for attachment of the four acyl chains. Without prior knowledge, determination of such an arrangement can be challenging.

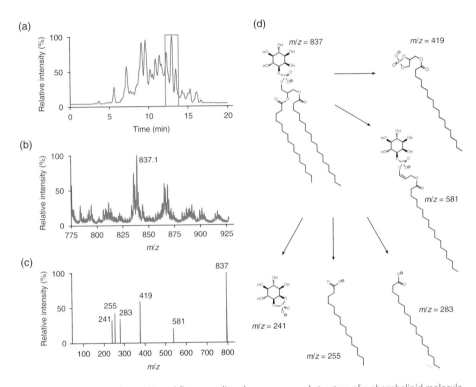

Figure 22.10 An example LC-MS workflow revealing the presence and structure of a phospholipid molecule in a sample. (a) The LC-MS chromatogram shows the total ion abundance detected over the course of the chromatography run in negative ion mode (in this case 20 minutes). From previous experiments or standards, the boxed area is known to putatively contain the phospholipid phosphatidylinositol (PI) amongst other molecules. (b) The mass spectrum derived from the boxed area of (a). Many clusters of peaks are observed and within each cluster the predominant peaks are 2 Da apart, indicative of the number of double bonds in a lipid moiety. For example, a species with one double bond will have a mass 2 Da less than a fully saturated species, since two hydrogen atoms will be lost to form the double bond. In this way, ions to the left of a cluster will generally have more double bonds than those on the right. The differences between each cluster will vary, with differences of 28 Da indicative of a C_2H_4 unit, i.e. the pairwise addition of carbon units seen in fatty acid biosynthesis. (c) The fragmentation spectrum of the ion with m/z = 837.1 observed in (b). Here, collision-induced dissociation has produced a number of fragment ions of the molecular ion. Note that the ion with m/z = 241 is diagnostic of the inositol headgroup of PI. (d) A cartoon of the putative assignments of the molecular and fragment ions observed in (c). Collectively, the fragments indicate a likely structure of PI containing C16:0 and C18:0 fatty acid moieties. The data shown in this figure do not represent a real experiment and have been compiled for illustrative purposes only.

One of the greatest strengths of NMR is to provide unambiguous positional information in a detectable molecule and so this approach can be more definitive than MS in lipidomics studies. In addition, since lipids constitute a large proportion of the biomass of a given biological system, the relatively poor sensitivity of NMR often does not present an issue in the context of lipidomics studies. However, the complexity of lipid structures is still a concern and so separation techniques such as solid-phase extraction (SPE) can be used to help simplify analysis by fractionating at least some lipid classes.

22.6.2 Metabolite Footprinting

Metabolite footprinting is a branch of metabolomics that is often incorporated into both untargeted and targeted metabolomics studies. Metabolite footprinting seeks to measure the uptake and excretion of metabolites by a cell system. Most commonly, this is performed in vitro, where the medium from cells grown in culture is analysed in order to determine nutrient usage (e.g. Figure 22.11). Since culture medium is usually readily available for analysis, both MS and NMR approaches are generally equally applicable to these analyses (metabolite dependent). Metabolite footprinting can be regarded as essential for establishing a full picture of the physiological state of the biosystem. When observing carbon/energy metabolism, for example, measurement of uptake of carbon sources (e.g. glucose, glutamine) from the medium and excretion of waste products (e.g. lactate, HCO_3^-) is required to enable a more complete assessment of metabolic flux and/or respiration.

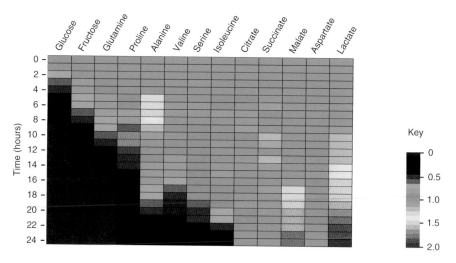

Figure 22.11 Footprinting analysis. The figure shows a heatmap depicting metabolite abundance changes in the medium of a cell culture over 24 hours. Cooler colours represent decreased abundance, green colours signify no change, while warmer colours represent increased abundance. In this example, glucose is the primary carbon source and, after this is used up, the organism preferentially uses fructose then glutamine, proline and so on. Warmer colours show that malate and lactate appear to be metabolic waste products of the cell, while alanine is at first secreted as waste when glucose is replete, but can also be used as a carbon source when primary carbon sources are depleted.

Metabolic footprinting also has a number of other applications. For example, footprinting is widely used in toxicology (including environmental toxicology) and pharmacological studies in order to determine uptake and half-lives of drugs or toxins in a biosystem and the resultant physiological effects of these.

22.6.3 Mass Spectrometry Imaging (MSI)

MSI is a powerful technique used to visualise the spatial distribution of metabolites in a tissue section or cell. While not a true metabolomics platform (as well as metabolites, MSI can be used to analyse drugs, proteins and peptides), MSI is often considered an off-shoot of metabolomics due to the thousands of metabolites and lipids that can be potentially observed in a sample. The main techniques in MSI are MALDI imaging, ambient mass spectrometry (AMS) imaging, and secondary ion mass spectrometry (SIMS) imaging.

The process of MALDI imaging is discussed in Section 15.2.5 and essentially relies on the process of spraying a matrix onto a thin tissue section and using a laser to produce ions. This approach can enable detection of a broad range of metabolites/lipids (in terms of both chemical properties and size) with an optimal resolution around 20 μm. This allows metabolite mapping of tissues, but finer detail is as yet not possible.

AMS imaging often uses desorption electrospray ionisation (DESI) to allow analysis of a sample under atmospheric or ambient conditions. Here, highly charged aqueous droplets touch the sample surface, causing molecules to desorb and ionise. This soft ionisation technique often gives rise to the molecular ion (i.e. little fragmentation) of a metabolite, which is subsequently analysed by electrospray MS. This technique is sensitive and can be very useful for analysis of volatiles or labile molecules, although robustness and resolution are not as good as for other methods.

SIMS imaging uses application of a primary high-energy ion beam onto a sample surface, resulting in ejection of secondary ions, which are subsequently collected and analysed by MS. This process, termed sputtering, is more sensitive than MALDI and results in elemental, molecular and fragment ions. Moreover, since the ion beam is more focussed than the laser of MALDI MSI, SIMS imaging has a better resolution, with surface imaging of < 200 nm resolution possible. However, SIMS is limited in its detectable mass range compared to MALDI MSI, with only fragment (rather than molecular) ions of larger, more complex molecules being detected. This can lead to challenges in mass spectral interpretation, although fragmentation libraries are always expanding to aid this. While both MALDI and SIMS imaging by definition degrade the sample being analysed, this can sometimes be of benefit to the analyst. For example, sequential SIMS sputtering over an entire surface will remove the top layer from a given sample, revealing a sublayer. This sublayer can be analysed in the same way, and so on in a layer-by-layer manner, ultimately resulting in a 3D image metabolite map.

22.6.4 Headspace Analysis

Headspace analysis is almost always associated with GC-MS, since it involves the analysis of volatiles present in the (local) atmosphere. Headspace analysis has a number of applications, both in basic biological research and in industry. In biology, examples of headspace analysis include pheromone study, toxin release, and organism to organism signalling/interaction. In food technology, headspace analysis is instrumental in quality control, including identification of factors controlling fragrance and flavours, as well as shelf-life and degradation.

A common headspace analytical set up is shown in Figure 22.12. The sample – for instance a cell culture or aliquot of wine – is introduced to a vial and conditioned (e.g. heated or agitated until an equilibrium of volatiles between the sample and the headspace is reached) as shown in Step 1. A microfiber (Step 2) coated in an adsorbent porous polymer is inserted into the headspace of the vial for a set amount of time to allow collection (trapping) and concentration of the volatile analyte (Step 3). This process is known as solid–phase microextraction (SPME). The fibre is then removed from the vial and inserted into the inlet port of a GC-MS instrument (Step 4). Since the inlet is held at a high temperature, the analytes revolatilise and enter the GC-MS for analysis. The fibre may then be reused, providing all analytes have been desorbed into the GC-MS inlet. The nature of the polymer can be controlled to give different volatile specificities. In addition to coating SPME fibres, the adsorbent porous polymer resin can be packed into flow-through cartridges. By drawing gas through the cartridge, volatiles stick to the resin, which can then be introduced directly into the GC-MS by thermal desorption (as is the case for SPME), or can be extracted and further concentrated using suitable chemicals. In both SPME and cartridge assemblies, analyte adsorption and subsequent analysis is highly sensitive and robust.

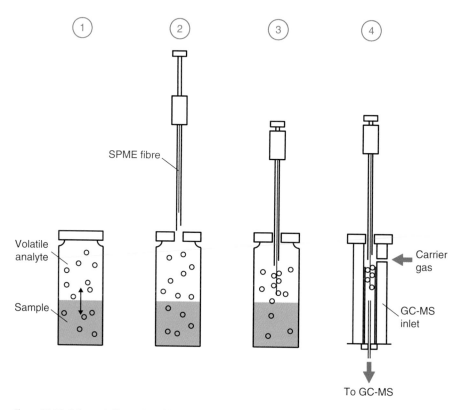

Figure 22.12 Schematic illustration of headspace analysis. For explanation of each step, see the main text.

22.7 SUGGESTIONS FOR FURTHER READING

22.7.1 Experimental Protocols

Creek D.J., Chokkathukalam A., Jankevics A., *et al.* (2012) Stable isotope-assisted metabolomics for networkwide metabolic pathway elucidation. *Analytical Chemistry* **84**, 8442–8447.

MacRae J.I., Sheiner L., Nahid A., *et al.* (2012) Mitochondrial metabolism of glucose and glutamine is required for intracellular growth of *Toxoplasma gondii*. *Cell Host Microbe* **12**, 682–692.

Metallo C.M., Walther J.L. and Stephanopoulos G. (2009) Evaluation of ^{13}C isotopic tracers for metabolic flux analysis in mammalian cells. *Journal of Biotechnology* **144**, 167–174.

Norris J.L. and Caprioli R.M. (2013) Analysis of tissue specimens by matrix-assisted laser desorption/ionization imaging mass spectrometry in biological and clinical research. *Chemical Reviews* **113**, 2309–2342.

Saunders E.C., MacRae J.I., Naderer T., *et al.* (2012) *LeishCyc: a guide to building a metabolic pathway database and the visualization of metabolomic data.* In *Microbial Systems Biology: Methods in Molecular Biology*, Vol. 881, Humana Press, Totowa, NJ, USA, pp 505–529.

Tymoshenko S., Oppenheim R.D., Agren R., *et al.* (2015) Metabolic needs and capabilities of *Toxoplasma gondii* through combined computational and experimental analysis. *PLoS Computational Biology* **11**, e1004261.

Zomorrodi A.R. and Maranas C.D. (2010) Improving the iMM904 *S. cerevisiae* metabolic model using essentiality and synthetic lethality data. *BMC Systems Biology* **4**, 178.

22.7.2 General Texts

Fiehn O. (2016) Metabolomics by gas chromatography-mass spectrometry: combined targeted and untargeted profiling. *Current Protocols in Molecular Biology*, 30.4.1–30.4.32.

Grimm F., Fets L. and Anastasiou D. (2016) *Gas Chromatography Coupled to Mass Spectrometry (GC-MS) to study metabolism in cultured cells.* In *Tumor Microenvironment: Study Protocols*, Koumenis C., Coussens L.M., Giaccia A. and Hammond E., Eds., Springer, New York, USA.

Weckwerth W. (2007) *Metabolomics: Methods and Protocols: Methods in Molecular Biology*, Vol. 358, Humana Press, Totowa, NJ, USA.

22.7.3 Review Articles

Breitling R., Pitt A.R. and Barrett M.P. (2006) Precision mapping of the metabolome. *Trends in Biotechnology* **24**, 543–548.

Brown M., Wedge D.C., Goodacre R., *et al.* (2011) Automated workflows for accurate mass-based putative metabolite identification in LC/MS-derived metabolomic datasets. *Bioinformatics* **27**, 1108–1112.

Fan T.W. and Lane A.N. (2016) Applications of NMR spectroscopy to systems biochemistry. *Progress in Nuclear Magnetic Resonance Spectroscopy* **92–93**, 18–53.

Fiehn O. (2002) Metabolomics : the link between genotypes and phenotypes. *Plant Molecular Biology* **48**, 155–171.

Orth J.D., Thiele I. and Palsson B.Ø. (2010) What is flux balance analysis? *Nature Biotechnology* **28**, 245–248.

Sumner L.W., Amberg A., Barrett D., *et al.* (2007) Proposed minimum reporting standards for chemical analysis Chemical Analysis Working Group (CAWG) Metabolomics Standards Initiative (MSI) *Metabolomics* **3**, 211–221.

Zamboni N. (2011) ^{13}C metabolic flux analysis in complex systems. *Current Opinion in Biotechnology* **22**, 103–108.

Zamboni, N., Saghatelian, A. and Patti, G. J. (2015) Defining the metabolome: size, flux and regulation. *Molecular Cell* **58**, 699–706.

22.7.4 Websites

The Human Metabolome Database
www.hmdb.ca/ (accessed May 2017)

KEGG Pathway Database
www.genome.jp/kegg/pathway.html (accessed May 2017)

LipidBlast – an MS/MS database for lipid identification
fiehnlab.ucdavis.edu/projects/LipidBlast/ (accessed May 2017)

MapMan – Conversion of large datasets into metabolic pathway diagrams
mapman.gabipd.org (accessed May 2017)

MBROLE – Metabolites Biological Role
csbg.cnb.csic.es/mbrole2/ (accessed May 2017)

MetaboAnalyst
www.metaboanalyst.ca/ (accessed May 2017)

MetaboLights – Metabolomics experiments and derived information
www.ebi.ac.uk/metabolights/ (accessed May 2017)

MS-DIAL – MS/MS deconvolution for comprehensive metabolome analysis
prime.psc.riken.jp/Metabolomics_Software/MS-DIAL/ (accessed May 2017)

XCMS (Scripps Center for Metabolomics)
www.xcmsserver.com/ (accessed May 2017)

Yeast Metabolome Database
www.ymdb.ca/ (accessed May 2017)

23 | Enzymes and Receptors

MEGAN CROSS AND ANDREAS HOFMANN

23.1 DEFINITION AND CLASSIFICATION OF ENZYMES

Catalytic reactions in biological processes are facilitated by two types of catalysts, enzymes and ribozymes. Whereas enzymes are proteins, ribozymes consist of ribonucleic acid (RNA). Most enzymes are much larger than the substrates they process, but this is not a requirement (for example, restriction enzymes that cleave DNA). The catalytic features of enzymes arise through a particular three-dimensional arrangement of functional groups in a small number of amino acids (see Section 23.4.2) in the active site of the enzyme. The geometrical arrangement of such groups enables productive interactions with the bound substrate and leads to formation of a transition state for which the energy barrier (activation energy) is significantly reduced as compared to the non-catalysed reaction (see Figure 23.1). Consequently, the reaction rate is increased by several orders of magnitude relative to the non-catalysed reaction. Importantly, enzymes do not alter the position of equilibrium of the reversible reactions they catalyse, rather they accelerate establishment of the position of equilibrium for the reaction.

Many enzymes are the key players in metabolic or signalling pathways. In these coordinated pathways, enzymes are collectively responsible for maintaining the metabolic needs of cells under varying physiological conditions (Section 23.5). Through their individual catalytic activities, they control the rate of a particular metabolic or signalling pathway. A range of regulatory mechanisms operates to allow short-, medium- and long-term changes in activity (Section 23.5.1). Therefore, the over- or under-expression of an enzyme can lead to cell dysfunction, which may manifest itself

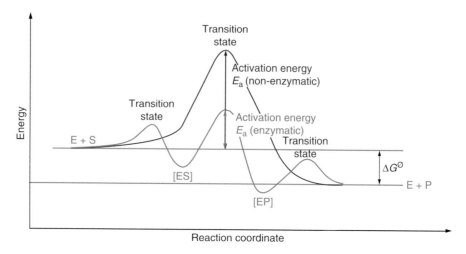

Figure 23.1 Comparison of the energy profiles of non-catalysed and catalysed reactions. The formation of the enzyme–substrate ([ES]) and enzyme–product ([EP]) complexes, as well as the release of the formed product P proceeds via several transition states. The activation energy E_a for the overall reaction is dictated by the initial free energy of enzyme and substrate, and the transition state with the highest energy. The activation energy for the non-catalysed reaction is substantially higher than that of the catalysed reaction.

as a particular disease. Hence, enzymes have been the most important target in the development of therapeutics, and typical drugs for the treatment of pathological conditions act as inhibitors of particular enzymes (Sections 23.5.3). Other clinically relevant applications include monitoring of enzyme levels in assessment of disease states. For example, damage to the heart muscle as a result of oxygen deprivation following a heart attack results in the release of cellular enzymes into extra-cellular fluids and eventually into the blood. Such release can be monitored to aid diagnosis of the organ damage and to make a prognosis for the patient's future recovery (Section 10.3).

23.1.1 Specificity and Nomenclature

Individual enzymes are characterised by their **specificity** for a particular type of chemical reaction, but generally it is observed that enzymes involved in biosynthetic or signalling reactions show a higher specificity than others involved in degradation reactions. Different types of specificity can be distinguished (Table 23.1).

By international convention, Enzyme Commission (EC) rules (www.chem.qmul .ac.uk/iubmb/enzyme/) have been established, that allow assignment of a unique four-figure code and an unambiguous systematic name to each individual enzyme, based on the reaction catalysed. The EC rules classify enzymes into six groups on the basis of the type of chemical reaction that they catalyse (Table 23.2). Each group is divided into subgroups according to the nature of the chemical group and coenzymes involved in the reaction (see Example 23.1).

The catalytic properties of an enzyme are often dependent upon the presence of non-peptide molecules, which are called cofactors or coenzymes. If a cofactor

Table 23.1 **Different types of enzyme specificities**

Type of specificity	Mechanism	Examples
Bond specificity	Specificity of these enzymes is determined by the presence of specific functional groups within the substrate adjacent to the bond to be cleaved	Peptidases and esterases
Group specificity	Enzymes that promote a particular reaction on a structurally related group of substrates	Kinases catalyse the phosphorylation of substrates that have a common structural feature such as a particular amino acid (e.g. the tyrosine kinases) or sugar (e.g. hexokinase)
Stereospecificity	Distinguish between optical and geometrical isomers of substrates	

Table 23.2 **The six different groups of enzymes**

Group	Enzymes	Reaction	Examples
EC 1	Oxidoreductases	Transfer hydrogen or oxygen atoms or electrons from one substrate to another	Dehydrogenases, reductases, oxidases, dioxidases, hydroxylases, peroxidases and catalase
EC 2	Transferases	Transfer chemical groups between substrates	Kinases, aminotransferases, acetyltransferases and carbamyltransferases
EC 3	Hydrolases	Catalyse the hydrolytic cleavage of bonds	Peptidases, esterases, phosphatases and sulfatases
EC 4	Lyases	Catalyse elimination reactions resulting in the formation of double bonds	Adenylyl cyclase, enolase and aldolase
EC 5	Isomerases	Inter-convert isomers of various types by intramolecular rearrangements	Phosphoglucomutase and glucose-6-phosphate isomerase
EC 6	Ligases	Catalyse covalent bond formation with the concomitant breakdown of a nucleoside triphosphate, commonly ATP	Carbamoyl phosphate synthase and DNA ligase

is tightly bound to the enzyme, it is commonly referred to as a **prosthetic group**. Examples of coenzymes include NAD^+, $NADP^+$, FMN and FAD, whilst examples of prosthetic groups include haem and oligosaccharides, and divalent metal ions such as Mg^{2+}, Fe^{2+} and Zn^{2+}. DNA and RNA polymerases, as well as many nucleases, for example, require two divalent cations for their active site. The cations correctly orientate the substrate and promote acid–base catalysis.

The activity of an enzyme is defined by the speed of the chemical reaction it catalyses. Therefore, **enzyme activity** is measured by means of a kinetic rate describing the molar amount of product formed (or substrate consumed) per unit time:

$$\text{Enzyme activity} = \frac{\Delta n}{\Delta t} \qquad \text{(Eq 23.1)}$$

typically with units of $\mu mol \ min^{-1}$.

The SI unit for enzyme activity is called the **katal**: $1 \ kat = 1 \ mol \ s^{-1} = 60 \ mol \ min^{-1}$.

Since the rate of the chemical reaction catalysed by an enzyme depends on the amount of enzyme present, a more useful quantity in this context is the **specific**

Example 23.1   OPERATION OF ENZYME CLASSIFICATION RULES

As an example of the EC rules, consider the enzyme alcohol dehydrogenase which catalyses the reaction:

alcohol + NAD$^+$ $\rightleftharpoons$ {aldehyde or ketone} + NADH$^+$ + H$^+$

It has the systematic name alcohol:NAD oxidoreductase and the classification number 1:1:1:1. The first '1' indicates that it is an oxidoreductase, the second '1' that it acts on a CH-OH donor, the third '1' that NAD$^+$ or NADP$^+$ is the acceptor and the fourth '1' that it is the first enzyme named in the 1:1:1 subgroup.

Systematic names tend to be user-unfriendly and for day-to-day purposes recommended trivial names are preferred. When correctly used, the trivial names give a reasonable indication of the reaction promoted by the enzyme in question, but fail to identify fully all the reactants involved. For example, the name 'glyceraldehyde-3-phosphate dehydrogenase' fails to identify the involvement of orthophosphate and NAD$^+$. Its EC number is 1.2.1.12: '1' oxidoreductase, '2' acting on an aldehyde, '1' NAD$^+$ or NADP$^+$ as acceptor, '12' glyceraldehyde-3-phosphate dehydrogenase.

The name 'phosphorylase kinase' fails to convey the information that it is the b form of phosphorylase that is subject to phosphorylation involving ATP. Its EC number is 2.7.11.19: '2' transferase, '7' transfer of phosphorus-containing groups, '11' serine/threonine kinase, '19' phosphorylase kinase.

enzyme activity, which relates the measured activity to the amount of enzyme present. Logically, this normalisation should be carried out per molar amount of enzyme; however, the traditionally used biochemical quantity used to measure the amount of protein is the mass. Therefore, the specific activity is expressed in units of mmol min^{-1} mg^{-1}.

23.1.2 Isoenzymes and Multi-Enzyme Complexes

Isoenzymes

Some enzymes exist in multiple forms that differ in amino-acid sequence, but catalyse the same chemical reaction. Such enzymes are called isoenzymes or isozymes. A prominent example is lactate dehydrogenase (LDH; EC 1:1:1:27), which exists in five isoforms. LDH is a tetramer that can be assembled from two subunits, H (for heart) and M (for muscle). The five observed forms are LDH_1 (H_4), LDH_2 (H_3M), LDH_3 (H_2M_2), LDH_4 (HM_3) and LDH_4 (M_4) which can be separated by electrophoresis and have different tissue distributions. Importantly, the kinetic properties differ among the different isoforms of isoenzymes, and the five LDH forms thus have different affinities for their substrates, lactate and pyruvate, and different K_M and v_{max} values (see Section 23.2). Consequently, isoenzymes are important in diagnostic enzymology (see Section 10.3).

Multi-Enzyme Complexes

Some enzymes that promote consecutive reactions in a metabolic pathway associate to form a multi-enzyme complex. Examples include the pyruvate dehydrogenase complex consisting of three enzymes that convert pyruvate into acetyl-CoA: pyruvate dehydrogenase (EC 1.2.4.1), dihydrolipoyl transacetylase (EC 2.3.1.12) and dihydrolipoyl dehydrogenase (EC 1.8.1.4). Other examples are the fatty-acid synthase system, consisting of two identical monomers with multiple active sites, as well as RNA polymerase, where five subunits combine together with the initiation factor σ to form the holoenzyme consisting of six subunits.

There are a number of advantages of multi-enzyme complexes over individual enzymes: the transit time for the diffusion of the product of one catalytic site to another is reduced, there is a lower possibility of the product of one step being acted upon by another enzyme not involved in the pathway and of one enzyme directly activating an adjacent enzyme in the pathway (Section 23.5.2).

23.2 ENZYME KINETICS

Enzyme-catalysed reactions proceed via the formation of an enzyme–substrate complex [ES] in which the substrate is non-covalently bound to the active site of the enzyme (see also Figure 23.1). The formation of this complex is rapid and reversible, and hence can be described by forward (k_1) and reverse (k_{-1}) rate constants. The conversion of [ES] to product P is most simply represented by an irreversible reaction with the rate constant k_2 (also called k_{cat}). This yields the following reaction pathway:

$$E + S \underset{k_{-1}}{\overset{k_1}{\rightleftharpoons}} [ES] \xrightarrow{k_2 = k_{cat}} E + P \qquad \text{(Eq 23.2)}$$

When an enzyme is mixed with an excess of substrate there is an initial short period of time (a few hundred microseconds) during which intermediates leading to the formation of the product gradually build up (Figure 23.2). This so-called pre-steady state requires special techniques for study and these are discussed in Section 23.3.3. After this pre-steady state, the reaction rate and the concentration of intermediates change relatively slowly with time with so-called steady-state kinetics.

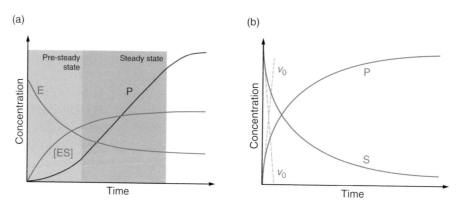

Figure 23.2 (a) Concentrations of the free enzyme (E), the enzyme–substrate complex ([ES]) and the product (P) during a two-step mechanism. (b) Determination of the initial rate v_0 from the time-dependent change in the concentration of substrate (S) or product (P). The numerical value of v_0 is independent of the choice of observable (S or P).

The most common approach to analyse enzyme kinetics uses the initial rate. In a plot of substrate/product concentration versus time, tangents are drawn through the steady-state curves in the origin; their slopes yield the initial rate v_0. This is the fastest rate for a given concentration of enzyme and substrate under the defined experimental conditions. Its numerical value is influenced by many factors, including substrate and enzyme concentration, pH, temperature and the presence of activators or inhibitors.

For many enzymes, the initial rate varies hyperbolically with substrate concentration for a fixed concentration of enzyme (Figure 23.3). This relationship is expressed by the Michaelis–Menten equation:

$$v_0 = \frac{v_{max} \times c(S)}{K_M + c(S)}$$

(Eq 23.3)

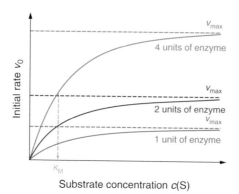

Substrate concentration $c(S)$

Figure 23.3 The effect of substrate concentration $c(S)$ on the initial rate v_0 of an enzyme-catalysed reaction in the presence of different concentrations of enzyme. Doubling the enzyme concentration doubles the maximum initial rate v_{max} but has no effect on the Michaelis constant K_M.

which is a deterministic rate equation, assuming that the number of molecules in the sample under investigation is large. For single-molecule observations, which are increasingly becoming technologically possible, deterministic rate equations cannot be used to describe the reaction dynamics. This is because the inherent noise due to random timing of biochemical reactions dominates such observations. In such circumstances, stochastic rate equations need to be considered.

Figure 23.3 illustrates that at low substrate concentrations the occupancy of the active sites on the enzyme molecules is low and the reaction rate is directly related to the number of sites occupied. This approximates to first-order kinetics in that the rate is proportional to substrate concentration. At high substrate concentrations effectively all of the active sites are occupied and the reaction becomes independent of the substrate concentration, since no more enzyme–substrate complex [ES] can be formed and zero-order or saturation kinetics are observed. Under these conditions the reaction rate is only dependent upon the conversion of the enzyme–substrate complex to products and the diffusion of the products from the enzyme.

In Equation 23.3, v_{max} is the limiting value of the initial rate that is observed when all active sites are occupied. K_M is called the Michaelis constant and carries the units of mol dm^{-3}; its numerical value equals the substrate concentration at which the initial rate is one-half of the maximum rate v_{max}. Values of K_M are usually in the range of 10^{-2} to 10^{-5} M; knowledge of the K_M value is important since it enables calculation of the substrate concentration required to saturate all enzyme active sites in the assay. From Equation 23.3, it can easily be deduced that at substrate concentrations considerably larger than K_M, the observed initial rate v_0 approximates v_{max}. More specifically, when $c(S) = 10 \times K_M$, $v_0 = 0.9 \times v_{max}$, but when $c(S) = 100 \times K_M$, $v_0 = 0.99 \times v_{max}$.

As mentioned above (see Equation 23.2), a key feature of enzyme-catalysed reactions is the formation of an enzyme–substrate complex [ES]; this is a reversible reaction and hence treated as a chemical equilibrium. When a chemical reaction has reached equilibrium, its forward and reverse rates are the same, which leads to the law of mass action:

$$k_1 \times c(E) \times c(S) = k_{-1} \times c([ES]) \quad \Rightarrow \quad K_d = \frac{c(E) \times c(S)}{c([ES])} = \frac{k_{-1}}{k_1} = \frac{1}{K_a} \qquad \text{(Eq 23.4)}$$

where K_d is the dissociation constant and K_a the association (or affinity) constant of the enzyme–substrate complex. Obviously, when K_d takes a large numerical value, the equilibrium is in favour of the unbound enzyme and substrate, whilst when K_d is numerically small, the equilibrium rests on the side of the enzyme–substrate complex.

The formation of product is typically an irreversible reaction, characterised by the first-order rate constant k_2 which is also called k_{cat}. This second reaction generally proceeds slower than the pre-equilibrium ($k_{cat} < k_1$, k_{-1}), which makes the conversion of the enzyme–substrate complex to product the rate-limiting step, such that the concentration of [ES] remains essentially constant. Under these conditions, the Michaelis constant K_M is given by:

$$K_M = \frac{k_{cat} + k_{-1}}{k_1} = K_d + \frac{k_{cat}}{k_1} \qquad \text{(Eq 23.5)}$$

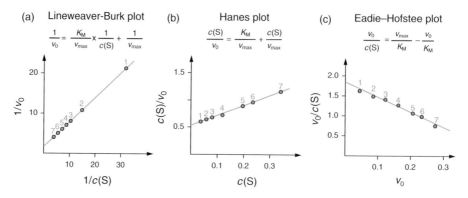

Figure 23.4 (a) Lineweaver–Burk, (b) Hanes and (c) Eadie–Hofstee plots for the same set of experimental data of the effect of substrate concentration on the initial rate of an enzyme-catalysed reaction.

It is evident that under these circumstances, K_M must be numerically larger than K_d, and only when k_{cat} is very small, are K_M and K_d approximately equal.

Although the Michaelis–Menten equation (Equation 23.3) can be used to calculate K_M and v_{max}, its use has historically been subject to the difficulty of experimentally measuring initial rates at high substrate concentrations, as well as fitting the hyperbolic curve to give accurate values of K_M and v_{max}. Linear transformations of the Michaelis–Menten equation have therefore been commonly used alternatives. From Figure 23.4 it can be seen that the Lineweaver–Burk equation gives an unequal distribution of points and there is greater emphasis on the points at low substrate concentration that are subject to the greatest experimental error, whereas the Eadie–Hofstee equation and the Hanes equation give a better distribution of points. In the case of the Hanes plot, greater emphasis is placed on the experimental data at higher substrate concentrations and on balance it is the statistically preferred plot. In spite of their widespread use, these linear transformations of enzyme kinetic data are subject to error. Specifically, they assume that the scatter of points around the line follows a Gaussian distribution and that the standard deviation of each point is the same. In practice this is rarely true. With the advent of widely available non-linear regression software packages, there are now strong arguments for their preferential use in cases where accurate kinetic data are required.

It is important to appreciate that whilst K_M is a characteristic of an enzyme for its substrate and is independent of the amount of enzyme used for its experimental determination, this is not true of v_{max}. It has no absolute value, but varies with the amount of enzyme used. This is illustrated in Figure 23.3 and is discussed further in Example 23.2. A characteristic parameter for a particular enzyme is the rate constant for the conversion of the enzyme–substrate complex to product, k_{cat}, which is also called the turnover number; it describes the number of catalytic cycles performed in a given time interval, divided by that time interval. From the general observations with enzymatic reactions it follows that k_{cat} can be calculated from the maximum initial rate v_{max} and the total concentration of enzyme $c_0(E)$:

$$k_{cat} = k_2 = \frac{v_{max}}{c_0(E)}$$ (Eq 23.6)

Values for k_{cat} range from 1 to 10^7 s^{-1}; one of the most efficient enzymes known, catalase, has a turnover number of 4×10^7 s^{-1}.

The catalytic potential of high turnover numbers can only be realised at high (saturating) substrate concentrations; this is rarely achieved under normal cellular conditions. Thus, to capture how efficiently an enzyme converts substrate to product at low substrate concentrations, a measure of efficiency is required. The substrate concentration required to achieve 50% of the maximum rate of an enzymatic reaction is described by the Michaelis–Menten constant K_M. Therefore, a lower K_M will result in higher efficiency. Likewise, the efficiency will be higher if the turnover number (k_{cat}) is high. The **catalytic efficiency** (or **specificity constant**) η of an enzyme is thus defined as:

$$\eta = \frac{k_{cat}}{K_M} = \frac{k_1 \times k_{cat}}{k_{-1} + k_{cat}} \qquad \text{(Eq 23.7)}$$

with units of M^{-1} s^{-1}.

For a substrate to be converted to product, molecules of the substrate and of the enzyme must first collide by random diffusion and then combine in the correct orientation. Diffusion and collision have a theoretical limiting rate constant value of about 10^9 M^{-1} s^{-1} and yet many enzymes, including acetylcholine esterase, carbonic anhydrase, catalase, β-lactamase and triosephosphate isomerase, have catalytic efficiencies approaching this value, indicating that they have evolved to almost **maximum kinetic efficiency**. Notably, when enzymes are embedded in an organised assembly, such as in cellular membranes, the product of one enzyme is channelled to the next enzyme in the pipeline. In such cases, the rate of catalysis is not limited by diffusion and can exceed the diffusion-controlled limit.

Multi-enzyme complexes (Section 23.1.2) overcome some of the diffusion and collision limitations to specificity constants. The product of one reaction is passed directly by a process called channelling to the active site of the next enzyme in the pathway as a consequence of its juxtaposition in the complex, thereby eliminating diffusion limitations.

Since specificity constants are a ratio of two other constants, enzymes with similar specificity constants can have widely different K_M values. The enzyme catalase, mentioned earlier, for example, possesses a catalytic efficiency of 4×10^7 M^{-1} s^{-1} with a K_M of 1.1 M (very high), whereas fumarase with a similar catalytic efficiency of 3.6×10^7 M^{-1} s^{-1} possesses a K_M of 2.5×10^{-5} M (very low).

23.2.1 Burst Kinetics

In practical settings, enzymatic reactions are often monitored by measurements of an observable (e.g. difference in light absorbance) in a time-dependent manner. According to Michaelis–Menten kinetics, a linear increase of the observable with time is expected to intercept with zero at time $t = 0$. However, with some enzymes, one may observe that the plot of the raw data against time does not proceed through the origin

Example 23.2   **PRACTICAL ENZYME KINETICS**

Question   The enzyme α-D-glucosidase isolated from *Saccharomyces cerevisiae* was studied using the synthetic substrate *p*-nitrophenyl-α-D-glucopyranoside (PNPG), which is hydrolysed to release *p*-nitrophenol (PNP) which is yellow in alkaline solution. A 3 mM solution of PNPG was prepared and portions used to study the effect of substrate concentration on initial rate using a fixed volume of enzyme preparation. The total volume of each assay mixture was 10 cm³. A 1 cm³ sample of the reaction mixture was withdrawn after 1 min, and quenched with 4 cm³ borate buffer pH 9.0 to stop the reaction and develop the yellow colour. The change in absorbance at 400 nm was determined and used as a measure of the initial rate. The following results were obtained:

Volume (PNPG) in cm³	0.1	0.2	0.3	0.4	0.6	0.8	1.2
Initial rate in AU min⁻¹	0.055	0.094	0.130	0.157	0.196	0.230	0.270

What kinetic constants can be obtained from these data?

Answer   Subject to the calculation of the molar concentration of PNPG in each reaction mixture, it is possible to construct Lineweaver–Burk, Hanes and Eadie–Hofstee plots to obtain the values of K_M and v_{max}. The fact that a 1 cm³ sample of the reaction mixture was used to measure the initial rate is not relevant to the calculation of the molar concentration of the substrate, c(PNPG). Lineweaver–Burk, Hanes and Eadie–Hofstee plots derived from these data are shown in Figure 23.4 in which v_0 measurements are expressed simply as the increase in absorption at 400 nm.

In order to calculate numerical values of v_{max} in terms of the molar amount of product formed in unit time, it is necessary to convert absorbance units (AU) to the amount of product (PNP). This requires knowledge of the molar absorption coefficient ε of PNP, which can be derived from a plot of c versus A, where ε represents the slope according to the Beer–Lambert law (Section 13.2.2). Data for such a PNP dilution series are given in Table A.

Table A

c(PNP) in µM	2.0	4.0	6.0	8.0	12.0	16.0	24.0
Absorbance at 400 nm	0.065	0.118	0.17	0.23	0.34	0.45	0.65

A plot of these data confirms that the Beer–Lambert law is obeyed and thus enables the amount of product to be calculated. From this, v_0 values in units of µmol min⁻¹ can be calculated. The data for the three linear plots are presented in Table B.

Table B

c(S) in mM	0.030	0.060	0.090	0.12	0.18	0.24	0.36
v_0 in AU min^{-1}	0.055	0.094	0.13	0.16	0.20	0.23	0.27
v_0 in µmol min^{-1}	0.078	0.15	0.22	0.27	0.34	0.40	0.48
$1/c(S)$ in mM^{-1}	33	17	11	8.3	5.6	4.2	2.8
$1/v_0$ in (µmol min^{-1})$^{-1}$	13	6.7	4.6	3.7	2.9	2.5	2.1
$c(S)/v_0 \times 10^{-3}$ in min dm^{-3}	0.38	0.40	0.41	0.45	0.53	0.59	0.75
$v_0/c(S) \times 10^3$ in dm^3 min^{-1}	2.6	2.5	2.4	2.2	1.9	1.7	1.3

Data derived from the three linear plots are presented in Table C.

Table C

Plot	Regression coefficient	Slope	Intercept	K_M (mM)	v_{max} (µmol min^{-1})
Lineweaver–Burk	0.998	0.352	0.909	0.39	1.1
Hanes	0.988	1.15	0.325	0.28	0.87
Eadie–Hofstee	0.955	−6.11	3.09	0.16	0.50

The agreement between the three plots for the values of K_M and v_{max} was good, but the quality of the fitted regression line for the Lineweaver–Burk plot was noticeably better. However, the distribution of the experimental points along the line is the poorest for this plot (Figure 23.4). The value for v_{max} indicates the amount of product released per minute, but of course this is for the chosen amount of enzyme and is for 10 cm^3 of reaction mixture. For the value of v_{max} to have an absolute value, the amount of enzyme and the volume of reaction mixture have to be taken into account. The volume can be adjusted to 1 dm^3 by multiplication of above value of v_{max} by 100, thus yielding a v_{max} value of 50 µmol min^{-1} dm^{-3} (when using the result from the Eadie–Hofstee analysis). However, it is only possible to correct for the amount of enzyme, if it was pure and the molar amount present was known. α-D-glucosidase has a molecular mass of 68 kDa; so if there was 3 µg of pure enzyme in each 10 cm^3 reaction mixture, its molar concentration would be 4.4×10^{-3} µM. This allows the value of the turnover number k_{cat} to be calculated (see Equation 23.6):

$$k_{cat} = \frac{v_{max}}{c_t(E)} = \frac{50 \ \mu M \ min^{-1}}{4.4 \times 10^{-3} \ \mu M} = 11 \times 10^3 \ min^{-1} = 1.8 \times 10^2 \ s^{-1}$$

k_{cat} is a measure of the number of molecules of substrate (PNPG in this case) converted to product (PNP) per second by the enzyme under the defined experimental conditions. The value of 180 s^{-1} is in the mid-range for the majority of enzymes.

It is also possible to calculate the specificity constant η (Equation 23.7), which is a measure of the efficiency with which the enzyme converts substrate to product at low substrate concentrations ($c(S) \leq K_M$):

$$\eta = \frac{k_{cat}}{K_M} = \frac{1.8 \times 10^2 \, s^{-1}}{0.16 \, mM} = 11 \times 10^2 \, mM^{-1}s^{-1} = 11 \times 10^5 \, M^{-1}s^{-1}$$

Note that the units of the specificity constant are that of a second-order rate constant, effectively for the conversion of the enzyme–substrate complex [ES] to the enzyme–product complex [EP]. Its value in this case is typical of many enzymes and is lower than the limiting value.

when extrapolated (see Figure 23.5). Instead, the data depict a biphasic reaction with an initial burst of product, followed by a slower steady-state reaction.

Indeed, such reactions proceed via an intermediate [ES'] and release two different products, P_1 and P_2:

$$E \; + \; S \; \underset{k_{-1}}{\overset{k_1}{\rightleftharpoons}} \; [ES] \; \overset{k_2}{\longrightarrow} \; [ES'] + P_1 \; \overset{k_3 \, = \, k_{cat}}{\longrightarrow} \; E \; + \; P_2 \qquad \text{(Eq 23.8)}$$

This introduces a further rate constant, since the product formation proceeds in two steps; as the rate-determining step in that mechanism is the one leading to product P_2, its rate constant is k_{cat}. Importantly, since the initial velocity v_0 of Michaelis–Menten kinetics refers to the beginning of the steady-state region, its value needs to be determined from the slope of a linear fit of appropriate data points.

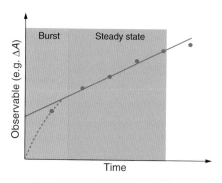

Figure 23.5 Enzymatic reactions involving a covalent enzyme–substrate intermediate often display burst kinetics. Extrapolation of the observable to time $t = 0$ shows 1:1 stoichiometry between the amounts of product produced and enzyme used.

23.2.2 Environmental Parameters

Effect of Temperature on Enzymatic Catalysis

The rate v of a chemical reaction varies with temperature according to the **Arrhenius** equation:

$$v = A \times e^{\frac{-E_a}{R \times T}} \qquad \text{(Eq 23.9)}$$

where A is a constant known as the **pre-exponential factor**, and related to the frequency with which two reactants (enzyme and substrate) collide in the correct orientation to produce the enzyme–substrate complex. E_a is the activation energy (see

Figure 23.1; units: J mol^{-1}), R is the gas constant (8.2 J mol^{-1} K^{-1}), and T is the absolute temperature (units: K).

Equation 23.9 explains the sensitivity of enzyme reactions to temperature: the relationship between reaction rate and absolute temperature is exponential. A plot of the natural logarithm of the initial rate v_0 against the reciprocal of the absolute temperature T allows for determination of the activation energy E_a. The rate of most enzyme reactions approximately doubles for every 10 K rise in temperature.

The so-called **optimum temperature**, at which the enzyme appears to have maximum activity, arises from a combination of environmental parameters, as well as the thermal stability of the enzyme, and is not normally chosen for the study of enzyme activity. Enzyme assays are routinely carried out at 30 or 37 °C. Obviously, the optimum temperature has to lie below the temperature at which the enzyme denatures, i.e. typically below 40–70 °C. However, enzymes from mesophiles (growing at moderate temperature) and thermophiles (growing best at high temperature) seem to possess genuine temperature optima and become reversibly less active above the temperature optimum, but not as a consequence of denaturation.

We discussed earlier (Section 23.1) that enzymes work by facilitating the formation of a transition state, which is a transient intermediate in the formation of the product from the substrate. The transition state of a catalysed reaction has a lower energy barrier (decreased activation energy E_a) than that of the non-catalysed reaction (Figure 23.1). A decrease in the activation energy of as little as 5.7 kJ mol^{-1}, equivalent in energy terms to the strength of a hydrogen bond, will result in a 10-fold increase in reaction rate. Notably, the energy barrier is, of course, lowered equally for both the forward and reverse reactions, so that the position of equilibrium is unchanged. As an extreme example of the effect of an enzyme in this context, the enzyme catalase decomposes hydrogen peroxide 10^{14} times faster than occurs in the non-catalysed reaction!

The intermediates occurring in the transition state cannot be isolated, since they are species in which covalent bonds are being made and broken. For the majority of enzyme-catalysed reactions, the major energy barrier is the formation of one or more intermediates in which covalent bonds are being made and broken, as illustrated in Figure 23.1. In these cases, the formation of intermediates dictates the activation energy for the overall reaction and hence its rate.

However, for a few enzymes, e.g. ATP synthase, the energy-requiring step is not the formation of the intermediates, but the initial binding of the substrate and the subsequent release of the product. The thermodynamic constants ΔG^0, ΔH^0 and ΔS^0 for the binding of substrate to the enzyme can be calculated from knowledge of the binding constant, K_a (= $1/K_d$). Specifically, ΔG^0 can be obtained from the equation:

$$\Delta G^0 = -R \times T \times \ln K_a \qquad \text{(Eq 23.10)}$$

which can be transformed to

$$\ln K_a = -\frac{\Delta H^0}{R \times T} + \frac{\Delta S^0}{R} \qquad \text{(Eq 23.11)}$$

If K_a is measured at two or more temperatures, a plot of $\ln K_a$ versus $1/T$, known as the van't Hoff plot, will give a straight line, slope $-\Delta H^0/R$, with an intercept on the y-axis of $\Delta S^0/R$.

A small number of enzymes appear to operate by a mechanism that does not rely on the formation of a transition state. Studies with the enzyme methylamine dehydrogenase, which promotes the cleavage of a C-H bond, have shown that the reaction is independent of temperature and hence is inconsistent with **transition-state theory**.

The observation is explained in terms of **enzyme-catalysed quantum tunnelling**. Under this mechanism, rather than overcoming the potential energy barrier, the reaction proceeds through the barrier (hence 'tunnelling') at an energy level near that of the ground state of the reactants. Concerted enzyme and substrate vibrations are coupled in such a way as to reduce the width and height of the potential energy barrier and facilitate the cleavage of the C-H bond by the process of quantum mechanical tunnelling. This phenomenon is known to occur with some chemical reactions, but only at low temperatures.

Effect of pH on Enzymatic Catalysis

The state of **ionisation** of amino-acid residues in the catalytic site of an enzyme is pH-dependent (see Section 2.3.2). Many catalytic mechanisms involve the transfer of a proton from or to a protein residue, a nucleophilic attack, as well as stabilisation of the intermediate by a protein residue with a free lone pair or a negative charge. Therefore, the ionisation state of active-site residues, indicated by their pK_a **values**, are of crucial importance to the catalytic mechanism. This is reflected in the variation of the Michaelis–Menten parameters K_M and v_{max} (and thus k_{cat}, Equation 23.6) with pH. If the enzyme mechanism involves a single ionisable amino-acid residue, a plot of k_{cat} against pH has the shape of a titration curve and shows a plateau. However, if two ionisable residues are involved, the plot appears bell-shaped, thus yielding a narrow **pH optimum** (Figure 23.6). By studying the variation of K_M and k_{cat} with pH, it is thus possible to identify the pK_a values of key amino-acid residues involved in the binding and catalytic processes. This experimental approach is of particular importance, since

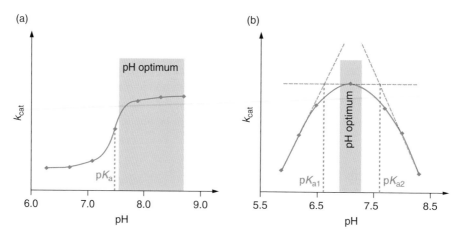

Figure 23.6 The effect of pH on k_{cat} of an enzyme-catalysed reaction involving one (a) or two (b) ionisable groups in the active site of the enzyme. Analysis of these data allows determination of the pK_a values of amino-acid residues involved in the mechanism.

the local environment of an amino-acid residue within the protein can give rise to deviations from the theoretical pK_a.

23.2.3 Inhibition of Monomeric Enzymes

The classification and mechanism of enzyme inhibition is of importance in a number of respects:

- It gives an insight into the mechanisms by which enzymes promote their catalytic activity (Section 23.4)
- It gives an understanding of the possible ways by which metabolic activity may be controlled in vivo (Section 23.5)
- It allows specific inhibitors to be synthesised and used as therapeutic agents to block key metabolic pathways underlying clinical conditions (see Sections 10.3 and 24.8.1).

In this section, we summarise the different inhibition mechanisms and their analysis for an enzyme with a single active site. The presence of multiple active sites, for example in oligomeric enzymes gives rise to particular phenomena, which are described in Section 23.2.4.

Competitive Reversible Inhibition

All **reversible inhibitors** combine non-covalently with the enzyme and can therefore be readily removed by dialysis. **Competitive reversible inhibitors** combine at the same site as the substrate and must therefore be structurally related to the substrate. An example is the inhibition of succinate dehydrogenase by malonate (Figure 23.7).

All types of reversible inhibitors are characterised by their dissociation constant K_i, called the **inhibition constant**. Note that in contrast to K_a and K_d, which have no units, K_i typically carries units of concentration. For competitive inhibition, this relates to the dissociation of the enzyme–inhibitor complex [EI]:

$$E + S \rightleftharpoons [ES] \longrightarrow E + P$$

$$-I \; \updownarrow \; +I$$

$$[EI] \xrightarrow{\;\;/\!\!/\;\;}$$

(Eq 23.12)

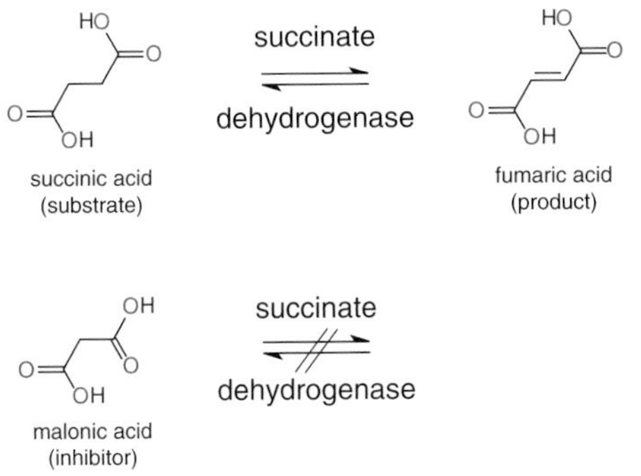

Figure 23.7 Succinate dehydrogenase can convert succinic to fumaric acid. Malonic acid shares structural similarity with the substrate succinic acid and can thus bind in the active site of the enzyme. However, since malonic acid does not possess an aliphatic group that can be oxidised, it acts as an inhibitor for the enzyme.

Since the binding of both substrate and inhibitor involves the same site, the effect of a competitive reversible inhibitor can be overcome by increasing the substrate concentration. The result is that v_{max} is unaltered, but the concentration of substrate required to achieve it is increased, so that when $v_0 = 0.5 \times v_{max}$ then:

$$c(S) = K_M \times \left(1 + \frac{c(I)}{K_i}\right)$$ (Eq 23.13)

where $c(I)$ is the concentration of inhibitor.

It can be seen from Equation 23.13 that K_i is equal to the concentration of inhibitor that apparently doubles the value of K_M. In the presence of a competitive inhibitor, the Lineweaver–Burk equation (Figure 23.4) becomes:

$$\frac{1}{v_0} = \frac{K_M}{v_{max}} \times \left(1 + \frac{c(I)}{K_i}\right) \times \frac{1}{c(S)} + \frac{1}{v_{max}}$$ (Eq 23.14)

Application of this equation allows the diagnosis of competitive inhibition (Figure 23.8). The reaction is carried out for a range of substrate concentrations in the presence of a series of fixed inhibitor concentrations, and a Lineweaver–Burk plot for each inhibitor concentration constructed. The numerical value of K_i can be calculated from Lineweaver–Burk plots for the uninhibited and inhibited reactions. In practice, however, a more accurate value is obtained from a secondary plot where the slope from Equation 23.14: $\dfrac{K_M}{v_{max}} \times \left(1 + \dfrac{c(I)}{K_i}\right)$ is plotted against the inhibitor concentration. Alternatively, the apparent Michaelis constant: $K_M \times \left(1 + \dfrac{c(I)}{K_i}\right)$ calculated from the reciprocal of the negative intercept on the $1/c(S)$ axis, can be plotted against the inhibitor concentration. In both cases, the intercepts on the $c(I)$ axis will be $-K_i$.

Sometimes it is possible for two molecules of inhibitor to bind at the active site. In these cases, although all the primary double reciprocal plots are linear, the secondary plot is parabolic. This is referred to as parabolic competitive inhibition to distinguish it from normal linear competitive inhibition.

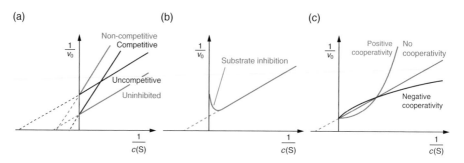

Figure 23.8 Lineweaver–Burk plots showing (a) the effects of three types of reversible inhibitor, (b) substrate inhibition and (c) homotropic cooperativity (see Section 23.2.4).

Non-Competitive Reversible Inhibition

A non–competitive reversible inhibitor binds to a site distinct from that for the substrate. Since the substrate can still bind to the catalytic site, the inhibited enzyme may now comprise either the inhibitor-bound enzyme [EI] or a ternary complex consisting of enzyme, substrate and inhibitor [ESI]. This complex is unable to convert the substrate to product and is referred to as a **dead-end complex**.

$$E + S \rightleftharpoons [ES] \longrightarrow E + P$$

$$-I \Updownarrow +I \qquad -I \Updownarrow +I \qquad \text{(Eq 23.15)}$$

$$[EI] \underset{-S}{\overset{+S}{\rightleftharpoons}} [ESI] \not\longrightarrow$$

It is obvious from Equation 23.15, that there are now two different inhibition constants, $K_{[EI]}$ and $K_{[ESI]}$. In the case of non-competitive inhibition (a special case of mixed inhibition, see below), the two constants are numerically equal and K_i is therefore obtained as per:

$$K_i = K_{[EI]} = K_{[ESI]} \qquad \text{(Eq 23.16)}$$

Since this inhibition involves a site distinct from the catalytic site, the inhibition cannot be overcome by increasing the substrate concentration. The consequence is that v_{max}, but not K_M, is reduced, because the inhibitor does not affect the binding of substrate, but it does reduce the amount of productive enzyme–substrate complex [ES] that can proceed to the formation of product. In this case, the Lineweaver–Burk equation becomes:

$$\frac{1}{v_0} = \frac{K_M}{v_{max}} \times \frac{1}{c(S)} + \frac{1}{v_{max}} \times \left(1 + \frac{c(I)}{K_i}\right) \qquad \text{(Eq 23.17)}$$

Once non-competitive inhibition has been diagnosed (Figure 23.8), the K_i value is best obtained from a secondary plot of either the slope of the primary plot (K_M/v_{max}) or of the $1/v_0$ intercept: $\dfrac{1}{v_{max}} \times \left(1 + \dfrac{c(I)}{K_i}\right)$ against inhibitor concentration. Both secondary plots will have an intercept of $-K_i$ on the inhibitor concentration axis (Figure 23.9).

Uncompetitive Reversible Inhibition
An **uncompetitive reversible inhibitor** can bind only to the enzyme–substrate complex [ES] and not to the free enzyme, so that inhibitor binding must be located either at a site created by a conformational change induced by the binding of the substrate to the catalytic site or directly to the substrate molecule. The resulting ternary complex [ESI] is also a dead-end complex.

$$E + S \rightleftharpoons [ES] \longrightarrow E + P$$

$$-I \Updownarrow +I \qquad \text{(Eq 23.18)}$$

$$[ESI] \not\longrightarrow$$

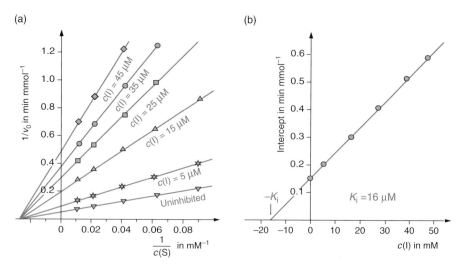

Figure 23.9 (a) Primary Lineweaver–Burk plots showing the effect of a simple linear non-competitive inhibitor at a series of concentrations and (b) the corresponding secondary plot that enables the inhibitor constant K_i to be calculated.

As with non-competitive inhibition, the effect cannot be overcome by increasing the substrate concentration, but in this case both K_M and v_{max} are reduced by a factor of: $\dfrac{1}{v_{max}} \times \left(1 + \dfrac{c(I)}{K_i}\right)$. An inhibitor concentration equal to K_i will therefore halve the values of both K_M and v_{max}. The Lineweaver–Burk equation here becomes:

$$\frac{1}{v_0} = \left(\frac{K_M}{v_{max}} \times \frac{1}{c(S)} + \frac{1}{v_{max}}\right) \times \left(\frac{1 + c(I)}{K_i}\right) \tag{Eq 23.19}$$

The value of K_i is best obtained from a secondary plot of either the slope: $\dfrac{K_M}{v_{max}} \times \left(1 + \dfrac{c(I)}{K_i}\right)$ or the y-intercept: $\dfrac{1}{v_{max}} \times \left(1 + \dfrac{c(I)}{K_i}\right)$ of the primary plot against inhibitor concentration. Both secondary plots will have an intercept of $-K_i$ on the x-axis.

Mixed Reversible Inhibition

With some inhibitors, it is observed that their mechanism of enzyme inhibition does not fit any of the above inhibition mechanisms. Rather, it is possible that either the complex [ESI] has some catalytic activity or the two different inhibition constants $K_{[EI]}$ and $K_{[ESI]}$ are numerically not equal.

$$\text{E + S} \rightleftharpoons \text{[ES]} \longrightarrow \text{E + P}$$

$$-\text{I} \big\updownarrow +\text{I} \qquad\qquad -\text{I} \big\updownarrow +\text{I} \tag{Eq 23.20}$$

$$\text{[EI]} \underset{-S}{\overset{+S}{\rightleftharpoons}} \text{[ESI]} \longrightarrow \text{[EI] + P}$$

These cases are called mixed inhibition, and are characterised by linear Lineweaver–Burk plots that do not fit any of the patterns shown in Figure 23.8a. The plots for the uninhibited and inhibited reactions may intersect either above or below the $1/c(S)$ axis. The associated K_i can be obtained from a secondary plot of the slope, either of the primary plot or from the slope of a secondary plot graphing $1/v_{max}$ against inhibitor concentration. In both cases, the intercept on the inhibitor concentration axis is $-K_i$. Non-competitive inhibition may be regarded as a special case of mixed inhibition.

Substrate Inhibition

For a number of enzymes and their substrates, it is observed that the substrate acts as an uncompetitive inhibitor and forms a dead-end complex. Thus, the substrate is binding to a second, non-catalytic site on the enzyme whereby it modulates the activity, hence the term substrate inhibition. This phenomenon manifests itself substantially at high substrate concentrations where the enzymatic rate decreases with increasing substrate concentrations. The graphical diagnosis of this situation is shown in Figure 23.8b.

Product Inhibition

Similar to substrate inhibition, the end-product of an enzymatic reaction might modulate the enzyme activity in a process called product inhibition. Typically, the first enzymes in an unbranched metabolic pathway are regulated by such a mechanism. Here, the final product of the pathway acts as an inhibitor of the first enzyme in the pathway, thus switching off the whole pathway when the final product begins to accumulate. The inhibition of aspartate carbamoyltransferase (EC 2.1.3.2) by cytosine triphosphate (CTP) in the CTP biosynthetic pathway is an example of this form of regulation. In branched pathways, product inhibition usually operates on the first enzyme after the branch point.

Irreversible Inhibition

Irreversible inhibitors, such as the organophosphorus and organomercury compounds, cyanide, carbon monoxide and hydrogen sulfide, combine with an enzyme to form a covalent bond. The extent of their inhibition of the enzyme is dependent upon the rate constant (and hence time) for covalent bond formation and upon the amount of inhibitor present. The effect of irreversible inhibitors, which cannot be removed by simple physical techniques such as dialysis, is to reduce the amount of enzyme available for reaction. The inhibition involves reactions with a functional group, such as a hydroxyl or sulfydryl group, or with a metal atom in the active site or a distinct allosteric site (allostery is explained in Section 23.2.4). The irreversible inhibitor diisopropylfluorophosphate, for example, reacts with a serine group in the active site of esterases such as acetylcholinesterase (EC 3.1.1.7). Other examples include the organomercury compound p-hydroxymercuribenzoate which reacts with cysteine groups in cysteine proteases such as papain (EC 3.4.22.2) and acetylcholinesterase.

Not all irreversible inhibitors form covalent adducts with their enzyme targets. Some reversible inhibitors bind so tightly to their target enzyme that they are essentially irreversible. These tight-binding inhibitors possess so-called slow-binding kinetics, which appears similar to covalent irreversible inhibitors. Several clinically important drugs fall into this category, including acyclovir, a guanosine analogue used for the treatment of herpes simplex virus infections, chickenpox and shingles. For the so-called suicide inactivation of herpes simplex virus DNA polymerase by acyclovir, apparent inhibition constants of 4–6 nM are observed.

23.2.4 Oligomeric Enzymes

Allostery and Cooperativity

The discussion of the mechanism of enzyme action so far has been based on the assumption that there exists one particular active site on the protein molecule. However, some enzymes, for example β-carbonic anhydrases (EC 4.2.1.1), are homo-oligomers and therefore possess multiple identical active sites. In such cases, successive substrate molecules can bind to the enzyme with either progressively greater ease (affinity) or with reduced ease. If this is the case, these enzymes are said to be under allosteric control (in this context, see also Section 14.5.3).

Not all homo-oligomeric enzymes possess allostery; they might well display simple Michaelis–Menten kinetics in which case all substrate molecules bind with the same affinity, as is the case with the tetramer lactate dehydrogenase (Section 23.1.2). However, if the Michaelis–Menten plot takes a sigmoid shape (Figure 23.10), then this is indicative of an allosteric enzyme.

Examples of such enzymes are aspartate carbamoylase and phosphofructokinase. Progressive binding of the substrate molecules to the subunits of an allosteric enzyme may result in either increased (positive cooperativity) or decreased (negative cooperativity) activity towards the binding of further substrate molecules. In such cases, the substrate molecules are said to display a homotropic effect. Changes in catalytic activity towards the substrate may also be brought about by the binding of molecules other than the substrate at distinct allosteric binding sites on one or more subunit (cf. non-competitive, uncompetitive and mixed reversible inhibition!).

Compounds that induce such changes are referred to as heterotropic effectors. They are commonly key metabolic intermediates such as ATP, ADP, AMP and inorganic phosphate. Heterotropic activators increase the catalytic activity of the enzyme, making the curve less sigmoid and moving it to the left, whilst heterotropic inhibitors cause a decrease in activity, making the curve more sigmoid and moving it to the right (Figure 23.10). The diagnosis of cooperativity by use of the Lineweaver–Burk plot is

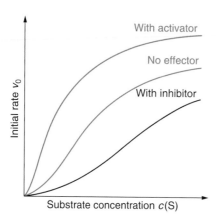

Figure 23.10 Effect of activators and inhibitors on the sigmoid kinetics of an enzyme subject to allosteric control.

shown in Figure 23.8c. The operation of cooperative effects may be confirmed by a Hill plot, which is based on the equation:

$$\log \frac{v_0}{v_{max} - v_0} = h \times \log c(S) + \log K \qquad \text{(Eq 23.21)}$$

where h is the Hill coefficient, and K is an overall binding constant related to the individual binding constants for n sites. The Hill coefficient, which is equal to the slope of the Hill plot, is a measure of the cooperativity between the sites such that:

- if $h = 1$, binding is non-cooperative and normal Michaelis–Menten kinetics exist
- if $h > 1$, binding is positively cooperative
- if $h < 1$, binding is negatively cooperative.

At very low substrate concentrations that are insufficient to fill more than one site, as well as at high substrate concentrations at which most of the binding sites are occupied, the slopes of Hill plots tend to a value of 1. The Hill coefficient is therefore taken from the linear central portion of such a plot. One of the problems with Hill plots is the difficulty in estimating v_{max} accurately.

The Michaelis constant K_M is not used with allosteric enzymes. Instead, the term $S_{0.5}$, which is the substrate concentration required to produce 50% saturation of the enzyme, is used. From a practical perspective, it is important to carefully appraise the sample under investigation. If the sample contains more than one enzymatic species capable of acting on the substrate, the kinetic data will also appear sigmoid. In such cases, the presence of more than one enzyme needs to be established, for example by checking for a discrepancy between the amount of substrate consumed and the expected amount of product produced.

Non-Enzymatic Proteins

Importantly, allostery is not confined to enzymes; the phenomenon might be observed with any protein that possesses multiple binding sites and also includes monomeric proteins, for example haemoglobin (tetramer) and many cell membrane receptors especially those of the G-protein coupled receptors (GPCRs; e.g. human dopamine D_3 receptor may exist as monomer, dimer and tetramer) type (see Section 23.6).

Molecular Mechanisms of Allostery

Historically, two models have been proposed to interpret allosteric regulation of oligomeric proteins. Both are based on the assumption that the allosteric enzyme consists of a number of subunits (called protomers), each of which can bind substrate and exist in two conformations referred to as the R (relaxed) and T (tense) states. In these models, the substrate binds more tightly to the R form.

The first model was introduced by Monod, Wyman and Changeux, and is referred to as the symmetry model. It assumes that conformational changes between the R and T states of the individual protomers is highly coupled, so that all subunits must exist in the same conformation. Thus, binding of substrate to a T-state protomer, causing it to change conformation to the R state, will automatically switch the other protomers to the R form, thereby enhancing reactivity (Figure 23.11a).

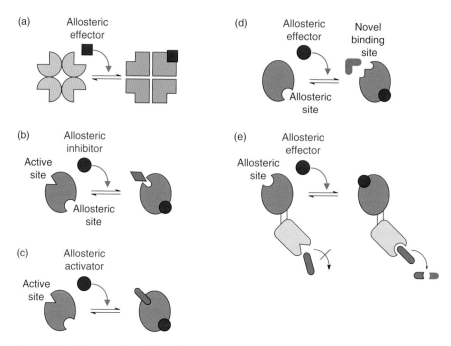

Figure 23.11 Different modes of allosteric behaviour. (a) A representation of the Monod–Wyman–Changeux model of allosteric transitions. A symmetric, multimeric protein can exist in one of two distinct conformational states – the active and inactive conformations. Each subunit has a binding site for an allosteric effector, as well as an active site or binding site. Upon binding of the first ligand to one monomer, the other monomers change their states to readily accept a ligand. (b) A monomeric, allosterically inhibited protein. The binding of an allosteric inhibitor alters the active site or binding site geometry in an unfavourable way, thereby decreasing affinity or catalytic efficiency. (c) A monomeric, allosterically activated protein. The binding of an allosteric activator results in an increased affinity or activity in the second site. (d) The binding of an allosteric effector might introduce a new binding site to a protein. Binding of a ligand to this new binding site could lead to changes in active-site geometry, providing an indirect mechanism of allosteric control. This type of effect is of great interest in the design of allosteric drugs and can be considered as a subset of the example in (c). (e) The fusion of an enzyme to a protein under allosteric control. This type of construct can act as an allosteric switch because the activity of the enzyme is indirectly under allosteric control via the bound protein with an allosteric site. Such constructs are both present in nature and the target of protein engineering studies.

The second model of Daniel Koshland, known as the induced-fit or sequential model, does not assume the tightly coupled concept and hence allows protomers to exist in different conformations, but in such a way that binding to one protomer modifies the reactivity of others. Importantly, induced fit is a substrate specificity effect not a catalytic mode.

However, we have mentioned before that allostery is not only observed with oligomeric, but also with monomeric proteins (see Figure 23.11b–e). It is generally believed that allostery is a consequence of the flexibility of proteins whereby the continuous folding and unfolding in localised regions of the protein gives rise to a population of conformations that interconvert on various timescales. The interconverting conformations have similar energies and their mixture constitutes the native state of the protein. However, the different conformations differ in their affinity for certain ligands and, in the case of enzymes, in their catalytic activity. The binding of an allosteric effector at its distinct site results in the redistribution of the conformational ensembles as a result of the alteration of their rates of interconversion, and, as a consequence, the conformation and activity of the active site is modified. Integral to this redistribution of protein conformations is the existence of amino–acid networks that facilitate communication between the different sites and that are linked to the mechanism of catalysis.

NMR (Section 14.3.2) and isothermal titration calorimetry (Sections 14.2, 23.3.4) studies have shown that the timescale of linked amino-acid networks is milliseconds to microseconds and the timescale of binding site change is microseconds to nanoseconds. These values compare with the very fast rate of atomic fluctuations (nanoseconds to picoseconds).

23.3 ANALYTICAL METHODS TO INVESTIGATE ENZYME KINETICS

Enzyme assays are undertaken for a variety of reasons, but the most common are:

- To determine the amount (or concentration) of enzyme present in a particular preparation, which is particularly important in diagnostic enzymology (Section 10.3)
- To gain an insight into the kinetic characteristics and enzyme mechanisms by determining a range of kinetic constants such as K_M, v_{max} and k_{cat}
- To study the effect of environmental parameters and inhibitors on the enzyme in the elucidation of enzyme mechanisms and determine involvement in a metabolic or signalling pathway. The study of enzyme inhibition is fundamental to the development of new drugs.

Analytical methods for enzyme assays may be classified as either continuous (kinetic) or discontinuous (fixed-time). Continuous methods monitor some property change (e.g. absorbance or fluorescence) in the reaction mixture, whereas discontinuous methods require samples to be withdrawn from the reaction mixture and analysed by some convenient technique. The inherent greater accuracy of continuous methods commends them whenever they are available.

23.3.1 Fixed-Time Assays

For simplicity, initial rates are sometimes determined experimentally on the basis of a single measurement of the amount of substrate consumed or product produced in a given time rather than by the tangent method. This approach is valid only over the short period of time when the reaction is proceeding effectively at a constant rate (steady state). This linear rate section comprises at most the first 10% of the total possible change and the error is smaller the earlier the rate is measured. In such cases, the rate is proportional either to the reciprocal of the time to produce a fixed change (fixed–change assays) or to the amount of substrate reacted in a given time (fixed-time assays).

Figure 23.12 illustrates the caution required when conducting fixed-time assays, which represents the effect of enzyme concentration on the progress of the reaction in

(a)

(b)

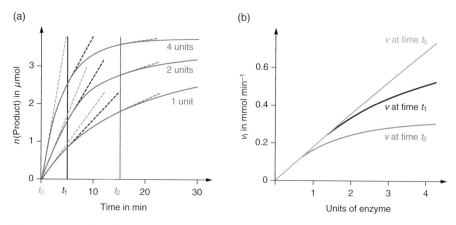

Figure 23.12 Initial and fixed-time rate determination in enzyme assays. (a) Time-dependent variation in the concentration of products in the presence of 1, 2 and 4 units of enzyme; (b) variation of reaction rate with enzyme concentration using the true initial rate v_0 (time t_0) and two fixed-time assays (times t_1 and t_2).

the presence of a constant initial substrate concentration (Figure 23.12a). Measurement of the rate of the reaction at time t_0 (by the tangent method) to give the true initial rate or at two fixed times, t_1 and t_2, gives the relationship between initial rate and enzyme concentration shown in Figure 23.12b. Clearly, only the initial rate method results in a linear relationship. Since the correct determination of initial rate means that the observed changes in the concentration of substrate or product are relatively small, it is inherently more accurate to measure the increase in product concentration because the relative increase in its concentration is significantly larger than the corresponding decrease in substrate concentration.

23.3.2 Steady State

The importance of choosing the right time at which the enzymatic rate is determined has been highlighted in the previous section. Experimental methods used to observe steady-state kinetics include UV/vis spectrometry, fluorimetry, luminescence and ITC, as well as immunochemical methods.

Many substrates and products absorb light in the visible or ultraviolet region and the change in the absorbance during the reaction can be used as the basis for the enzyme assay (see Table 13.1). It is essential that the substrate and product do not absorb at the same wavelength and that the Beer–Lambert law (Section 13.2.2) is obeyed for the chosen analyte. A large number of common enzyme assays are based on the interconversion of $NADP^+$ and NADPH, either directly or via a coupled reaction. Both of these nucleotides absorb light at a wavelength of 260 nm, but only the reduced form absorbs at 340 nm. The scope of visible spectrophotometric enzyme assays can be extended by the use of synthetic substrates that release a coloured product. Many such artificial substrates are available commercially, particularly for the assay of hydrolytic enzymes. Commonly used products are phenolphthalein and p-nitrophenolate, both of which are coloured in alkaline solution.

The particular advantage of fluorimetric enzyme assays is that they are highly sensitive and can therefore detect and measure compounds at low concentrations. Since NAD(P)H is fluorescent, it can not only be assayed by the absorption at 340 nm, but also by emission at 378 nm (excitation wavelength 340 nm). Synthetic substrates that release a fluorescent product are also available for the assay of some enzymes. The large commercial interest in the development of inhibitors of kinases, phosphatases and proteases for their therapeutic potential has stimulated the development of assays for these enzymes in vivo using the principle of fluorescence resonance energy transfer (FRET; see Sections 13.5.3 and 14.4.2). The most commonly used fluorophores in this respect are cyan, red and yellow fluorescent protein.

The bioluminescence reaction of luciferase, which converts luciferin and O_2 to oxyluciferin is a commonly used assay for ATP since during this reaction one ATP molecule is consumed and a photon is released. As discussed in Section 13.5.3, bioluminescence assays are extremely sensitive and therefore frequently used for enzyme assays. Moreover, any enzymatic reaction that utilises ATP can be assayed with this approach by means of **coupled reactions**.

It is also possible to monitor enzyme activity based on highly specific immunochemical assays (ELISA). Since monoclonal antibodies raised against a particular enzyme can distinguish between isoenzyme forms, such assays are attractive for diagnostic purposes. An important clinical example is creatine kinase (EC 2.7.3.2) (Section 10.3.2), a dimer based on two different subunits, M (muscle type) and B (brain type). Therefore, three different isoenzymes exist: CK-MM, CK-BB and CK-MB. CK-MB is important in the diagnosis of myocardial infarction (heart attack) and an immunological assay is important in its assay.

23.3.3 Pre-Steady State

The experimental techniques discussed in the previous section are not suitable for the study of the progress of enzyme reactions in the short period of time (milliseconds) before steady-state conditions, with respect to the formation of enzyme–substrate complex, are established. Figure 23.2 shows the progress curves for this pre-steady state initial stage of an enzymatic reaction. Typically, methods to monitor and analyse pre-steady-state kinetics are also required with rapidly proceeding enzymatic reactions such as carbonic anhydrases. Two main methods are available for the study of fast kinetics: rapid-mixing methods and relaxation methods.

The **rapid-mixing methods** are carried out either by means of stopped flow or continuous flow; both approaches are described in Section 13.8. These rapid mixing methods are limited only by the time required to mix the two reactants. Absorption (UV/vis absorption, CD spectroscopy) or emission (fluorescence spectroscopy, light scattering) spectrometers can be coupled to the mixing apparatus and therefore a variety of different assay types can be implemented with this technique. The stopped-flow method can be varied and used as a **quenching method**, whereby a quenching agent (e.g. trichloroacetic acid) is injected into the reaction mixture. This stops the primary reaction and allows for analysis of intermediates that might have built up.

Example 23.3   **ENZYME ASSAYS BY COUPLED REACTIONS**

In many cases, enzymes that cannot be assayed directly may be monitored by a coupled reaction that involves two enzyme reactions linked by means of common intermediates. The assay of 6-phosphofructokinase (PFK; EC 2:7:1:11) coupled to fructose-bisphosphatase aldolase (FBPA; EC 4.1.2.13) and glyceraldehyde-3-phosphate dehydrogenase (G3PDH; EC 1:2:1:12) illustrates the principle:

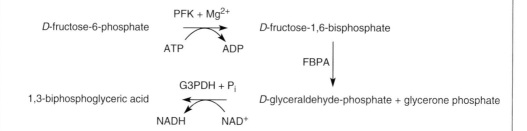

The assay mixture would contain D-fructose-6-phosphate, ATP, Mg^{2+}, FBPA, G3PDH, NAD^+ and phosphate (P_i), all in excess, so that the reaction would proceed to completion and the rate of reduction of NAD^+ and the production of NADH – and hence the increase in absorbance at 340 nm – would be determined solely by the activity of PFK. In principle, there is no limit to the number of reactions that can be coupled in this way, provided that the enzyme under investigation is always present in limiting amounts.

The number of units of enzyme in the test preparation can be calculated by applying the Beer–Lambert law (Section 13.2.2) to calculate the molar amount of product formed per second (= katal):

$$\text{enzyme units} = \frac{\left(\dfrac{\Delta A_{340}}{\Delta t}\right)}{\varepsilon_{340}} \times \frac{V_{mix}}{V_{test}}$$

where $\Delta A_{340}/\Delta t$ is the control-corrected change in the absorbance at 340 nm per second, V_{mix} is the total volume of reaction mixture (in cm^3) in the cuvette of 1 cm light path, V_{test} is the volume of test solution (in mm^3) added to the reaction mixture, and ε_{340} is the molar extinction coefficient for NADH at 340 nm (6.3×10^3 M^{-1} cm^{-1}). By dividing the above equation by the total concentration of enzyme under investigation, the specific activity (katal kg^{-1}) of the preparation can be calculated.

A limitation of the rapid-mixing methods is the **dead time** during which the enzyme and substrate are mixed and observation of the reaction is not possible due to instrumental constraints. In the relaxation methods, a reaction is allowed to establish its equilibrium and then the position of the equilibrium is altered by a change in environmental conditions. Most commonly, this is achieved by the **temperature jump** (T-jump) technique in which the reaction temperature is raised rapidly by 5–10 K by means of the discharge of a capacitor or by an infrared laser. The rate at which the reaction mixture adjusts to its new equilibrium (**relaxation time** t, generally a few microseconds) is inversely related to the rate constants involved in the reaction. This return to equilibrium is monitored by one or more suitable spectrophotometric methods. From the recorded data, the number of intermediates can be deduced and the various rate constants calculated from the relaxation times.

These pre-steady-state techniques have shown that the enzyme and its substrate(s) associate very rapidly, with rate constants for the formation of the enzyme–substrate complex [ES] in the range 10^6 to 10^8 M^{-1} s^{-1} (note: [ES] formation follows second-order kinetics) and for the dissociation of [ES] in the range 10 to 10^4 s^{-1} (dissociation of [ES] follows first-order kinetics). The upper limit of these values is such that for some enzymes virtually every interaction between an enzyme and its substrate leads to the formation of a complex. The rapid-mixing methods have also been used to study other biochemical processes that are kinetically fast and may involve transient intermediates, in particular protein folding, protein conformational changes, receptor–ligand binding and the study of second-messenger pathways.

23.3.4 Isothermal Titration Calorimetry

The method of isothermal titration calorimetry (ITC) was introduced in Section 14.2. With ITC detecting heat changes during physico-chemical processes occurring in the sample cell, it can be used to directly monitor enzymatic reactions based on the enthalpy change, which is a characteristic feature of any reaction. In this methodology, the heat generated as the enzymatic reaction proceeds is the directly observable quantity.

The observed heat change during the reaction is correlated with the apparent enthalpy ΔH_{app} as per:

$$Q = c(P) \times V \times \Delta H_{app} \qquad \text{(Eq 23.22)}$$

where $c(P)$ is the molar concentration of product, V is the volume of the ITC sample cell and ΔH_{app} the apparent enthalpy, which comprises the enthalpy of reaction as well as any enthalpy changes due to additional processes happening during the reaction (e.g. buffer ionisation). Since the compensation power $P(t)$ of the calorimeter – which is the directly observed quantity in ITC experiments – equals the accumulated heat, it follows that

$$P(t) = \frac{dQ}{dt} = \frac{dc(P)}{dt} \times V \times \Delta H_{app} \qquad \text{(Eq 23.23)}$$

which allows calculation of the rate of the enzymatic reaction, $dc(P)/dt$, from the determined enthalpy change ΔH_{app} and the compensation power $P(t)$. Experimentally, the enzyme kinetics can be assessed either by titrating substrate into enzyme solution or vice versa. The two different approaches are described below. Both require that dilution experiments are performed, whereby identical solutions of the titrant are injected into buffer lacking the other component. These data represent the 'baseline' and are subtracted from the data obtained in the presence of all components.

The Enzyme is Titrated into the Substrate Solution

A complicating issue arises from the fact that the enzymatic reaction in the sample cell starts during injection of the enzyme into the sample solution containing the substrate (Figure 23.13). However, using a slightly more expanded mathematical treatment of the ITC data that includes a function correcting for the change in enzyme concentration, an improved model of the ITC data can be established that allows approximation of enzyme kinetics data from ITC using Michaelis–Menten theory.

If the enzyme solution is titrated into the sample solution immediately at the beginning of the experiment, as illustrated in Figure 23.13, the resulting data can be fitted computationally and the apparent reaction enthalpy ΔH_{app}, k_{cat} and K_M can be obtained from the fit parameters.

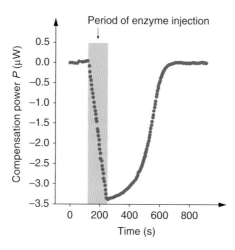

Figure 23.13 Scheme of an isothermal titration calorimetric observation of an enzymatic reaction where the enzyme is titrated into the substrate solution at once (grey shaded area). Before injection, the baseline stability is checked (~2 min). Then, enzyme solution is injected. Negative compensation power is required to adjust the sample cell for the exothermic process, which comprises the enzymatic reaction as well as dilution. The maximum rate is observed at ~250 s when all the enzyme had been added. The compensation power then decreases and returns to baseline level when all substrate has been processed.

The Substrate is Titrated into the Enzyme Solution

In this case, one needs to carry out two types of experiments when assessing enzyme kinetics with ITC:

- In the first experiment, the total molar enthalpy ΔH_{app} is determined with a relatively large concentration of enzyme in the cell, and low amounts of substrate in the injection syringe. The gaps between the individual injections should be sufficiently long to allow for all substrate to be converted to product.
- A second experiment is then carried out where rather large amounts of substrate are injected into the sample cell that now contains enzyme at a rather low concentration. The gaps between individual injections should now be short, with the aim to only have a small part of the substrate being converted.

With ΔH_{app} determined in the first experiment, the data from the second experiment allows plotting reaction rates against substrate concentration from which the Michaelis–Menten parameters k_{cat}, K_M and v_{max} can then be obtained.

23.3.5 Substrate Assays

The amount of a compound present in a sample can be determined by means of a substrate assay, if that compound is a substrate for a particular enzyme or a reaction of that compound can be coupled to an enzymatic reaction (see examples in Table 23.3). Such substrate assays are of great importance for many clinical biochemistry applications (Section 10.3), and also form the basis of biosensors (Sections 10.2.2, 14.6), which afford a simple means for the fast measurement of these substrates.

In general, substrate assays use an excess of enzyme, making sure that every substrate molecule can rapidly be converted to product. *In praxi*, the substrate concentration should thus be less than K_M so that the observed initial rate is directly related to the substrate concentration. If the reaction is freely reversible, then it is necessary to change the experimental conditions (e.g. pH) or to chemically trap the product, so that an individual substrate molecule only undergoes conversion to product once. The sensitivity of the initial rate methodology depends upon the value of the molar extinction coefficient of the analyte being assayed and also on K_M for the substrate. Since this puts practical constraints on the analytes that can be monitored, two approaches have been developed to overcome this problem:

- The end-point technique avoids the measurement of initial rate by converting all the substrate to product and then computing the amount present by correlating it with the total change in a parameter such as absorbance or fluorescence.
- In the method of enzymatic cycling, the substrate is regenerated by means of a coupled reaction and the total change in the observed quantity is measured in a given

Table 23.3 **Examples of substrate assays for cholesterol and serum glucose**

Compound	Enzyme	Reaction	Type of assay	Analyte
cholesterol	Cholesterol oxidase Peroxidase	Reaction produces H_2O_2 H_2O_2 is used for enzymatic oxidation of dye	Coupled	Oxidised dye
β-D-glucose	Glucose dehydrogenase	Reduction of NAD to NADH, oxidation of glucose	Direct	NADH
D-glucose	Glucose oxidase Peroxidase	Reaction produces H_2O_2 H_2O_2 is used for enzymatic oxidation of dye	Coupled	Oxidised dye
D-glucose	Hexokinase Glucose-6-phosphate dehydrogenase	Synthesis of glucose-6-phosphate Reduction of NAD to NADH, oxidation of glucose-6-phosphate	Coupled	NADH

time. Pre-calibration using a range of substrate concentrations with all the other reactants in excess allows the substrate concentration in a test solution to be computed. Due to the amplification of the signal, this method has a 10^4- to 10^5-fold increase in sensitivity relative to the end-point technique.

23.4 MOLECULAR MECHANISMS OF ENZYMES

23.4.1 Active Sites

As pointed out, enzymes are characterised by their high specificity, catalytic activity and capacity for regulation. These properties are imprinted into the specific three-dimensional interactions between the enzyme and its substrate and ligands. A complete understanding of the way enzymes work must therefore include the elucidation of the mechanism underlying the binding of a substrate(s) to the enzyme catalytic site and the subsequent conversion of the substrate(s) to product(s). The mechanism must include details of the nature of the binding and catalytic sites, the nature of the enzyme–substrate complex [ES], and the associated electronic and stereochemical events that result in the formation of the product. A wide range of strategies and analytical techniques has been adopted to gain such an understanding.

X-Ray Crystallographic Studies

The technique is described in Section 14.3.4 and enables determination of the three-dimensional atomic positions of enzymes in their free and ligand-bound forms. From the three-dimensional structures, it is possible to deduce how the substrate binds and undergoes a reaction. The Protein Data Bank (www.rcsb.org/pdb) as a repository of three-dimensional structures of proteins and the Catalytic Site Atlas (see Section 23.8.4) as a database of the active sites and catalytic residues of enzymes are valuable resources that collect results from such studies and often allow inference of mechanisms for enzymes whose three-dimensional structures have not yet been determined or for computational screening of small-molecule libraries in the drug-discovery process (Section 24.5).

Site-Directed Mutagenesis Studies

The investigation of mutant enzymes where a particular amino-acid residue, thought to be involved in substrate binding and catalysis, has been replaced by another amino acid allows for a verification of hypothesised molecular mechanisms. One or more mutations are introduced by means of molecular biology and cloning approaches (Chapter 4) and the mutant enzyme is subjected to the same characterisation as the wild-type enzyme. By studying the impact of the replacement of an ionisable or nucleophilic amino acid with a non-reactive one on the catalytic properties of the enzyme, conclusions can be drawn about the role of the amino-acid residue that has been replaced.

Such studies assume that the impact of the single amino-acid replacement is confined to the active site and has not affected other aspects of the enzyme.

Irreversible Inhibitor and Affinity Label Studies

Irreversible inhibitors act by forming a covalent bond with the enzyme. By locating the site of the binding of the inhibitor, the identity of specific amino acids in the catalytic or binding site can be deduced. A similar approach uses photoaffinity labels that structurally resemble the substrate, but contain a functional group, such as an azo group (-N=N-), which, on exposure to light, is converted to a reactive functional group that forms a covalent bond with a neighbouring functional group in the active site. It is common practice to tag the inhibitor or photoaffinity label with a radioisotope (Chapter 9) so that its location in the enzyme protein can easily be established experimentally.

Kinetic Studies

This approach is based on the use of a range of substrates and/or competitive inhibitors and the determination of the associated K_M, k_{cat} and K_i values. These allow correlations to be drawn between molecular structure and kinetic constants, and hence deductions about the structure of the active site. Further information about the structure of the active site can be gained by studying the influence of pH on the kinetic constants. Specifically, the effect of pH on K_M (i.e. on the enzyme–substrate binding) and on v_{max} or k_{cat} (i.e. conversion of [ES] to products) is studied. By plotting the variation of either K_M, v_{max} or k_{cat} with pH, indications of the pK_a values of ionisable groups involved in the active site can be obtained (Figure 23.6). These are then compared with the pK_a values of the ionisable groups known from other proteins. For example, pH sensitivity around the range 6–8 could reflect the importance of one or more imidazole side chains of a histidine residue in the active site (see also Section 23.2.1).

Isotope Exchange Studies

The replacement of the natural isotope of an atom in the substrate by a different isotope of the same element and the study of the impact of the isotope replacement on the observed rate of enzymatic reaction and its associated stereoselectivity, often enables deductions to be made about the mechanism of the reaction. For example, alcohol dehydrogenase oxidises ethanol to ethanal using NAD+:

$$\text{(Eq 23.24)}$$

The two hydrogen atoms on the methylene (CH_2) group of ethanol are chemically indistinguishable, but if one is replaced by a deuterium (D) or tritium isotope, the methylene carbon atom becomes a chiral centre (stereogenic centre) and the resulting molecule can be identified as adopting either the R or the S configuration according to the Cahn–Ingold–Prelog rules for defining the stereochemistry of asymmetric centres. Studies have shown that alcohol dehydrogenase exclusively removes the hydrogen atom in the pro-R configuration, i.e. (R)-CH_3CHDOH loses the D isotope in its conversion to ethanal, but (S)-CH_3CHDOH retains it. Such a finding can only be interpreted in terms of the specific orientation of the ethanol molecule in the binding site such that the two hydrogen atoms are effectively not equivalent.

23.4.2 Catalytic Mechanisms

The application of the various strategies outlined above to a wide range of enzymes has enabled mechanisms to be deduced for many of them. The most commonly occurring amino-acid residues in enzyme catalytic sites are the side-chain residues of histidine (imidazole), arginine (guanidinium group), glutamate and aspartate (carboxylate group), lysine (amino group), serine, threonine and tyrosine (hydroxyl group), and cysteine (thiol group). These specific amino-acid side chains commonly occur in pairs or triplets within catalytic sites.

The most commonly observed residue side chain involved in catalytic reactions is the imidazole group of histidine. In lipases and peptidases, for example, the catalytic triad comprising serine, aspartate and histidine occurs very frequently. The catalytic triad of the serine protease chymotrypsin is formed by His-57, Asp-102 and Ser-195. In such catalytic triads, the basic nitrogen of the histidine imidazole group is used to abstract a proton from either serine (as in the case of chymotrypsin), threonine or cysteine, which activates the latter amino-acid side chains as a nucleophile. In other catalytic mechanisms, histidine may act as a proton shuttle or coordinating residue for a catalytic metal ion.

The roles of the catalytic residues include:

- Activation of the substrate by forming a hydrogen bond, thereby lowering the energy barrier (E_a) for the reaction. Such a hydrogen bond is referred to as a low-barrier hydrogen bond
- Provision of a nucleophile to attack the substrate
- Acid–base catalysis of the substrate similar to the mechanisms well known in conventional organic chemistry
- Stabilisation of the transition state of the reaction.

23.5 REGULATION OF ENZYME ACTIVITY

23.5.1 Molecular Mechanisms of Control

The activity of an enzyme can be regulated in two basic ways:
1. by alteration of the kinetic conditions under which the enzyme is operating
2. by alteration of the amount of the active form of the enzyme present by promoting enzyme synthesis, enzyme degradation or the chemical modification of the enzyme.

The latter option is a long-term control, exerted in hours, and operates at the level of enzyme synthesis and degradation. Whereas many enzymes are synthesised at a virtually constant rate and are said to be constitutive enzymes, the synthesis of others is variable and is subject to the operation of control mechanisms at the level of gene transcription and translation. The metabolic degradation of enzymes is a first-order process characterised by a half-life. The half-life of enzymes varies from a few hours to many days. Importantly, enzymes that exert control over pathways have relatively short half-lives. Metabolic enzyme degradation operates via poly-ubiquitinylation, which comprises C-terminal 'tagging' of a protein destined for proteolytic degradation with multiple **ubiquitin** monomers. Poly-ubiquitinylated proteins are then subjected to proteolytic degradation by the **proteasome**.

Dysfunction of the ubiquitin–proteasome pathway has been implicated in a number of disease states. For example, there is evidence that the accumulation of abnormal or damaged proteins due to impairment of the pathway contributes to a number of neurodegenerative diseases, including Alzheimer's. In contrast, deliberately blocking the pathway in cancer cells could lead to a disruption of protein regulation that in turn could cause the apoptosis of the malignant cells. Accordingly, proteasome inhibitors have been developed for evaluation as anti-tumour agents against selected cancers. **Bortezomib** is one such inhibitor that targets the 26S proteasome, and in combination with other chemotherapeutic agents has been shown to have therapeutic potential.

In contrast to the long-term control, there are several mechanisms by which the activity of an enzyme can be altered almost instantaneously:

- Reversible inhibition by small molecules (drugs) is of eminent importance for many therapeutic applications.
- **Product inhibition:** Here the product produced by the enzyme acts as an inhibitor of the reaction so that unless the product is removed by further metabolism the reaction will cease. This process is of relevance for negative-feedback control (in contrast, positive-feedback control stimulates the formation of product). An example is the inhibition of **hexokinase** (EC 2.7.1.1) by glucose-6-phosphate. Hexokinase exists in four isoenzyme forms (I–IV). The first three isoforms are distributed widely and have a low K_M for glucose (about 10–100 µM), commensurate with the glucose concentrations they encounter, and are inhibited by glucose-6-phosphate. Isoform IV (also called **glucokinase**) is confined to the liver where its higher K_M allows it to deal with high glucose concentrations following a carbohydrate-rich meal.
- **Allosteric regulation:** Here a small molecule that may be a substrate, product or key metabolic intermediate such as ATP or AMP alters the conformation of the catalytic site as a result of its binding to an allosteric site. A good example is the regulation of 6-phosphofructokinase discussed earlier (Section 23.2.4).
- Reversible covalent modification: This may involve adenylation of a Tyr residue by ATP (e.g. glutamine synthase) or the ADP-ribosylation of an arginine residue by NAD^+ (e.g. nitrogenase), but most frequently involves the phosphorylation of specific Tyr, Ser or Thr residues by a protein kinase. During phosphorylation, the γ-phosphate group of ATP is transferred to the target enzyme where it might induce conformational changes in the enzyme structure such as to either activate or deactivate the enzyme. Reversible covalent modification is quantitatively the most important of the three mechanisms.

23.5.2 Implications for Control of Pathways

In a metabolic pathway, individual enzymes combine to produce a flow of substrates and products through the pathway. This flow is referred to as the metabolic flux. Its value is determined by factors such as the availability of starting substrate and cofactors, but above all by the activity of the individual enzymes. The different enzymes in a particular pathway do not all possess the same activity. As a consequence, one or at most a small number of enzymes with the lowest activity determine the overall flux through the pathway. In order to identify these rate-controlling enzymes three types of study need to be carried out:

1. In vitro kinetic studies of each individual enzyme conducted under experimental conditions as near as possible to those found in vivo and such that the enzyme is saturated with substrate (i.e. such that $c(S) > 10 \times K_M$)
2. Studies to determine whether or not each individual enzyme stage operates at or near equilibrium in vivo
3. Studies to determine the flux control coefficient, C, for each enzyme.

The flux control coefficient is a property of the enzyme that expresses how the flux of reactants through a pathway is influenced by a change in the activity of the enzyme under the prevailing physiological conditions, assuming constant enzyme concentrations. Such a change may be induced by allosteric activators or inhibitors, or by feedback inhibition. Values for C can vary between 0 and 1. A flux control coefficient of 1 means that the flux through the pathway varies in proportion to the increase in the activity of the enzyme, whereas a flux control coefficient of 0 means that the flux is not influenced by changes in the activity of that enzyme. The sum of the C values for all the enzymes in a given pathway is 1 so that the higher a given C value, the greater the impact of that enzyme on the flux through the pathway. The C values are therefore highest for the rate-determining enzymes.

23.5.3 Enzyme Regulation by Therapeutics

Whereas there are examples of irreversible inhibitors that covalently bind to the enzyme and thus affect its catalytic activity (e.g. afatinib, a cancer therapeutic targeting receptor tyrosine kinases, and ornithine-analogue inhibitors of polyamine biosynthesis used to treat the sleeping sickness African trypanosomiasis), the most frequently exploited mechanism of action of drugs is the reversible inhibition of an enzyme.

Reversible Inhibitors
Reversible inhibitors attach to enzymes by non-covalent interactions and this process can thus be described quantitatively as binding of the inhibitor to the enzyme or the enzyme–substrate complex, depending on the type of inhibition. Thermodynamically, a binding process is typically characterised by a dissociation constant K_d; when applied to enzyme inhibition, the dissociation constant becomes an inhibition constant K_i.

The dissociation constant K_d is a true thermodynamic equilibrium constant for ligand binding to a receptor, and describes the concentration required to achieve half-maximum binding (albeit used without units). Smaller values indicate stronger binding. Accordingly, the inhibition constant K_i (used with units of concentration) describes the concentration required to achieve half-maximum inhibition.

Whereas the dissociation constant K_d is directly measured, the inhibition constant K_i is indirectly measured by observing the inhibitory effect on enzyme activity (which requires another ligand – the substrate – to bind). Since the values of inhibition constants can range from 10^{-15} M to 10^{-3} M, it is often more convenient to express the results as $pK_i = -\lg K_i$.

EC_{50} and IC_{50} (sometimes also called $I_{0.5}$) values describe the concentrations that are required to achieve half-maximum binding (inhibition) under a defined set of conditions. EC_{50} and IC_{50} values may thus be the same as the values of the dissociation (K_d) or inhibition (K_i) constant for a ligand, but they do not have to be.

Values for the inhibition constant K_i can be estimated using the Michaelis–Menten parameters in the presence (v'_{max}, K'_M) and absence (v_{max}, K_M) of inhibitor:

- for competitive inhibition :
$$K_i = \frac{K_M \times c(I)}{K'_M} - K'_M \qquad \text{(Eq 23.25)}$$

- for non-competitive inhibition:
$$K_i = \frac{v'_{max} \times c(I)}{v_{max}} \qquad \text{(Eq 23.26)}$$

Here, $c(I)$ is the molar concentration of the inhibitor. Whereas K_i is a thermodynamic quantity, the frequently used IC_{50} is a phenomenological measure that describes the concentration of inhibitor required to lower the enzyme activity to 50%. Relationships between IC_{50} and K_i values have been investigated for different inhibition mechanisms and are summarised in Table 23.4.

Table 23.4 Relationships between IC_{50} and K_i

Non–competitive inhibition	$IC_{50} = K_i$
Uncompetitive inhibition	$IC_{50} = K_i$
Competitive inhibition	$IC_{50} = K_i \times \left(1 + \frac{c(S)}{K_M}\right)$
If $c(S) \ll K_M$	$IC_{50} \approx K_i$
If $c(S) = K_M$	$IC_{50} \approx 2 \times K_i$

Note: $c(S)$ is the molar concentration of substrate

Irreversible Inhibitors

Inhibitors that covalently bind to the enzyme are irreversible inhibitors and cannot be characterised by an IC_{50} value. Since the covalent modification of the enzyme by the inhibitor is a chemical reaction, characterised by a reaction rate constant, the amount of active enzyme at a particular concentration of inhibitor varies with the incubation time. The incubation time therefore becomes a parameter that needs to be captured in the characterisation of the irreversible inhibition reaction. Therefore, the ratio

$$\frac{k_{obs}}{c(I)} \qquad \text{(Eq 23.27)}$$

is used to characterise irreversible inhibition. Here, k_{obs} is the observed pseudo-first-order rate of inactivation, which is obtained from a plot of the enzyme activity (expressed as a percentage) versus time. This ratio is valid until all enzyme molecules have reacted with the inhibitor (saturation).

23.6 RECEPTORS

Besides enzymes, receptors play an important role for the healthy organism, since they enable communication between cells, initiate a cellular response to external stimuli and coordinate activities to achieve homeostasis. Like enzymes, receptors have specific sites for substrates; upon substrate binding to a receptor, a cellular response is elicited.

The communication between cells is called **inter-cellular signalling** and is achieved by:

- The release of **signalling molecules** by the 'signalling' cells, referred to as **endogenous agonists or first messengers**.
- The specific recognition and binding of these agonists by receptor molecules, simply referred to as receptors, located either in the cell membrane or in the cytoplasm of the 'target' cell. Each cell membrane contains between 10^3 and 10^6 molecules of a given receptor. Binding of the agonist to a specific binding domain on the receptor changes the receptor from its inactive, **resting state**, to an active state. Exceptionally, a receptor may possess activity in the absence of agonist. Such receptors are said to possess **constitutive activity**.
- The initiation of a sequence of molecular events commonly involving interaction between the active receptor and other, so-called effector molecules, the whole process being referred to as intra-cellular signal transduction, which terminates in the final cellular response.

Agonist signalling molecules can be either lipophilic or hydrophilic and range from the gas nitric oxide, amines, amino acids, nucleosides, nucleotides and lipids to hormones, growth factors, interleukins, interferons and cytokines. The receptors for inter-cellular signalling are embedded in the cell membrane and possess an agonist-binding domain exposed on the extra-cellular side.

Cell membrane receptor proteins possess three distinct domains:

- **Extra-cellular domain**: This protrudes from the external surface of the membrane and contains all or part of the agonist-binding domain known as the **orthosteric agonist-binding site**.
- **Transmembrane domain**: This is inserted into the phospholipid bilayer of the membrane and may consist of several regions that loop repeatedly back and forth across

the membrane. In some cases these loops form a channel for the 'gating' (hence the channel may be open or closed) of ions across the membrane, whilst in other receptors the loops create part of the orthosteric site.

- Intra-cellular domain: This region of the protein has to respond to the extra-cellular binding of the agonist to initiate the transduction process. In some cases it is the site of the activation of enzyme activity within the receptor protein, commonly kinase activity, or is the site that interacts with effector proteins.

The existence of three domains within receptor proteins reflects their amphipathic nature in that they contain regions of 19 to 24 amino-acid residues possessing polar groups that are hydrophilic, and similar-sized regions that are rich in non-polar groups that are hydrophobic and hence lipophilic. The hydrophobic regions, generally in the form of α-helixes, are the transmembrane regions that are inserted into the non-polar, long-chain fatty acid portion of the phospholipid bilayer of the membrane. Superfamilies of receptor proteins can be recognised from the precise number of transmembrane regions each possesses. In contrast, the hydrophilic regions of the receptor are exposed on the outside and inside of the membrane, where they interact with the aqueous, hydrophilic environment.

There is evidence that many receptors exist in multiple isoforms that have subtly different physiological roles and that many receptors form homo- and/or heterodimers or oligomers that are sensitive to allosteric regulation and function as partners to initiate interactions between the downstream signalling molecules triggered by each receptor. This cross-talk between receptors allows cells to integrate signalling information originating from various external sources and to respond to it with maximum regulatory efficiency.

23.7 CHARACTERISATION OF RECEPTOR–LIGAND BINDING

23.7.1 Dose–Response Curves

The response of membrane receptors in their resting inactive state to exposure to an increasing concentration (dose) of agonist is a curve that has three distinct regions:

1. An initial threshold below which little or no response is observed
2. A slope in which the response increases rapidly with increasing dose
3. A declining response with further increases in dose and a final maximum response.

Since such plots commonly span several-hundred-fold variations in agonist concentration, they are best expressed in semi-logarithmic form (Figure 23.14).

Types of Agonists

Molecules that bind to a receptor (ligands) can typically be classfied into one of the following:

- Full agonists: These ligands increase the activity of the receptors and produce the same maximal response, but they differ in the dose required to achieve it (Figure 23.14).
- Partial agonists: These ligands also increase the activity of the receptors, but do not produce the maximal response shown by full agonists even when present in large excess such that all the receptors are occupied (Figure 23.14).

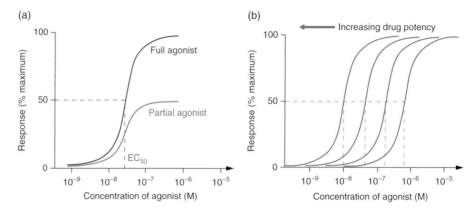

Figure 23.14 Dose–response curves for receptor agonists. (a) The biological effect (percentage maximum response) and the concentration of a full agonist are plotted on a logarithmic scale. An equipotent partial agonist has a lower efficacy than a full agonist – it cannot achieve the maximum response, even when all the receptors are occupied. EC_{50} is the concentration of agonist that produces 50% maximum effect. (b) Dose–response curves for four full agonist drugs of different potencies, but equal efficacy.

- Inverse agonists: These agonists decrease the activity of constitutively active receptors to their inactive state. A dose–response curve for an inverse agonist (Section 23.7.2) acting on a receptor with constitutive activity would be a mirror image of that shown in Figure 23.14, resulting in a progressive decrease in receptor activity.
- Partial inverse agonists: These agonists decrease the activity of constitutively active receptors, but not to their inactive state.
- Protean agonists: These agonists acting on receptors possessing constitutive activity display any response ranging from full agonism to full inverse agonism, depending on the level of constitutive activity in the system and the relative efficacies of the constitutive activity and that induced by the agonist.
- Biased agonists: This form of agonist behaviour is found with receptors that can couple to two or more different G-proteins and as a consequence the agonist preferentially selects one of them, thus favouring one specific transduction pathway.
- Antagonists: In the absence of agonists these ligands produce no change in the activity of the receptors. Three subclasses have been identified using the antagonist in the presence of an agonist and using receptors not possessing constitutive activity:

  (i) Competitive reversible antagonists: The antagonist competes with the agonist for the orthosteric sites so that the effect of the antagonist can be overcome by increasing the concentration of agonist (see Figure 23.15)
  (ii) Non-competitive reversible antagonists: The antagonist binds at a different site on the receptors to that of the orthosteric site so that the effect of the antagonist cannot be overcome by increasing the concentration of agonist.
  (iii) Irreversible competitive antagonists: The antagonist competes with the agonist for the orthosteric site, but the antagonist forms a covalent bond with the site so that its effect cannot be overcome by increasing the concentration of agonist.

- **Allosteric modulators:** These bind to a site distinct from that of the orthosteric site and can only be detected using functional, as opposed to ligand-binding, assays.

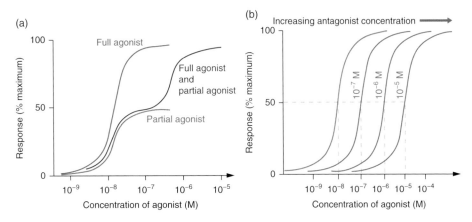

Figure 23.15 Dose–response curves for receptor agonists. (a) In the presence of a partial agonist, the dose–response curve of the full agonist is shifted to the right. (b) In the presence of a reversible antagonist, the dose–response curve for a full agonist is shifted to the right because the full agonist competes for receptor binding. The higher the concentration of antagonist, the greater the shift, but the maximum response remains unchanged. The yellow curve shows the receptor response to agonist binding in the absence of antagonist.

Types of Agonist Effects

As a result of the above agonist classification, ligand action on receptors can be characterised by a number of parameters:

- **Intrinsic activity:** This is a measure of the ability of an agonist to induce a response by the receptors. It is defined as the maximum response to the test agonist relative to the maximum response to a full agonist acting on the same receptors. All full agonists, by definition, have an intrinsic activity of 1, whereas partial agonists have an intrinsic activity of less than 1.
- **Efficacy:** This is a measure of the inherent ability of an agonist to initiate a physiological response following binding to the orthosteric site. The initiation of a response is linked to the ability of the agonist to promote the formation of the active conformation of the receptors, whereas for inverse agonists it is linked to their ability to promote the formation of the inactive conformation. While all full agonists must have a high efficacy, their values will not necessarily be equal, in fact they have no theoretical maximum value. Partial agonists have a low efficacy, antagonists have zero efficacy and inverse agonists have negative efficacy.
- **Collateral efficacy:** This relates to the ability of the agonist to preferentially select one of the two or more possible transduction pathways displayed by the binding receptors. It primarily relates to G-protein-coupled receptors that can bind to two or more G-proteins.
- **Potency:** This is a measure of the concentration of agonist required to produce the maximum effect; the more potent the agonist the smaller the concentration required.

The potency of an agonist is related to the position of the sigmoidal curve on the log dose axis. It is expressed in a variety of forms, including the effective dose or concentration for 50% maximal response, ED_{50} or EC_{50}. On a semi-logarithm plot, the value emerges as pED_{50} or pEC_{50} values [i.e. $-\log_{10} ED_{50}/(1\ M)$]. Thus an agonist with an EC_{50} of 3×10^{-5} M would have a pEC_{50} of 4.8. The potency of a reversible antagonist is expressed by its pA_2 value, defined as negative logarithm of the concentration of antagonist that will produce a two-fold shift in the concentration–response curve for an agonist.

- Affinity: This is a measure of the concentration of agonist required to produce 50% binding. As will be shown in the following section, affinity is a reflection of both the rate of association of the ligand with the receptors and the rate of dissociation of the resulting complexes. The rate of association is a reflection of the three-dimensional interaction between the two and the rate of dissociation a reflection of the strength of binding within the complexes. Affinity of an agonist can be expressed by an affinity or binding constant, K_a, but is more commonly expressed as a dissociation constant, K_d, of the receptor–ligand complex, where K_d is equal to the reciprocal of K_a. The affinity of receptors for an antagonist is expressed by the corresponding dissociation constant K_b.

- Selectivity: This is a measure of the ability of an agonist to discriminate between receptor subtypes. This is particularly important from a therapeutic perspective.
- Functional selectivity: This is a measure of the ability of the agonist to induce selective response from receptors capable of promoting more than one transduction activity.

23.7.2 Mechanisms of Receptor–Agonist Interactions

The long-held view of receptor–agonist interaction was based on a two-state model that visualised that the binding of the agonist (A) by the receptor to form a receptor–agonist complex triggers a conformational change in the receptor that converts it from a dormant or resting inactive state (R) to an active state (R*):

$$R + A \longrightarrow AR \longrightarrow AR^* \longrightarrow \text{transduction via effector} \quad \text{(Eq 23.28)}$$

The formation of the active state of the receptor initiates a transduction (linking) process in which the receptor activates an effector protein. The effector protein may be the receptor itself or a distinct protein that is either attached to the inside of the membrane or free in the cytoplasm. This activated effector either allows the passage of selected ions across the membrane, thereby changing the membrane potential, or it produces a second messenger, which initiates a cascade of molecular events, involving molecules located on and/or at the internal surface of the cell membrane or within the cell cytoplasm, that terminate in the final target cell response. Examples of second messengers are Ca^{2+}, cAMP (cyclic adenosine monophosphate), cGMP (cyclic guanosine monophosphate), 1,2-diacylglycerol and inositol-1,4,5-trisphosphate.

However, in 1989 it was discovered that opioid receptors in NG108 cells possessed activity in the absence of agonist. This receptor activity in the absence of agonist was termed **constitutive activity**, and synthetic ligands were identified that could bind to the receptor and decrease its constitutive activity in the absence of physiological agonist. Such ligands were termed **inverse agonists**. Subsequent in vivo and in vitro investigations with a wide range of receptor types, many of which were G-protein-coupled receptors (GPCRs), identified other examples of receptors with constitutive activity and showed that this activity may have a physiological role, thereby confirming that constitutive activity is not solely a consequence of a mutation or over-expression of a receptor gene. Of particular interest was the observation that certain receptor mutations (known as **constitutively active mutants**) were associated with such clinical disorders as retinitis pigmentosa, hyperthyroidism and some autoimmune diseases.

The **conformational selection model** of receptor action was formulated to rationalise the concomitant existence of active and inactive conformations. The model envisages that in the absence of agonist, receptors exist as an equilibrium mixture of inactive (R) and active (R*) forms, and that the relative proportion of the two forms is determined by the associated equilibrium constant. An introduced ligand will preferentially bind to one conformation, thereby stabilising it, and causing a displacement of the equilibrium between the two forms. Agonists will preferentially bind to the R* state, displacing the equilibrium to increase the proportion of the R* form. Partial agonists are deemed to have the ability to bind to both forms with a preference for the R* form, again resulting in an increase in the R* form, but by a smaller amount than that produced by full agonists. Inverse agonists preferentially bind to the R conformation, displacing the equilibrium and increasing the proportion of R. Partial inverse agonists can bind to both states, with a preference for the R form, resulting in decrease in the proportion of the R* form. Unlike the two-state model, the conformational selection model does not require the binding ligand to cause a conformational change in the receptor in order to alter the activity of the receptor.

Prior to the discovery of constitutive activity in some receptors, ligands were classified either as agonists or antagonists. This classification formed the basis of the understanding of the pharmacological action of many therapeutic agents and the development of new ones by the pharmaceutical industry. However, retrospective evaluation of ligands previously classified as antagonists, but using receptors produced by cloning techniques and shown to possess constitutive activity revealed that many were actually inverse agonists and possessed negative efficacy, whilst others were **neutral antagonists** in that they neither increased nor decreased the receptor activity. To date approximately 85% of all antagonists that have been re-evaluated have been shown to be inverse agonists. These observations can be rationalised in that:

- **Agonism** is a behaviour characteristic of a particular ligand and can be demonstrated in the absence of any other ligand for the receptor
- **Antagonism** can only be demonstrated in the presence of an agonist

- **Inverse agonism** can only be demonstrated when the receptor possesses constitutive activity that can be reduced by the agent. In the absence of constitutive activity, inverse agonists can only demonstrate simple competitive antagonism of a full or partial agonist.

Further understanding of the nature of constitutive activity and the mode of action of different agonists on a given receptor has come from studies on the histamine H_3 receptor (H_3R). This is a G-protein-coupled receptor, specifically coupling to G_i/G_o proteins. Constitutive activity has been found in both rat and human brain in which the activity inhibits histamine release from synaptosomes. Studies using the ligand proxyfan, previously classified as an antagonist, have assigned to it a spectrum of activities ranging from full agonist through partial agonist to partial inverse agonist and full inverse agonist. Such behaviour by a ligand has been classified as **protean agonism** and proxyfan as a **protean agonist**. The precise behaviour of proxyfan in a given study correlated with the level of constitutive activity of the system and the relative efficacy of the constitutively active state R* and that induced by the ligand, AR*. Thus, in the absence of any R* or in the presence of AR* with a lower efficacy than that of R*, the ligand will display agonist activity. When both R* and AR* states are present with equal efficacy the ligand will display neutral antagonism and when the AR* state has a higher efficacy than R* the ligand will display inverse agonism. Such behaviour can only be explained by a **multi-state model** in which multiple R* and AR* states of the receptor can be formed. Furthermore, studies using a range of agonists on the histamine H_3 receptor indicated that different ligands could promote the creation of distinct active states that can display differential signalling. A wider understanding of the mechanism by which receptors are activated by agonists is linked to the discovery, initially made with the α_{1B} adrenergic receptor, that certain mutations in the sequence of G-protein-coupled receptors caused a large increase in the constitutive activity of the receptor. This observation had the implication that there may be domains in the receptor that are crucial to the conservation of a receptor not displaying constitutive activity, and that the action of agonists was to release these constraints creating the active receptor. In mutant receptors possessing constitutive activity these constraints have been released as a result of the mutation. Studies have shown that the activation of inactive receptors by agonists proceeds by a series of conformational changes. The question as to whether agonists and inverse agonists switch the receptor on a linear 'on–off' scale or whether they operate by different mechanisms has been studied using the α_{2A} adrenergic receptor and a FRET-based approach (see Section 13.5.3). Differences in the kinetics and character of the conformational changes induced by these two classes of agonists provided clear evidence for distinct types of molecular switch. Moreover, full agonists and partial agonists also showed distinct differences, indicating that receptors do not operate by a simple 'on–off' switch, but rather that they have several distinct conformational states and that these states can be switched with distinct kinetics by the various classes of agonist.

As previously pointed out, the pharmaceutical industry seeks to identify receptor agonists or antagonists that can be used for the treatment of specific clinical conditions. The potential role of inverse agonists in this respect remains to be fully evaluated. However, it is apparent that clinical conditions caused by a mutant receptor that has constitutive activity, in contrast to the normal receptor that is only active in the presence of the physiological agonist, could be treated with an inverse agonist that would eliminate the constitutive activity. Equally, use of an inverse agonist may be advantageous in conditions resulting from the over-stimulation of the receptor due to the over-production of the signalling agonist. At the present time many inverse agonists are used clinically, although at the time of their development they were believed to be competitive antagonists.

23.7.3 Ligand-Binding Studies

Receptor Preparations

Preparations of receptors for ligand-binding studies may either leave the membrane intact or involve the disruption of the membrane and the release of the receptor with or without membrane fragments, some of which could form vesicles with variable receptor orientation and control mechanisms. Membrane receptor proteins show no or very little ligand-binding properties in the absence of phospholipid, so that if a purified receptor protein is chosen, it must be introduced into a phospholipid vesicle for binding-study purposes. The range of receptor preparations available for binding studies is shown in Table 23.5.

Kinetic studies aimed at the determination of individual rate constants are best carried out using isolated cells, whilst studies of the number of receptors in intact tissue are best achieved by labelling the receptors with a radiolabel, preferably using an irreversible competitive antagonist and applying the technique of quantitative autoradiography (see Section 9.3.3).

Ligands

A common technique for the study of ligand–receptor interaction is the use of a radiolabelled ligand with isotopes such as ^{3}H, ^{14}C, ^{32}P, ^{35}S and ^{135}I (see Section 9.3.3). Generally, a high-specific-activity ligand is used as this minimises the problem of non-specific binding (see below). If a large number of ligands are being studied, such as in the screening of potential new therapeutic agents, the cost and time of producing the radiolabelled forms becomes virtually prohibitive and experimental techniques such as fluorescence spectroscopy and surface plasmon resonance spectroscopy have to be used. However, the use of radiolabelled ligands remains attractive as a means of distinguishing between ligands targeting the main binding site (orthosteric ligands) and allosteric ligands.

The technique of using radiolabelled ligand generally requires the separation of bound and unbound ligand once equilibrium has been achieved. This is most commonly achieved by techniques such as equilibrium dialysis and ultrafiltration, exploiting the inability of receptor-bound ligand to cross a semi-permeable membrane, and by simple centrifugation, exploiting the ability of the receptor-bound ligand to be pelleted by an applied small centrifugal field.

For the study of ligand–receptor interactions that occur on a submillisecond timescale, special approaches such as stopped-flow and quench-flow methods (see Section 13.8) need to be adopted to deliver the ligand to the receptor. An alternative approach is the use of so-called **caged compounds**. These possess no inherent ligand properties, but on **laser flash photolysis** with light of a specific wavelength, a protecting group masking a key functional group is instantaneously cleaved, releasing the active ligand.

Table 23.5 **Receptor preparations for the study of receptor–ligand binding**

Receptor preparation	Comments
Tissue slices	5–50 μm thick, generally adhered to gelatine-coated glass slide. Good for study of receptor distribution.
Cell membrane preparation	Disrupt cells (from tissue or cultures) by sonication and isolate membrane fraction by centrifugation. Lack of cytoplasmic components may compromise receptor function. Used to study ligand binding and receptor distribution in lipid rafts and caveolae. Increasingly used with cell lines transfected with human receptor genes.
Solubilised receptor preparation	Disrupt membrane with detergents and purify receptors by affinity chromatography using an immobilised competitive antagonist. Isolation from other membrane components may compromise studies.
Isolated cells	Release cells from tissue by mechanical or enzymatic (collagenase, trypsin) means. Cells may be in suspension or monolayers. May be complicated by the presence of several cell types. Widely used for the study of a range of receptor functions. Allows ligand binding and cellular functional responses to be studied under the same experimental conditions.
Cultured cell lines	Very popular. Has the advantage of cell homogeneity and ease of replication.
Recombinant receptors	Produced by cloning or mutagenesis techniques and inserted into specific cultured cell line, including ones of human origin. Popular for the study of the effect of mutations on receptor function, such as constitutive activity and cell signalling. Care needed to ensure that receptors have the same functional characteristics (e.g. post-translational modifications) as native cells.

Experimental Procedures for Ligand-Binding Studies

The general experimental approach for studying the kinetics of receptor–ligand binding and hence to determine the experimental values of the binding constants and the total number of binding sites is to incubate the receptor preparation with the ligand under defined conditions of temperature, pH and ionic concentration for a specific period of time that is sufficient to allow equilibrium to be attained. The importance of allowing the system to reach equilibrium cannot be overstated as Equations 23.31 to 23.42 below do not hold if **equilibrium** has not been attained. Using an appropriate analytical procedure, the bound and unbound forms of the ligand are then quantified or some associated change measured. This quantification may necessitate the separation of the bound and unbound fractions. The study is then repeated for a series of ligand concentrations to cover 10–90% of maximum binding at a fixed receptor concentration. The binding data are then analysed using Equations 23.36 to 23.42, often in the form of a computer program, many of which are available commercially. Most perform linear or non-linear least-squares regression analysis of the experimental data. Measurements may be made on a **steady-state** (single measurement) basis or a **time-resolved** (multiple measurements over a period of time) basis, most commonly by stopped-flow or quench-flow procedures (see Section 13.8).

Non-Specific Binding

A general problem in the study of receptor–ligand binding is the non-specific binding of the ligand to sites other than the orthosteric or allosteric binding sites. Such non-specific binding may involve the membrane lipids and other proteins either located in the membrane or released by the isolation procedure. The characteristic of non-specific binding is that it is non-saturable, but is related approximately linearly to the total concentration of the ligand. Thus the observed ligand binding is the sum of the saturable (hyperbolic) specific binding to the receptor and the non-saturable (linear) binding to miscellaneous sites:

$$B = B_s + B_{ns} \tag{Eq 23.29}$$

The specific binding component is usually obtained indirectly, either by carrying out the binding studies in the presence of an excess of non-labelled ligand (agonist or antagonist) if a labelled ligand is being used, or by using a large excess of an agonist or competitive antagonist in other studies. The presence of the excess unlabelled or competitive ligand will result in the specific binding sites not being able to bind the ligand under study and hence its binding would be confined to non-specific sites (Figure 23.16). In practice, a concentration of the competitive ligand of at least 1000 times its K_d or K_b must be used and confirmation that under the conditions of the experiment, non-specific binding was being studied would be sought by repeating the study using a range of different and structurally dissimilar competitive ligands that should give consistent estimates of the non-specific binding.

23.7.4 Quantitative Characterisation of Receptor–Ligand Binding

Note: In contrast to the previous sections, the association and dissociation constants K_a and K_d are used with units in the following section.

The previous discussion has shown that ligands capable of binding to a receptor may be of a variety of types. To fully characterise any ligand it is essential that its binding be expressed in quantitative terms. This is achieved by binding studies. If under the conditions of the binding studies the total concentration of ligand is very much greater than that of receptor (so-called **saturation conditions**), changes in ligand concentration due to receptor binding can be ignored, but changes in the free (unbound) receptor concentration cannot. Hence if:

$c(R_t)$ is the total concentration of receptor that determines the maximum binding capacity for the ligand

$c(L)$ is the free ligand concentration

$c(RL)$ is the concentration of receptor–ligand complex

then $c(R_t) - c(RL)$ is the concentration of free receptor.

At equilibrium, the forward and reverse reactions for ligand binding and dissociation will be equal:

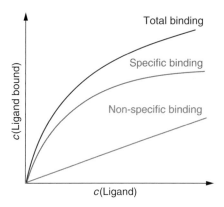

Figure 23.16 Specific and non-specific binding of a ligand to a membrane receptor. Specific binding is normally hyperbolic and shows saturation. Non-specific binding is linear and is not readily saturated.

$$k_{+1} \times [c(R_t) - c(RL)] \times c(L) = k_{-1} \times c(RL) \qquad \text{(Eq 23.30)}$$

where k_{+1} is the association rate constant and k_{-1} is the dissociation rate constant. Therefore:

$$\frac{k_{-1}}{k_{+1}} = K_d \times 1\,M = \frac{1}{K_a} \times 1\,M = \frac{\left[c(R_t) - c(RL)\right] \times c(L)}{c(RL)} \qquad \text{(Eq 23.31)}$$

where K_d is the dissociation constant for the complex [RL] and K_a is the association or affinity constant. Rearranging gives:

$$c(RL) = \frac{c(L) \times c(R_t)}{K_d \times 1\,M + c(L)} \qquad \text{(Eq 23.32)}$$

Note that the factor of 1 M appears because K_a and K_d are thermodynamic equilibrium constants and thus have no units. Importantly, when the receptor sites are half saturated, then $c(L) = K_d \times 1$ M.

Determination of Dissociation Constants

Equation 23.32 is of the form of a rectangular hyperbola which predicts that ligand binding will reach a limiting value as the ligand concentration is increased and therefore that receptor binding is a saturable process. The equation is of precisely the same form as Equation 23.3 that defines the binding of the substrate to its enzyme in terms of K_M and v_{max}. For the experimental determination of K_d, Equation 23.32 can be used directly by analysing the experimental data by non-linear regression curve-fitting software. However, several linear transformations of Equation 23.32 have been developed, including the **Scatchard equation**:

$$\frac{c(RL)}{c(L)} = \frac{c(R_t)}{K_d \times 1\,M} - \frac{c(RL)}{K_d \times 1\,M} \qquad \text{(Eq 23.33)}$$

This equation predicts that a plot of $c(RL)/c(L)$ against $c(RL)$ will be a straight-line slope $-1/K_d$ allowing for K_d to be calculated. However, in many studies the relative molecular mass of the receptor protein is unknown so that the concentration term $c(RL)$ cannot be calculated in molar terms. In such cases, it is acceptable to express the extent of ligand binding in any convenient unit (B), e.g. pmol per 10^6 cells, pmol per mg protein or more simply as an observed change, for example in fluorescence emission intensity (ΔF), under the defined experimental conditions. Since maximum binding (B_{max}) will occur when all the receptor sites are occupied, i.e. when:

$$c(R_t) = B_{max},$$

(Eq 23.34)

Equation 23.33 can be written in the form:

$$\frac{B}{c(L)} = \frac{B_{max}}{K_d \times 1\,M} - \frac{B}{K_d \times 1\,M}$$

(Eq 23.35)

Hence a plot of $B/c(L)$ against B will be a straight line, slope $-1/K_d$ and intercept on the y-axis of B_{max}/K_d (Figure 23.17). In cases where the relative molecular mass of the receptor protein is known, the Scatchard equation can be expressed in the form:

$$\frac{B}{c(L) \times B_{max}} = \frac{n}{K_d \times 1\,M} - \frac{B}{K_d \times 1\,M \times B_{max}}$$

(Eq 23.36)

where n is the number of independent ligand-binding sites on the receptor. The expression B/B_{max} is the molar amount of ligand bound to one mole of receptor. If this expression is defined as r, then:

$$\frac{r}{c(L)} = \frac{n}{K_d \times 1\,M} - \frac{r}{K_d \times 1\,M}$$

(Eq 23.37)

In this case, a plot of $r/c(L)$ against r will again be linear with a slope of $-1/K_d$ but in this case the intercept on the x-axis will be equal to the number of ligand-binding sites, n, on the receptor.

Alternative linear plots to the Scatchard plot are:

Lineweaver–Burk plot

$$\frac{1}{B} = \frac{1}{B_{max}} + \frac{K_d \times 1\,M}{B_{max} \times c(L)}$$

(Eq 23.38)

Hanes plot

$$\frac{c(L)}{B} = \frac{K_d \times 1\,M}{B_{max}} + \frac{c(L)}{B_{max}}$$

(Eq 23.39)

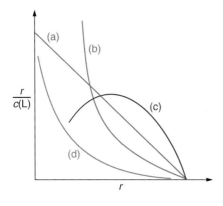

Figure 23.17 Scatchard plot for (a) a single set of sites with no cooperativity, (b) two sets of sites with no cooperativity, (c) a single set of sites with positive cooperativity and (d) a single set of sites with negative cooperativity.

In practice, Scatchard plots are commonly used although, statistically, they are prone to error since the experimental variable B occurs in both the x and y terms so that linear regression of these plots over-estimates both K_d and B_{max}. There is a view that linear transformations of the three types above are all inferior to the non-linear regression analysis of Equation 23.32 since they all distort the experimental error. For example, linear regression assumes that the scatter of experimental points around the line obeys Gaussian distribution and that the standard deviation of the points is the same. However, in practice, this is rarely true and as a consequence values of the slope and intercept are not the 'best' value.

The derivation of Equation 23.32 is based on the assumption that there is a single set of homogeneous receptors and that there is no cooperativity between them in the binding of the ligand molecules. In practice, two other possibilities arise, namely that there are two distinct populations of receptors, each with different binding constants, and secondly that there is cooperativity in binding within a single population. In both cases, the Scatchard plot will be curvilinear (Figure 23.17). If cooperativity is suspected, it should be confirmed by a Hill plot which, in its non-kinetic form, is:

$$\log\left(\frac{Y}{1-Y}\right) = h \times \log\frac{c(L)}{1\,M} - \log K_d \qquad \text{(Eq 23.40)}$$

or

$$\log\left(\frac{B}{B_{max} - B}\right) = h \times \log\frac{c(L)}{1\,M} - \log K_d \qquad \text{(Eq 23.41)}$$

where Y is the fractional saturation of the binding sites (and assumes values between 0 and 1) and h is the Hill coefficient. Note that because the argument of a logarithm cannot include units, $c(L)$ needs to be divided by the standard molar concentration ($c^0 = 1$ M). For a receptor with multiple binding sites that function independently $h = 1$, whereas for a receptor with multiple sites that are inter-dependent, h is either greater than 1 (positive cooperativity) or less than 1 (negative cooperativity). Scatchard plots that are biphasic due to ligand multivalence (i.e. multiple binding sites) rather than receptor cooperativity, are sometimes taken to indicate that the two extreme, and approximately linear, sections of the curvilinear plots represent high affinity (high bound/free ratio at low bound values) and low affinity (low bound/free ratio at high bound values) sites and that tangents drawn to these two sections of the curve can be used to calculate the associated K_d and B_{max} values. This is incorrect, and the correct values can only be obtained from the binding data by means of careful mathematical analysis, generally undertaken by the use of special computer programs, many of which are commercially available.

Determination of Rate Constants

To determine the dissociation rate constant, k_{-1} (in units of s^{-1}) of any ligand for the receptor, some of the receptor–ligand complex (B_0) is allowed to form, usually using

Example 23.4   **ANALYSIS OF LIGAND-BINDING DATA**

Question   The extent of the binding of an agonist to its membrane-bound receptor on intact cells was studied as a function of ligand concentration in the absence and presence of a large excess of unlabelled competitive antagonist.

c(ligand) (nM)	40	60	80	120	200	500	1000	2000
Total ligand bound (pmol per 10^6 cells)	0.284	0.365	0.421	0.547	0.756	1.269	2.147	2.190
Ligand binding in presence of competitive antagonist (pmol per 10^6 cells)	0.054	0.068	0.084	0.142	0.243	0.621	1.447	1.460

In all cases, the extent of total ligand binding was such that there was no significant change in the total ligand concentration. What quantitative information about the binding of the ligand to the receptor can be deduced from these data?

Answer   To address this problem, it is first necessary to calculate the specific binding of the ligand to the receptor (B_s). The use of a large excess of unlabelled competitive antagonist enables the non-specific binding to be measured. The difference between this and the total binding gives the specific binding. Once this is known, various graphical options are open to evaluate the data. The simplest is a plot of the specific ligand binding as a function of the total ligand binding. More accurate methods are those based on linear plots such as a Scatchard plot (Equation 23.33) and a Lineweaver–Burk plot (Equation 23.38). In addition, it is possible to carry out a Hill plot (Equation 23.40 or 23.41) to obtain an estimate of the Hill constant, h. The derived data for each of these three plots are shown in the following table:

c(L) (nM)	40	60	80	120	200	500	1000	2000
Total bound ligand B (pmol per 10^6 cells)	0.284	0.365	0.421	0.547	0.756	1.269	2.147	2.190
Non-specific binding B_{ns} (pmol per 10^6 cells)	0.054	0.068	0.084	0.142	0.243	0.621	1.447	1.460
Specific binding B_s (pmol per 10^6 cells)	0.230	0.297	0.337	0.405	0.513	0.648	0.700	0.730
$B_s/c(L) \times (10^{-3}\ dm^3)$	5.75	4.95	4.21	3.37	2.56	1.30	0.70	0.43

$1/B_s$ (10^6 cells pmol^{-1})	4.35	3.37	2.97	2.47	1.95	1.54	1.43	1.37
$1/c(L)$ (nM^{-1})	0.0250	0.0170	0.0125	0.0083	0.0050	0.0020	0.0010	0.0005
$(B_{max} - B_s)$ (pmol per 10^6 cells)	0.52	0.45	0.413	0.345	0.237	0.102	0.050	0.020
$B_s/(B_{max} - B_s)$	0.44	0.66	0.816	1.174	2.164	6.35	14.00	36.50
lg $[B_s/(B_{max} - B_s)]$	−0.356	−0.180	−0.088	0.070	0.335	0.803	1.146	1.562
B_s/B_{max}	0.3067	0.3960	0.4493	0.5400	0.6840	0.8640	0.9333	0.9733
$B_s/[B_{max} \times c(L)]$	0.0077	0.0066	0.0056	0.0045	0.0034	0.0017	0.0009	0.0005

The hyperbolic plot allows an estimate to be made of the maximum ligand binding, B_{max}. It is approximately 0.75 pmol per 10^6 cells. An estimate can then be made of K_d by reading the value of $c(L)$ that gives a ligand binding value of 0.5 B_{max} (0.375 pmol per 10^6 cells). It gives an approximate value for K_d of 100 nM.

A Scatchard plot obtained by regression analysis gives a correlation coefficient of $R = 0.992$, B_{max} of 0.786 pmol per 10^6 cells and K_d of 97.3 nM.

A Lineweaver–Burk plot gives a correlation coefficient of $R = 0.997$, B_{max} of 0.746 pmol per 10^6 cells and K_d of 90.5 nM. Note that there is some variation between these three sets of calculated values; the ones given by the Lineweaver–Burk plot are more likely to be correct since, as previously pointed out, the Scatchard plot over-estimates both values when the binding data are subjected to linear regression analysis.

The Hill plot based on a value of B_{max} of 0.75 pmol per 10^6 cells, gave a correlation coefficient of $R = 0.996$ and a value for the Hill coefficient of $h = 1.13$.

ligand labelled with a radioactive isotope. The availability of the remaining unoccupied receptors to the labelled ligand is then blocked by the addition of at least 100-fold excess of the unlabelled ligand or competitive antagonist and the rate of release (B_t) of the radiolabelled ligand from its binding site monitored as a function of time. This generally necessitates the separation of the bound and unbound fractions. The rate is given by the expression:

$$\frac{dB_0}{dt} = -k_{-1} \times B_0 \qquad\qquad \text{(Eq 23.42)}$$

and the equation governing the release by the expression:

$$B_t = B_0 \times e^{-k_{-1} \times t} \qquad\qquad \text{(Eq 23.43)}$$

hence:

$$\lg B_t = \lg B_0 - 2.303 \times k_{-1} \times t \qquad\qquad \text{(Eq 23.44)}$$

Thus, a plot of lg B_t against time will give a straight line with a slope of $-2.303 \times k_{-1}$, allowing for estimation of the dissociation rate constant k_{-1}.

The association rate constant, k_{+1} (units: $M^{-1}\,s^{-1}$), is best estimated by the approach to equilibrium method by which the extent of agonist binding is monitored continuously until equilibrium is reached under conditions that are such that $c(L) >> c(R_t)$ – this yields pseudo first-order conditions rather than second order; under these conditions $c(R_t)$ decreases with time but $c(L)$ remains constant. Ligand binding increases asymptotically such that:

$$\lg\left(\frac{B_{eq}}{B_{eq} - B_t}\right) = 2.303 \times \left[k_{+1} \times c(L) + k_{-1}\right] \times t \qquad \text{(Eq 23.45)}$$

Thus, a plot of log $B_{eq}/(B_{eq} - B_t)$ against time will be linear with a slope of $2.303 \times (k_{+1} \times c(L) + k_{-1})$, where B_{eq} and B_t are the ligand binding at equilibrium and time t respectively. From knowledge of k_{-1} (obtained by the method discussed above) and $c(L)$, the value of k_{+1} can be calculated from the slope.

The K_d values observed for a range of receptors binding to their physiological agonist are in the range 10^{-6}–10^{-11}, which is indicative of a higher affinity than is typical of enzymes for their substrates. The corresponding k_{+1} rate constants are in the range 10^5–$10^8\,M^{-1}\,min^{-1}$ and k_{-1} in the range 0.001–$0.5\,min^{-1}$. Studies with G-protein-coupled receptors that form a tertiary complex (AR*G) have shown that the tertiary complex has a higher affinity for the agonist than has the binary complex (AR*). Receptor affinity for its agonist is also influenced by receptor interaction with various adaptor protein molecules present in the intra-cellular cell membrane.

It is relatively easy to calculate the number of receptors on cell membranes from binding data. The number is in the range 10^3–10^6 per cell. Although this may appear large, it actually represents a small fraction of the total membrane protein. This partly explains why receptor proteins are sometimes difficult to purify. From knowledge of receptor numbers and the K_d values for the agonist, it is possible to calculate the occupancy of these receptors under normal physiological concentrations of the agonist. In turn it is possible to calculate how the occupancy and the associated cellular response will respond to changes in the circulating concentration of the agonist. The percentage response change will be greater the lower the normal occupancy of the receptors. This is seen from the shape of the dose–response curve within the physiological range of the agonist concentration. It is clear that if the normal occupancy is high, the response to change in agonist concentration is small. Under such conditions, the response is likely to be larger if the receptor–agonist binding is a positively cooperative process.

Quantitative Characterisation of Competitive Antagonists

The ability of a competitive antagonist to reduce the response of a receptor to a given concentration of agonist can be quantified in two main ways:

1. IC_{50} value: The antagonist concentration that reduces the response of the receptor, in the absence of the antagonist, to a given concentration of agonist by 50%.

2. K_b value: The dissociation constant for the binding of the antagonist.

To determine an IC_{50} value, the standard procedure is to study the effect of increasing concentrations of antagonist on the response to a fixed concentration of agonist. In the absence of antagonist the response will be a maximum. As the antagonist concentration is increased, the response will decrease in a manner that is a mirror image of a dose–response curve (Figure 23.14). From the curve the antagonist concentration required to reduce the response by half (IC_{50}) can be determined. If this study is repeated for a series of increasing fixed agonist concentrations it will be evident that the IC_{50} value is critically dependent on the agonist concentration used, i.e. it is not an absolute value. In spite of this, it is commonly used, because of its simplicity, particularly in the screening of potential therapeutic agents. From knowledge of the IC_{50} value, in principle it is possible to calculate the K_b value using the **Cheng–Prusoff** equation:

$$K_b = \frac{IC_{50}}{1 + \dfrac{c(L)}{K_d \times 1M}}$$ (Eq 23.46)

where $c(L)$ is the concentration of agonist and K_d is its dissociation constant. It is evident from this equation that IC_{50} only approximates to K_b when $c(L)$ is very small and the denominator approaches 1. Although this equation is commonly used to calculate K_b values, its application is subject to reservations, primarily because inhibition curves do not reveal the nature of the antagonism. Furthermore, the numerical values of IC_{50} depend on the concentration of agonist used. Antagonist equilibrium constants, K_b, are thus best determined by application of the **Schildt equation**:

$$r = 1 + \frac{c(B)}{K_b \times 1M}$$ (Eq 23.47)

or in its logarithmic form:

$$\log(r - 1) = \log \frac{c(B)}{1M} - \log K_b$$ (Eq 23.48)

where $c(B)$ is the concentration of antagonist. The **dose ratio** r measures the amount by which the agonist concentration needs to be increased in the presence of the antagonist to produce the same response as that obtained in the absence of antagonist. This assumes that the same fraction of receptor molecules needs to be activated in the presence and absence of the antagonist to produce a given response. Experimentally, the dose ratio is equal to the dose of agonist required to give 50% response in the presence of the given antagonist concentration divided by the EC_{50} value.

To generate a Schild plot, the receptor response to increasing concentrations of agonist for a series of fixed concentrations of antagonist is studied and the dose factor for each concentration of antagonist calculated. The maximum response in all cases should be the same. Equation 23.49 then predicts that a plot of the $\log(r - 1)$

against log $[c(B)/(1\ M)]$ should be a straight line of slope unity with an intercept on the abscissa equal to log K_b. If the slope is not unity then either the antagonist is not acting competitively or more complex interactions are occurring between the antagonist and the receptor, possibly allosteric in nature. It is important to note from the Schild equation that the value of K_b, unlike that of IC_{50}, is independent of the precise agonist used to generate the data and is purely a characteristic of the antagonist for the specific receptor. A limitation of a Schild plot is that any error in measuring EC_{50} (i.e. the response in the absence of antagonist) will automatically influence the value of all the derived dose ratios. The intercept of a Schild plot on the y-axis also gives the pA_2 value for the antagonist (pA_2 is the dose of antagonist that requires a two-fold increase in agonist concentration). The pA_2 will be equal to the value of $-\log K_b$ since at an antagonist concentration that gives a dose ratio of $r = 2$, the Schildt equation reduces to log $[c(B)/(1\ M)] = \log K_b$. pA_2 is a measure of the potency of the antagonist.

Example 23.5   **SCHILD PLOT: CALCULATION OF A K_b AND A pA_2 VALUE**

Question   Use the data in Figure 23.15b to construct a Schild plot. The left-hand curve is the receptor response to agonist binding in the absence of antagonist. The next three responses are in the presence of 10^{-7} M, 10^{-6} M and 10^{-5} M antagonist. Read off from the graph the concentration of agonist required to produce 50% maximum response in the absence and presence of the antagonist and calculate the dose ratio (r) at each of the three antagonist concentrations. Then plot a graph of lg(r – 1) against lg[c(antagonist) / (1 M)] and hence calculate both pA_2 and K_b.

Answer

$c(B)$ (M)	10^{-7}	10^{-6}	10^{-5}
EC_{50} (M)	10^{-7}	10^{-6}	10^{-5}
r	10	100	1000
$r - 1$	9	99	999
lg $(r - 1)$	0.954	1.9956	2.999
lg $(c(B) / (1\ M))$	-7	-6	-5

You will see that the Schild plot is linear (correlation coefficient $R = 0.9999$) and that it has a slope of 1.02, confirming the competitive nature of the antagonist. The extrapolation of the line to the y-axis yields a value of -8.12. This is equal to $-\lg K_b$, hence $K_b = 7.58 \times 10^{-9}$ and $pA_2 = 8.12$.

23.8 SUGGESTIONS FOR FURTHER READING

23.8.1 Experimental Protocols

Ali H. and Haribabu B. (2006) *Transmembrane Signalling Protocols*, 2nd Edn., Methods in Molecular Biology, vol. 332, Humana Press, New York, USA.

Cortés A., Cascante M., Cárdenas M. and Cornish-Bowden A. (2001) Relationships between inhibition constants, inhibitor concentrations for 50% inhibition and types of inhibition: new ways of analysing data. *Biochemical Journal* 357, 263–268.

Karim N. & Kidokoro S. (2006) Precise evaluation of enzyme activity using isothermal titration calorimetry. *Netsu Sokutei* 33, 27–35.

Lorsch J.R. (2014) Practical steady-state enzyme kinetics. *Methods in Enzymology* 536, 3–15.

Williams M. and Daviter T. (2013) *Protein–Ligand Interactions: Methods and Applications*, 2nd Edn., Methods in Molecular Biology, vol. 1008, Humana Press, New York, USA.

Willars G.B. and Challiss R.A.J. (2016) *Receptor Signal Transduction Protocols*, 3rd Edn., Methods in Molecular Biology, vol. 746, Humana Press, New York, USA.

23.8.2 General Texts

Cornish-Bowden A. (2012) *Fundamentals of Enzyme Kinetics*, 4th Edn., Wiley-Blackwell, Weinheim, Germany.

Creighton T.E. (2010) *The Biophysical Chemistry of Nucleic Acids and Proteins*, Helvetian Press, Eastbourne, UK.

23.8.3 Review Articles

Arrang J.-M., Morisset S. and Gbahou F. (2007) Constitutive activity of the histamine H$_3$ receptor. *Trends in Pharmacological Sciences* 28, 350–357.

Baker J.G. and Hill S.J. (2007) Multiple GPCR conformations and signalling pathways: implications for antagonist affinity estimates. *Trends in Pharmacological Sciences* 28, 374–380.

Barglow K. T. and Cravatt B. F. (2007) Activity-based protein profiling for the functional annotation of enzymes. *Nature Methods*, 4, 822–827.

Furnham N., Holliday G.L., de Beer T.A., *et al.* (2014) The Catalytic Site Atlas 2.0: cataloging catalytic sites and residues identified in enzymes. *Nucleic Acids Research* 42, D485–489.

Grima R., Walter N.G. and Schnell S. (2014) Single-molecule enzymology a la Michaelis–Menten. *FEBS Journal* 281, 518–530.

Matsuo T. and Hirota S. (2014) Artificial enzymes with protein scaffolds: structural design and modification. *Bioorganic and Medicinal Chemistry* 22, 5638–5656.

Mross S., Pierrat S., Zimmermann T. and Kraft M. (2015) Microfluidic enzymatic biosensing systems: a review. *Biosensors and Bioelectronics* 70, 376–391.

Raynal M., Ballester P., Vidal-Ferran A. and van Leeuwen P.W. (2014) Supramolecule catalysis. Part 2: artificial enzyme mimics. *Chemical Society Reviews* **43**, 1734–1787.

Schnell S. (2014) Validity of the Michaelis–Menten equation – steady-state or reactant stationary assumption: that is the question. *FEBS Journal* **281**, 464–472.

Schwartz T.W. and Holst B. (2007) Allosteric enhancers, allosteric agonists and ago-allosteric modulators: where do they bind and how do they act? *Trends in Pharmacological Sciences* **28**, 366–372.

Seibert E. and Tracy T.S. (2014) Different enzyme kinetic models. *Methods in Molecular Biology* **1113**, 23–35.

Truhlar, D.G. (2015) Transition state theory for enzyme kinetics. *Archives of Biochemistry and Biophysics* **582**, 10–17.

Wyllie D.J.A. and Chen P.E. (2007) Taking the time to study competitive antagonism. *British Journal of Pharmacology* **150**, 541–551.

23.8.4 Websites

CalcuSyn: software for the analysis of ligand-binding data
www.biosoft.com/ (accessed May 2017)

Catalytic Site Atlas
www.ebi.ac.uk/thornton-srv/databases/CSA/ (accessed May 2017)

Enzyme Commission (EC) rules
www.chem.qmul.ac.uk/iubmb/enzyme/ (accessed May 2017)

GPCRdb: database of G protein-coupled receptors
gpcrdb.org/ (accessed May 2017)

Origin: data analysis and graphing software
www.originlab.com/ (accessed May 2017)

Prism: data analysis and graphing software
www.graphpad.com/ (accessed May 2017)

Protein Data Bank
www.rcsb.org/pdb/ (accessed May 2017)

R: software for statistical computing and graphics
www.r-project.org/ (accessed May 2017)

SDAR: data analysis and graphing software
www.structuralchemistry.org/pcsb/sdar.php (accessed May 2017)

SigmaPlot: data analysis and graphing software
www.systat.com/products/sigmaplot (accessed May 2017)

Tutorial on enzyme catalysis
bcs.whfreeman.com/webpub/Ektron/pol1e/Animated%20Tutorials/at0302/at_0302_enzyme_catalysis.html (accessed May 2017)

24 | Drug Discovery and Development

DAVID CAMP

24.1 INTRODUCTION

An enormous cache of human biology remains to be explored as discoveries and new insights emerge from the '-omics' sciences. The goal for many pharmaceutical companies (and increasingly also for academic organisations) is to find selective small molecules that can modulate these newly identified targets (mainly proteins), so that they can ultimately be developed into therapeutics. In this context, the term **small molecules** is used for organic compounds, typically with a molecular weight < 1000 dalton (Da) comprised mainly of carbon, hydrogen, nitrogen and oxygen atoms, but frequently also sulfur and fluorine, and less often chlorine, bromine and phosphorus. The modulation of new biological targets may require novel structural frameworks (called **scaffolds**), which contain various functional groups in strategic positions. The correlation between the chemical structure of compounds and their biological activities constitutes the so-called **structure–activity relationships** and their analysis is an integral part of drug development. Indeed, the identification of new structural motifs of small molecules is one of the many drivers *en route* to understanding biological systems and developing innovative, safer therapeutics with novel modes of action.

The task of finding a selective molecule is truly daunting considering there is an estimated 10^{20}–10^{200} compounds with a molecular weight below 500 Da comprised of the atoms that make up current small-molecule therapies. To put this into perspective, approximately 2.7×10^7 molecules have been reported to date, which equates roughly to the ratio of the mass of the Sun compared to the mass of a proton, if the Bohecek number (6×10^{62}) is used to represent the set of small molecules with drug-like properties (see Section 24.2.2). Moreover, it would be impossible to screen this large number of molecules across multiple assays if we were limited to using the available resources on our planet, given the Earth is estimated to contain 'only' 10^{51} atoms.

The real challenge, which will be the primary focus of this chapter, lies in:

- the initial selection of small molecules that are the most biologically relevant for the assays they will subsequently be tested in
- the identification of the best compounds following screening (called **hit compounds**)
- the downstream development of hits into a useful therapeutic.

To meet these challenges, a variety of strategies have been employed in contemporary drug discovery, which began around the 1990s when more modern approaches had to be developed to keep abreast of the massive amount of data generated by screening so-called mega-libraries (libraries in excess of 10^6 compounds) against a biomolecular target or cell line. The resulting paradigm shift was underpinned by advances in molecular biology to aid the creation of biomolecular targets, robust automation that facilitated the synthesis and screening of molecular libraries, and increasingly reliable informatics platforms to analyse data. Indeed, the **high-throughput screening** (HTS) of mega-libraries to identify compounds that impede the function of specific proteins is now commonplace in the pharmaceutical industry. The modern drug-discovery pipeline is outlined in Figure 24.1, and this chapter will concentrate on the stages from hit identification to lead optimisation. The different, often complementary, approaches to current drug discovery will also be explored in this chapter.

24.2 MOLECULAR LIBRARIES AND DRUG-DISCOVERY STRATEGIES

24.2.1 The Random Approach

When pharmaceutical companies first began preparing compound libraries for downstream high-throughput screening (HTS), it was thought that biological activity would be evenly distributed through chemistry space. Using this premise, the likelihood of a hit being found is increased if the molecular library is comprised of compounds with diverse chemical motifs and/or appendages. Vast libraries with assorted structural features were subsequently produced using established synthetic methodologies by a process known as **combinatorial chemistry**. Here, hundreds of thousands to millions of compounds could be synthesised in a highly efficient manner using different 'building blocks' that contained chemical functional groups that could react cleanly with each other. By way of example, eight different scaffolds (base structures) containing a carboxylate group could each be reacted with 12 different amines to afford 96 unique amides. The synthesis would be undertaken on a robotic platform in a 96-well plate, and the products subsequently purified, typically via automated high-performance liquid chromatography.

Unfortunately, it did not take long to realise that this hypothesis needed refinement, given the unacceptably low hit rates that were obtained following HTS. Additionally, attempts to progress many of the hit compounds to the next stage of the

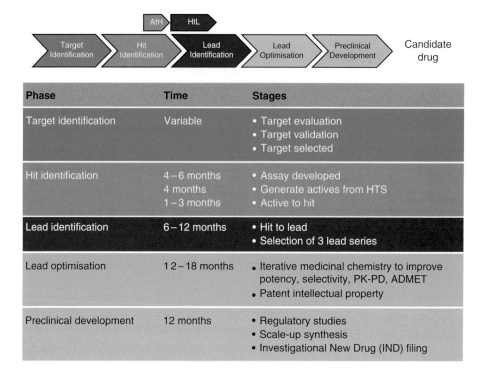

Figure 24.1 A generic timeline for the pre-clinical drug discovery process.

Phase	Time	Stages
Target identification	Variable	• Target evaluation • Target validation • Target selected
Hit identification	4–6 months 4 months 1–3 months	• Assay developed • Generate actives from HTS • Active to hit
Lead identification	6–12 months	• Hit to lead • Selection of 3 lead series
Lead optimisation	12–18 months	• Iterative medicinal chemistry to improve potency, selectivity, PK-PD, ADMET • Patent intellectual property
Preclinical development	12 months	• Regulatory studies • Scale-up synthesis • Investigational New Drug (IND) filing

drug-discovery process, i.e. lead identification (see Figure 24.1), were unsuccessful. Indeed, the observed activity for a substantial number of campaigns was caused by artefacts. This rather sad state of affairs prompted one of the pioneers who advocated improving the quality of compounds that made up screening libraries, Chris Lipinski, to comment that, 'The combinatorial libraries in the early years were so flawed that if you took the libraries across Pharma from 1992 to 1997 and stored them in dumpsters you would have improved productivity.'

Clearly, the random approach to drug discovery was not a successful strategy. If nothing else, the early data from HTS indicated that biological activity was not dispersed evenly through chemistry space and, moreover, suggested it was most likely confined to specific regions (i.e. areas of biologically relevant chemical space). Nevertheless, it must be remembered that biological activity also accounts for compounds that can have a deleterious effect on the body, such as poisons. Useful therapeutics, therefore, would occupy only a small portion of all biologically relevant chemical space, and an even smaller portion of the total space.

Contemporary drug discovery is now concerned with how biologically relevant chemistry space is defined and how libraries can be designed that effectively probe that space. The following sections describe some of the more well-known strategies. Although there are a number of complementary approaches, most of these mainly comprise variation of physico-chemical properties, structure or biological activity in a certain target class. Some, like the privileged structure approach (see Section 24.2.7) combine two of the three overarching themes.

24.2.2 Lipinski's Rule of Five and Drug-Like Molecules

Combinatorial chemistry was attractive at the onset of the high-throughput paradigm given its perceived promise to deliver large numbers of novel compounds in an effort to discover novel effective compounds (called new chemical entities or NCEs) more efficiently. However, as already noted, the early combinatorial chemistry libraries failed to live up to the hyperbole. This untenable position forced the pharmaceutical industry to critically examine the role of chemistry, since the vast numbers of compounds that could be efficiently synthesised and screened was ultimately deceptive – a false sense of security was created that, in turn, led to unrealistic expectations being placed upon nascent technologies to reverse the lack of new chemical entities emerging from company pipelines.

Physico-chemical properties drive a drug's efficacy, safety and metabolism by controlling binding affinities for macromolecules, typically proteins. The processes of desolvation, diffusion and resolvation required for passive permeability across a cell membrane are largely governed by a few fundamental physico-chemical properties, i.e. size, polarity, lipophilicity and conformational dynamics. A constructive interplay between these properties is necessary for a compound to be orally available. Highly polar molecules, for example, fail to desolvate and enter the cell membrane, while highly lipophilic compounds may not dissolve in the gastrointestinal tract or remain bound to the membrane.

Absorption is a complex process and in 1997 an examination of drug–likeness in the context of oral bioavailability was first reported by Lipinski and coworkers in their pivotal analysis that identified four physico-chemical properties common to the 90th percentile of approximately 2250 drugs and candidate drugs believed to have entered

Table 24.1 **Physico-chemical profiles for drugs, leads and fragments**

	Drugs	Leads		Fragments
Physico-chemical property	Rule of five	Lead-like	Reduced complexity	Rule of three
Molecular weight (Da)	≤ 500	≤ 460	≤ 350	≤ 300
$\log P$ or $cLogP$	≤ 5	$(-4.6) - (+4.2)$	≤ 2.2	≤ 3
H-bond donors	≤ 5	≤ 5	≤ 3	≤ 3
H-bond acceptors	≤ 10	≤ 9	≤ 8	≤ 3
Rotatable bonds	–	≤ 10	≤ 6	≤ 3
Polar surface area	–	–	–	≤ 60 Å^2
Heavy atoms	–	–	≤ 22	–
Number of rings	–	≤ 4	–	–
$\log S_w{}^a$	–	≥ -5	–	–

Note: ^{a}Logarithm of water solubility

phase II clinical trials. Poor absorption or permeation is more likely to occur if two or more physico-chemical properties (represented by so-called molecular descriptors, see also Section 17.3) exceed the parameters shown in Table 24.1.

Today, their analysis is better known as the **rule of five**, so-called because the maximum value for each parameter is a multiple of five. In essence, the rule of five is a set of guidelines that identifies key properties for *orally* administered drugs that permeate through a cell membrane via passive diffusion. Molecules obeying the rule of five are commonly referred to as drug-like; however, this is not really the case since the distribution of properties is similar in drug and non-drug libraries. The rule of five should be viewed more as a measure of the quality of bioactive molecules that highlights potential bioavailability issues if two or more violations occur. A brief description for each of the properties follows.

Molecular Weight (MW)

AstraZeneca scientists reported that the molecular weight (= molecular mass) of leads and drugs is linked to lipophilicity, as shown by the fact that as molecules are made larger then they also need to be made more lipophilic in order to successfully permeate through cell membranes. In Lipinski's rule of five, the cut-off of 500 Da for molecular weight is consistent with the experimentally observed upper limit of permeability of molecules through membranes via passive diffusion.

Calculated Logarithm of the 1-Octanol/Water Partition Coefficient (cLogP)

This property is a measure of lipophilicity (see also Section 24.2.4) and reflects the key event of desolvation as a molecule passes from an aqueous environment through a cell membrane, which is typically hydrophobic in nature, before binding to a protein. A highly lipophilic molecule will be less water soluble, which tends to decrease oral bioavailability from the onset. In addition, as lipophilicity is increased, there is a higher probability that the molecule will become less selective and attach to hydrophobic protein targets other than the one intended. This promiscuous behaviour carries the risk of greater toxicity as, for example, lipophilic compounds can undesirably bind to the active sites for P450 inhibition and the human ether-a-go-go-related gene (hERG) which codes for a potassium ion channel (see Section 24.8.5). Interestingly, this gene received its name following a mutagenesis experiment involving *Drosophila* flies that, after being anaesthised with ether, responded with a frenetic shaking of the legs and occasional twitching of the abdomen. This was apparently reminiscent of a then popular dancing style at the Whisky A Go-Go nightclub in West Hollywood in the 1960s when the discovery was made.

Hydrogen-Bond Donors (HBDs)

H-bond donors are defined in the rule of five as the sum of -OH and -NH moieties. Hydrogen bonding capacity is detrimental for the transport of a molecule into the non-polar environment of the cell membrane. Reducing the number of HBDs effectively lowers the desolvation energy penalty that occurs upon absorption into a lipid bilayer on account of the loss of hydrogen bonds to bulk solvent (water), thus facilitating permeability across the membrane.

Example 24.1 **APPRAISAL OF COMPOUNDS FOR COMPLIANCE WITH THE RULE OF FIVE PARAMETERS**

Question Molecular weight, H-bond donors and H-bond acceptors are three properties that can be gauged from examining the structure of a molecule. Lipophilicity is either determined experimentally or calculated (yielding cLogP).
Appraise the following compounds with respect to compliance with Lipinski's rule of five: aspirin (cLogP = 1.70), penicillin (cLogP = 1.82), doxorubicin (cLogP = −1.33), erythromycin (cLogP = −0.14).

Answer

Aspirin

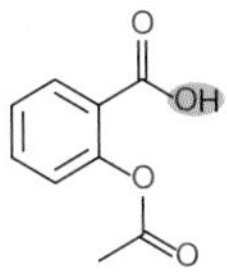

		Rule of five	
Molecular mass	180 g mol^{-1}	< 500 g mol^{-1}	✓
cLogP	1.70	≤ 5	✓
HBD = sum of OH and NH groups	1	≤ 5	✓
HBA = sum of O and N atoms	4	≤ 10	✓
Rule of five violations	0		

Penicillin

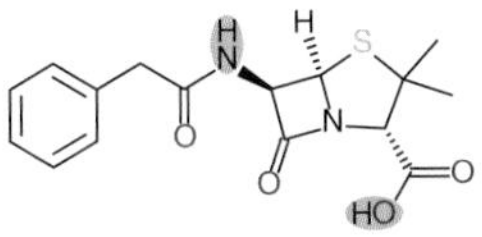

		Rule of five	
Molecular mass	334 g mol^{-1}	< 500 g mol^{-1}	✓
cLogP	1.82	≤ 5	✓
HBD = sum of OH and NH groups	2	≤ 5	✓
HBA = sum of O and N atoms	6	≤ 10	✓
Rule of five violations	0		

Doxorubicin

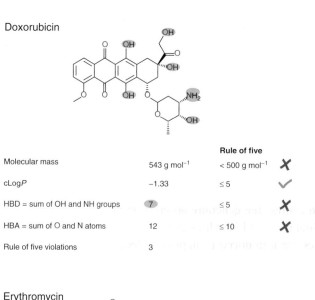

		Rule of five	
Molecular mass	543 g mol^{-1}	< 500 g mol^{-1}	✗
cLogP	−1.33	≤ 5	✓
HBD = sum of OH and NH groups	7	≤ 5	✗
HBA = sum of O and N atoms	12	≤ 10	✗
Rule of five violations	3		

Erythromycin

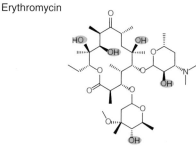

		Rule of five	
Molecular mass	734 g mol^{-1}	< 500 g mol^{-1}	✗
cLogP	−0.14	≤ 5	✓
HBD = sum of OH and NH groups	5	≤ 5	✓
HBA = sum of O and N atoms	14	≤ 10	✗
Rule of five violations	2		

Hydrogen-Bond Acceptors (HBAs)

H-bond acceptors are defined as the sum of oxygen and nitrogen atoms. Various publications have reported an overall increase in the mean MW of drugs launched prior to 1983, compared with those launched afterwards. Interestingly though, the mean cLogP has not changed in a statistically significant manner. This has been attributed, in part, to the interplay that exists when additional polarity or H-bonding is incorporated into a molecule. A study by Leeson suggested this was largely confined to oxygen and nitrogen atoms, and thus HBAs, but not HBDs.

Importantly, the rule of five also contains a significant caveat, i.e. 'compound classes that are substrates for biological transporters are exceptions to the rule'. The authors observed that certain classes, like antibiotics, antifungals, vitamins and car-diac glycosides, tended to violate the rule of five. Undeniably being highly effective drugs, these instances demonstrate that the rule of five should be seen as a guide rather than dogma. Further, there is currently considerable debate as to whether drugs permeate a cell exclusively via a carrier-mediated mechanism, or if this coexists with passive diffusion. Finally, while the rule of five is a valuable tool that helps define the end-point of a drug-discovery programme, it does not describe the physico-chemical properties required for chemical starting points.

24.2.3 Lead-Like Libraries

Lead-like molecules are smaller and more polar than drug-like molecules and were proposed to address the generally observed trend within the pharmaceutical industry that the initial hit from high-throughput screening tends to increase in mass as chem-ical moieties are introduced to improve selectivity, potency and bioavailability. For this reason, it makes sense to start below the drug-like profile so that additional mass and functionality can be added *en route* to the candidate drug without potentially suffering from 'molecular obesity'.

Exactly how far below the drug-like profile a chemical starting point should reside is open to interpretation (see Table 24.1). On the one hand, lead-like molecules have been defined by employing a property-based analysis. On the other hand, molecules with **reduced complexity** that have even more stringent physico-chemical profiles have also been advocated. Significantly, although the actual guidelines may dif-fer, the physico-chemical profiling of compound libraries that now occurs in the pharmaceutical industry has become an intrinsic part of the design and selection process.

The emphasis now placed on quality starting points reflects the simple fact that a project is less likely to succeed without a suitable molecular platform that lays the foundation for future development. This trend is also echoed by the progressively more stringent definition that has evolved for lead-like molecules over the past two decades as the cost to develop a new therapeutic has exceeded US$ 1 billion in some companies. Accordingly, the physico-chemical properties of the screening library are addressed early in contemporary drug discovery.

24.2.4 Congreve's Rule of Three and Fragment-Based Drug Discovery

Fragment libraries are designed for structure-based screening (see Section 24.5.2) and are conceptually no different from traditional compound libraries except that the mol-ecules are smaller. Congreve's rule of three (see Table 24.1) gives an idea of how the physico-chemical properties compare with drug-like, lead-like and reduced-complex-ity sets. Because the molecules comprising fragment libraries are more restricted in terms of their physico-chemical properties, are less structurally complex and occupy a smaller region of chemical space, it is somewhat counter-intuitive to learn that they are also more likely to bind to a greater number of targets and increase the probability of finding a hit. Indeed, a far greater sampling of chemical space can theoretically be achieved with a fragment-like library compared with a lead-like or drug-like library, even if there are many more orders of magnitude of compounds available in the latter two cases.

The theoretical basis lies in the observation that the probability of binding to a biological target decreases with increasing complexity of a small molecule because of the way chemical space grows exponentially with the number of atoms. Smaller molecules have fewer possible sites to interact with and, consequently, are more likely to match the corresponding elements on the target. As a result, however, fragments tend to have a low affinity towards the target and do not bind as strongly as larger compounds. Therefore, testing must be done at high, often millimolar, concentrations. In contrast to hits from biological screening (see Section 24.5.3), an acceptable hit from a fragment library could have an activity of 1 mM, i.e. a 1000 times less than a conventional hit (see Section 24.6). The IC_{50} value represents the concentration of a small molecule that is required to inhibit 50% of a biological process. The high concentrations required for fragment screening make many biochemical assays impractical, which means that binding must typically be observed directly using either Fourier transform mass spectrometry (Section 15.3.4), nuclear magnetic resonance (NMR) spectroscopy or X-ray crystallography (see Chapter 14).

Unlike high-throughput screening, which screens larger lead- or drug-sized molecules for their ability to modulate a protein target, fragment-based screening identifies potential inhibitors in a more calculated, knowledge-based approach by assembling the leads and, ultimately the drugs, piece by piece as the project progresses. Thus, while a single fragment with 1 mM binding is not a particularly attractive proposition, it can act as a starting point to further explore the binding site through attachment of other appropriately shaped fragments, so that a more strongly bound ligand is obtained; this technique is sometimes referred to as 'growing' (Figure 24.2). Alternatively, if multiple different fragments are also found to bind near the same site on a target, then they can be linked together to form more elaborate analogues with increased potency and selectivity. In both examples, the tether between the fragments is of critical importance. Structural information about the ligand-target binding can be acquired from X-ray crystallography and used to guide the positioning and length of the tether between individual fragments.

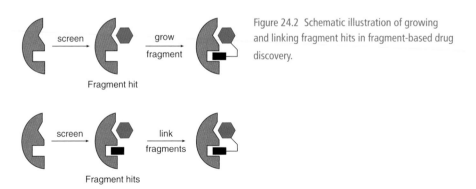

Figure 24.2 Schematic illustration of growing and linking fragment hits in fragment-based drug discovery.

Ligand Efficiency Metrics

The often low-affinity binding that is generally observed for fragments calls into question which hits should be selected for downstream optimisation. A useful tool used in fragment-based drug discovery to help filter hits is given by various **ligand efficiency** (LE) metrics. The two most widely used are LE and lipophilic ligand efficiency (LLE). LE is defined as the free energy of binding (ΔG) of the ligand for a specific target averaged for each non-hydrogen ('heavy') atom; for N heavy atoms this yields:

$$\text{LE} = -\frac{\Delta G}{N} \qquad \text{(Eq 24.1)}$$

Because the free energy of binding can be expressed as $\Delta G = \text{R} \times T \times \ln K_d$ (see also Equation 23.10), where K_d is the (unit-less) dissociation constant, R is the ideal gas constant, and T is the temperature, this can be rewritten as:

$$\text{LE} = -\frac{\text{R} \times T \times \ln K_d}{N} \qquad \text{(Eq 24.2)}$$

However, in practice, IC_{50} values can be substituted for K_d to gain a relative comparison. In analogy to pH and pK (see Section 2.3.2), the parameter pIC_{50} is defined as the negative decadic logarithm of the binding affinity of a hit compound:

$$pIC_{50} = -\lg\left(\frac{IC_{50}}{1\,\text{M}}\right) \qquad \text{(Eq 24.3)}$$

Equation 24.2 can therefore be transformed to yield an expression that is based on pIC_{50}:

$$\text{LE} = -\frac{\text{R} \times T \times \ln\left(\dfrac{IC_{50}}{1\,\text{M}}\right)}{N} = -\frac{\text{R} \times T \times \dfrac{\lg\left[IC_{50}/(1\,\text{M})\right]}{\lg e}}{N} = \frac{2.303 \times \text{R} \times T \times pIC_{50}}{N} \qquad \text{(Eq 24.4)}$$

In this sense, LE can be thought of as the average binding affinity per heavy atom.

Lipophilic ligand efficiency (LLE) extends the concept of LE by also taking into account **lipophilicity** (Equation 24.5). This recognises the trade-off that exists between activity and lipophilicity when selecting hits following screening. Thus, LLE is defined as the difference between pIC_{50}, and the logarithm of the partition coefficient (log$P$, or cLog$P$ if the value is calculated).

$$\text{LLE} = pIC_{50} - \text{cLog}P \qquad \text{(Eq 24.5)}$$

It has been suggested that acceptable values for drug candidates are LE > ~1.3 kJ mol^{-1} and LLE > ~5 based on an IC_{50} value of 10 nM, molecular weight of 500 Da (typically containing ~38 heavy atoms) and a cLog$P$ < 3. In the development of novel orally available drug leads, compounds are thus modified with the goal to achieve highest possible activity (large LE) and maximum bioavailability (large LLE).

Example 24.2   **CALCULATION OF LIGAND EFFICIENCY AND LIPOPHILIC LIGAND EFFICIENCY**

Cholesteryl ester transfer protein (CETP) is an important target for atherosclerosis as its inhibition raises levels of high-density lipoprotein and concomitantly lowers that of the low-density lipoprotein. CETP has proved a challenging target when it comes to controlling the physico-chemical properties of potential candidate drugs, with 53% of ligands violating the rule of five. Boehringer Ingelheim has nevertheless managed to increase the potency and reduce the lipophilicity of a hit that was active in inhibiting CETP by

● replacing carbon atoms with heteroatoms
● removing non-polar groups and adding hydrophilic groups
● keeping the molecular weight under control.

This strategy is reflected in the ligand efficiency (LE) and lipophilic ligand efficiency (LLE) values of the initial hit and lead compound developed from it. In developing the lead, LE was kept constant, but LLE has been vastly improved.

Question   Appraise the screening hit and the lead compound shown above with respect to LE and LLE. Note that, usually, the temperature chosen for these calculations is 27 °C (i.e. $T = 300$ K).

Answer

Appraisal of the initial screening hit

IC_{50}	250 nM
pIC_{50} (Equation 24.3)	$-\lg (250 \times 10^{-9}) = 6.6$
Number of heavy atoms	30
cLog$P$	7.6
LE (Equation 24.4)	$\dfrac{2.303 \times 8.3144 \text{ J K}^{-1} \text{ mol}^{-1} \times 300 \text{ K} \times 6.6}{30} = 1.3 \text{ kJ mol}^{-1}$
LLE (Equation 24.5)	$6.6 - 7.6 = -1.0$

Appraisal of the lead compound

IC_{50}	20 nM
pIC_{50} (Equation 24.3)	$-\lg (20 \times 10^{-9}) = 7.7$
Number of heavy atoms	34
$cLogP$	4.6
LE (Equation 24.4)	$\dfrac{2.303 \times 8.3144 \text{ J K}^{-1} \text{ mol}^{-1} \times 300 \text{ K} \times 7.7}{34} = 1.3 \text{ kJ mol}^{-1}$
LLE (Equation 24.5)	$7.7 - 4.6 = 3.1$

24.2.5 Other Important Physico-Chemical Indices

Since publication of the seminal paper by Lipinski and coworkers, drug hunters have become increasingly mindful of the causal relationships between physico-chemical properties of potential drug candidates and models that predict absorption, distribution, metabolism, excretion, and toxicity (ADMET) in humans. Subsequent studies have identified additional indices considered to be useful predictors of oral bioavailability. The more commonly employed ones are outlined below.

Number of Rotatable Bonds and Polar Surface Area (Veber's Rule)

Rotatable bonds are defined as any single bond, not in a ring, bound to a non-terminal (i.e. non-hydrogen) atom. C–N bonds are excluded due to their high rotational energy barrier. Increased molecular flexibility can reduce the ability of a compound to cross a membrane. The polar surface area is the sum of surfaces of polar atoms in a molecule and is typically calculated from its three-dimensional structure. The upper value for the polar surface area is equivalent to 12 H-bond donors and/or H-bond acceptors, which correlates with Lipinski's rules. Based on an analysis of 1100 drug candidates it appears that compounds with 10 or less rotatable bonds and a polar surface area less than or equal to 140 Å^2 are more likely to possess good oral bioavailability in rats.

Aromatic Ring Count

The number of aromatic rings in molecules being considered for drug discovery and development has received greater attention after an analysis of GlaxoSmithKline compounds was published that outlined how increasing the number of aromatic rings, and particularly benzenoid rings, negatively impacted on properties like aqueous solubility, serum albumin binding and hERG inhibition (see Section 24.8.5). Aqueous solubility was shown to decrease, even when $cLogP$ was kept relatively constant, suggesting aromatic ring count is influencing solubility independently of lipophilicity. Since higher numbers of aromatic rings increase lipophilicity (and often potency), but at the expense of solubility, it has been suggested that their numbers in an orally administered drug be limited to three.

Number of Rings

The number (and type) of ring systems affords some idea of the rigidity and complexity of the compound. Flexible ligands with identical H-bond and hydrophobic interactions generally show weaker binding affinity towards a protein compared with a more rigid analogue. This can be attributed to a greater loss of **entropy** upon binding in the case of more flexible molecules. The incorporation of simple or more complex fused, bridged and spiro ring systems are employed to increase rigidity.

Tetrahedral Carbon Atom Shape-Based Descriptors

Around the same time that the deleterious effects of increasing aromatic ring count were published, it was reported that two simple shape-based descriptors, namely the fraction of tetrahedral carbon atoms in a molecule and the number of **stereogenic carbon atoms** (sometimes incorrectly referred to as 'chiral carbons'), correlated with aqueous solubility and the successful passage of compounds from discovery, through clinical testing, to (orally available) drugs. Indeed, the shift to combinatorial chemistry libraries comprised of predominately achiral, aromatic compounds may have been a contributing factor behind the high failure rate observed at the onset of the high-throughput era. As a consequence, contemporary drug design includes efforts to decrease the aromatic ring count and, concomitantly, increase the proportion of tetrahedral carbon atoms, which often leads to increased aqueous solubility.

Calculated Logarithm of the Distribution Coefficient (cLog D)

In contrast to cLogP, the calculated logarithm of the 1-octanol/water distribution coefficient cLogD takes into account both the ionised and neutral (non-ionised) forms of a compound at a fixed pH, typically the physiological pH = 7.4, is another important parameter, given that the degree of ionisation of a compound can dramatically effect crucial physico-chemical properties such as solubility, lipophilicity and permeability. According to **pH–partition theory**, it is the neutral species that preferentially permeates through the gastrointestinal and other lipid membranes via passive diffusion. Additionally, the theory correlates lipophilicity to both the rate and the degree to which a compound is absorbed and permeates a membrane.

Aqueous Solubility

Aqueous solubility, often expressed as $\text{Log}S_w$, is an important property associated with oral drug absorption and distribution. More than 80% of marketed drugs have a $\text{Log}S_w$ value greater than −4. Generally speaking, poor absorption can be traced back to a compound that has low solubility. Improved formulation strategies can assist in making insoluble compounds more bioavailable, but it is likely that the body must then work harder to eliminate them. Ultimately, this can place more demands on the medicinal chemist during lead optimisation to make molecules that are more metabolically stable and thus prolong their duration of action.

Finally, in a somewhat more lateral approach to identifying the most critically important physico-chemical properties in drug discovery, it has been proposed that those properties that remained consistent over time (when comparing orally administered drugs) were the criteria to consider when constructing a screening library or undertaking downstream optimisation of leads. Examination of drug approvals prior to, and after, 1983 indicated that the median cLog$P$, percentage polar surface area and the number of HBDs were virtually unaffected, while other physico-chemical properties like the molecular weight increased steadily over time in a statistically significant manner. Of the three most constant physico-chemical properties identified, lipophilicity, as determined experimentally or through computation is considered the 'Lord of the rules' for drug discovery and development.

24.2.6 Diverse Libraries

Diversity remains a key aspect in the design and production of screening libraries; however, the emphasis once placed on producing large numbers of compounds as exemplified by the random method (see Section 24.2.1) has been transformed into a more nuanced approach. Nowadays, smaller libraries are built around a larger number of scaffolds that align with current wisdom on physico-chemical profiles. A scaffold is a structural framework, most often a ring system, that contains various functional groups strategically positioned around the periphery. These functional groups can react with complementary chemical moieties on other molecules in a selective manner to afford a small-molecule library.

Structural novelty is an especially important consideration in the pharmaceutical industry, as the possibility of patenting lead molecules must stand up to the rigours of a patent application in the highly competitive world of drug discovery. Building novelty into the screening library addresses this concern early in the process. The novelty may come from new approaches like diversity-oriented synthesis, that utilise known chemical reactions to deliver, for example, skeletal diversity through strategically placed 'branching' points in the starting material when reacted with different reagents or catalysts.

24.2.7 Focussed, Targeted and Privileged Structure Libraries

Screening diverse libraries is ideal when little is known about the function or the binding site of the biological target. However, given the vastness of chemical space, there is always a real danger that biologically irrelevant regions will be populated. Because screening a large library is costly, there has been a shift towards focussed libraries (also called biased or directed libraries). Here, the screening library can be focussed either toward the chemistry or the biology. In the first instance (chemistry), the library is focussed in some way on a particular theme used to guide assembly of the screening set, e.g. molecules with drug-like properties, known drugs or natural products. In the second approach (biology), the screening set is focussed around a limited number of scaffolds for which binding interactions with the target are generally known. Peripheral functionality is added, removed or modified in some way in an effort to improve affinity and selectivity.

Targeted libraries refine the concept of focussed libraries to concentrate on specific chemotypes (chemical motifs) for target clusters and families. Because the active site of each protein in a target class would be expected to vary in some way, targeted libraries would ideally contain a diverse set of small molecules that have the potential to probe as many members of the protein family as possible. Clearly, the successful design of targeted libraries is dependent on prior knowledge of any ligands that bind to the target protein in the active site or indeed the three-dimensional structure of the active site. Ultimately, the type of information available for a particular target class will determine whether ligand-based (i.e. interrogating the protein family via small molecules) or structure-based approaches can be employed. A ligand-based approach is most useful for target families with minimal structural data such as G-protein-coupled receptors (GPCRs). In contrast, a structure-based approach could be used to design a targeted library of kinase inhibitors, given the significant amount of structural information already available for this family of proteins.

Another popular strategy has been to exploit privileged structures, with known drug-like properties. Privileged-structure screening essentially relies on chemotypes such as benzodiazepines or opioids that are known to target specific gene families. Conceptually, the privileged structure approach lies somewhere between focussed and targeted library strategies, i.e. modifications on a known motif (focussed approach) are undertaken to increase the library size for subsequent biological evaluation against other targets from a gene cluster or family (targeted strategy).

24.2.8 Bio-Isosteres and Scaffold Hopping

The idea behind bio-isosterism and scaffold hopping is that single atoms, substructures or even entire molecules that have similar volume, shape and/or physico-chemical properties to a known biologically active (bioactive) molecule can produce a similar biological response. The process begins through isosteric substitution of an atom or group of atoms in the parent molecule with other atoms/groups having similar electronic and steric configurations, so that the core structure morphs into a novel chemotype, e.g. the replacement of the oxygen atom in an analogue of the analgesic pethidine initially with an -NH-, then with the -CH$_2$- group (Figure 24.3). Importantly, if the resultant isosteres can elicit a broadly comparable biological profile, then they are referred to as bio-isosteres (Figure 24.4).

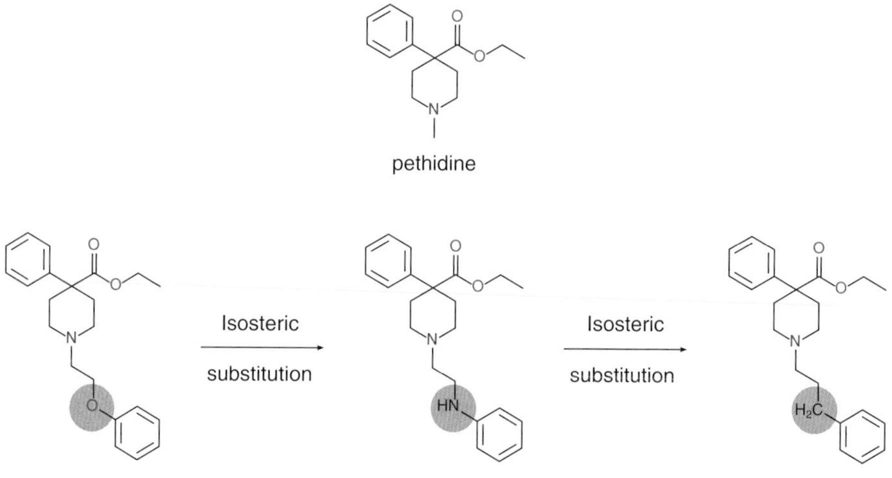

pethidine

Figure 24.3 Example of isosteric substitution. The pethidine analogue on the left with an oxygen atom in the linker is replaced first by a nitrogen atom, then a carbon atom. Relative to pethidine, the analgesic activity of the oxygen analogue is 12× more potent, the nitrogen analogue 80×, and the carbon analogue 20×.

The concept of bio-isosterism can be extended to the complete overhaul of core structures. This strategy, more commonly known as **scaffold hopping**, aims to explore new areas of chemical space by modifying the central motif of a known bioactive, whilst maintaining a similar biological response, as shown by the non-steroidal anti-inflammatory drugs (NSAIDs) in Figure 24.5. Scaffold hopping is a particularly attractive option to industry as it provides a mechanism that can address novelty and achieve a favourable intellectual-property position.

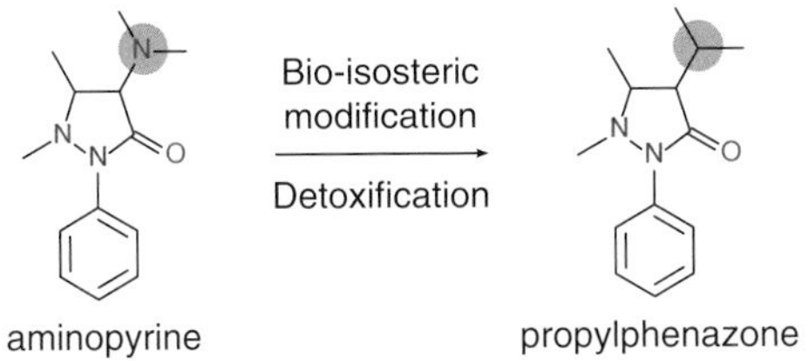

aminopyrine

Bio-isosteric modification

Detoxification

propylphenazone

Figure 24.4 Example of bio-isosteric substitution. Initially marketed in 1896 as an analgesic and anti-inflammatory, it was later revealed in 1922 that aminopyrine was a carcinogen. Propylphenazone was developed by Roche in 1951. Bio-isosteric modification of the dimethylamino group removed the carcinogenic properties.

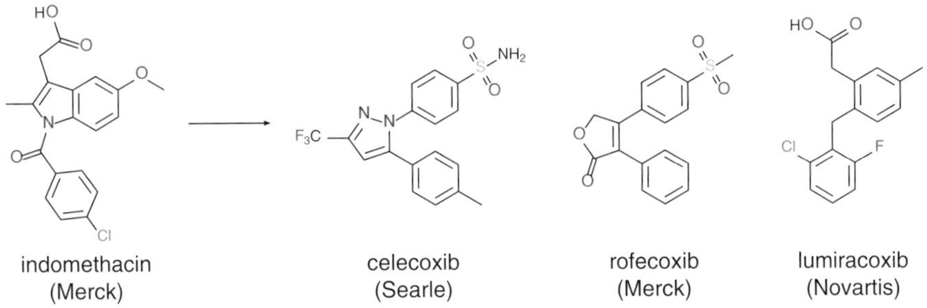

indomethacin (Merck)

celecoxib (Searle)

rofecoxib (Merck)

lumiracoxib (Novartis)

Figure 24.5 An example of scaffold hopping to yield NSAIDs from the lead molecule, indomethacin.

24.2.9 Drug Repurposing

Another technique used for the identification of bioactive molecules is **drug repurposing** (also called redirecting, repositioning or reprofiling). This concept was introduced in 2004 and was initially limited to the process of finding new uses for existing pharmaceuticals. The original idea was that drug repurposing offered greater benefits over *de novo* drug discovery (the traditional approach that searches for a new bioactive compound having a therapeutic value) as the development risks would be reduced given there was already a wealth of ADME and toxicity information about the drug. Over time, however, this definition has been expanded to include the 'rescue' of drug candidates that industry had de-prioritised for a variety of reasons (sometimes colloquially referred to as 'fallen angels'). This strategy is particularly attractive to universities and publicly funded research organisations that have strengths in the discovery of new targets that may be implicated in a particular disease, but not the finances to collate a significant

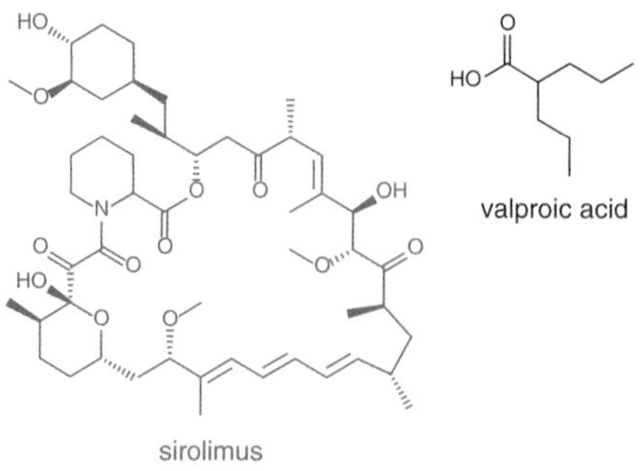

Figure 24.6 Two examples of drug repurposing for the treatment of cancer.

valproic acid

sirolimus

compound library or develop a hit into a drug. This approach has led to the unexpected discovery of anticancer activity in compounds such as sirolimus (an immunosuppressant also known as rapamycin) and the antiepileptic agent, valproic acid (Figure 24.6).

While there are generally no issues for programs that screen drugs that have come off patent, issues may arise around intellectual property for drugs that are still under some sort of patent protection. In this scenario, completely novel therapeutic applications must generally be pursued. By way of example, Celgene successfully repurposed thalidomide (an antiemetic once used to treat morning sickness in pregnant women, but later discovered to be a teratogen) for leprosy; however, due to its teratogenic effects, use of this drug is not recommended by the WHO.

24.2.10 Probe Compounds

At the other extreme, a compound library could be comprised entirely of **probe compounds** or chemical 'tools'. Semantically, probes and tools are molecules in which the rules for lead- and drug-likeness are relaxed, as the ultimate goal is not drug discovery, but rather chemical biology. Chemical biology examines biological systems through the application of chemical techniques and tools; its goal is to use small-molecule probes to discover specific protein targets and pathways that are modulated by the particular compound. This contrasts with classical molecular biology techniques that permits specific proteins to be eliminated by 'knocking out' genes; increasing the concentrations of particular proteins by increasing the number of copies of the corresponding genes or by using a more active gene promoter; altering the function of a protein by introducing specific mutations in the corresponding gene. Although these methods are powerful in model organisms such as *Saccharomyces cerevisiae* and *Drosophila melanogaster*, mammals represent a significant experimental challenge for molecular biology because of slower rates of reproduction, larger sizes and larger genomes. The chemical biology approach avoids these problems by studying the effect of small molecules on the mammalian proteome.

Small molecules typically modulate protein function by inducing conformational changes or by competing for endogenous protein–ligand or protein–protein interaction sites, resulting in altered activity (see also Chapter 23). This allows a temporal study of signalling pathways and the ability to wash out probes to study reversible inhibition. In this respect, probes are chemical tools used to validate targets and interrogate function to further our understanding of biological processes.

The important point to note is that although drug discovery and probing biological function are equally important, the types of compounds used in drug discovery are more limited. Target validation and interrogating biological function are not limited to molecules obeying lead- and drug-like rules. However, the two goals are not necessarily exclusive. A hit from a library based on natural product probes that does not comply with Lipinski's rule of five may provide a foundation for more focussed libraries that explore structure–function relationships within a particular protein family. Ultimately, enough information may be gained to progress the original chemical starting point into a molecule with a therapeutic application.

24.3 ASSEMBLING A MOLECULAR LIBRARY

Assembling libraries for subsequent screening was historically done in-house by many pharmaceutical firms, and typically involved the screening of purified natural products or their derivatives, prepared one at a time, and then screened in whole-animal systems. As understanding of the scientific disciplines that underpin modern drug discovery grew, it was found that the proteins were often the actual targets of many drugs. Indeed, by this time, the monopoly of natural products was starting to weaken, as more drugs with synthetic (non-natural) motifs were approved. Advances in automation facilitated the screening of tens of thousands of compounds in days, compared to years. By the early 1990s, high-throughput screening was heralded (somewhat prematurely as it turned out) as the pinnacle of drug discovery, if only there were enough compounds to assay. The following acquisition strategies were prosecuted in an effort to grow screening libraries.

24.3.1 Corporate Collections

At the outset of high-throughput screening, a company's compound collection became its screening library. In the major pharmaceutical companies at the time, these corporate collections could number between 50 000 and 100 000 compounds and, not surprisingly, reflected the areas of past research. As a result, the level of chemical diversity was not high, prompting aggressive augmentation strategies in an effort to explore a much wider area of chemical space in line with the conventional wisdom of the time. As noted earlier, combinatorial chemistry and parallel synthesis quickly became the methods of choice for producing large numbers of molecules in short time frames. New compounds would also enter corporate collections, albeit at a much slower rate, as they were produced following a screening campaign during the lead identification and optimisation phases, where structure–activity relationships (SAR) around a hit or lead structure were explored in greater detail (see Sections 24.6 and 24.7).

24.3.2 Vendor Libraries

Another approach to augmenting a corporate collection is to purchase compounds, as there is a significant cost saving compared with producing new compounds in-house. Somewhat fortuitously for the pharmaceutical industry, the appetite for large amounts of new compounds coincided with the dissolution of the former Soviet Union in

1991. Many laboratories that had previously been funded by the Soviet government suddenly found themselves without a benefactor. Some of these laboratories formed companies (e.g. ChemBridge, ChemDiv, Enamine) and sold compounds (typically in the range of 0.5–1.0 mg) that they held in stock.

A concern if purchasing compounds from several suppliers is the amount of overlap between collections from different vendors. To this end, a recent publication by Günther and coworkers analysed the catalogues of 115 suppliers and the ZINC database. Of the approximately 125 million compounds available from these combined sources, only about half (68 million) were found to be unique. Just as importantly, their analysis also revealed that a significant proportion of the total set (13 million, 10%) did not pass recognised medicinal chemistry filters like **pan-assay interference compounds** (PAINs) or reactive group alerts (see Section 24.3.5). From this, a general caveat was broadcast by the authors: 'unsupervised use of commercial catalogues for screening is [to be] discouraged.'

On the plus side, though, a detailed analysis of scaffold diversity revealed that there were thousands of scaffolds in commercially available compounds unique to each dataset. Moreover, there was a wide range of molecular complexity among the combined datasets, ranging from very simple motifs to more complex natural-product-like structures with high scores in the fraction of tetrahedral carbon atoms, including stereogenic carbon atoms.

24.3.3 Outsourced Libraries

Large pharmaceutical companies have traditionally undertaken all the different aspects of drug discovery in-house. However, given the increasing cost of successfully launching a new drug on the market, currently estimated to be over US$1 billion in some companies, there has been an appreciable rise in outsourcing various components, including access to compound libraries. Thus, instead of preparing or purchasing a library, companies can effectively hire one. Here, the owner of the library typically enters into an agreement that allows a drug company to access either a subset or the whole library upon payment of an up-front fee. If the outsourced library produces a hit in the screening campaign, then the owner could also receive milestone payments as the compound progresses through the drug-discovery pipeline. Additional revenue in the form of royalties is also possible if the original hit is ultimately developed into a marketed product.

Outsourcing libraries has increasingly been taken up by academic organisations over the past decade. By way of example, the Eskitis Institute at Griffith University has produced a screening set of approximately 200 000 natural-product extracts, known as Nature Bank. The extracts have been derived from plants and marine invertebrates and, importantly, have: (i) removed various nuisance compounds that can interfere with some assay technologies and (ii) been enriched with components that comply with drug-like Log$P$ scores.

24.3.4 Natural Products

It has been estimated that up to half of all drugs on the market originated from natural products, either directly with no modifications, as an analogue based on a natural pharmacophore, or otherwise inspired by them in some way. Indeed, the vast majority of antibiotics (the penicillins, cephalosporins, tetracyclines and vancomycin to name a few), and anticancer compounds (paclitaxel, the anthracyclines, ixabepilone, trabectedin and eribulin, Figure 24.7) were all derived from natural sources, as was the blockbuster cholesterol-lowering statin class of compounds that originated from the fungal-derived natural product, lovastatin. Despite this undeniable track record, their popularity has waned considerably over the past 20–30 years. The decline can be traced back to the paradigm shift noted earlier, that coalesced advances in molecular biology, combinatorial chemistry and high-throughput screening. At the same time access to the world's biodiversity became harder to legally acquire and use for commercial research due to lack of certainty and clarity over access and benefit-sharing requirements.

Clearly, natural-product drug discovery would not be possible without access to the world's biodiversity, of which greater than 80% of terrestrial organisms are estimated to be spread across a mere 17 so-called megadiverse countries. A country's riches in biodiversity have, for the most part, been inversely proportional to wealth and scientific capacity to not only undertake biodiscovery, but also commercialise outcomes. As a consequence, megadiverse countries from the developing world became suppliers of biodiversity that the developed world translated into commercial outcomes. By the late 1980s, many governments of developing countries effectively lobbied the United Nations that the situation was inequitable, which ultimately ushered in the Convention on Biological Diversity.

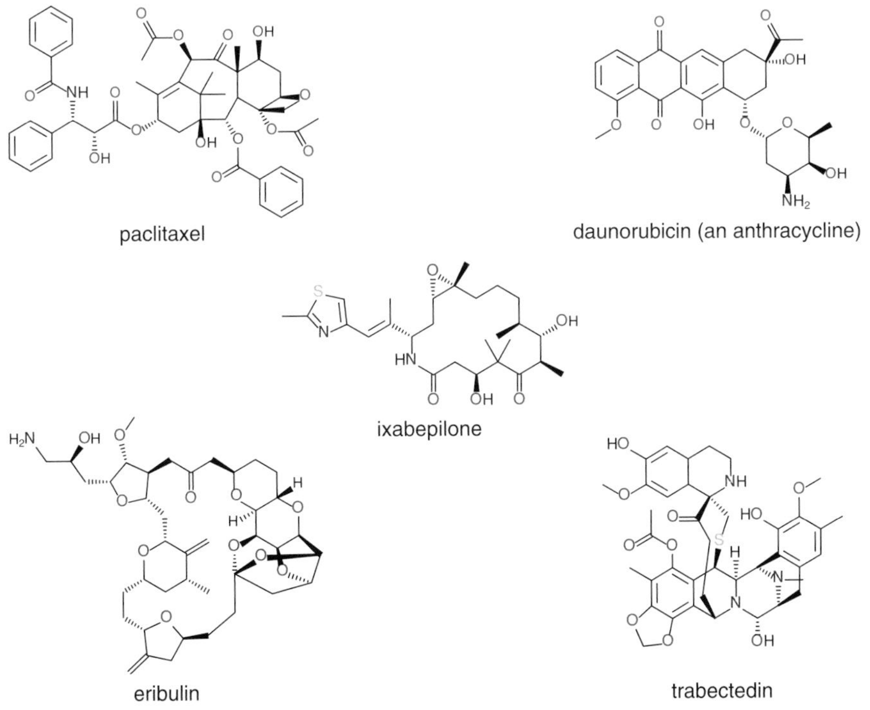

paclitaxel

daunorubicin (an anthracycline)

ixabepilone

eribulin

trabectedin

Figure 24.7 Anticancer drugs where the lead molecule originally came from nature.

The synergies between new technologies, and the changing landscape for bio-discovery, led many pharmaceutical companies to either cut back or disband their natural-product drug-discovery programmes and shift resources into the new high-throughput paradigm. To maintain the forward momentum of projects performed in industry that utilised both pure compound library and natural-product extract screening, hits from pure compound libraries were often progressed to the hit-to-lead phase in six months, while it took one or two full-time employees in this time frame to complete a reasonable number of **bioassay-guided fractionations** (say 20) to arrive at the hit stage. Bioassay-guided fractionation involves the large-scale extraction of an organism (or culture in the case of many bacteria or fungi), followed by iterative rounds of chromatographic fractionation and biological testing, to home in on the pure compound responsible for the desired activity. This disadvantage increasingly marginalised natural-product drug discovery and it was often viewed as a last resort by project teams, only to be considered when screening the compound library did not yield any leads. Put simply, classical bioassay-guided fractionation so engrained in natural-product drug discovery was not competitive with the high-throughput screening timelines of pure compound libraries. This attitude has changed in recent years, as the output from pharma pipelines has slowed. Proponents of natural-product drug discovery have argued that natural products are already biologically validated, having been produced by an enzyme to interact with other proteins, either within the same organism or another. Because there are a limited number of ways in which a protein can fold, it has been suggested that so long as the topology of a protein cavity is conserved in terms of the positioning of polar and lipophilic residues, then natural products that bind strongly to one protein have a high probability of binding to a protein with a similar cavity.

More recent attempts to better integrate natural products into the high-throughput paradigm have included the preparation of screening libraries containing a diverse set of pure natural compounds that are drug-like in their physico-chemical parameters. However, it was noted that although this strategy can compete with the screening of combinatorial chemistry libraries, it nevertheless fails to deliver a comprehensive coverage of natural-product chemical space. The exposure of a larger range of natural product scaffolds can only be achieved via a comprehensive sampling of the chemical diversity found in nature by screening extracts. Here, a number of strategies have been devised that enrich the extracts with components containing desirable chemical motifs (e.g. alkaloids) or desirable physico-chemical properties.

24.3.5 Structural Alerts and Pan-Assay Interference Compounds (PAINs)

Because the original strategy to increase the number of new chemical entities (NCEs) flowing through company pipelines was to screen every available compound, it was not uncommon to find that synthetic intermediates, reagents and even catalysts were incorporated into the early screening collections. Inevitably, many of the hits from these sets contained compounds that were false positives (compounds that interfere with the assay technology in a manner that does not involve specific binding to the target protein). There are a number of reasons why a compound could be regarded as a false positive, but the three most common are:

- The compound reacts in some way with the protein target
- Some compounds can form aggregates that then interfere with binding interactions
- Compounds can directly interfere with the assay signal.

Removal of suspected false positives is critical to ensure the quality of high-throughput screening. The latter two reasons will be discussed in more detail in Section 24.5.3.

In the case of protein-reactive compounds, it was realised that not every imaginable structure was suitable for screening; *in silico* (cheminformatics) methods were developed to filter out compounds containing undesirable substructures. One of the most common causes of false positives at the onset of high-throughput screening was the presence of functional groups that could irreversibly react with proteins and consequently interfere with the readout in the assay. Thus, **structural alerts** were developed that initially included alkylating and acylating reagents; such molecules can react with a particular amino-acid-residue side chain in a protein. Structural alerts also had to account for reactive groups present in synthetic intermediates that were traditionally produced (and submitted to the corporate collection) by medicinal chemists *en route* to the desired target molecule. Such functional groups flagged by these structural alerts included aldehydes, epoxides and alkyl halides, among others.

Structural alerts could also include molecular features that would make compounds unattractive as potential starting points in the hit-to-lead phase of drug discovery. Thus, molecules that contained too many halogen atoms or those considered too structurally complex, as judged by the number of stereo centres and fused ring systems, could be discounted. Additionally, structural alerts were coded for functional groups known to be or implicated as the cause of toxicity, such as aromatic amines, hydrazines and diazo groups. Some of the most widely applied structural alerts are listed in Figure 24.8.

Recently, the idea of structural alerts was extended by Baell and coworkers, who provided evidence that many hits from high-throughput screening were more likely to be caused by artefacts rather than acting like a true drug that inhibits or activates a protein by fitting into a binding site. Several compound classes were identified that lacked any obvious reactive groups yet ostensibly exhibited drug-like binding across a variety of assays and targets classes. Taken one screen at a time, it would be difficult to identify these 'serial offenders' (named **pan assay interference compounds** or PAINs), but when multiple screens were analysed concurrently, their subversive reactivity was brought to light. The issue at stake here is that a massive amount of resources have already been spent on developing hits from these classes that are essentially chemical and biological dead ends. Indeed, patents have even been filed before the full extent of PAINs was realised.

PAINs can act in a variety of ways, which makes their detection in any single assay format problematic; however, most PAINs function as reactive chemicals rather than discriminating drugs. Some are fluorescent or strongly coloured so that a positive signal is obtained, even though no protein-related activity may be present. Others can bind to metals, typically used in catalysts to synthesise molecules in a medicinal chemistry programme. These metals then produce a signal that has no relationship with the small-molecule–protein interaction. Yet others can coat a protein or sequester

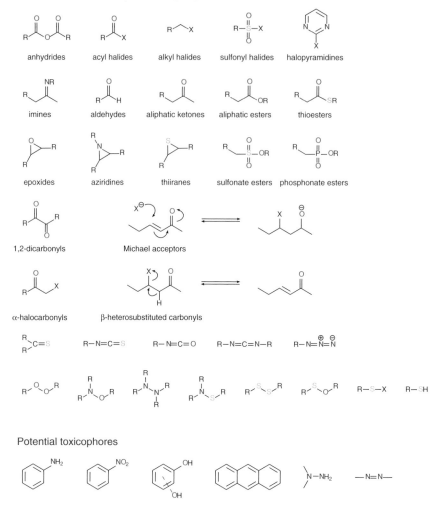

Figure 24.8 Common structural alerts in screening libraries.

metal ions that are essential to the protein functioning properly. Some classes of PAINs such as the toxoflavins, quinones and catechols act as redox cyclers that can produce hydrogen peroxide, an antiseptic that is produced by certain immune cells, under some assay conditions. In the assays of a drug-discovery programme, hydrogen peroxide can inactivate the protein of interest, thus making the PAINs appear to be good inhibitors.

It has been estimated that 400 structural classes could be considered PAINs, although this number is much smaller in a typical screening library. Some of the more insidious PAINs that may still be present in many screening libraries are shown in Figure 24.9.

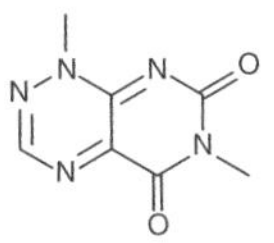

Toxoflavin: a redox cycler that can produce hydrogen peroxide, which can activate or deactivate proteins.

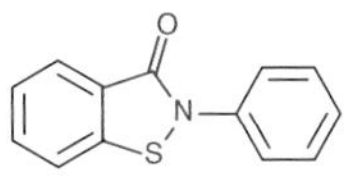

Isothiazolones: covalent modifiers that react with proteins in a non-specific, non-drug-like manner.

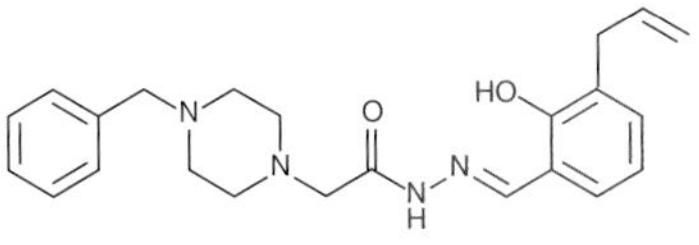

Hydroxyphenyl hydrazones: metal complexers that bind to metal ions required by some proteins to function.

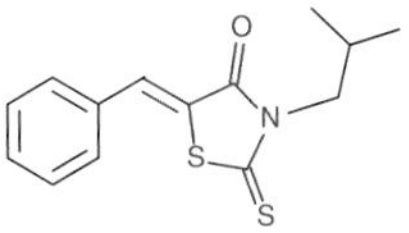

Ene-rhodanines: covalent modifiers and metal complexers.

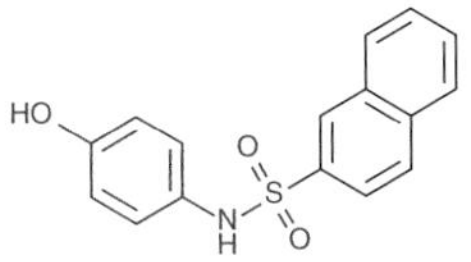

Phenol sulfonamides: unstable compounds that break down into molecules that give false signals.

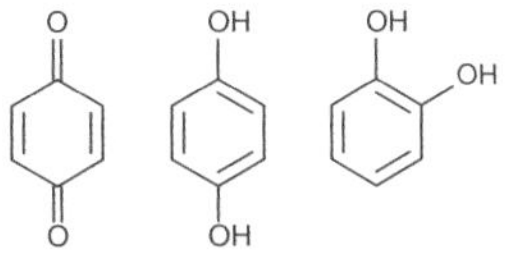

Quinones and catechols: redox cyclers, metal complexers and covalent modifiers.

Figure 24.9 Examples of some PAINs often encountered in screening libraries and their effect upon an assay.

24.4 COMPOUND MANAGEMENT

Compound management is the logistical nexus between chemistry and biology. Until the late 1990s, the handling of newly synthesised compounds was a largely manual and resource-intensive process. With the introduction of high-throughput screening, the need to weigh hundreds and then thousands of samples per week forced a radical rethink of how compound management could better support the technological changes taking place in both the chemistry and bioscience laboratories. Within ten years of high-throughput screening being introduced, the major players in the pharmaceutical industry had all implemented automated compound management facilities. Compound management was typically centralised, although many companies also had local stores if their operations extended across different continents. The central store held all 'dry' (non-solubilised) samples produced in-house as well as a 'liquid' store that maintained samples prepared as **dimethyl sulfoxide** (DMSO) solutions at known concentrations. The establishment of liquid stores greatly facilitated rapid production and delivery of screening sets to the high-throughput screening sites.

DMSO was universally adopted as a solvent of choice for several reasons. First, it solubilises the highest proportion of test compounds compared with dimethyl formamide, ethanol or methanol. Second, it is well tolerated in both cellular (0.1–0.2% v/v) and biochemical assays (1–5% v/v). Third, DMSO has a melting point of 18 °C, so samples stored below this temperature (provided there is no uptake of adventitious moisture) can be stored as solids, theoretically making them less prone to decomposition. Fourth, DMSO has a boiling point of 189 °C, rendering it less volatile than many other organic solvents. This helps maintain samples at a known concentration over prolonged periods of time; something that is critically important when following up hits, given that activity is related to compound concentration. Clearly, if the concentration of the stock solution is not accurate, then any subsequent assessment of activity will also be incorrect.

However, there are nevertheless some serious disadvantages to using DMSO. As alluded to earlier, there are issues related to the hygroscopic nature of DMSO, which can absorb up to 10% of its weight in water in as little as five hours under normal laboratory conditions. Indeed, there is anecdotal evidence of collections made up in DMSO housed in boiler rooms that, due to water uptake, were never depleted! In addition to altering the concentration, the ingress of adventitious water can significantly affect solubility, and also accelerate degradation.

The solubility can be affected through successive freeze–thaw cycles if the liquid stock is maintained at temperatures between −20 °C and 4 °C. When samples are removed from cold storage, there is a danger that they absorb moisture during the thawing process. The situation is exacerbated the more often this is repeated. As more water enters into the DMSO solution, the two solvents can form an ice-like crystal lattice that effectively excludes the compound of interest, causing it to precipitate. If not detected, it is possible that subsequent screens will not test the compound at all. Fortunately, precipitated components can be forced back into solution by the use of ultrasonic devices. Degraded samples, in contrast, cannot be rescued and must be discarded. Here, it is important to note that strong acids, such as **trifluoroacetic acid** (TFA), are often used to facilitate the purification of compounds. TFA can form salts with functional groups containing a basic nitrogen atom, and thus be a causative agent of degradation if DMSO becomes more aqueous and the solution thus becomes appreciably more acidic. Acoustic auditing devices are able to concomitantly measure the concentration of water in DMSO stocks and the volume of liquid. Early detection of aqueous contamination can flag a sample for potential errors in high-throughput screening and replenishment. Because of these considerations, most liquid stores are maintained at a low relative humidity, or under a dry nitrogen atmosphere. Under these more rigorous storage regimes, it is possible to avoid freeze–thaw cycles altogether and keep the stock solution at ambient temperature so that it is always in its liquid state to facilitate reformatting into microtitre plates.

Two popular ways of achieving long-term storage that maintains the integrity of a sample in DMSO are microtubes and minitubes. Microtubes are available in a range of volumes from 500 to 1000 μl and can be stored in a format-free manner within modular or purpose-built stores. All tracking is done through on-board software with the aid of 2D barcodes that provide additional assurance. Screening sets are prepared in microtitre plates from the microtube stock solutions and accessed multiple times to prepare source plates for primary screening, retest, dose response, etc.

An alternative storage mechanism is 'one-shot' minitubes. The essential difference when compared to microtubes is that individual tubes, typically with a capacity of < 50 μl, are arrayed into 384-well plates. Solubilised sample is introduced from freshly prepared stock into a number of plates containing the minitubes. These are subsequently sealed and placed in long-term storage. Whole plates or specific tubes can then be accessed for primary screening, secondary screening or retest, as needed. As the name suggests, they are used once and discarded.

Finally, more well-resourced compound management groups will also analyse a subset of the collection to ensure the integrity of samples. This is aided by automated high-performance liquid chromatography (HPLC) coupled to a photo diode array (PDA) detector that can scan a range of wavelengths in the ultraviolet to visible wavelengths. Sometimes other detection methods are necessary if, for example, selected compounds do not have a chromophore. In these cases, an evaporative light scattering detector (ELSD) is able to quantify the amount of sample and any degradation product. A mass spectrometer (MS) is often linked to HPLC, with the resulting combination often abbreviated to LC-MS (see Chapter 21), to provide structural information related to the molecular weight.

24.5 SCREENING STRATEGIES USED IN HIT DISCOVERY

24.5.1 *In Silico* Methods

Because drug discovery is a time-consuming and expensive endeavour that can cost many millions of dollars for each campaign, the methods used throughout the entire process are continuously refined and optimised in an effort to reduce costs. Efficiency gains can be achieved during the initial search for compounds that can interact with a biological target and modulate its action (i.e. hit identification) by screening large databases of compounds against a protein target by **automated docking** (virtual screening) or by screening a smaller number of compounds against a range of targets (inverse virtual screening). Setting up a computer-based screening resource is less costly to establish and maintain than in vitro approaches like high-throughput screening or high-content screening (HCS). *In silico* methods are typically employed as preliminary filters of corporate collections so that only the most promising of the predicted hits proceed to experimental testing.

Virtual Screening

The process of prioritising specific compounds from a large number of molecules using computational methods according to their potential binding affinity for a certain protein target is generally referred to as **virtual screening** (also called virtual ligand screening or *in silico* screening). This strategy is increasingly used to identify small molecules that can potentially bind to a protein target before committing resources towards more expensive empirical methods. Importantly, virtual libraries need not be limited to known compounds. A greater exploration and assessment of chemical space can be achieved if, for example, novel motifs, such as those generated via scaffold hopping (see Section 24.2.8) were included. Understandably, any molecules submitted to virtual screening must be able to be synthesised for this approach to be successful. Ordinarily, molecules would also be filtered for physico-chemical properties, structural alerts and PAINs prior to being screened *in silico*.

The technique relies on computational modelling of the interactions (see Section 17.4) between the binding site on a target and a small molecule. Clearly, detailed structural information must first be obtained of the target protein before a virtual screen can even commence. Such information is generally obtained from protein X-ray diffraction data, but also from nuclear magnetic resonance (NMR) experiments or comparative modelling using a previously determined structure. A simplified computational approach for the binding of small molecules in the active site considers the ligand as a series of interacting groups (i.e. van der Waals interactions, electrostatic interactions, H-bond donors and acceptors) in three-dimensional space. Binding can occur when these groups are aligned in the correct orientation so as to make energetically favourable contacts with the amino-acid backbone and/or side-chain groups. The compound is manipulated so as to maximise these complementary interactions. Stronger interactions are indicative of potentially higher-affinity binding.

The above process is complicated by the flexibility inherent in both the small molecule ligand and the protein target. Because ligands can adopt different molecular configurations, depending on the protein they interact with, *in silico* methods must address conformational flexibility, and other factors, such as tautomers, pH-dependent ionisations and stereochemistry. The different conformations and other features may be generated prior to virtual screening, so that each conformer is essentially treated as a unique ligand. Alternatively, the ligand can be allowed to retain a degree of flexibility and the energy of interaction is calculated, while incremental changes are made to the conformation of the small molecule in the binding site. Some virtual screening methods also take into account the flexibility of the target protein, which may also change conformation as a small-molecule ligand is introduced into a binding site. This is a much more difficult proposition than that for small molecules, given the larger size of the target protein and secondary structure the amino-acid residues may adopt. Various methods have been employed to simulate protein flexibility, including multiple crystal or NMR structures and molecular dynamics. Energy minimisation is also applied with conformational analysis to optimise the ligand–protein interaction. Because molecular dynamics methods do not explicitly include electrons, more recent hybrid approaches incorporating quantum mechanics and molecular mechanics have recently been introduced.

Once the starting conformation for both the target and each ligand has been determined, every small molecule in the database is evaluated at the binding site. There are two steps to this process:

1. Docking, which is the search for the conformation, position and orientation of the ligand in the active site
2. Scoring, which is an evaluation of the interaction energy between the target and the ligand.

Scoring functions approximate, and then sum, the various forces that contribute towards binding. It is important to note, however, that scoring methods tend to over-estimate the binding affinity for many ligands that show little or no activity during follow-up in vitro validation. Fortunately, the number of false positives from virtual screening can be reduced through consensus scoring. In consensus scoring, several different scoring algorithms are used to calculate the interaction energy of the same ligand to protein binding. Compounds that score highly in more than one method are prioritised for further investigation. This would entail screening in a biochemical assay, but may also involve *de novo* synthesis if the ligand existed only *in silico* up to this point.

Inverse Virtual Screening

While finding out which ligands from a large virtual library are predicted to bind most effectively to a specific protein has become more routine, the opposite question of which target(s) a known ligand has affinity for is still largely unknown. Such knowledge may help identify new biological activities for existing therapeutics and bioactives, the goal of drug repurposing, explain off-target activity (side effects) in cases where a candidate drug lacks specificity or ascertain previously overlooked proteins that may in fact be 'druggable'.

To help answer this question, a relatively new computational tool known as **inverse virtual screening** (also called reverse docking) was recently developed. Inverse virtual screening involves the *in silico* interrogation of multiple protein targets described in protein databases (up to 2000) by a relatively small set of compounds (to date generally < 100 compounds). Although reverse docking also works on the prediction of the most likely ligand–target interactions following docking and scoring, the evaluation of a single ligand against a panel of protein targets is presently more difficult to set up and time-consuming to undertake, given the large number of heterogeneous targets that are investigated. Currently, inverse virtual screening facilitates the identification of potentially druggable targets rather than being able to accurately predict a range of activities for one or more compounds.

24.5.2 Structure-Based Methods

Structure-based methods encompass approaches that are informed by knowledge of the three-dimensional structure of the biological target derived from protein nuclear magnetic resonance (NMR) or X-ray crystallographic experiments (see Chapter 14). The biological target is typically a protein, and approaches mentioned earlier that utilise biostructural information include fragment-based drug discovery and *in silico* methods. When structural information for the biological target is not available, **comparative modelling** can provide three-dimensional models (also called homology models), based on an experimentally determined structure of a related protein. This is currently an active area of research, as automated methods of comparative modelling can be inaccurate, since even minor errors can make the homology model useless for drug-discovery applications.

Originally, **structure-based drug design** was employed during lead optimisation to refine single compounds prior to selecting a candidate drug for clinical trials. The scope has since significantly broadened to now include many earlier stages of the drug-discovery process, including the design of focussed libraries. The design of such libraries for kinase inhibition, for example, has received much attention of late, given the crucial role kinases play in the regulation of a wide range of cellular processes. However, despite their attractiveness as drug targets, their relatively conserved ATP binding site presents a significant challenge in the discovery of non-promiscuous inhibitors. Given the nucleotide binding sites of kinases are the most prominent, but also most highly conserved sites, structure-based methods are used to develop compounds that selectively bind to a specific nucleotide binding site rather than any of the other >500 kinases currently identified in the human genome. Additionally, compounds must possess very high potency in order to compete with endogenous ATP concentrations at the millimolar level. Here, other structure-based methods like docking and scoring are employed to select suitable compounds for synthesis from an enumerated virtual library. As noted above, three-dimensional structures are generated in the computer, passed through appropriate property filters, aligned into the active site of the target protein using automated docking algorithms, and then prioritised on the basis of their docking scores.

24.5.3 Biomolecular Methods

Biomolecular methods encompass the screening of pure compound or natural-product-extract libraries for their effects on relevant biological targets and typically fall under one of two broad categories: **target-based** biochemical assays and **phenotypic screens** employing cell-based or whole organism assays. These two principal approaches differ in emphasis. Phenotypic screens start with the observation of a cell or organism displaying a certain phenotype. Compounds are then screened to find those that can alter the phenotype. For phenotypic screens, the emphasis is on function. In contrast, for target-based screens, the emphasis is placed on the isolated protein produced by a specific gene. Compounds are screened to find those that bind with high affinity. Downstream assays are then conducted to confirm the primary activity.

Target-based assays can be subdivided into assays conducted on isolated molecular targets such as a purified enzyme like a protease or kinase, or cell-free multi-component assays associated with cell extracts, membranes or reconstituted signalling cascades. Similarly, cellular assays can be further categorised into 'reporter gene'-type assays or phenotypic assays that measure outputs resulting from intact cellular processes. Assays employing whole organisms are typically used in drug discovery against parasitic worms (helminths) and either conducted as survival or phenotypic assays. Compounds are screened in an effort to identify a hit that could potentially be developed into a lead or tool. Leads, as already noted, may ultimately be developed into drugs.

High-Content Screening

High-content screening (HCS) is a term used to describe automated image capture and analysis technologies that monitor cellular responses in the presence of a small molecule, e.g. cell death, growth or change in morphology. The ability to collect information from individual cells and subpopulations, rather than an average readout of the total population, is what differentiates HCS from most other cellular assay formats. Until relatively recently, image-based assays were very low-throughput and required significant amounts of time to be dedicated to the manual collection and visual inspection of images. The extensive information that could nevertheless be collected from individual cells at these lower throughputs was termed **high-content analysis**. Subsequent technological advances in both automated microscopy and analysis facilitated the high-throughput screening to be applied to image-based assays and the term HCS was coined to describe image-based screening performed at the same rate as the high-throughput screening of target-based assays. HCS underpins newer inter-facial sciences like chemical biology and is also known as cell-based or phenotypic screening.

Imaging screens typically involve cells first being plated into optical bottom microtitre plates, then treated with small molecules and incubated for an appropriate period of time. Proteins of interest are then made to fluoresce by using a combination of small-molecule fluorescent probes, immunofluorescence with antibodies or expression of green fluorescence protein (GFP)-tagged proteins in cells. GFP-tagged proteins can be utilised in both live and fixed cells. Images are captured by automated microscopy and subsequently analysed to determine the type and extent of any phenotypic change. In small-molecule screening, it is eventually necessary to identify the biochemical target causing the phenotypic change. This can be challenging, especially if phenotypic change is caused by the perturbation of more than one protein.

High-Throughput Screening

High-throughput screening (HTS) is a drug-discovery process that is underpinned by various synergistic technologies to facilitate the rapid and reproducible testing of a biochemical or cellular event, generally taken to be between 10 000 and 100 000 compounds per day. While HTS really only refers to the number of assays that can be performed each day, the term has become synonymous with **target-based screening**. Campaigns delivering throughputs of between 1000 and 10 000 data points per day have been described as medium-throughput, while those affording less than 1000 compounds are considered to be low-throughput. At the other end of the spectrum, ultra HTS (uHTS) implies that more than 100 000 data points are generated per day.

Like any biological assay, those adapted for HTS incorporate positive and negative controls in an effort to determine the assay 'window' and validate the biological response. An **assay window** is typically measured using the signal-to-background-noise ratio ($S{:}B$), defined as the mean signal (μ_S) divided by the mean background (μ_B) (Equation 24.6) which, for a good assay, is greater than 5.

$$S{:}B = \frac{\mu_S}{\mu_B} \qquad\qquad\qquad \text{(Eq 24.6)}$$

Reference compounds are used as positive controls to produce the same result sought from an active compound in the screening set. Negative controls, on the other hand, are usually associated with additives demonstrated to have no activity on the assay at the concentrations being used, e.g. DMSO, the solvent used to solubilise compounds. Because large numbers of compounds are screened in an HTS campaign, it is also important to consider controls between different plates to facilitate on-the-fly assessment, so potential corrective action can be taken if the biological response changes over time. Intra-plate controls are also essential to assess, for example, edge effects, which can affect the uniformity of the biological response; over the course of a screen they permit the analysis of the uniformity of the biological response.

Because HTS of a large corporate collection is expensive, positive and negative controls are also used to determine the robustness of an assay, described by the Z'-factor, to help predict if useful data could be obtained if the assay was scaled up to 10^4–10^6 compounds. The Z'-factor measures the quality of the assay itself in the *absence* of test compounds. Four parameters are required for the determination of Z' (Equation 24.7): the mean (μ) and standard deviation (σ) of both the positive and negative controls.

$$Z' = 1 - \frac{3 \times \left(\sigma_{positive} + \sigma_{negative}\right)}{\mu_{positive} - \mu_{negative}}$$ (Eq 24.7)

The performance of an assay during the screen, when test compounds are assayed, is commonly reported as the Z-factor (Equation 24.8). The Z-factor is a statistical parameter that takes into account the signal-to-background and assay signal variation. Here the mean (μ) and standard deviation (σ) of the samples and controls are used to derive the value. Values for the Z-factor ranging from 0.5 to 1.0 are considered good, while anything below 0.5 is marginal.

$$Z = 1 - \frac{3 \times \left(\sigma_{sample} + \sigma_{control}\right)}{\left| \mu_{sample} - \mu_{control} \right|}$$ (Eq 24.8)

Both control (Z'-) and sample (Z-) factors are typically reported. Comparing the values of Z'- and Z-factors of the same assay under the same conditions gives an idea of how the compound library being screened affects the assay. By way of example, if the Z'-factor is large, but the value of the Z-factor is relatively small (and where both values are greater than 0), it could indicate that the compound library and/or compound concentration needs to be optimised further.

Inter-plate controls containing vehicle only are usually distributed uniformly throughout the screen at ten-plate intervals to monitor systematic variation in background. These controls are used to assess and correct, when possible, systematic variations in the biological response during the course of the screening campaign (e.g. spurious results caused by a faulty dispenser tip).

Many current screening technologies rely heavily on light-based detection methods, such as fluorescence, to quantify the effect of a compound on a biological target. While light-based detection technologies achieve a desirable balance between sensitivity and ease of automation, as noted earlier in the discussion of PAINs, they are also susceptible to different types of assay interference. Thus, in addition to both positive and negative controls that are built into any assay, it is not uncommon for HTS campaigns to also incorporate a range of secondary screens to filter out false positives that in some way interfere with the assay technology. Their aim is to differentiate between compounds that generate false positives from those compounds that are genuinely active against the target. Here, a different assay format is used to test actives from the primary assay in an effort to confirm that activity is directed toward the biological target of interest. A negative result indicates that the activity observed in the primary screen was most likely due to the assay format and not specific to the biology. Inactive compounds are removed from further consideration. Secondary assays are also used to confirm the estimated IC_{50} value of actives in the primary screen, selectivity (e.g. in the case of kinases) to minimise off-target effects, cytotoxicity, permeability, metabolism and P450 inhibition.

24.6 ACTIVE-TO-HIT PHASE

In the context of high-throughput screening, the terms 'active' and 'hit' can be used inter-changeably or have different meanings, depending on the author or group reporting the biological data. In this chapter, an active is defined as a compound that shows reproducible activity above a specified threshold at a tested concentration while a hit is defined as a compound that has:

- reproducible activity in a relevant bioassay
- a structure that has been confirmed from a high purity sample
- selectivity data generated
- analogues tested
- a chemically tractable structure
- potential for novelty.

A hit molecule would normally have a potency of between 100 nM and 5 µM at the drug target. Because many high-throughput screening campaigns can deliver large numbers of compounds with confirmed activity, a key goal of the active-to-hit (AtH) phase is to deliver three- to five-hit series to take forward to the next (hit-to-lead) stage. To do this, the actives are typically grouped together into clusters, based on their structural similarity. Consideration is given to the properties of each cluster, particularly as to whether an identifiable structure–activity relationship can be discerned from a subset of compounds (known as a series) that have a common substructure or chemical motif. The synthetic tractability of a potential series will also be examined at this point. Representative examples from the various series will be subjected to various in vitro assays designed to provide early data on the absorption, distribution, metabolism and elimination (ADME), as well as physico-chemical and pharmacokinetic (PK) measurements, of the compound. Selectivity profiling is also undertaken during the AtH stage, especially if the target is a kinase. While the actual number of series taken forward can vary, depending on resources, more than one is preferred to allow for attrition at later stages.

24.7 HIT-TO-LEAD PHASE

Once hits have been established, the next step is to find the best possible leads that can be progressed to lead optimisation (see Section 24.9) in the shortest period of time. The process of identifying a lead is known as the hit-to-lead phase, often abbreviated as HtL or H2L, and also referred to as lead generation or lead discovery (Figure 24.1). Importantly, while hit criteria are aimed at providing evidence that the activity is not associated with impurities or false positives (after the initial high-throughput screening), and that the issues to be addressed in HtL have at least been evaluated, lead criteria tend to build on the hit criteria to establish sufficient scope and robustness before moving on to the even more resource-intensive lead optimisation phase. In essence, a lead can be thought of as a hit with a genuine structure–activity relationship; this suggests that more potent and selective compounds can be found.

The HtL process typically addresses five general objectives. The first is the validation of structure and purity. This may seem obvious but, as noted above at the onset of the high-throughput screening paradigm, many of the screening samples in corporate collections were already decades old. This often meant they were prepared and purified prior to the widespread use of high-field NMR spectroscopy for structure determination and chromatographic techniques for purity assessment. Moreover, many samples simply decomposed during long-term storage, especially in DMSO if traces of a strong acid were present. The second objective is to find the minimum active fragment (i.e. the **pharmacophore**) if the hit is a complex molecule, so that an atom-efficient series can be developed with demonstrable structure–activity relationships. The third objective is to enhance selectivity in an effort to minimise activity at closely related targets. The fourth is to exclude compounds with inappropriate modes of action, such as non-specific binding to the molecular target, while the fifth is to increase potency.

24.8 ADMET

The most significant additions to the five objectives of the hit-to-lead phase noted above are the characterisation of hits and leads in terms of their physico-chemical and metabolic properties. Early lead generation efforts tended to focus on confirming the identity of a hit, expansion of the structure–activity relationships and increasing the potency to a point where confidence to enter lead optimisation was obtained. However, the importance placed on pursuing potency alone has been downplayed in many companies after it was realised that the late-stage attrition of compounds in development is highly costly. To minimise later-stage failure, chemical entities are also assessed for their pharmacodynamic and pharmacokinetic properties. **Pharmacodynamics** relates to the effect a drug has on a biological system, e.g. activity or toxicity, while **pharmacokinetics** describes how the body reacts to a drug, i.e. the way the drug is absorbed, distributed, metabolised and eliminated, and its toxicity (hence the term ADMET). An early awareness of the selectivity, solubility, permeation, metabolic stability, etc., has facilitated the selection and prioritisation of compound series with the best development potential. This also meant that leads not reaching pharmacokinetic, pharmacological and toxicological targets were identified early in the process, typically described as 'failing fast and failing cheap'.

24.8.1 Absorption

The first process that is generally considered for a drug or bioactive molecule that has been administered to the body is **absorption**. Most drugs are administered by the oral route, but the inhalation (via the lungs), sublingual (beneath the tongue), transdermal (across the skin) and subcutaneous injection (beneath the skin) routes are more appropriate for the administration of certain drugs due to their particular physico-chemical properties.

For drugs administered orally, the overall rate-limiting factor is often their solubility in the hydrophilic intestinal milieu. However, after the initial absorption, which requires hydrophilic properties of the compound, the majority of drugs need to be absorbed from the gastrointestinal tract in order to enter the circulatory system; this typically occurs by passive diffusion, and thus requires that the drug is lipophilic, i.e. it possesses adequate lipid solubility and therefore is non-ionised. Many drugs are weak bases and as such exist in an ionised and hence hydrophilic state at the pH of 1 prevailing in the stomach. Thus, generally speaking, the extent of absorption of drugs from the stomach is low. In contrast, in the small intestine, with a pH of 7, most drugs are non-ionised and hence suitable for passive diffusion across the gut wall. In these cases, the gastric emptying rate is normally the limiting factor for absorption and this is influenced by the food content of the stomach. On entering the circulatory system from the gut, the drug is taken by the hepatic portal vein to the liver, which is the major site of **drug metabolism**. Some drugs are readily metabolised by hepatic enzymes and therefore subject to significant inactivation by metabolism before they have entered the general circulatory system. Loss of active drug at this stage is referred to as the **first-pass effect** – hence the importance of assessing the susceptibility of a candidate drug to hepatic metabolism in the early stages of drug discovery. The proportion of an oral dose of drug reaching the systemic circulation from its site of administration is referred to as its **bioavailability** (β). It is possible to avoid the first-pass effect by administering the drug by routes such as sublingual and transdermal.

For an orally administered drug, its concentration in the blood increases as absorption from the gut continues. It eventually reaches a plateau, at which point the rate of absorption and the rate of loss of the drug are equal, and then declines as the drug is eliminated from the body (Figure 24.10). For a drug to exert its desired pharmacological effect, its concentration in the blood needs to exceed a threshold value referred to as the **minimum effective concentration**. At a higher concentration the drug may begin to display toxic side effects, referred to as the **toxicity threshold**. The ratio of these two thresholds is referred to as the **therapeutic index** or therapeutic ratio. The closer the index to 1, the more problematic the drug is in clinical use. Drugs such as the antiepileptics phenytoin, carbamazepine and phenobarbitone, with an index in the region of 1, are often subject to therapeutic drug monitoring to ensure that the patient is not exposed to potential toxic effects. The aim of repeat dosing with any drug is to maintain the concentration of the drug in the therapeutic range or window in which toxic effects are not observed. If the dosing interval is adjusted correctly in relation to the plasma half-life (see below), the drug plasma concentration will oscillate within the therapeutic range (Figure 24.10). Once in the general circulation, the drug may bind to a plasma protein, especially albumin, and if the extent of binding is high (> 90%) the ability of the drug to cross membranes to reach its site of action and exert a pharmacological effect may be impaired. There are a number of experimental methods to facilitate the rapid determination of membrane permeability. For example, a monolayer of caco-2 human epithelial cells can be grown in a multi-well filter plate; the elution of compounds through the monolayer can be measured to give a sense of their permeability. However, because cell lines are prone to experimental variability, there has been a shift toward testing against a **parallel artificial membrane permeability assay** (PAMPA). Since the membranes are artificial, and the filter plates can be prepared with automated platforms in multi-well formats, there is a decreased opportunity for variability between plates.

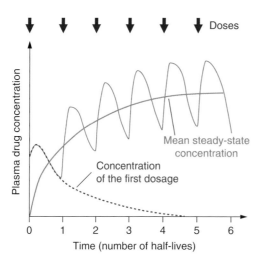

Figure 24.10 Plasma drug concentration following repeated oral dosing. Following each dose, the plasma concentration increases, reaches a peak and declines normally at an exponential rate. When the next dose is administered, the plasma concentration profile is superimposed on the existing profile but if the dosing is given after one half-life, the peak plasma concentration gradually reaches a maximum and thereafter remains at this level following further repeated dosing. This maximum is generally achieved after five doses.

The pharmacokinetic parameters of a drug quantify the non-pharmacological behaviour of the drug from the time of its administration to the time of its removal from the body. There are three main parameters: intrinsic clearance, apparent volume of distribution, and plasma or elimination half-life.

Intrinsic Clearance (Cl_{int})

This is defined as the volume of plasma apparently cleared of drug per unit time by all routes. It has units of cm^3 min^{-1}. Drugs are removed from the body by two main routes:

1. Metabolism, normally by hepatic enzymes, to one or more polar metabolites that generally lack pharmacological activity and which are readily excreted by the kidneys
2. Renal excretion of unchanged drug.

Intrinsic clearance is therefore the sum of hepatic clearance Cl_{hep} (the volume of plasma apparently cleared of drug in unit time by hepatic metabolism) and renal clearance Cl_r (the volume of plasma apparently cleared of drug in unit time by renal excretion of unchanged drug). The value of Cl_{int} can be calculated (Equation 24.9) by administering a dose (units: mg) of the drug by the oral route and dividing the dose by the area under the resulting plasma concentration/time curve (AUC) (units: mg min cm^{-3}) making an allowance for the bioavailability of the drug:

$$Cl_{int} = \frac{Dose \times \beta}{AUC}$$

(Eq 24.9)

The bioavailability (β) is calculated from the ratio of the AUC values for an oral dose and for the same dose administered intravenously, and which is not subject to first-pass loss. It is expressed as a percentage and can vary from 0 to 100%.

Apparent Volume of Distribution (V_d)

This is defined as the volume of body fluid in which the drug appears to be distributed (Equation 24.10); it has units of dm^3. In an adult body, the total volume of water is about 42 dm^3, and is made up of 3 dm^3 plasma water, 14 dm^3 extra-cellular water and 25 dm^3 intra-cellular water. The value of V_d is measured by administering a dose by bolus (fast) intravenous injection (to avoid the first-pass loss) and using the equation:

$$V_d = \frac{\text{Dose}}{\text{Peak plasma concentration}} \qquad \text{(Eq 24.10)}$$

Many drugs have V_d values in the range of 42 dm^3, indicating that they are fully distributed in body water. Abnormally high values are the result of a low plasma concentration caused by the deposition of the drug in some particular tissue, most commonly fat tissue for highly lipophilic drugs. Generally speaking, this is an undesirable property of a drug. Equally, an abnormally low V_d value due to a high plasma concentration is indicative of a poor ability of the drug to penetrate lipid barriers.

Plasma or Elimination Half-Life ($t_{1/2}$)

This is defined as the time required for the plasma concentration to decline by 50% following its intravenous administration. It is a so-called hybrid constant, as its value is linked to both Cl_{int} and V_d by Equation 24.11:

$$t_{1/2} = \frac{0.693 \times V_d}{Cl_{int}} \qquad \text{(Eq 24.11)}$$

Values of $t_{1/2}$ normally range from 1 to 24 hours. The clinical importance arises from the fact that the value of $t_{1/2}$ determines the frequency with which the drug needs to be administered in order to maintain the plasma concentration in the therapeutic range. Thus, drugs with a short half-life need to be administered more frequently, whereas drugs with a long half-life can be given on a daily basis.

24.8.2 Distribution

Following absorption, the next step in the pharmacokinetic process is the distribution of a bioactive compound from the bloodstream to the intended site of interest. Orally administered compounds must pass through a number of membranes before they can enter, and then also leave, the bloodstream. Membrane permeability assays have therefore become important tools to guide the selection of compounds for further development that can be expected to have favourable distribution properties. Permeability assays can be tailored to consider the target area of interest. By way of example, the **blood–brain barrier** protects the brain and the central nervous system (CNS) from penetration by a number of larger biomolecules and microorganisms. It also prevents many smaller molecules from entering. To gain access to the CNS, small-molecule drugs should be sufficiently permeable to have a passive diffusion that is not significantly decreased by any efflux transport. This difference largely accounts for CNS drugs having more restricted physicochemical properties than orally administered drugs; indeed it has been suggested that the upper limit of the polar surface area should be around 75 Å^2 to ensure sufficient permeability. **Parallel artificial membrane permeability assays** (PAMPAs) have been used to provide an in vitro simulation of the blood–brain barrier to facilitate the selection of compounds with the best chance of in vivo activity.

Distribution can also be examined via permeability through modified cell lines. For example, the caco-2 cell line (see above) can be engineered to over-express p-glycoprotein (Pgp), which actively transports many different compounds across membranes that would not pass through in passive mode. A reasonable approximation of the overall distribution of a compound, whether an active or passive mechanism is operating, can be determined by performing assays under standard conditions, as well as in the presence of Pgp inhibitors such as ketoconazole or cyclosporin A used at concentrations of 50 μM.

Membrane permeability is not the only mechanism by which small molecules can exit the bloodstream. Fenestrated capillaries contain small pores that permit low-molecular-weight compounds, but not larger proteins, to pass through and enter the interstitial fluid surrounding cells. Proteins retained in the blood plasma can therefore end up playing a critical role in the distribution of potential drugs. This is particularly important if compounds associate with plasma proteins; in this situation, the therapeutic compound is unlikely to permeate through capillary pores to reach the desired target. One of the most abundant proteins in human blood is albumin. A primary function of albumin is to transport lipophilic compounds such as fatty acids and hormones. However, given it is a fairly non-selective transporter, albumin also binds many other lipophilic small molecules, and ultimately impedes their distribution out of the circulatory system. For this reason, an albumin protein binding assay is routinely performed during drug discovery. Although high protein binding (> 99%) does not automatically disqualify a compound from being a viable drug, it can dramatically reduce its distribution and pharmacodynamic effects.

24.8.3 Metabolism

Most drugs are sufficiently lipophilic to be poorly excreted by the kidneys and hence would be retained by the body for very long periods of time were it not for the intervention of metabolism, mainly in the liver. Metabolism occurs in two phases (Figure 24.11). Phase I mainly involves oxidation reactions and Phase II conjugation of either the drug or its Phase I metabolites with glucuronic acid, sulfate or glutathione to increase the polarity of the drug or its metabolite(s) and hence ease of renal excretion. The oxidation reactions are carried out by a group of haem-containing enzymes collectively known as cytochrome P450 monooxygenases (CYP), so-called because of their absorption maximum at 450 nm when combined with CO. These membrane-bound enzymes are associated with the endoplasmic reticulum and operate in conjunction with a single NADPH-cytochrome P450 reductase, and are capable of oxidising drugs at C, N and S atoms. More than 50 CYP human genes have been sequenced and divided into four families (CYP1–4) of which CYP2 is the largest.

Five CYPs appear to be responsible for the metabolism of the majority of drugs in humans: CYP1A2, CYP2C9, CYP2C19, CYP2D6, CYP2E1 and CYP3A4 (Table 24.2). The

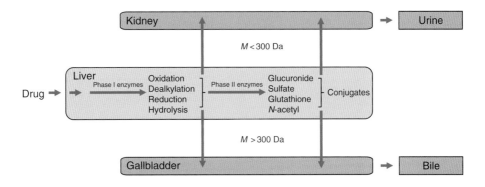

Figure 24.11 Drug metabolism. Phase I enzymes catalyse the modification of existing functional groups in drug molecules (oxidation reactions). Conjugating enzymes (Phase II) facilitate the addition of endogenous molecules such as sulfate, glucuronic acid and glutathione to the original drug or its Phase I metabolites.

genes for these cytochromes have been cloned and expressed in cell lines suitable for drug metabolism studies. Several of these cytochromes are expressed polymorphically (i.e. they exist in several forms due to their expression by related genes) in humans resulting in considerable inter-individual variation in the rate of metabolism of some drugs. There are also ethnic variations in the expression of some of these isoforms. This can be critically important in the use of drugs that have a narrow therapeutic index. In principle, it is possible to genotype individual patients for their CYP activity and hence to 'personalise' drug dosage, an approach known as pharmacogenomics (Section 4.20), but in practice this has yet to be put into widespread clinical practice. One other problem associated with the clinical use of some drugs is that they either inhibit or induce one or more of the cytochrome P450s (Table 24.2). Inhibition means that the intrinsic clearance of the drug and that of other concomitantly adminis-tered drugs is impaired and this may have toxicological consequences, whilst enzyme induction results either in increased clearance of the drug that may render its therapy ineffective and/or in the production of toxic metabolites.

24.8.4 Excretion

By the time a drug has undergone metabolism to the form that is eliminated from the body, it has often outlived its original purpose. For this reason, excretion tends to be the least considered of the four processes that make up the pharmacokinetic profile of a compound during the lead optimisation phase of drug discovery. However, there are in vitro assays that help provide some insights into the excretion process; for example the multi-drug resistance-associated protein 2 (mrp2) transport pathway is used to predict biliary excretion.

Largely ignored to date, drugs eliminated from the body may have further effects on ecological systems, and, ultimately, the health of humans and animals. Due to extensive use in the past 100 years, eliminated pharmaceuticals (as well as personal care products, summarised as **PCPs**) are present in water bodies and soil throughout

Table 24.2 **Action of some drugs on the major human cytochrome P450 isoforms**

P450 isoform	Substrate	Inhibitor	Inducer
CYP1A2	Imipramine, oestradiol, paracetamol, verapramil, propranolol	Fluvoxamine,[a] cimetidine, ciprofloxacin	Omeprazole, cigarette smoke
CYP2C9	Fluvastatin, ibuprofen, phenytoin, amitriptyline, tamoxifen	Fluconazole,[a] fluvastatin, lovastatin, sulfaphenazole, phenylbutazone	Rifampicin, secobarbital
CYP2C19	Diazepam, propranolol, amitriptyline, omeprazole, lansoprazole	Lansoprazole, omeprazole, cimetidine	Carbamazepine, rifampacin, prednisone
CYP2D6	Amitriptyline, imipramine, propranolol	Quinidine,[a] bupropion,[a] cimetidine, ranitidine	Rifampicin, dexamethazone
CYP2E1	Paracetamol, theophylline, ethanol	Cimetidine, disulfiram	Ethanol
CYP3A4	Indinavir, diazepam, lansoprazole, saquinavir, lovastatin	Ketoconazole, indinavir,[a] nelfinavir,[a] ritonavir[a]	Carbamazepine, nevirapine, phenytoin

Note: [a]Strong inhibitors that cause at least 80% decrease in clearance.

the world. For example, the increased levels of oestrogen and other synthetic hormones in waste water due to birth control and hormonal therapies have been linked to increased feminisation of aquatic organisms. There are further concerns of inducing antibiotic resistance through accumulation of antibiotic substances in waste water. New research disciplines such as **pharmacoenvironmentology** are now concerned with the effects of drugs in the environment after elimination from humans and animals post-therapy.

24.8.5 Toxicity

Toxicity is the largest cause of pre-clinical attrition and, as a consequence, compounds are generally evaluated at the same time as in vitro pharmacokinetic optimisation of the four other processes such that ADMET (or ADME/Tox) are now common terms. Indeed, there are a number of **cytotoxicity assays** that facilitate the rapid assessment of compounds so that a selectivity profile is easily generated.

However, although there are a number of assays that measure toxicity in whole cells, there are not as many where the mechanism of toxicity has been fully developed. One exception is the **hERG ion channel** which regulates potassium ion concentration across muscle cells and is particularly important in the heart muscle, where it helps maintain proper intra-cellular potassium levels during ventricular repolarisation. Compounds that block this channel, such as the cyclooxygenase 2 (COX-2) inhibitor rofecoxib (Vioxx, Figure 24.5), which was withdrawn from the market in 2004, can produce cardiac arrhythmia (measured as so-called QT interval prolongation in an electrocardiogram) and thus potentially be fatal. To complicate matters, the hERG ion channel has been observed to bind a number of different drug candidates, especially those that are lipophilic or contain an aromatic amine group. Because the hERG channel is somewhat promiscuous, an electrophysiology patch-clamp assay is now routinely employed to assess compounds for hERG liability as part of the lead optimisation process.

The Ames test is another frequently employed toxicity assay that is used to assess compounds for mutagenicity. Here, several bacterial strains are utilised to determine the capacity of a compound to induce mutations and, ostensibly, their likelihood of also being potential carcinogens.

24.9 LEAD OPTIMISATION

The object of this final drug-discovery phase is to maintain favourable properties in the lead molecule while, at the same time, improving on any deficiencies. In general, molecules that have entered lead optimisation need to be examined in models of genotoxicity such as the Ames test. High-dose pharmacology, together with dose linearity, drug-induced metabolism and metabolic profiling are investigated during this stage. Consideration is also given to any chemical stability issues and salt selection for the developing lead compound. All the information gathered about the molecule during this phase will be used to prepare a profile of the candidate drug which, when combined with toxicological and chemical manufacture and control considerations, will underpin a regulatory submission to allow commencement of clinical testing.

Once a candidate is selected, the attrition rate of compounds entering clinical trials is very high, with only one in ten candidates typically reaching the market. However, the financial consequences of failure at this point are much higher and it can become increasingly problematic to terminate the project. By the time a candidate has reached testing in humans, the project has become public knowledge and killing it can erode confidence in the company and decrease its value to shareholders.

24.10 SUGGESTIONS FOR FURTHER READING

24.10.1 Experimental Protocols

Burke M.D. and Schreiber S.L. (2004) A planning strategy for diversity-oriented synthesis. *Angewandte Chemie International Edition* 43, 46–58.

Hopkins A.L., Keseru G.M., Leeson P.D., Rees D.C. and Reynolds C.H. (2014) The role of ligand efficiency metrics in drug discovery. *Nature Reviews Drug Discovery* 13, 105–121.

Inglese J., Shamu C.E. and Guy R.K. (2007) Reporting data from high-throughput screening of small-molecule libraries. *Nature Chemical Biology* 2007, 438–441.

Sarnpitak P., Mujumdar P., Taylor P., *et al.* (2015) Panel docking of small-molecule libraries: prospects to improve efficiency of lead compound discovery. *Biotechnology Advances* 33, 941–947.

Teague S.J., Davis A.M., Leeson P.D. and Oprea T. (1999) The design of leadlike combinatorial libraries. *Angewandte Chemie International Edition* 38, 3743–3747.

Zhang J.H., Chung T.D. and Oldenburg K.R. (1999) A simple statistical parameter for use in evaluation and validation of high throughput screening assays. *Journal of Biomolecular Screening* 4, 67–73.

24.10.2 General Texts

Leahy W.L. (2016) Medicinal chemistry and lead optimization of marine natural products. In: *Marine Biomedicine: From Beach to Bedside*, Baker B.J., Ed., CRC Press, Boca Raton, Florida, USA.

Rankovic Z. and Morphy R. (2010) *Lead Generation Approaches in Drug Discovery*, Wiley, Hoboken, New Jersey, USA.

Rouhi A.A. (2003) Rediscovering natural products. *Chemical and Engineering News* 81, 77–91.

Sneader W. (2005) *Drug Discovery: A History*, John Wiley & Sons, Chichester, West Sussex, UK.

24.10.3 Review Articles

Baell J. and Walters M. A. (2014) Chemical con artists foil drug discovery. *Nature* 513, 481–483.

Davis A.M., Keeling D.J., Steele J., Tomkinson N.P. and Tinker A.C. (2005) Components of successful lead generation. *Current Topics in Medicinal Chemistry* 5, 421–439.

Ekins S., Mestres J. and Testa, B. (2007) *In silico* pharmacology for drug discovery: methods for virtual ligand screening and profiling. *British Journal of Pharmacology* 152, 9–20.

Gleeson M.P., Hersey A., Montanari D. and Overington J. (2011) Probing the links between in vitro potency, ADMET and physicochemical parameters. *Nature Reviews Drug Discovery* 10, 197–208.

Hughes J.P., Rees S., Kalindjian S.B. and Philpott K.L. (2011) Principles of early drug discovery. *British Journal of Pharmacology* 162, 1239–1249.

Leeson P.D. and Davis A.M. (2004) Time-related differences in the physical property profiles of oral drugs. *Journal of Medicinal Chemistry* 47, 6338–6348.

Lipinski C.A., Lombardo F., Dominy B. and Feeney, P J. (1997) Experimental and computational approaches to estimate solubility and permeability in drug discovery and development settings. *Advanced Drug Delivery Reviews* 23, 3–25.

Lucas X., Grüning B.A., Bleher S. and Günther S. (2015) The purchasable chemical space: a detailed picture. *Journal of Chemical Information and Modeling* 55, 915–924.

Macarron R. (2006) Critical review of the role of HTS in drug discovery. *Drug Discovery Today* 11, 277–279.

Rees D.C., Congreve M., Murray C.W. and Carr, R. (2004) Fragment-based lead discovery. *Nature Reviews Drug Discovery* 3, 660–672.

Renaud J.-P., Chung C.-W., Danielson U.H., *et al.* (2016) Biophysics in drug discovery: impact, challenges and opportunities. *Nature Reviews Drug Discovery* 15, 679–698.

Ritchie T.J., Macdonald S.F., Young R.J. and Pickett S.D. (2011) The impact of aromatic ring count on compound developability: further insights by examining carbo- and hetero-aromatic and aliphatic ring types. *Drug Discovery Today* **16**, 164–171.

Thorne N., Auld D.S. and Inglese J. (2010) Apparent activity in high-throughput screening: origins of compound-dependent assay interference. *Current Opinion in Chemical Biology* **14**, 315–324.

Veber D.F., Johnson S.R., Cheng, H.-Y., *et al.* (2002) Molecular properties that influence the oral bioavailability of drug candidates. *Journal of Medicinal Chemistry* **45**, 2615–2623.

Vieth M., Siegel M.G., Higgs R.E., *et al.* (2004) Characteristic physical properties and structural fragments of marketed oral drugs. *Journal of Medicinal Chemistry* **47**, 224–232.

Wenlock M.C., Austin R.P., Barton P., Davis A.M. and Leeson P.D. (2003) A comparison of physiochemical property profiles of development and marketed oral drugs. *Journal of Medicinal Chemistry* **46**, 1250–1256.

Wermuth C.G. (2006) Similarity in drugs: reflections on analogue design. *Drug Discovery Today* **11**, 348–354.

24.10.4 Websites

ChEMBL database
www.ebi.ac.uk/chembldb/ (accessed May 2017)

CDD Collaborative Drug Discovery
www.collaborativedrug.com/ (accessed May 2017)

PubChem database
pubchem.ncbi.nlm.nih.gov/ (accessed May 2017)

ZINC database
zinc.docking.org/ (accessed May 2017)

INDEX